A Specialist Periodical Report

# Inorganic Chemistry of the Main-group Elements

Volume 4

A Review of the Literature Published between October 1974 and September 1975

Senior Reporter
**C. C. Addison**

Reporters
**M. G. Barker**
**G. Davidson**
**M. F. A. Dove**
**P. G. Harrison**
**P. Hubberstey**
**N. Logan**
**R. J. Pulham**
**D. B. Sowerby**

All of: Department of Chemistry, University of Nottingham

© Copyright 1977

The Chemical Society
Burlington House, London W1V 0BN

**ISBN**: 0 85186 782 0
**ISSN**: 0305–697X
Library of Congress Catalog Card No. 72-95098

*Filmset in Northern Ireland at The Universities Press, Belfast, and printed by photolithography and bound in Great Britain at The Pitman Press, Bath*

## Preface

On the basis of the published reviews of the first three volumes of this Series, and the many written and verbal comments received by the Reporters, it does appear that our general plan for presenting the chemistry of the Main-Group elements is now accepted as the best pattern, and it has therefore been continued in Volume 4. Each of the eight Chapters is concerned with one of the Main Groups of the Periodic Table, and the lengths of the Chapters give a very rough measure of the quantity of research work carried out on each Group. As the volume has again been written entirely by the inorganic chemists in the University of Nottingham, there have been day-to-day discussions on the contents, so that internal overlap between the various Chapters is reduced to a minimum.

To an increasing extent, it is becoming possible to discuss the chemistry of a compound against the background of its known structure; this adds to the relevance and the interest of the chemistry involved, and helps us to concentrate on chemical, rather than purely physical, properties. However, there is then a danger of overlap with other Reviews which deal specifically with molecular structures. In principle, this is acceptable, since a reader does not welcome too much cross-reference between different Reports, but in view of the economic pressures we are attempting to minimise overlap by discussion with Senior Reporters of other related Reports.

Volume 4 is approximately the same length as was Volume 3, and the authors have again found it difficult to do justice to the large quantity of worth-while research which has been published during the year 1974—75.

C. C. Addison

# Contents

**Chapter 1  Elements of Group I**    1
*By R. J. Pulham*
1. The Alkali Metals    1
2. Alloys and Intermetallic Compounds    4
3. Solvation of Alkali-metal Ions    7
4. Compounds containing Organic Molecules or Complex Ions    12
5. Alkali-metal Oxides    20
6. Alkali-metal Halides    21
7. Lithium Compounds    25
8. Sodium Compounds    28
9. Potassium Compounds    32
10. Rubidium Compounds    35
11. Caesium Compounds    36

**Chapter 2  Elements of Group II**    37
*By R. J. Pulham*
1. Beryllium    37
2. Magnesium    42
3. Calcium    47
4. Strontium    52
5. Barium    56

**Chapter 3  Elements of Group III**    60
*By G. Davidson*
1. Boron    60
     General    60
     Boranes    61
     Borane Anions and Their Metallo-derivatives    65

| | | |
|---|---|---|
| | Carbaboranes | 74 |
| | Metallo-carbaboranes | 81 |
| | Compounds containing B—C Bonds | 87 |
| | Amino-boranes and other Compounds containing B—N Bonds | 89 |
| | Compounds containing B—P or B—As Bonds | 92 |
| | Compounds containing B—O Bonds | 95 |
| | Compounds containing B—S or B—Se Bonds | 99 |
| | Boron Halides | 101 |
| | Boron-containing Heterocycles | 105 |
| | Boron Nitride, Metal Borides, and Related Species | 112 |
| **2** | **Aluminium** | **113** |
| | General | 113 |
| | Aluminium Hydrides | 114 |
| | Compounds containing Al—C Bonds | 116 |
| | Compounds containing Al—N Bonds | 117 |
| | Compounds containing Al—O Bonds | 119 |
| | Aluminium Halides | 125 |
| **3** | **Gallium** | **129** |
| | General | 129 |
| | Gallium Hydrides | 129 |
| | Compounds containing Ga—C Bonds | 130 |
| | Compounds containing Ga—N, Ga—As, or Ga—Sb Bonds | 130 |
| | Compounds containing Ga—O, Ga—S, or Ga—Se Bonds | 132 |
| | Gallium Halides | 134 |
| | Other Gallium Compounds | 135 |
| **4** | **Indium** | **136** |
| | Compounds containing Bonds between Indium and the Elements of Group VI | 136 |
| | Indium Halides | 138 |
| | Other Indium Compounds | 140 |
| **5** | **Thallium** | **140** |
| | Thallium(III) Compounds | 140 |
| | Thallium(I) Compounds | 142 |
| | Other Thallium Compounds | 144 |

## Chapter 4  Elements of Group IV  145

*By P. G. Harrison and P. Hubberstey*

| | | |
|---|---|---|
| **1** | **Carbon** | **145** |
| | Carbon Allotropes | 146 |
| |   Synthetic Studies | 146 |
| |   Structural Studies | 147 |
| |   Chemical Studies | 148 |

| | |
|---|---:|
| Graphite Intercalation Compounds | 151 |
|    Metals | 152 |
|    Halides and Acids | 155 |
| Methane and its Substituted Derivatives | 156 |
|    Theoretical Studies | 157 |
|    Structural Studies | 157 |
|    Spectroscopic Studies | 159 |
|    Chemical Studies | 159 |
| Formaldehyde and its Substituted Derivatives | 167 |
|    Formaldehyde, Carbonyl Halides, *etc.* | 167 |
|    Formic Acid, Formates, *etc.* | 169 |
| Derivatives of Group VI Elements | 169 |
|    Oxides, Sulphides, and Related Species | 169 |
|    Carbonates, Thiocarbonates, and Related Anions | 173 |
| Derivatives of Group V Elements | 178 |
|    Cyanogen, Cyanides, Cyanates, and Related Species | 178 |

## 2 Silicon, Germanium, Tin, and Lead     182

| | |
|---|---:|
| Hydrides of Silicon, Germanium, and Tin | 182 |
| Silicon(IV) Oxide and Related Silicates | 184 |
|    Silicon Dioxide | 185 |
|    Silicates | 190 |
| Germanium(IV), Tin(IV), and Lead(IV) Oxide Phases | 199 |
| Molecular Silicon(IV)–, Germanium(IV)–, Tin(IV)–, and Lead(IV)–Oxygen Compounds | 203 |
|    Oxides | 203 |
|    Alkoxides and Related Derivatives | 203 |
| Halides of Silicon, Germanium, Tin, and Lead | 211 |
| Pseudohalide Derivatives of Silicon and Tin | 222 |
| Sulphur and Selenium Derivatives of Silicon, Germanium, Tin, and Lead | 224 |
| Nitrogen and Phosphorus Derivatives of Silicon, Germanium, and Tin | 226 |
| Derivatives of Silicon, Germanium, Tin, and Lead containing Bonds to Main-Group Elements | 230 |
| Derivatives of Silicon, Germanium, Tin, and Lead containing Bonds to Transition Metals | 234 |
| Bivalent Derivatives of Silicon, Germanium, Tin, and Lead | 240 |
|    Silylenes | 240 |
|    Germylenes | 241 |
|    Tin(II) and Lead(II) Halide Systems | 242 |
|    Oxides and Molecular Oxygen Derivatives of Bivalent Tin and Lead | 244 |
|    Sulphur and Selenium Derivatives of Bivalent Germanium, Tin, and Lead | 248 |
|    Nitrogen and Phosphorus Derivatives of Bivalent Tin and Lead | 249 |
|    Bivalent Organo-tin and -lead Derivatives | 249 |
|    Miscellaneous Studies | 251 |

|   |   |   |
|---|---|---|
| **3** | **Intermetallic Phases** | 251 |
|   | Binary Systems | 251 |
|   | Ternary Systems | 254 |

## Chapter 5   Elements of Group V    258
*By N. Logan and D. B. Sowerby*

|   |   |   |
|---|---|---|
| **1** | **Nitrogen** | 258 |
|   | Elementary Nitrogen | 258 |
|   | $N_2$ Complexing | 260 |
|   | Bonds to Hydrogen | 261 |
|   | $NH_3$ | 261 |
|   | $NH_4^+$ | 263 |
|   | $NH_2OH$ and Derivatives | 264 |
|   | Bonds to Nitrogen | 265 |
|   | $N_2H_2$ | 265 |
|   | $N_2H_4$ | 265 |
|   | $N_3H$ | 266 |
|   | $N_4H_4$ | 266 |
|   | Bonds to Oxygen | 266 |
|   | General | 266 |
|   | $N_2O$ and Other Nitrogen(I) Species | 267 |
|   | NO and Other Nitrogen(II) Species | 268 |
|   | Nitrogen(III) Species | 269 |
|   | $NO_2$–$N_2O_4$ | 270 |
|   | Nitric Acid | 271 |
|   | Nitrates | 272 |
|   | Other Nitrogen(V) Species | 274 |
|   | Bonds to Fluorine | 275 |
|   | Bonds to Chlorine and Iodine | 276 |
| **2** | **Phosphorus** | 277 |
|   | Phosphorus and Phosphines | 277 |
|   | Phosphides | 278 |
|   | Compounds containing P—P Bonds | 279 |
|   | Bonds to Boron | 281 |
|   | Bonds to Carbon | 283 |
|   | Phosphorus(III) Compounds | 283 |
|   | Phosphorus(V) Compounds | 286 |
|   | Bonds to Silicon or Germanium | 288 |
|   | Bonds to Fluorine | 290 |
|   | Phosphorus(III) Compounds | 290 |
|   | Phosphorus(V) Compounds | 291 |
|   | Bonds to Chlorine | 293 |
|   | Bonds to Bromine or Iodine | 295 |
|   | Bonds to Nitrogen | 296 |
|   | Phosphorus(III) Compounds | 296 |
|   | Phosphorus(V) Compounds | 298 |
|   | Compounds containing $P_nN_n$ Rings | 304 |

|  |  |
|---|---|
| Bonds to Oxygen | 309 |
| Lower Oxidation States | 309 |
| Phosphorus(v) Compounds | 311 |
| Monophosphates | 316 |
| Apatites | 320 |
| Diphosphates | 321 |
| Meta- and poly-phosphates | 321 |
| Phase Studies | 322 |
| Bonds to Sulphur or Selenium | 322 |
| **3 Arsenic** | **326** |
| Arsenic and Arsenides | 326 |
| Bonds to Carbon | 326 |
| Bonds to Halogens | 328 |
| Bonds to Nitrogen | 329 |
| Bonds to Oxygen | 330 |
| Bonds to Sulphur or Selenium | 332 |
| **4 Antimony** | **333** |
| Antimonides | 333 |
| Bonds to Carbon or Nitrogen | 333 |
| Bonds to Halogens | 334 |
| Antimony(III) Compounds | 334 |
| Antimony(v) Compounds | 335 |
| Bonds to Oxygen | 337 |
| Antimony(III) Compounds | 337 |
| Antimony(v) Compounds | 337 |
| Bonds to Sulphur or Selenium | 339 |
| **5 Bismuth** | **340** |

# Chapter 6  Elements of Group VI  342

*By M. G. Barker*

| | |
|---|---|
| **1 Oxygen** | **342** |
| The Element | 342 |
| Oxides, Peroxides, and Superoxides | 344 |
| Oxygen Halides | 346 |
| **2 Sulphur** | **346** |
| The Element | 346 |
| Sulphur–Halogen Compounds | 348 |
| Sulphur–Oxygen–Halogen Compounds | 353 |
| Sulphur–Nitrogen Compounds | 354 |
| Linear Compounds | 354 |
| Cyclic Compounds | 358 |
| Other Sulphur-containing Ring Compounds | 363 |

|   |   |   |
|---|---|---|
| | Sulphur–Oxygen Compounds | 365 |
| |    Sulphates | 367 |
| |    Thiosulphates and Thionates | 369 |
| |    Sulphuric Acid and Related Compounds | 370 |
| | Sulphides | 371 |
| |    Preparation and Properties of Sulphides | 372 |
| | Hydrogen Sulphide | 375 |
| | Polysulphides | 376 |
| | Other Sulphur-containing Compounds | 376 |
| **3** | **Selenium** | **378** |
| | The Element | 378 |
| | Selenium–Halogen Compounds | 379 |
| | Selenium–Oxygen Compounds | 379 |
| |    Selenates | 380 |
| |    Selenites | 382 |
| | Selenides | 383 |
| | Other Compounds of Selenium | 386 |
| **4** | **Tellurium** | **388** |
| | The Element | 388 |
| | Tellurium–Halogen Compounds | 388 |
| | Tellurium–Oxygen Compounds | 390 |
| | Tellurides | 392 |
| | Other Compounds of Tellurium | 392 |
| **5** | **Polonium** | **393** |

## Chapter 7   The Halogens and Hydrogen   394
*By M. F. A. Dove*

|   |   |   |
|---|---|---|
| **1** | **Halogens** | **394** |
| | The Elements | 394 |
| | Halides | 400 |
| | Interhalogens and Related Species | 401 |
| | Oxide Halides | 407 |
| | Oxides and Oxyanions | 408 |
| | Hydrogen Halides | 413 |
| **2** | **Hydrogen** | **415** |
| | Hydrogen-bonding | 415 |
| | Miscellaneous | 416 |

## Chapter 8   The Noble Gases   417
*By M. F. A. Dove*

**1  The Elements**   417

| | | |
|---|---|---|
| 2 | **Krypton, Xenon, and Radon(II)** | 418 |
| 3 | **Xenon(IV)** | 421 |
| 4 | **Xenon(VI)** | 421 |
| 5 | **Xenon(VIII)** | 422 |

# Author Index 423

# 1
# Elements of Group I

BY R. J. PULHAM

## 1 The Alkali Metals

The high-resolution Auger spectrum, in undifferentiated form, of polycrystalline lithium shows a single peak in the region of 50 eV, which is identified with $KVV$ Auger transitions, and has a profile qualitatively similar to the $L_{2,3}VV$ spectrum of aluminium; there is a broad peak which falls steeply from the maximum to a high-energy threshold, generally consistent with a band-like picture for the conduction electrons of the metal, and on the low-energy side of the peak there is an intense plasmon tail. Two free-electron models have been developed, one being a bulk model, the other including an idealized surface; the latter gives better agreement with the major features of the spectrum, suggesting that the surface environment may be important in determining the shape of the profile.[1] The $NaK_\alpha$-line shift during the transition from Na to $Na^+$ has been calculated as 0.21 eV. Analysis of the electron densities during formation of $X$-ray vacancies shows that even in ionic compounds of sodium, the sodium valence electron is not completely removed from the atom.[2] Photoelectron and Auger electron measurements have been made on polycrystalline films of sodium evaporated in vacuum, and on sodium monoxide produced *in situ* by oxidation with dry oxygen. Core- and valence-electron binding energies, photoionization cross-sections relative to Na1s, $KLL$ and $KLV$ Auger energies, and transition probabilities are reported. Valence band influences on the $KLV$ Auger spectra are discussed with references to the $X$-ray photoelectron spectrum and other sources of band information. Unexpected structure was found in the $KLV$ spectra of the metal which, pending thorough interpretation, offsets the sensitivity and resolution advantages which these spectra otherwise offer for valence band studies.[3]

Photodetachment of alkali-metal negative ions by an argon-ion laser (4880 Å) has been studied by the crossed-beam technique. Analysis of the energy of the photodetached electrons yields the electron affinities of the alkali-metals shown in Table 1. The values are relative to the independent measurement of the electron affinity of potassium.[4] Included in the table for comparison are values from another investigation using a similar technique;[5] these are refined values of data reported in Vol. 3, Chapter 1.

---

[1] A. J. Jackson, C. Tate, T. E. Gallon, P. J. Bassett, and J. A. D. Matthew, *J. Phys. (F)*, 1975, **5**, 363.
[2] L. N. Mazalov, V. U. Murakhtanov, and T. I. Guzhavina, *Zhur. strukt. Khim.*, 1975, **16**, 49.
[3] A. Barrie and F. J. Street, *J. Electron Spectroscopy Related Phenomena*, 1975, **7**, 1.
[4] A. Kasdan and W. C. Lineberger, *Phys. Rev. (A)*, 1974, **10**, 1658.
[5] H. J. Kaiser, E. Heinicke, R. Rackwitz, and D. Feldmann, *Z. Phys.*, 1974, **270**, 259.

**Table 1** Electron affinities/kJ of alkali-metal atoms

| | | |
|---|---|---|
| Li | 59.75±0.67 | 58.88±1.93 |
| Na | 52.82±0.39 | 51.88±1.93 |
| K  | 48.31±0.05 | 47.90±1.93 |
| Rb | 46.84±0.28 | 47.23±1.93 |
| Cs | 45.30±0.28 | 44.34±1.93 |

The chemistry of liquid metals, particularly the alkali metals, has been reviewed in an article which covers the physical properties, the solution and solvation ability for both metallic and non-metallic solutes, chemical interactions in solution, surface reactions, corrosion aspects, and techniques.[6] The need for high-purity liquid metals is clear and in the case of lithium, this is conveniently achieved by gettering. High-purity liquid lithium has been prepared by gettering with yttrium and cerium. The metal was subsequently used to prepare lithium hydride which contained only 30 p.p.m. oxygen. Yttrium decreased the oxygen concentration of the metal to 15 p.p.m. and yttrium was more suitable than cerium which corroded the molybdenum container more rapidly. The carbon concentration in liquid lithium, however, was not affected by exposure to either yttrium or cerium.[7] The purity of liquid lithium influences its electrical conductivity or resistivity. The resistivity of liquid and solid lithium contained in a stainless steel capillary has been determined using a Kelvin–Wheatstone bridge. The values are given by the equations:

$$\rho \Omega \, m = 16.476 \times 10^{-8} + 4.303 \times 10^{-10} \, \theta_c - 2.297 \times 10^{-13} \, \theta_c^2$$
$$180 \leq \theta_c \leq 460 \,°C$$

and

$$\rho \Omega \, m = 8.685 \times 10^{-8} + 3.261 \times 10^{-10} \, \theta_c + 1.821 \times 10^{-13} \, \theta_c^2$$
$$15 \leq \theta_c \leq 180 \,°C$$

Lithium which had been purified by filtration followed by gettering with titanium and yttrium at 753 K had a lower resistivity than metal purified by other methods. For the liquid, $d\rho/d\theta_c$ was positive but decreased with increasing temperature, whereas for the solid, the value increased with increasing temperature.[8] The resistance of dissolved oxide and hydride impurities in eutectic alloys of sodium and potassium appears to be a complex function of the concentration of each impurity, which can be attributed to chemical interaction in the metal to form hydroxide. Dissolved hydride causes a considerable increase in the resistance of the alloy but hydroxide has a much smaller effect. Dissolved lithium hydride affects the resistance of the alloy more than does sodium or potassium hydride but, again, hydroxide, as lithium hydroxide, has a smaller effect.[9] Information on the solubility of lithium salts in liquid lithium has been critically reviewed. Recommended solubilities are provided for solutions of oxide and nitride as

---

[6] C. C. Addison, *Chem. in Britain*, 1974, **10**, 331.
[7] H. C. Weed and O. H. Krikorian, *J. Nuclear Mater.*, 1974, **52**, 142.
[8] G. K. Creffield, M. G. Down, and R. J. Pulham, *J. C. S. Dalton*, 1974, 2325.
[9] M. N. Arnol'dov, M. N. Ivanovskii, A. D. Pleshivtsev, V. I. Subbotin, and B. A. Shmatko, *Fluid Mech. Soviet Res.*, 1974, **3**, 18.

follows:

$$\log_{10}(\text{mol\% O}) = 2.778 - 2740/T \qquad 523 \leq T \leq 908 \text{ K}$$
and
$$\log_{10}(\text{mol\% N}) = 5.088 - 3540/T \qquad 523 \leq T \leq 673 \text{ K}$$

No overall solubility trend could be discerned; at temperatures >673 K, solubilities decrease in the order: N>H>C>O>F, I>Cl>F and Si>C>Ge. The solubilities are compared with those for the corresponding salts in sodium. Carbon, nitrogen, and hydrogen are more soluble in lithium; oxygen solubility is similar; fluorine and germanium are less soluble in lithium.[10] A thermodynamic analysis has been made of the interaction of lithium vapour with the refractory oxides BeO, CaO, MgO, $Al_2O_3$, $SiO_2$, $Sc_2O_3$, and $Y_2O_3$ at 1800—2300 K and at vapour pressures of lithium from 0.01 to 760 Torr. With $Al_2O_3$, the ternary oxide $LiAlO_2$ was taken into account; the most resistant refractory oxides were BeO and $Y_2O_3$.[11] Lithium isotope ratios can be determined by atomic absorption by using $^6Li$ and $^7Li$ hollow cathode lamps and an air–acetylene flame. The method is based on the isotope shift of the line at 670.8 nm, which amounts to 0.015 nm and gives a precision better than 0.5 mol% $^6Li$.[12] Physical properties such as electrical, thermal, mechanical, and magnetic properties of the alkali metals are covered in a review which also discusses corrosion by these metals.[13] This is taken up again for liquid sodium in a review of corrosion and mass transport and their effects on the mechanical properties of structural materials.[14] Corrosion by oxygen of austenitic stainless steel is enhanced by caesium. In caesium vapour at between $10^{-2}$ and 1 atm pressure, the enhanced attack takes the form of surface oxidation, intergranular oxidation, and deep penetration to produce local high caesium concentrations.[15] The corrosion resistance of several metals, Ag, Cu, Be, Ti, Zr, Nb, V, Cr, Mo, W, Fe, and Ni towards caesium has been estimated from their solubilities, determined by weight changes, in the liquid alkali metal.[16]

Methods of analysing for dissolved salts in the liquid alkali metals continue to be of interest. A variation of the electrochemical oxygen meter is described which employs a solid oxide electrolyte ($ThO_2$–10% $Y_2O_3$) and a reference gas oxygen or oxygen–argon mixture and which allows continuous measurement of oxygen concentrations in liquid sodium.[17] Carbon in sodium may be determined by gamma-photon activation analysis in which the sample is $\gamma$-irradiated for ca. 5 min at 35 MeV and 50 $\mu$A. Carbon-11 was separated by heating the irradiated sample in oxygen with sodium carbonate as carrier followed by treatment with 5M sulphuric acid. The evolved gases were passed through zirconium oxychloride in 4M sulphuric acid to remove interfering $^{18}F$, and the carbon dioxide was trapped out. Recovery of $^{11}C$ was 86.6% on 0.22—1.2 $\mu$g of carbon per g sodium.[18] The solubility of carbon in sodium has been determined by

[10] P. F. Adams, P. Hubberstey, and R. J. Pulham, *J. Less-Common Metals*, 1975, **42**, 1.
[11] B. F. Yudin, S. A. Lapshin, and Yu. A. Polonskii, *Zhur. priklad. Khim.*, 1975, **48**, 305.
[12] H. S. Raede, *Atomic Absorption Newsletter*, 1974, **13**, 81.
[13] R. Fromageau, 'Monographe Metal Haute Purete', ed. G. Chaudron. Masson, Paris, 1972, Vol. 1, p. 39.
[14] R. S. Fidler and M. J. Collins, *Atomic Energy Rev.*, 1975, **13**, 8.
[15] J. E. Antill, K. A. Peakall, and E. F. Smart, *J. Nuclear Mater.*, 1975, **56**, 47.
[16] M. M. Godneva, N. D. Sedel'nikova, and E. S. Geizler, *Zhur. priklad. Khim.*, 1974, **47**, 2177.
[17] J. Jung, *J. Nuclear Mater.*, 1975, **56**, 213.
[18] F. Nordmann, C. Tinelli, and G. Engelmann, *Analysis*, 1974, **2**, 739.

exposing the liquid metal to $^{14}C$ dissolved as an unsaturated solution in nickel and counting the $^{14}CO_2$ produced by combustion of sodium samples in an oxygen–air mixture. The solubility is given by the equation:

$$\log(\text{wt. p.p.m.}) = 7.646 - 5970/T \qquad 763 \leq T \leq 1105 \text{ K}$$

The carbon is present in sodium probably as the acetylide anion, $C_2^{2-}$, except if oxygen or nitrogen is added when the amount of dissolved carbon is increased due to formation of carbonate and cyanide, respectively.[19] Methods of determining hydrogen in the liquid alkali metals are varied. Thus the determination of hydrogen in lithium is achieved by fusion extraction under vacuum in a bath of tin at 623—673 K. The amalgamation method can be used for hydrogen in sodium, but the vacuum distillation method is more appropriate for determining hydrogen (and oxygen) in the more volatile caesium.[20]

The rate of reaction of liquid sodium with hydrogen is affected by metal solutes. Although the rates of reaction of solutions of lithium (up to 8.6 mol%), barium (up to 6.1 mol%), strontium (up to 4.0 mol%), and mercury (up to 50 mol%) in sodium with hydrogen were all directly proportional to pressure up to 35 kN m$^{-2}$ and from 390 to 565 K, reaction was slower with mercury solutions and more rapid with lithium, strontium, and barium solutions than with sodium alone. The pre-exponential factor varied with the solute and its concentration, and the rate-determining step was considered to involve the hydrogen molecule. The differences in behaviour were related to possible changes in the energy profiles of conduction band electrons in the solutions.[21] The reaction of hydrogen with potassium has been studied by continuous injection of a jet of clean molten metal, circulated by a miniature electromagnetic pump, into hydrogen. Changes in hydrogen pressure over the range 0.3—22.2 kN m$^{-2}$ at 483—606 K showed that the reaction obeyed first-order kinetics with an apparent activation energy of 66.5 kJ mol$^{-1}$, which is slightly lower than the corresponding value for sodium. At higher temperatures the process was complicated by the gradually increasing dissociation pressure of potassium hydride.[22]

## 2 Alloys and Intermetallic Compounds

The crystal structure of LiAl has been confirmed by neutron diffraction to be of the NaTl-type with $a = 6.37$, $c = 1.76$ Å and $Z = 8$.[23] The crystal structure of LiGaGe belongs to the hexagonal system with $a = 4.175$, $c = 6.78$ Å, space group $P6_3mc$, $d(\text{calc.}) = 4.84$, $d(\text{expt.}) = 4.55$, and $Z = 2$. In addition to LiGaGe, the fcc phases $Li_2GaGe_{0.8}$, $Li_2GaGe_{0.2}$ and $LiGaGe_{0.2}$ with larger homogeneity ranges have been found by X-ray diffraction studies. The homogeneity range of LiGaGe is limited at the composition $Li_{1.4}GaGe$; the Ga and Ge atoms form a wurtzite lattice, the octahedral sites of which are occupied by the Li atoms.[24] The

---

[19] R. Ainsley, L. P. Hartlib, P. M. Holroyd, and G. Long, J. Nuclear Mater., 1974, **52**, 255.
[20] E. D. Malikova, L. L. Kunin, and Kh. N. Evzhanov, 'Internationale Congresse Hydrogene Metal', Paris, 1972 (Pub. 1973), Vol. 1, p. 184.
[21] M. R. Hobdell and A. C. Whittingham, J.C.S. Dalton, 1975, 1591.
[22] G. Parry and R. J. Pulham, J.C.S. Dalton, 1975, 446.
[23] K. Kuriyama and N. Masaki, Acta Cryst., 1975, **B31**, 1793.
[24] W. Bockelmann and H. U. Schuster, Z. anorg. Chem., 1974, **410**, 233.

structure of $Li_{13}Si_4$, earlier reported as $Li_7Si_2$, has been redetermined and the structural relationship between $Li_{13}Si_4$, $Li_7Si_2$, and $Li_3Sb$ discussed.[25] A new lithium germanide, $Li_{11}Ge_6$, has been reported: it crystallizes in the orthorhombic space group *Cmcm* with $a = 4.38$, $b = 24.45$, and $c = 10.64$ Å. The germanium sub-lattice consists of isolated, planar, five-membered germanium rings.[26] Information concerning compounds of tin with the alkali and alkaline-earth metals has been published in handbook form.[27] The new compound $Li_{13}Sn_5$ has been reported to have the trigonal structure, space group *P3m1* and lattice parameters $a = 4.70$, $c = 17.12$ Å.[28] Also $Li_7Sn_3$ has been synthesized by melting the stoicheiometric proportions of the elements in a tantalum crucible at 1173 K under argon and annealing at 753 K. This intermetallic compound is monoclinic with space group $P2_1/m$ and $a = 9.45$, $b = 8.56$, $c = 4.72$ Å, $\gamma = 105.95°$, $d$(expt.) $= 3.72$, and $d$(calc.) $= 3.67$.[29] Other lithium–tin compounds reported are $Li_5Sn_2$ and $Li_7Sn_2$. The former is trigonal/rhombohedral with space group $R\bar{3}m$, $a = 4.74$, $c = 19.83$ Å, $d$(expt) $= 3.5$, and $d$(calc.) $= 3.54$ for $Z = 3$, and the latter is orthorhombic with space group *Cmmm* and $a = 9.80$, $b = 13.8$, $c = 4.75$ Å, $d$(expt.) $= 2.99$, and $d$(calc.) $= 2.96$ for $Z = 4$. Structural relationships between $Li_5Sn_2$ and $Li_2Si$ and $Li_9Ge_4$ have been discussed, together with the close relationship between the structures of $Li_7Sn_2$ and $Li_7Pb_2$.[30,31] The high stability of many intermetallic gaseous compounds has been explained in terms of the Pauling concept of the polar bond. No intermetallic gaseous compounds of lithium, however, have yet been observed although stable molecules might be expected between lithium and Group IB elements owing to the comparatively large difference in electronegativity. Such molecules have been investigated by mass spectrometry and their dissociation energies determined. For CuLi, AgLi, and AuLi, the dissociation energies, $D_0^0$, are 189.3, 173.6, and 280.8 kJ mol$^{-1}$. These values are much lower than those calculated using the Pauling model.[32]

The sodium–lithium phase diagram has been redetermined over the entire composition range by a combination of resistance and thermal methods. Each method is particularly effective for specific parts of the diagram. Two liquid phases separate below the new consolute temperature 578 K and composition 63 mol% Li. The two-liquid immiscibility boundary extends from 10.1 to 97.0 mol% Li at the monotectic temperature 443.90 K. The eutectic occurs at 3.0 mol% Li and 365.25 K. Positive deviation from ideality is demonstrated for both sodium- and lithium-rich solutions.[33] The tendency to separate into two immiscible liquids makes itself apparent above 578 K in the temperature coefficient of resistivity of these solutions. Above 623 K, $d\rho/dT$ changes smoothly from sodium to lithium, but at 580—588 K the coefficient shows a maximum at *ca.* 68 mol% Li which is attributed to incipient immiscibility. A capillary method was

[25] U. Frank, W. Mueller, and H. Schaefer, *Z. Naturforsch.*, 1975, **30b**, 10.
[26] U. Frank and W. Mueller, *Z. Naturforsch.*, 1975, **30b**, 313.
[27] Gmelin's Handbook of Inorganic Chemistry, System No. 46: Tin, Pt. C, Sect. 3: Compounds with Alkali and Alkaline Earth Metals. 8th Edn.
[28] U. Frank and W. Mueller, *Z. Naturforsch.*, 1975, **30b**, 316.
[29] W. Mueller, *Z. Naturforsch.*, 1974, **29b**, 304.
[30] U. Frank, W. Mueller, and H. Schaefer, *Z. Naturforsch.*, 1975, **30b**, 1.
[31] U. Frank, W. Mueller, and H. Schaefer, *Z. Naturforsch.*, 1975, **30b**, 6.
[32] A. Neubert and K. F. Zmbov, *J.C.S. Faraday I*, 1974, **70**, 2219.
[33] M. G. Down, P. Hubberstey, and R. J. Pulham, *J.C.S. Dalton*, 1975, 1490.

used to determine the resistivity from 373 to 723 K. A parabolic relationship was obtained over the full concentration range apart from a small deviation near the sodium axis. The excess resistivity of the solutions over that of the linear interpolation between the pure metals was small compared with other mixtures of alkali metals with sodium.[34] These phase relationships and resistivities are largely confirmed by a parallel study[35] which is augmented by Knight shift measurements of $^{23}$Na and $^{7}$Li nuclei in liquid sodium–lithium solutions.[36] Measurements of the absorption and velocity of ultrasound (15—105 MHz) in liquid Na–K and Na–Rb alloys from 268 K (or from the liquidus temperature) to 523 K has disclosed a slight peak in the plot of the excess absorption *versus* concentration for Na–K and a corresponding substantial peak for Na–Rb similar to that found previously for Na–Cs. These peaks, like those for $d\rho/dT$ in Na–Li, are attributed to a forewarning of immiscibility in the single-phase region.[37] A new phase, $Na_2Cs$, which has a hexagonal Laves-type structure with $a = 7.861$ and $c = 13.062$ Å at 223 K, has been detected in the Na–Cs system. The radius ratio (calculated as $r_{Cs} : r_{Na} = 1.43$) deviates considerably from the ideal value, 1.225, leading to a shortened Cs—Cs distance of 4.86 Å.[38]

The electrical resistivities of solutions of germanium (up to 1.0 mol%), tin (up to 6.9 mol%), and lead (up to 7.8 mol%) in liquid sodium have been determined over the temperature ranges 473—723, 473—673, 373—573 K respectively. All three solutes, especially tin, increase the resistivity of sodium by a relatively large amount (Table 2). The data in Table 2 are of value in the study of chemical reactions in sodium solutions by resistometric methods, *e.g.* a high value enhances the accuracy when determining the rate of depletion of solute from the alloy during a selective reaction with a gas. As might be expected from the general chemistry of germanium and tin, dilute solutions containing both elements in liquid sodium show additive resistivities consistent with lack of association between the two solutes in this medium.[39] Resistivity has been further exploited to determine the solubilities of germanium (up to 1.0 mol%) and tin (up to 6.9 mol%) in liquid sodium. Tin (and lead) are much more soluble than germanium. At 563 K, for example, the solubilities are 6.55, 2.05, and 0.28 mol% for lead, tin, and germanium, respectively. Partial molar enthalpies and entropies of solution in sodium for tin (with respect to $Na_{15}Sn_4$) and germanium (with respect to NaGe) are 46.8 and 44.3 kJ mol$^{-1}$ and 50.7 and 29.7 J K$^{-1}$ mol$^{-1}$, respectively. Use of a simple thermochemical cycle demonstrates that metallic tin

**Table 2** *The increase in the resistivity* ($\rho/\Omega$ m $\times 10^8$) *of sodium caused by one atom percent of solute at 573 K*

| | | | | | | |
|---|---|---|---|---|---|---|
| Li | 0.20 | | | | | |
| K | 1.4 | | | | | Ge 8.60 (673 K) |
| Rb | 3.4 | Sr 0.20 | Ag 2.8 | Cd 5.4 | | Sn 11.85 |
| Cs | 4.2 | Ba 2.47 | Au 5.0 | Hg 4.1 | | Pb 11.20 |

[34] M. G. Down, P. Hubberstey, and R. J. Pulham, *J.C.S. Faraday I*, 1975, **71**, 1387.
[35] P. D. Feitsma, J. J. Hallers, F. B. D. Werff, and W. Van der Lugt, *Physica (B, C)*, 1975, 35.
[36] P. D. Feitsma, G. K. Slagter, and W. Van der Lugt, *Phys. Rev. (B)*, 1975, **11**, 3589.
[37] J. E. Amarat and S. V. Letcher, *J. Chem. Phys.*, 1974, **61**, 92.
[38] A. Simon and G. Ebbinghaus, *Z. Naturforsch.*, 1974, **29b**, 616.
[39] P. Hubberstey and R. J. Pulham, *J.C.S. Faraday I*, 1974, **70**, 1631.

and lead dissolve exothermically but that carbon dissolves endothermically. The cycle is also used to derive solvation enthalpies for the elements Pb, Sn, and C of $-248$, $-306$, and $-607$ kJ mol$^{-1}$, respectively.[40] The crystal structure of $Na_{3.7}Sn$, earlier reported as $Na_{15}Sn_4$ and before that as $Na_4Sn$, has now been determined. The crystals are orthorhombic, with space group *Pnma*, and $a = 9.82$, $b = 5.57$, and $c = 22.79$ Å. The tin atoms are isolated with respect to each other and the packing of atoms is less dense than in $Na_{15}Pb_4$.[41]

An e.m.f. technique incorporating the $\beta$-alumina electrolyte has been used to measure the thermodynamics of Na–Pb solutions from 623 to 773 K and from mol fraction $x_{Na} = 0.044$ to 0.78. The results are good agreement with many previous determinations.[42] Activity measurements for potassium and mercury in Na–Hg solutions from 473 to 573 K by a similar e.m.f. method provide a dissociation constant for $KHg_2$ of $2.75 \times 10^{-8}$ at $5 \times 10^{-4}$ mol fraction potassium and a free energy of formation of $-19.7$ kcal mol$^{-1}$.[43]

## 3 Solvation of Alkali-metal Ions

The theory of hydrogen bonding in water and ion hydration has been the topic for a conference in which the structures and energies of $M(H_2O)^+$ (M = Li, Na, or K) and $X(H_2O)^-$ (X = F or Cl) were discussed.[44] Electric permittivity, dipole moments, and structure in solutions of ions and ion pairs have also been reviewed.[45]

The hydration energies of lithium chloride, sodium chloride, and potassium chloride in their saturated aqueous solutions at 298 K are reported to be 12.1, 19.4, and 22.8 kcal mol$^{-1}$, respectively. These relatively low values are attributed to ion pairing in the saturated solution rather than dissociation into separate ions which occurs on dilution. The values are about an order lower than the additive value of a metal cation and the chloride anion hydration energies.[46] Calculations on lithium hydrates predict that the preferential first co-ordination sphere of $Li^+$ is tetrahedral.[47] The phase diagrams of each of the nitrites and nitrates of lithium, sodium, and potassium with water have been determined. The characteristics of singular points of each diagram were correlated with the properties of each electrolyte. The disordering effect of electrolyte on the structure of free water appears to increase in the order $KNO_3 < NaNO_3 < LiNO_3$; and $NaNO_2 \leqslant KNO_2 \ll LiNO_2$. The effect of any cation on the water structure also depends on the nature of the anion and account must be taken of ion pairing which is a function of both the nature of the ions and their attraction for water molecules.[48] By measuring the spin–lattice relaxation times of protons in aqueous solutions of sodium and potassium hydroxides, the hydration numbers ($n$) of $Na^+$, $K^+$, and $OH^-$, and their

---

[40] P. Hubberstey and R. J. Pulham, *J.C.S. Dalton*, 1975, 1541.
[41] W. Mueller and K. Volk, *Z. Naturforsch.*, 1975, **30b**, 494.
[42] D. J. Fray and B. Savory, *J. Chem. Thermodynamics*, 1975, **7**, 485.
[43] L. F. Kozin, M. B. Dergacheva, and N. G. Almazova, *Izvest. Akad. Nauk Kaz. S.S.R., Ser. khim.*, 1974, **24**, 68.
[44] P. Schuster, 'Structure of Water in Aqueous Solutions, Proceedings International Symposium', 1973 (Pub. 1974), p. 141.
[45] E. Grunwald, S. Highsmith, and T.-P. I, *Ions, Ion Pairs, Org. Reactions*, 1974, **2**, 447.
[46] Ya. G. Goroshchenko, *Zhur. fiz. Khim.*, 1974, **48**, 1256.
[47] P. A. Kollman and I. D. Kuntz, *J. Amer. Chem. Soc.*, 1974, **96**, 4766.
[48] A. A. Ennan and V. A. Lapshin, *Zhur. fiz. Khim.*, 1975, **49**, 119.

mobilities have been determined for various OH⁻ concentration. For $Na^+$, $n=6$ and 18; for K, $n=8$ and 24, depending on OH⁻ concentration. The anomalous mobility and chemical shift of $H^+$ and $OH^-$ are explained in terms of the structure of their hydration complexes.[49] The solvation of an ion in a binary solvent mixture has been treated in terms of consecutive equilibria in which the free energy of solvent exchange varied as the amount of one component in the solvation shell increased. Derived equations were used to fit observed n.m.r. chemical shifts and u.v. peak shifts in binary solvents such as DMSO–H₂O and MeCN–MeOH. The observed behaviour was classified in terms of two parameters, one of which was dependent on solvation number. It was found that in DMSO–H₂O mixtures, the ions $Cs^+$, $NO_3^-$, and $I^-$ were preferentially solvated by DMSO but $Li^+$ was slightly selectively solvated by H₂O. In MeCN–H₂O, the ions $Na^+$, $Cl^-$, $Br^-$, and $I^-$ were all preferentially hydrated.[50] The ultrasonic relaxation observed in solutions of lithium chloride in N,N-dimethylacetamide (DMA) has been attributed to the formation of solvated ion pairs. On introducing water, the ternary system generated an additional equilibrium involving hydrated $Li^+$ and $Cl^-$ ions. No relaxation was observed in either system on replacing LiCl with $LiClO_4$.[51] Interactions between water at low concentrations and 1:1 electrolytes have been studied in dipolar aprotic solvents of increasing basicity *i.e.* sulpholane (TMS) < MeCN < propylene carbonate < DMSO, using n.m.r., calorimetry, and vapour pressure measurements. Results are interpreted in terms of hydration constants and hydration enthalpies for the 1:1 complexes with $Li^+$, $Na^+$, $Ag^+$, $H^+$, $Cl^-$, and $NO_3^-$ in these solvents. The hydration constants are rather insensitive to the nature of the ion but sensitive to the nature of the solvent. For DMSO, all hydration constants are low, thus accounting for the ability of this solvent to retain its dipolar aprotic character in the presence of some water. The enthalpy changes are low and nearly equal for replacement of one MeCN or PC molecule by one H₂O molecule in the solvation shells of $Li^+$ and $Cl^-$ ions. The calculated solvation enthalpies, $\Delta H_s$(kcal mol⁻¹), in PC are $Li^+(H_2O)$, −101; $Na^+(H_2O)$, −83; and $K^+(H_2O)$, −75 (estimated); compared with those in water of $Li^+$ −124 and $Na^+$ −97.[52]

The proceedings of a conference on metal–ammonia solutions have been published, featuring reviews of the physical properties of dilute and concentrated solutions, electrical, n.m.r., i.r., and Raman spectroscopic studies of diffusion, the solvated electron, kinetics, and solution structure.[53] Electron spin resonance in metallic Li–NH₃ systems has been investigated from 12° to 296 K. In the liquid solutions and in the cubic phase of $Li(NH_3)_4$ the conduction e.s.r. lineshapes are in agreement with theory. To a good approximation the solvated ions are the only spin scatterers in the liquid state. The paramagnetic susceptibility of liquid $Li(NH_3)_4$ indicates that the concentration of localized moments is low and they order antiferromagnetically below 20 K.[54]

[49] L. V. Bertyakova, A. M. Polyakov, and L. G. Romanov, *Izvest. Akad. Nauk S.S.R., Ser. Fiz-Mat.*, 1975, **13**, 7.
[50] A. K. Covington and J. M. Thain, *J.C.S. Faraday I*, 1974, **70**, 1879.
[51] M. J. Adams, C. B. Baddiel, R. J. Jones, and A. J. Matheson, *J.C.S. Faraday II*, 1974, **70**, 1114.
[52] R. L. Benoit and S. Y. Lam, *J. Amer. Chem. Soc.*, 1974, **96**, 7385.
[53] 'Electrons in Fluids: the nature of metal-ammonia solutions'. Colloque Weyl 3rd 1972, ed. J. Joshua, Springer, New York, 1973.
[54] W. S. Glaunsinger and M. J. Sienko, *J. Chem. Phys.*, 1975, **62**, 1873.

The vapour pressures of solutions of lithium in methylamine have been determined from 218 to 278 K. The activity of the amine showed positive deviations from Raoult's law at low metal concentrations and negative deviation with increasing concentration as in metal-ammonia solutions. The system was considered to contain solvated metal atoms coexisting with free $MeNH_2$ molecules.[55]

An apparatus has been developed for the preparation of solutions of lithium, sodium, and potassium in HMPA, and their conductivities have been measured as a function of time and concentration. The conductivities appear to reflect the balance between the rates of metal dissolution and decomposition of the solvated electron and, in the case of lithium and potassium, electron pairing giving enhanced conductivity by electron drift. Spin pairing and the existence of ion trios of one $Na^+$ ion with two solvated electrons was indicated by e.s.r. measurements.[56] Aggregation of the solvated electron with $Li^+$ ions in THF forms an ion pair ($Li^+$, $e_s^-$) which has been investigated by pulse radiolysis. The ion pair has a near-i.r. absorption band with $\lambda_{max} = 1180$ nm and molar extinction coefficient of $2.28 \times 10^4$ l mol$^{-1}$ cm$^{-1}$. Rates of reaction of this species with anthracene, biphenyl, and dibenzyl mercury have been determined which are compared with those for the reactions of ($Na^+$, $e_s^-$) and $e_s^-$ in THF.[57] The optical absorption maximum in the spectrum of the solvated electron at room temperature in amides and amines is reported as follows: DMF, 1680 nm; DMA, 1800 nm; $NN$-diethylformamide, 1775 nm; dimethylamine, 1950 nm; and 1,2-diaminopropane, 1500 nm. The rate constants for the reactions in ethanol of $e_s^-$ with formamide, methylformamide, and DMF are $1.8 \times 10^8$, $2.2 \times 10^8$, and $1.1 \times 10^9$ l mol$^{-1}$ s$^{-1}$, respectively.[58] On dissolving the alkali metals Na, K, Rb, and Cs in binary solutions of ethylenediamine and HMPA at room temperature, the optical absorption spectrum shows a maximum for $e_s^-$ between the maxima reported for the single solvents. The peak position for the $M^-$ species was dependent on both the metal and solvent composition. The spectra of $Na^-$, $K^-$, $Rb^-$, and $Cs^-$ obey criteria for a CTTS transition.[59] Photoionization cross-sections have been determined, probably for the first time, of solvated electrons (at photon energies of 1.2—4.1 eV) and the $Na^-$ species (at photon energies of 2.8—4.1 eV) in HMPA.[60] The $^{23}Na$ n.m.r. spectra of solutions of $NaL^+,Na^-$ (L = 4,7,13,16,21,25-hexaoxa-1,10-diazabicyclo[8,8,8]hexacosane) in ethylamine from 271.45 to 274.45 K show a signal which is unshifted from one solvent to another and which indicates the presence of a genuine $Na^-$ anion and not a complex of the solvated cation and an electron pair. The $Na^-$ anion probably exists in solution as a large solvated anion with two electrons in the 3s orbital. The solutions employed were made directly from the ligand and an excess of sodium in ethylamine, and also by dissolving $NaL^+,L^-$ crystals in the appropriate solvent. Both methods gave identical n.m.r. spectra.[61]

---

[55] Y. Nakamura, Y. Horie, and M. Shimoji, *J.C.S. Faraday I*, 1974, **70**, 1376.
[56] N. Gremmo and J. E. B. Randles, *J.C.S. Faraday I*, 1974, **70**, 1480.
[57] B. Bockrath and I. M. Dorfman, *J. Phys. Chem.*, 1975, **79**, 1509.
[58] J. F. Gavlas, F. Y. Jou, and L. M. Dorfman, *J. Phys. Chem.*, 1974, **78**, 2631.
[59] P. Childs and R. R. Dewald, *J. Phys. Chem.*, 1975, **79**, 58.
[60] H. Aulich, L. Nemec, and P. Delahay, *J. Chem. Phys.*, 1974, **61**, 4235.
[61] J. M. Ceraso and J. L. Dye, *J. Chem. Phys.*, 1974, **61**, 1585.

Evidence for ion pairing and solvation in liquid ammonia comes from Raman spectra of solutions of potassium, sodium, ammonium, calcium, and silver nitrate solutions. Digital Raman spectra of solutions of $NaNO_3(1—5.5\ mol\ l^{-1})$, $KNO_3(1—3\ mol\ l^{-1})$, $NH_4NO_3(6.4, 8.7\ mol\ l^{-1})$, $Ca(NO_3)_2(1—2.6\ mol\ l^{-1})$ and $AgNO_3(1—4\ mol\ l^{-1})$ have been analyzed by least-squares curve fitting. The perturbations of the $NO_3^-$ ion spectra clearly indicate the presence of ion pairs in all the solutions. The effects are similar to but simpler than those observed for concentrated aqueous solutions, indicating that only one environment is important in all except $Ca(NO_3)_2$ solutions. Direct cation–anion contact is indicated and the splitting of $\nu_4$ of $NO_3^-$ correlates well with the ionic potential of the cations. The failure to observe $\nu_4$ splitting for $K^+$, $NH_4^+$, and $Ag^+$ cations is attributed to their low polarizing power. The splitting of $NO_3^-$ $\nu_3$ by ion pairing in ammonia is considerably less than that observed in water. The $\nu_3$ splittings of the univalent electrolytes show an inverse correlation with the heats of ammoniation of the cations except in the case of the small $Li^+$ ion, suggesting that charge transfer from ammonia to metal ion decreases the cation polarizing power.[62] A comparison of the i.r. spectra in the $(\nu_1, \nu_3)$ region for $M^+ClO_3^-$ ion pairs (M = Na, K, or Li) in xenon, glassy water, or ammonia matrices shows that association with the water matrix reduces the $\nu_3$ splitting, a measure of the anion distortion, whereas the $\nu_1$ frequency increases, the magnitude of both effects being ca. 15 cm$^{-1}$. Both observations are definite indications of a reduced charge density at the cation, and the decrease in the $\nu_3$ splitting suggests that this reduction is extensive. The $\nu_3$ splitting of $K^+ClO_3^-$ is nearly eliminated in the ammonia matrix.[63]

Chemical shifts in the n.m.r. spectrum of $^7Li$ in solutions of the salts $LiClO_3$, LiCl, LiBr, LiI, LiI$_3$, and LiBPh$_4$ in several solvents have been measured using aqueous 4M-LiClO$_4$ as reference. The shifts range from +2.80 p.p.m. in acetonitrile to −2.54 p.p.m. in pyridine. In DMSO and DMF, no evidence for contact ion pairing is observed. Formation of ion pairs is particularly evident, however, in THF, nitromethane, and tetramethylguanidine. The large broadening of the $^{35}Cl$ resonance of the perchlorate anion in these solvents is in agreement with these deductions. In contrast to $^{25}Na$ n.m.r., no correlation was found between limiting chemical shifts in different solvents and the Gutmann donor numbers of these solvents.[64] The variations in the n.m.r. chemical shifts of $^{23}Na$ in sodium tetraphenylborate, with mole fraction, of the binary solvent mixtures of any two of nitromethane, acetonitrile, DMSO, tetramethylurea, HMPA, and pyridine, indicated preferential solvation of the $Na^+$ ion. In DMSO and in HMPA, introduction of pyridine destroyed the polymeric structure by a dipole interaction to give enhanced donicity of DMSO and HMPA.[65] The electrical conductivity of the salts LiCl, LiBr, LiNO$_3$ and LiClO$_4$ in methanol has been determined. A quasi-crystalline diffuse lattice model of solvated ions was combined with a method of calculating the electrostatic contribution to the activity coefficient to derive solvation numbers of 10.7, 10.8, 10.8, and 9.2 for LiCl, LiBr, LiNO$_3$, and LiClO$_4$, respectively.[66]

[62] J. W. Lundeen and R. S. Tobias, *J. Chem. Phys.*, 1975, **63**, 924.
[63] N. Smyrl and I. P. Devlin, *J. Chem. Phys.*, 1974, **61**, 1596.
[64] Y. M. Cahen, P. R. Handy, E. T. Roach, and A. I. Popov, *J. Phys. Chem.*, 1975, **79**, 80.
[65] M. S. Greenberg and A. I. Popov, *Spectrochim. Acta*, 1975, **31A**, 697.
[66] P. A. Skabichevskii, *Zhur. fiz. Khim.*, 1975, **49**, 181.

The solvation of the salts LiI, NaI, NH$_4$I, KI, NaBr, and NaCl in methanol at 298 K has been investigated using ultrasound from infinite dilution to saturation. The solvation numbers decreased with increasing salt concentration, and the rate of change in the solvation numbers increased with increasing cation mass. At infinite dilution, the solvation numbers depend linearly on the molecular weights of the salts.[67] I.r. absorption spectra (frequency 1000—1500 and 1650—3000 cm$^{-1}$) of solutions of lithium, sodium, and magnesium perchlorates in acetone at 298 K have been measured and used to determine the association constants and solvation numbers of the cations. Interactions between acetone and cations are similar to those between acetonitrile and the same cations. The solvation numbers are 4, 4, and 6 for Li$^+$, Na$^+$, and Mg$^{2+}$, respectively. The association constants increase in the order NaClO$_4$ < LiClO$_4$ < Mg(ClO$_4$)$_2$, being considerably smaller (0.31, 0.16, and 1.31 l mol$^{-1}$, respectively,) than in acetonitrile.[68] The association constant and the enthalpy, $\Delta H$, and entropy, $\Delta S$, of ion-pair formation have been determined from shifts in the i.r. absorption bands displayed by solutions of sodium and lithium iodides in acetone from 183 to 318 K. Association constants increased with temperature for both salts. Values of $\Delta H$ and $\Delta S$ were $-0.25$, $-0.13$ kcal mol$^{-1}$ and 0.31, 0.09 cal mol$^{-1}$ deg$^{-1}$, respectively, for LiI and NaI. The association is less than in acetone–perchlorate and acetonitrile–iodide solutions.[69] I.r. spectra of solutions of alkali-metal cyanides (0.1 mol l$^{-1}$) show that in DMF and DMSO, the salts exist uniquely as associated ion pairs. The cations in the ion pair are solvated, and the anions exist in the non-solvated free ion state. No appreciable amounts of free ions are detectable other than in a solution of sodium cyanide containing a polycyclic ether and in a solution of tetraphenylarsonium cyanide.[70] Ebullioscopic molecular weight and conductance measurements have been carried out on the compounds LiAlH$_4$ and LiBH$_4$ and their alkoxy-derivatives in THF and diethyl ether. In THF the results are consistent with the formation of ion pairs and triple ions for LiAlH$_4$, NaAlH$_4$, LiBH$_4$, and for the larger alkoxy-derivatives of LiAlH$_4$ and LiBH$_4$. The alkoxy-group is believed to encourage formation of linear aggregates *via* oxygen-bridge bonds. Larger aggregates than the triple ion appear to be formed for LiAlH$_4$ and LiBH$_4$ in diethyl ether. The higher stereoselectivity of LiAlH$_4$ in the reduction of cyclic ketones is attributed to formation of a complex between lithium and the carbonyl group as the initial step in the reduction process.[71] Solvation of the Li$^+$ cation in several solvents has been studied by means of cation-sensitive glass electrodes. Mono- and di-solvated species were indicated in methanol, pyridine, formamide, methylformamide, and HMPA. Tri-solvated species were also formed in DMF, DMA, and DMSO. The complexing of the Li$^+$ ion was more pronounced than that of the Na$^+$ ion. When water was used, the equilibrium potential could not be established.[72] The free energies of solvation of the alkali-metal chlorides in unit molality AlCl$_3$–propylene carbonate (PC) solution have been determined

---

[67] A. M. Beznoshchenko, E. F. Ivanova, L. A. Petrenko, and Yu. A. Petrenko, *Zhur. fiz. Khim.*, 1975, **49**, 1351.
[68] I. S. Perelvgin and M. A. Klimchuk, *Zhur. fiz. Khim.*, 1975, **49**, 138.
[69] I. S. Perelvgin and M. A. Klimchuk, *Zhur. fiz. Khim.*, 1975, **49**, 164.
[70] A. Loupy and J. Corset, *Compt. rend.*, 1974, **279**, C, 713.
[71] E. C. Ashby, F. R. Dobbs, and H. P. Hopkins, *J. Amer. Chem. Soc.*, 1975, **97**, 3158.
[72] T. Nakamura, *Bull. Chem. Soc. Japan*, 1975, **48**, 1447.

**Table 3** *Free energies of solvation/kcal mol$^{-1}$ of the alkali-metal chlorides, $\Delta G°_{solv}$ (MCl), in 1M-AlCl$_3$–PC, and of alkali-metal ions, $\Delta G°_{solv}$(M), in 1M-AlCl$_3$–PC, PC, and H$_2$O at 298 K*

| Salt | $\Delta G°_{solv}$(MCl) | $\Delta G°_{solv}$(M)$_{AlCl_3-PC}$ | $\Delta G°_{solv}$(M)$_{PC}$ | $\Delta G°_{solv}$(M)$_{H_2O}$ |
|---|---|---|---|---|
| LiCl | −188.43 | −98.4 | −95.0 | −97.8 |
| NaCl | −166.64 | −76.6 | −71.9 | −72.4 |
| KCl | −149.60 | −59.6 | −56.6 | −54.9 |
| RbCl | −143.78 | −53.8 | −54.5 | −50.7 |
| CsCl | −141.70 | −51.7 | −52.7 | −46.7 |

from e.m.f. measurements. Although the chlorides, with the exception of LiCl, are insoluble in PC, addition of AlCl$_3$ forms a complex thereby rendering the halide soluble:

$$MCl + AlCl_3 = M^+ + AlCl_4^-$$

The free energies of solvation of the alkali-metal chlorides and the derived solvation enthalpies of the individual ions are given in Table 3. The free energies of transfer, $\Delta G°_t(M^+)$, estimated by comparison with aqueous data, are −0.6, −4.2, −4.7, −3.1, and −5.0 kcal mol$^{-1}$ for Li$^+$, Na$^+$, K$^+$, Rb$^+$, and Cs$^+$, respectively. The negative sign indicates that the cations are on a lower energy level and strongly solvated by the local negative charge of the solvent dipole.[73] The behaviour of the lithium salts of the tetraphenylene radical anion (T$^-$Li$^+$) and dianion (T$^{2-}$, 2Li$^+$) in THF has been investigated by spectroscopy. A strong absorption band occurs at 385 nm in the spectrum of T$^{2-}$, 2Li$^+$ which does not appear in the spectrum of T$^{2-}$, 2Na$^+$. This is tentatively attributed to a change-transfer transition involving a partially desolvated and tightly bound Li$^+$ cation. The dissociation of the radical anion and dianion and the disproportionations:

$$2T^-, Li^+ = T + T^{2-}, 2Li^+$$

and

$$T^- + T^-, Li^+ = T + T^{2-}, Li^+$$

have been investigated conductometrically and spectrophotometrically, respectively, over the range 203—293 K.[74]

### 4 Compounds containing Organic Molecules or Complex Ions

Co-ordination compounds of the alkali- and alkaline-earth metals with covalent characteristics have been reviewed.[75]

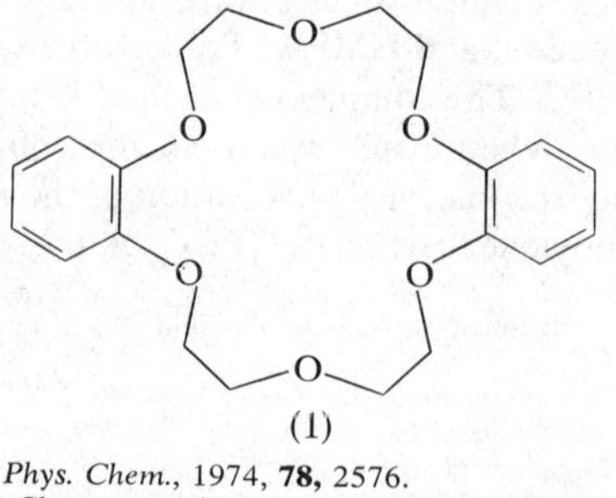

(1)

---
[73] J. Jorne and C. Tobias, *J. Phys. Chem.*, 1974, **78**, 2576.
[74] B. Lundgren, G. Levin, S. Claesson, and M. Szwarc, *J. Amer. Chem. Soc.*, 1975, **97**, 262.
[75] P. N. Kapoor and R. C. Mehrotra, *Coordination Chem. Rev.*, 1974, **14**, 1.

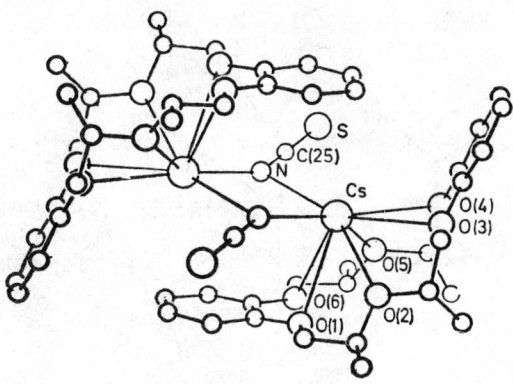

**Figure 1** *The centrosymmetric complex molecule of* (II)
(Reproduced from *J.C.S. Perkin II*, 1975, 261)

The crystal structures of complexes between alkali-metal salts and cyclic polyethers have been studied as models of the transport mechanism for alkali-metal cations across biological membranes.[76] Complex formation tends to enclose the cation in a lipophilic shell involving a change in conformation of the ligand. Changes in composition of the ligand may influence the ability to achieve the required conformation. Modification of dibenzo-18-crown-6 (1) by methyl substitution of four of the aliphatic carbon atoms gives five isomers (I), two of which have been designated F and G. Crystal structure determinations show that the caesium thiocyanate, CsNCS, complexes of F and G, designated (II) and (III), respectively, have methyl groups in the *cis,anti,cis,cis*-configuration relative to the benzene ring, and the *trans,anti,trans,trans*-configuration, respectively. The crystal structure of the F compound (II), (7R,9R,18S,20S)-tetramethyldibenzo-18-crown-6 caesium thiocyanate ($C_{24}H_{32}O_6$,CsNCS), is triclinic, with space group $P\bar{1}$ and $a = 11.12$, $b = 11.87$, $c = 13.46$ Å, $\alpha = 107.75$, $\beta = 95.25°$, $\gamma = 121.9°$, and $Z = 4$. This is a dimeric, centrosymmetric 1:1 complex with Cs—N—Cs bridges, and further Cs co-ordination by six ether-oxygen atoms. The complex molecule is shown in Figure 1, and bond lengths are shown in Figure 2. The crystal structure of the G compound, (III), bis-(18R,20R)-tetramethyldibenzo-18-crown-6 caesium thiocyanate, $C_{48}H_{64}O_{12}CsNCS$, is tetragonal with space group $P4_2/n$ and $a = 15.22$, $c = 11.14$ Å, and $Z = 2$. Compound (III) is a 1:2 complex consisting of [Cs(isomer G)$_2$]$^+$ cations with $\bar{4}$ symmetry as shown in Figure 3, and with 12 Cs—O contacts (3.12—3.36 Å), and disordered thiocyanate anions.[76]

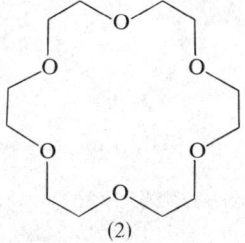

(2)

[76] P. R. Mallinson, *J.C.S. Perkin II*, 1975, 261.

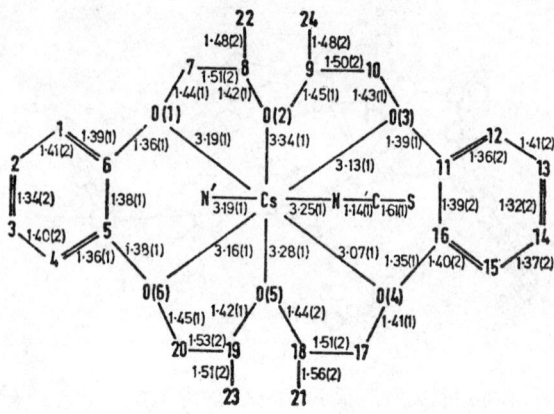

**Figure 2** Bond lengths/Å in (II)
(Reproduced from *J.C.S. Perkin II*, 1975, 261)

The structures of 1,4,7,10,13,16-hexaoxacyclooctadecane ($C_{12}H_{24}O_6$) (2) and its complexes with the thiocyanates of sodium, potassium, rubidium and caesium (and calcium) have been discussed. In the $K^+$, $Rb^+$, $Cs^+$, and $Ca^{2+}$ complexes, this unsubstituted hexaether adopts nearly $D_{3d}$ symmetry although the larger $Rb^+$ and $Cs^+$ ions are displaced by $>1$ Å from the mean plane of the ligand. In the $Na^+$ complex, the ring is more strongly distorted to accommodate the smaller cation. The shortening of the C—C bonds found in these and related complexes is judged to be mainly an artificial effect arising from inadequate treatment of curvilinear vibrations.[77] The structures of the complex thiocyanates

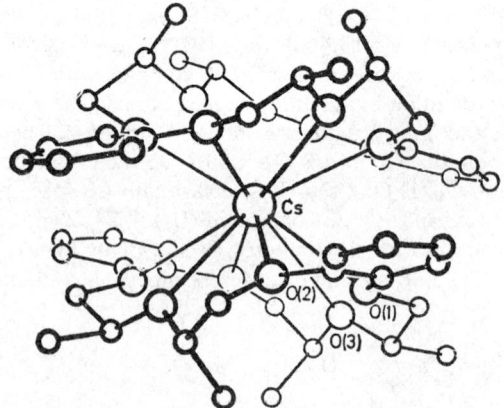

**Figure 3** *The complex cation of* (III), *with symmetry* $\bar{4}$. *The two ligands are enantiomerically related*
(Reproduced from *J.C.S. Perkin II*, 1975, 261)

[77] J. D. Dunitz, M. Dobler, P. Seiler, and R. P. Phizackerley, *Acta Cryst.*, 1974, **B30**, 2733.

# Elements of Group I

$C_{12}H_{24}O_6$,MCNS (M = K, Rb, or Cs) are monoclinic with space group $P2_1/c$ and values of $a$, $b$, $c$, $\beta$, and $Z$ of 8.190, 11.982, 12.063; 14.285, 8.347, 8.409; 7.775, 21.996, 22.370 Å; 99.19, 125.16, 125.50°; and 2, 4, 4, respectively.[78—80] Whereas in these structures the six oxygen atoms lie alternately above and below their mean plane so that the co-ordination polyhedron around the cation approximates to a very flat octahedron, in the hydrated sodium complex, $C_{12}H_{24}O_6$,NaNCS,$H_2O$, the hexaether adopts a highly irregular conformation in which five approximately coplanar atoms surround the sodium cation. A distorted pentagonal-bipyramidal co-ordination of the cation is completed by the remaining ether-oxygen and by a water molecule. Crystals are monoclinic, with space group $P2_1/c$ and $a = 12.316$, $b = 13.737$, $c = 11.215$ Å, $\beta = 105.32°$ and $Z = 4$, and are built from dimeric units formed by hydrogen bonding between the water molecules and the nitrogen atoms of the thiocyanate ions.[81]

Previous investigations have shown that 1,4,7,10-tetraoxacyclododecane complexes with alkali-metal halide salts possess $D_4$ symmetry, where the oxygen atoms of two cyclomer molecules occupy the vertices of a square antiprism centred on the cation. An X-ray diffraction study has established that the 2:1 octahydrate complex of 1,4,7,10-tetraoxacyclododecane with sodium hydroxide, $2C_8H_{16}O_4$,NaOH,$8H_2O$, retains to a large extent the structural principles of the halide series. The crystals have space group $Pbcn$ with $a = 14.794$, $b = 15.845$, $c = 23.960$ Å and $Z = 8$. The sodium ion forms an eight-co-ordinate sandwich complex of ca. $D_4$ symmetry with two polyether rings each with ca. $C_4$ symmetry. The oxygen co-ordination round sodium, shown in Figure 4, has two Na—O(10A and 10B) distances (2.441 and 2.459) significantly shorter than the other six

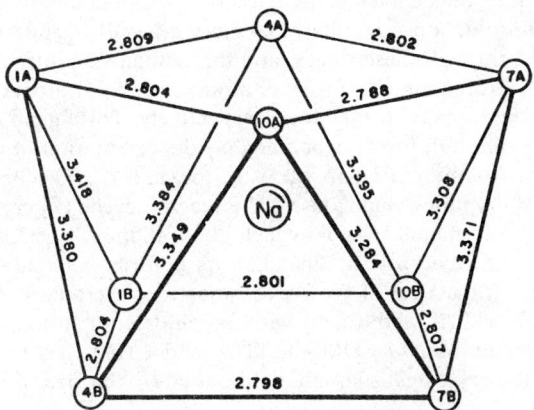

**Figure 4** *Dimensions of the square antiprism formed by the oxygen atoms in the* $[Na^+(C_8H_{16}O_4)_2]$ *unit*
(Reproduced by permission from *Inorg. Chem.*, 1974, **13**, 2826)

[78] P. Seiler, M. Dobler, and J. D. Dunitz, *Acta Cryst.*, 1974, **B30**, 2744.
[79] M. Dobler and R. P. Phizackerley, *Acta Cryst.*, 1974, **B30**, 2746.
[80] M. Dobler and R. P. Phizackerley, *Acta Cryst.*, 1974, **B30**, 2748.
[81] M. Dobler, J. D. Dunitz, and P. Seiler, *Acta Cryst.*, 1974, **B30**, 2741.

(average 2.497 Å). The hydroxide ion is associated with the water molecules to form infinite sheets of linked four- and five-membered rings. The configurations of the [Na$^+$(C$_8$H$_{16}$O$_4$)$_2$] units are virtually identical with those in [Na$^+$(C$_8$H$_{16}$O$_4$)$_2$][Cl$^-$,5H$_2$O], and the unusual feature of a crystal separating into alternating organic and aqueous layers, containing the positive and negative ions, respectively, is again apparent.[82] Complexes of sodium, potassium, and caesium tetraphenylborates with a range of polyethers have been prepared. Whereas some complexes have the usual metal co-ordination, the non-co-ordination of the BPh$_4^-$ ion opens up the synthesis of new products when conditions are varied. Thus the sandwich compound Na(benzo-15-crown-5)$_2$BPh$_4$ crystallizes in high yield under the appropriate conditions and dibenzo-30-crown-10 co-ordinates either one or two sodium ions depending on the solvent used.[83] Twelve new complexes of the type MX-crown-HL (M = K or Na, X = Cl, I or NCS, HL = PhCO$_2$H, 2-HOC$_6$H$_4$CO$_2$H, 2-nitro-, 2,4-dinitro-, or 2,4,6-trinitro-phenol) have been prepared. Potassium formed a complex cation with two molecules of benzo-15-crown-5 or one of dibenzo-30-crown-10, and these cations were protonated by HL. Sodium formed a monocation with one molecule of benzo-15-crown-5 but a dication with one molecule of dibenzo-30-crown-10.[84]

The distribution of alkali-metal picrate 1:1 complexes of dibenzo-18-crown-6 between water and benzene has been investigated. The formation of 2:1 complexes was recognized for Rb$^+$ and Cs$^+$ ions in a large excess of polyether. The extractability of complex cation–picrate ion pairs decreased in the order K > Rb > Cs > Na > Li.[85] This order was confirmed in a parallel investigation,[86] and bis-(3,5-di-t-butylbenzo)-18-crown-6 also has good extraction ability for alkali-metal ions.[87]

Macrocyclic ethers define a circular cavity for alkali-metal ions and macrobicyclic ligands an ellipsoidal one. A spheroidal macrotricyclic ligand has been successively synthesized through these stages, and this contains an intramolecular cavity lined with ten co-ordination sites; four N atoms and six O atoms situated at the corners of a tetrahedron and octahedron, respectively. This ligand (3) may exist in five topologically different forms depending on the orientation of the N pyramids into (i) or out of (o) the cavity; i$_4$, i$_3$o, i$_2$o$_2$, io$_3$, and o$_4$. The ligand forms 1:1 macrotricyclic [3] cryptates analogous to the macrobicyclic [2] cryptates (see vol. 2, p. 26) with alkali-metal salts in which the cation K$^+$ or Cs$^+$ (or Ba$^{2+}$) is contained inside the cavity of the ligand in its i$_4$ form.[88] A $^7$Li n.m.r. study of lithium complexes (cryptates) of the polyoxadiamine macrobicyclic ligands (cryptands) (4), C222, C221, and C211 with perchloate or triiodide salts in the non-aqueous solvents MeNO$_2$, DMSO, THF, and CHCl$_3$ has been carried out. The stability of the cryptates is largely determined by the size of the crypt cavity

---

[82] F. P. Boer, M. A. Neuman, F. P. Van Remoortere, and E. C. Steiner, *Inorg. Chem.*, 1974, **13**, 2826.
[83] D. G. Parsons, M. R. Truter, and J. N. Wingfield, *Inorg. Chim. Acta*, 1975, **14**, 45.
[84] N. S. Poonia, *J. Inorg. Nuclear Chem.*, 1975, **37**, 1855.
[85] A. Sadakane, T. Iwachido, and K. Toei, *Bull. Chem. Soc. Japan*, 1975, **48**, 60.
[86] P. R. Danesi, H. Meider-Gorican, R. Chiarizia, and G. Scibona, *J. Inorg. Nuclear Chem.*, 1975, **37**, 1479.
[87] L. Tusek, P. R. Danesi, and R. Chiarizia, *J. Inorg. Nuclear Chem.*, 1975, **37**, 1538.
[88] E. Graf and J. M. Lehn, *J. Amer. Chem. Soc.*, 1975, **97**, 5022.

# Elements of Group I

and by the solvating ability of the solvent used. Crypts 222 and 221, which have a large cavity (2.8 and 2.2 Å, respectively), do not form a complex with $Li^+$ in DMSO solutions since solvent molecules have a strong solvating ability and compete successfully with the cryptands, which do not form strong cryptates with $Li^+$ ions. In nitromethane, however, which is a poorly solvating solvent, the respective cryptates are readily formed as shown by the chemical shift of the $^7Li$ resonance upon the addition of the ligand. Not surprisingly, the chemical shift of the $Li^+$ ion when complexed by C211 is essentially independent of the solvent; the $^7Li$ chemical shift depends exclusively on nearest neighbours and is insensitive both to solvent molecules surrounding the crypt and to the counter ion.[89] The kinetics of complexation of $Li^+$ ion with C211 in pyridine, water, DMSO, DMF, and formamide, and with C221 in pyridine, have also been measured, and provide activation energies for the release of the cation from the $Li^+$-C211 complex which increase with increasing Gutmann donicity of the solvent. This is explained by invoking a transition state involving substantial ionic solvation. The formation constants of the $Li^+$-C211 cryptate in water were used to deduce the rate constant for the forward reaction as $0.98 \times 10^3 \, s^{-1}$.[90]

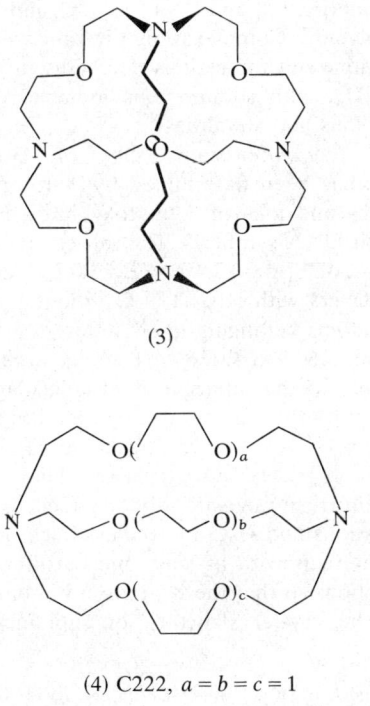

(3)

(4) C222, $a = b = c = 1$
C221, $a = b = 1, c = 0$, C211, $a = b = c = 0$

---

[89] V. M. Cahen, J. L. Dye, A. I. Popov, and I. Alexander, *Inorg. Nuclear Chem. Letters*, 1974, **10**, 899.
[90] V. M. Cahen, J. L. Dye, and A. I. Popov, *J. Phys. Chem.*, 1975, **79**, 1292.

The preparation and crystal structure of a compound which contains the cryptated sodium cation separated from the sodium anion has been reported. Crystals of $Na_2C_{18}H_{36}N_2O_6$ form as thin hexagonal plates when a saturated solution of sodium in ethylamine is cooled in the presence of the bicyclic polyoxadiamine (2,2,2-crypt). The compound crystallizes in the rhombohedral system ($R32$) with three molecules per hexagonal unit cell and $a = 8.83$, $c = 29.26$ Å. One sodium species is trapped in the crypt at distances from the N and O atoms which are characteristic of a $Na^+$ ion. The other sodium species is outside the crypt at a large distance from all other atoms and has all the indications of being the $Na^-$ ion.[91] E.s.r. spectroscopy has been used to investigate solutions of potassium in benzene or toluene mixtures with macrocyclic ether or 2,2,2-crypt complexing agents. These results indicate the formation of radical anions of $C_6H_6$ and PhMe as counter ions to the complexed $K^+$ cation.[92] The macrocyclic antibiotic, monensin, has been incorporated as a carrier into a membrane which can selectively move $Na^+$ ions from a region of low to high concentration. The energy for this movement comes from the simultaneous countertransport of protons. The transport is dominated by the chemical characteristics of the mobile carrier and is unaffected by electrostatics and osmotic effect.[93] This drug has been investigated by potentiometric, i.r., and $^{23}Na$, $^7Li$, $^{13}C$, and $^1H$ n.m.r. techniques in several non-aqueous solvents. Complexation titration with other cations shows that the selectivity of monensin for cations decreases in the order $Ag^+ > Na^+ > K^+ > Rb^+ > Cs^+ > Li^+ = NH_4^+$, with sodium ions monensin forms two complexes with different configurations and stabilities.[94]

The structure of the potassium salt of 3-hydroxybiuret, (1:2) $C_2H_4KN_3O_3,C_2H_5N_3O_3$, has been determined by $X$-ray methods and shown to consist of potassium ions and ions of 3-hydroxybiuret, $H_2NCON(OH)CONH_2$, which is known to inhibit DNA synthesis. The salt crystallizes in the space group $P2_1/c$ with $a = 3.86$, $b = 20.38$, $c = 12.40$ Å, $\beta = 90.38°$, and $Z = 4$. The $K^+$ ions form slightly puckered layers with $H(C_2H_4N_3O_3)^-$ ions between. Each $K^+$ ion is surrounded by eight O atoms belonging to six different 3-hydroxybiuret residues with K—O distances from 2.663 to 3.026 Å. There is no regular geometry around potassium probably owing to the intensive intermolecular hydrogen bonding in this structure.[95] Potassium 5-ethylbarbiturate $\frac{5}{3}$hydrate is triclinic with space group $P\bar{1}$ and $a = 12.276$, $b = 10.843$, $c = 15.800$ Å, $\alpha = 103.88$, $\beta = 124.42$, $\gamma = 106.78°$, and unit cell $K_6[C_6H_7N_2O_3]_6,10H_2O$. The anions are linked by NH $\cdots$ OC bonds to form an unusual ribbon stacked to make channels to accommodate the $K^+$ cations and $H_2O$ molecules. Each of three crystallographically distinct $K^+$ ions interacts with at least one barbiturate oxygen atom from each of two different ribbons so that the $K^+$ ions aid in binding together separate H-bonded ribbons.[96] The crystal structure of rubidium hydroxylamine NN-

---

[91] F. Tehan, B. L. Barnett, and J. L. Dye, *J. Amer. Chem. Soc.*, 1974, **96**, 7203.
[92] B. Kaempf, S. Raynal, A. Collet, F. Schue, S. Boileau, and J. M. Lehn, *Angew. Chem.*, 1974, **86**, 670.
[93] E. M. Choy, D. F. Evans, and E. L. Cussler, *J. Amer. Chem. Soc.*, 1974, **96**, 7085.
[94] P. G. Gertenbach and A. I. Popov, *J. Amer. Chem. Soc.*, 1975, **97**, 4738.
[95] I. K. Larsen, *Acta Chem. Scand.* (A), 1974, **28**, 787.
[96] G. L. Gartland, B. M. Gatehouse, and B. M. Craven, *Acta Cryst.*, 1975, **B31**, 208.

disulphonate, $Rb_5\{[ON(SO_3)_2H]_2\},3H_2O$, has been determined as triclinic. Crystals belong to the $P\bar{1}$ space group with $a = 9.357$, $b = 11.109$, $c = 11.206$ Å, $\alpha = 102.03$, $\beta = 99.09$, and $\gamma = 115.53°$ with $Z = 2$. The $\{[ON(SO_3)_2]_2H\}^{5-}$ anions are separated by $Rb^+$ ions and water molecules, but the cations do not possess well-defined oxygen co-ordination spheres.[97] Crystals of lithium hydrogen maleate dihydrate are monoclinic, space group $P2_1/C_1$, with $a = 5.860$, $b = 6.070$, $c = 19.690$ Å, $\beta = 106.3°$, $d$(expt.) = 1.54, $d$(calc.) = 1.56 for $Z = 4$. The hydrogen maleate anions are arranged in layers which are H-bonded through water molecules. The $Li^+$ ion is surrounded by four oxygen atoms at an average distance of 1.99 Å and is sandwiched between the layers of anions.[98]

The phase diagrams of rubidium chloroacetates–chloroacetic acids–water reveal the formation of the compounds $CH_2ClCO_2Rb,CH_2ClCO_2H$, $CH_2ClCO_2$-$Rb,2CH_2ClCO_2H$, $CHCl_2CO_2Rb,CHCl_2CO_2H$, and $CCl_3CO_2Rb,CCl_3CO_2H$.[99] In the corresponding caesium systems the compounds formed are $CH_2ClCO_2$-$Cs,CH_2ClCO_2H$, $3CH_2ClCO_2Cs,5CH_2ClCO_2H$, $CHCl_2CO_2Cs,CHCl_2CO_2H$, and $CCl_3CO_2Cs,CCl_3CO_2H$.[100] The structure of potassium hydrogen bis(trichloroacetate), $KH(CCl_3CO_2)_2$, has space group $I\bar{4}2d$ with lattice parameters $a = 13.060$, $c = 14.449$ Å with $d$(expt.) = 1.95 and $d$(calc.) = 1.97 for $Z = 8$. Two crystallographically equivalent trichloroacetate groups are bonded by a strong symmetrical H bond (2.46 Å). The $K^+$ ion is surrounded by four oxygen-atoms and two chlorine-atoms.[101]

The crystal structure of potassium and rubidium thioacetates have been determined. The compounds are orthorhombic with space group $Pnma$ and lattice constants $a = 9.702$, $10.008$, $b = 4.212$, $4.423$, $c = 11.756$, $11.823$ Å, $d$(expt.) = 1.58, 2.08 and $d$(calc.) = 1.57, 2.02 for $Z = 4$, for the potassium and rubidium salts, respectively. In each case, the alkali-metal ion is surrounded by three oxygen and four sulphur atoms, to form a trigonal prism with extra oxygen atom displaced outward from a rectangular face.[102,103] Potassium dithioacetate is also orthorhombic, with space group $Cmcm$ and lattice parameters $a = 7.688$, $b = 10.565$, $c = 6.490$ Å, $d$(expt.) = 1.65 and $d$(calc.) = 1.63 for $Z = 4$.[104] The type of co-ordination at potassium seen in the thioacetate also appears in potassium tetraacetatoborate, $K[B(OAc)_4]$, which is monoclinic with space group $P2_1/n$ and $a = 17.665$, $b = 12.456$, $c = 6.090$ Å, $\beta = 100.11°$ and $Z = 4$. The structure consists of slightly distorted boron–oxygen tetrahedra held together by potassium ions. Four acetoxy-groups with only one B—O bond form each boron–oxygen tetrahedron of the anionic unit $B(OAc)_4^-$. The $K^+$ ion is co-ordinated to seven oxygen atoms, as shown in Figure 5, with K—O distances 2.68—3.02 Å.[105]

---

[97] R. J. Guttormson, J. S. Rutherford, B. E. Robertson, and D. B. Russell, *Inorg. Chem.*, 1974, **13**, 2062.
[98] M. P. Gupta, S. M. Prasad, and T. N. P. Gupta, *Acta Cryst.*, 1975, **B31**, 37.
[99] J. Pokorny, *Z. Chem.*, 1975, **15**, 197.
[100] J. Pokorny, *Z. Chem.*, 1975, **15**, 238.
[101] L. Golic and F. Lazarini, *Cryst. Structure Comm.*, 1974, **3**, 645.
[102] M. M. Borel, G. Dupriez, and M. Ledesert, *J. Inorg. Nuclear Chem.*, 1975, **37**, 1533.
[103] M. M. Borel and M. Ledesert, *Acta Cryst.*, 1974, **B30**, 2777.
[104] M. M. Borel and M. Ledesert, *Z. anorg. Chem.*, 1975, **415**, 285.
[105] A. D. Negro, G. Rossi, and A. Perotti, *J.C.S. Dalton*, 1975, 1232.

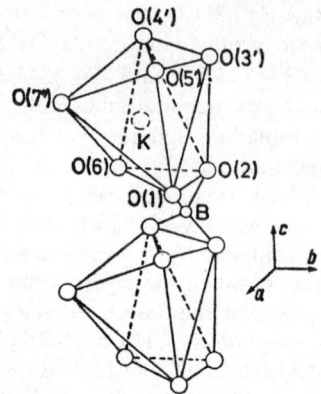

**Figure 5** *Potassium co-ordination pattern in* K[B(OAc)₄]
(Reproduced from *J.C.S. Dalton*, 1975, 1232)

## 5 Alkali-metal Oxides

The standard enthalpy of formation of lithium monoxide, $\Delta H^{\ominus}_{298}$, has been determined as $-142.902$ kcal mol$^{-1}$; the measured enthalpy of reaction of Li$_2$O with water of $-31.488$ kcal mol$^{-1}$ was combined with auxiliary thermochemical data to yield this result.[106] The Raman spectra of the peroxides Li$_2$O$_2$, Na$_2$O$_2$, K$_2$O$_2$, Rb$_2$O$_2$, Cs$_2$O$_2$, MgO$_2$, CaO$_2$, SrO$_2$, BaO$_2$, ZnO$_2$, and CdO$_2$ have been measured and show O—O stretching frequencies below and above 800 cm$^{-1}$ for the alkali-metal peroxides and bivalent metal peroxides, respectively, as expected due to the increasing polarization of the electrons away from the anion by the cation. In the hydrates of the alkali-metal (M) peroxides, this absorption moves to higher wavenumbers indicating that the structure of the hydrated species is of the type MOOH. In contrast, the O—O stretching frequency moves to lower wavenumbers on hydration of the alkaline earth peroxides. This is attributed to a decrease in the cationic field due to co-ordination of water to the cation.[107] The reaction of potassium peroxide with the molten alkali-metal nitrates has been studied by d.t.a. The peroxide anion is oxidized to the superoxide anion, O$_2^-$, which dissolves to give a stable solution in the molten nitrates. This oxidation is accompanied by reduction of nitrate ion.[108] Potassium superoxide, KO$_2$, reacts rapidly with nitric oxide initially but thereafter reaction comes to a complete stop. At pressures between 100 and 280 Torr, the major and minor products are KNO$_2$ and KNO$_3$, respectively, which are produced by a diffusion controlled mechanism.[109] Many other aspects of the synthesis, properties, and reactions of the alkali- and alkaline-earth metal peroxides, superoxides, and ozonides have also been described.[110] The e.s.r. spectrum at 4 K of sodium ozonide, NaO$_3$, prepared from

---

[106] G. K. Johnson, R. T. Grow, and W. N. Hubbard, *J. Chem. Thermodynamics*, 1975, **7**, 781.
[107] H. H. Eysel and S. Thym, *Z. anorg. Chem.*, 1975, **411**, 97.
[108] V. Bruner, A. Salta, and Dz. Peica, *Latv. PSR Zinal. Akad. Vestis, Khim. Ser.*, 1974, 651.
[109] T. P. Firsova and E. Ya. Filatov, *Izvest. Akad. Nauk S.S.S.R., Ser. khim.*, 1974, 1416.
[110] *Tezisy Dokl. Vses. Soveshch. Khim. Neorg. Perekisnykh Soedin*, ed. I. I. Vol'vov and A. Ya. Blum, Rizh. Politekh. Inst., Riga, U.S.S.R., 1973.

the reaction of sodium with ozone diluted with argon, indicates that the sodium atom is equidistant from the terminal oxygen atoms of the non-planar ozonide ion.[111] The potassium analogue, when dissolved in liquid ammonia, absorbs strongly in the u.v. from 300 nm in addition to strong absorption at longer wavelengths. Irregular spacing of the maximum in the absorption in the visible region points to some interference from a second vibrational mode of the excited state. Conflict between previously reported spectral data is attributed to interference by other species such as $O_3$, $O_2^-$, and $HO_2$.[112] A theoretical study reports the assignment of a charge of $-0.61$ proton units to each of the oxygen atoms in the chromate ion, $CrO_4^{2-}$. This is used to derive an enthalpy of formation of $-7055$ kJ mol$^{-1}$ for the $CrO_4^{2-}$ ion and total lattice potential energies for $M_2CrO_4$ of 1836, 1714, 1653, and 1596 kJ mol$^{-1}$ for M = Na, K, Rb, and Cs, respectively.[113]

### 6 Alkali-metal Halides

The standard X-ray diffraction patterns for 57 compounds have been collected. The data comprise Miller indices, interplanar spacings, and lattice constants for inorganic compounds, solid solutions of alkali-metal halides and of alkaline-earth metal nitrates.[114]

The mutual solubilities of Li–LiCl and Li–LiCl–KCl systems have been determined both in liquid–liquid and solid–liquid equilibrium regions. For Li–LiCl, the consolute temperature could not be detected despite raising the temperature to 1623 K. This is higher than the boiling point of the metal. Solubilities increase from 0.007 (567 K) to 0.23 (923 K) mol% LiCl in Li, and from 0.66 (935 K) to 1.22 (1123 K) mol% Li in LiCl. The mutual solubilities for Li–LiCl are considerably smaller than those for Na–NaX or K–KX (X = halogen) and accordingly their temperature coefficients are also very small. The enthalpy of solution of LiCl(s) in Li(l) was estimated to be ca. 14 kcal mol$^{-1}$.[115] The fumes liberated when lithium dissolves in liquid LiCl–KCl mixtures have been identified as potassium vapour liberated by the exchange reaction of lithium with potassium chloride. The rate of this reaction has been measured using a simple electrochemical half-cell containing two lithium electrodes. The amount of unchanged lithium was determined by electrochemical transport between the electrodes. The reaction rate was independent of KCl concentration, current density, and electrode-melt surface area, and proceeded with an activation energy of 111 kcal mol$^{-1}$ of Li which is nearly equal to the enthalpy of vaporization of potassium, 115 kcal mol$^{-1}$, from the melt. Thus the rate of reaction of lithium with potassium chloride under these conditions is probably controlled by the rate of potassium vaporization.[116]

Lithium and magnesium bromides and iodides which have been crystallized from organic solvents have been examined by X-ray diffraction methods. A new

---

[111] F. J. Adrian, V. A. Bowers, and E. L. Cochran, *J. Chem. Phys.*, 1974, **61**, 5463.
[112] P. A. Giguere and K. Herman, *Canad. J. Chem.*, 1974, **52**, 3941.
[113] H. D. B. Jenkins, A. Winsor, and T. C. Waddington, *J. Phys. Chem.*, 1975, **79**, 578.
[114] H. F. McMurdie, M. C. Morris, L. E. Evans, B. Paretzkin, J. H. De Groot, C. R. Hubbard, and S. J. Carmel, *Nat. Bur. Stand. (U.S.A.), Monographs*, 1975, **25**, Sect. 12.
[115] T. Nakajima, R. Minami, K. Nakanishi, and N. Watanabe, *Bull. Chem. Soc. Japan*, 1974, **47**, 2071.
[116] R. N. Seefurth and R. A. Sharma, *J. Electrochem. Soc.*, 1975, **122**, 1049.

cubic modification of lithium bromide with the caesium chloride-type structure ($a = 1.014$ Å) was observed.[117] The He$^I$ photoelectron spectra of the gaseous chlorides, bromides, and iodides of the metals Na, K, Rb, and Cs have been obtained using a molecular beam technique. The spectra are consistent with an ionic model, but it is necessary to postulate polarization of the halogen atoms by the alkali-metal ions to explain features of the spectra for the lighter molecules, thereby making the lighter molecules more covalent. The proportion of dimers to monomers in the gas falls as the molecules become more ionic.[118]

Metal-gas reactions as studied by matrix-isolation i.r. and Raman spectroscopic techniques have been reviewed.[119] This technique has been applied to several alkali-metal halide systems. Thus the i.r. spectra (900—4000 cm$^{-1}$) of vapours over LiF—NaF mixtures at 1173—2272 K have been obtained by condensation. Five of the six i.r.-active frequencies of lithium sodium difluoride, LiNaF$_2$, were observed and assigned on the basis of a planar $C_{2v}$ structure. Harmonic force constants are available for Li$_2$F$_2$, Na$_2$F$_2$, and LiNaF$_2$.[120] Matrix-isolated sodium and potassium chlorides and cyanides have also been examined by i.r. spectroscopy. Temperature cycling was pursued in an attempt to differentiate between bands due to monomers, dimers, and higher polymers. Symmetry force constants, calculated on the basis of a cyclic orthorhombic structure where CN is equivalent to halogen, are available for the dimers of the alkali-metal fluorides, chlorides, and cyanides.[121] Laser-Raman and i.r. spectroscopy have been employed to study alkali-metal matrix reactions with fluorine. Raman signals appropriate for the $\nu_1$ intraionic $(F \leftrightarrow F)^-$ mode in the M$^+$F$_2^-$ species were observed and showed an alkali-metal effect due to interaction with the $\nu_2$ interionic (M$^+ \leftrightarrow$ F$_2^-$) mode. The i.r. spectra revealed several bands assigned to $\nu_2$ for the M$^+$F$_2^-$ species and intense sharp bands due to the MF and M$_2$F$_2$ species.[122] The M$^+$Cl$_3^-$ species is formed by condensing alkali-metal chloride vapour with argon-containing chlorine and the $\nu_3$ absorptions for the Cl$_3^-$ anion appear from 375 cm$^{-1}$ for M = Na to 327 cm$^{-1}$ for M = Cs. The analogous hydrogen chloride-salt reaction forms the M$^+$HCl$_2^-$ species which has $\nu_3$ absorptions ranging from 658 to 736 cm$^{-1}$ depending upon M; the isolated HCl$_2^-$ ion absorption assignment is at 696 cm$^{-1}$.[123] Matrix-isolation i.r. spectra (15 K) have been obtained for complexes of lithium chloride and lithium bromide with several electron-donor bases. The complexes studied and the absorption frequencies assigned as the lithium–halogen stretching frequency (cm$^{-1}$) are: (Me$_3$N)$^6$LiCl, 632; H$_3$N$^6$LiCl, 538; Me$_2$O$^6$LiCl, 524; H$_2$O$^6$LiCl, 553; Me$_3$N$^7$LiBr, 554; H$_3$N$^7$LiBr, 584, 479; and H$_2$O$^7$LiBr, 460. The relative frequency shifts (related to the average isolated monomer frequency) vary in the same way as the hydrogen-bond shifts when plotted in a vibrational correlation diagram. Such evidence indicates that these complexes have geomet-

---

[117] B. M. Bulychev, K. N. Semenenko, V. N. Verbetskii, and K. B. Bitsoev, *Vestnik. Moskov. Univ., Khim.*, 1975, **16**, 45.
[118] A. W. Potts, T. A. Williams, and W. C. Price, *Proc. Roy. Soc.*, 1974, **A341**, 147.
[119] G. A. Ozin and A. Van der Voet, *Progr. Inorg. Chem.*, 1975, 105.
[120] A. Snelson, B. N. Cyvin, and S. J. Cyvin, 'Molecular Structure Vibrations', ed. S. J. Cyvin, Elsevier, Amsterdam, 1972, p. 246.
[121] Z. K. Ismail, R. H. Hauge, and J. L. Margrave, *J. Mol. Spectroscopy*, 1975, **54**, 402.
[122] W. F. Howard and L. Andrews, *Inorg. Chem.*, 1975, **14**, 409.
[123] B. S. Ault and L. Andrews, *J. Amer. Chem. Soc.*, 1975, **97**, 3824.

# Elements of Group I

ries similar to those found in hydrogen-bonded structures, hence the interaction may reasonably be called a lithium bond.[124] The argon matrix reactions of iodine (Ar:I = 500:1) with alkali-metal atoms have been examined by Raman spectroscopy which revealed a resonance Raman progression for each of the five alkali-metal $M^+I_2^-$ species. The dissociation energy of $I_2^-$ ($I_2^- \rightarrow I + I^-$) in the $M^+I_2^-$ species is estimated to be 21($LiI_2$), 20($NaI_2$), and 18($KI_2$) kcal mol$^{-1}$. The average value of 20 kcal mol$^{-1}$ is in agreement with the thermodynamic value 24.5 ± 2 obtained from a Hess's Law calculation.[125] The i.r. and Raman spectra of the trihalides $CsI_3$, $RbI_3$, $CsIBr_2$, and $RbIBr_2$ have been obtained. The triiodides probably have the same space group $Pnma$ with the same slightly dissymmetric structure for $I_3^-$.[126,127] Complex halides that have been investigated are collected in Table 4.

**Table 4** *Complex halides that have been investigated*

| Compound | | Ref. |
|---|---|---|
| $KCaCl_3$ | Dissociates on melting | 128 |
| $KMgCl_3$ | $K_2MgCl_4$ dissociates into KCl and $KMgCl_3$ on freezing | 128 |
| $LiBeF_3$, $(LiBeF_3)_2$, $Li_2BeF_4$ | Matrix isolation i.r. spectroscopy; gaseous $BeF_2$ reaction with $^6LiF$ and $^7LiF$ | 129 |
| $Na_2Li(BeF_4)_2$ | Monoclinic, space group $P2_1/c$, $a = 6.542$, $b = 9.634$, $c = 9.939$ Å, $\beta = 93.83°$, and $Z = 4$. Rings of eight or four $F_4$ tetrahedra linked by Li and Na polyhedra | 130 |
| $KBeF_2$ | Mass spectrometry of $KF-BeF_2$ melt vapours. Enthalpy, entropy of dissociation and dimerization are 67.1, 38.5 kcal mol$^{-1}$ and 34.2, 29.8 e.u., respectively | 131 |
| $Cs_2BeCl_4$, $CsBe_2Cl_5$, $CsBeCl_3$ | $CsCl-BeCl_2$ phase diagram. $Cs_2BeCl_4$ and $CsBe_2Cl_5$, congruent melting points at 881 and 587 K, respectively. $CsBeCl_3$ incongruent melting point at 575 K | 132 |
| $MBCl_4$ (M = Rb or Cs) | Thermochemistry. Standard enthalpy of formation = $-220.7$ (M = Rb), and $-225.0$ kcal mol$^{-1}$ (M = Cs) | 133 |
| $KAlF_4$, $(KAlF_4)_2$, $K_2AlF_5$, $NaAlF_4$, $(NaAlF_4)_2$ | Mass spectroscopy of $AlF_3-MF$ (M = K or Na) melt vapours. The dissociation enthalpies, $\Delta H°$ kcal mol$^{-1}$, are | 134, 135 |

$$KAlF_4 = KF + AlF_3 \quad \Delta H°_{1020K} = 83.4$$
$$(KAlF_4)_2 = 2KAlF_4 \quad \Delta H°_{838K} = 38.6$$
$$K_2AlF_5 = KF + KAlF_4 \quad \Delta H°_{1054K} = 46$$

---

[124] B. S. Ault and G. C. Pimentel, *J. Phys. Chem.*, 1975, **79**, 621.
[125] W. F. Howard and L. Andrews, *J. Amer. Chem. Soc.*, 1975, **97**, 2956.
[126] J. P. Coignac and M. Debeau, *Spectrochim. Acta*, 1974, **30A**, 1551.
[127] J. P. Coignac and M. Debeau, *Spectrochim. Acta*, 1974, **30A**, 1365.
[128] B. F. Markov, T. A. Tishura, and A. N. Budarina, *Rev. Roumaine Chim.*, 1975, **20**, 597.
[129] A. Snelson, B. N. Cyvin, and S. J. Cyvin, *J. Mol. Structure*, 1975, **24**, 165.
[130] J. Vicat, Tran Qui Duc, S. Aleonard, and P. Richard, *Acta Cryst.*, 1974, **B30**, 2678.
[131] A. N. Rykov, Yu. M. Korenev, A. F. Loshin, and A. V. Novoselova, *Zhur. neorg. Khim.*, 1974, **19**, 1923.
[132] B. P. Podafa, P. G. Dubovoi, and V. Barchuk, *Ukrain. khim. Zhur.*, 1974, **40**, 1211.
[133] A. Finch, P. L. Gardner, N. Hill, and N. Roberts, *J.C.S. Dalton*, 1975, 357.
[134] E. N. Kolosov, T. N. Tuvaeva, and L. N. Sidorov, *Zhur. fiz. Khim.*, 1975, **49**, 805.
[135] E. N. Kolosov, V. B. Shol'ts, and L. N. Sidorov, *Zhur. fiz. Khim.*, 1974, **48**, 2199.

**Table 4** (continued)

| | | |
|---|---|---|
| KAlF$_4$, NaAlF$_4$ | The enthalpies of (NaAlF$_4$)$_2$ dissociation and NaAlF$_4$ sublimation are 45.0 and 62.0 kcal mol$^{-1}$, respectively Raman spectroscopy of AlF$_3$–MF(M = K or Na) melts. Evidence for tetrahedral AlF$_4^-$ ion with some distortion by cation | 136 |
| Caesium chlorogallates | Raman spectroscopy of solid and molten mixtures of CsCl, GaCl$_3$, and CsAlGaCl$_7$. Successive formation of GaCl$_4^-$, Ga$_2$Cl$_7^-$, Ga$_n$Cl$_{3n+1}$ ($n \geq 3$), and Ga$_2$Cl$_6$ is proposed on addition of GaCl$_3$ to CsCl. Evidence for bridge bonds in Ga$_2$Cl$_7^-$ and Al$_2$Cl$_7^-$ | 137 |
| Potassium chloroindates | Raman spectroscopy of KCl–InCl$_3$ melts. Identification of InCl$_4^-$, In$_2$Cl$_7^-$ and In$_2$Cl$_2$ with structures similar to corresponding Al and Ga species | 138 |
| KTlF$_4$ | Trigonal, space group $P3_1$, $a = 8.025$, $c = 10.16$ Å $Z = 6$. New superstructure variation of the CaF$_2$-type. Isostructural with other A$^+$B$^{3+}$F$_4^-$ compounds (A = Na, K, or Ag; B = In, Tl, or Ln) | 139 |
| MTlF$_4$ (M = Rb or Cs) | Space group $Pb2_1a$, $Z = 4$. For M = Rb, Cs, respectively, $a = 8.525$, 8.405, $b = 8.359$, 8.484, $c = 6.244$, 6.566 Å; $d$(expt.) = 5.77, 6.09, $d$(calc.) = 5.64, 5.86. RbTlF$_4$ is isostructural with TlF$_2$ and contains layers of two-dimensional corner-linked (TlF$_4^1$F$_2$) octahedra connected by Rb$^+$ ions | 140 |
| MSn$_2$F$_5$ (M = K, Rb, NH$_4$, or Tl) | Crystallographic parameters of MSn$_2$F$_5$ are (M, system, space group, $a$, $b$, $c$, $\beta$, and $Z$): K, monoclinic, $C2/m$, $C2$, or $Cm$, 9.860, 4.208, 7.286 Å, 90.09°, 2; Rb, monoclinic, $C2m$, $C2$, or $Cm$, 10.124, 4.272, 7.401 Å, 90.07°, 2; Tl, monoclinic, $C2m$, $C2$, or $Cm$, 10.247, 4.263, 7.404 Å, 90.07°, 2; Tl, orthorhombic, $C222$, $Cmm2$, or $Cmmm$, 13.92, 7.109, 6.385 Å, —, 4; NH$_4$, orthorhombic, —, 13.33, 10.233, 7.661 Å, —, 6 | 141 |
| Rb$_2$PbI$_4$ | Orthorhombic, space group $Pnam$, with $a = 10.275$, $b = 17.381$, and $c = 4.773$ Å, $d$(calc.) = 5.24 for $Z = 4$. Pb is nearly octahedrally surrounded by I; octahedra held together by Rb | 142 |
| MSb$_2$F$_7$ (M = Rb, NH$_4$, or Tl) | Crystallographic parameters are [M, system, space group, $a$, $b$, $c$, $\beta$, $d$(expt.), $d$(calc.), and $Z$]: Rb, monoclinic, $P2_1/c$, 12.393, 6.568, 8.768 Å, 96.67°, 4.328, 4; NH$_4$, monoclinic, $P2_1/c$, 12.377, 6.587, 8.684 Å, 95.71°, 3.737, 3.719, 4; Tl, monoclinic $P2_1/c$, 24.79, 19.683, 8.870 Å, 97.72°, 5.401, 5.397, 4. Effect of cation size on structure of anion shown | 143 |
| Na$_2$SbF$_5$ | Orthorhombic, space group $P222$, with $a = 5.451$, $b = 11.234$, $c = 8.083$ Å, $Z = 4$, $d$(expt.) = 3.540 and $d$(calc.) = 3.541. Structure contains isolated Na$^+$ and distorted square-pyramidal SbF$_5^-$ ions | 144 |
| Cs$_2$SbBr$_6$, Cs$_2$SbBr$_5$, CsBr | Standard enthalpies, $\Delta H_{298}^\ominus$, and energies $\Delta G_{298}^\ominus$, of formation (kcal mol$^{-1}$), respectively, for Cs$_2$SbBr$_6$, $-268.3$, $-253.3$; Cs$_2$SbBr$_5$, $-264.5$, 251.4. For CsBr, $\Delta H_{298}^\ominus = -94.64$, $\Delta H_{298}^\ominus$(Cs$^+$, aq) = $-59.35$. $S_{298}^\ominus$(Cs$^+$, aq) = 31.5 cal mol$^{-1}$ K$^{-1}$ | 145 |

*Elements of Group I*

## 7 Lithium Compounds

A book on the chemistry of the alkali metals (lithium to francium) has been published[146] and the chemistry of the typical elements has been reviewed.[147]

The equilibrium pressures (0.5—760 Torr) of hydrogen existing above mixtures of lithium with lithium hydride (0.5—99 mol% LiH) sealed in iron capsules have been measured from 983 to 1176 K. The isotherms confirm the phase diagram to consist of two immiscible liquid phases with boundaries 25.2 and 98.4 mol% LiH at 983 K and 45.4 and 85.8 mol% LiH at 1176 K. For dilute solutions of lithium hydride in liquid lithium, the relationship between the mole fraction in solution, $x_{\text{LiH}}$, and the equilibrium pressure, $(p_{H_2})^{\frac{1}{2}}$, at $T(K)$ is given by

$$x_{\text{LiH}}/(p_{H_2})^{\frac{1}{2}} = \exp(-6.498 + 6182/T) \text{ atm}^{-\frac{1}{2}}$$

The standard free energy of formation (kcal mol$^{-1}$) of liquid lithium hydride is given by $(13.47 \times 10^{-3}T - 16.55)$.[148] The thermodynamic properties of the Li–LiH system have been determined by a mass spectrometric Knudsen effusion method in the plateau region from 973 to 1146 K. The enthalpy, $\Delta H^{\ominus}_{1023}$, of the reaction

$$\text{LiH(solid, sat. with Li)} = \text{Li(liquid, sat. with LiH)} + \tfrac{1}{2}H_2(g)$$

was calculated to be $24.0 \pm 0.5$ kcal mol$^{-1}$. The enthalpies of formation of lithium hydride calculated from the second Law ($\Delta H^{\ominus}_{298} = -21.7 \pm 0.5$ kcal mol$^{-1}$) and from the third Law ($\Delta H^{\ominus}_{298} = -22.0 \pm 1$ kcal mol$^{-1}$) are in good agreement. The corresponding free energy of formation, calculated from the hydrogen partial pressure which was extrapolated to 298 K, was $-16.77 \pm 0.5$ kcal mol$^{-1}$. The 'plateau' pressures (*i.e.* the dissociation pressure of hydrogen above non-stoicheiometric metal-rich lithium hydride) are considerably lower than those for the other alkali metals and are given in Table 5. These low values reflect the

**Table 5** *Variation in the partial pressure of hydrogen in the Li–LiH system with temperature in the 'plateau' region*

| T/K | $p_{H_2}$/Torr | T/K | $p_{H_2}$/Torr |
|---|---|---|---|
| 725 | 0.0694 | 763 | 0.3839 |
| 730 | 0.0894 | 773 | 0.5176 |
| 740 | 0.1447 | 790 | 0.9535 |
| 745 | 0.1745 | 800 | 1.0301 |

[136] B. Gilbert, G. Mamantov, and G. M. Begun, *Inorg. Nuclear Chem. Letters*, 1974, **10**, 1123.
[137] H. A. Oeye, and W. Bues, *Acta Chem. Scand. (A)*, 1975, **29**, 489.
[138] H. A. Oeye, E. Rytter, and P. Klaeboe, *J. Inorg. Nuclear Chem.*, 1974, **36**, 1925.
[139] C. Hebecker, *Z. Naturforsch.*, 1975, **30b**, 305.
[140] C. Hebecker, *Z. anorg. Chem.*, 1975, **412**, 37.
[141] L. L. Abbas, G. Jourdan, C. Avinens, and L. Cot, *Compt. rend.*, 1974, **279**, C, 307.
[142] H. J. Haupt, F. Huber, and H. Preut, *Z. anorg. Chem.*, 1974, **408**, 209.
[143] N. Habibi, B. Ducourant, R. Fourcade, and G. Mascherpa, *Bull. Soc. chim. France*, 1974, 2320.
[144] R. Fourcade, G. Mascherpa, E. Philippot, and M. Maurin, *Rev. Chim. Minerale.*, 1974, **11**, 481.
[145] S. H. Lee, R. M. Murphy, and C. A. Wulff, *J. Chem. Thermodynamics*, 1975, **7**, 33.
[146] 'The Chemistry of Lithium, Sodium, Potassium, Rubidium, Caesium and Francium', S. A. Hart, O. Beumel, and T. P. Whaley, Pergamon, New York, 1975.
[147] D. Millington, J. M. Winfeld, and M. G. H. Wallbridge, *Ann. Reports (A)*, 1974, **70**, 279.
[148] E. Veleckis, E. H. Van Deventer, and M. Blander, *J. Phys. Chem.*, 1974, **78**, 1933.

difficulty in removing hydrogen isotopes from liquid lithium by vacuum methods.[149] The ionization potential of the molecule LiD is reported as $7.7 \pm 0.1$ eV, giving a binding energy $D_0^0(\text{LiD}) = 0.14 \pm 0.1$ eV. From measurements of the gaseous equilibrium

$$\text{Li}_2(g) + \text{D}_2(g) = 2\text{LiD}(g)$$

over dilute solutions of lithium deuteride in liquid lithium, a value of $D_0^0(\text{LiD}) = 2.49 \pm 0.04$ eV was obtained for the dissociation energy of the deuteride.[150]

Lithium peroxocarbonate hydrate, $\text{Li}_2\text{CO}_4,\text{H}_2\text{O}$, has been prepared by treating lithium hydroxide with hydrogen peroxide in a 2:1 ratio at 253—273 K. The salt was characterized by n.m.r. spectroscopy, and thermally decomposes at 353 K to carbonate, oxygen, and water.[151]

Equilibrium constants for the reaction,

$$M + HCN = MCN + H$$

where M = alkali metal, have been determined by measuring the effect of added cyanogen on the intensity of emission of metallic resonance lines from flames at 1500 K. The yield of HCN from the cyanogen was measured by withdrawing samples through a quartz micro-probe for mass spectrometric analysis. The dissociation energies for the metal cyanide molecules, $D(\text{M}-\text{CN})$ are 497, 474, 497, 487, and 501 (all $\pm 22$) kJ mol$^{-1}$ for M = Li, Na, K, Rb, and Cs, respectively.[152] Some structural studies of the pseudohalides of $s$- and $p$-block elements have been reviewed.[153]

The standard enthalpy of formation of lithium nitride has been determined by solution calorimetry. The enthalpies of reaction of Li$_3$N with water and hydrochloric acid are $-581.62 \pm 1.42$ and $-803.50 \pm 1.26$ kJ mol$^{-1}$, respectively. These values are combined with auxiliary thermochemical data to derive a weighted mean value of $-164.93 \pm 1.09$ kJ mol$^{-1}$ for the standard enthalpy of formation of lithium nitride. This value differs by about 8 kcal mol$^{-1}$ from previous determinations.[154] Lithium nitride reacts in the solid state with M (M = Ca, Th, or Hf) to form MN as intermediate, and with MN to form Li$_2$MN$_2$. These reactions have been studied using d.t.a. and t.g.a. X-ray powder diffraction investigation shows that Li$_2$HfN$_2$ is isostructural with the known Li$_2$ZrN$_2$.[155] Lithium nitride reacts with Ge$_3$N$_4$ to form at least four phases: Li$_8$GeN$_4$, cubic, space group $Im3m$, $a = 9.619$ Å; Li$_2$GeN$_2$, probably isostructural with Li$_2$SiN$_2$; LiGe$_2$N$_3$, isotypic with LiSi$_2$N$_3$, orthorhombic, $a = 9.523$, $b = 5.497$, $c = 5.041$ Å, average Li—N, Ge—N distances of 2.19, 1.82 Å. The fourth phase has a variable composition between Li$_5$GeN$_3$ and Li$_{12}$Ge$_3$N$_8$.[156]

Lithium phosphide reacts with MP (M = Ce or Pr) in the presence of phosphorous vapour to form the ternary phosphides Li$_2$MP$_2$. X-Ray investigation showed

---

[149] H. R. Ihle and C. H. Wu, *J. Inorg. Nuclear Chem.*, 1974, **36**, 2167.
[150] H. R. Ihle and C. H. Wu, *J. Chem. Phys.*, 1975, **63**, 1605.
[151] T. P. Firsove, V. I. Kvlividze, A. N. Molodkina, and T. G. Morozova, *Izvest. akad. Nauk S.S.S.R., Ser. khim.*, 1975, 1424.
[152] J. N. Mulvihill and L. F. Phillips, *Chem. Phys. Letters*, 1975, **33**, 608.
[153] C. Glidewell, *Inorg. Chim. acta*, 1974, **11**, 257.
[154] P. A. G. O'Hare and G. K. Johnson, *J. Chem. Thermodynamics*, 1975, **7**, 13.
[155] M. G. Barker and I. C. Alexander, *J.C.S. Dalton*, 1974, 2166.
[156] J. David, J. P. Charlot, and J. Lang, *Rev. Chim. Minerale*, 1974, **11**, 405.

that the compounds are hexagonal with $d$(expt.) = 3.4, $d$(calc.) = 3.45 for $Z = 1$ and lattice constants $a = 4.189$, 4.190, and $c = 6.834$, 6.835 Å, respectively, for $Li_2CeP_2$ and $Li_2PrP_2$. In $Li_2CeP_2$, cerium is at the centre of a deformed tetrahedron of phosphorus, and lithium is contained within a similarly deformed tetrahedron with shortest Li—P distances of 2.44 Å.[157]

A new lithium lead phosphate has been reported, single crystals of which analyse to $LiPbPO_4$. The compound crystallizes in the space group $Pna2$, with $a = 7.080$, $b = 18.64$, and $c = 4.938$ Å.[158]

Structural and i.r. spectral analyses have been carried out on lithium, rubidium, and caesium hydroxide hydrates. The i.r. spectrum of $LiOH,H_2O$ shows seven bands in the 1000—300 cm$^{-1}$ region, in addition to the stretching and bending modes of OH and $H_2O$ groups. Four of these bands give characteristic shifts upon D substitution and confirm the presence of co-ordinated water and hydroxide ion. Analysis indicates that unlike other alkali-metal hydroxide hydrates, the water molecules and hydroxide ions form discrete planar hydrogen-bonded $[(OH^-)_2(H_2O)_2]$ anionic units rather than extended chains in the monohydrate. The mono- and hemi-hydrates of rubidium and caesium hydroxides are similar.[159] X-ray analysis of $2LiOH,KOH$ has shown it to be monoclinic, with space group $P2_1/m$, $a = 6.134$, $b = 5.810$, $c = 5.197$ Å, $\gamma = 103.12°$, $d$(expt.) = 1.85 and $d$(calc.) = 1.89 for $Z = 2$. The structure consists of $Li(OH)_4$ tetrahedra (Li—O distances 1.907—2.044 Å) and $K(OH)_{12}$ polyhedra (K—O distances from 2.826—3.767 Å).[160]

I.r. spectra of matrix-isolated alkali-metal perchlorate ion pairs have been obtained. The vapours over alkali-metal perchlorate salts at 673—723 K were condensed as molecular beams in argon and other matrices at 12 K. The spectra indicated that the dominant vapour species were the monomeric ion pairs, $M^+ClO_4^-$, though for the lithium salt, in particular, a large percentage of dimers was indicated. The monomer ion-pair spectra show strong splitting of $\nu_3$ and $\nu_4$ modes, with the magnitude of the $\nu_3$ splittings in the same range (200 cm$^{-1}$) as previously reported for $M^+NO_3^-$ monomers and, therefore, suggestive of strong distortion of the tetrahedral perchlorate structure. The perchlorate ion distortion is reduced by as much as a factor of six in water or ammonia matrices, apparently as a result of diminution of cation charge by electron transfer from the matrix molecule.[161] X-Ray and neutron diffraction studies of lithium perchlorate trihydrate, $LiClO_4,3H_2O$, show that it crystallizes in the space group $P6_3mc$ with $a = 7.719$ and $c = 5.455$ Å and consists of columns of —Li—$(H_2O)_3$—Li— entities which are hydrogen-bonded to regular tetrahedral $ClO_4^-$ ions with Cl—O = 1.440 Å. The Li atoms are co-ordinated by an almost regular octahedron of water molecules (average Li—O = 2.133 Å). Each H atom forms a weak bond (H—O = 2.044 Å) with one perchlorate oxygen atom and a very weak (H—O = 2.617 Å), but structurally significant bond with a second perchlorate oxygen atom.[162]

---

[157] A. El Maslout, J. P. Motte, A. Courtois, and C. Gleitzer, *Compt. rend.*, 1975, **280**, C, 21.
[158] L. H. Brixner and C. M. Foris, *Mater. Res. Bull.*, 1975, **10**, 31.
[159] I. Gennick and K. H. Harmon, *Inorg. Chem.*, 1975, **14**, 2214.
[160] N. V. Rannev, T. A. Demidova, and V. E. Zavodnik *Kristallografiya*, 1974, **19**, 998.
[161] G. Ritzhaupt and J. P. Devlin, *J. Chem. Phys.*, 1975, **62**, 1982.
[162] A. Sequeira, I. Bernal, I. D. Brown, and R. Faggiani, *Acta Cryst.*, 1975, **B31**, 1735.

An n.m.r. survey has been made of the properties of lithium intercalation compounds of layered transition-metal chalcogenides. The inclusion of a guest Li atom between the host disulphides and diselenides of Group IVB and VB is accompanied by donation of ca. one electron to the host.[163] Butyl-lithium in hexane serves as an excellent reagent to effect this intercalation. Under mild conditions, highly crystalline, uncontaminated stoicheiometric products are formed. In addition the course of the reaction can be followed by titration of samples of the supernatant solution.[164] Stable stoicheiometric compounds prepared by this method have been characterized by $X$-ray analysis. Intercalation proceeds by simple expansion of the lattice at the van der Waals gap. The Group VIB chalcogenides (with the exception of $MoS_2$ which forms a metastable compound), $SnS_2$, and $ReS_2$ do not form intercalation compounds.[165]

## 8 Sodium Compounds

The crystal structure of hydrated sodium borate, $Na_3[B_3O_5(OH)_2]$ (space group $Pnma$, $a = 8.923$, $b = 7.152$ and $c = 9.548$ Å) has been determined from single-crystal diffractometer $X$-ray data. The structure shows that the compound, usually labelled in the literature as 1:1:1, actually has a composition corresponding to $Na_2O : B_2O_3 : H_2O$ molar proportions of 3:3:2. The basic structural unit is the isolated polyion $[B_3O_5(OH)_2]^{3-}$ formed by corner sharing among a tetrahedron and two triangles. Of the four attached oxygen atoms, only the two linked to the tetrahedral boron exist as hydroxy-groups. Sodium polyhedra form a tight two dimensional network by the sharing of corners, edges, and faces. All oxygen atoms linked to the sodium atoms belong at the same time to the B—O polyions and *vice versa*. A high cohesion in all directions is the result of the perpendicular arrangement between B—O rings and Na—O sheets.[166] The vapours at 935—1100 K above mixtures of molten borates, $MBO_2$, with fluorides MF (M = Na, K, or Rb) have been shown by mass spectrometry to contain the associated molecules $M_2BO_2F$ with enthalpies of sublimation 77.0, 78.3, and 73.2 kcal mol$^{-1}$ for M = Na, K, and Rb, respectively. The dissociation process:

$$M_2BO_2F(g) = MF(g) + MBO_2(g)$$

proceeds with enthalpies of 57.6, 51.5, and 48.3 kcal mol$^{-1}$ for $Na_2BO_2F$, $K_2BO_2F$, and $Rb_2BO_2F$, respectively.[167] Detailed methods of preparing sodium salts containing the octahydrotriborate anion, $(B_3H_8)^-$, have been described. In the first, $NaBH_4$ reacts with iodine in anhydrous diglyme to give $NaB_3H_8$. This compound will react with $(Bu_4N)I$ to produce $(Bu_4N)B_3H_8$.[168] In the second method, reaction of $NaBH_4$ with $BF_3OEt_2$ in diglyme, produces $NaB_3H_8$ mixed with $NaBF_4$, which is removed by filtration. The octahydrotriborate is extracted with ether and isolated as $NaB_3H_8,3C_4H_8O_2$. Treatment with caesium bromide produces the moderately stable $CsB_3H_8$.[169] The preparation and reagent proper-

---

[163] B. G. Silbernagel, *Solid State Comm.*, 1975, **17**, 361.
[164] M. B. Dines, *Mater. Res. Bull.*, 1975, **10**, 287.
[165] M. S. Whittingham and F. R. Gamble, *Mater. Res. Bull.*, 1975, **10**, 363.
[166] E. Corazza, S. Menchetti, and C. Sabelli, *Acta Cryst.*, 1975, **B31**, 1993.
[167] V. E. Shevchenko, O. T. Nikitin, and L. N. Sidorov, *Vestnik Moskov. Univ., Khim.*, 1974, **15**, 756.
[168] G. E. Ryschkewitsch and K. C. Nainan, *Inorg. Synth.*, 1974, **15**, 113.
[169] W. J. Dewkett, M. Grace, and H. Beall, *Inorg. Synth.*, 1974, **15**, 115.

ties of sodium tetrakis($m$-chlorophenyl)borate have been described. The compound was prepared by Grignard reaction of $NaBF_4$ and $m$-bromochlorobenzene followed by reaction of the product with triethylamine to give the intermediate $Et_3NB(PhCl)_4$, which reacted with NaOMe to give $NaB(PhCl)_4$. This compound was similar to $NaBPh_4$ with regard to selectivity in reaction with large cations but was superior with regard to stability and precipitation. The ions $Cs^+$, $Rb^+$, $K^+$, and $NH_4^+$ give heavy dense precipitates, and the compound can be used for the gravimetric determination of $K^+$ ions where recovery can be 97.4—99.7%. The solubility of the potassium salt, $KB(PhCl)_4$, is $0.0992 \times 10^{-3}\,g\,cm^{-3}$.[170] Monosodium cyanamide, NaNHCN, has been prepared by treatment of $H_2NCN$ with NaOH in propan-2-ol at 298 K.[171]

The kinetics of the reactions of sulphur dioxide with sodium and with potassium have been investigated at 383—493 and 298—403 K, respectively. The only product identified from the former reaction was sodium sulphide, $Na_2S$. A diffusion-controlled process was indicated. The reaction with potassium appeared to be simpler since only potassium dithionate was formed:

$$2K + 2SO_2 \rightarrow K_2S_2O_4$$

A mechanism based on adsorption-reaction is in general agreement with the observed kinetics.[172] New solid-state syntheses of crystalline selenates $M_2SeO_4$ (M = K, Rb, and Cs) and antimonates, $MSbO_3$ (M = K or Rb), $M_3SbO_4$ (M = K, Rb, and Cs) involve treatment of the powdered superoxides, $MO_2$, with $SeO_2$ and $Sb_2O_3$, respectively. These ternary oxides have been characterized by their X-ray diffraction patterns.[173]

A structural investigation of compounds with anionic groups of the pyro-type $Na_6M_2S_7$ (M = Ge or Sn) and $Ba_3Sn_2S_7$ shows that $Na_6Ge_2S_7$ and $Na_6Sn_2S_7$ are isotypic and crystallize in the monoclinic system, space group $C2/c$ with lattice parameters, respectively, $a = 9.094$, 9.395; $b = 10.437$, 10.719; $c = 15.464$, 15.671 Å; $\beta = 109.49$, 109.97°. $Ba_3Sn_2S_7$ crystallizes in the monoclinic system, space group $P2_1/c$ with $a = 11.073$ $b = 6.771$, $c = 18.703$ Å and $\beta = 100.77°$. In these compounds, the crystal structure is built up from $Na^+$ or $Ba^{2+}$ cations and $M_2X_7^{6-}$ anions. The $M_2X_7^{6-}$ anion results from the condensation of two $MX_4$ tetrahedra with one common apex.[174]

The crystal structures of the high-temperature modifications of $\beta$-sodium antimony sulphide and $\beta$-sodium antimony selenide, determined by X-ray methods, are isotypic and possess a fcc lattice of the NaCl type with a statistical distribution of (Na + Sb) atoms in the $Na^+$ positions and of S (or Se) atoms in the $Cl^-$ positions. Crystal data are space group $Fm3m$ for each, $a = 5.7748$, 5.965 Å; $d(expt.) = 3.54$, 4.68; $d(calc.) = 3.60$, 4.73; $Z = 2$, 2; for $\beta$-$NaSbS_2$ and $\beta$-$NaSbSe_2$, respectively. The low-temperature modifications $\alpha$-$NaSbS_2$ and $\alpha$-$NaSbSe_2$ are not isostructural and exhibit lower symmetry. Atoms of Na and Sb in

---

[170] F. Jarzembowski, F. Cassaretto, H. Posvic, and C. E. Moore, *Analyt. Chim. Acta*, 1974, **73**, 409.
[171] S. Weiss and H. Krommer, *Angew. Chem.*, 1974, **86**, 590.
[172] P. Touzain, H. F. Avedi, and J. Besson, *Bull. Soc. chim. France*, 1974, 421.
[173] G. Duquenoy, *Rev. Chim. Minerale*, 1974, **11**, 474.
[174] J. C. Jumas, J. Olivier-Fourcade, F. Vermor-Gaud-Daniel, M. Ribes, E. Philippot, and M. Maurin, *Rev. Chim. Minerale*, 1974, **11**, 13.

these structures, in contrast to the β-analogues, are arranged in an orderly manner which provides the lower symmetry.[175]

The decomposition of sodium oxyhyponitrite [sodium trioxodinitrate(II)], $Na_2N_2O_3$, has been investigated in both the solid and solution phases. In the solid state the major products are $N_2O$, $NaNO_2$, and $Na_2O$, with minor products of nitrogen and oxygen. Secondary reactions produce $NaNO_3$ and NO. Each mole of oxyhyponitrite produces approximately one mole of nitrite and half a mole of oxide. The mechanism proposed envisages that the salt decomposes primarily according to:

$$Na_2N_2O_3 \rightarrow N_2O + Na_2O_2$$

but this is accompanied by some disproportionation:

$$2Na_2N_2O_3 \rightarrow Na_2N_2O_2 + 2NaNO_2$$

The hyponitrite is responsible for the production of nitrogen according to:

$$3Na_2N_2O_2 \rightarrow 2Na_2O + 2NaNO_2 + 2N_2$$

The main source of nitrite and oxide is from the process:[176]

$$Na_2N_2O_3 + Na_2O_2 \rightarrow 2NaNO_2 + Na_2O$$

In aqueous solution, the decomposition has been studied over a range of pH values by kinetic measurements based upon u.v. absorption and tracer experiments using the labelled form $Na_2(O^{15}NNO_2)$. The reaction products are exclusively $N_2O$ and $NO_2^-$, and the N atoms in the $N_2O$ come only from the N atom bound to a single O atom at pH ≥ 3. The rate of decomposition of the $N_2O_3^{2-}$ ion is low and decreases with increasing pH. Much of the process centres round the decomposition of $H_2N_2O_3$, $HN_2O_3^-$, and $HNO_2$ species. A very early hypothesis of HNO as intermediate in the decomposition and direct precursor of $N_2O$ was confirmed and it is postulated that cleavage of the N=N bond to produce HNO and $NO_2^-$ is the primary controlling process in acidic as well as in basic solutions.[177] The equilibrium gaseous species generated by liquid sodium tripolyphosphate have been investigated by mass spectrometry at 1270—1567 K. Third Law enthalpies and, where possible, Second Law enthalpies were measured for the following gas-phase reactions:

| | $\Delta H/kJ =$ |
|---|---|
| NaPO = Na + PO | 268 |
| NaOP = Na + PO | 281 |
| $NaPO_2$ = Na + $PO_2$ | 333 |
| $2NaPO_2$ = $NaPO_3$ + NaOP | 40 |

The derived atomization and standard enthalpies of formation of these gaseous molecules NaPO, NaOP, $NaPO_2$, and $NaPO_3$ are 861 ±30, −176 ±30; 874 ±30, −191 ±30; 1419 ±20, −192 ±20, and 1925 ±45 and −753 ±45 kJ mol$^{-1}$, respectively.[178] Sodium potassium trimetaphosphate, $Na_2KP_3O_9$, has been re-

---

[175] V. G. Kuznetsov, A. S. Kanishcheva, and A. V. Salov, Zhur. neorg. Khim., 1974, **19**, 1280.
[176] T. M. Oza and N. L. Dipali, Indian J. Chem., 1975, **13**, 178.
[177] F. T. Bonner and B. Ravid, Inorg. Chem., 1975, **14**, 558.
[178] K. A. Gingerich and F. Miller, J. Chem. Phys., 1975, **63**, 1211.

ported to be triclinic, space group $P\bar{1}$, with $a = 6.886$, $b = 9.494$, $c = 6.797$ Å, $\alpha = 110.07$, $\beta = 104.69$, $\gamma = 86.68°$, and $Z = 2$.[174] The sodium tri- and tetra-metaphosphates $(NaPO_3)_n$ ($n = 3$ and 4) undergo cyanamidolysis reactions with sodium cyanamide, $Na_2NCN$, to give $Na_3PO_3NCN$ and $Na_3PO_2(NCN)_2$ for $n = 3$ and 4, respectively. In alkaline aqueous solutions, parallel reactions of hydrolysis and cyanamidolysis occur.[179]

The structure of sodium chlorite trihydrate, $NaClO_2,3H_2O$, has been determined as triclinic, space group $P\bar{1}$, with $a = 5.492$, $b = 6.412$, $c = 8.832$ Å, $\alpha = 72.06$, $\beta = 87.73$, $\gamma = 70.88°$ and $Z = 2$. The bond distances and angles of the $ClO_2$ anion are Cl—O 1.564 and 1.557 Å, and $\angle OClO = 180.23°$. One of the oxygens belonging to the $ClO_2^-$ ion together with the oxygen atoms of the water molecules form a distorted octahedron around the sodium ion. This is shown in Figure 6. These octahedra form zigzag columns by sharing edges and the columns are linked by hydrogen bonds.[180]

The $MX-BiX_3-H_2O$ systems (M = Na, K, Rb, or Cs; X = Cl or Br) have been investigated at 298 K. The compounds which are formed include $Na_2BiCl_5,5H_2O$, $Na_3BiBr_6,12H_2O$, $Na_5Bi_2Br_{11},12H_2O$, $Na_2BiBr_5,4H_2O$, $Cs_3Bi_2X_9$ (X = Cl or Br), $Rb_7Bi_3Cl_{16}$, and $Rb_3Bi_2Br_9$. In these solutions, the CsX and RbX salts are significantly less soluble than NaX and KX, so that caesium and rubidium ions can be separated from mixtures of the alkali-metal ions by precipitation of their less soluble bismuth halogeno complexes.[181]

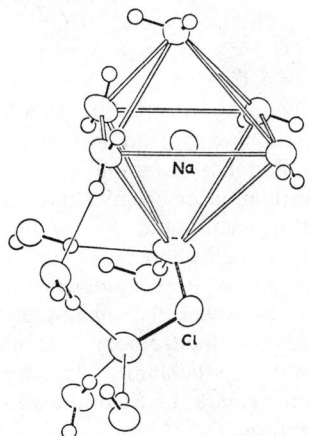

**Figure 6** *The structure of $NaClO_2,3H_2O$. The water oxygens and one of the oxygens of $ClO_2^-$ form a distorted octahedron around the sodium ion. Thin lines represent hydrogen bonds, heavy solid lines the O—H and O—Cl chemical bonds, and double lines describe the octahedron of oxygens in van der Waals contact with the sodium*
(Reproduced by permission from *Acta Cryst.*, 1975, **B31**, 2146)

---

[179] H. Koehler, U. Lange, and R. Uebel, *Z. anorg. Chem.*, 1975, **413**, 119.
[180] C. Tarimci, E. Schempp, and S. C. Chang, *Acta Cryst.*, 1975, **B31**, 2146.
[181] V. E. Plyushchev, N. M. Septsova, S. B. Stepina, and I. V. Vlasova, *Zhur. priklad. Khim.*, 1975, **48**, 637.

## 9 Potassium Compounds

Potassium hydride has more potential value in organic and organometallic synthesis than is commonly realized. The reactivity and applications of this salt in these areas have been investigated and show, for example, that reaction occurs rapidly (<1 min) at 293—298 K with excess t-butyl alcohol to yield the alkoxide in quantitative yield in contrast to the reactions of metallic potassium, NaH, and LiH, which are only 80, 5, and 0 percent complete, respectively, after 20 min. Potassium hydride also rapidly metallates unhindered amines (*e.g.* pyrrolidine) and DMSO. A suspension in THF rapidly metallates highly hindered weak acids such as $(Me_3Si)_2NH,N$-isopropylaniline, t-butyl alcohol, and 2,6-di-t-butylphenol. Under comparable conditions only the latter reacted significantly with NaH, and none reacted with LiH. Ketones are metallated to potassium enolates in high quantitative yield; no reduction of carbonyl groups by hydride is observed. Also KH reacts rapidly with weak and hindered Lewis acids under conditions where NaH is very sluggish and LiH is inert. Hindered trialkylboranes are readily converted into the corresponding borohydrides ($K^+HBR_3^-$) at room temperature. Reactions of KH appear to be entirely heterogeneous, occurring at the crystal surface. The salt is reasonably thermally stable but needs handling as a dispersion in mineral oil for protection from the atmosphere.[182] Complex hydrides containing the alkali metals with germanium, $MGeH_3$ (M = Li, K, Rb, or Cs), have been prepared and their structures investigated. Preparation was from the metals and $GeH_4$ in dimethoxyethane. The compounds $KGeH_3$ and $RbGeH_3$ have the NaCl-type structure with $a = 7.245$ and $7.518$ Å, respectively. The TlI-structure is adopted by $CsGeH_3$ with $a = 5.1675$, $b = 14.435$, $c = 6.9664$ Å and $Z = 4$.[183]

A development in the chemistry of graphite–alkali-metal intercalation compounds is the reaction of graphite with liquid metal alloys to produce ternary compounds K–Na–graphite and Na–Cs–graphite. Excess liquid is removed by centrifuge rather than by distillation or chemical means. Sodium and the heavier alkali-metal are entrained in metastable phases but prolonged exposure of graphite to the liquid alloy causes almost complete preferential extraction of sodium into the graphite lattice. X-Ray examination shows that the metals are inserted as solid solutions; the lower the alkali-metal content of the ternary compounds, the higher becomes the sodium content. Caesium enhances the uptake of sodium more than does potassium.[184] Ternary compound formation has also been achieved through the vapour phase of the alkali metals. An alloy of two metals heated to a temperature $T_0$ at one end of a tube produced a ternary compound with graphite heated to $T \geqslant T_0$, at the other end. For $T < 573$ K, the compounds had the formula $M^1_x M^2_{1-x} C_8$ where $M^1$ and $M^2$ are K, Rb, and Cs. A thermodynamic study revealed that the compounds $K_x Rb_{1-x} C_8$ and $Rb_x Cs_{1-x} C_8$ were essentially ideal solutions of $KC_8$ with $RbC_8$, and of $RbC_8$ with $CsC_8$. The calculated enthalpies of formation of $RbC_8$ and $CsC_8$ are $-11.9$, $-17.5$ kcal mol$^{-1}$ and $-3.9$, $-5.8$ cal mol$^{-1}$, respectively. Compounds of $^{85}Rb^{87}Rb$–graphite were prepared to examine possible isotopic enrichment. An enrichment of the heavier

---

[182] C. A. Brown, *J. Org. Chem.*, 1974, **39**, 3913.
[183] G. Thirase and E. Weiss, *Z. Naturforsch.*, 1974, **29b**, 800.
[184] D. Billaud and A. Hérold, *Bull. Soc. chim. France*, 1974, 2715.

isotope of ca. 1.2% was observed.[185] As with Na–K–graphite these ternary compounds contain solid solutions of the two alkali metals. X-Ray analysis showed that the interplanar spacings do not vary in a linear fashion with composition, which precludes a purely ionic model, but dilation measurements indicate that the attraction between the layers of inserted metals and graphite increases from compounds of K to those of Cs, corresponding to the ionic character of the metal–graphite bond.[186]

A new preparative use for the $MC_8$ species has been found in transition-metal carbonyls synthesis. $KC_8$ has been found to react with $M(CO)_6$ (M = Cr, Mo, or W), $Fe(CO)_5$, and $Co_2(CO)_8$ to give the dinuclear carbonyls $K_2[M_2(CO)_{10}]$ and $K_2[Fe_2(CO)_8]$, and $K[Co(CO)_4]$, respectively.[187] The catalytic effect of alkali metals adsorbed on active carbon is illustrated by the $H_2$–$D_2$ exchange reaction at 273 K which has been studied by e.s.r. Exchange occurs readily with first-order kinetics with respect to pressure. The activation energy is lower on K–C (1—2 kcal mol$^{-1}$) than on Na–C (2—7 kcal mol$^{-1}$). The catalytic activity parallels the electron spin density of the catalyst as the alkali content is increased. The electron spins originate from donation of electrons to carbon by metal. The observed spins are those of the donated electron. The active site appears to be located near the alkali ion. Hydrogen adsorption occurs by dissociation into atoms which react with the electron spin causing a decrease in spin density by electron pairing. The difference between K and Na is ascribed to a loss of K into the graphite structure. Although the active site on the carbon surface is formed by adsorbed alkali-metal atoms, aggregation of atoms into a metallic phase reduces the number of sites.[188] The characteristics of the adsorption at 473—573 K of both nitrogen and hydrogen on alkali-metal-promoted ruthenium on active carbon has also been studied. Extensive nitrogen adsorption occurred on K–Ru and K–Ru–C but not on Ru–C or K–C. The nitrogen was probably initially chemisorbed on the ruthenium and then migrated to the alkali-metal which was partly transformed into amide, and, under the ammonia synthesis condition, delayed the presence of ammonia in the gas phase.[189]

Potassium perthiocarbonate, $K_2CS_4$, and its adducts $K_2CS_4, nL$ (L = MeOH or $Me_2NH$, $n = 1$; L = $H_2O$, $n = 0.5$) have been prepared and characterized by t.g.a., tensiometry, X-ray diffraction, and i.r. spectroscopy. The adducts have the characteristics of inclusion compounds in which the geometry and arrangement of the $CS_4^{2-}$ ions in the crystal allow the inclusion of MeOH but not the larger PrOH molecules.[190]

The crystal structure of potassium orthostannate, $K_4SnO_4$, has been determined as triclinic, with space group $P\bar{1}$ and $a = 6.48$, $b = 6.51$, $c = 9.70$ Å, $\alpha = 71.82$, $\beta = 99.89$, and $\gamma = 113.13°$. The compound consists of discrete tetrahedral $SnO_4$ groups which are bounded by $K^+$ cations.[191]

---

[185] D. Billaud, D. Balesdent, and A. Hérold, *Bull. Soc. chim. France*, 1974, 2402.
[186] D. Billaud and A. Hérold, *Bull. Soc. chim. France*, 1974, **11**, 2407.
[187] C. Ungurenasu and M. Palie, *J.C.S. Chem. Comm.*, 1975, 388.
[188] M. Ishizuka and A. Ozaki, *J. Catalysis*, 1974, **35**, 320.
[189] K. Aika and A. Ozaki, *J. Catalysis*, 1974, **35**, 61.
[190] M. Abrouk, *Rev. Chim. Minerale*, 1974, **11**, 726.
[191] R. Marchand, Y. Piffard, and M. Tournoux, *Acta Cryst.*, 1975, **B31**, 511.

An oxothioantimonite, $K_3SbS_3,3Sb_2O_3$, with a tube structure has been prepared from alkaline aqueous solutions of $K_2S$ and $Sb_2O_3$. The hexagonal structure, space group $P6_3$, with $a = 14.256$, $c = 5.621$ Å, $d(\text{expt.}) = 4.16$, $d(\text{calc.}) = 4.24$ for $Z = 2$, consists of $Sb_2O_3$ tubes which contain the $K^+$ cations. Their charge is neutralized by $SbS_3^{3-}$ pyramids which occupy positions between the tubes.[192] The thermodynamic properties (enthalpies and entropies of formation) of the four phases $K_3As$, $K_5As_4$, KAs, and $KAs_2$ have been investigated by vapour pressure measurements and e.m.f. methods. The standard enthalpies of formation of $K_3As$, $K_5As_4$, KAs, and $KAs_2$ appear to be $-10.0$, $-12.0$, $-12.4$, and $-9.9$ kcal mol$^{-1}$, respectively. The arsenic compounds in the series $M_3B$ (M = Na or K; B = As or Sb) appear to be more thermodynamically stable than the antimony analogues.[193] The potassium arsenate–alkaline earth arsenate–water system at 295 K contains hydrated double arsenates. With strontium and barium the only double arsenates formed are $KM(AsO_4),8H_2O$. The compound $KSr(AsO_4),8H_2O$ decomposes in moist air and at 593 and 1128 K to $\alpha$-$KSr(AsO_4)$ and $\beta$-$KSr(AsO_4)$, respectively. The decomposition of $KBa(AsO_4)8H_2O$ occurs at 298 K, and $\alpha$- and $\beta$-$KBa(AsO_4)$ form at 523 and 1253 K, respectively.[194]

Procedures for the synthesis of potassium polysulphides in non-aqueous solvents such as EtOH and liquid $NH_3$ have been reported. The addition of sulphur to KHS produced $K_2S_5$. The reaction of sulphur with EtOK yielded impure products, $K_2S_4$ or $K_2S_6$, depending on the S:EtOK ratio. The tri- or pentasulphides were obtained by the reaction of sulphur with an equimolar mixture of EtOK and KHS. The reaction of potassium with sulphur furnished all the existing polysulphides including the hexasulphide.[195] Crystallographic data for the higher polysulphides $K_2S_n$ ($3 \leq n \leq 6$) are [compound, $a$, $b$, $c$, $\alpha$, $\gamma$, $d(\text{expt.})$, $d(\text{calc.})$, and $Z$] $K_2S_3$ 6.42, 12.40, 12.52 Å, 106°6′, 99°9′, 86°4′, 2.14, 2.146, 7; $K_2S_4$, 6.88, 12.284, 13.837 Å, 114°30′, 90°7′, 87°46′, 2.01, 2.087, 7; $K_2S_5$, 7.40, 13.550, 13.975 Å, 118°55′, 85°16′, 94°45′, 2.15, 2.272, 7; and $K_2S_6$, 7.93, 14.266, 14.917 Å, 123°51′, 79°36′, 99°13′, 2.21, 2.287, 7.[196] The enthalpies of fusion (kcal mol$^{-1}$) of the polysulphides $K_2S_3$, $K_2S_4$, $K_2S_5$, and $K_2S_6$ are 3.5, 3.4, 6.0, and 6.3, respectively.[197] Preparation of potassium antimony disulphide, $KSbS_2$, from an aqueous solution of KHS and $Sb_2S_3$ under mild hydrothermal conditions provides a monoclinic form, space group $C2/c$, with $a = 8.75$, $b = 8.98$, $c = 6.84$ Å, $\beta = 121.6°$, $d(\text{expt.}) = 3.19$, $d(\text{calc.}) = 3.26$ for $Z = 4$. The structure consists of $SbS_2^-$ chains built up from trigonal bipyramids which share edges. The $K^+$ ions between these chains have nearly octahedral co-ordination.[198]

The crystal structure of mixed potassium iodate–iodic acid, $KIO_3,HIO_3$, has been determined. There are two types of $K^+$ ion, $K_1$ and $K_2$, each surrounded by nine oxygen atoms ($K_1$—O = 2.70—3.13 and $K_2$—O = 2.77—3.35 Å). The basic structure is composed of two columns of $K_1$ and $K_2$ polyhedra, parallel to [100], which are joined along the edges. The $IO_3^-$ ions adopt the umbrella-shaped

---

[192] H. A. Graf and H. Schaefer, *Z. anorg. Chem.*, 1975, **414**, 220.
[193] G. F. Voronin and L. N. Bludova, *Vestnik Moskov. Univ., Khim.*, 1974, **15**, 433.
[194] N. Ariguib-Kbir and R. Stahl-Brasse, *Bull. Soc. chim. France*, 1974, 2343.
[195] J. M. Letoffe, J. M. Blanchard, and J. Bousquet, *Bull. Soc. chim. France*, 1975, 485.
[196] J. M. Letoffe, J. M. Blanchard, M. Prost, and J. Bousquet, *Bull. Soc. chim. France*, 1975, 148.
[197] J. Bousquet, J. M. Letoffe, and M. Diot, *J. Chim. phys.*, 1974, **71**, 1180.
[198] H. A. Graf and H. Schaefer, *Z. anorg. Chem.*, 1975, **414**, 211.

configuration. The crystals are orthorhombic, space group $Pca2_1$, with $a = 8.604$, $b = 7.508$, $c = 19.548$ Å, $d(\text{calc.}) = 4.10$ for $Z = 8$.[199] In $K_2H(SO_4)(IO_3)$, two independent $K^+$ ions are surrounded by eight oxygen atoms ($K_1$—O = 2.75—3.02, and $K_2$—O = 2.74—3.13 Å). The basic structural motif is a sinusoidal wave-shaped column of K polyhedra joined along the faces and apices parallel to [110]. Within each column, two polyhedra are offset at half wavelengths. The $IO_3^-$ ions are almost of regular pyramidal form, and the second complex anion comprises sulphur tetrahedrally co-ordinated by oxygen. In the monoclinic cell, the crystallographic parameters are $a = 13.94$, $b = 7.41$, $c = 7.25$ Å, $\gamma = 93°$, and $Z = 4$.[200]

## 10 Rubidium Compounds

A New Treatise of Inorganic Chemistry has been published which is a sequel to Volume 1 and encompasses the chemistry of rubidium, caesium, and francium.[201] The standard enthalpies of formation of rubidium and caesium ethoxides have been determined from calorimetric measurements on their reactions with sulphuric acid. A similar value of 96.5 ±1 cal mol$^{-1}$ was obtained for both substances.[202] Single crystals of $RbBO_2, \frac{4}{3}H_2O$ have assigned the structural formula $Rb_3[B_3O(OH)_4], 2H_2O$. X-Ray examination showed that the crystals belonged to space group $Pna2_1$ with $a = 7.90$, $b = 13.95$, $c = 9.28$ Å, $d(\text{calc.}) = 2.97$ for $Z = 12$, and contained eight-cornered polyhedra of oxygen atoms round the cations. The compound is isostructural with the potassium salt.[203] As part of a study of the oxoindates of the alkali-metals, the crystal structure of rubidium oxoindate, $Rb_2In_4O_7$, has been determined as trigonal, space group $P\bar{3}1m$ with $a = 5.62$, $c = 7.35$ Å, $d(\text{expt.}) = 6.04$, $d(\text{calc.}) = 6.13$ for $Z = 1$. The structure is derived from ccp oxygen atoms with indium in $\frac{1}{9}$ of the tetrahedral and $\frac{2}{9}$ of the octahedral interstices. Rubidium replaces $\frac{2}{3}$ of the oxygen atoms of each third layer.[204] Mass spectrometric and Knudsen effusion methods have been used to study the vapour above rubidium metaphosphate. Between 1142 and 1308 K, the vapour contains only $RbPO_3$. The enthalpy and entropy of sublimation at 1000 K are 66.3 kcal mol$^{-1}$ and 36.4 e.u. The standard enthalpy and entropy of formation of $RbPO_3$ are $-227$ ±3.9 kcal mol$^{-1}$ and 70.1 ±1.1 e.u.[205]

The reactions of rubidium and caesium sulphates with barium halides in the solid state and under anhydrous conditions have been studied. With $BaCl_2$, $BaBr_2$, or $BaI_2$, reaction proceeds in two stages in the absence of air to produce $BaSO_4$ and the alkali-metal halide at 828—963 K. Conductivity and thermal measurements showed that the exothermic first stage involved formation of a ternary eutectic, whereas the second stage was endothermic.[206] A combination of

---

[199] V. I. Vavilin, V. M. Ionov, V. V. Ilyukhin, and N. V. Belov, *Doklady Akad. Nauk S.S.S.R.*, 1974, **219**, 1108.
[200] V. I. Vavilin, V. V. Ilyukhin, and N. V. Belov, *Doklady Akad. Nauk S.S.S.R.*, 1974, **219**, 1352.
[201] 'Nouveau Traite de Chimie Minerale' Complement, Vol. 1. ed. A. Pacault and G. Pannetier, Masson, Paris, 1974.
[202] J. Bousquet, J. M. Blanchard, R. D. Joly, J. M. Letoffe, G. Perachon, and J. Thourey, *Bull. Soc. chim. France*, 1975, 478.
[203] I. Zviedra and A. Ievins, *Latv. PSR Ainat. Akad. Vestis. Khim. Ser.*, 1974, 395.
[204] D. Fink and R. Hoppe, *Z. anorg. Chem.*, 1974, **409**, 97.
[205] J. A. Ratkovskii, V. A. Ashuiko, V. A. Urikh, and V. A. Sinyaev, *Zhur. fiz. Khim.*, 1974, **48**, 2887.
[206] L. G. Berg, N. P. Burmistrova, N. I. Lisov, and N. G. Sabadash, *Izvest. V. U. Z., Khim. ikhim. Tekhnol.*, 1974, **17**, 1119.

X-ray, thermal, spectroscopic, and chemical techniques have been employed in investigating the thermal behaviour of rubidium selenite up to 1273 K. At 453 K some small decomposition to selenium occurs; oxidation in air begins at 583 K producing $Rb_2SeO_3,3Rb_2SeO_4$. This melts at 1263 K without decomposition.[207]

## 11 Caesium Compounds

Caesium thiogallate has been prepared by the reaction of $Cs_2CO_3$ with $Ga_2O_3$ at 1123 K in $H_2S$. The crystals are monoclinic, space group $C2/c$, with $a = 7.425$, $b = 12.21$, $c = 5.907$ Å, $\beta = 113.1°$, and $Z = 4$. The atomic arrangement is isotypic with $RbFeS_2$. Structural similarities between alkali thio-gallates and -ferrates can be used to study the magnetic interactions between Fe atoms within the sequence of mixed crystals of $AFe_{1-x}Ga_xS_2$, where A is an alkali-metal atom.[208]

I.r. studies have been made at 10 and 310 K of caesium cyanate and of $NCO^-$ isolated in caesium iodide by employing the alkali-metal halide pressed disc technique. On replacing CsCNO by KCNO, heating for several hours at 673 K was necessary to encourage diffusion away of $K^+$ ions and prevent absorptions due to KCNO appearing with aging.[209]

The reaction of caesium and lead metaphosphates, $CsPO_3$ and $Pb(PO_3)_2$, produces $CsPb(P_3O_9)$ containing a trimetaphosphate anion. This compound has a melting point of 814 K and was identified by means of d.t.a. and X-ray methods.[210] Tricaesium cyclotri-$\mu$-oxotris(tetrafluoroantimonate), $Cs_3(Sb_3F_{12}O_3)$, has been shown to possess an orthorhombic structure, space group $Cmca$, with $a = 14.260$, $b = 10.793$, $c = 20.656$ Å and $Z = 8$. The compound does not form a dimeric unit in the solid state such as that previously ascribed to $(As_2F_8O_2)^{2-}$, but the structure contains the anion $(Sb_3F_{12}O_3)^{3-}$ which has symmetry $m(C_s)$ giving a boat-like six-membered $(Sb_3O_3)$ ring.[211]

The crystal structure of caesium triselenocyanate, $Cs(SeCN)_3$, has been determined as monoclinic, space group $C2/c$, with $a = 7.969$, $b = 21.156$, $c = 5.593$ Å, $\beta = 98.84°$ and $Z = 4$. The selenium sequence in the triselenocyanate anion is very nearly linear, with $\angle SeSeSe = 178.31°$ and Se—Se bond length 2.650 Å, which is 0.31 Å longer than single covalent Se—Se bonds. The middle selenocyanate group is exactly linear and, within experimental error, the terminal selenocyanate groups are also linear.[212]

---

[207] T. V. Klushina, O. N. Evstaf'eva, N. M. Selivanova, Yu. M. Khozhainov, and A. Y. Monosova, *Zhur. neorg. Khim.*, 1975, **20**, 291.
[208] D. Schmitz and W. Bronger, *Z. Naturforsch.*, 1975, **30b**, 491.
[209] D. J. Gordon and D. F. Smith, *Spectrochim. Acta*, 1974, **30A**, 2047.
[210] V. M. Shpakova, I. V. Mardirosova, and G. A. Bukhalova, *Izvest. Akad. Nauk S.S.S.R., neorg. Materialy*, 1974, **10**, 2184.
[211] W. Haase, *Acta Cryst.*, 1974, **B30**, 2465.
[212] S. Hauge, *Acta Chem. Scand. (A)*, 1975, **29**, 163.

# 2
# Elements of Group II

BY R. J. PULHAM

## 1 Beryllium

The physicochemical properties of beryllium compounds and alloys have been reviewed; subjects include phase diagrams, crystal structure, and density data on alloys and compounds, with a special section devoted entirely to halides and chalcogenides.[1]

The optical spectra of beryllium atoms and diatomic beryllium molecules have been studied at 4 K, trapped in solid matrices of the elements Ne, Ar, and Kr. A strong absorption near 235 nm was observed which was assigned to the resonance transition $^1P_1 \leftarrow {}^1S_0$ that appears in the gas phase at 234.86 nm. The absorption was dependent on the matrix, appearing at 232.0, 236.0, and 240.5 nm in neon, argon, and krypton, respectively. In concentrated matrices, absorption by the $Be_2$ molecule yielded a progression of bands. The bands at 350.15 and 361.30 nm in neon and argon, respectively, are assigned to the $A^1\Sigma_u^+ \leftarrow X^1\Sigma_g^+$ transition by analogy with the similar van der Waals molecules $Mg_2$ and $Ca_2$. The vibrational frequency in the upper state is 489 and 474 cm$^{-1}$ in neon and argon matrices, respectively. No other molecular electronic transitions were detected up to 50 000 cm$^{-1}$.[2]

The crystal and molecular structures of dichlorobis(acetonitrile)beryllium have been determined by X-ray crystallography and compared with MO calculations. Beryllium occupies a distorted tetrahedral environment in $BeCl_2,2MeCN$ consisting of two Cl atoms and two N atoms. A novel feature lies in the slight 'crab-claw' configuration adopted by the nitrile ligands, as shown in Figure 1. The bond angles Cl—Be—Cl and N—Be—N are 116.8 and 100.5°, respectively, and are typical for tetrahedral distortion, i.e. as in the other beryllium compounds $BeMe_2(NC_7H_{13})_2$ and $[BeMe(C\equiv CMe)NMe_3]_2$. The Be—Cl distances (1.970 and 1.985 Å) are longer than in monomeric $BeCl_2$(1.75 Å), and, as expected, the N—C triple-bond length of 1.131 Å in the complex is 0.03 Å shorter than in free methyl cyanide.[3] X-Ray, electron, and i.r. spectroscopic studies of the nitrilotriacetates $KMN(CH_2CO_2)_3$, where M = $K_2$, Sr, Ca, Mg, Be, or KH, have been made which show that the N(1s) binding energies increase systematically from 399.8 to 402.6 eV in this series. The i.r. absorption attributed to the C—H vibrations also increased in this order, from 2835 to 3002 cm$^{-1}$. The O(1s) and C(1s) line

---
[1] Atomic Energy Reviews, 1973, Special Issue 4, p. 45.
[2] J. M. Brom, W. D. Hewett, and W. Weltner, J. Chem. Phys., 1975, 62, 3122.
[3] C. Chavant, J. C. Daran, Y. Jeannin, G. Kaufmann, and J. MacCordick, Inorg. Chim. Acta, 1975, 14, 281.

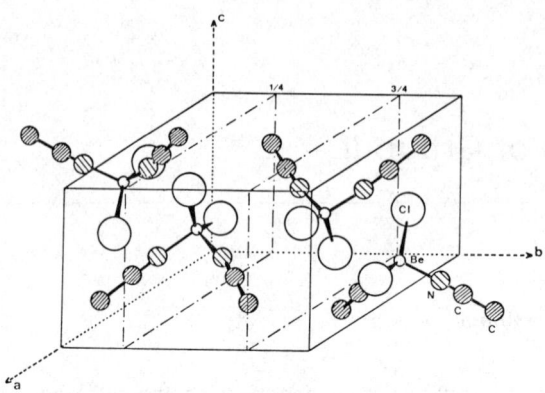

**Figure 1** *Occupancy of the unit cell of* $BeCl_2(NCMe)_2$
(Reproduced by permission from *Acta Cryst.*, 1975, **B31**, 1164)

energies and wavenumbers, however, attributed to the C—O symmetric and asymmetric vibrations, did not change significantly. The character of the M—O and M—N bonds changes from ionic for M = Sr to covalent for M = Be.[4] The i.r. and $^1$H n.m.r. spectra of beryllium oxoacetate, $Be_4O(OAc)_6$, in the presence of limited amounts of water in ethanol, indicate that contraction of the polymer network occurs, forming $[Be(OEt)OAc]_n$. Thermal dissociation of $[Be(OEt)OAc]_n$ in vacuum below 200 °C consists of cleavage of terminal groups with the simultaneous loss of alcohol, forming volatile $Be_4O(OAc)_5OEt$.[5] The hydroxyacetate. Be(OAc)OH reacts with liquid ammonia and propylamine to give the complexes $Be_2(OAc)_2(OH)_2NH_3$ and $Be_2(OAc)_2(OH)_2PrNH_2$, respectively. Spectral data indicate that the complexes exist either as trimers or polymers, containing bridging OH and OAc groups.[6]

The crystal structure of bis(pentane-2, 4-dionato)beryllium, $Be(acac)_2$, has been refined by direct X-ray methods. The compound crystallizes in the space group $P2_1$ with cell parameters $a = 13.537$, $b = 11.378$, $c = 7.762$ Å, $\beta = 100.76°$ and $Z = 4$, with two molecules in the asymmetric unit see Figure 2). The average interatomic separation for Be—O is 1.62 Å and the average O—Be—O chelate ring angle is 107° (see Figure 3). In one molecule, the Be atom is located on each of the least-squares planes through the ring atoms of the two different chelate ligands, while in the other molecule the Be atom is 0.20 Å removed from either of the two least-squares planes.[7] Chelation of beryllium ions by anthranilic acid, A, is reported to occur in 0.5 M sodium perchlorate solutions at 25 °C to give several complexes with the following formation constants, $K$; $BeA^+$, 91; $Be_2(OH)A_2^+$, 32; $Be_3(OH)_3A_2^{2+}$, $4.6 \times 10^{-8}$; and $Be_3(OH)_3A_2^+$, $4.6 \times 10^{-6}$.[8] A gravimetric method for

---

[4] A. I. Grigor'ev, V. I. Nefedov, N. I. Voronezheva, and Ya. V. Salyn, *Zhur. neorg. Khim.*, 1974, **19**, 2576.
[5] A. V. Sipachev and A. I. Grigor'ev, *Zhur. neorg. Khim.*, 1975, **20**, 25.
[6] A. I. Grigor'ev and L. N. Reshetova, *Zhur. neorg. Khim.*, 1974, **19**, 2612.
[7] J. M. Stewart and B. Morosin, *Acta Cryst.*, 1975, **B31**, 1164.
[8] G. Duc, F. Bertin, and G. Thomas-David, *Bull Soc. chim. France*, 1975, 495.

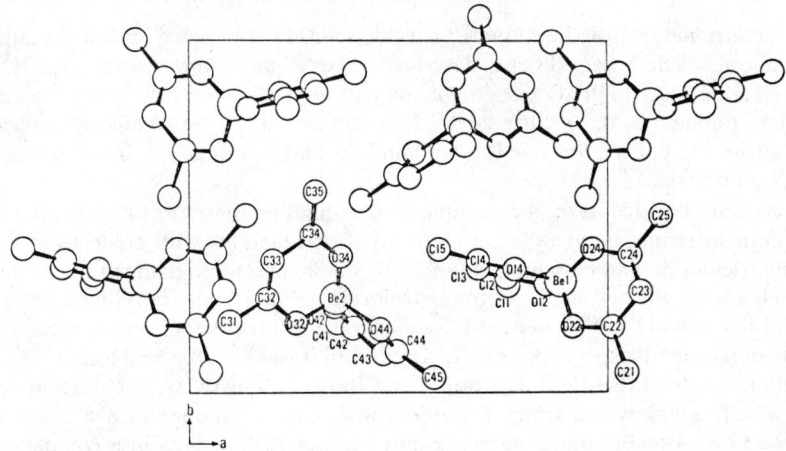

**Figure 2** *Projection along the c axis of the crystal structure of* Be(acac)$_2$. *The labelling scheme used in the original text is shown.*
(Reproduced by permission from *Acta Cryst.*, 1975, **B31**, 1164)

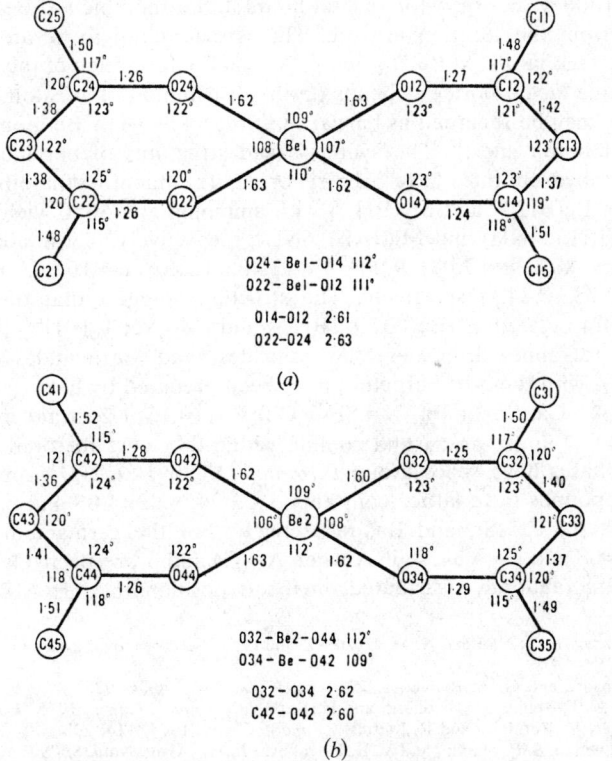

**Figure 3** *Interatomic separations and angles for the two molecules in* Be(acac)$_2$.
(Reproduced by permission from *Acta Cryst.*, 1975, **B31**, 1164)

the determination of beryllium in acid solution by using nitrilotrimethylphosphoric acid has been described. Beryllium ions react to form $Be_3[NH(CH_2PO_3)_3],10H_2O$ but do not complex with the necessary buffer reagents, glycine, phthalic acid, or citric acid. There appears to be no significant interference from $Mg^{2+}$, $Ca^+$, $Sr^{2+}$, or $Ba^{2+}$ in hundred-fold amounts, nor from $NO_3^-$, $Cl^-$, $SO_4^{2-}$, nitrilotriacetic acid, or edta.[9]

Beryllium borides have been synthesized at high temperatures (>1000 °C) and the enthalpies measured by calorimetry of their reactions with chlorine to form boron trichloride and beryllium chloride. By combining these data with the known enthalpies of formation of boron trichloride (-98.5) and beryllium chloride (-123.0 kcal mol$^{-1}$), the standard free energies of the beryllium borides have been determined as -18.3, -16.7, 15.5, -26.0, and -38.6 kcal mol$^{-1}$ for the compounds $Be_4B$, $Be_2B$, $BeB_2$, $BeB_6$, and $BeB_{12}$, respectively.[10] The compound $Al_{0.06}BeB_{3.05}$ has space group $P6/mmm$, with lattice parameters $a = 9.800$ and $c = 9.532$ Å. The Be and B atoms occupy the polyhedral units in a complex $B_{12}$ icosahedral structure where the linkage between polyhedra is related to that of $B_{12}$ in $\beta$-rhombohedral boron. Aluminium atoms occupy interstitial sites.[11] Tetragonal $AlBeB_{24}$ [or $\beta$-(Al, Be)$B_{12}$] has been prepared by heating a mixture of the elements Be, Al, and B with Si, C, Nb, or Mg at 500 °C in vacuum and then at 1400—1500 °C in argon for several hours. Orthorhombic (Al,Be)$B_{12}$ was also separated from the reaction mixture. This is considered to be an intermediate between tetragonal $\beta$-(Al,Be)$B_{12}$ and $\beta$-$AlB_{12}$. The formation of the tetragonal or orthorhombic mixed boride depends on the distribution of Al atoms; the degree of ordering may be regarded as partial substitution of Al or Be atoms on or near icosahedral boron sites.[12] The synthesis and structures of beryllium and magnesium tetrahydroborates have been reported. Treatment of the etherates of the compounds $Be(BH_4)_2$ and $Mg(BH_4)_2$ with ammonia at −50 °C yielded the complexes $Be(BH_4)_2,4NH_3$ and $Mg(BH_4)_2,6NH_3$, respectively. These adducts are orthorhombic, with $a = 7.07$, 9.00; $b = 12.40$, 12.40; $c = 10.00$, 10.78 Å, and $d$(expt) = 0.78, 0.84, respectively. The structures suggest that the compounds should be formulated as $[Be(NH_3)_4](BH_4)_2$ and $[Mg(NH_3)_6](BH_4)_2$.[13]

Ternary alkaline-earth beryllium silicides and germanides of formula $ABe_{0.75}X_{1.75}$ with the $AlB_2$ structure have been prepared by heating the elements A:Be:X (A = Ca, Sr, or Ba; X = Si or Ge) in 1:0.75:1.25 ratio for an hour at 1300—1350 °C under argon and cooling within 24 hours to room temperature. The new phases have space group $P_6/mmm$-$D_{6h}^4$ or $P6m2$-$D_{3h}^1$ and $Z = 3$. The silicon compounds have lattice constants $a = 3.94$, 4.03, 4.05 and $c = 4.38$, 4.68, 5.05 Å, for A = Ca, Sr, and Ba, respectively. For the germanium compounds, $a = 4.00$, 4.08 and $c = 4.34$, 4.66 Å when A = Ca and Sr, respectively. The Be and X atoms are randomly distributed on the B positions of this $AlB_2$-type structure.[14]

[9] S. S. Morozova, L. V. Nikitina, N. M. Dyatlova, and G. V. Serebryakova, *Zhur. analit. Khim.*, 1974, **29**, 1323.
[10] M. P. Morozova and G. A. Rybakova, *Zhur. fiz. Khim.*, 1974, **48**, 1608.
[11] R. Mattes, K. F. Tebbe, H. Neidhard, and H. Rethfeld, *Z. anorg. Chem.*, 1975, **413**, 1.
[12] H. J. Becher, H. Rethfeld, and R. Mattes, *Z. anorg. Chem.*, 1975, **414**, 203.
[13] K. N. Semenenko, S. P. Shilkin, and V. B. Polyakova, *Izvest. Akad. Nauk S.S.S.R., Ser, khim.*, 1975, 735.
[14] N. May, W. Mueller, and H. Schaefer, *Z. Naturforsch.*, 1974, **29b**, 325.

The crystal structure of sodium beryllium orthosilicate, $Na_2BeSiO_4$, has been determined as monoclinic, space group $P2_1/b$, with $a = 7.022$, $b = 9.933$, $c = 4.960$ Å, $\gamma = 90.03°$, $Z = 4$. The structure contains discrete $SiO_4$ tetrahedra (Si—O = 1.61—1.65 Å), as in the compound $Na_2Zn(SiO_4)$, which are joined by $Be_2O_6$ groups in corrugated sheets. The $Be_4O$ tetrahedra (Be—O = 1.61—1.67 Å) are in dinuclear $Be_2O_6$ groups with Be—Be distances of 2.33 Å. There are two crystallographically different sodium ions, which are surrounded by six oxygen atoms at 2.32—2.67 Å and 2.31—2.65 Å, respectively.[15] The crystal structure of high-temperature lithium beryllium silicate, $\gamma$-$Li_2BeSiO_4$, has been determined on crystals quenched to room temperature from 1100 °C. The structure is directly related to the structures of the oxides $LiAlO_2$, $NaAlO_2$, and $NaFeO_2$ (high temperature, $\gamma$-forms). The orthorhombic symmetry and cell of the silicate $\gamma$-$Li_2BeSiO_4$ ($a = 6.853$, $b = 6.927$, $c = 6.125$ Å, space group $C222_1$), as distinct from the tetragonal symmetry of the oxide $\gamma$-$LiAlO_2$, is due to systematic, ordered replacement of aluminium by silicon and beryllium. The O atoms form a distorted hexagonal close-packed arrangement; the cations are distributed over half the available tetrahedral sites on either side of the oxygen layers. Mean bond distances for Si—O and Be—O are 1.635 and 1.647 Å, respectively. A comparison of Li—O distances in $LiO_4$ tetrahedra which share a common edge with the polyhedra $BeO_4$, $SiO_4$(in $\gamma$-$Li_2BeSiO_4$), and $AlO_4$(in $\gamma$-$LiAlO_2$) shows that the size and the distortion of the $LiO_4$ tetrahedron is similar in all these structures but that the Li atom is displaced progressively further from the centre of its tetrahedron and away from the common edge, as the cation in the other tetrahedron changes from Be to Al to Si.[16]

A structural study of potassium and rubidium amidoberyllates, prepared in a reaction between a solution of potassium or rubidium in liquid ammonia with metallic beryllium, reveals that the compounds belong to the space group $Pbca$, with $a = 12.88$, $b = 11.21$, $c = 13.19$ Å for $KBe(NH_2)_3$, and $a = 12.96$, $b = 11.55$, and $c = 13.48$ Å for $RbBe(NH_2)_3$. The structures are composed of trigonal groups of $[Be(NH_2)_3]^-$ and $M^+$ ions. I.r. and Raman spectral data are also available.[17,18]

The crystal structure of sodium oxopyroberyllate, $Na_6Be_2O_5$, has been determined as tetragonal, space group $P4_2/nm$, $P4n2$, or $P4_2/mnm$, with $a = 5.67$, $c = 10.21$ Å, $d$(expt.) = 2.38, $d$(calc.) = 2.45 for $Z = 2$. This structure contains $Be_2O_5$ units which are formed from $BeO_3$ triangles sharing a corner; the angle at the bridging oxygen atom is 167.3°. The $Be^{2+}$ ion lies 0.17 Å outside the plane of the three nearest oxygen atoms.[19] The oxoberyllates $K_2Be_2O_3$, $K_4BeO_3$, $Rb_2Be_3O_4$, $Rb_2Be_2O_3$, $Rb_4BeO_3$, $Cs_2Be_3O_4$, and $Cs_2BeO_2$ have been shown to exist, by means of X-rays. For single crystals the crystallographic data are (compound, structure, $a$, $b$, $c$, $\alpha$, $\beta$, $\gamma$): $Li_4BeO_3$, triclinic, 5.49, 9.53, 4.93 Å, 114, 105, 106°; $Na_2Be_2O_3$, triclinic, 6.32, 8.53, 2.71 Å, 107, 105, 78°; $Na_2Be_2O_3$,

---

[15] G. F. Plakhov, M. A. Simonov, and N. V. Belov, *Doklady Akad. Nauk S.S.S.R.*, 1974, **218**, 335.
[16] R. A. Howie and A. R. West, *Acta Cryst.*, 1974, **B30**, 2434.
[17] M. G. B. Drew, J. E. Goulter, L. Guemas-Brisseau, P. Palvadeau, J. Rouxel, and P. Herpin, *Acta Cryst.*, 1974, **B30**, 2579.
[18] L. Guemas, P. Palvadeau, J. Rouxel, G. Lucazeau, D. Bougeard, and A. Novak, *Spectrochim. Acta*, 1975, **31A**, 421.
[19] M. Jansen, *Naturwiss.*, 1975, **62**, 236.

**Table 1** Standard enthalpies/kcal(mol dimer)$^{-1}$ of dimerization of gaseous chlorides

| | | | | | |
|---|---|---|---|---|---|
| $MgCl_2$ | $-40.5^a$ | $ZnCl_2$ | $-26.5$ | $LiCl$ | $-49.5^a$ |
| $CaCl_2$ | $-50.3$ | $CdCl_2$ | $-27.3$ | $NaCl$ | $-48.6^a$ |
| $SrCl_2$ | $-51.2$ | $HgCl_2$ | $-13.0$ | $KCl$ | $-45.0^a$ |
| $BaCl_2$ | $-47.1$ | | | $RbCl$ | $-42.9^a$ |
| | | | | $CsCl$ | $-42.9^a$ |

$^a$ Previously reported values.

tetragonal, 5.67, —, 10.2 Å; $K_2BeO_2$, monoclinic, 7.09, 10.57, 5.70 Å, $\gamma = 131°$, and $Rb_2BeO_2$, monoclinic, 7.46, 11.17, 5.88 Å, $\gamma = 131°$.[20]

The reaction of elemental beryllium with gaseous molecules $BeF_2$ and $Cl_2$ has been studied by effusion mass spectrometry at 1415—1592 K. The standard enthalpy of formation of the chlorofluoride BeClF(g) was calculated as $-139.0 \pm 4$ and $-140.5 \pm 2$ kcal mol$^{-1}$ by second- and third-law thermodynamics, respectively. The dissociation energy of 104 and 144 kcal mol$^{-1}$ calculated by the same method for the species BeF agrees with values reported previously.[21] Mass spectrometry has also been used to study gaseous dimeric molecules. Standard enthalpies of dimerization for the chlorides $CaCl_2$, $SrCl_2$, and $BaCl_2$ are larger than those for the corresponding chlorides $ZnCl_2$, $CdCl_2$, and $HgCl_2$. Values are shown in Table 1, together with previously reported values for magnesium chloride and the alkali-metal halides for comparison.[22] Gaseous $BeCl_2$ molecules have been shown to react with $HClO_4$ vapour at 250 °C to form $Be_4O(ClO_4)_6$. This compound, characterized by its i.r. spectrum and by X-ray diffraction, contains bidentate bridging ligands similar to other known beryllium oxy-salts.[23]

## 2 Magnesium

The preparation and electrical properties of magnesium mercury, $Mg_5Hg_3$, have been reported. The intermetallic compound was prepared by heating the elements together above 562 °C, at which temperature the compound melts, in an evacuated silica tube. Electrical resistivity and Hall-effect measurements were made from 2.4 to 297 K. No phase transition was observed over this range. The Hall constant was low and the ideal resistivity (the difference between total and residual resistivity) had the form $\rho \propto T^\alpha$, where $\alpha = 1.9$ and 1.0 at 11.50 K and 150—300 K, respectively.[24] The solubility of mercury in magnesium has been determined by measuring the saturated pressure of mercury vapour above the alloys under conditions of non-equilibrium solidification. The solubility increases from 0.71% at 298 K to 0.93% at 721 K. The pressures were used to determine the activity of mercury in the alloys. The activity was characterized by a net concentration independence in the heterogeneous region ($Mg + Mg_3Hg$), but in the solid-solution region the activity increased monotonically with mercury con-

---

[20] P. Kastner and R. Hoppe, Z. anorg. Chem., 1975, **415**, 249.
[21] M. Farber and R. D. Srivastava, J.C.S. Faraday I, 1974, **70**, 1581.
[22] H. Schaefer and M. Binnewies, Z. anorg. Chem., 1974, **410**, 251.
[23] Z. K. Nikitina and V. Ya. Rosolovskii, Izvest. Akad. Nauk S.S.S.R., Ser khim., 1975, 1173.
[24] E. Cruceanu, D. Niculescu, and O. Ivanciu, J. Materials Sci., 1974, **9**, 1389.

## Elements of Group II

**Table 2** Core level binding energies for some magnesium compounds measured by XPS

| Peak (core level assignment) | Binding energy/eV with respect to Fermi level | | | | |
|---|---|---|---|---|---|
| | Mg | $Mg_2Cu$ | $Mg_3Au$ | $Mg_3Bi_2$ | MgO |
| $L_{2,3}(2p)$ | 49.9 | 49.8 | 49.7 | 50.6 | 51.6 |
| $L_1$ | 89.0 | 88.8 | not observed | 89.6 | 90.4 |
| $K(1s)$ | 1303.4 | 1302.9 | 1303.0 | 1303.9 | 1305.3 |

centration.[25] The KLL and KLM Auger and X-ray photoelectron spectra for the species Mg, $Mg_2Cu$, $Mg_3Au$, $Mg_3Bi_2$, and MgO have been reported. The binding energies of magnesium core levels measured with respect to the Fermi level for magnesium and the intermetallics, together with the binding energies for magnesium oxide measured with respect to the pure metal level, are given in Table 2. Interpretation of the chemical shifts is not straightforward because, rigorously, these should be measured with respect to the vacuum level. As expected, however, there is a general shift to higher binding energies when the magnesium atom is bonded to a more electronegative atom. The magnesium KLL Auger spectra energies and assignments are given in Table 3. Several of the peaks are given assignments analogous to those made earlier for magnesium oxide.[26] The reactions of clean surfaces of the metals Mg, Al, Mn, and Cr with oxygen and water vapour have been followed by X-ray photoelectron spectroscopy. With exposures up to $10^{-2}$ Torr s, the intensity of the $O(1s)$ peak was approximately proportional to the quantity of oxygen adsorbed on the metals Mg, Al, and Cr at room temperature. The spectra could distinguish between surface oxides, hydroxides, and adsorbed water. The metals principally give oxides in their reaction with water at room temperature, but hydroxides commonly form at lower temperatures. With oxygen, the surfaces of the metals Mg, Al, and Cr become protected by the first layers of oxide which are formed, but this is not the case for manganese.[27] The reaction of magnesium with mixtures of nitrogen and oxygen has been studied and shown to proceed with irregular kinetics. Initially the metal reacted simultaneously with the gases to form a nitride layer next to the metal and an oxide layer adjacent to the gas. The double layer was temporarily protective, but cracking accelerated both oxidation and nitriding.[28]

**Table 3** X-Ray-excited KLL Auger spectrum (eV) of Mg, $Mg_2Cu$, $Mg_3Au$, $Mg_3Bi_2$, and MgO

| Assignment | Mg | $Mg_2Cu$ | $Mg_3Au$[a] | $Mg_3Bi_2$ | MgO |
|---|---|---|---|---|---|
| $KL_2L_3(2s^22p^4; {}^1D_2)$ | 1185.5 | 1185.7 | 1185.4 | 1184.6 | 1180.0 |
| $KM-L_{23}L_{23}M$ | 1183.5 | | | 1183.9 | |
| $KL_2L_2(2s^22p^4; {}^1S)$ | 1180.5 | 1180.6 | 1179.9 | 1178.9 | 1175.4 |
| $KL_1L_{23}(2s^12p^5; {}^3P)$ | 1154.1 | 1154.0 | 1153.8 | 1153.8 | 1148.8 |
| $KL_1L_{23}(2s^12p^5; {}^1P)$ | 1140.1 | 1139.9 | 1139.8 | 1138.9 | 134.4 |
| $KL_1L_1(2s^02p^6; {}^1S)$ | 1106.2 | 1106.1 | 1106.0 | 1105.2 | 1101.2 |

[a] The composition of this alloy is uncertain

[25] M. Ya. Vyazner, I. P. Vyatkin, and S. V. Mushkov, *Izvest. Akad. Nauk S.S.S.R., Met.*, 1975, 237.
[26] J. C. Fuggle, L. M. Watson, D. J. Fabian, and S. Affrossman, *J. Phys. (F)*, 1975, **5**, 375.
[27] J. C. Fuggle, L. M. Watson, D. J. Fabian, and S. Affrossman, *Surface Sci.*, 1975, **49**, 61.
[28] B. Dupré and R. Streiff, *J. Nuclear Materials*, 1975, **56**, 195.

The crystal structure of the 1:1 complex magnesium chloride hexahydrate with 1,4,7,10-tetraoxacylododecane, $Mg(H_2O)_6Cl_2,C_8H_{16}O_4$, has been determined by single-crystal $X$-ray diffraction methods. The crystal is monoclinic, space group $C2/c$, with cell constants $a = 16.406$, $b = 8.443$, $c = 12.729$ Å, and $\beta = 93°18'$, with $d(\text{calc}) = 1.432$ for $Z = 4$. The structure contains octahedral $Mg(H_2O)_6^{2+}$ units, very similar to those in the hydrate $MgCl_2, 6H_2O$, with Mg—O distances of 2.054 of 2.054—2.082 Å. Eight of the twelve water hydrogens form H-bonds to ether oxygens from four different cyclomer molecules. The $Mg(H_2O)_6^{2+}$ and Cl⁻ ions form two-dimensional layers, as do the twelve-membered rings. The tetraoxacyclododecane molecule has virtually normal bond distances and angles.[29] In the series of paper on structures of divalent metal complexes with pyridinecarboxylic acids, the structure of tetra-aquo(isonicotinato-$N$)(isonicotinato-$O$)magnesium has been determined. The crystal has space group $P2_2/n$, with lattice parameters $a = 6.38$, $b = 36.69$, $c = 7.11$ Å, $\beta = 113.5°$, $d(\text{expt.}) = 1.48$, $d(\text{calc.}) = 1.48$ for $Z = 4$. The compound consists of discrete complex molecules with two organic and four water molecules co-ordinated octahedrally to magnesium. One of the two isonicotinate ligands co-ordinates through the carboxylic oxygen while the other co-ordinates through the pyridine nitrogen.[30] Complexing between metals and organic acids has been studied by pH–potentiometric titrations. The cations $Mg^{2+}$, $Ca^{2+}$, $Sr^{2+}$, and $Ba^{2+}$ all form a monoprotonated and a normal complex with 2-hydroxy-1,3-diaminopropane-$NN'$-dimalonic acid. In addition, the $Mg^{2+}$ ion also forms a dinuclear complex. The stability constants $K$ for the complexes of $N$-(carboxymethyl)aspartic acid ($H_3L$) with ions $Mg^{2+}$, $Ca^{2+}$, $Sr^{2+}$, and $Ba^{2+}$ decrease with decreasing cation radium (ML⁻, log $K$; 4.57, 3.71, 3.32, and 3.21, respectively). The first and second acid dissociation constants are much larger than the third ($pK_1 = 2.58$, $pK_2 = 2.85$, and $pK_3 = 9.65$), indicating a betaine-type of structure for the molecule in solution.[31,32] The i.r. and Raman spectra of solid magnesium dibromide etherate, $MgBr_2,2Et_2O$, have been studied from 90 to 300 K in the 30—4000 cm⁻¹ range. The crystal exhibits a reversible phase change which involves primarily a conformational change in the ether molecules and, much less, the arrangement of the ligands around the central Mg atom.[33]

The standard free energies of formation have been calculated from vapour-pressure data for the compounds $Mg_2Si$, $Mg_2Ge$, $Mg_2Sn$, $Mg_2Pb$, $MnSi$, $MnSi_{1.73}$, and $CoSi$; values are $-12.2$, $-27.4$, $-7.2$, $-1.2$, $-22.9$, $-22.3$, and $-22.3$ kcal mol⁻¹, respectively.[34]

The preparation and structure of magnesium polyphosphide, $MgP_4$, have been described. The compound was prepared by the reaction of gaseous phosphorus with the phosphide $Mg_3P_2$ at 600 °C in a sealed silica tube. Evidence for a primitive monoclinic cell was obtained from electron microdiffraction. Refinement of $X$-ray powder diffraction data showed that the compound is isostructural with

---

[29] M. A. Neuman, E. C. Steiner, F. P. Van Remoortere, and F. P. Boer, *Inorg. Chem.*, 1975, **14**, 734.
[30] C. M. Biagini, V. A. Chiesi, C. Guastini, and D. Viterbo, *Gazzetta*, 1974, **104**, 1087.
[31] A. I. Kapustnikov and P. Gorelov, *Zhur neorg Khim.*, 1975, **20**, 904.
[32] I. P. Gorelov and V. M. Nikol' skii, *Zhur. neorg. Khim.*, 1975, **20**, 1722.
[33] J. Kress and A. Novak, *J. Mol. Structure*, 1974, **23**, 215.
[34] N. Rozyev and Ch. Agabaev, *Izvest. Akad. Nauk Turkm. S.S.R., Ser. fiz-tekh., khim., geol. Nauk.*, 1974, 43.

## Elements of Group II

CdP$_4$, with $a = 5.15$, $b = 5.10$, $c = 7.50$ Å, and $\beta = 81°$.[35] The crystal structure of the compound $\alpha$-Mg$_3$Sb$_2$ has been determined. The compound crystallizes in the space group $P\bar{3}m1$, with $a = 4.568$, $c = 7.229$ Å, and $d$(calc.) $= 4.01$ for $Z = 1$. The structure resembles that of the oxide La$_2$O$_3$.[36]

Crystals of the oxide K$_6$MgO$_4$ have been prepared by the reaction of melted potassium and magnesium monoxides. The symmetry is hexagonal, space group $P6_3mc$, with cell parameters $a = 8.47$, $c = 6.58$ Å, and $Z = 2$. The structure is made up of two types of triangular based antiprisms, A and B (Figure 4), linked by short K—O bonds. Antiprism A consists of three atoms of K(2) and three atoms of O(1) superimposed. This antiprism is capped by a Mg and an O(2) atom, which form a near regular tetrahedron Mg(3)O(1)O(2) [Mg—O(1) = 2.03, Mg—O(2) = 1.99 Å]; this assembly has the formula K$_3$MgO$_4$, with K(2)—O(1) = 2.73, K(2)—K(2) = 3.34, and O(1)—O(1) = 3.33 Å. Antiprism B is composed exclusively of six K atoms (Figure 4), with basal K(1)—K(1) = 3.66 Å and vertical K(1)—K(1) = 3.91 Å. The overall structure is then synthesized by assembling the capped and normal antiprisms, A and B respectively, as shown in Figure 5.[37]

The Mg$K\beta$ emission band has been measured in single crystals of magnesium hydroxide. For magnesium and other third-row elements, the outermost electron

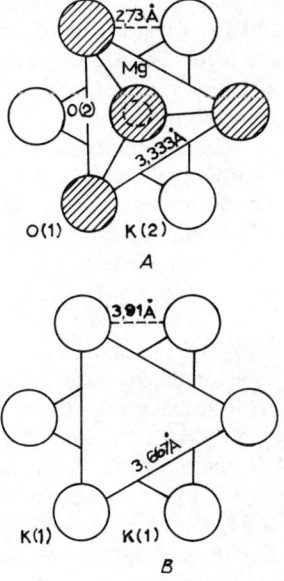

**Figure 4** *Capped antiprism A (K$_3$O$_3$) and antiprism B (K$_6$) in K$_6$MgO$_4$* (Reproduced by permission from *Acta Cryst.*, 1974, **B30**, 2667)

---

[35] A. E. Maslout, M. Zanne, F. Jeannot, and C. Gleitzer, *J. Solid State Chem.*, 1975, **14**, 85.
[36] M. Martinez-Ripoll, A. Haase, and G. Brauer, *Acta Cryst.*, 1974, **B30**, 2006.
[37] B. Darriet, M. Devalette, F. Roulleau, and M. Avallet, *Acta Cryst.*, 1974, **B30**, 2667.

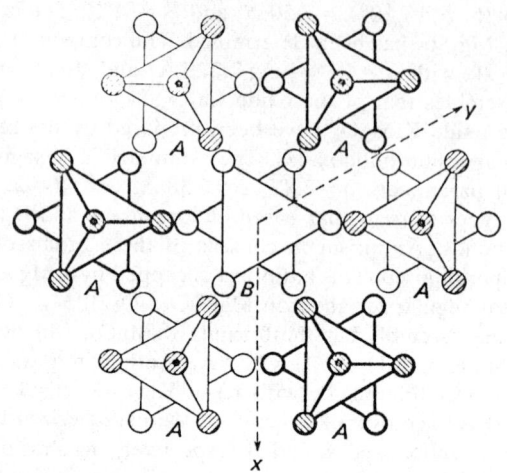

**Figure 5** *Assembly of antiprisms A and B in* $K_6MgO_4$
(Reproduced by permission from *Acta Cryst.*, 1974, **B30**, 2667)

levels which from the valence band consist of a mixed $3s$, $3p$, and $3d$ symmetries. The $K\beta$-emission band results from transitions of these outermost electrons to a vacancy created in the $1s$ core level. The $MgK\beta$ band shows two maxima at 1297 and 1292 eV, having an intensity ratio of about 2:1, which are assigned to $3T_{1u} \rightarrow 1s$ and $2T_{1u} \rightarrow 1s$ transitions involving $1\pi$ and $3\sigma$ molecular orbitals of the OH⁻ groups, respectively. The assignment is based on octahedral symmetry, $O_h$, since each Mg atom is co-ordinated octahedrally by six O atoms, even though, when H atoms are included, the structure becomes trigonal. The MO energy levels are shown in Figure 6. The atomic orbitals on the cation side are $3s(A_{1g})$, $3p(T_{1u})$, and $3d(T_{2g} + E_g)$ with rising energy. On the anion side, the atomic orbitals are $2\sigma$, $3\sigma$, and $1\pi$, originating from the oxygen $2s$, $2p_z$, and $2p_xp_y$, orbitals, respectively. The bands are assigned as shown, and the 2:1 intensity reflects the occupancy of the respective states $3T_{1u}$ and $2T_{1u}$.[38]

Detailed X-ray powder data for the hydrate $MgBr_2,6H_2O$ have been indexed on a monoclinic unit cell with space group $C2/m$ and lattice parameters $a = 10.290$, $b = 7.334$, $c = 6.211$ Å, $\beta = 93°25'$, and $Z = 2$.[39] Formation of complex ions has been investigated in melts of $MgCl_2$ + alkali-metal chloride by X-ray analysis. The compound $K_2MgCl_4$ forms in these mixtures and is tetragonal, with space group $I4/mmm$, $a = 4.94$, $c = 15.58$ Å, and $Z = 2$. The compound is isostructural with the fluoride $K_2NiF_4$ and contains an infinite two-dimensional array of $MgCl_6$ octahedra, sharing edges. The compound $Cs_2MgCl_4$ is orthorhombic, with space group $Pnma$, $a = 9.777$, $b = 7.514$, $c = 13.234$ Å, and $Z = 4$. This compound is isostructural with $\beta$-$K_2SO_4$ and contains discrete $MgCl_4^{2-}$ ions, which probably persist in the melt at temperatures near to the melting point.[40] The

---

[38] F. Freund, *Phsy. Status Solidi (B)*, 1974, **66**, 271.
[39] C. A. Sorrell and R. R. Ramey, *J. Chem. and Eng. Data*, 1974, **19**, 307.
[40] C. S. Gibbons, V. C. Reinsborough, and W. A. Whitla, *Canad. J. Chem.*, 1975, **53**, 114.

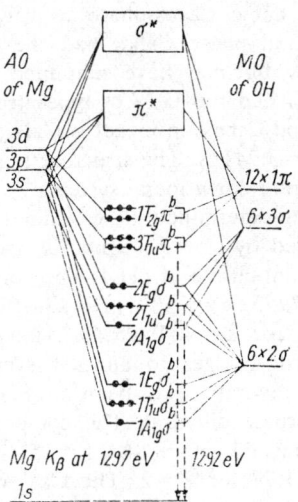

**Figure 6** *MO Scheme for* $Mg(OH)_2$ *assuming octahedral symmetry, with tentative assignment of the main Mg Kβ band maxima*
(Reproduced by permission from *Phys. Status Solidi (B)*, 1974, **66**, 271)

enthalpies of solution of the hydrazine complexes of magnesium, calcium, strontium, and barium perchlorates have been determined in hydrochloric acid, together with the enthalpies of solution of the non-solvated perchlorates in water. The standard enthalpies of all these compounds were calculated from the results and are given in Table 4.[41]

## 3 Calcium

The electrical resistivity and thermopower of metallic calcium, strontium, and barium have been measured from room temperature to near their melting points. From discontinuities observed in these parameters as functions of temperature, the f.c.c.–b.c.c. phase transition was determined in calcium at 428 ± 2 °C and in strontium at 542 ± 2 °C, both at ambient pressure.[42] Four compounds have been identified in the Ca–Ni system by means of X-ray methods. The intermetallic

**Table 4** *Standard enthalpies of formation*/kcal mol$^{-1}$ *of* $M(ClO_4)_2$ *and* $M(ClO_4)_2, xN_2H_4$.

| M | | x | | x | |
|---|---|---|---|---|---|
| Mg | −134.49 | — | −154.39 | — | |
| Ca | −175.56 | 2 | −178.17 | 1 | −184.86 |
| Sr | −184.19 | 2 | −183.23 | 1 | −187.89 |
| Ba | −187.93 | 2 | −179.99 | 1 | −180.96 |

[41] N. V. Krivtsov, Zh. G. Sakk, and V. Ya. Rosolovskii, *Zhur. neorg. Khim.*, 1974, **19**, 3213.
[42] J. Katerberg, S. Neimeyer, D. Penning, and J. B. Van Zytveld, *J. Phys. (F)*, 1975, **5**, L74.

compound $CaNi_2$ had a cubic Laves-phase $MgCu_2$-type structure, with $a = 7.239$ Å. The rhombohedral phase $CaNi_3$ had the $PuBi_3$-type structure, with $a = 5.030$ and $c = 24.27$ Å, but may have contained small amounts of adjacent phases. The phase $Ca_2Ni_7$ had the $Gd_2Co_7$-type structure, with $a = 5.009$ and $c = 36.06$ Å. The compound richest in nickel is $CaNi_5$, with a $CaCu_5$-type structure and $a = 4.055$ and $c = 3.941$ Å. The structure types of these compounds are usually found in calcium–rare-earth metal systems.[43]

The structure has been determined of calcium hydrogen citrate trihydrate $Ca(C_6H_6O),3H_2O$, prepared by slow precipitation from an aqueous mixture of calcium chloride and sodium citrate at pH 4. The compound is monoclinic, with space group $P2_1/c$, $a = 8.7955$, $b = 5.5891$, $c = 23.817$ Å, $\beta = 116.77°$, $d$(expt.) = 1.79, and $d$(calc.) = 1.806 for $Z = 4$. Each $Ca^{2+}$ ion is surrounded by seven O atoms, two of which form bridges to another symmetry-related $Ca^{2+}$ ion.[44] Calcium binding to carbohydrates has been investigated through the crystal structure of calcium ascorbate dihydrate. The compound $Ca(C_6H_7O_6)_2,2H_2O$ is monoclinic, with space group $P2_1$, $a = 8.842$, $b = 15.77$, $c = 6.364$ Å, $\beta = 115.88°$, $d$(expt.) = 1.76, $d$(calc.) = 1.77 for $Z = 2$. The $Ca^{2+}$ ion is bonded to two $H_2O$ molecules and three ascorbate anions. One ascorbate ion chelates the $Ca^{2+}$ ion through the two OH groups of its glycerol side-chain; a second ascorbate anion chelates the cation through a pair of glycerol OH groups in concert with the ionized O atom; a third co-ordinates to the cation through the carbonyl oxygen atom of the lactone moiety, as shown in Figure 7. The eight O atoms surrounding the $Ca^{2+}$ ion form a slightly distorted square antiprism, with Ca—O distances ranging from 2.415 to 2.530 Å. The two ascorbate anions assume different conformations about the C—C bond of the glycerol side-chain.[45] Another determination of the structure of this compound verifies the square antiprismatic co-ordination about potassium, and gives the K—O distances of from 2.409 to 2.520 Å. In this investigation the cell parameters of the monoclinic cell were $a = 8.335$, $b = 15.787$, $c = 6.360$ Å, and $\beta = 107.48°$.[46] Binding of calcium to D-glucuronate residues of oligo- and poly-saccharides has been investigated through the determination of the crystal structure of the hydrated calcium bromide salt of D-glucuronic acid, $CaBr(C_6H_9O_7),3H_2O$. Crystals of the salt are monoclinic, with space group $P2_1$, $a = 6.410$, $b = 10.784$, $c = 8.879$ Å, $\beta = 92.07°$, and $Z = 2$. An outstanding feature of the crystal packing is the interaction of D-glucuronate anions with $Ca^{2+}$ cations. The cation is co-ordinated to three symmetry-related D-glucuronate anions and to two $H_2O$ molecules. These anions bind $Ca^{2+}$ through three chelation sites; one involving a carboxyl-O atom. The cation is thereby surrounded by a co-ordination polyhedron of eight oxygen atoms forming a slightly distorted square antiprism. The Ca—O distances range from 2.38 to 2.57 Å. This co-ordination is similar to that found in several other calcium–carbohydrate salts and complexes.[47] The crystal structure has been determined for the isomorphous pair calcium sodium galacturonate and strontium

---

[43] K. H. J. Buschow, *J. Less-Common Metals*, 1974, **38**, 96.
[44] B. Sheldrick, *Acta Cryst.*, 1974, **B30**, 2056.
[45] R. A. Hearn and C. E. Bugg, *Acta Cryst.*, 1974, **B30**, 2705.
[46] J. Hvoslef and K. E. Kjellevold, *Acta Cryst.*, 1974, **B30**, 2711.
[47] L. DeLucas, C. E. Bugg, A. Terzis, and R. Rivest, *Carbohydrate Res.* 1975, **41**, 19.

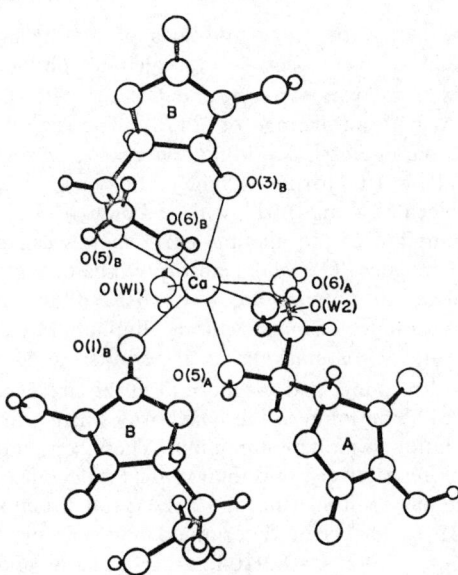

**Figure 7** *Environment of the calcium ion in calcium ascorbate dihydrate. Atoms from ascorbate ions A and B are represented by the respective letter. O(W1) and O(W2) are oxygen atoms from water molecules.*
(Reproduced by permission from *Acta Cryst.*, 1974, **B30**, 2705)

sodium galacturonate, these being models for the biologically important interactions between $Ca^{2+}$ ions and the pyranose forms of sugar carboxylic acids. The space group is $P6_3$ and the cell dimensions are $a = 13.56$, $c = 9.70$ Å for $NaCaC_{18}C_{39}O_{27}$, and $a = 13.50$, $c = 9.63$ Å for $NaSrC_{18}H_{39}O_{27}$. There are six galacturonate anions, twelve $H_2O$ molecules, two $Na^+$ cations, and two alkaline-earth cations in the hexagonal unit cell. Each bivalent cation is co-ordinated by three sugar anions and three $H_2O$ molecules in a nine-co-ordinate trigonal prism arrangement. The sugar O atoms co-ordinating to the cation are those of the carboxylate and sugar ring. It is suggested that sugar anions will generally co-ordinate to $Ca^{2+}$ through the carboxylate oxygen and the oxygen function on the α-carbon atom, with these two oxygen atoms being in an eclipsed or nearly eclipsed conformation.[48]

A disordered structure has been found for crystals of the calcium thiocyanate complex of 1,4,7,10,13,16–hexaoxacyclo-octadecane, $Ca(NCS)_2,C_{12}H_{24}O_6$, which are monoclinic, with space group $C2/c$, $a = 13.34$, $b = 10.69$, $c = 15.39$ Å, $\beta = 106.2°$, and $Z = 4$. Each $Ca^{2+}$ ion is surrounded by a nearly planar hexagon of O atoms in at least two orientations that are related by rotation in the mean plane. The Ca—O distances are 2.56—2.74 Å. Hexagonal-bipyramidal co-ordination is completed by the N atoms (Ca—N = 2.35 Å) of the thiocyanate

---

[48] S. E. B. Gould, R. O. Gould, D. A. Rees, and W. E. Scott, *J.C.S. Perkin II*, 1975, 237.

anions, which point along the three-fold axis of the hexagon.[49] Six-fold co-ordination of calcium also occurs in calcium phthalate monohydrate, $CaC_8H_4O_4,H_2O$. Six oxygen atoms surround $Ca^{2+}$, at Ca—O distances ranging from 2.31 to 2.60 Å, with an average of 2.47 Å. The crystals have space group $P2_1$, with lattice parameters $a = 11.28$, $b = 6.67$, $c = 11.91$ Å, $\beta = 99.3°$, $d(\text{expt.}) = 1.67$, $d(\text{calc.}) = 1.68$ for $Z = 4$. The carboxylate groups in the crystal are linked by water molecules along [001], with H-bonds of 3.12 and 2.96 Å.[50] The photometric determination of the alkaline-earth metals can be achieved by the use of Arsenazo III reagent. The molar absorptivities of the calcium, strontium, and barium complexes at 650 nm are $4.40 \times 10^4$, $4.00 \times 10^4$, and $3.65 \times 10^4$, respectively. The interference of magnesium is eliminated by analysis at pH 5—6. The use of sodium sulphate as masking agent permits the determination of small amounts of calcium, strontium, and magnesium in the presence of barium.[51]

Single crystals (1 to 5 mm long) of the hexaborides $CaB_6$, $SrB_6$, and $BaB_6$ have been grown from solution in molten aluminium. These compounds are black, dark blue, and dark green, respectively, and their optical reflectance in the 1.5—5.5 eV photon energy range has been measured.[52] The crystal structure of calcium tetra-fluroroborate, $Ca(BF_4)_2$, has been determined. Crystals are orthorhomic, with space group $Pbca$, $a = 9.2792$, $b = 8.9103$, $c = 13.3719$ Å, and $Z = 8$. The structure consists of columns of $BF_4^-$ ions and columns of alternating $Ca^{2+}$ and $BF_4^-$ ions, all parallel to [010]. There are twice as many $[Ca^{2+}, BF_4^-]$ columns as $BF_4^-$ columns. All columns are linked together through Ca---F bonds. The $Ca^{2+}$ ion is co-ordinated by a square antiprism of F atoms, each from a different $BF_4^-$ ion. Each F atom is bonded to a $Ca^{2+}$ ion, and four $Ca^{2+}$ ions form an approximate tetrahedron about the $BF_4^-$ ion.[53]

The compounds $Ca_5Si_3$ and $Ca_5Ge_3$ have been prepared by cooling molten mixtures of composition 1.8—1.3 : 1 Ca : Si(Ge). The crystals have a tetragonal structure of the $Cr_5B_3$-type, space group $I4/mcm$, with cell parameters $a = 7.64$, 7.74, $c = 11.64$, 14.66 Å, $d(\text{calc.}) = 2.22$, 3.16, for $Z = 4$, respectively, for the silicon and germanium compounds.[54] The dehydration of the compound $CaPb(OH)_6$ is reported to lead to the formation of three previously unknown compounds of tervalent lead. In the presence of water vapour, orthorhombic $Ca_2Pb_2O_5(OH)_2$ is formed, with lattice parameters $a = 5.661$, $b = 6.133$, and $c = 15.250$ Å. If dehydrated in air, it loses water to form an oxygen-deficient oxide $CaPbO_{3-x}$, with a defect disordered fluorite structure and lattice parameter $a = 5.321$ Å. When dehydrated in an atmosphere of oxygen, $CaPb(OH)_6$ loses three molecules of water and forms the stoichiometric oxide $CaPbO_3$, with the ilmenite structure.[55]

The electrical conductivity of solutions of the alkaline-earth metals in liquid ammonia has been studed as a function of time, temperature, and metal concentration. The conductivity is relatively high and similar to that observed for

---

[49] J. D. Dunitz and P. Seiler, *Acta Cryst.*, 1974, **B30**, 2750.
[50] M. P. Gupta, S. M. Prasad, and R. P. Sinha, *Current Sci.*, 1974, **43**, 509.
[51] V. Mikhailova and N. Koleva, *Talanta*, 1974, **21**, 523.
[52] S. Muranaka and S. Kawai, *J. Crystal Growth*, 1974, **26**, 165.
[53] T. H. Jordan, B. Dickens, L. W. Schroeder, and W. E. Brown, *Acta Cryst.*, 1975, **B31**, 669.
[54] B. Eisenmann and H. Schaefer, *Z. Naturforsch.*, 1974, **29b**, 460.
[55] Cl-Levy Clement and A. Michel, *Ann. Chim. (Paris).* 1975, **10**, 63.

metal-salt melts. The solutions appear metastable and spontaneously separate into a more stable two-phase equilibrium.[56] The preparation and properties of calcium phosphide chloride and strontium phosphide chloride have been reported. Brown $Ca_2PCl$ and brown-violet $Sr_2PCl$ are formed from the reactions of metal chloride, metal, and red phosphorus in the ratio 1:3:2 at 890—950 °C in quartz vessels under argon, followed by prolonged heating at 1000—1100 °C. The compounds are cubic, with $a = 5.84$ and 6.18 Å for $Ca_2PCl$ and $Sr_2PCl$, respectively.[57] Alkaline-earth iodopnictides have been prepared in the systems M–M′–I (M = Ca, Sr, or Ba; M′ = P, As, or Sb) which can be classed as pseudochalcogenides in which a Group VI element is replaced by a pair of Group V + VII elements. The phases $M_{3+x}M'_{1+x}I_{3-x}$ are cubic for M = Ca and M′ = Sb. For the calcium compounds the composition range of the phase is extended on passing from P to As to Sb. The compound $Ca_2PI$ crystallizes in the trigonal system, space group $R3m$, and $Ca_3PI_3$ is cubic, space group $I4_132$. The compounds possess sodium chloride type superstructures. The iodide $Ca_2PI$ has the anti-$\alpha$-$NaFeO_2$ structure. The compounds $M_2M'$ exist in all cases, and crystallize in the rhombohedral system. Their cell angle increases linearly with the ionic radii of M′.[58,59] The thermal decompositon of the phase $Ca_2Sb_{1-x}I_x$ leads to the formation of the antimonide $Ca_2Sb$, which has space group $I4mm$, with $a = 4.69$ and $c = 16.39$ Å. This structure can be built up from cubes of Sb atoms with Ca atoms in each face.[58—60.] In a separate investigation, the arsenide $Ca_2As$ is reported to form from the elements above 700 °C and to crystallize in the tetragonal space group $I4/mmm$, with $a = 4.63$, $c = 15.56$ Å, and $Z = 4$. The compound is isostructural with the antimonide $Sr_2Sb$ and has shortest distances As—As = 4.20, Ca—As = 3.0, and Ca—Ca = 3.27 Å, which, when compared with the ionic radii, indicate that the compound $Ca_2As$ is a typical Zintl phase.[61] When calcium and arsenic are melted together at 900—1500 °C, in the ratio two to three parts Ca to one of As, the compound $Ca_5As_3$ results, having the $Mn_5Si_3$-type structure, with space group $P6/mcm$, $a = 8.43$, $c = 6.75$ Å, and $d$(calc.) = 3.42 for $Z = 2$.[62]

Studies have continued on the intercalation of the alkaline-earth metals into lattices of the transition-metal sulphides. Calcium and strontium are intercalated into $MoS_2$ from liquid ammonia solution. X-Ray data reveal a lowering of the crystal symmetry of the sulphide and an increase in the complexity of the structure on intercalation. The intercalation compounds begin to superconduct at $ca$ 4 K (for Ca) and 5.6 K (for Sr), and they show considerable anisotropy with respect to the critical magnetic field.[63] Calcium in liquid ammonia also intercalates with $TiS_2$. Two $Ca_xTiS_2$ phases have been identified. The limits of the first are $0.03 \le x \le 0.50$, for which a relationship between $x$ and cell parameters $a$ and $c$ was derived; for $Ca_{0.5}TiS_2$, $a = 3.44$ and $c = 20.06$ Å. The second phase is less

[56] M. M. Tarnorutskii, V. N. Epimakhov, N. V. Konyasheva, and Yu. M. Mishenov, *Zhur. fiz. Khim.*, 1974, **48**, 3048.
[57] C. Hadenfeldt, *Z. Naturforsch.*, 1975, **30b**, 165.
[58] C. Hamon, R. Marchand, and J. Lang, *Rev. Chim. minérale*, 1974, **11**, 287.
[59] C. Hamon, R. Marchand, Y. Laurent, and J. Lang, *Bull. Soc. France minérale Crist.*, 1974, **97**, 6.
[60] C. Hamon, R. Marchand, P. L'Haridon, and Y. Laurent, *Acta Cryst.*, 1975, **B31**, 427.
[61] A. Huetz and G. Nagorsen, *Z. Metallkunde*, 1974, **65**, 618.
[62] A. Huetz and G. Nagorsen, *Z. Metallkunde*, 1975, **66**, 314.
[63] R. B. Somoano, V. Hadek, R. Rembaum, S. Samson, and J. A. Woollam, *J. Chem. Phys.*, 1975, **62**, 1068.

precisely defined, being between $0.13 \leq x \leq 0.25$. The influence of electronic factors is emphasized by the maximum Ca content allowed in the first phase, $Ca_{0.5}TiS_2$, and that achieved using a univalent metal, as in $NaTiS_2$.[64]

## 4 Strontium

Although the position of the strontium $K$ absorption edge has been located in a few of its compounds, viz $SrCl_2$, $SrBr_2$, and SrO, the position of the $K$ absorption edge for the metal had not been measured until recently. It has now been found that the $K$ edge of strontium is shifted towards the high-energy side with respect to the pure metal in all of its compounds. The $K$ absorption edge is assumed to arise from the electronic transition $1s \to 5p$, and the energy of the edge is almost equal to the binding energy of the $1s$ electrons, which is affected by chemical bonding. The shifts are related to interatomic distances and percentage ionic character of the bond; in the former, a linear relationship exists for crystals of the same structure, whereas the latter shows an exponential variation, with maximum charge transfer to fluorine, when the degree of ionic character is derived by Pauling's method. The shifts $\Delta eV$ are given in Table 5, and are used to determine the ionic character of the bonds in the polyatomic compounds where Pauling's formula is not applicable.[65]

The Sr–Al and Ba–Al phase diagrams have been examined by thermal micrographic and X-ray methods. The compounds which were observed are shown in Table 6, together with their lattice parameters, derived from several sources, and their melting points. Eutectics were observed at 81.75 mol% Sr (590 °C) and 71.5 mol% Ba (538 °C) in the respective systems.[66] A further investigation of $BaAl_2$ confirms the trigonal space group $P\bar{3}m1$, with lattice parameters $a = 6.099$ and $c = 17.269$ (see Table 6) but reveals that the stoicheiometry is more properly described as $Ba_7Al_{13}$. The atomic arrangement appears closely related to that of

**Table 5** Shift ($\Delta$) in K absorption edge of strontium in strontium compounds

| Substances | Crystal form | $\Delta eV^a$ | Interatomic distances /Å | % Ionic character | Ref. 65 |
|---|---|---|---|---|---|
| Sr | Cubic | — | — | — | — |
| SrS | Cubic | 0.85 | 2.93 | 32 | — |
| $SrBr_2,6H_2O$ | Rhombohedral | 1.48 | 2.82 | 41 | — |
| $SrCl_2,6H_2O$ | Hexagonal | 1.90 | 2.67 | 47 | — |
| SrO | Cubic | 3.60 | 2.57 | 65 | — |
| $SrF_2$ | Cubic | 5.88 | 2.20 | 85 | — |
| $SrCO_3$ | Rhombohedral | 6.56 | — | — | 88.5 |
| $Sr(NO_3)_2,4H_2O$ | Cubic | 7.13 | — | — | 91.0 |
| $SrC_2O_4,5H_2O$ | Tetragonal | 7.55 | — | — | 92.5 |

$^a$ $\Delta eV = E_{compd} - E_{metal}$.

[64] A. Le Blanc-Soreau and J. Rouxel, *Compt. rend.*, 1974, **279**, C, 303.
[65] M. K. Gupta, *J. Phys. (F)*, 1975, **5**, 359.
[66] G. Bruzzone and F. Merlo, *J. Less-Common Metals*, 1975, **39**, 1.

# Elements of Group II

**Table 6** Structural data for the Sr–Al and Ba–Al intermetallic compounds

| Compound | Structure type | Space group | Lattice parameters/Å | | | m.pt./°C |
|---|---|---|---|---|---|---|
| | | | a | b | c | |
| SrAl$_4$ | BaAl$_4$ | $D_{4h}^{17}$-I4/mmm | 4.463 | | 11.203 | 1040 |
| | | | 4.46 | | 11.07 | |
| | | | 4.462 | | 11.197 | |
| SrAl$_2$ | CeCu$_2$ | $D_{2h}^{28}$-Imma | 4.793 | 7.922 | 7.937 | 936 |
| | | | 4.84 | 7.92 | 7.99 | |
| Sr$_3$Al$_2$ | cubic | P23 or Pm3 | 12.753 | | | 666 |
| BaAl$_4$ | BaAl$_4$ | $D_{4h}^{17}$-I4/mmm | 4.566 | | 11.278 | 1104 |
| | | | 4.540 | | 11.16 | |
| | | | 4.566 | | 11.250 | |
| BaAl$_2$ | trigonal | P321 or P3m1 | 6.100 | | 17.25 | 914 |
| | | or P$\bar{3}$m1 | 6.06 | | 17.25 | |
| | | | 6.099 | | 17.269 | |
| BaAl | hexagonal | P6$_3$mc or P$\bar{6}$2c | 6.103 | | 17.80 | 730 |
| | | or P6$_3$/mmc | 6.01 | | 17.78 | |

the Laves phase MgNi$_2$.[67] Two further phases in Sr–Al have been detected at concentrations below 45 mol% Sr. These were SrAl$_4$ + α + SrAl$_4$ eutectic and SrAl$_4$ + β + SrAl$_4$ eutectic, where α and β are the solid solutions of aluminium in strontium and of strontium in aluminium, respectively. X-Ray diffraction confirmed the tetragonal structure of SrAl$_4$, with $a = 4.462$ and $c = 11.198$ Å (see Table 6). The compound SrAl could not be detected. In Sr–Pb systems, the region up to 25 mol% Sr consists of a mixture of SrPb$_3$ crystals and β + SrPb$_3$ eutectic (β is the solid solution of Sr in Pb). The congruently melting compound Sr$_2$Pb was observed, melting at 970 °C on increasing the strontium concentration. D.t.a. revealed a new phase, SrPb, formed by a peritectic reaction at 900 °C.[68] The variation in vapour pressure, measured by the Knudsen effusion method, with temperature over Sr—Al alloys suggests that the phase SrAl$_4$ is the most stable intermetallic compound existing in this system between 827 and 1117 °C. The partial pressure of strontium is expressed by:

$$\log(p/\text{Torr}) = A - 12\,000/(T/\text{K})$$

and the values of $A$ are 8.12 and 9.36 for alloys containing 45.8 and 76.6 wt.% Sr, respectively. The enthalpy of formation of the aluminium-rich compound SrAl$_4$ is $-20.4$ kcal mol$^{-1}$.[69]

The crystal structure has been determined of tetrakis(biuret)strontium perchlorate, [Sr{NH$_2$CO)$_2$NH}$_4$] (ClO$_4$)$_3$. The compound is monoclinic, with space group P2/c and $a = 11.21$, $b = 7.30$, $c = 14.52$ Å, $\beta = 98.2°$, $d(\text{expt.}) = 1.976$, and $d(\text{calc.}) = 1.973$ for $Z = 2$. Biuret, NH$_2$CONHCONH$_2$, complexes with transition metals as a neutral bidentate ligand *via* oxygen atoms, a dianionic bidentate ligand *via* nitrogen atoms, or as a unidentate ligand *via* one of two oxygen atoms. In the strontium perchlorate complex, the biuret moieties bond to strontium as bidentate ligands *via* the carbonyl oxygen atoms. The molecular unit

---

[67] M. L. Mornasini and G. Bruzzone, *J. Less-Common Metals*, 1975, **40**, 335.
[68] A. V. Vakhobov, T. D. Dzhuraev, B. A. Bardin, and G. A. Zademidko, *Izvest. Akad. Nauk S.S.S.R., Met.*, 1975, 194.
[69] A. V. Vakhobov, T. D. Dzhuraev, and V. N. Vigdorovich, *Zhur. fiz Khim.*, 1974, **48**, 2204.

[Sr(NHCONHCONH$_2$)$_4$] (ClO$_4$)$_2$ shows $C_2$ symmetry. The oxygen co-ordination polyhedron round strontium is best described as square antiprismatic with ligands spanning opposite edges of rectangular faces, as expected for the neutral bidentate ligand. The compound presents an example of eight-co-ordination of a non-transition-metal complex where all the ligands are non-ionic, chemically identical, bidentate oxygen donors.[70] Crystals of strontium acetate thioacetate, Sr(MeCO$_2$)(MeCOS),4H$_2$O, are monoclinic, with space group $P2_1/c$, $a$ = 12.72, $b$ = 709, $c$ = 12.97, $\beta$ = 111.13°, and $Z$ = 4. In this structure, strontium is co-ordinated by nine atoms; one S (Sr—S = 3.224), one O [Sr—O(thioacetate) = 2.69], for O [Sr—O(acetate) = 2.50—2.72], and three O [Sr—O(H$_2$O) = 2.62—2.75 Å]. The polyhedron is similar to that found in hydrated strontium salts and consists of a distorted trigonal prism in which each lateral face is surmounted by a pyramid.[71] The extraction of strontium ions by monocarboxylic acids of differing structure has been investigated. The length of the $n$-alkyl chain had very little effect on the extractive ability of the acids, but branched or unsaturated chains inhibited extraction, though the effect diminished if these features were further from the carboxy-group. The extraction process can be described by:[72]

$$Sr^{2+}(aq) + 4H_2O + 3(HA)_2(org) \rightarrow SrA_2,4HA,4H_2O(org) + 2H^+(aq)$$

The compounds SrSi and SrGe$_{0.76}$ have been prepared from mixtures of the elements. Their structures are orthorhombic, space group $D_{2h}^{25}$-$Immm$, with parameters $a$ = 12.98, 13.38, $b$ = 4.89, 4.84, $c$ = 28.03, 18.52 Å, $d$(calc.) = 3.29 and 3.95, for $Z$ = 20, respectively.[73]

The vapour pressure of strontium metaphosphate has been determined by the Knudsen effusion method. The vapour contains P$_4$O$_{10}$ molecules due to the process:

$$4[Sr(PO_3)_2] \rightarrow 2Sr_2P_2O_7 + P_4O_{10}$$

for which the enthalpy and entropy change at 1327 °C are 35 kcal mol$^{-1}$ and 14.6 e.u., respectively. The standard enthalpy and entropy of formation of the metaphosphate Sr(PO$_3$)$_2$ are calculated to be $-587.5$ kcal mol$^{-1}$ and 38.5 e.u., respectively.[74] The Sr–Sb phase diagram shows considerable complexity and contains several compounds. Thus the compound SrSb is formed at 935 °C, and Sr$_2$Sb melts incongruently at 840 °C; Sr$_3$Sb$_2$ and SrSb$_3$ form by peritectic reactions at 880 and 680 °C, respectively. The elements, having different crystal lattices, show little mutual solubility, and reaction between them becomes increasingly exothermic with increasing antimony concentration.[75] The new compounds Sr$_2$Bi, Ba$_2$Sb, Sr$_5$Bi$_3$, Ba$_5$Sb$_3$, and Ba$_5$Bi$_3$ have been prepared and their structures determined. The isotypic M$_2$B compounds (M = alkaline-earth metal, B = Sb or B) have the body-centred tetragonal structure, as illustrated for Sr$_2$Sb in Vol. 2, of these Reports p. 94. The cell constants for this family of compounds are collected

---

[70] S. Haddad and P. S. Gentile, *Inorg. Chim. Acta*, 1975, **12**, 131.
[71] M. M. Borel and M. Ledesert, *Acta Cryst.*, 1975, **B31**, 725.
[72] A. I. Mikhailichenko, M. A. Klimenko, and T. V. Fedulova, *Zhur. neorg. Khim.*, 1974, **19**, 3344.
[73] B. Eisenmann, H. Schaefer, and K. Turban, *Z. Naturforsch.*, 1974, **29b**, 464.
[74] V. A. Ashuiko, L. N. Urchsovskaya, and I. A. Ra'kovskii, *Zhur. fiz. Khim.*, 1975, **49**, 812.
[75] A. V. Vakhobov, Z. U. Niyazova, and B. N. Polev, *Izvest. Akad. Nauk S.S.S.R., Neorg. Materialy*, 1975, **11**, 363.

in Table 7. The $M_5B_3$ (M = Sr or Ba, B = Sb or Bi) compounds possess the $Mn_5Si_3$-type structure. In both $M_2B$ and $M_5B_3$ structures, the B atoms are co-ordinated only to M atoms. The structures of the calcium compounds $Ca_5Sb_3$ and $Ca_5Bi_3$ were reported in Vol. 3, p. 82. Data for all $M_5B_3$ compounds are also collected in Table 7.[76]

The phase diagrams of strontium with selenium and with tellurium have been investigated. The compound SrSe melts congruently at 1000 °C, $Sr_2Se_3$ forms by a peritectic reaction at 910 °C, and $SrSe_2$ and $SrSe_3$ form at 820 and 730 °C, respectively. The eutectic on the selenium side is close to the melting point (217 °C) of selenium, and indicates an insignificant solid solubility of strontium in selenium. In Sr–Te, no compounds were observed up to 50 mol% Te. The telluride SrTe melts at 1490 °C without decomposition, and two compounds form by peritectic reactions; $SrTe_3$ at 850 and $Sr_2Te_3$ at 920 °C.[77]

Strontium, barium, or lead tin fluorides with the $PbCl_2$-type structure have been reported. The fluorides $BaSnF_4$, $SrSnF_4$, and $PbSnF_4$ were prepared by the reaction of tin difluoride with the fluorides $BaF_2$, $SrF_2$, and $PbF_2$, respectively. These compounds crystallize in the tetragonal system, space group $P4/nmm$, with $Z = 2$. Lattice parameters are $a = 4.356$, 4.175, 4.220, $c = 11.289$, 11.448, 11.415 Å, $d$(expt.) = 5.10, 4.80, 6.50, and $d$(calc.) = 5.15, 4.20, and 6.57, respectively, for the ternary fluorides $BaSnF_4$, $SrSnF_4$, and $PbSnF_4$.[78] The crystal structures of the halides SrFCl and BaFCl have been refined. The compounds belong to the tetragonal PbFCl-structure type, space group $P4/nmm$, with $Z = 2$ and $a = 4.1259$, 4.3939, $c = 6.9579$, 7.2248 Å for the strontium and barium compounds, respectively.[79]

**Table 7** Structural data for $M_2B$ and $M_5B_3$ (M = Ca, Sr, or Ba; B = Sb or Bi) compounds.

| Compound | Structure | Cell parameters/Å | | |
|---|---|---|---|---|
| | | a | b | c |
| $Ca_2Sb$ | Tetragonal $I4/mmm$-$D_{4h}^{17}$ | 4.67 | — | 12.28 |
| $Ca_2Bi$ | $I4/mmm$-$D_{4h}^{17}$ | 4.72 | — | 16.54 |
| $Sr_2Sb$ | $I4/mmm$-$D_{4h}^{17}$ | 5.00 | — | 17.41 |
| $Sr_2Bi$ | $I4/mmm$-$D_{4h}^{17}$ | 5.01 | — | 17.68 |
| $Sr_2Sb$ | $I4/mmm$-$D_{4h}^{17}$ | 5.22 | — | 18.46 |
| $Ca_5Sb_3$ | Orthorhombic, $Pnma$ | 12.502 | 9.512 | 8.287 |
| $Ca_5Bi_3$ | Orthorhombic, $Pnma$ | 12.722 | 9.666 | 8.432 |
| $Sr_5Sb_3$ | Hexagonal, probably $P6_3/mcm$ | 9.496 | — | 7.422 |
| $Sr_5Bi_3$ | Hexagonal, probably $P6_3/mcm$ | 9.63 | — | 7.67 |
| $Ba_5Sb_3$ | Hexagonal, probably $P6_3/mcm$ | 9.97 | — | 7.73 |
| $Ba_5Bi_3$ | Hexagonal, probably $P6_3/mcm$ | 10.13 | — | 7.79 |

[76] B. Eisenmann and K. Deller, *Z. Naturforsch.*, 1975, **30b**, 66.
[77] Yu. B. Lyskova and A. V. Vakhobov, *Izvest. Akad. Nauk S.S.S.R., Neorg. Materialy*, 1975, **11**, 361.
[78] G. Denes, J. Pannetier, and J. Lucas, *Compt. rend.*, 1975, **280**, C, 831.
[79] M. Sauvage, *Acta Cryst.* 1974, **B30**, 2786.

## 5 Barium

The electrical resistivity of liquid barium has been measured. The value of $314 \times 10^{-8}\ \Omega$ m at the melting point is one of the largest for liquid metals. The temperature coefficient of the electrical resistivity is negative, as is characteristic for liquid zinc. The effect of impurity barium oxide appears to be minimal.[80] The Ba–Cd phase diagram has been determined from the results of d.t.a. and metallographic examination of thirty alloys. In addition to the known compounds $BaCd_{11}$, $BaCd_2$, and BaCd, the new compound $Ba_2Cd$, with an incongruent melting point of 411 °C, was found.[81] The Ba–Hg phase diagram has been investigated using thermal and X-ray methods. Three compounds melt congruently; BaHg, 822; $BaHg_2$, 726; and $BaHg_4$, 517 °C. Five other compounds melt peritectically: $Ba_2Hg$, 434; $Ba_2Hg_9$, 505; $BaHg_6$, 410; $BaHg_{11}$, 255; and $BaHg_{13}$, 160 °C Three eutectics occur at 372 (83.0 mol% Ba), 660 (40.0 mol% Ba), and 490 °C (24.0 mol% Ba). No solid solubility was observed for the intermediate phases.[82]

The enthalpies of solution of the trifluoromethanesulphonates, of barium, zinc, and cadmium have been shown to decrease in their strong degree of exothermicity in the solvent order $DMF > DMSO > MeOH > HCONH_2 > H_2O > MeCN$. Single-ion enthalpies of transfer from water to the solvents, based on the assumption that $\Delta H_{tr}(Ph_4As^+) = \Delta H_{tr}(Ph_4B^+)$ have been discussed in terms of the properties of the ions and of the solvents. Transfer of $Ba^{2+}$ and other bivalent cations from water to MeOH, DMF, or DMSO is strongly exothermic, but the transfer of many anions is endothermic. Transfer and solvation enthalpies for the cations are provided in Table 8.[83]

The compound $BaB_{12}H_{12},6H_2O$ has been investigated by means of X-rays. The structure is orthorhombic, with space group $Cmcm$, $a = 11.880$, $b = 9.185$, $c = 14.020$ Å, $d$(expt.) = 1.675, and $d$(calc.) = 1.690, for $Z = 4$. The crystals are composed of $[Ba(H_2O)_6]^{2+}$ cations and $B_{12}H_{12}^{2-}$ anions, which have a distorted icosahedron structure with B—B bonds of 1.70—1.85 Å. All six $H_2O$ molecules are in the first co-ordination sphere and form a boat-like shape which covers only one side of the cation; the opposite side contacts the anion. On heating, the compound loses four $H_2O$ molecules at 140 and two more at 180 °C.[84]

The vapour above barium metaborate has been shown to consist of molecules of $BaBO_2$ and $BaB_2O_4$ at 1470—1550 °C. The enthalpies of formation of these gaseous species were deduced to be −161.4 and −321.3 kcal mol$^{-1}$, respectively.[85]

The reaction of barium with graphite is reported to form a mixture of insertion phases and $BaC_2$. The insertion compound $BaC_6$ crystallized in the space group $P6_3/mmc$, with $a = 4.302$ and $c = 10.51$ Å. The compound is less reactive than alkali-metal graphite compounds.[86]

The crystal structure of the glaserite form of barium sodium phosphate has been determined. The compound is trigonal, with space group $P\bar{3}m1$, $a = 5.622$

---

[80] H. J. Guentherodt, E. Hauser, and H. U. Kuenzi, *J. Phys.* (F), 1975, 889.
[81] R. T. Dirstine, *J. Less-Common Metals*, 1975, **39**, 271.
[82] G. Bruzzone and F. Merlo, *J. Less-Common Metals*, 1975, **39**, 271.
[83] G. R. Hedwig and A. J. Parker, *J. Amer. Chem. Soc.*, 1974, **96**, 6589.
[84] K. A. Solntsev, N. T. Kuznetsov, and N. V. Rannev, *Doklady Akad. Nauk S.S.S.R.*, 1975, **221**, 1378.
[85] M. K. Ll'in, A. V. Makarov, and O. T. Nikitin, *Vestnik Moskov Univ., Khim.*, 1974, **15**, 436.
[86] D. Guerard and A. Herold, *Compt. rend.*, 1974, **279**, C, 455.

**Table 8** Single ion enthalpies of transfer/kJ mol$^{-1}$ from water to non-aqueous solvents and enthalpies of solvation$^{a,b}$/kJ mol$^{-1}$ of some uni- and bi-valent ions at 298 K

| Ion | H$_2$O $\Delta H_{tr}$ | H$_2$O $\Delta H_s$ | MeOH $\Delta H_{tr}$ | MeOH $\Delta H_s$ | HCONH$_2$ $\Delta H_{tr}$ | HCONH$_2$ $\Delta H_s$ | DMF $\Delta H_{tr}$ | DMF $\Delta H_s$ | DMSO $\Delta H_{tr}$ | DMSO $\Delta H_s$ | MeCN $\Delta H_{tr}$ | MeCN $\Delta H_s$ |
|---|---|---|---|---|---|---|---|---|---|---|---|---|
| Na$^+$ | 0 | −418.0 | −20.5 | −438.5 | −16.3 | −434.3 | −33.1 | −451.1 | −27.6 | −445.6 | −13.0 | −431.0 |
| K$^+$ | 0 | −333.5 | −18.4 | −351.9 | −16.7 | −350.2 | −39.3 | −372.8 | −34.7 | −368.2 | −22.6 | −356.1 |
| Ag$^+$ | 0 | −487.9 | −20.9 | −508.8 | −22.6 | −510.5 | −38.5 | −526.4 | −54.8 | −542.7 | −52.7 | −540.6 |
| Ba$^{2+}$ | 0 | −1328 | −59.2 | −1387 | −40.2 | −1368 | −85.5 | −1413 | −78.5 | −1406 | −8.5 | −1336 |
| Zn$^{2+}$ | 0 | −2069 | −45.6 | −2115 | −23.9 | −2093 | −62.7 | −2132 | −62.2 | −2132 | +20.1 | −2049 |
| Cd$^{2+}$ | 0 | −1831 | −40.4 | −1871 | −27.5 | −1858 | −63.3 | −1894 | −70.8 | −1902 | +8.2 | −1823 |

$^a$ The hydration enthalpies were calculated from the conventional standard enthalpies of hydration using $\Delta H_{hyd}(H^+) = -1103.3$ kJ mol$^{-1}$;
$^b$ For the non-aqueous solvents, the solvation enthalpies were calculated from the hydration enthalpies and the single ion enthalpies of transfer i.e., $\Delta H_s(M^{n+}) = \Delta H_{hyd}(M^{n+}) + \Delta H_{tr}(M^{n+})$ from H$_2$O → solvent

and $c = 7.259$ Å. There are three different cation sites; Na and Ba atoms lie at sites of $3m$ symmetry and are co-ordinated to six and twelve O atoms, respectively. The Na—O distances are 2.344 Å and the two unique Ba—O distances are 2.788 and 3.247 Å. The remaining cation site has $3m$ symmetry, a co-ordination number of ten, and contains equal amounts of Na and Ba atoms, with metal-oxygen bond lengths of from 2.548 to 3.017 Å. The phosphate ion also has $3m$ symmetry.[87]

The vapour species above barium oxide has been investigated by means of effusion mass spectrometry from 1643 to 1803 K. Reaction enthalpies yield second- and third-law standard heats of formation of $-26.5$ and $-28.3$ kcal mol$^{-1}$, respectively, for BaO (g). A third-law dissociation energy of 129.7 kcal mol$^{-1}$ was derived for BaO.[88] Another determination gives a value of $D_0^\circ(\text{BaO}) = 133.5$ kcal mol$^{-1}$.[89] A mass-spectrometric effusion study has been made of the evaporation of barium aluminium oxide, $BaAl_2O_4$, prepared by the reaction of sub- and super-stoicheiometric mixtures of barium carbonate with alumina. The mixed oxide evaporates incongruently at 1840—2086 K by dissociating into the solid compound $BaAl_{12}O_{19}$ and gaseous molecules of BaO, according to:

$$6BaAl_2O_4(s) \rightarrow 5BaO(g) + BaAl_{12}O_{19}(s)$$

for which the reaction enthalpy is 645 kcal mol$^{-1}$ at 1963 K. The dissociation pressure of barium monoxide over the mixed oxide $BaAl_2O_4$ is given by:[90]

$$\log(p/\text{atm.}) = 6.50 - 2.814 \times 10^4/(T/K); \quad 1840 \leq T/K \leq 2086$$

Further effusion mass spectrometry work has been carried out on barium monoxide from a variety of Knudsen cells made of molybdenum, or graphite, or alumina, or platinum to study the reaction of the monoxide with the cell material. The ionic species $Ba^+$, $BaO^+$, and $Ba_2O_2^+$ were detected in the vapour, but, contrary to the results of other investigations, no $Ba_2O_3^+$ ions were observed. The vapour pressure of barium monoxide is given by:

$$\log(p/\text{atm.}) = 7.19 - 2.173/(T/K); \quad 1332 \leq T/K \leq 1681$$

For the sublimation:

$$BaO(s) \rightarrow BaO(g)$$

the standard enthalpy is 105 kcal mol$^{-1}$. The corresponding energy/kcal mol$^{-1}$ and entropy/kcal mol$^{-1}$ deg$^{-1}$ are given by:[91]

$$\Delta G^\ominus = 99 - 0.03289(T/K); \quad 1332 \leq T/K \leq 1681$$

and

$$\Delta S^\ominus = 0.027 + 7.59/(T/K); \quad 1332 \leq T/K \leq 1681$$

The i.r. spectra have been measured of BaO molecules in nitrogen matrices,

[87] C. Calvo and R. Faggiani, *Canad. J. Chem.*, 1975, **53**, 1849.
[88] M. Farber and R. D. Srivastava, *High Temp. Sci.*, 1975, **7**, 74.
[89] P. J. Dagdigian, H. W. Cruse, A. Schultz, and R. Zare, *J. Chem. Phys.*, 1974, **61**, 4450.
[90] K. Hilpert, A. Naoumidis, and G. Wolff, *High Temp. Sci.*, 1975, **7**, 1.
[91] K. Hilpert and H. Gerads, *High Temp. Sci.*, 1975, **7**, 11.

formed by the reactions of alkaline-earth metal atoms with ozone. The fundamental frequencies of the molecules BaO, SrO, and CaO are 613, 620, and 707 cm$^{-1}$, respectively.[92] Using oxygen produces metal superoxide and metal oxide dimeric species. The $\nu_2$ interionic modes of the peroxides MO$_2$ (M = Ba, Sr, or Ca) were located at 541, 474, and 498 cm$^{-1}$, respectively, in a nitrogen matrix. Frequencies observed for the antisymmetric modes $\nu_6(B_{3u})$ and $\nu_5(B_{2u})$ of (BaO)$_2$ at 487 and 393 cm$^{-1}$ indicate a $D_{2h}$ structure, with OBaO angle ca. 102°. A mechanism has been proposed for the formation of the oxide molecules MO and MO$_2$ which features MO$_2$M and (MO)$_2$ as intermediates. Nitrogen proves to be more effective as a matrix material in stabilizing the ionic compounds than argon.[93] The compound BaCl$_2$,Ba(OH)$_2$,4H$_2$O is reported to be the only barium hydroxide chloride to separate from aqueous solution. On dehydration, the sole product is BaCl$_2$,Ba(OH)$_2$. The standard enthalpies of formation of the hydrated and anhydrous compounds are −3037.6 and −1824.4 kJ mol$^{-1}$, respectively.[94] The crystal structures of the compounds Ba$_2$S$_3$ and BaS$_3$ have been determined as tetragonal, with space group $I4_1md$, $a$ = 6.112, $c$ = 15.950 Å, $Z$ = 4, and space group $P42_1m$, with $a$ = 6.871, $c$ = 4.168 Å, and $Z$ = 2, respectively. The polysulphide Ba$_2$S$_3$ contains a sulphide ion as well as an S$_2^{2-}$ polysulphide ion. One Ba atom is in the centre of a distorted trigonal prism whose corners are occupied by the S$_2^{2-}$ ions. The distances between Ba$^{2+}$ ions and the nearest sulphur of the dumbell-shaped anions are 3.11 and 3.91 Å, with three additional S$^{2-}$ ions capping the rectangular faces at 3.15 and 3.24 Å. A second Ba$^{2+}$ ion is surrounded by an irregular polyhedron of nine S atoms; three of the vertices are occupied by S$_2^{2-}$ ions at 3.14 and 3.71 Å, and six vertices are occupied by S$_2^{2-}$ ions. The nearest S atoms of the polysulphide ion are at 3.21 and 3.42 Å from the cation. In the compound BaS$_3$, the polysulphide S$_3^{2-}$ ion has S—S distances 2.074 Å and the SSS angle is 114.8°. Barium ions are twelve-co-ordinated, with Ba—S distances of from 3.204 to 3.541 Å.[95] A neutron-diffraction study of barium thiosulphate monohydrate has shown that the crystal is made up of Ba$^{2+}$ cations, S$_2$O$_3^{2-}$ anions, and H$_2$O molecules. The Ba—S distances are 3.355 and 3.424 Å, and the Ba—O distances range from 2.775 to 3.428 Å. In the tetrahedral S$_2$O$_3^{2-}$ anion the S—S and S—O bond lengths are 1.979 and 1.472—1.483 Å, respectively. The co-ordination number of barium is eleven, and it can be described as a distorted octahedron with three vertices occupied by O—O edges, and another two by S—O edges, of five different S$_2$O$_3^{2-}$ tetrahedra. The octahedron is completed by the O atom of the H$_2$O molecule.[96]

The preparation and properties of barium monofluorotrioxiodate, BaIO$_3$F, have been described. This sparingly soluble salt was prepared by the reaction of the hydrated hydroxide Ba(OH)$_2$,2H$_2$O with the compound KIO$_2$F$_2$ in alcohol, and was characterized by $X$-ray diffraction and i.r. and Raman spectroscopy. The IO$_3$F$^{2-}$ ion has three oxygen atoms equidistant from iodine, and, for this reason, easily breaks down in aqueous solution to iodate and fluoride.[97]

---

[92] B. S. Ault and L. Andrews, *J. Chem. Phys.*, 1975, **62,** 2320.
[93] B. S. Ault and L. Andrews, *J. Chem. Phys.*, 1975, **62,** 2312.
[94] O. A. Markova, *Zhur. fiz. Khim.*, 1975, **49,** 38.
[95] S. Yamaoka, J. T. Lemley, J. M. Jenks, and H. Steinfink, *Inorg. Chem.*, 1975, **14,** 129.
[96] L. Manojlović-Muir, *Acta Cryst.*, 1975, **B31,** 135.
[97] S. Okransinski, R. Jost, R. Rakshapal, and G. Mitra, *Inorg. Chim. Acta*, 1975, **12,** 247.

# 3
# Elements of Group III

BY G. DAVIDSON

## 1 Boron

**General.**—A new extended Gaussian set of atomic orbitals for use in MO calculations involving boron has been published.[1] It is described as the 6-31G basis set, and gives a better description of nuclear regions, and energies nearer the Hartree–Fock limit than does the previously reported 4-31G set.

*Ab initio* MO calculations have been made on 17 neutral one- and two-heavy-atom, B-containing molecules, with C, N, O, or F also present.[2] At the STO-3G level, geometries and conformational preferences agree with those previously calculated for analogous $C^+$ species. The B atom was a stronger $\sigma$-donor and weaker $\pi$-acceptor than $C^+$, however. At the 6-31G level, B may be stabilized by attachment of $\pi$-donors, by hyperconjugation, by dimerization or by complexation with Lewis bases. BH, BCH, and $B_2H_2$ were calculated to have triplet ground states. $H_2BNH_2$ and $H_2BOH$ were predicted to be planar, with barriers to non-rigid rotation of 29 and 14 kcal mol$^{-1}$, respectively. The barrier in $H_2BCH_3$, on the other hand, appeared to be negligible. $H_2BBH_2$ seems to prefer the perpendicular, $D_{2d}$, conformation by 10 kcal mol$^{-1}$ with respect to the planar one.

Equilibrium geometries and spin properties have been calculated by the INDO method for a number of B-containing radicals (BO, BS, $BH_2$, $BF_2$, $BH_3^-$, $BMe_3^-$, and $\cdot CH_2BMe_2$). In general the results were in good agreement with such experimental data as are available. Thus for BO the experimental bond length is 1.204 Å, the calculated 1.3 Å. The equivalent figures for BS are 1.609, 1.63 Å, and for $BH_2$ 1.18, 1.184 Å.[3]

Boron may be extracted from aqueous boric acid solutions rapidly and effectively (97% of B in one extraction) using napththalene-2,3-diol–diphenylguanidine–butanol.[4]

Boron-10 concentrations of *ca.* 29% may be obtained (*cf.* about 18% initially) by $CO_2$ TEA-laser-induced photochemical reaction of $BCl_3$–$H_2S$ mixtures using 10.55 $\mu$m radiation, *i.e.* the P(16) line of the 001-100 band of $CO_2$. The $^{10}B$ concentration may be reduced, on the other hand, to *ca.* 14%, by using 10.18 $\mu$m radiation [R(30) line of the same $CO_2$ vibrational band].[5]

---

[1] J. D. Dill and J. A. Pople, *J. Chem. Phys.*, 1975, **62**, 2921.
[2] J. D. Dill, P. von R. Schleyer, and J. A. Pople, *J. Amer. Chem. Soc.*, 1975, **97**, 3402.
[3] A. Hudson R. F. Treweek, and J. T. Wiffen, *Theor. Chim. Acta*, 1975, **38**, 355.
[4] F. Vlačil and K. Drbal, *Coll. Czech. Chem. Comm.*, 1975, **40**, 2792.
[5] S. M. Freund and J. J. Ritter, *Chem. Phys. Letters*, 1975, **32**, 255.

Elements of Group III

**Boranes.**—Muetterties et al. have summarized the existing state of knowledge concerning intramolecular rearrangements in boron clusters (in borane derivatives), with particular reference to expected rearrangement barriers in a number of different geometries.[6]

The topological approach recently devised for the determination of allowed transition states for nucleophilic and electrophilic reactions of boranes has been revised so as to exclude $sty(-1)$ valence structures.[7] These are the structures in which one B has no terminal hydrogen, and they should be excluded because of the 'awkward' hybridizations involved.

A series of empirical rules has been presented for helping in the assignment of $^{11}$B signals in the $^{11}$B n.m.r. spectra of nido-boranes.[8]

Extensive series of CI (configuration interaction) MO calculations on BH have yielded energy values and spectroscopic data for the five lowest singlet states: $X^1\Sigma^+$, $A^1\Pi$, $C^1\Delta$, $B^1\Sigma^+$, and $C^1\Sigma^+$.[9]

An *ab initio* direct calculation method has been used to evaluate the ionization potential of reaction (1) and the electron affinity of reaction (2) over a range of internuclear distances.[10]

$$BH(^1\Sigma^+) \rightarrow BH^+(^2\Sigma^+) \qquad (1)$$

$$BH(^1\Sigma^+) \rightarrow BH^-(^2\Pi) \qquad (2)$$

MO calculations have been reported for the BH and BH$_3$ molecules, using the 'pair natural orbital configuration interaction' (PNO—CI) and 'coupled electron pair approximation with natural orbitals' (CEPA—PNO) methods.[11] The force constant and equilibrium distance of BH agreed very well with experimental values.

Some exploratory MO calculations have been made, using CI–INDO methods, for borane adducts—chiefly BH$_3$CO.[12] The $\pi$-type interactions of the BH$_3$ may be described best in terms of hyperconjugation. It was suggested that $^2J(H,H)$ coupling constants and $^{11}$B quadrupole coupling constants are likely to be useful probes for investigating the binding of BH$_3$ to donors.

The electronic structure of B$_2$H$_6$ has been calculated within a simple group-function model, where the group functions are products of singly occupied non-orthogonal orbitals. The calculation produces a molecular wavefunction which incorporates a considerable amount of chemically significant electron correlation.[13]

An SCF calculation using a large Gaussian basis set yields a value of $\Delta E_f = -19.9$ kcal mol$^{-1}$ for reaction (3):

$$2BH_3 \rightarrow B_2H_6 \qquad (3)$$

this is larger than most previous results (using more restricted basis sets) but still

---

[6] E. L. Muetterties, E. L. Hoel, C. G. Salentine, and M. F. Hawthorne, *Inorg. Chem.*, 1975, **14**, 950.
[7] R. W. Rudolph and D. A. Thompson, *Inorg. Chem.*, 1974, **13**, 2779.
[8] S. Heřmánek and J. Plešek, *Z. anorg. Chem.*, 1974, **409**, 115.
[9] S. A. Houlden and I. G. Csizmadia, *Theor. Chim. Acta*, 1974, **35**, 173.
[10] K. M. Griffing and J. Simons, *J. Chem. Phys.*, 1975, **62**, 535.
[11] R. Ahlrichs, F. Driessler, H. Lischka, V. Staemmler, and W. Kutzelnigg, *J. Chem. Phys.*, 1975, **62**, 1235.
[12] K. F. Purcell and R. L. Martin, *Theor. Chim. Acta*, 1974, **35**, 141.
[13] S. Wilson and J. Gerratt, *Mol. Phys.*, 1975, **30**, 765.

some way from the experimental value of $-35$ kcal mol$^{-1}$.[14] Another calculation on this system yields a Hartree–Fock limit of $-20.7$ kcal mol$^{-1}$ (giving a value of $-36.6$ kcal mol$^{-1}$ on taking account of correlation) for $\Delta E_f$.[15]

Pyrolysis of $B_2H_6$ at atmospheric pressure by a filament heated to 240 °C produces a 25% yield of $B_{10}H_{14}$ after purification. The necessary apparatus is particularly simple.[16]

Monochromatic low-power radiation from a $CO_2$ laser (ca. 1.5 W) initiates a chain reaction in $B_2H_6$ (gas, 200 Torr).[17] The chief product is $B_{20}H_{16}$, with smaller amounts of $B_5H_9$ and $B_{10}H_{14}$.

Dipolar coupling tensors of the B, $H_{br}$, and $H_t$ atoms have been calculated for the neutral diboranyl radical $\cdot B_2H_5$, from e.s.r. data.[18]

$^1$H and $^{11}$B n.m.r. parameters of monoiodoborane have been reported.[19] The direct $J$(BH) coupling constants remain very similar throughout the series $B_2H_5X$, even though the chemical shifts change markedly.

(2-Biphenylyl)diethylborane and ethyldiborane (6) react to give ca. 80% of (1), which on treatment with $BF_3,Et_2O$ gives (2), and with diborane gives (3).[20]

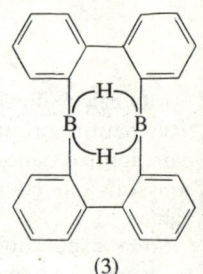

The action of Lewis acids (HCl, $B_2H_6$, $BF_3$, or MeOH) upon the adducts $B_5H_9,2L$ (L = NMe$_3$ or PMe$_3$; 2L = Me$_2$NC$_2$H$_4$NMe$_2$) has been discussed.[21] A number of different reaction types are found in these series.

Electron diffraction measurements on 1- and 2-silylpentaboranes confirmed the conclusion of an EHMO calculation that the former should be more stable. Thus, $r$(B—Si) in the former is $1.981 \pm 0.005$ Å, in the latter $2.006 \pm 0.004$ Å.[22] Thus there appears to be greater $\pi$-interaction for the apically substituted isomer. The analogous methyl derivatives (where a similar, but much smaller stability difference was calculated) give $r$(B—CH$_3$) values which are very similar ($1.592 \pm 0.005$, $1.595 \pm 0.005$ Å).

The previously reported positions of two- and three-centre bonds in the

---

[14] D. S. Marynick, J. H. Hall, and W. N. Lipscomb, J. Chem. Phys., 1974, **61**, 5460.
[15] R. Ahlrichs, Theor. Chim. Acta, 1974, **35**, 59.
[16] J. Dazord, G. Gullevic, and H. Mongest, Bull. Soc. chim. France, 1975, 981.
[17] H. R. Bachmann, H. Nöth, R. Rinck, and K. L. Kompa, Chem. Phys. Letters, 1974, **29**, 627.
[18] O. Edlund and J. Sohma, Mol. Phys., 1975, **29**, 1229.
[19] J. E. Drake and B. Rapp, J. Inorg. Nuclear Chem., 1974, **36**, 2611.
[20] R. Köster and H.-G. Willemsen, Annalen, 1974, 1843.
[21] A. B. Burg and L. Maya, Inorg. Chem., 1975, **14**, 942.
[22] D. Wieser, D. C. Moody, J. C. Huffman, R. L. Hildebrandt, and R. Schaeffer, J. Amer. Chem. Soc., 1975, **97**, 1074.

topological structures of $B_5H_9(PMe_3)_2$ (A. V. Fratini et al., J. Amer. Chem. Soc., 1974, **96**, 3013) are erroneous, and should be reversed.[23]

$^{11}$B N.m.r. has been used to monitor reactions of $B_5H_{11}$ with ethers: $Me_2S$ and $Et_2S$ gave the symmetrical cleavage products $R_2S,BH_3$ and $R_2S,B_4H_8$; oxoethers gave products which depended on the base strength of the ether. Thus THF (very strongly basic) produced $H_2B(THF)_2^+(B_4H_9)^-$: $Me_2O$ and $Et_2O$ (moderately basic), $B_5H_{11},OR_2$; weakly basic ethers such as $Pr_2^iO$ did not react.[24] No symmetrical cleavage was found for any of the oxoethers.

$B_6H_{10},2PMe_3$ forms monoclinic crystals, belonging to the space group $P2_1/c$.[25] The bond distances within the $B_6H_{10}$ fragment are as expected, with $r(BB)$ lying in the range 1.745(7)—1.841(9) Å. The structure is the first to be confirmed in which the boron arrangement is a fragment of the equatorial belt of an icosahedron. The P-1—B-3 and P-2—B-6 distances are 1.890(6) and 1.884(6) Å, respectively, compared with 1.90(1) and 1.98(1) Å in the analogous $B_5H_9$ adduct.

A new investigation of the electron diffraction behaviour of decaborane(14) gave the following values for bond lengths: $r(BH_t)$ 1.18(2) Å, $r(BH_{br})$ 1.34(2) Å, and $r(BB)$ 1.78(1) Å.[26] These are closer to the X-ray parameters for the solid than those obtained previously by electron diffraction.

The experimental Compton scattering profile of decaborane(14), using $^{241}$Am γ-rays, is in excellent agreement with that calculated using localized MO's transferred from smaller boranes.[27]

The $B_{10}H_{14}$-benzene system is a simple eutectic (eutectic point: 20 mol% $B_{10}H_{14}$, −7.0 °C).[28] In $B_{10}H_{14}$-toluene an incongruently melting compound is formed. Another group of workers have reported equilibrium diagrams for the $B_{10}H_{14}$-arene systems (arene = benzene, durene, naphthalene, or biphenyl).[29]

An acid-base reaction sequence of $B_7H_{12}^-$ with $B_6H_{10}$, reactions (4) and (5)

$$KB_7H_{12} + HCl(liq) \xrightarrow{Me_2O} B_7H_{11},OMe_2 + H_2 \quad (4)$$

$$B_7H_{11},OMe_2 + BF_3 \xrightarrow{B_6H_{10}} BF_3,OMe_2 + B_{13}H_{19} + H_2 \quad (5)$$

gives the new hydride $B_{13}H_{19}$. Its presumed structure, based on n.m.r. evidence, is shown in Figure 1.[30]

Controlled hydrolysis of $B_{16}H_{20}$ produces the new species $B_{14}H_{18}$ (preliminary report: S. Heřmánek et al., Chem. and Ind., 1972, 606).[31] The structure proposed for this is consistent with the 70.6 MHz $^{11}$B and 33.7 MHz $^2$H n.m.r. spectra of $B_{14}H_{18-x}D_x$ (Figure 2) and is built up by an edge fusion of decaborane and hexaborane frameworks (similar to that found in $i$-$B_{18}H_{22}$).

Tetradecaborane(20), $B_{14}H_{20}$, has been prepared for the first time, by the reaction of excess $B_8H_{12}$ with $KB_6H_9$ in $Et_2O$.[32] The compound forms crystals

---

[23] A. V. Fratini, G. W. Sullivan, M. L. Denniston, R. K. Hertz, and S. G. Shore, J. Amer. Chem. Soc., 1974, **96**, 6819.
[24] G. Kodama and D. J. Saturnino, Inorg. Chem., 1975, **14**, 2243.
[25] M. M. Mangion, J. R. Long, W. R. Clayton, and S. G. Shore, Cryst. Structure Comm., 1975, **4**, 501.
[26] V. S. Mastryukov, O. V. Dorofeeva, and L. V. Volkov, J. Struct. Chem., 1975, **16**, 110.
[27] I. R. Epstein, P. Pattison, M. G. H. Wallbridge, and M. J. Cooper, J.C.S. Chem. Comm., 1975, 567.
[28] K. G. Myakishev, I. S. Posnaya, and V. V. Volkov, Russ. J. Inorg. Chem., 1974, **19**, 761.
[29] V. A. Kuznetsov, N. D. Golubeva, and K. N. Semenenko, Russ. J. Inorg. Chem., 1974, **19**, 778.
[30] J. Rathke, D. C. Moody, and R. Schaeffer, Inorg. Chem., 1974, **13**, 3040.
[31] S. Heřmánek, K. Fetter, J. Plešek, L. J. Todd, and A. R. Garber, Inorg. Chem., 1975, **14**, 2250.
[32] J. C. Huffman, D. C. Moody, and R. Schaeffer, J. Amer. Chem. Soc., 1975, **97**, 1621.

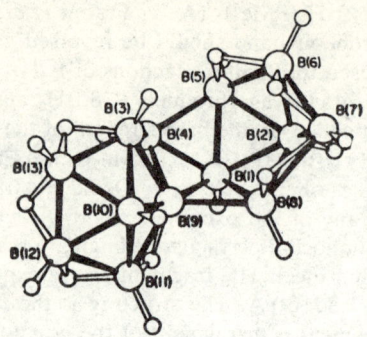

**Figure 1** *Presumed structure of* $B_{13}H_{19}$
(Reproduced by permission from *Inorg. Chem.*, 1974, **13**, 3040)

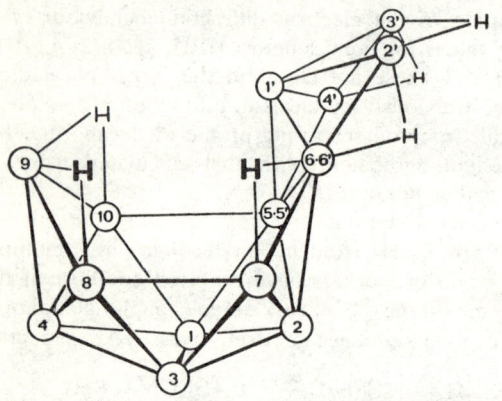

**Figure 2** *Proposed structure of* $B_{14}H_{18}$
(Reproduced by permission from *Inorg. Chem.*, 1975, **14**, 2250)

belonging to the space group $P2_12_12_1$, and the molecular structure is as shown in Figure 3. There are two $B_8H_{12}$ fragments, fused at B-7—B-12, and the bond lengths are as expected. However, the open faces are *cis* to each other, the first time such a feature has been observed, while the B-3—B-7—B-2—B-12 framework is approximately planar.

$B_6H_{10}$ reacts with $B_8H_{12}$ and $B_9H_{13}$ to give the new boranes $B_{14}H_{22}$ and $B_{15}H_{23}$,

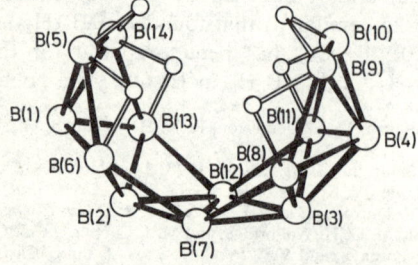

**Figure 3** *Molecular structure of* $B_{14}H_{20}$ *(terminal hydrogens omitted)*
(Reproduced by permission from *J. Amer. Chem. Soc.*, 1975, **97**, 1621)

respectively; the proposed structure for the latter is shown in Figure 4.[33] These and other $B_6H_{10}$ reactions are thought to arise from a rather unusual mechanism, namely formation of a three-centre bond by reaction of a hydride containing a B—B two-centre bond with an acidic hydride having a readily available empty B orbital.

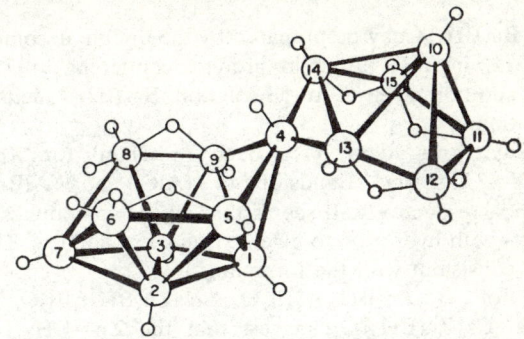

**Figure 4** *Proposed structure of* $B_{15}H_{23}$
(Reproduced by permission from *Inorg. Chem.*, 1974, **13**, 3008)

**Borane Anions and their Metallo-derivatives.**—EHMO calculations and a consideration of the Jahn–Teller theorem may be used to rationalize the relative stereochemical rigidities of some *closo*-borane anions.[34] Thus, in $B_8H_8^{2-}$ there is a very small energy gap between the highest occupied and lowest unoccupied MO's, and this species is very non-rigid. Relatively large barriers to rearrangements, on the other hand, are predicted for $D_{3h}$ $B_5H_5^{2-}$ and $B_9H_9^{2-}$, since these are degenerate in the likely transition states or reactive intermediates, which are of $C_{4v}$ symmetry.

Rate measurements on the hydrolysis of $BH_4^-$ in DMSO have led to the suggestion that proton transfer from hydronium ion to $BH_4^-$ in water uses a water molecule as a proton relay. In DMSO, the $H^+$ rate coefficient is much smaller, because DMSO cannot act in this way, and DMSO$\cdots$H$\cdots$HBH$_3^+$ is very unstable.[35]

General analytical methods have been developed for determining the composition of Li or Al borohydride complexes with substituted hydrazines.[36] Thus, it was shown that the action of hydrazine or *NN*-dimethylhydrazine on $LiBH_4$ in $Et_2O$ produces $LiBH_4,N_2H_4$, $LiBH_4,2N_2H_4$, and $LiBH_4,NH_2NMe_2$.[37]

Phase relationships have been elucidated in the $LiBH_4$–$M(BH_4)_2$–THF systems (M = Sr or Ba).[38] No solid solutions or double salts are formed.

Kinetic data have been given for reaction (6) (in liquid ammonia), but no mechanism was proposed.[39]

$$(NH_4)_2SO_4 + 2NaBH_4 \rightarrow Na_2SO_4 + 2H_2 + 2BNH_6 \qquad (6)$$

---

[33] J. Rathke and R. Schaeffer, *Inorg. Chem.*, 1974, **13**, 3008.
[34] E. L. Muetterties and B. F. Beier, *Bull. Soc. chim. belges*, 1975, **84**, 397.
[35] L. M. Abts, J. T. Langland, and M. M. Kreevoy, *J. Amer. Chem. Soc.*, 1975, **97**, 3181.
[36] J. Samanos and S. J. Teichner, *Bull. Soc. chim. France*, 1975, 77.
[37] J. Samanos and S. J. Teichner, *Bull. Soc. chim. France*, 1975, 81.
[38] V. I. Mikheeva, L. N. Tolmacheva, and A. S. Sizareva, *Russ. J. Inorg. Chem.*, 1974, **19**, 622.
[39] T. S. Briggs and W. L. Jolly, *Inorg. Chem.*, 1975, **14**, 2267.

Phase diagrams have been published for the systems $NaBH_4$–$NaClO_4$–$H_2O$[40] and $Mg(BH_4)_2$–THF.[41]

I.r. bands due to $BH_4^-$ in CsCl lattices have been assigned. It appears that in the presence of the $BH_4^-$ the CsCl is present both as simple cubic and as f.c.c. crystals.[42]

A solution of $Ba(BH_4)_2$ may be prepared by the double decomposition of $BaI_2$ with excess $LiBH_4$ in THF. A solid product containing >97% $Ba(BH_4)_2$ is obtained by the addition of $Et_2O$ to the solution. $Ba(BH_4)_2$ melts at 385 °C with slight decomposition.[43]

Almost unassigned i.r. data have been presented for $K[Al(BH_4)_4]$ and $K[Al(BH_4)_3X]$ (X = Cl or Br).[44] Bands due to $\nu$(AlCl) and $\nu$(AlBr) were found at 430, and 336 cm$^{-1}$, respectively. It seems that all the compounds are polymeric.

$Al(BH_4)_3$ reacts with hydrazine to give 1:1 and 1:4 adducts. The i.r. spectrum of the former is consistent with the formula (4).[45]

I.r. spectra of $Li_2Zn(BH_4)_4,3Et_2O$, $NaZn(BH_4)_3Et_2O$, $K_2Zn_3(BH_4)_8$, $RbZn(BH_4)_3$ and $Cs_nZn(BH_4)_{2+n}$ suggest that the Zn—$BH_4$ bonding is very similar to that found in tetrahydroboratoaluminates.[46]

When $CoCl_2,6H_2O$ and $P(C_6H_{11})_3$ are dissolved in EtOH–toluene, and treated with $NaBH_4$ under an atmosphere of $N_2$, the complex $[CoH(BH_4)\{P(C_6H_{11})_3\}]$ is formed.[47] This contains a bidentate $BH_4$ ligand, and the co-ordination at the Co is as shown in (5).

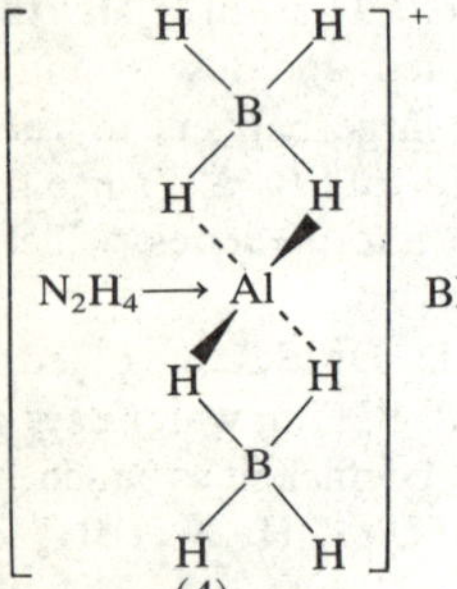

(4)

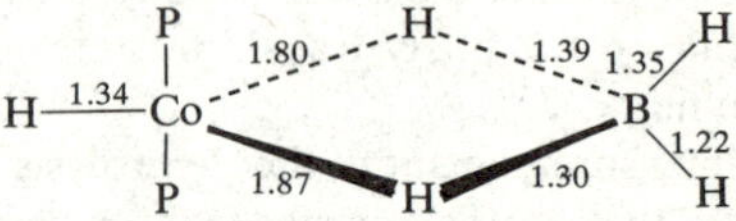

(5) bond lengths/Å

Convenient, reliable, high-yield processes have been developed for the preparation of $Cp_2Zr(BH_4)_2$ (i.e. $Cp_2ZrCl_2 + LiBH_4$ in benzene solution) and $Zr(BH_4)_4$ (a similar reaction of $ZrCl_4 + LiBH_4$).[48] No vacuum sublimation is needed, as the products are precipitated cleanly from solution.

Analysis of the Raman and i.r. wavenumbers of $Hf(BH_4)_4$ and $Hf(BD_4)_4$, together with intensity data from the Raman spectra, suggests that the $BH_4$ unit is

---

[40] V. I. Mikheeva and K. V. Titova, *Russ. J. Inorg. Chem.*, 1974, **19**, 1586.
[41] V. N. Konoplev and T. A. Silina, *Russ. J. Inorg. Chem.*, 1974, **19**, 1383.
[42] W. C. Schutte and H. Coker, *J. Chem. Phys.*, 1974, **61**, 2808.
[43] V. I. Mikheeva and L. N. Tolmacheva, *Russ. J. Inorg. Chem.*, 1974, **19**, 665.
[44] K. N. Semenenko, V. B. Polyakova, O. V. Kravchenko, S. P. Shilkin, and Yu. Ya. Kharitonov, *Russ. J. Inorg. Chem.*, 1975, **20**, 173.
[45] J. Samanos and S. J. Teichner, *Bull. Soc. chim. France*, 1975, 87.
[46] N. N. Mal'tseva, N. S. Kedrova, V. V. Klinkova, and N. A. Chumaevskii, *Russ. J. Inorg. Chem.*, 1975, **20**, 339.
[47] M. Nakajima, H. Moriyama, A. Kobayashi, T. Saito, and Y. Sasaki, *J.C.S. Chem. Comm.*, 1975, 80.
[48] B. D. James and B. E. Smith, *Synth. React. Inorg. Metal-Org. Chem.*, 1974, **4**, 461.

bonded to the hafnium *via* the B atom rather than H-bridges.[49] Thus, a strong, polarized Raman band at 480 cm$^{-1}$ in Hf(BH$_4$)$_4$ [398 cm$^{-1}$ in Hf(BD$_4$)$_4$] appears to be due mainly to the HfB$_4$ breathing mode. Normal-co-ordinate analyses were reported, but these gave rather poor agreement between the observed and calculated wavenumbers.

The vibrational spectra of Cp$_3$UH$_3$Br (R = H, Et, or Ph) are consistent with the presence of a triple bridge, as in (6).[50] The alkyl derivatives were formed by the action of BR$_3$ on the tetrahydroborate. The $^{11}$B-decoupled n.m.r. spectrum of Cp$_3$UH$_3$BH exhibits a collapse of the BH$_4$ resonances at low temperatures, since the paramagnetism has induced sufficient energy separation between exchanging sites (bridge and terminal) to slow down the dynamic intramolecular rearrangement process.

$^{11}$B Nuclear quadrupole coupling parameters have been obtained for Me$_4$N$^+$B$_3$H$_8^-$ and KB$_3$H$_8$.[51] The B$_3$H$_8^-$ ions, whose static structure, (7), contains two types of B and three of H, undergo motions such that all like nuclei become equivalent at 298 K.

One of the impurities in Be(BH$_4$)$_2$ prepared from BeCl$_2$ and LiBH$_4$ has been identified as Be(B$_3$H$_8$)$_2$.[52] The $^{11}$B n.m.r. spectrum of this indicates that all hydrogen atoms in each B$_3$H$_8$ unit are exchanging, as are all three B atoms. The postulated static structure is given in Figure 5.

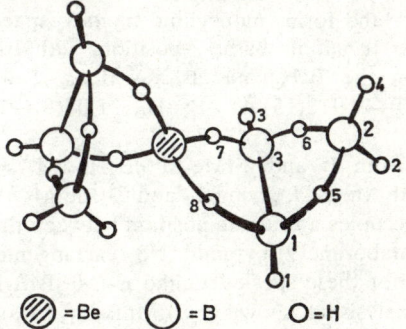

**Figure 5** *Proposed static structure of* Be(B$_3$H$_8$)$_2$
(Reproduced from *J.C.S. Chem. Comm.*, 1975, 626)

[49] T. A. Keiderling, W. T. Wozniak, R. S. Gay, D. Jurkowitz, E. R. Bernstein, S. J. Lippard, and T. G. Spiro, *Inorg. Chem.*, 1975, **14**, 576.
[50] T. J. Marks and J. R. Kolb, *J. Amer. Chem. Soc.*, 1975, **97**, 27.
[51] R. T. Baine, E. Fukushima, and S. B. W. Roeder, *Chem. Phys. Letters*, 1975, **32**, 566.
[52] D. F. Gaines and J. H. Morris, *J.C.S. Chem. Comm.*, 1975, 626.

KH or $NH_3$ can both act as deprotonating agents towards $B_4H_{10}$, producing $B_4H_9^-$ in good yields (over 90%). Addition of $BH_3$ to the latter gave a polyhedral expansion reaction, forming $B_5H_{12}^-$, $B_6H_{11}^-$, and $B_7H_{12}^-$. The $B_5H_{12}^-$ ion is the first reported binary hydride species belonging to the class of *hypho*-boranes, *i.e.* those containing $2n+8$ framework electrons. All of the ions made are stereochemically non-rigid.[53]

Reaction of $CuCl(PPh_3)_3$ with $K[RB_5H_7]$ (R = H, 1-Me, or 4-Me) yields the bridging derivatives (8; R = H or Me) and (9), respectively.[54] These structures are based on the usual spectroscopic observations, which also show that none of them is subject to fluxional behaviour.

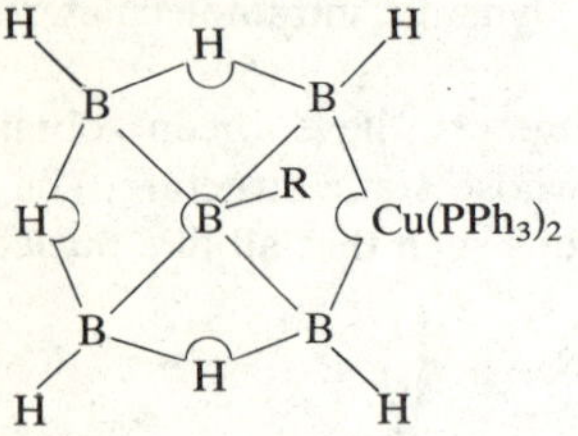

(8) 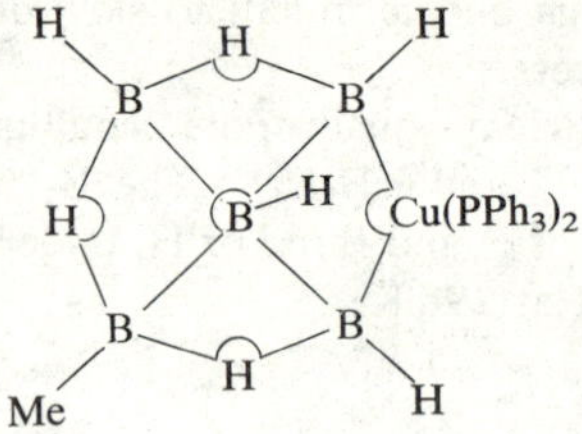       (9)

A number of bis(pentaboranyl)–Group IV compounds have been isolated from reactions of $LiB_5H_8$ with $RMeM^{IV}Cl_2$ ($M^{IV}$ = Si, R = H or Me; $M^{IV}$ = Ge, R = Me).[55] The chief products were $(B_5H_8)_2M^{IV}RMe$. When M = Si and R = H, two isomers are formed, in both of which the $B_5H_8$ units are bound at their basal terminal positions B-2 to the Si (Figure 6). For $(B_5H_8)_2SiMe_2$, one $B_5H_8$ is bonded at a bridging site ($\mu$) to Si, with the other bonded *via* B-2. In the Ge analogue, both $\mu,2'$ and $2,2'$ isomers are formed, however. Other species isolated were 2-$HMeClSi(B_5H_8)$, $Me_2HSi(B_5H_8)$, and $Me_2HGe(B_5H_8)$.

$Ir(B_5H_8)Br_2(CO)(PMe_3)_2$ is formed by the oxidative addition of 1- or 2-$BrB_5H_8$ to $IrCl(CO)(PMe_3)_2$,[56] and forms monoclinic crystals (space group $P2_1/c$). The $B_5H_8$ unit is linked to Ir *via* its basal 2-position, with Ir—B-2 = 2.071(14) Å. Bond distances within the $B_5H_8$ are: $B_{apical}$—$B_{basal}$, 1.641(21)—1.691(18) Å; $B_{basal}$—$B_{basal}$, 1.805(20)—1.912(15) Å; B—$H_t$, 1.08(8)—1.55(11) Å; B—$H_{br}$, 1.03(14)—1.45(10) Å.

Proton abstraction from 1- and 2-$(Me_3M^{IV})B_5H_8$ ($M^{IV}$ = Si or Ge), leads to anions which react with $Me_2BCl$, giving 1- and 2-$(Me_3M^{IV})$-$\mu$-$(Me_2B)B_5H_7$. The $BMe_2$ unit therefore occupies a bridging position between the two B atoms in the basal plane of the pentaborane(9) pyramid. No rearrangement to a hexaborane-(10) derivative occurs for these species, unlike $\mu$-$(Me_2B)B_5H_8$ itself.[57]

X-Ray diffraction analysis has shown $(CO)_3MnB_8H_{13}$ to contain a terdentate $B_8$ ligand, bonded to the Mn by two Mn—H—B bridges from borons bordering the open face of the boron cage, and one Mn—H—B bond from an adjacent B in the

---

[53] R. J. Remmel, H. D. Johnson, I. S. Jaworiwsky, and S. G. Shore, *J. Amer. Chem. Soc.*, 1975, **97**, 5395.
[54] V. T. Brice and S. G. Shore, *J.C.S. Dalton*, 1975, 334.
[55] D. F. Gaines and J. Ulman, *Inorg. Chem.*, 1974, **13**, 2792.
[56] M. R. Churchill and J. J. Hackbarth, *Inorg. Chem.*, 1975, **14**, 2047.
[57] D. F. Gaines and J. Ulman, *J. Organometallic Chem.*, 1975, **93**, 281.

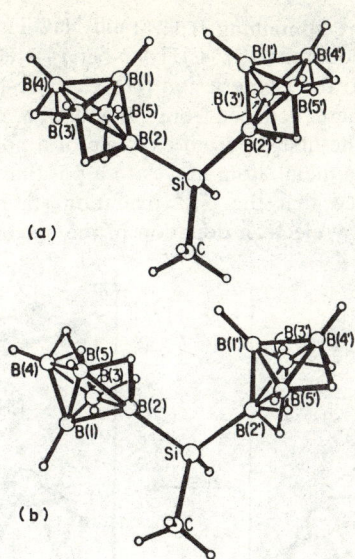

**Figure 6** Suggested structures for the two least sterically-hindered conformers of 2,2'-$(B_5H_8)_2$SiHMc
(Reproduced by permission from *Inorg. Chem.*, 1974, **13**, 2792)

base of that cage (Figure 7).[58] I.r., Raman, n.m.r., u.v., and mass spectra were also reported.

Reaction of 2-THF-6-$(CO)_3$-6-$MnB_9H_{12}$ with $NEt_3$ leads to the formation of 8-[$Et_3N(CH_2)_4O$]-6-$(CO)_3$-6-$MnB_9H_{12}$ *via* an unusual internal rearrangement of the metalloborane cage. An X-ray study of the structure shows that it is similar to $B_{10}H_{14}$, with the B-6 replaced by the Mn of the $Mn(CO)_3$ group. The Mn co-ordination is almost octahedral, while the $O(CH_2)_4NEt_3$ side-chain is attached to the metalloborane cage at either B-8 or B-10.[59]

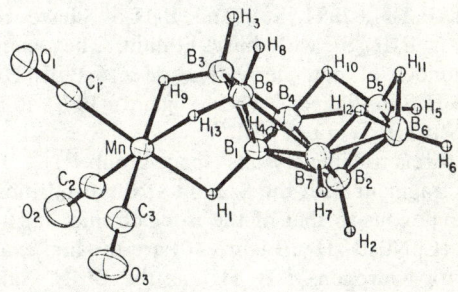

**Figure 7** *The structure of* $(OC)_3MnB_8H_{13}$
(Reproduced by permission from *J. Amer. Chem. Soc.*, 1974, **96**, 6318)

[58] J. C. Calabrese, M. B. Fischer, D. F. Gaines, and J. W. Lott, *J. Amer. Chem. Soc.*, 1974, **96**, 6318.
[59] D. F. Gaines, J. W. Lott, and J. C. Calabrese, *Inorg. Chem.*, 1974, **13**, 2419.

Treatment of a mixture containing $B_9H_{12}^-$ and $NaC_5H_5$ with sodium amalgam produces the new metalloborane $[(\eta^5-C_5H_5)-2-Ni(\eta^5-B_9H_9)]^-$. This may be converted quantitatively into the isomeric ion $[(\eta^5-C_5H_5)-1-Ni(\eta^5-B_9H_9)]^-$ by heating.[60] The structures of these, deduced from $^{11}B$ and $^1H$ n.m.r., are shown in Figure 8. The isomerization is the first reported example of a polyhedral metalloborane rearrangement in which a metal atom moves to a position of lower co-ordination number. It was suggested that the $Ni^{IV}$ oxidation state in the latter may be preferentially stabilized by electron donation in the apical position.

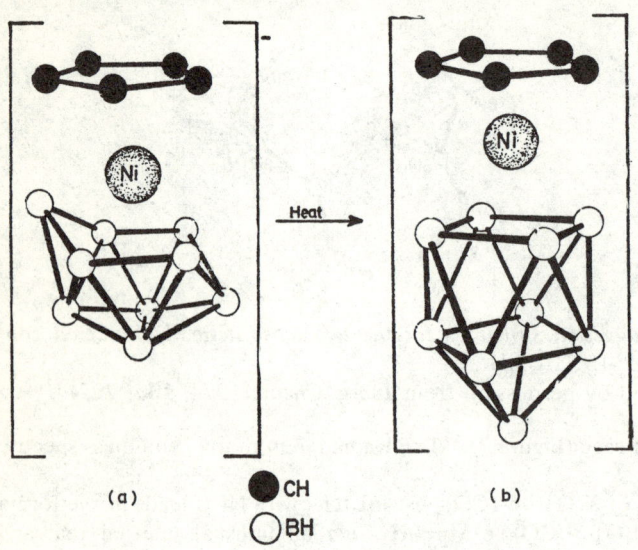

**Figure 8** *The proposed structures of* (a) $[(\eta^5-C_5H_5)-2-Ni(\eta^5-B_9H_9)]^-$ *and* (b) $[(\eta^5-C_5H_5)-1-Ni(\eta^4-B_9H_9)]^-$
(Reproduced from *J.C.S. Chem. Comm.*, 1975, 310)

The reactive ions $B_9H_{13}^{2-}$, $B_9H_{11}S^{2-}$, and $B_9H_9S^{2-}$ have been prepared by reactions of $B_9H_{14}^-$ or $B_9H_{12}S^-$ with butyl-lithium. They can be used for the preparation of a number of new derivatives, *e.g.* $6-PhB_{10}H_{13}$, $(B_9H_{11}S)_2M^{2-}$, (M = Ni or Pd), $(B_9H_9S)Pd(ligand)_x^n$ where ligand = $PPh_3$ ($x = 2$, $n = 0$), phen ($x = 1$, $n = 0$), or $C_2S_2(CN)_2$ ($x = 1$, $n = 2$).[61]

$(Ph_3P)_3AuB_9H_{12}S$ forms triclinic crystals (space group $P\bar{1}$).[62] The $B_9H_{12}S^-$ ion is an open icosahedral fragment with the S atom at the 6-position; the structure is indeed completely analogous to that of the isoelectronic $B_{10}H_{14}$ (Figure 9).

$CpNi(B_{11}H_{11})$ and $(CpNi)_2B_{10}H_{10}$ (Figure 10) are the first examples of metalloboranes without 'extra hydrogen'.[63] $B_{11}H_{11}^{2-}$ as the $Bu_4N^+$ salt and $(CpNiCO)_2$ react in refluxing THF to give the former, and $(CpNiCO)_2$ and $B_{10}H_{10}^{2-}$ as the

---

[60] R. N. Leyden and M. F. Hawthorne, *J.C.S. Chem. Comm.*, 1975, 310.
[61] A. R. Siedle, D. McDowell, and L. J. Todd, *Inorg. Chem.*, 1974, **13**, 2735.
[62] L. J. Guggenberger, *J. Organometallic Chem.*, 1974, **81**, 271.
[63] B. P. Sullivan, R. N. Leyden, and M. F. Hawthorne, *J. Amer. Chem. Soc.*, 1975, **97**, 455.

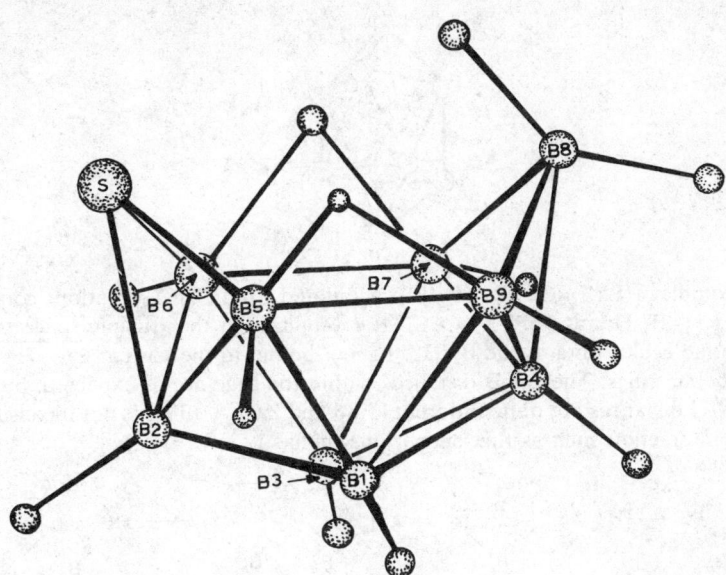

**Figure 9** *The proposed structure of* $B_9H_{12}S^-$
(Reproduced by permission from *J. Organometallic Chem.*, 1974, **81**, 271)

Et$_3$NH$^+$ salt give the latter. Both are stable to air and water, and provide examples of the predictive value of polyhedral electron-counting rules.

A reconsideration of X-ray results for $Cu_2^IB_{10}H_{10}$ (10), together with $\nu(BH)$ and $\nu(BD)$ wavenumbers for the species and its perdeuterio-analogue, suggest that there are both Cu—B and Cu—H—B interactions. The latter are less than full-bridge bonds, but they do modify the i.r. spectra by comparison with true terminal BH units.[64]

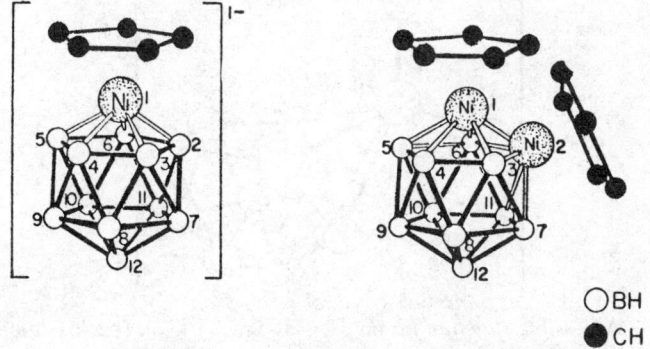

**Figure 10** *Proposed structures for* $[(B_{11}H_{11})NiCp]^-$ *and* $B_{10}H_{10}(NiCp)_2$
(Reproduced by permission from *J. Amer. Chem. Soc.*, 1975, **97**, 455)

[64] T. E. Paxson, M. F. Hawthorne, L. D. Brown, and W. N. Lipscomb, *Inorg. Chem.*, 1974, **13**, 2772.

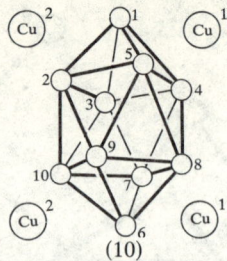

(10)

The complex $[(Ph_3P)_2Cu]_2B_{10}H_{10}$ is precipitated from $CHCl_3$ solutions containing $(NEt_3H)_2B_{10}H_{10}$ and $(Ph_3P)_3CuCl$. It crystallizes in the triclinic space group $P\bar{1}$. The molecule contains the $B_{10}H_{10}^{2-}$ ligand, bound to the Cu via Cu—H—B—B—H chelate rings. The B—B distances within the cage are as expected, but the two Cu—H distances are quite different (1.86 and 2.08 Å). This is not inconsistent with an interaction such as that shown in Scheme 1.[65]

**Scheme 1**

$K_2B_{10}H_{10}$ reacts with $L_2PtCl_2$ [L = $PPh_3$, $L_2$ = 1,2-bis(diphenylphosphino)-ethane, (diphos)] in EtOH–$CHCl_3$ to give isomeric, ethoxy-*nido*-metalloboranes: $L_2PtB_{10}H_{11}(OEt)$. Two isomers are formed for L = $PPh_3$, three for $L_2$ = diphos.[66]

$B_{10}H_{12}^{2-}$ (as $Ph_4As^+$ or $Ph_3MeP^+$ salts) reacts with $SnCl_2$ to give, among other products, the *nido*-stannadecaborate ion, $B_{10}H_{12}SnCl_2^{2-}$ (Figure 11). With $Me_2SnCl_2$ the chelating derivative (11) is formed.[67]

PRDDO (partial retention of diatomic differential overlap) calculations have

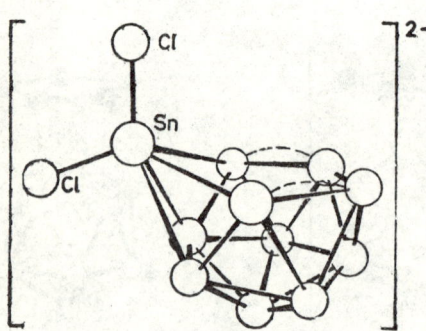

**Figure 11** *A possible structure for the* $[B_{10}H_{12}SnCl_2]^{2-}$ *ion (each B atom is bonded to a terminal H; bridging H atoms are indicated by broken lines)* (Reproduced by permission from *J. Organometallic Chem.*, 1975, **86**, 243)

[65] J. T. Gill and S. J. Lippard, *Inorg. Chem.*, 1975, **14**, 751.
[66] T. E. Paxson and M. F. Hawthorne, *Inorg. Chem.*, 1975, **14**, 1604.
[67] N. N. Greenwood and B. Youll, *J.C.S. Dalton*, 1975, 158.

## Elements of Group III

(11)

been reported for $B_8H_{12}$, $B_8H_{14}$, $B_8H_{13}^-$, $B_9H_{15}$, $B_9H_{14}^-$, $B_{10}H_{14}$, $B_{10}H_{14}^{2-}$, $B_{10}H_{13}^-$, $B_{11}H_{13}^{2-}$, $C_2B_7H_{13}$, $C_2B_9H_{12}^-$, and $C_2B_{10}H_{13}^-$.[68] Localized molecular orbitals, LMO, may be derived for all of these, and the molecules may be grouped into three families (prototypes $B_9H_{12}$, $B_{10}H_{14}$, and $B_{11}H_{13}^{2-}$). Within each group, differences in geometrical structure and charge distribution may be correlated with differences in the LMO structure.

$[(\eta^5\text{-}C_5H_5)Fe(CO)_2(\text{cyclohexene})]^+PF_6^-$ reacts with $B_{10}H_{13}^-$ and $7,8\text{-}B_9C_2H_{12}^-$ to give $6\text{-}[(\eta^5\text{-}C_5H_5)Fe(CO)_2]B_{10}H_{13}$ and $(\eta^5\text{-}C_5H_5)Fe(CO)_2(7,8\text{-}B_9C_2H_{12})$.[69] These have single Fe—B bonds, and the proposed structure for the former is given in Figure 12.

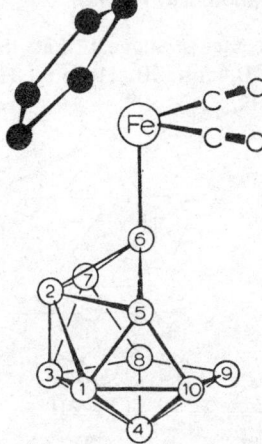

**Figure 12** *Proposed structure for* $6\text{-}[\eta^5\text{-}C_5H_5Fe(CO)_2]B_{10}H_{13}$ (H *atoms omitted*).
(Reproduced by permission from *J. Organometallic Chem.*, 1975, **86,** 243)

Deprotonation of decaborane(14) with $Me_3M$ (M = In or Tl) forms $(Me_2M)^+$ $[B_{10}H_{12}MMe_2]^-$ (M = In or Tl), $(B_{10}H_{12}InMe)$, and $(Me_2Tl)^+(B_{10}H_{13})^-$.[70] The first type of species, containing two distinct $Me_2M$ groups is given the structure shown in Figure 13, on the basis of analytical and spectroscopic results.

Thia-*closo*-dodecaborane(11), $SB_{11}H_{11}$, can be prepared by reaction (7).

$$7\text{-}SB_{10}H_{12} + Et_3N,BH_3 \xrightarrow{120-190\,°C} SB_{11}H_{11} + S\bar{B}_{10}H_{11}\text{—}\overset{+}{N}HEt_3 + NEt_3 \quad (7)$$
$$(35\%) \quad\quad (50\%)$$

---

[68] J. H. Hall, jun., D. A. Dixon, D. A. Kleier, T. A. Halgren, L. D. Brown, and W. N. Lipscomb, *J. Amer. Chem. Soc.*, 1975, **97,** 4202.
[69] F. Sato, T. Yamamoto, J. R. Wilkinson, and L. J. Todd, *J. Organometallic Chem.*, 1975, **86,** 243.
[70] N. N. Greenwood, B. S. Thomas, and D. W. Waite, *J.C.S. Dalton*, 1975, 299.

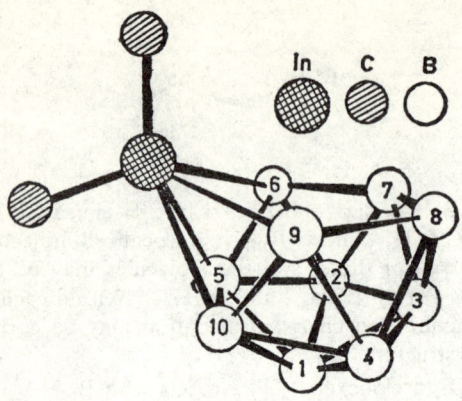

**Figure 13** *Heavy-atom structure of* $[B_{10}H_{12}InMe_2]^-$, (*Numbered as for* $B_{10}H_{14}$; *bridging H atoms between* B—6—B—7 *and* B—8—B—9)
(Reproduced from *J.C.S. Dalton*, 1975, 299)

The usual physical measurements suggest that this new compound has the structure (12).[71] Action of $Br_2$ on $SB_{11}H_{11}$ in $CH_2Cl_2$ in the presence of Al powder forms 12-Br-1-$SB_{11}H_{10}$.

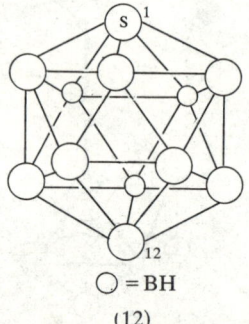

○ = BH

(12)

Reactions of $B_{10}H_{10}^{2-}$ and $B_{12}H_{12}^{2-}$ with $FeCl_3$ in the presence of nitriles produce nitrile adducts as well as $B_{20}H_{18}^{2-}$, $B_{24}H_{23}^{3-}$, $B_{12}H_{11}Cl^{2-}$, $B_{24}H_{22}Cl^{3-}$, and $B_{12}H_{11}NH_3^-$). These adducts could be isolated in the partially hydrolysed form $B_nH_{n-a}(NH_2CO_2R)_a^{a-2}$ ($n = 10$ or 12, $a = 1$ or 2, R = Me, $CH_2CO_2Et$, $CH_2CONH_2$, or $CH_2OSO_2Ph$).[72]

Studies on the $MnO_4^-$ oxidation of $B_{12}H_{12}^{2-}$ in neutral or alkaline aqueous solutions have provided definite evidence for the presence of $B_{12}H_{11}(OH)^{2-}$ and $B_{12}H_{10}(OH)_2^{2-}$ as intermediates, and there are good grounds for believing that more highly hydroxylated species must also be present.[73]

**Carbaboranes.**—He I photoelectron spectra of the *closo*-carbaboranes 1,5-$C_2B_3H_5$, 1,6-$C_2B_4H_6$, 2,4-$C_2B_5H_7$, and 1,7-$C_2B_{10}H_{12}$ indicate that the theoretical

---
[71] J. Plešek and S. Heřmánek, *J.C.S. Chem. Comm.*, 1975, 127.
[72] A. H. Norman and A. Kaczmarczyk, *Inorg. Chem.*, 1974, **13**, 2316.
[73] A. Kaczmarczyk and M. Collins, *Inorg. Chem.*, 1975, **14**, 207.

# Elements of Group III

separation of the MO's of these species into *endo*- and *exo*-polyhedral types is experimentally justified. Data for $B_5H_9$ show that this separation can also be applied to boranes.[74]

Flash thermolysis of 1,2-bis(trimethylsilyl)pentaborane(9) apparently gives rise to carbon insertion producing 2- and 4-methyl derivatives of $CB_5H_7$ and some *C*-silyl derivatives of the smallest known carbaborane: $1\text{-}H_3Si\text{-}1,5\text{-}C_2B_3H_3$, (13), $1\text{-}MeH_2Si\text{-}1,5\text{-}C_2B_3H_4$, $2\text{-}Me\text{-}1\text{-}(H_3Si)\text{-}1,5\text{-}C_2B_3H_3$.[75] Variable-temperature n.m.r. measurements reveal a fast bridge-hydrogen tautomerization at *ca.* 100 °C.

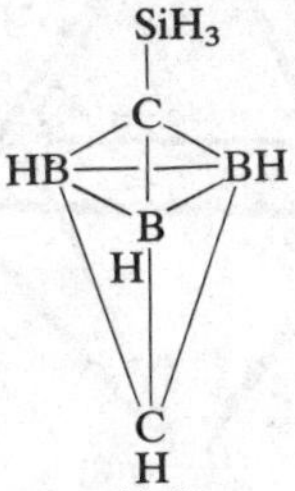

```
        SiH₃
         |
         C
  HB─────┼─────BH
         B
         H|
         
         C
         H
        (13)
```

EHMO calculations have been used to determine the natures of allowed rearrangements in the reduced carbaboranes $C_2B_3H_5^{2-}$ and $C_2B_4H_6^{2-}$.[76]

Gas-phase electron diffraction patterns of 1,2-dicarba-*closo*-hexaborane(6) (Figure 14) and carbahexaborane(7), (Figure 15), yielded a comprehensive set of structural parameters.[77]

The rearrangement of 1,2- to $1,6\text{-}C_2B_4H_6$ is known to occur, but a suggested co-operative twist ('diamond-square-diamond') mechanism did not appear (by a CNDO/2 calculation) to give an energetically accessible pathway. However, PRDDO calculations using a modified type of this mechanism imply that there exists a third stable geometry on the $C_2B_4H_6$ geometry surface. This is a possible intermediate in the interconversion, and has the distorted trigonal prismatic form (14).[78]

Use of computer line-narrowing techniques in the $^{11}B$ FT n.m.r. spectra of small *nido*-carbaboranes $C_2R_2B_4H_6$ has resulted in a higher resolution being achieved.[79] Some previously unresolved examples of $^{11}B\text{-}^{11}B$ coupling were noted, *e.g.* $J(^{11}B^{11}B)$ in $C_2Me_2B_4H_6$ is 26.0 Hz.

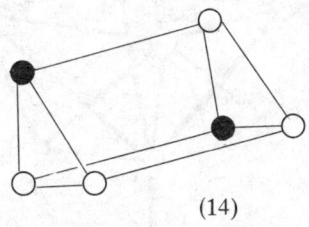

(14)

---

[74] T. P. Fehlner, *Inorg. Chem.*, 1975, **14**, 934.
[75] J. B. Leach, G. Oates, S. Tang, and T. Onak, *J.C.S. Dalton*, 1975, 1018.
[76] B. J. Meneghelli and R. W. Rudolph, *Inorg. Chem.*, 1975, **14**, 1429.
[77] E. A. McNeill and F. R. Scholer, *Inorg. Chem.*, 1975, **14**, 1081.
[78] T. A. Halgren, I. M. Pepperberg, and W. N. Lipscomb, *J. Amer. Chem. Soc.*, 1975, **97**, 1248.
[79] J. W. Akitt and C. G. Savory, *J. Magn. Resonance*, 1975, **17**, 122.

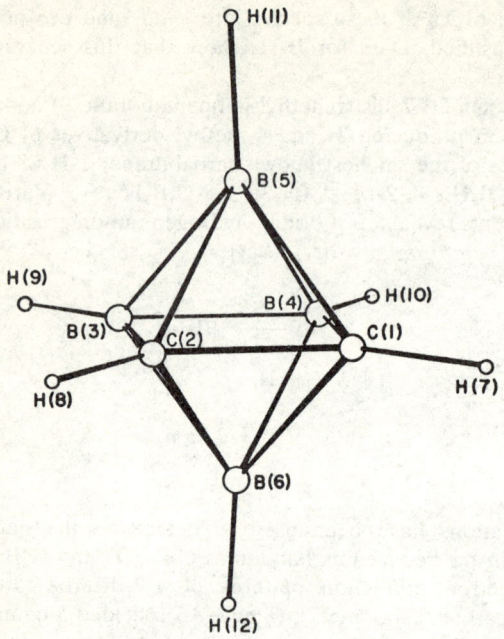

**Figure 14** *Molecular structure and numbering scheme for* $1,2\text{-}B_4C_2H_6$
(Reproduced by permission from *Inorg. Chem.*, 1975, **14**, 1081)

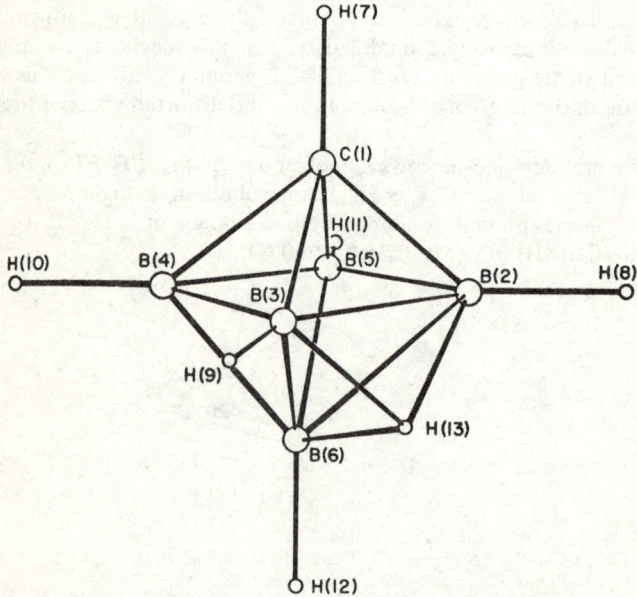

**Figure 15** *Molecular structure and numbering scheme for* $CB_5H_7$
(Reproduced by permission from *Inorg. Chem.* 1975, **14**, 1081)

The microwave spectrum of $CB_5H_9$ shows that this compound has a distorted structure one face of which contains three long B—B bonds.[80] This suggests that an H atom is located in or above that face, as previously postulated on the basis of n.m.r. data (T. Onak, R. Drake, and G. Dunks, *J. Amer. Chem. Soc.*, 1965, **87**, 2505).

The molecular structure of 2,4-dicarba-*closo*-heptaborane, $C_2B_5H_7$, has been determined by electron diffraction (Figure 16). The C atoms are in the pentagonal belt, separated by one B atom; the pentagonal belt is planar.[81]

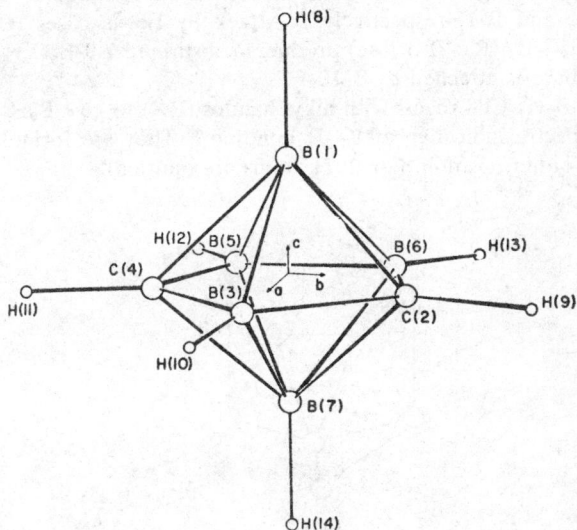

**Figure 16** *The molecular structure of* 2,4-$C_2B_5H_7$
(Reproduced by permission from *J. Mol. Structure*, 1975, **27**, 151)

Controlled reaction of $C_2B_5H_7$ with $F_2$ leads to the following new compounds: $BF_2CH_2BF_2$, 3-$FC_2B_5H_6$, 5-$FC_2B_5H_6$, 1,5-$F_2C_2B_5H_5$, 1,3-$F_2C_2B_5H_5$, and 5,6-$F_2C_2B_5H_5$.[82] These are the first fluorine-containing compounds derived from small carboranes. All of them were characterized by $^{11}B$ n.m.r., i.r., and mass spectrometry.

Electrochemical reduction of the bis(oxy) derivatives: 4,7-(ORO)-2,3-$(Me)_2$-2,3-$C_2B_9H_7$ (R = 1,2-ethanediyl, 1,2-phenyldiyl, or 1,3-dimethyl-1,3-propanediyl) takes place *via* two one-electron processes. The first is electrochemically reversible, and gives a stable radical-anion, detectable by e.s.r.[83]

4,7-$(OH)_2$-2,3-$C_2B_9H_7$(11) undergoes reaction (8) with glycols; $^{18}O$ enrichment of the glycol with R = H shows that the carbaborane hydroxy-groups are displaced.[84]

[80] G. L. McKown, B. P. Don, R. A. Beaudet, P. J. Vergamini, and L. H. Jones, *J.C.S. Chem. Comm.*, 1974, 765.
[81] E. A. McNeill and F. R. Scholer, *J. Mol. Structure*, 1975, **27**, 151.
[82] N. J. Maraschin and R. J. Lagow, *Inorg. Chem.*, 1975, **14**, 1855.
[83] G. D. Mercer, J. Lang, R. Reed, and F. R. Scholer, *Inorg. Chem.*, 1975, **14**, 761.
[84] D. Gladkowski and F. R. Scholer, *J. Organometallic Chem.*, 1975, **85**, 287.

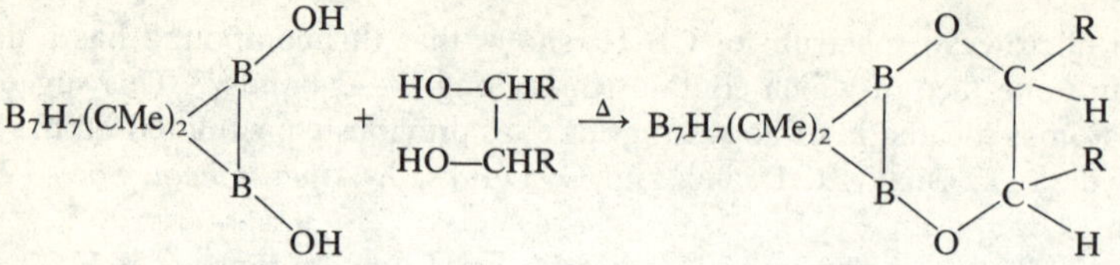

$$B_7H_7(CMe)_2\begin{matrix}B\diagdown OH\\ \diagup\\ B\diagdown OH\end{matrix} + \begin{matrix}HO-CHR\\ |\\ HO-CHR\end{matrix} \xrightarrow{\Delta} B_7H_7(CMe)_2\begin{matrix}B\diagdown O\diagup C\diagdown R\\ \diagup \diagdown H\\ B\diagdown O\diagup C\diagdown R\\ H\end{matrix}\qquad(8)$$

The $^{11}B$ n.m.r. spectrum of $7,9\text{-}B_9C_2H_{12}^-$ shows six doublets, relative intensities $2:1:2:2:1:1$, reading to higher field. these can be assigned to B-2,5, B-3,4, B-10,11, B-6, and B-1, respectively.[85] Attack by Lewis bases (OEt⁻ etc.) on closo-$2,3\text{-}B_9H_9C_2R_2$ (R = H or Me) produces substituted $7,9\text{-}B_9C_2H_{12}^-$ derivatives in which the base is attached to B-10.

7,8- or $7,9\text{-}B_9H_{11}CP^-$ reacts with alkyl halides (RX) to give $B_9H_{11}CPR$, which gave n.m.r. spectra indicative of P—R bonding.[86] They are formulated as (15), with the C at either position 8 or 9 (H atoms are omitted).

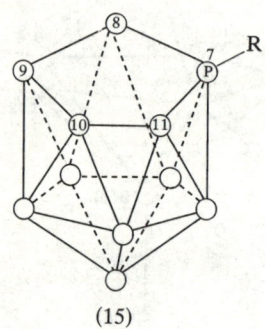

(15)

Reaction of $Tl_2(7,8\text{-}C_2B_9H_{11})$ with $RAsX_2$ (R = Me, X = Br; R = Ph or Bu$^n$, X = Cl) in $Et_2O$ solution produces closo-icosahedral structures. The 3-Ph-3-As-$1,2\text{-}C_2B_9H_{11}$ reacted with $BBr_3$ to give the 3-Br analogue. If $Tl_2(C_2B_9H_{11})$ is treated with $Me_2AsBr$ in a 1:2 ratio, $(Me_2As)_2C_2B_9H_{11}$ is formed, which is thought to have a nido-, 12-vertex structure (Figure 17).[87]

9,10-Dichloro-1,7-phosphacarbaborane (9,10-$Cl_2$-1,7-$CHPB_{10}H_8$) forms monoclinic crystals, space group $P2_1/n$. The 9 and 10 positions for the Cl atoms were confirmed, and it was shown that the presence of the P atom led to considerable distortion of the icosahedral framework. The same paper reported the isolation of six 1,7- and two 1,12-monochlorophosphacarbaborane isomers: the isomers were identified on the basis of the above crystal structure, v.p.c. retention times, and $^{11}B$ n.m.r. spectra.[88]

A new, four-carbon carbaborane, $Me_4C_4B_8H_8$, is a by-product in the formation of the metallo-derivatives $(2,3\text{-}Me_2C_2B_4H_4)_2Co^{III}H$ and $(2,3\text{-}Me_2C_2B_4H_4)_2Fe^{II}H_2$

---

[85] L. J. Todd, A. R. Siedle, F. Sato, A. R. Garber, F. R. Scholer, and G. D. Mercer, *Inorg. Chem.*, 1975, **14**, 1249.
[86] B. N. Storhoff and A. J. Infante, *J. Organometallic Chem.*, 1975, **84**, 291.
[87] H. D. Smith, jun. and M. F. Hawthorne, *Inorg. Chem.*, 1974, **13**, 2312.
[88] H. S. Wong and W. N. Lipscomb, *Inorg. Chem.*, 1975, **14**, 1350.

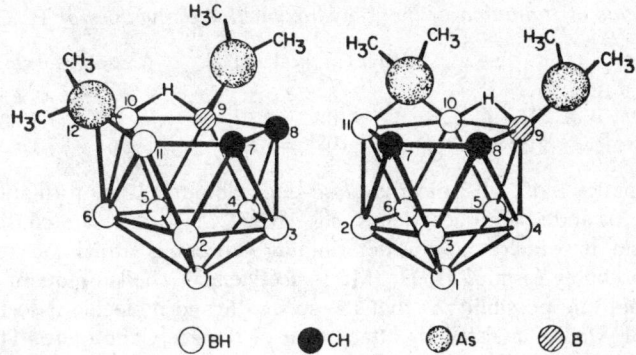

**Figure 17** *Possible structures for* $(Me_2As)_2C_2B_9H_{11}$
(Reproduced by permission from *Inorg. Chem.*, 1974, **13**, 2312)

from $Na^+(2,3-Me_2C_2B_4H_5)^-$ in THF. The initial product isomerizes in solution, and the n.m.r. spectra of the resulting species suggests that it has the structure shown in (16). It acts as a ligand towards transition metals, forming $(OC)_3Mo(Me_4C_4B_5H_5)$ on treatment with $Mo(CO)_6$. This, and a tungsten analogue, are the first metallocarbaboranes containing an electrically neutral ligand.[89]

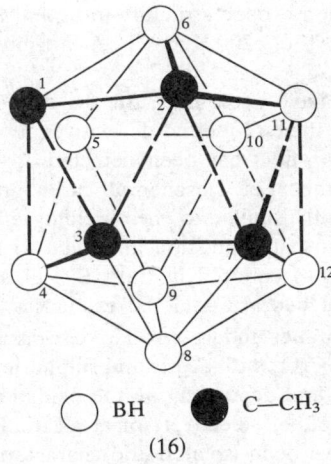

(16)

Heats of combustion have been used to determine the heats of formation of the 1-hydroxymethyl derivatives of the three isomers of $B_{10}C_2H_{12}$ (Table 1).[90]

A general discussion has been given of the bond lengths involving six-co-ordinate carbon in carbaboranes $C_2B_{10}H_{12}$, and their derivatives.[91]

[89] W. M. Maxwell, V. R. Miller, and R. N. Grimes, *J. Amer. Chem. Soc.*, 1974, **96**, 7116.
[90] G. L. Gal'chenko, V. K. Pavlovich, Yu. V. Gol'tyapin, and V. I. Stanko, *Doklady Chem.*, 1974, **216**, 344.
[91] V. S. Mastryukov, L. V. Vilkov, and O. V. Dorofeeva, *J. Mol. Structure*, 1975, **24**, 217.

**Table 1** *Heats of formation of the 1-hydroxymethyl derivatives of* $B_{10}C_2H_{12}$

| | $\Delta H°_{f(cryst)}$/kcal mol$^{-1}$ | $\Delta H°_{f(gas)}$/kcal mol$^{-1}$ |
|---|---|---|
| 1-($CH_2OH$)-$o$-$B_{10}C_2H_{12}$ | $-82.7 \pm 3.5$ | $-61.8 \pm 3.5$ |
| 1-($CH_2OH$)-$m$-$B_{10}C_2H_{12}$ | $-108.2 \pm 3.3$ | $-87.7 \pm 3.3$ |
| 1-($CH_2OH$)-$p$-$B_{10}C_2H_{12}$ | $-109.6 \pm 3.3$ | $-87.1 \pm 3.3$ |

The kinetics of 1,2-dicarba-*closo*-dodecaborane(12) formation from $B_{10}H_{12}(Me_2S)_2$ and several acetylenes, *e.g.* HC≡CCH$_2$Br *etc.*, are consistent with a mechanism in which the rate-determining process is attack on the borane substrate, probably forming $B_{10}H_{12}(Me_2S)$,acetylene.[92] The subsequent stages are not clear, but one possibility is that the second ligand molecule is lost from the 9-boron, followed or assisted by attachment of the acetylene there. The C$_2$ unit would then be suitably placed at the open face of the $B_{10}$ fragment to yield the *closo*-carbaborane cage after loss of H$_2$ and slight rearrangement of the C and B atoms.

The crystal structure of 1,12-dibromo-1,2-dicarba-*closo*-dodecaborane, 1,12-Br$_2$-1,2-C$_2$B$_{10}$H$_{10}$, has been determined. The crystals have a tetragonal unit cell, space group $P4_2$. The Br atoms are substituted in *para*-positions in the *o*-carbaborane cage, one on B-12, the other on C-1.[93]

Listings, and tentative assignments, of proton chemical shifts due to BH and CH groups in *o*- and *m*-carbaboranes have been given.[94] Solutions in C$_6$D$_6$, CCl$_4$, CDCl$_3$, (CD$_3$)$_2$CO, and [$^2$H$_6$]-DMSO, were used.

A sudden change in the i.r. band structure in *o*- and *m*-carbaboranes and their derivatives in the range 170—200 K has been attributed to an order–disorder phase transition.[95]

Some $^{11}$B n.m.r. data have been given for the dianions derived from *o*-, *m*-, and *p*-carbaboranes and their *C*-methyl derivatives.[96]

A long-range shielding effect has been detected in the $^1$H, $^{11}$B, $^{13}$C, and $^{31}$P n.m.r. spectra of a number of icosahedral carbaboranes and their metalloderivatives. It is apparently similar to the shielding effect previously found for B$_5$H$_9$ derivatives; it occurs at a position antipodal to the point of substitution, leading to a net shielding of endopolyhedral $^{11}$B, $^{13}$C, and $^{31}$P resonances, and a net deshielding of exopolyhedral C—H $^1$H resonances.[97]

*o*-, *m*-, and *p*-Carbaboranes undergo oxidative hydroxylation in the presence of KMnO$_4$ in acetic acid or CrO$_3$ with acetic and sulphuric acids.[98] The products are *B*-hydroxycarbaboranes, and depending on the oxidation potential of the system the reaction takes place either selectively or at all the B—H bonds. All possible hydroxylated isomers have been isolated and characterized.

Although previous work indicated that 1,2-dicarbaundecaborane derivatives

---

[92] W. E. Hill, F. A. Johnson, and R. W. Novak, *Inorg. Chem.*, 1975, **144**, 1244.
[93] V. Šubrtová, A. Línek, and C. Novák, *Coll. Czech. Chem. Comm.*, 1975, **40**, 2005.
[94] M. Selim, B. Barlet, R. Freymann, F. Mathey, G. Rabilloud, and B. Sillion, *Compt. rend.*, 1974, **279**, B, 593.
[95] M. Selim, G. Capderroque, and R. Freymann, *Compt. rend.*, 1975, **281**, B, 33.
[96] A. M. Alymov, A. M. Vassilyev, and S. P. Knyaznev, *J. Organometallic Chem.*, 1974, **78**, 313.
[97] A. R. Siedle, G. M. Bodner, A. R. Garber, D. C. Beer, and L. J. Todd, *Inorg. Chem.*, 1974, **13**, 2321.
[98] V. I. Stanko, V. A. Brattsev, N. N. Dvsyannikov, and T. P. Klimova, *J. Gen. Chem. (U.S.S.R.)*, 1974, **44**, 2441.

were produced when nucleophiles such as ammonia and amines reacted with 1-(halogenomethyl)-o-carbaboranes, a reinvestigation has shown that some dialkylamines bring about the smooth replacement of halogen giving the corresponding dialkylaminomethyl-o-carbaboranes. The reaction with $Et_2NH$ is particularly easy, proceeding at 20 °C in benzene solution, with yields >70%.[99]

Isomerization of the o-carbaborane nucleus is known to occur via dianions to the meta-structure, but similar reactions with the para-structure have not been reported[100] It has now been found that dianions obtained from 12-chloro-1-methyl-p-carbaborane and sodium in liquid ammonia, after oxidation, give a mixture of 1-methyl-o- (35%), 1-methyl-m- (55%), and 1-methyl-p-carbaboranes (2—3%). The exact product ratio depends upon the ratio of reactants and the rate of sodium addition.

1,7-Bis(hydroxydimethylsilyl)-m-carbaborane is the starting material for a new synthesis of difunctional 1,7-bis(disiloxanyl)-m-carbaboranes with the formula $[XR(Me)SiO(Me)Si]_2C_2B_{10}H_{10}$ (R = Me, Ph, H or $CH_2{=}CH$, X = Cl, H, MeO, $MeCO_2$, or OH).[101]

A study has been made of the rates of base cleavage of o-, m-, and p-carbaboranyl phenyl ketones.[102]

Phase effects in the oligomerization of 1-allyl-1,2-dicarba-closo-dodecaborane by electron radiolysis have been investigated.[103] Liquid state radiolysis indicates an activation energy of ca. 5 kcal mol$^{-1}$ for the formation of oligomer, and ca. 0 kcal mol$^{-1}$ for the formation of unsaturated dimers. For the plastic crystalline phases, negative activation energies were observed, indicating the existence of a complex reaction sequence.[103]

E.s.r. data for γ-irradiated carbaboranes in organic glass-forming solvents at 77 K suggest the formation of cage-centred o-carbaborane anions.[104]

**Metallo-carbaboranes.**—$Li^+$ and $Na^+$ salts of $2,3$-$C_2B_4H_7^-$ react with organometallic species containing Al, Ga, Rh, Au, or Hg with insertion of the metal atom into a bridging position on the base of the carbaborane pyramidal skeleton (17).[105] The metal is thought to be attached to the cage via a B—M—B three-centre, two-electron bond. $\mu$-$Me_2Ga$-$2,3$-$nido$-dicarbahexaborane(8) is thermally quite stable, but it reacts rapidly with HCl (giving $Me_2GaCl$ and $C_2B_4H_8$).

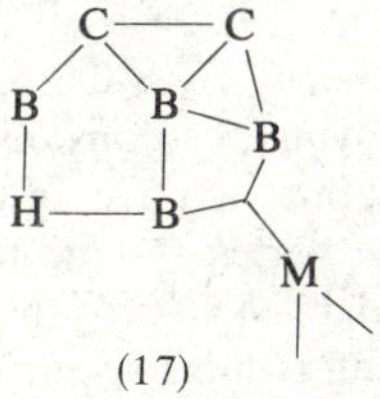

```
       C——C
    B  B  \
    |  |   B
    H——B   |
          \|
           M
           |\
(17)
```

---

[99] L. I. Zakharkhin, V. S. Kozlova, and S. A. Babich, *J. Gen. Chem. (U.S.S.R.)*, 1974, **44**, 1858.
[100] V. I. Stanko and G. A. Anorova, *J. Gen. Chem. (U.S.S.R.)*, 1974, **44**, 2074.
[101] V. V. Korol'ko, E. G. Kagan, Yu. A. Yuzhelevskii, and E. I. Sokolov, *J. Gen. Chem. (U.S.S.R.)*, 1974, **44**, 1501.
[102] V. I. Stanko, T. V. Klimova, and I. P. Beletskaya, *Doklady Chem.*, 1974, **216**, 322.
[103] T. J. Klingen and D. R. Hepburn, jun., *J. Inorg. Nuclear Chem.*, 1975, **37**, 1343.
[104] R. M. Thibault and T. J. Klingen, *J. Inorg. Nuclear Chem.*, 1974, **36**, 3667.
[105] C. P. Magee, L. G. Sneddon, D. C. Beer, and R. N. Grimes, *J. Organometallic Chem.*, 1975, **86**, 159.

The crystal structure of $(Me_4N)_2[(1,6-C_2B_{10}H_{10}Me_2)_2Ti]$ has been determined.[106] The molecular structure is similar to that found for $CpCoC_2B_{10}H_{12}$, but the Ti—C (2.181,2.468 Å) and Ti—B (2.399 Å) bonds are much longer than the equivalent ones in the Co species. This complex is the most electron-deficient metallocarbaborane yet investigated, but there are insufficient data as yet to discern any regular trends in molecular parameters of such complexes.

A preparative route for the species in this series, *i.e.* $[M(C_2R_2B_{10}H_{10})_2]^{2-}$ (M = Ti, R = H or Me; M = Zr, R = Me; M = V, R = H), involves the reaction of $1,2\text{-}C_2R_2B_{10}H_{10}$ (R = H or Me) with $TiCl_4$, $ZrCl_4$, or $VCl_3$. The compounds are significantly more stable than the cyclopentadienyl derivatives of these metals.[107]

Polyhedral expansion of $4,5\text{-}C_2B_7H_9$ with $FeCl_2$ and NaCp gives two isomers (one para-, one dia-magnetic) of $CpFe_2C_2B_6H_8$.[108] The diamagnetic form was shown by X-ray diffraction to be $1,6\text{-}(\eta^5\text{-}C_5H_5)_2\text{-}1,6,2,3\text{-}Fe_2C_2B_6H_8$ (Figure 18), *i.e.* it is a ten-vertex species derived from a bicapped square antiprism. It is not easy to decide on the relationship between this structure and that of the paramagnetic form.

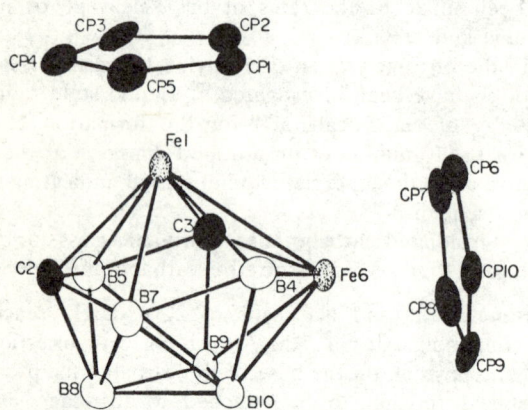

**Figure 18** Structure of $1,6\text{-}(\eta^5\text{-}C_5H_5)_2\text{-}1,6,2,3,\text{-}Fe_2C_2B_6H_8$
(Reproduced by permission from *J. Amer. Chem. Soc.*, 1975, **97,** 296)

The molecular structure of $8\text{-}\eta\text{-}C_5H_5\text{-}8\text{-}Co\text{-}6,7\text{-}C_2B_7H_{11}$ is decaborane-like with the three heteroatoms occupying positions on the open face.[109] The C atoms are adjacent, one in a four-, the other in a five-co-ordinate position. The Co atom occupies another five-co-ordinate vertex (bound to one C, three B's and the Cp ring), being further co-ordinated by a hydrogen atom bridging the Co and an adjacent B atom. Another bridging H links a four-co-ordinate B and the adjacent B on the open face.

---

[106] F. Y. Lo, C. E. Strouse, K. P. Callahan, C. D. Knobler, and M. F. Hawthorne, *J. Amer. Chem. Soc.*, 1975, **97,** 428.
[107] C. G. Salentine and M. F. Hawthorne, *J. Amer. Chem. Soc.*, 1975, **97,** 426.
[108] K. P. Callahan, W. J. Evans, F. Y. Lo, C. E. Strouse, and M. F. Hawthorne, *J. Amer. Chem. Soc.*, 1975, **97,** 296.
[109] K. P. Callahan, F. Y. Lo, C. E. Strouse, A. L. Sims, and M. F. Hawthorne, *Inorg. Chem.*, 1974, **13,** 2842.

# Elements of Group III

The non-icosahedral metallocarbaborane 1,2,4-$(C_5H_5)Co(CMe)_2B_8H_8$ rearranges on heating to give, first, 10,2,3-$(C_5H_5)Co(CMe)_2B_8H_8$, and finally, the thermodynamically stable isomer 1,2,3-$(C_5H_5)Co(CMe)_2B_8H_8$ (Scheme 2).[110] The intermediate is the second known example of monometallocarbaborane with five B atoms at the bonding face. It should be noted that the first stage in the rearrangement does not fit the recently developed rules for monometallocarbaborane isomerizations.

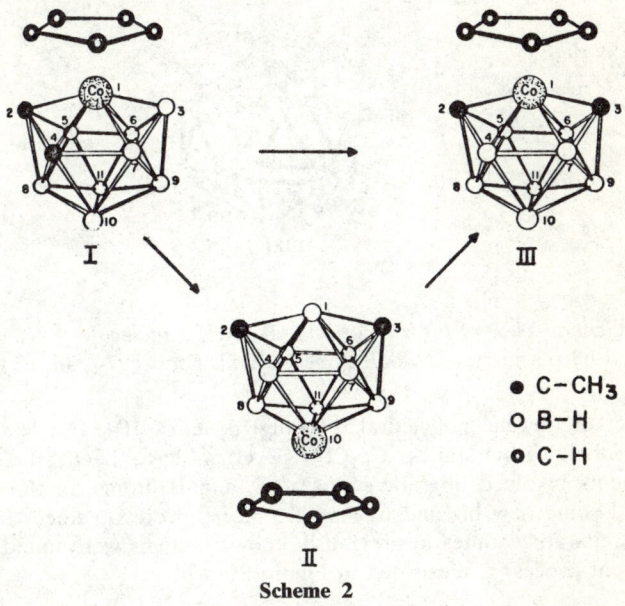

Scheme 2

The species $(\pi\text{-}C_5H_5)Co(\pi\text{-}C_5H_4B_9C_2H_{11})$ is zwitterionic, being derived from a cobalticenium ion and $B_9C_2H_{12}^-$, linked by a C—C bond, with loss of a terminal H from each species.[111] The carbaborane fragment is linked to the cation via atom C-1 (Figure 19). The remaining 10 atoms of the $B_9C_2$ framework bear apical H atoms [C-2—$H_t$ = 0.991(14), B—$H_t$ = 1.045(14)—1.152(15) Å]. The 'extra' facial H atom lies above the open pentagonal face C-1—B-4—B-8—B-7—C-2 (18), the largest interactions being with B-4 and B-7, ($H_{fac}$-B approx. 1.63 Å).

(18)

---

[110] G. D. Mercer, M. Tribo, and F. R. Scholer, *Inorg. Chem.*, 1975, **14**, 764.
[111] M. R. Churchill and B. G. DeBoer, *J. Amer. Chem. Soc.*, 1974, **96**, 6310.

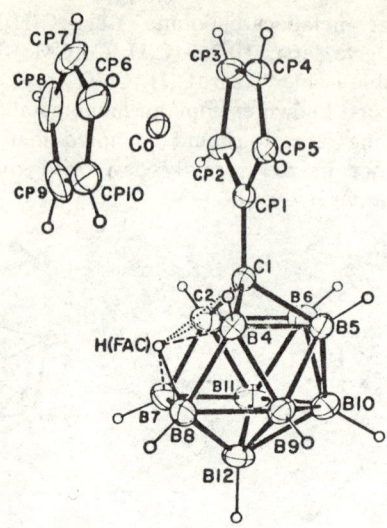

**Figure 19** *General view of the* $CpCo(C_5H_4\text{-}B_9C_2H_{11})$ *molecule*
(Reproduced by permission from *J. Amer. Chem. Soc.*, 1974, **96**, 6310)

A number of thermal polyhedral rearrangements of 10-, 11-, and 12-vertex bimetallic cobaltocarbaboranes, $Cp_2Co_2C_2B_nH_{n+2}$, have been studied.[112] The rearrangements involved migration of Co, C, and B atoms on the polyhedral surface, and some new bi- and tri-metallic species were obtained. In addition, improved preparative routes to previously known systems were found. A typical rearrangement process is illustrated in Figure 20.

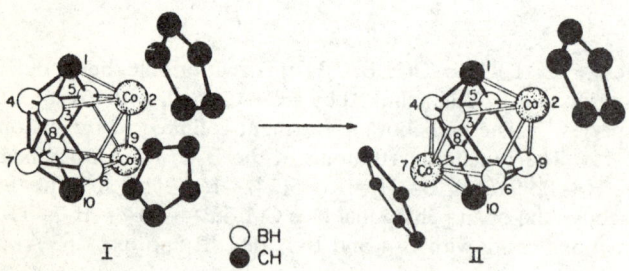

**Figure 20** *The rearrangement of* 2,6,1,10-$Cp_2Co_2(C_2B_6H_8)$, I, *to the* 2,7,1,10-*isomer*, II
(Reproduced by permission from *J. Amer. Chem. Soc.*, 1974, **96**, 7405)

Metal complexes with the *nido*-dicarbaborane structure, *e.g.* ($\pi$-$B_{10}H_{10}C_2R^1R^2)_2^{n-}$ (M = Co, Ni, or Fe), are readily oxidized by $CuCl_2$ and other reagents to carbaboranes with the *closo*-structure. The study reported here

---

[112] W. J. Evans, C. J. Jones, B. Štíbr, R. A. Grey, and M. F. Hawthorne, *J. Amer. Chem. Soc.*, 1975, **96**, 7405.

involves a variety of substituted o- and m-carbaboranes; a low-temperature process for the isomerization of o- to m- and p-carbaboranes has been identified, via oxidation of metal complexes of the dianions.[113]

7,8-$B_9C_2H_{11}^{2-}$ reacts with $(Ph_3P)_2Rh(CO)Cl$, forming $(B_9C_2H_{11})RhCl(PPh_3)_2$.[114] The new species $[(B_9C_2H_{11})Rh(PPh_3)(C_6H_6)_2]_2$ and $(B_9C_2H_{11})Rh(H)PPh_3$ were obtained from the same species reacting with $(Ph_3P)_3RhCl–C_6H_6$.

An important new series of B-σ-metallocarbaboranes has been reported by Hoel and Hawthorne.[115] Thus, 3-$[(PPh_3)_2IrHCl]$-1,2-$C_2B_{10}H_{11}$ is prepared in good yield (ca. 84%) and rapidly by treating $(PPh_3)_2IrCl$ (prepared in situ from $[Ir(C_8H_{14})_2Cl]_2$ and two equivalents of $PPh_3$) with 1,2-$C_2B_{10}H_{12}$ in benzene. This and related compounds are good models for intermediates in the transition-metal-catalysed exchange reactions; they are presumably formed by an oxidative-addition mechanism.

The reaction of $(Ph_3P)_3IrCl$ with 1-Li-2-R-1,2-$B_{10}C_2H_{10}$ in $Et_2O$ leads to the formation of a carbaborane-iridium(I) complex, but this is unstable, and subsequent rapid oxidative-addition and reductive-elimination reactions occur (Scheme 3).[116]

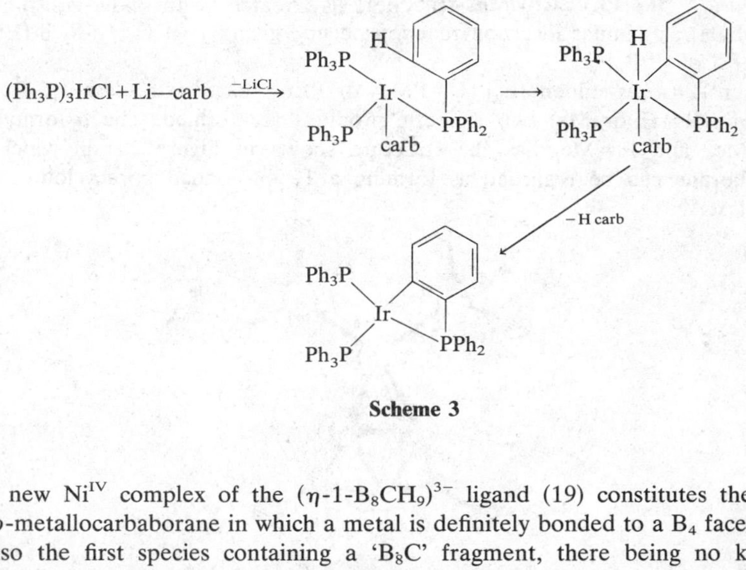

**Scheme 3**

A new $Ni^{IV}$ complex of the $(\eta$-1-$B_8CH_9)^{3-}$ ligand (19) constitutes the first closo-metallocarbaborane in which a metal is definitely bonded to a $B_4$ face.[117] It is also the first species containing a '$B_8C$' fragment, there being no known monocarbon carbaboranes (neutral or anionic) of this type. The compound was prepared, in very small amounts, as a side-product from the reaction of $Na_3B_{10}CH_{11}$ with NaCp and $NiBr_2,2C_2H_4(OMe)_2$.

$Pt(PEt_3)_3$, $[Pt(PMe_3)_2(trans$-stilbene)]$, $Pt(PMe_2Ph)_4$, $[Ni(1,5-C_8H_{12})_2]$, $[Ni(PEt_3)_2,(1,5-C_8H_{12})]$, and $[Pd(CNBu^t)_2]$ react with closo-2,3-$Me_2$-2,3-$C_2B_9H_9$ to

---

[113] L. I. Zakharkin, V. N. Kalinin, and N. P. Levina, *J. Gen. Chem. (U.S.S.R.)*, 1974, **44**, 2437.
[114] A. R. Siedle, *J. Organometallic Chem.*, 1975, **90**, 249.
[115] E. L. Hoel and M. F. Hawthorne, *J. Amer. Chem. Soc.*, 1974, **96**, 6770.
[116] B. Longato, F. Morandini, and S. Bresadola, *J. Organometallic Chem.*, 1975, **88**, C7.
[117] C. G. Salentine, R. R. Rietz, and M. F. Hawthorne, *Inorg. Chem.*, 1974, **13**, 3025.

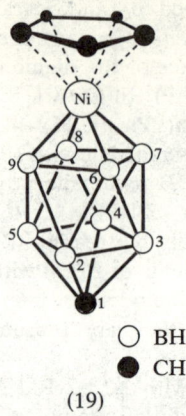

○ BH
● CH

(19)

give the *closo*-metallocarbaboranes: 1,1-L$_2$-2,4-Me$_2$-1,2,4-MC$_2$B$_9$H$_9$ (M = Pt, L = PEt$_3$, PMe$_3$, or PMe$_2$Ph; M = Ni, L$_2$ = 1,5-C$_8$H$_{12}$ or L = PEt$_3$; M = Pd, L = Bu$^t$NO).[118] If [Pt(PEt$_3$)$_2$(*trans*-stilbene)] is treated with 1-($\eta$-C$_5$H$_5$)-1,2,4-CoC$_2$B$_8$H$_{10}$, a similar insertion reaction occurs, giving 1-($\eta$-C$_5$H$_5$)-8, 8-(Et$_3$P)$_2$-1,2,7,8-CoC$_2$PtB$_8$H$_{10}$.

When [Pt(*trans*-stilbene)L$_2$] (L = PMe$_3$ or PEt$_3$) reacts with *closo*-1,6-R$_2$-1,6-C$_2$B$_6$H$_6$ (R = H or Me) two isomeric products are formed. The α-form, with L = PMe$_3$ and R = Me, has the structure shown in Figure 21, in which the carbaborane can be regarded as forming a 1–4-$\eta$-bonded boracyclobutadiene complex.[119]

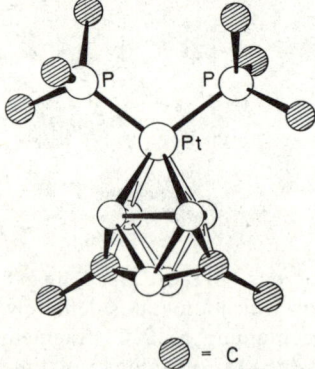

⊘ = C

**Figure 21** *The structure of closo*-1,1-(Me$_3$P)$_2$-6,8-PtC$_2$B$_6$H$_6$
(Reproduced from *J.C.S. Chem. Comm.*, 1974, 794)

*closo*-2,4-C$_2$B$_5$H$_7$ reacts with [Pt(styrene)(PEt$_3$)$_2$] to give *closo*-2,3-[(Et$_3$P)$_2$]$_2$-1,2,3,6-CPt$_2$CB$_5$H$_7$.[120] [Pt(1,5-C$_8$H$_{12}$)(PMe$_3$)$_2$], on the other hand, forms with

---

[118] M. Green, J. L. Spencer, F. G. A. Stone, and A. J. Welch, *J.C.S. Dalton*, 1975, 179.
[119] M. Green, J. L. Spencer, F. G. A. Stone, and A. J. Welch, *J.C.S. Chem. Comm.*, 1974, 794.
[120] G. K. Barker, M. Green, J. L. Spencer, F. G. A. Stone, B. F. Taylor, and A. J. Welch, *J.C.S. Chem. Comm.*, 1975, 804.

1,6-$C_2B_8H_{10}$ the complex nido-$\mu$(4,8)-[$(Me_3P)_2Pt$]-8,8-[$(Me_3P)_2$]-7,8,10-$CPtCB_8H_{10}$. The structures were determined by $X$-ray diffraction.

Detailed $X$-ray crystallographic results have been reported for 1,1-bis-(dimethylphenylphosphine) - 2, 4 - dimethyl - 2, 4 - dicarba - 1 - platina - closo - dodecaborane [$Pt(Me_2C_2B_9H_9)(PMe_2Ph)_2$].[121] The carbaborane unit has distorted icosahedral geometry, and carbon atoms at the 2- and 4-positions, with the Pt at the 1-position. The Pt atom is ca. 1.81 Å above the $C_2B_3$ face.

Consideration of approximate MO calculations on metallocarbaboranes with $d^8$ and $d^9$ metals and 1,2- and 1,7-$C_2B_9H_{11}^{2-}$ ligands has suggested that the unsymmetrical co-ordination to the metal ('slipped sandwich' structure) is related to the presence of two or three electrons in the (antibonding) $2e_{1g}$ orbitals. It was not clear, however, why $d^8$ and $d^9$ metal complexes with $C_5H_5^-$ should not also be distorted.[122]

New six-membered cyclic compounds (20; Z = $SnMe_2$, $GeMe_2$, PPh, or AsMe) can be prepared when the di-lithium derivative of the di-o-carbaboranyldimethylsilane reacts with the dihalogeno-compounds $ZCl_2$.[123] Similar compounds containing Ge in place of Si were also obtained.

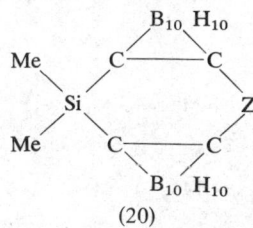

(20)

**Compounds containing B—C Bonds.**—High-resolution i.r. spectra of $^{10}BD_3CO$ and $^{11}BD_3CO$ were obtained in the region of the $K = 0$ sub-bands of the $\nu_2$ and $\nu_2 + \nu_8 - \nu_8$ bands.[124] It was not possible, however, to determine the borine carbonyl structure accurately using only the $B_0$ rotational constants.

Exchange reactions of $HBr + BMe_3$, $BMe_2Br$, or $BMeBr_2$ yield single specific products under the influence of $CO_2$ laser radiation.[125] Thus, reaction (9) is the only one to occur when irradiation is at 970.5 cm$^{-1}$.

$$BMe_3 + HBr \rightarrow BMe_2Br + CH_4 \quad (9)$$

If 1039.4 cm$^{-1}$ radiation is used, however, the only process to be activated is reaction (10).

$$BMe_2Br + HBr \rightarrow BMeBr_2 + CH_4 \quad (10)$$

$Me_2B(CH_2Cl)$ forms two crystalline modifications, belonging to the space groups *Pbcm* and *Pnma*. The chief molecular parameters are shown in (21).[126]

[121] A. J. Welch, *J.C.S. Dalton*, 1975, 1473.
[122] P. A. Wegner, *Inorg. Chem.*, 1975, **14**, 212.
[123] L. I. Zakharkin and N. F. Shemyakin, *J. Gen. Chem. (U.S.S.R.)*, 1974, **44**, 1043.
[124] C. Pépin, L. Lambert, and A. Cabana, *J. Mol. Spectroscopy*, 1974, **53**, 120.
[125] H. R. Bachmann, H. Nöth, R. Rinck, and K. L. Kompa, *Chem. Phys. Letters*, 1975, **33**, 261.
[126] J. C. Huffman, *Cryst. Structure Comm.*, 1974, **3**, 649.

The $^7$Li chemical shift data for Et$_2$O, THF, or dimethoxyethane solutions of LiMMe$_4$ (M = B, Al, Ga, or Tl) are consistent with the formation of ion-pairs *via* the equilibria (11), except when M = B. In this case there is evidence for an interaction of the type (22a) or (22b).[127]

$$\text{Li}^+,\text{MMe}_4^- \rightleftharpoons [S_n\text{Li}]^+[\text{MMe}_4]^- \rightleftharpoons [S_n\text{Li}]^+ \| [\text{MMe}_4]^- \qquad (11)$$

unsolvated solid        'tight' ion-pair        solvent-separated ion-pair

$J(B, C)$ coupling constants for a number of organoboranes [e.g. 1-MeB$_5$H$_8$, B(CH=CH$_2$)$_3$, BH$_3$CN$^-$] are in good agreement with values calculated from an INDO-SCF finite perturbation calculation. This only took into account the Fermi contact contribution to $J(B, C)$.[128]

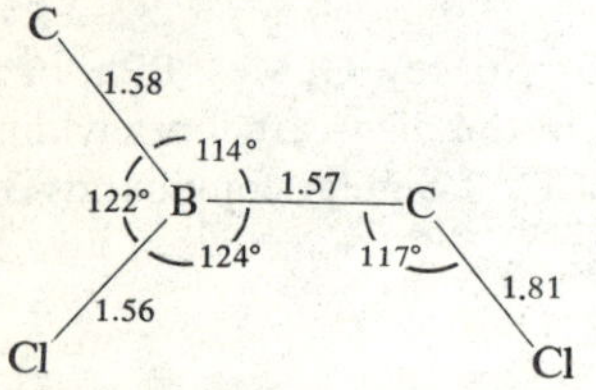

(21) 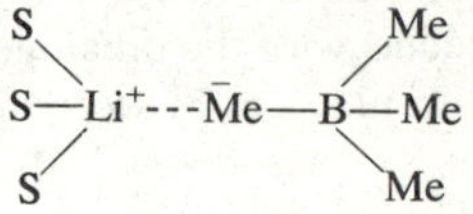     (22a)

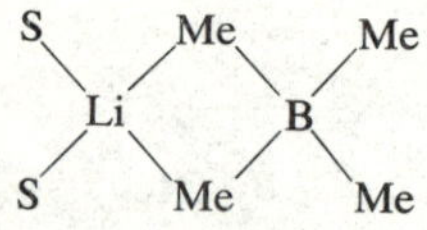

(22b)

A number of molecular structural parameters for trivinylborane, B(CH=CH$_2$)$_3$, are shown in Table 2.[129] The most acceptable molecular model is a planar dynamic one, in which steric strain is reduced by distortion of the vinyl groups. Since the molecule is probably planar, some C··· B π-electron delocalization was postulated; this was corroborated by the value for the C=C bond length.

$^{13}$C and $^{11}$B n.m.r. data have been presented for trivinylborane, six vinylhalogenoboranes [including (CH$_2$=CH)$_2$BBr, prepared for the first time] and some Lewis acid–base adducts of these. The ability of the vinyl substituent to provide π-electron density for B is relatively insensitive to the nature of the halogen present.[130]

**Table 2** *Structural parameters for trivinylborane* B(CH=CH$_2$)$_3$

| | | | |
|---|---|---|---|
| $r$(C—H) | 1.092 ± 0.003 Å | ∠BCH | 116.5 ± 0.9° |
| $r$(C—C) | 1.370 ± 0.006 Å | ∠BCC | 122.4 ± 0.9° |
| $r$(B—C) | 1.558 ± 0.003 Å | ∠CCH (*trans* to B) | 124.0 ± 1.6° |
| | | ∠CCH (*cis* to B) | 132.2 ± 2.3° |

[127] R. J. Hogans, P. A. Scherr, A. T. Weibel, and J. P. Oliver, *J. Organometallic Chem.*, 1975, **85**, 265.
[128] L. W. Hall, D. W. Lowman, P. D. Ellis, and J. D. Odom, *Inorg. Chem.*, 1975, **14**, 580.
[129] A. Foord, B. Beagley, W. Reade, and I. A. Steer, *J. Mol. Structure*, 1975, **24**, 131.
[130] L. W. Hall, J. D. Odom, and P. D. Ellis, *J. Amer. Chem. Soc.*, 1975, **97**, 4527.

*Elements of Group III*

Further evidence of significant $\pi$-character in a B—C bond of vinylboranes was produced by $^{13}$C n.m.r. spectra of $\alpha,\beta$-unsaturated alkenylboranes such as $CH_2$=$CHB(OBu^n)_2$. Thus the $\beta$-carbon is strongly deshielded.[131] Exactly analogous results and conclusions followed a study of the $^{13}$C n.m.r. spectra of boron-substituted alkynes, such as $HC\equiv CB(OBu^n)_2$.[132]

**Amino-boranes and other Compounds containing B—N Bonds.**—Trimethylamine-borane, on exposure to $^{60}$Co $\gamma$-rays at 77 K, forms the radical-anion $Me_3\dot{N}$—$BH_3^-$. This has a four-line e.s.r. spectrum, with a large hyperfine coupling to the boron atom. Annealing this species at temperatures >77 K gives $Me_2\dot{N}$—$BH_3$.[133]

Fluorescence spectra of 10 aminoborane analogues of stilbenes and styrenes were interpreted in terms of $\pi\pi^*$ emission from a polar excited state to a less polar, more rigidly planar ground state.[134]

A direct comparison has been made of the properties of a number of amino-boranes and related borazines.[135] Thus, $HgX_2$ (X = halogen) are more reactive towards $H_3BNMe_3$ than towards $H_3B_3N_3H_3$. The decreased reactivity of the latter appears to be due to the $\pi$-bonding in the ring, which reduces the hydridic character of the H attached to the boron, and the partial positive charge on the boron itself, by comparison with $H_3BNMe_3$ [the mode of attack is thought to be as in (23)]. Purely steric effects would have led to the opposite reactivity sequence. $^1$H N.m.r. data on the amino-boranes $H_2XBNMe_3$ and $H_2XBNMe_2H$, together with those on $H_2XB_3N_3Me_3$ and $H_2XB_3N_3H_3$ (X = Cl, Br, Me, or CN), support the idea that a resonance interaction between the substituent and the borazine ring is responsible for the changes in the relative chemical shifts of the o- and p-protons in $H_2XB_3N_3H_3$.

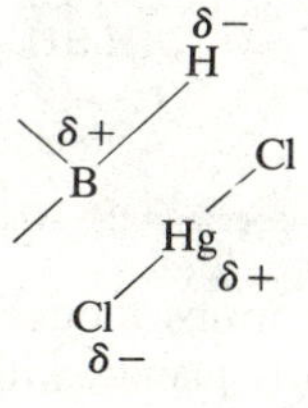

(23)

Dialkylaminohydridophenoxyboranes [$HB(OPh)NR_2$] may be formed by the three-stage process (12)—(14)

$$4BF_3,OEt_2 + 3NaBH_4 \rightarrow 3NaBF_4 + 2B_2H_6(g) + 4Et_2O \quad (12)$$

$$\tfrac{1}{2}B_2H_6(g) + HNR_2 \rightarrow H_3BNHR_2 \quad (13)$$

$$H_3BNHR_2 + HOPh \xrightarrow{\Delta} 2H_2 + HB(OPh)NR_2 \quad (14)$$

---

[131] Y. Yamamoto and I. Moritani, *Chem. Letters*, 1975, 57.
[132] Y. Yamamoto and I. Moritani, *Chem. Letters*, 1975, 439.
[133] T. A. Claxton, S. A. Fieldhouse, R. E. Overill, and M. C. R. Symons, *Mol. Phys.*, 1975, **29**, 1453.
[134] K. G. Hancock, Y. Ko, D. A. Dickinson, and J. D. Kramer, *J. Organometallic Chem.*, 1975, **90**, 23.
[135] O. T. Beachley, jun. and B. Washburn, *Inorg. Chem.*, 1975, **14**, 120.

[R = Me, Et, Pr$^i$, Bu$^n$, or CH$_2$Ph; R$_2$ = (CH$_2$)$_4$ or (CH$_2$)$_5$]. The products are planar and monomeric, and there is apparently significant B---N $\pi$-interaction.[136]

Amorphous poly(aminoborane)s are formed in high yield by reaction (15).

$$B_2H_6 + LiNH_2 \xrightarrow{Et_2O} 1/n(BH_2NH_2)_n + LiBH_4 \qquad (15)$$

They are not highly polymeric, $n$ being mainly 3, 4, or 5, and they were much more highly reactive than cyclic borazanes. They were, therefore, thought to be linear.[137]

LiNMe$_2$ reacts with HB(NMe$_2$)$_2$, LiMe$_2$N,BH$_3$, and [Me$_2$NBH$_2$]$_2$ at decreasing rates to give B(NMe$_2$)$_3$ + LiH.[138] Reactions (16) and (17) have been proposed to account for the behaviour of the last-named.[139]

$$[Me_2NBH_2]_2 + LiH \xrightarrow{Et_2O} LiMe_2BH_2NMe_2,BH_3 \qquad (16)$$

$$LiMe_2BH_2NMe_2,BH_3 + LiH \xrightarrow{Et_2O} 2LiMe_2N,BH_3 \qquad (17)$$

Analysis of the microwave spectra of 10 isotopic species of H$_2$N,B$_2$H$_5$ gave the substitution values of the structural parameters shown in Table 3.[140] Unlike bromodiborane, aminodiborane has a structure significantly different from diborane itself.

**Table 3** *Structural parameters for $\mu$-amino-diborane, H$_2$N—B$_2$H$_5$*

| | | | |
|---|---|---|---|
| $r$(B—B) | 1.916 ± 0.002 Å | ∠BNB | 75.9 ± 0.1° |
| $r$(B—N) | 1.558 ± 0.001 Å | ∠BH$_{br}$B | 90.0 ± 0.6° |
| $r$(B—H$_{br}$) | 1.355 ± 0.005 Å | ∠H$_t$BH$_t$ | 121.0 ± 0.3° |
| $r$(B—H$_t$) | 1.193 ± 0.001 Å | $\varepsilon^a$ | 16.8 ± 0.1° |
| $r$(N—H) | 1.005 ± 0.006 Å | ∠HNH | 111.0 ± 1.2° |

$^a$ Defined as the angle between the BH$_2$ planes and the symmetry axis.

Dichloro(diphenylamino)borane forms monoclinic crystals (space group $P2_1$). The Cl$_2$BNC$_2$ skeleton is very nearly planar, with the Ph groups tilted out of this plane. The B—N distance is 1.38 Å, suggestive of B⇌N bonding.[141]

Reaction of AgNCO with liquid BCl$_3$ gives gaseous mixtures of BCl$_3$ and BCl$_2$(NCO). The latter was characterized by i.r. and mass spectrometry. The i.r. spectrum suggested the presence of a B—N bond. Liquid mixtures of this composition are unstable, with the BCl$_2$(NCO) giving polymeric decomposition products. BBr$_3$ + AgNCO gave only very small amounts of BBr$_2$(NCO).[142]

Analysis of microwave spectra due to eight isotopic species of Me$_3$NBH$_3$

---

[136] R. A. Kovac and G. G. Waldvogle, *Inorg. Chem.*, 1975, **14**, 2239.
[137] D. L. Denton, A. D. Johnson, C. W. Hickam, R. K. Bunting, and S. G. Shore, *J. Inorg. Nuclear Chem.*, 1975, **37**, 1037.
[138] P. C. Keller, *Inorg. Chem.*, 1975, **14**, 438.
[139] P. C. Keller, *Inorg. Chem.*, 1975, **14**, 440.
[140] K.-K. Lau, A. B. Burg, and R. A. Beaudet, *Inorg. Chem.*, 1974, **13**, 2787.
[141] F. Zettler and H. Hess, *Chem. Ber.*, 1975, **108**, 2269.
[142] H. Mongeot, J. Dazord, H. Atchekzal, and J. Carré, *Compt. rend.*, 1975, **280**, C, 1109.

gave the following structural parameters: $d(BN) = 1.638 \pm 0.01$ Å, $d(CN) = 1.483 \pm 0.01$ Å, and $\angle CNB = 109.9 \pm 1.0°$.[143]

Pyrolysis of $H_3NBF_3$ at 185 °C produces aminodifluoroborane, $BF_2NH_2$, as a primary product. This new compound has an ionization potential of $12.4 \pm 0.4$ eV. $\Delta H_f(BF_2NH_2)$ was calculated to be $-255$ kcal mol$^{-1}$.[144]

The i.r. spectra of the adducts $Me_3NBH_2X$ and $Me_3NBHX_2$ were analysed using the group-vibration approximation.[145]

$^{13}C$ and $^{11}B$ chemical shifts have been tabulated for the adducts $Me_3N,BX_2Y$ ($X = Y = H$, F, Cl, or Br; $X = H$, $Y = F$; $X = F$, $Y = H$). It is suggested that the adduct bond strengths are in the order $BH_3 > BH_2F > BF_3 > BF_2H$. This ordering is not absolutely certain, but $BH_3$ certainly forms the strongest adduct.[146]

The adducts $PhMe_2N,BX_3$ ($X = Cl$, Br, or I) show three-bond coupling between methyl protons and $^{11}B$ at room temperature (*ca.* 3 Hz). The fine structure due to this coupling collapses at higher temperatures, the activation energy for the process being in the order $X = Cl < Br < I$.[147] This order follows the likely ease of ionization of the halogen, and not the order of the B—N bond energies. Therefore, B—X fission is more likely in the rate-determining step than B—N fission.

The exchange reaction (18) ($X = Cl$, Br, F, CN, NCS, NCO, or $BH_3CN$) proceeds satisfactorily in THF, but not in monoglyme or MeCN as solvent.[148] The adducts with $X = NCS$ and NCO have not previously been described.

$$Me_3N,BH_2I + MX \rightarrow Me_3N,BH_2X + MI \qquad (18)$$

Direct reaction of $BX_3$ ($X =$ halogen) with secondary amines produces amino-dihalogenoboranes, ammonium salts, and bis(amine) boronium salts as well as the simple adducts. This complication may be avoided by the halogenation of amino-boranes, and the syntheses of $R_2NH,BX_3$ were described for $X = Cl$ or Br, $R = Me$ or Et.[149]

Piperazine- and triethylenediamine-monoborane (but not diborane) adducts of $BH_2F$, $BHF_2$, and $BF_3$ are produced by displacement reactions between the diamine and the $Me_3N$-fluoroboranes. $^1H$ and $^{11}B$ n.m.r., and i.r. data were given for these new species.[150]

Approximate vibrational assignments have been given for $[(R_3N)_2BH_2]^+I^-$ ($R = H$ or Me) using the group frequency approach.[151] Bands due to the $\nu(BH_2)$, $\delta(BH_2)$, and $\nu(BN_2)$ modes appeared to shift very little on going to the ionic system from $R_3N,BH_2X$.

$^1H$ N.m.r. spectroscopy has been used to follow the reactions between $H_3B,Me_2NCH_2NMe_2,BH_3$, *i.e.* $TMED,2BH_3$, and the hydrogen halides HX ($X = F$, Cl, or Br) or $Br_2$. With the latter, symmetrical substitution takes place and

---

[143] P. Cassoux, R. L. Kuczkowski, P. J. Bryan, and R. C. Taylor, *Inorg. Chem.*, 1975, **14**, 126.
[144] E. F. Rothgery, H. A. McGee, and S. Pusatcioglu, *Inorg. Chem.*, 1975, **14**, 2236.
[145] P. J. Bratt, K. R. Seddon, and I. A. Steer, *Inorg. Chim. Acta*, 1974, **10**, 191.
[146] R. A. Geanangel, *Inorg. Chem.*, 1975, **14**, 696.
[147] J. R. Blackborow, *J. Magn. Resonance*, 1975, **18**, 107.
[148] P. J. Bratt, M. P. Brown, and K. R. Seddon, *J.C.S. Dalton*, 1974, 2161.
[149] G. E. Ryschkewitsch and W. H. Myers, *Synth. React. Inorg. Metal-Org. Chem.*, 1975, **5**, 123.
[150] J. M. Van Paaschen and R. A. Geanangel, *Canad. J. Chem.*, 1975, **53**, 723.
[151] P. J. Bratt, M. P. Brown, and K. R. Seddon, *J. Inorg. Nuclear Chem.*, 1975, **37**, 887.

TMED,2BH$_2$Br and TMED,2BHBr$_2$ were identified. Similar mono-fluorinated and-chlorinated compounds TMED,2BH$_2$X could be isolated, but further addition of HF led to [TMEDH$_2^{2+}$][BF$_4^-$]$_2$. The symmetrical substitution of the BH$_3$ groups in each case is considered to be evidence that the inductive effect of the halogen makes the remaining hydrogens less susceptible to electrophilic attack.[152]

A number of methylidyne cluster compounds have been prepared, *i.e.* Co$_3$(CO)$_9$COMX$_2$,NMe$_3$ (M = B, X = F or I; M = Al, X = Cl).[153] In addition, the new compound Co$_6$(CO)$_{18}$B results from the reaction of Co$_2$(CO)$_8$ with BBr$_3$ at 60 °C, or from Co$_2$(CO)$_8$ + B$_2$H$_6$ at 8—10 atm. If Co$_2$(CO)$_8$ is replaced by Co(CO)$_4^-$ in the former reaction, the product is Co$_3$(CO)$_9$B,NEt$_3$. This was characterized by i.r. and $^{11}$B n.m.r. spectroscopy (Figure 22).

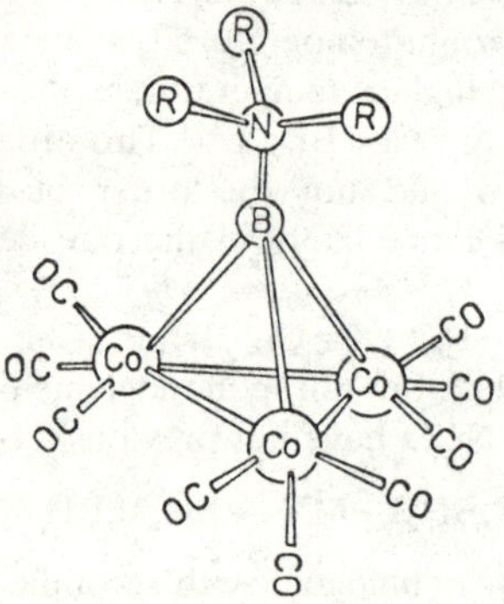

**Figure 22** *Proposed structure for* Co$_3$(CO)$_9$B,NEt$_3$
(Reproduced by permission from *J. Organometallic Chem.*, 1975, **86,** 257)

**Compounds containing B—P or B—As Bonds.**—Detailed vibrational spectroscopic studies have been reported for H$_3$P,BX$_3$ and D$_3$P,BX$_3$ (X = F, Br, or I).[154] The spectra may all be interpreted in terms of $C_{3v}$ symmetry. $\nu$(P—H) modes are at higher wavenumbers than in free phosphine, consistent with bond shortening on co-ordination. Normal-co-ordinate analyses were performed on each molecule. In H$_3$P,BF$_3$ there are are four lattice modes, suggesting that there are at least two molecules per unit cell. Similarly detailed i.r. and Raman spectra and assignments have been given for H$_3$P,BH$_3$, H$_3$P,BD$_3$, D$_3$P,BH$_3$, and D$_3$P,BD$_3$ in the solid state at 77 K. A normal-co-ordinate analysis yielded a value of 1.97 mdyn Å$^{-1}$ for the P—B stretching force constant. The number of observed lattice modes was inconsistent with the previously reported crystal structure.[155]

Durig's group has subjected MePH$_2$,BH$_3$, MePH$_2$,BD$_3$, Me$_2$DP,BH$_3$, and MePD$_2$,BD$_3$ to an analogous treatment.[156] The calculated P—B stretching force constant for this series of species was 2.44 mdyn Å$^{-1}$, *i.e.* significantly higher than in the previous case. Microwave data on the methyl-substituted adducts produced

[152] M. G. Hu and R. A. Geanangel, *Inorg. Chim. Acta*, 1974, **10,** 83.
[153] G. Schmid, V. Bätzel, G. Etzrodt, and R. Pfeil, *J. Organometallic Chem.*, 1975, **86,** 257.
[154] J. R. Durig, S. Riethmiller, V. F. Kalasinsky, and J. D. Odom, *Inorg. Chem.*, 1974, **13,** 2729.
[155] J. D. Odom, V. F. Kalasinsky, and J. R. Durig, *J. Mol. Structure*, 1975, **24,** 139.
[156] J. R. Durig, V. F. Kalasinsky, Y. S. Li, and J. D. Odom, *J. Phys. Chem.*, 1975, **79,** 468.

a value of $2.49 \pm 0.06$ kcal mol$^{-1}$ for rotation of the methyl group, and $1.57 \pm 0.06$ kcal mol$^{-1}$ for the BH$_3$ top.

The final paper in this sequence deals with Me$_2$HP,BH$_3$, Me$_2$DP,BH$_3$, and Me$_2$HP,BD$_3$. The i.r. and Raman spectra of all of these were assigned using $C_s$ symmetry as a basis. Analysis of the microwave spectra ($^{10}$B and $^{11}$B forms for each) gave the parameters listed in Table 4.[157]

**Table 4** *Some structural parameters for dimethylphosphino-borane,* Me$_2$HP,BH$_3$

| | | | |
|---|---|---|---|
| $r$(B—H) | $1.216 \pm 0.005$ Å | ∠CPB | $114.6 \pm 1.0°$ |
| $r$(P—B) | $1.898 \pm 0.010$ Å | ∠CPC | $105.4 \pm 1.0°$ |
| $r$(P—C) | $1.813 \pm 0.010$ Å | ∠PBH | $104.8 \pm 0.4°$ |
| $r$(P—H) | $1.414 \pm 0.009$ Å | ∠BPH | $118.1 \pm 4.0°$ |

Base-displacement reactions of substituted difluorophosphines established the following series of base strengths with respect to BH$_3$: PF$_2$Bu$^t$ ≳ PF$_2$Et > PF$_2$C≡CMe ≳ PF$_2$Me > PF$_2$NMe$_2$ > PF$_2$OPr$^i$ > PF$_2$OEt > PF$_2$OMe > PF$_2$OCH$_2$CF$_3$ ≳ PF$_2$SMe > PF$_3$ > PF$_2$Cl > PF$_2$Br. No overall correlations between $J$(P, B) and the base strengths could be found.[158]

$^{11}$B chemical shifts and the coupling constants $J$(P, B) and $J$(PH) have been reported for the adducts Me$_n$PH$_{3-n}$,BX$_2$Y, and Me$_n$PH$_{3-n}$,BXY$_2$ ($n = 0$ or 1; X,Y = Cl, Br, or I).[159] $J$(P,H) appears to be a useful internal test of the feasibility of using other parameters as measures of adduct stability or acceptor ability.

Decomposition of H$_3$B,PF$_3$ by irradiation with a CO$_2$ laser (10.6 μm band) is due to vibrational excitation, and not to net rise in temperature. No boron isotope effect was detected. The rate-determining step seems to be the reaction:[160]

$$H_3B + H_3B,PF_3 \xrightarrow{(v=2)} B_2H_6 + PF_3$$

Bis(ligand)-diborane(4) complexes of PF$_2$X (X = F, Cl, or Br) may be prepared from either Me$_2$O,B$_3$H$_7$ or B$_4$H$_{10}$ as shown in reactions (19) and (20);

$$3F_2XP + Me_2O,B_3H_7 \xrightarrow[5h]{0\,°C} (F_2XP)_2,B_2H_4 + F_2XP,BH_3 + Me_2O \qquad (19)$$

$$\text{excess } F_2XP + B_4H_{10} \xrightarrow[3-5d]{0\,°C} (F_2XP)_2,B_2H_4 + 2F_2XP,BH_3 \qquad (20)$$

mechanisms were proposed.[161]

Gas-phase calorimetry has produced values for the enthalpy of reaction, $\Delta H_{rxn}$, at 25 °C for the reaction (21) (Table 5).[162] The relative base strengths are Me$_3$P > Me$_3$As > Me$_3$Sb, and the relative acid strengths X = Br > Cl ≈ H > F > Me.

$$Me_3M(g) + BX_3(g) \rightarrow Me_3,M,BX_3(s) \qquad (21)$$

---

[157] J. R. Durig, B. A. Hudgens, Y. S. Li, and J. D. Odom, *J. Chem. Phys.*, 1974, **61**, 4809.
[158] E. L. Lines and L. F. Centofanti, *Inorg. Chem.*, 1974, **13**, 2796.
[159] J. E. Drake and B. Rapp, *J. Inorg. Nuclear Chem.*, 1974, **36**, 2613.
[160] E. R. Lory, S. H. Bauer, and T. Manuccia, *J. Phys. Chem.*, 1975, **79**, 545.
[161] R. T. Paine and R. W. Parry, *Inorg. Chem.*, 1975, **14**, 689.
[162] D. C. Mente, J. L. Mills, and R. E. Mitchell, *Inorg. Chem.*, 1975, **14**, 123.

**Table 5** Enthalpies of reaction/kcal mol$^{-1}$ MMe$_3$ and BX$_3$ at 25 °C

|    | F     | Cl    | Br     | H     | Me    |
|----|-------|-------|--------|-------|-------|
| P  | −45.5 | −68.6 | −122.3 | −79.9 | −41.0 |
| As | −20.4 | −46.2 | −81.2  | −49.6 |       |
| Sb |       | −26.8 | −19.8  |       |       |

The Raman and i.r. spectra of solid B$_2$Cl$_4$,2PX$_3$ (X = H or D) may be assigned in terms of $C_{2h}$ symmetry. $\nu_{sym}$(PB) wavenumbers are at 713 and 658 cm$^{-1}$ for the H and D species, respectively, while those due to $\nu_{asym}$(PB) are at 651 and 620 cm$^{-1}$, respectively. The changes in internal modes on co-ordination are as expected for P—B co-ordination.[163] The barriers to internal rotation about the P—B bonds are 2.92 ± 0.18 kcal mol$^{-1}$.

I.r. and Raman spectra of the Li$^+$ salts of X$_2$PBY$_3^-$, (X,Y = H, D) in the solid phase, and of H$_2$PBH$_3^-$ and H$_2$PBD$_3^-$ in Et$_2$O solutions can be assigned on the basis of $C_s$ molecular symmetry.[164] A normal-co-ordinate analysis gave a value for the P—B stretching force constant of *ca.* 2.4 mdyn Å$^{-1}$. This indicates a stronger B—P bond (and weaker B—H bonds) than in H$_3$BPH$_3$, which is the opposite trend to that seen in H$_3$CPH$_3^+$ and H$_3$CPH$_2$.

X-Ray diffraction studies on Me$_3$P,BX$_3$ (X = Cl, Br, or I) show that when X = Cl or Br the crystals are monoclinic (space group $P2_1/m$), while when X = I they are orthorhombic (space group *Pnma*).[165] All have the staggered conformation, with P—B distances of 1.957(5), 1.924(12), and 1.918(15) Å, for the Cl, Br, and I species respectively; B—X = 1.855(5) Å (Cl), 2.022(7) Å (Br), and 2.249(12) Å (I).

N.m.r. data, some vibrational assignments, and mass spectral information have been given for Me$_3$E,BX$_3$ (E = P, X = F, Cl, Br, H, or Me; E = As, X = F, Cl, Br, or H; E = Sb, X = F, Cl, Br, or H).[166] Adduct strength followed previously established orders. No salt formation or halogen exchange was found.

Microwave spectra of CH$_3$PF$_2$,XY$_3$, (X = $^{11}$B or $^{10}$B, Y = H or D) have been obtained.[167] No internal rotation splittings could be resolved, and hence the upper limit to the barrier to rotation for the BH$_3$ and CH$_3$ tops is *ca.* 2000 cal mol$^{-1}$. The P—B bond distance is 1.84 ± 0.02 Å.

Various new boron–phosphorus compounds have been reported in a very long paper.[168] Thus, BCl$_3$,OEt$_2$ reacts with LiPEt$_2$, in a 1:1 molar ratio, at −50 °C, to give BCl$_2$OEt,PEt$_3$, BCl$_3$,PEt$_3$ (main product), (Cl$_2$BPEt$_2$)$_2$, and B(OEt)$_3$. At +20 °C, (BCl$_2$,PEt$_2$)$_2$ and BCl$_3$PEt$_3$ are formed. A 1:2 molar ratio gives [ClB(PEt$_2$)$_2$]$_2$ while a 1:3 ratio yields [B(PEt$_2$)$_2$]$_2$, at −50 °C, or LiB(PEt$_2$)$_4$, at +20 °C. Similar series of reactions were reported for Cl$_2$B(OEt),PEt$_3$ *etc.*

Rather detailed vibrational assignments have been given for Me$_3$As–BH$_3$ and −BD$_3$.[169] These were backed up by a normal-co-ordinate analysis, which showed that the $\nu$(AsC) and $\nu$(AsB) stretches were extensively mixed, and that the As—B

---

[163] J. D. Odom, V. F. Kalasinsky, and J. R. Durig, *Inorg. Chem.*, 1975, **14**, 434.
[164] E. Mayer and H. Hofstötter, *J. Mol. Structure*, 1975, **27**, 309.
[165] D. L. Black and R. C. Taylor, *Acta Cryst.*, 1975, **B31**, 1116.
[166] D. C. Mente and J. L. Mills, *Inorg. Chem.*, 1975, **14**, 1862.
[167] R. A. Cresswell, R. A. Elzaro, and R. H. Schwendeman, *Inorg. Chem.*, 1975, **14**, 2256.
[168] G. Fritz and E. Sattler, *Z. anorg. Chem.*, 1975, **413**, 193.
[169] J. R. Durig, B. A. Hudgens, and J. D. Odom, *Inorg. Chem.*, 1974, **13**, 2206.

stretching force constant was 1.84 mdyn $Å^{-1}$. Microwave spectra of $^{11}B$ and $^{10}B$ isotopic forms of both adducts (making reasonable assumptions for the $BH_3$ and $CH_3$ bond lengths and angles) gave the following values for the adduct structural parameters: $r(As—B) = 2.035$ Å, $r(As—C) = 1.945$ Å, and $\angle CAsB = 113°$.

From a study of arsine-boranes it appears that $AsPh_3$ is a stronger base than $AsPh_2H$, while the Lewis-acidity of some boranes towards these are in the order[170] $BBr_3 > BH_3 > MeBBr_2 > Me_2BBr > Me_2B(SMe)$.

**Compounds containing B—O Bonds.**—Calculated bond lengths [$r(H—B) = 2.1913$ Bohr(1.1596 Å); $r(B—O) = 2.2284$ Bohr(1.1792 Å)] for the recently detected, linear HBO species were obtained from a restricted Hartree–Fock LCAO-MO-SCF study.[171]

Ab initio calculations on hydroxyborane, $H_2BOH$, reveal a $\pi$-bonding interaction between B and O. The barrier to internal rotation about this bond is computed to be 16.4 kcal $mol^{-1}$.[172]

Previous work on the $BF_3$–$H_2O$ system in acetone (L. Bernander and G. Olafsson, Acta Chem. Scand., 1973, **27,** 1034) had suggested that the '$BF_3,H_2O$' adduct in fact contained co-ordinated acetone, i.e. that it is $BF_3$,acetone,$H_2O$. Gillespie and Hartman, however, have shown that the equilibrium (22) lies very far over to the right. Thus acetone is a much weaker electron-pair donor to B than is $H_2O$.[173]

$$H_2O + acetone,BF_3 \rightleftharpoons acetone + H_2O,BF_3 \qquad (22)$$

Vibrational force constants have been calculated for the $XBF_3$ group in the adducts $Me_2X,BF_3$ (X = O, S, or Se).[174] Comparison with values for the separated fragments showed that the expected trends were occurring on adduct formation. Thus, in $Me_2O,BF_3$, $F(CO)$ is $4.54 \times 10^5$ dyn $cm^{-1}$ ($5.38 \times 10^5$ in free $Me_2O$), and $F(BF)$ is ca. $5 \times 10^5$ dyn $cm^{-1}$ (7.840 in free $BF_3$).

A vibrational force field was calculated for $Me_2CO,BF_3$, using as a basis force constants for free $Me_2CO$ and $BF_4^-$.[175] The B—O stretching force constant was ca. 2.1 mdyn $Å^{-1}$.

The $^{35}Cl$ n.q.r. spectrum of dimethylcarbamoyl-trichloroborane (24) shows that the C—Cl resonance is shifted very much to high field compared with the value for the uncomplexed $Me_2NC(O)Cl$.[176]

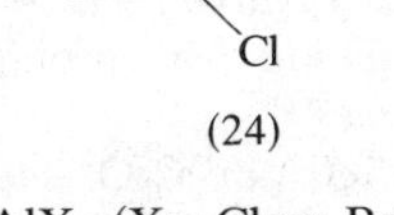

$$Me_2N—C\begin{matrix} O,BCl_3 \\ \\ Cl \end{matrix}$$

(24)

$BX_3$ (X = Cl, Br, or I) and $AlX_3$ (X = Cl or Br), react with $Co_2(CO)_8$ to form (25). $GaBr_3$ and $InBr_3$, on the other hand, give (26).[177]

---

[170] R. Goetze and H. Nöth, Z. Naturforsch., 1975, **30b,** 343.
[171] C. Thomson and B. J. Wishart, Theor. Chim. Acta, 1974, **35,** 267.
[172] O. Gropen and R. Johansen, J. Mol. Structure, 1975, **25,** 161.
[173] R. J. Gillespie and J. S. Hartman, Acta Chem. Scand. (A), 1974, **28,** 929.
[174] P. Labarbe and M. T. Forel, Spectrochim. Acta, 1975, **31A,** 525.
[175] M. Fouassier and M. T. Forel, J. Mol. Structure, 1975, **26,** 315.
[176] E. A. C. Lucken, A. Meller, and W. Gerger, Z. Naturforsch., 1975, **30b,** 286.
[177] G. Schmid and V. Bätzel, J. Organometallic Chem., 1974, **81,** 321.

```
        EX₃
        |
        O
        ‖
        C
       / \
(OC)₃ Co───Co(CO)₃
       \ /
        C
        ‖
        O
       (25)
```

```
              Br₂
              |
              M
             / \
(OC)₃ Co⇠Br   Co(CO)₃
           \ ⇢
            Br
           (26)
```

Large tables of vibrational wavenumbers have been assigned for the adducts of $HCO_2CH_3$, $HCO_2CD_3$, and $DCO_2CD_3$ with $BF_3$.[178] A force field was calculated for the $A'$ vibrations of the $HCO_2CH_3$–$BF_3$ molecule. The B—O stretching force constant appears to be 2.9 mdyn Å$^{-1}$.[179]

Cycloenones, on complexing with $BF_3$, show changes in their magneto-optical behaviour which can be ascribed to electronic delocalization over the whole C=C—C=O,$BF_3$ unit.[180]

The adducts $X_3PO,BY_3$ (X, Y = F, Cl, although $F_3PO,BF_3$ and $F_3PO,BCl_3$ were not isolated) can be prepared by the direct addition of the components. I.r. and Raman spectra were reported, and a number of bands assigned. $\nu$(P—O) of the P—O—B link was assigned to features in the range 1150—1270 cm$^{-1}$, while $\nu$(B—O) was between 1060 and 1140 cm$^{-1}$. The P—O—B bend was thought to occur at ca. 100 cm$^{-1}$.[181]

N.m.r. parameters have been listed for (27; R = Me or Et) and (28; R = Me, Et, or Ph).[182]

```
          R
          |
(RO)₂P
          \
           O→BF₃
          (27)
```

```
             R
             |
             O
            / \
(RO)₂P       BF₂
            \ /
             F
            (28)
```

Re-investigation of the $^{11}$B n.m.r. spectra of $B_3H_7$,$PEt_2$ and $B_3H_7$,THF shows that neither migration of the ether molecule nor base exchange is occurring on the n.m.r. time-scale. The observed spectra are, however, compatible with a structure analogous to that of $B_3H_7$,$NH_3$, with rapid H tautomerism taking place, to make two of the B atoms equivalent.[183]

Bubbling $BF_3$ into a $N_2O_5$–$CCl_4$ mixture at $-15\,°C$ gave the new nitryl salt $(NO_2)_2[(BF_3)_2O]$.[184] The formulation was proposed from an analysis of the vibrational spectrum of this species.[184]

$BF_3$ in pentane reacts with $SO_4^{2-}$ or $CrO_4^{2-}$ to give the salts $[XO_4(BF_3)_2]^{2-}$ [possible structure (29)] and $[XO_4(BF_3)_3]^{2-}$, (X = S or Cr).[185] $NO_3^-$ also forms adducts with $BF_3$, but $SO_3^{2-}$ and $CO_3^{2-}$ are decomposed.

---

[178] E. Taillandier and T. Ben Lakhdar, *Spectrochim. Acta*, 1975, **31A**, 541.
[179] E. Taillandier and T. Ben Lakhdar, *Spectrochim. Acta*, 1975, **31A**, 549.
[180] P. Castan, J.-P. Laurent, J. Torri, and M. Azzaro, *J. chim. Phys.*, 1975, **72**, 113.
[181] E. Payen, J. Ogil, and M. Migeon, *Compt. rend.*, 1975, **281**, C, 499.
[182] B. A. Arbuzov, Yu. V. Belkin, and N. A. Polezhaeva, *Doklady Chem.*, 1974, **216**, 367.
[183] G. Kodama, *Inorg. Chem.*, 1975, **14**, 452.
[184] J.-F. Herzog, B. Bonnett, and G. Mascherpa, *Compt. rend.*, 1975, **280**, C, 197.
[185] V. Gutmann, U. Mayer, and R. Krist, *Synth. React. Inorg. Metal-Org. Chem.*, 1974, **4**, 523.

Complex formation between boric acid and the terdentate ligands $MeC(CH_2OH)_3$ and $Me_2CHC(CH_2OH)_3$ has been investigated, and a number of products such as (30) were isolated ($M^I$ = Li, Na, K, or $NH_4$).[186] When absolute alcohol is the solvent, spiran-type products (31) result. DTA investigation of the thermal decomposition points to 'dimerization' of the former type, giving diborates (32), while those of the latter type give trimers.

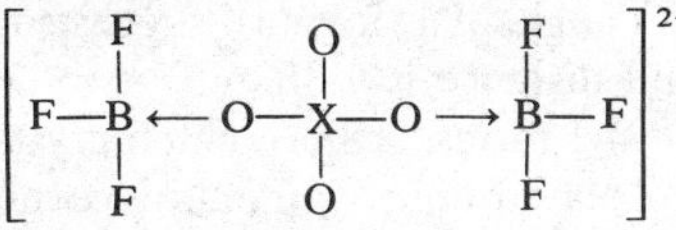

$$\begin{bmatrix} F & O & F \\ | & \| & | \\ F\!-\!B\leftarrow O\!-\!X\!-\!O\rightarrow B\!-\!F \\ | & \| & | \\ F & O & F \end{bmatrix}^{2-} \quad M\begin{bmatrix} & CH_2O \\ & \diagup \diagdown \\ Me\!-\!C\!-\!CH_2O\!-\!B\!-\!OH \\ & \diagdown \diagup \\ & CH_2O \end{bmatrix}$$

(29) 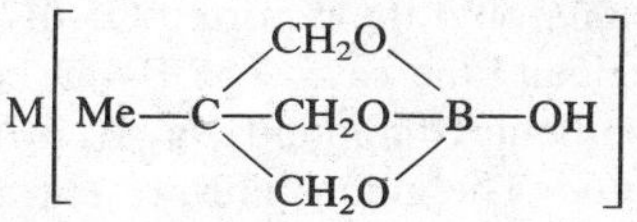            (30)

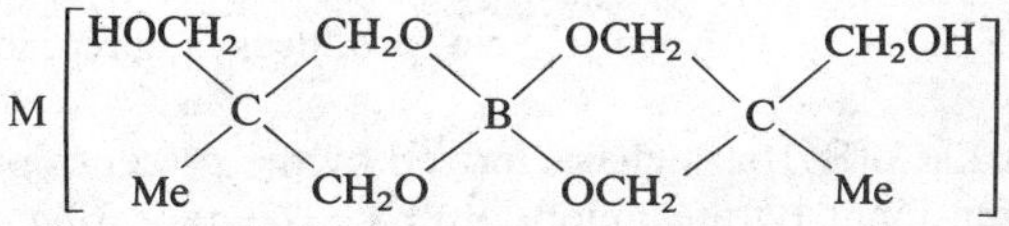

$$M\begin{bmatrix} HOCH_2 & CH_2O & OCH_2 & CH_2OH \\ \diagdown & \diagup \diagdown & \diagup \diagdown & \diagup \\ C & B & C \\ \diagup & \diagdown \diagup & \diagdown \diagup & \diagdown \\ Me & CH_2O & OCH_2 & Me \end{bmatrix}$$

(31)

$$\begin{bmatrix} & CH_2O & & OCH_2 & \\ & \diagup \diagdown & & \diagup \diagdown & \\ MeC\!-\!CH_2O\!-\!B\!-\!O\!-\!B\!-\!OCH_2\!-\!CMe \\ & \diagdown \diagup & & \diagdown \diagup & \\ & CH_2O & & OCH_2 & \end{bmatrix}^{2-}$$

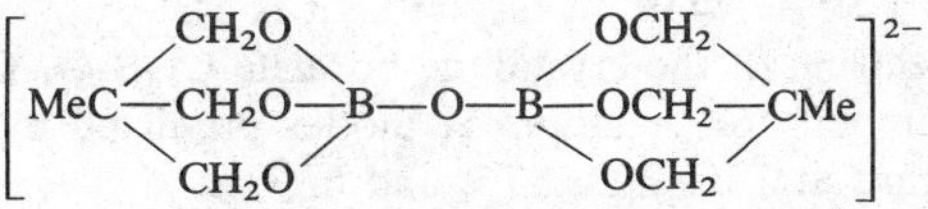

(32)

Ratios of the Larmor frequencies of $^{10}B$ and $^{2}H$, and of $^{11}B$ and $^{10}B$ have been measured by FT n.m.r. spectroscopy of aqueous solutions, and the magnetic moment of $^{10}B$ in the $B(OH)_4^-$ ion has been measured.[187]

Ab initio SCF calculations on the ground state of the $LiBO_2$ molecule show that the bent structure has a lower energy than the linear, in agreement with Walsh's rules.[188] Only calculations including polarization functions produced this result.

$Sr_{9.402}Na_{0.209}(PO_4)_6B_{0.996}O_2$ forms trigonal crystals of apatitic type, but belonging to the space group $P\bar{3}$. The B—O bond length is 1.253(10) Å.[189] A linear O—B—O group is found in some of the voids of the $PO_4$ framework structure.

The conversion of $BH_4^-$ into $B(OMe)_4^-$ by the action of methanol has been shown to proceed via $BH_3(OMe)^-$ and $BH(OMe)_3^-$, both of which could be isolated under appropriate conditions.[190]

Reaction of CsF with $B(OMe)_3$ in anhydrous EtOH forms $Cs[BF(OMe)_3]$, characterized by elemental analysis, $^{19}F$ n.m.r., and i.r. spectroscopy. It dissociates to the original components on heating above 115 °C.[191]

Potassium tetra-acetatoborate forms monoclinic crystals, space group $P2_1/n$.[192] The structure is made up of slightly distorted B—O tetrahedra, held together by

---

[186] A. A. Vegnere, E. M. Shvarts, and A. F. Ievin'sh, J. Gen. Chem. (U.S.S.R.), 1974, **44**, 2172.
[187] B. W. Epperlein, O. Lutz, and A. Schwenk, Z. Naturforsch., 1975, **30a**, 955.
[188] A. Dementjev and D. Kracht, Chem. Phys. Letters, 1975, **35**, 243.
[189] C. Calvo, R. Faggiani, and N. Krishnamachari, Acta Cryst., 1975, **B31**, 188.
[190] V. Kadlec and J. Hanzlik, Coll. Czech. Chem. Comm., 1974, **39**, 3200.
[191] V. N. Plakhotnik, N. G. Parkhomenko, and V. V. Evsikov, Russ. J. Inorg. Chem., 1974, **19**, 686.
[192] A. Dal Negro, G. Rossi, and A. Perotti, J.C.S. Dalton, 1975, 2132.

$K^+$ ions. The acetato-ions act as unidentate ligands in $B(OAc)_4^-$, and the B—O distances lie in the range 1.459—1.481 Å.

$Fe_3BO_6$ is orthorhombic, space group *Pnma*, and isostructural with mineral norbergite, $Mg_3SiO_4(OH)_2$. All of the B atoms are tetrahedrally co-ordinated.[193]

A redetermination of the crystal structure of calcium orthoborate, $Ca_3(BO_3)_2$, reveals that it crystallizes in the rhombohedral system, space group $R\bar{3}c$, and that there are six formula units in the hexagonal unit cell.[194] The $BO_3$ group is said to be non-planar, but the reported O—B—O angle of 119.95(4)° suggests that this is only marginally the case. The B—O bond distance is 1.3836(5) Å.

The synthetic compound $Na_3[B_3O_5(OH)_2]$ forms orthorhombic crystals, space group *Pnma*. The anions are present as isolated units, formed by corner-sharing among one tetrahedron and two triangles. The hydroxy-groups are attached to the four-co-ordinate boron.[195]

The reaction of $ZrOCl_2$ with boric acid produces a zirconium tetraborate, $ZrOB_4O_7,4H_2O$.[196]

The structure of $LaCo(BO_2)_5$ is characterized by the presence of the previously unknown two-dimensional borate anion, $B_5O_{10}^{5-}$. The repeating unit consists of three borate tetrahedra and two plane triangular units. In the former the mean B—O distance is 1.50 Å, while in the latter it is 1.38 Å.[197]

The Raman spectrum of the crystalline boracite $Cr_3B_7O_{13}Cl$ has been analysed to obtain wavenumbers for the four $A_1$ modes predicted by a group-theoretical analysis.[198] They are at 176, 209, 375, and 657 $cm^{-1}$.

Expressions have been derived for determining the content of tetrahedrally co-ordinated boron in sodium borosilicate glasses from their composition.[199]

An electrochemical method for synthesizing sodium tungstoborates with differing B:W ratios has been described.[200]

A large number of phases (barium silicates and borates, and borosilicates) have been identified from the hydrothermal crystallization of the $BaO-B_2O_3-SiO_2-H_2O$ system at 250 and 415 °C.[201] Investigations of the $CaO-B_2O_3-SiO_2-H_2O$ system at the same two temperatures showed that cristobalite, quartz, five identified silicates, six identified borates, and the borosilicates danburite and datolite were formed. Besides these at least 10 unidentified, and possibly new, borates were prepared.[202]

Removal of $H_2O$ from tungstoboric acid always leads to loss of *ca.* 23 mol of $H_2O$ per mole—irrespective of temperature and duration of dehydration. The associated thermal effect is *ca.* 5.9 kJ $mol^{-1}$.[203]

The $Li\|BO_2,SO_4-H_2O$ system has been studied at 25 and 40 °C.[204] Two

---

[193] R. Diehl and G. Brandt, *Acta. Cryst.*, 1975, **B31**, 1662.
[194] A. Vegas, F. H. Cano, and S. García-Blanco, *Acta Cryst.*, 1975, **B31**, 1416.
[195] E. Corazza, S. Menchetti, and C. Sabelli, *Acta Cryst.*, 1975, **B31**, 1993.
[196] Yu. A. Afanas'ev, A. I. Ryabinin, and V. P. Eremin, *Doklady Chem.*, 1974, **215**, 147.
[197] G. K. Abdullaev, Kh. S. Mamedov, and G. G. Dzhafarov, *J. Struct. Chem.* 1975, **16**, 61.
[198] D. J. Lockwood, *J. Raman Spectroscopy*, 1974, **2**, 555.
[199] S. P. Zhdanov, *Doklady Chem.*, 1974, **217**, 518.
[200] K. G. Burtseva and S. L. Saval'skii, *Russ. J. Inorg. Chem.*, 1974, **19**, 839.
[201] R. M. Barrer and E. F. Freund, *J.C.S. Dalton*, 1974, 2054.
[202] R. M. Barrer and E. F. Freund, *J.C.S. Dalton*, 1974, 2060.
[203] M. M. Sadykova, G. V. Kosmodem'yanskaya, and V. I. Spitsyn, *Russ. J. Inorg. Chem.*, 1974, **19**, 221.
[204] V. G. Skvortsov, *Russ. J. Inorg. Chem.*, 1974, **19**, 460.

crystallization branches are found in the solubility curve: $LiBO_2$ and the mixtures $mLiBO_2, nLi_2SO_4, xH_2O$.

Hydrothermal reactions of the system $CaO-B_2O_3-GeO_2-H_2O$ are very similar to those in the analogous system, with $GeO_2$ replaced by $SiO_2$.[205] The double borate of neodymium and strontium, $Nd_2Sr_3(BO_3)_4$, crystallizes from melts in the $Nd_2O_3-SrO-B_2O_3$ system at 1100 °C. The structure was determined by single-crystal X-ray diffraction; this was built up of isolated triangular $BO_3$ units and cations co-ordinated by oxygens. Two types of Nd are present, both eight-co-ordinate, and three types of Sr, two nine and one eight-co-ordinate.[206]

A number of other studies of phase relationships in boron–oxygen-containing systems have been reported.[207—215]

**Compounds containing B—S or B—Se Bonds.**—The electronic structure of $^2\Sigma^+$ BS has been calculated by *ab initio* SCF methods, with both GTO and STO basis sets.[216]

A high-resolution, gas-phase i.r. study has been made of $v_1$ of HBS.[217] Band centres and rotational constants were determined for the $10^00-00^00$ transitions of $H^{11}B^{32}S$, $H^{10}B^{32}S$, $H^{11}B^{34}S$, and $H^{10}B^{34}S$, and for the $11^10-01^10$ transitions of $H^{11}B^{32}S$ and $H^{10}B^{32}S$. Force constants were calculated, which yielded the wavenumber values shown in Table 6.

**Table 6** *Vibrational wavenumbers/*$cm^{-1}$ *for some isotopic variants of* HBS

|     | $H^{11}B^{32}S$ | $H^{10}B^{32}S$ | $H^{11}B^{34}S$ | $H^{10}B^{34}S$ | $D^{11}B^{32}S$ | $D^{10}B^{32}S$ |
|-----|---------|---------|---------|---------|---------|---------|
| $v_1$ | 2768.52 | 2784.21 | 2768.41 | 2784.08 | 2080 | 2106 |
| $v_2$ | 635     | 641     | 634     | 641     | 497  | 505  |
| $v_3$ | 1195    | 1234    | 1185    | 1225    | 1151 | 1184 |

*Ab initio* MO calculations on mercaptoborane, $H_2BSH$, suggest that there is an S—B π-bonding interaction, and yield values of *ca.* 20 kcal mol$^{-1}$ for the barrier to internal rotation about that bond.[218]

Thio- and seleno-boric acids, $R_2B(XH)$ (X = S or Se), are produced by the reaction of halogenoboranes *etc.* with $H_2X$, *e.g.* reaction (22).

$$Me_2B(SPh) + H_2S \xrightarrow{150°C} Me_2B(SH) + PhSH$$

[205] R. M. Barrer and E. F. Freund, *J.C.S. Dalton*, 1974, 2123.
[206] G. K. Abdullaev and Kh. S. Mamedov, *I. Struct. Chem.*, 1974, **15**, 145.
[207] T. O. Ashchyan, V. P. Danilov, I. N. Lepshkov, and O. F. Kosyleva, *Russ. J. Inorg. Chem.*, 1975, **20**, 24.
[208] I. Nowdjavan, P. Vitse, and J. Potier, *Compt. rend.*, 1975, **280**, C, 755.
[209] A. B. Zdanovskii, I. I. Strezhneva, and K. V. Tkachev, *Russ. J. Inorg. Chem.*, 1975, **20**, 147.
[210] A. V. Nikolaev, Yu. A. Afanas'ev, A. I. Ryabinin, V. P. Eremin, and N. B. Zernyakova, *Doklady Chem.*, 1974, **215**, 216.
[211] I. N. Belyaev and E. N. Efstifeev, *Russ. J. Inorg. Chem.*, 1974, **19**, 1705.
[212] R. A. Larina and V. A. Ocheretnyi, *Russ. J. Inorg. Chem.*, 1974, **19**, 1696.
[213] V. N. Pavlikov, V. A. Yurchenko, E. S. Lugovskaya, N. L. Korobanova, and S. G. Tresvyatskii, *Russ. J. Inorg. Chem.*, 1974, **19**, 869.
[214] V. T. Mal'tsev, P. M. Chobanyan, and V. L. Volkov, *Russ. J. Inorg. Chem.*, 1974, **19**, 879.
[215] V. G. Skvortsov, *Russ. J. Inorg. Chem.*, 1974, **19**, 1088.
[216] J. R. Ball and C. Thomson, *Chem. Phys. Letters*, 1975, **36**, 6.
[217] R. L. Sams and A. G. Maki, *J. Mol. Structure*, 1975, **26**, 107.
[218] O. Gropen, E. Wisløff Nilsen, and H. M. Seip, *J. Mol. Structure*, 1974, **23**, 289.

High temperatures lead to elimination of hydrocarbons from the thio-derivatives, and formation of borthiins, together with some loss of $H_2S$, giving $(R_2B)_2S$; the latter, however, tend to lose further $H_2S$, giving borthiins and tri-organyl-boranes [reactions (23)—(25)].[219]

$$2R_2BSH \xrightarrow{\Delta} \begin{cases} [(R_2BSBR_2)] + H_2 & (23) \\ \\ 2/3(RBS)_3 + 2RH & (24) \end{cases}$$

$$3(R_2BSBR_2) \xrightarrow{\Delta} (RBS)_3 + 3BR_3 \quad (25)$$

$Me_2S,BH_3$ reacts with $X_2$, HX, and $BX_3$ to produce $Me_2S,BH_2X$ and $Me_2S,BHX_2$.[220] $^1$H N.m.r. results suggest that there is an increase in the donor–acceptor bond strength as $n$ decreases in $Me_2S,BH_nX_{3-n}$ (X = Br or I).

$^1$H N.m.r. spectra of mixtures of $B(SMe)_3$ and $B(NMe_2)_3$ show that the SMe and $NMe_2$ substituents are able to undergo rapid interchange reactions.[221]

Tris (trifluoromethylthio)borane, $(CF_3S)_3B$, is prepared by reaction (26).

$$BBr_3 + 3Hg(SCF_3)_2 \xrightarrow{\text{2-methylbutane}} B(SCF_3)_3 + 3BrHg(SCF_3) \quad (26)$$

The new compound was characterized by i.r. and $^{19}F$ n.m.r. spectra. It decomposes slowly at room temperature to form $BF_3$ and $SCF_2$, and, unlike $B(SMe)_3$, it forms a 1:1 adduct with $NMe_3$.[222]

The ring-puckering vibrations of $\mu\text{-HS-B}_2H_5$ and $\mu\text{-DS-B}_2H_5$ have been assigned to bands at 306.5 and 306.3 cm$^{-1}$, respectively, for the $1 \leftarrow 0$ transition.[223] The values (and those for the $2 \leftarrow 1$ transitions *etc.*) can be rationalized on the assumption of a planar ring conformation and an essentially harmonic vibration.

Gas-phase electron diffraction has been used to determine the molecular structure of $Me_2BSSBMe_2$.[224] The B—S and S—S bond lengths are 1.805(5) and 2.078(4) Å respectively, while $\angle BSS$, $\angle SBC(1)$, and $\angle SBC(2)$ angles were 105.3(4), 114.0(6), and 123.4(4)°. The torsional angle $\phi(BSSB) = 120(6)°$; while the torsional angle $\phi(SSBC) = 179(4)°$. The B—S bonds are coplanar, suggesting some $\pi$-bond character.

The substituted boron halides, $BX_{3-n}(SeH)_n$ (X = Cl, Br, or I) are considered to be planar on the basis of vibrational spectroscopic data. These data are readily comparable with those of the sulphur analogues, but from force-field calculations it was shown that the B—Se bond is particularly sensitive to the electronegativity of the attached halogen atoms.[225] The shape of the $\gamma(SeH)$ band in $BCl_2SeH$ has been investigated in detail to show a tunnel effect for the hydrogen atom.

---

[219] W. Siebert, E. Gast, F. Regel, and M. Schmidt, *J. Organometallic Chem.*, 1975, **90**, 13.
[220] K. Kinberger and W. Siebert, *Z. Naturforsch.*, 1975, **30b**, 55.
[221] J.-P. Costes, G. Cros, and J.-P. Laurent, *Compt. rend.*, 1975, **280**, C, 665.
[222] M. L. Denniston and D. R. Martin, *J. Inorg. Nuclear Chem.*, 1975, **37**, 651.
[223] W. C. Pringle, C. J. Appeloff, and K. W. Jordan, *J. Mol. Spectroscopy*, 1975, **55**, 351.
[224] R. Johansen, H. M. Seip and W. Siebert, *Acta Chem. Scand.* (A), 1975, **29**, 644.
[225] J. Bouix, M. Fouassier and M. T. Forel, *Ann. Chim.* (France), 1975, **10**, 45.

A bis-borane adduct of selenium is formed in reaction (27) (in xylene as solvent).

$$KSeMe + B_2H_6 \rightarrow KMeSe(BH_3)_2 \quad (27)$$

This in turn reacts with $I_2$ to give the bridging derivative $\mu$-MeSeB$_2$H$_5$ [reaction (28)].

$$KMeSe(BH_3)_2 + \tfrac{1}{2}I_2 \rightarrow KI + \tfrac{1}{2}H_2 + \mu\text{-MeSeB}_2H_5 \quad (28)$$

This diborane derivative is thermally unstable [reaction (29)].

$$\mu\text{-MeSeB}_2H_5 \rightleftharpoons 1/n(MeSeBH_2)_n + 1/2 B_2H_6 \quad (29)$$

The reaction is reversible in toluene, however, so that it is stable in the presence of excess $B_2H_6$.[226]

Electron diffraction of gaseous B(SeMe)$_3$ yields the molecular parameters listed in Table 7.[227] The heavy atom skeleton is essentially planar, so that there may be some B—Se $\pi$-bonding. The bond length, however, is close to that predicted for a single bond. This prediction is not very accurate, and so some multiple-bond character may be ascribed to the B—Se bond.

**Table 7** *Structural parameters of gaseous* B(SeMe)$_3$

| | | | |
|---|---|---|---|
| $r$(B—Se) | 1.936(2) Å | ∠SeCH | 107.0(10)° |
| $r$(Se—C) | 1.954(4) Å | ∠BSeC | 102.5(5)° |
| $r$(C—H) | 1.102(15) Å | | |

Dimethyltin selenide reacts with diorganoiodoboranes to form the diborylselenanes $(R_2B)_2Se$ (R = Me, Bu, Ph, or $C_6H_{11}$). The compounds with R = Ph or $C_6H_{11}$ are stable, the others polymerize to $(RBSe)_n$. Reaction of elemental selenium with $R_2BI$ produces $(R_2B)_2Se_2$, which decomposes on heating [reaction (30)].[228]

$$2(R_2B)_2Se_2 \xrightarrow{\Delta} Se + 2BR_3 + R-B\begin{matrix}Se-Se\\ \\Se\end{matrix}B-R \quad (30)$$

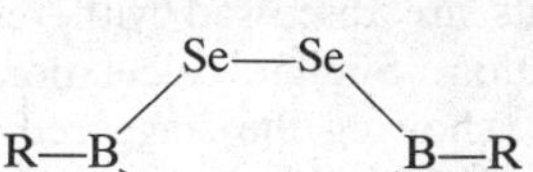

**Boron Halides.**—$^1$H N.m.r. data have been recorded for the adducts formed by $(XC_6H_4)_{3-n}PCl_n$ with boron trihalides.[229] A correlation was found between the shielding of the *ortho*-protons and the stability of the adducts.

$^{19}$F N.m.r. spectra of the adducts formed by *p*-F- and *m*-F-phenyl-phosphines and -arsines with boron halides reveal a correlation between the $^{19}$F chemical shift, $J$(F,H) and $J$(P,F), and the stabilities of the adducts formed.[230]

Mixed boron trihalide adducts of Me$_3$P, Me$_3$PO, and Me$_3$PS are formed by halogen redistribution from the BX$_3$ adducts themselves.[231] For the Me$_3$PS

---

[226] J. J. Mielcarek and P. C. Keller, *J. Amer. Chem. Soc.*, 1974, **96**, 7143.
[227] S. Lindøy, H. M. Seip, and W. Siebert, *Acta Chem. Scand. (A)*, 1975, **29**, 265.
[228] F. Riegel and W. Siebert, *Z. Naturforsch.*, 1974, **29b**, 719.
[229] E. Muylle, G. P. van der Kelen, and Z. Eeckhaut, *Spectrochim. Acta*, 1975, **31A**, 1039.
[230] E. Muylle and G. P. van der Kelen, *Spectrochim. Acta*, 1975, **31A**, 1045.
[231] M. J. Bula, J. S. Hartman, and C. V. Raman, *Canad. J. Chem.*, 1975, **53**, 326.

adducts only small amount of the F-containing adducts are present at equilibrium. $^1$H and $^{19}$F chemical shifts, and $^1$H–$^{31}$P, $^{11}$B–$^{19}$F, and $^{19}$F–$^{31}$P coupling constants, where appropriate, were tabulated.

The reaction with BF$_3$ of LiAlX$_4$ (X = SMe, SeMe, or NMe$_2$), which is generated *in situ*, by a reaction such as (31), provides a convenient synthetic route to BX$_3$.[232]

$$\text{LiAlH}_4 + 2\text{MeSSMe} \xrightarrow{-2\text{H}_2} \text{LiAl(SMe)}_4 \qquad (31)$$

LMO calculations have been reported for the boron fluorides BF, BH$_2$F, BHF$_2$, BF$_3$, BF$_2$NH$_2$, B$_4$F$_4$, and B$_2$F$_4$.[233] The LMO valence structures obtained by the Boys' procedure are typically (33) and (34), where the solid line originates at the atom donating an electron pair, and becomes dotted toward the atom which is electron deficient (indicating bond polarity). These are described as 'fractional bonds'.

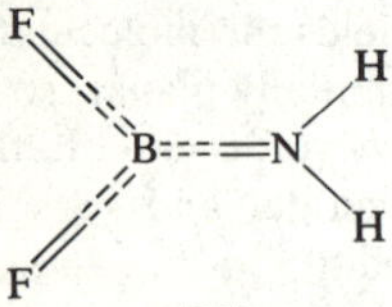

(33) 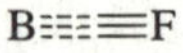        (34)

Ion–molecule studies reveal that HBF$_2$ and BF$_3$ are protonated by H$_3^+$, but not by CH$_5^+$. This sets the boundaries to their proton affinities as $5.10 \pm 0.40$ eV.[234]

The pure rotational Raman spectrum of $^{11}$BF$_3$ yields the set of molecular constants:[235] $B = 0.34502(\pm 3) \times 10^{-5}$ cm$^{-1}$; $D_J^0 = 4.38(\pm 0.10) \times 10^{-7}$ cm$^{-1}$, and $D_{JK} = -9.1(\pm 1.0) \times 10^{-7}$ cm$^{-1}$.

*Ab initio* MO calculations for molecules in the series BF$_n$(OH)$_{3-n}$ ($n = 0$—3) point to replacement of fluorine by hydroxide being energetically unfavourable, and the results are discussed with reference to the slow hydrolysis of BF$_3$ in aqueous solution. Similar calculations have been reported for the adducts NH$_3$,BH$_n$F$_{3-n}$, showing the staggered conformation to be more stable than the eclipsed, and the barrier to rotation about the B—N bond decreasing with increased fluorination.[236]

He I Photoelectron spectra have been obtained for the monomeric species BCl$_{3-n}$F$_n$ ($n = 0$—3) present in BCl$_3$—BF$_3$ mixtures.[237] In addition, microwave data were obtained for BClF$_2$ [$r$(BCl) = 1.71 Å, ∠FBF = 116.6°].

The interaction of BF$_3$ with n-butylamine is said to produce the compound [BF$_2$(NH$_2$C$_4$H$_9$)$_2$]BF$_4$.[238]

The Raman spectra of the low-temperature (orthorhombic) and high-temperature (cubic) phases of MBF$_4$ (M = K, Rb, or Cs), and that of molten LiBF$_4$

---

[232] W. Siebert, W. Ruf, and R. Full, *Z. Naturforsch.*, 1975, **30b**, 642.
[233] J. H. Hall, jun., T. A. Halgren, D. A. Kleier, and W. N. Lipscomb, *Inorg. Chem.*, 1974, **13**, 2520.
[234] R. C. Pierce and R. F. Porter, *Inorg. Chem.*, 1975, **14**, 1087.
[235] P. A. Freedman and W. J. Jones, *J. Mol. Spectroscopy*, 1975, **54**, 182
[236] D. R. Armstrong and P. G. Perkins, *Inorg. Chim. Acta*, 1974, **10**, 77.
[237] H. W. Kroto, M. F. Lappert, M. Maier, J. B. Pedley, and M. Vidal, *J.C.S. Chem. Comm.*, 1975, 810.
[238] V. I. Spitsyn, I. D. Kolli, E. A. Balabanova, and O. N. Mironenko, *Russ. J. Inorg. Chem.*, 1975, **20**, 29.

may be interpreted using a model involving an isolated $BF_4^-$ anion in a static field.[239]

$Ca(BF_4)_2$ crystallizes in the orthorhombic space group *Pbca*. The structure consists of columns of $BF_4^-$ ions, and columns of alternating $Ca^{2+}$ and $BF_4^-$ ions, all parallel to [010]. All of the columns are linked *via* Ca $\cdots$ F bonds.[240]

High yields of fluoroborates and fluoroaluminates can be obtained by the direct fluorination of $MBH_4$ (M = Li or K) and $LiAlH_4$.[241] The reactions are highly exothermic and must be strictly controlled. Solid $LiAlF_4$ has been obtained for the first time by this method.

*Ab initio* calculations, using moderately sized Gaussian basis sets, for $N_2O_4$ and the isoelectronic $B_2F_4$ confirm the greater stability of the planar conformation of the former, and the probable very low barrier to rotation about the B—B bond in the latter. The relative destabilization of the planar form of $B_2F_4$ was due to the small overlap of F orbitals and the greater electrostatic repulsion in that conformation.[242]

$^{19}$F N.m.r. spectra have been recorded at 118 K for a solution of $BF_3$ and $Bu_4^nN^+BF_4^-$ (in a 1:1 molar ratio); they may be analysed in terms of a $B_2F_7^-$ ion having a bridging F atom *i.e.* $F_3B$—F—$BF_3^-$. The B—F—B bond is apparently very weak, however. The *geminal* FBF(terminal) coupling constant is 95 Hz.[243]

Equilibrium constants have been reported for the redistribution equilibria involving $PhBX_2$ and $PhBY_2$ (X = F, Y = Cl or Br).[244] For $PhBX_2$–$BY_3$, a quantitative reaction gives $PhBCl_2$ or $PhBBr_2$ and $BF_3$ at 316.5 K.

Low-temperature n.m.r. studies of the $Me_3N$–$F_3P,B(BF_2)_3$ system suggest that adducts $(BF_2NMe_3)_n(BF_2)_{3-n}$, ($n = 1$—3) are formed; *e.g.* for $n = 3$ this is (35). At $-90\,°C$, $B_8F_{12}$ reacts vigorously with $NMe_3$. One possible product is $Me_3N,B(BF_2)_3$.[245]

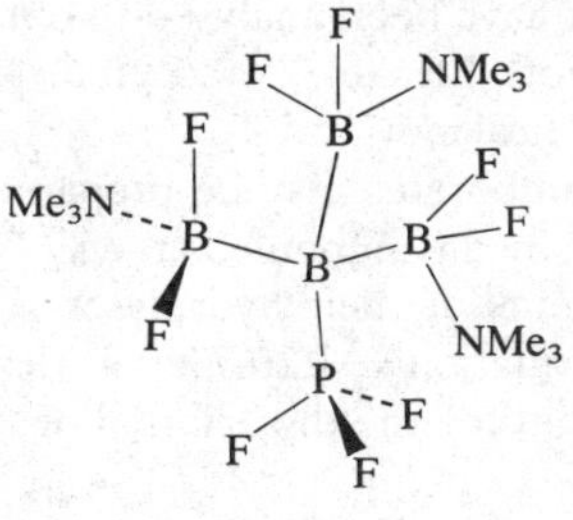

(35)

$BCl_3$–$H_2$ mixtures react under the influence of pulsed $CO_2$ laser irradiation to give $BHCl_2$ and HCl as the only products (50% conversion).[246]

The intensities of components of the Raman band of $BCl_3$ due to the symmetric B—Cl stretch ($\nu_1$) are not those expected from the isotopic composition of the

---

[239] J. B. Bates and A. S. Quist, *Spectrochim. Acta*, 1975, **31A**, 1317.
[240] T. H. Jordan, B. Dickens, L. W. Schroeder, and W. E. Brown, *Acta Cryst.*, 1975, **B31**, 669.
[241] R. J. Lagow and J. L. Margrave, *Inorg. Chim. Acta*, 1974, **10**, 9.
[242] J. M. Howell and J. R. Van Wazer, *J. Amer. Chem. Soc.*, 1974, **96**, 7902.
[243] J. S. Hartman and P. Stilbs, *J.C.S. Chem. Comm.*, 1975, 566.
[244] S. S. Krishnamurthy, M. F. Lappert, and J. B. Pedley, *J.C.S. Dalton*, 1975, 1214.
[245] J. S. Hartman and P. L. Timms, *J.C.S. Dalton*, 1975, 1373.
[246] S. D. Rockwood and J. W. Hudson, *Chem. Phys. Letters*, 1975, **34**, 542.

compound. The anomalies must be due to intermolecular interactions, which affect the different isotopic species differently.[247]

He I Photoelectron spectra have been obtained for the monomeric species $MX_3$, $MeMX_2$, $Me_2MX$, and $MMe_3$ (M = B, Al, Ga, or In; X = Cl, Br, or I).[248] The spectra for the boron trihalides, and those of the heavier Group III elements provided a basis for the assignments, being very similar. Some earlier interpretations could be modified in the light of these data, and among the conclusions reached were: (i) the position of the $a_2''$ level may give a measure of the X—M $\pi$-bonding (greatest for B); (ii) the relative Lewis acidities of the Group III elements are: Al > B ≈ Ga > In.

$BCl_3$ and $TeCl_4$ are almost completely mutually insoluble.[249]

$MBCl_4$ (M = Li, Na, or K) may be prepared by the reaction of MCl and $BCl_3$ in $SO_2$ at −30°C. Raman spectra showed bands due to $BCl_4^-$, as reported earlier (J. A. Creighton, *J. Chem. Soc.*, 1965, 6589).[250]

Reaction of $PF_5$ and $BCl_3$ constitutes a more convenient method than those used hitherto for the preparation of $PF_4Cl$.[251] The preparation of $B_2Cl_4$ from $BCl_3$ in a mercury discharge is thought to occur *via* formation of BCl, followed by attack at a second $BCl_3$ molecule, Scheme 4, rather than *via* intermediate formation of $BCl_2$ radicals.[252]

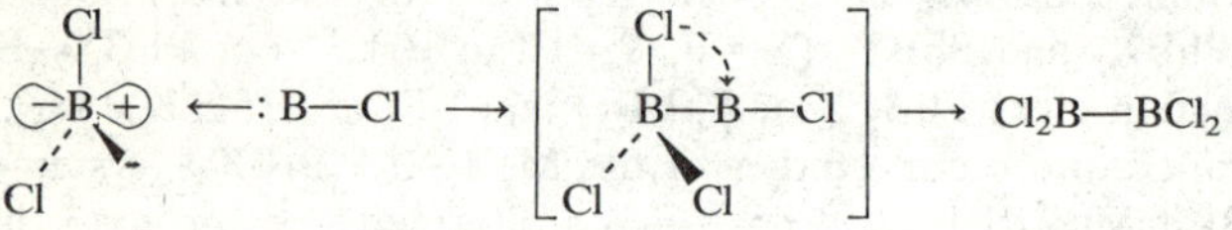

**Scheme 4**

Solution calorimetric data have been analysed to obtain the standard enthalpies of formation at 298.15 K of: $PBr_5(c)$ −60.5 ± 0.6, $PCl_4^+BCl_4^-(c)$ −223.9₅ ± 0.6, $PBr_4^+BBr_4^-$ (c) −135.5 ± 0.7 kcal mol$^{-1}$.[253]

Pure Rb and Cs tetrachloroborates may be prepared by heating RbCl or CsCl with $BCl_3$ in nitrobenzene in an ampoule, at *ca.* 7 atm and 90 °C for 4 h.[254] Thermochemical measurements of their hydrolyses gave values at 298.15 K for their standard heats of formation: −220.7 ± 0.4 (Rb), −225.0 ± 0.3 kcal mol$^{-1}$ (Cs). The equilibrium constant for the dissociation of $CsBCl_4$ (c) into CsCl(c) and $BCl_3(g)$ is *ca.* $8 \times 10^{-5}$ at 25 °C.

Pure dialkylbromoboranes may be obtained in yields greater than 75% by reaction (32) at −63 °C.[255]

$$2R_2^1B(SR^2) + Br_2 \rightarrow 2R_2^1BBr + R^2SSR^2 \quad (32)$$

---

[247] A. Loewenschuss, *Spectrochim. Acta*, 1975, **31A**, 679.
[248] G. K. Barker, M. F. Lappert, J. B. Pedley, G. J. Sharp, and N. P. C. Westwood, *J.C.S. Dalton*, 1975, 1765.
[249] A. V. Konov and V. V. Safonov, *Russ. J. Inorg. Chem.*, 1974, **19**, 619.
[250] M.-C. Dhamelincourt and M. Migeon, *Compt. rend.*, 1975, **281**, C, 79.
[251] R. H. Neilson and A. H. Cowley, *Inorg. Chem.*, 1975, **14**, 2019.
[252] A. G. Briggs, M. S. Reason, and A. G. Massey, *J. Inorg. Nuclear Chem.*, 1974, **37**, 313.
[253] A. Finch, P. J. Gardner, P. N. Gates, A. Hameed, C. P. McDermott, K. K. Sengupta, and M. Stephens, *J.C.S. Dalton*, 1975, 967.
[254] A. Finch, P. J. Gardner, N. Hill, and N. Roberts, *J.C.S. Dalton*, 1975, 357.
[255] A. Pelter, K. Rowe, D. N. Sharrocks, and K. Smith, *J.C.S. Chem. Comm.*, 1975, 531.

The reaction of $BBr_3$ with boron at $BBr_3$ pressures between $4.6 \times 10^{-3}$ and $20 \times 10^{-3}$ atm, in the temperature range 0—930 °C, involves physical adsorption (<75 °C), chemisorption (>75 °C), thermal dissociation of the $BBr_3$ into its elements (>400 °C), and formation of polymers $(BBr_{1-x})_n$.[256]

New 1:1 adducts of $BBr_3$ and $BI_3$ with $AsMe_3$ and $SbMe_3$ have been reported. The $SbMe_3$ adducts decompose slowly at room temperature to give $Me_3SbX_2$ (X = Br or I), but the $Me_3As$ adducts are stable indefinitely at such temperatures in the absence of $H_2O$.[257] The suggestion was made, on the basis of the $^{11}B$ chemical shifts, that these adducts may exist, in $CHCl_3$ solution, as ionic species $[Me_3EX]^+BX_4^-$, although the reactions to produce these were not explained.

If $BBr_3$ or $B_2Br_4$ is subjected to a silent electric discharge, the thermally stable species $B_9Br_9$ is produced.[258] Mixed halides $B_7ClBr_6$, $B_9ClBr_7$, $B_9Cl_nBr_{9-n}$ (n = 1—3), and $B_{10}ClBr_9$ are formed by the decomposition of $B_2Br_4$ containing a little chlorine. $B_9Br_nCl_{9-n}$ (n = 0—5) are formed when $B_9Br_9$ reacts with $TiCl_4$ or $SnCl_4$ at 250 °C.

**Boron-containing Heterocycles.**—Pyrolysis of $BMe_3$ produces yields of up to 25% of 2,4,6,8,9,10-hexamethyl-2,4,6,8,10-hexabora-adamantane (36).[259] This species is inert to $H_2O$, but breakdown of the cage structure occurs in the presence of $O_2$, $NH_3$, or propanoic acid. It is only a weak acceptor, and no complexes with $NMe_3$, $PMe_3$, or $PMe_2Ph$ could be isolated at room temperature.

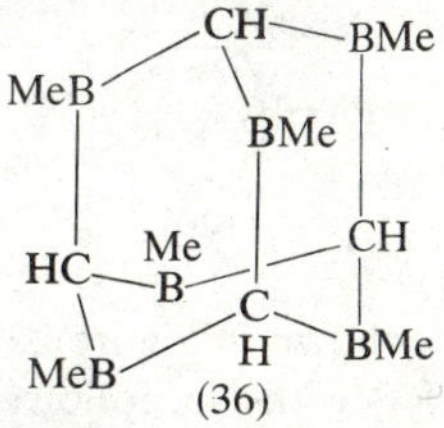

1,4-Difluoro-2,3,5,6-tetramethyl-1,4-diboracyclohexa-2,5-diene displaces CO from a number of metal carbonyls to form $\pi$-complexes.[260] The crystal structure of $[(C_4Me_4B_2F_2)_2Ni]$ (37) shows that all six ring atoms are attached to the nickel, and that the geometry is very like that of the isoelectronic duroquinone complex.

Both 1,3- and 1,4-di-iodobenzene react with $BI_3$ to give bis(di-iodoboryl)-benzenes, but a similar reaction with 1,2-di-iodobenzene produces the dibenzo-1,4-diborine derivative (38).[261]

The crystal structures of (39) and a related iron complex have been determined.[262] In each case the M—B bonds are longer than the M—C (ring) bonds. This non-centric co-ordination appears to be characteristic of $\left(\left\langle\bigcirc\right\rangle BR\right)$—M complexes.

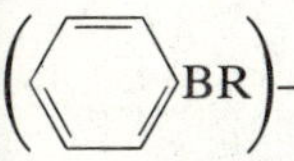

---

[256] B. A. Savel'ev and V. A. Krenev, *Russ. J. Inorg. Chem.*, 1974, **19**, 1288.
[257] M. L. Denniston and D. R. Martin, *J. Inorg. Nuclear Chem.*, 1974, **36**, 2175.
[258] M. S. Reason and A. G. Massey, *J. Inorg. Nuclear Chem.*, 1975, **37**, 1593.
[259] M. P. Brown, A. K. Holliday, and G. B. Way, *J.C.S. Dalton*, 1975, 148.
[260] P. S. Maddren, A. Modinos, P. L. Timms, and P. Woodward, *J.C.S. Dalton*, 1975, 1272.
[261] W. Siebert, K.-J. Schaper, and B. Asgarouladi, *Z. Naturforsch.*, 1974, **29b**, 642.
[262] G. Huttner and W. Gartzke, *Chem. Ber.*, 1974, **107**, 3786.

E.s.r. parameters for (40) suggest that the bonding is similar to that in ferrocene, with 0.7 of an electron back-donated to each borabenzene ring.[263]

The phenylborinato-complex shown in Figure 23 is formed by the reaction of 3-($\eta^5$-C$_5$H$_5$)-3,1,2-Co($\eta^5$-C$_2$B$_9$H$_{11}$) with Na, followed by PhBCl$_2$.[264]

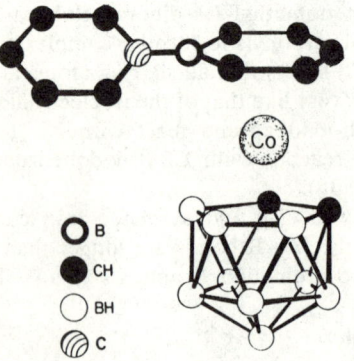

**Figure 23** *Proposed structure for* 3-[1-C$_6$H$_5$-($\eta^6$-C$_5$BH$_5$)]-3,1,2,—Co($\eta^5$-C$_2$B$_9$H$_{11}$) (Reproduced by permission from *Inorg. Chem.*, 1975, **14**, 2018)

---

[263] G. E. Herberich, T. Lund, and J. B. Raynor, *J.C.S. Dalton*, 1975, 985.
[264] R. N. Leyden and M. F. Hawthorne, *Inorg. Chem.*, 1975, **14**, 2018.

# Elements of Group III

Reaction of the bis(borinato)cobalt complexes $Co(C_5H_5BR)_2$ (R = Ph or Me) with iron carbonyls gives the complexes (41).[265] Pyrolysis of these at 230 °C produces (42).

The chemistry of inorganic polymers containing B—N and B—P bonds, other than the well-known cyclotriborazenes, has been reviewed.[266]

Cyclo(hexacyanoborane), $(BH_2CN)_6$, forms triclinic crystals, space group $P\bar{1}$. The macrocyclic ring is chair-like (Figure 24), and the mean dimensions were reported as follows: $r(B—C) = r(B—N) = 1.56$ Å, $r(C≡N) = r(B—H) = 1.14$ Å, $\angle BCN = 107.0°$ and the B—B—B—B dihedral angle = $59.6°$.[267]

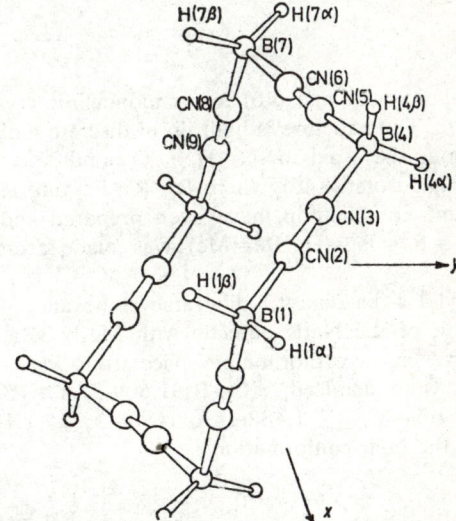

**Figure 24** *Molecular conformation and atom numbering scheme for* $(BH_2CN)_6$ (Reproduced from *J.C.S. Dalton*, 1975, 784)

Two 2-oxy-bis(1,3-dimethyl-2-diazaboracycloalkanes) (43); n = 2 or 3 may be prepared by the controlled hydrolysis of 1,3-dimethyl-2-methylthiodiazaboracycloalkanes with the stoicheiometric amounts of $H_2O$.[268]

---

[265] G. E. Herberich, H. J. Becker, and G. Greiss, *Chem. Ber.*, 1974, **107**, 3780.
[266] V. I. Spitzin, I. D. Colley, T. G. Sevastjanova, and E. M. Sadykova, *Z. Chem.*, 1974, **14**, 459.
[267] A. T. McPhail and D. L. McFadden, *J.C.S. Dalton*, 1975, 1784.
[268] W. R. Conway, P. F. Litz, and K. Niedenzu, *Synth. React. Inorg. Metal-Org. Chem.*, 1975, **5**, 37.

Several 1-imidazolylboranes have been prepared by reaction (33).[269]

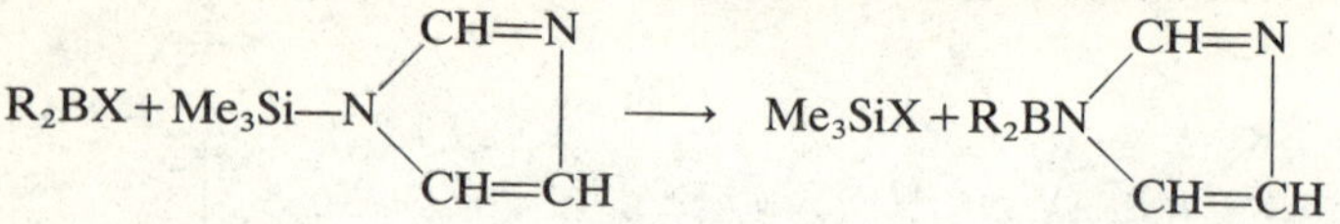

$$R_2BX + Me_3Si-N\begin{pmatrix}CH=N\\|\\CH=CH\end{pmatrix} \longrightarrow Me_3SiX + R_2BN\begin{pmatrix}CH=N\\|\\CH=CH\end{pmatrix} \quad (33)$$

Radical-initiated hydrophosphination of diethylaminodipropynylborane with $H_2PPh$ leads to the formation of derivatives of the new heterocyclic system 1-bora-4-phospha-cyclohexadiene (44).[270]

A large number of vibrational assignments, based on solid-phase data only, have been listed for adducts of the tetrazaborolines (45); R = Me, Ph, or Cl) with $SnCl_4$, $SbCl_5$, and $TiCl_4$ (in which it acts as a unidentate ligand).[271]

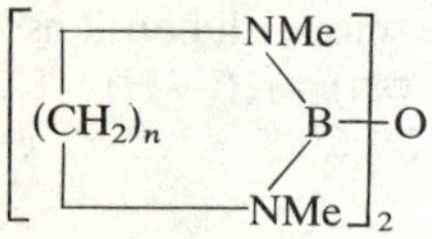

(43) 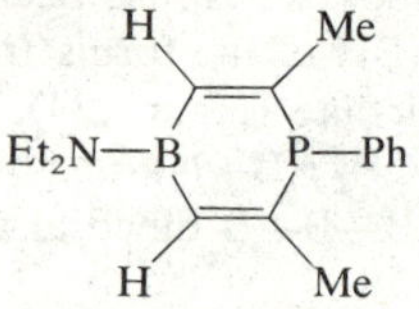    (44) 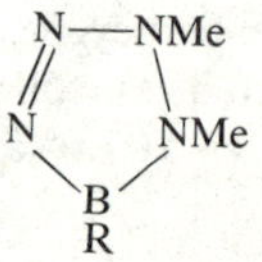    (45)

B-Phenyl-*diptych*-boroxazolidine (46) forms monoclinic crystals belonging to the space group $P2_1/c$. The structure is built up of discrete molecules linked into continuous spirals along the b axis by N—H···O bonds.[272]

2-Phenyl-1,3,2-oxazaborolans (47), where $R^1$, $R^2$, $R^3$, and $R^4$ are H or Me or Ph or $CH_2Ph$ in some combination, have been prepared and characterized;[273] Compound (47; $R^1 = R^3 = R^4 = H$, $R^2 = Me$) was made from $B(SEt)_2Ph$ and $H_2NCH_2CHMeOH$.

1,1,4,4-Tetramethyl-1,4-diazonia-2,5-diboratacyclohexane, (48), has been prepared by the reaction of the Na/K eutectic with $Me_3N,BBr_3$ in $NMe_3$ as the solvent.[274] It crystallizes in the orthorhombic space group *Pbca*, and the following stuctural parameters were deduced: r(C—B) 1.609(2) Å, r(B—N) 1.615(2) Å, r(N—C) 1.511(2) Å, $r(N-C_{Me;eq})$ 1.482(2) Å, $r(N-C_{Me,ax})$ 1.484(2) Å. The six-membered ring is in the boat conformation.

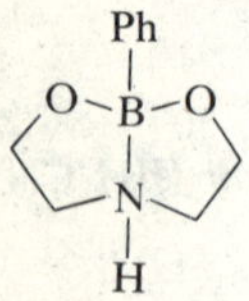

(46) 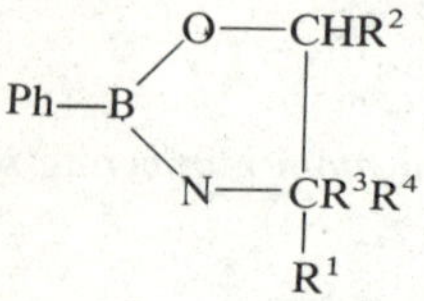    (47) 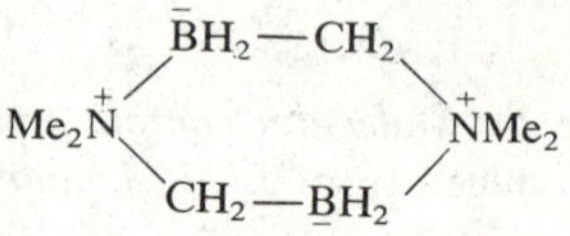    (48)

[269] I. A. Boenig, W. R. Conway, and K. Niedenzu, *Synth. React. Inorg. Metal-Org. Chem.*, 1975, **5**, 1.
[270] H.-O. Berger and H. Nöth, *Z. Naturforsch.*, 1975, **30b**, 641.
[271] J. H. Morris, *J. Inorg. Nuclear Chem.*, 1974, **36**, 2439.
[272] S. J. Rettig and J. Trotter, *Canad. J. Chem.*, 1975, **53**, 1393.
[273] R. H. Cragg and A. F. Weston, *J.C.S. Dalton*, 1975, 93.
[274] T. H. Hseu and L. A. Larsen, *Inorg. Chem.*, 1975, **14**, 330.

The bis-(3,5-dimethylpyrazolyl)borane dimer, $[(C_5N_2H_7)_2BH]_2$ contains a ring of the type (49), in a flattened-chair conformation (dihedral angle of 12.7° between the NBN plane and the central plane).[275] Steric effects of the bulky terminal pyrazolyl units force this structure to be adopted, by contrast to the boat conformation of pyrazolyldideuteriogallane.

$N$-Metallated derivatives of (50) react with $(ClR^1P)_2NR^2$ or $[Cl_2P(O)]_2NMe$ to give the 'phosphaborazines' (51); $R^2 = Me$, $R^1 = Me, Ph$, or $Cl$, no X; $R^2 = Ph$, $R^1 = Cl$, no X; $R^1 = Cl$, $R^2 = Me$, $X = O$).[276]

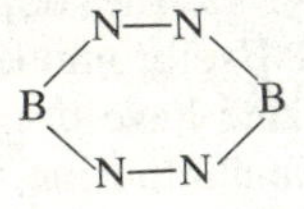

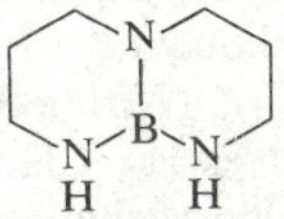

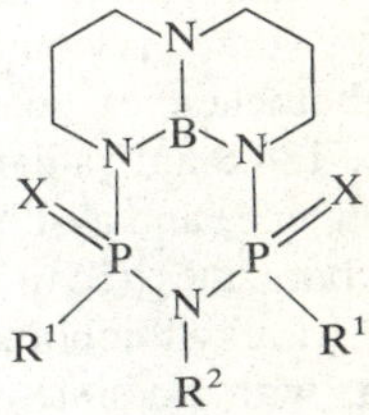

A new boron–nitrogen ring compound, a cyclohexadecaborazane containing in addition four six-membered borazine systems (52) has now been prepared from 1,2,3,5-tetramethyl-4,6-dichloroborazine and hexamethyldisilazane, in good yields.[277] A silylated derivative (53) is obtained in small yield, but the bis-(borazinyl)amine (54) accounts for ca. 25% of the reaction products.

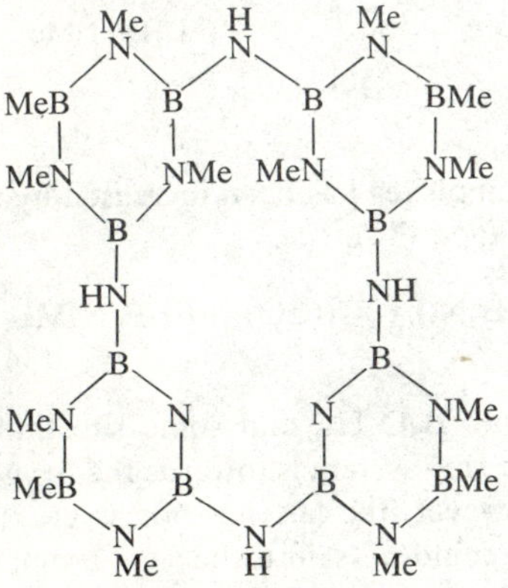

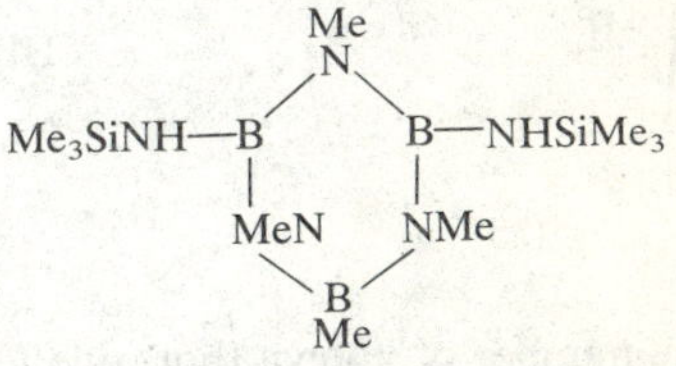

The electronic structures of borazine, $B_3N_3H_6$, and boroxine, $B_3O_3H_3$, have been compared, using *ab initio* SCF-LCAO-MO calculations.[278] It was found that

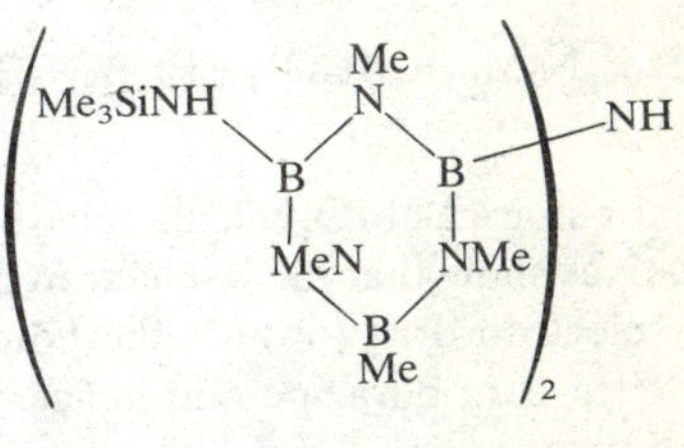

---

[275] N. W. Alcock and J. F. Sawyer, *Acta Cryst.*, 1974, **B30**, 2899.
[276] H. Nöth and W. Tinhof, *Chem. Ber.*, 1974, **107**, 3806.
[277] A. Meller and H.-J. Füllgrabe, *Angew. Chem. Internat. Edn.*, 1975, **14**, 359.
[278] A. Serafini and J.-F. Labarre, *J. Mol. Structure*, 1975, **26**, 129.

while π-delocalization occurs in both, the π-distribution in boroxine is much more polarized than in borazine.

An almost complete vibrational assignment has been proposed for N-trimethylborazine, (BHNMe)$_3$, this was achieved using several isotopically substituted species, and a normal-co-ordinate analysis of the molecule was carried out.[279] N-Trimethylborazine forms the adducts Me$_3$N$_3$B$_3$H$_3$,3HX with anhydrous hydrogen halides (some vibrational assignments were made).[280] At higher temperatures, (above ca. 100 °C) hydrogen–halogen exchange occurs to give B-halogenated derivatives of the original trimethylborazine. Reaction of the borazine with bromine gave the previously unknown N-trimethyl-B-dibromoborazine. A number of photochemical reactions of N-trimethylborazine have been reported. For example, 184.9 nm radiation gives H$_2$, CH$_4$, borazanaphthalene, N-methylborazanaphthalene, and N-dimethylborazanaphthalene.[281]

Polycyclic borazines (55); X = O, NH, or NMe, n = 2 or 3) result from the reaction of a trisalkylthioborane with hydroxyamines or amines.[282] Their i.r. and mass spectra, with possible mechanisms of formation, were reported. Photochemical reactions of borazine in the presence of (photochemically generated) radicals derived from hexafluoroacetone have led to the formation of B-(2H-hexafluoro-2-propoxy- and B-(perfluoro-butoxy)-borazine.[283]

Mercury-sensitized photolysis ($\lambda$ = 253.7 nm) of H$_2$ with N-trimethylborazine produces the new derivative 1,2-di-(3′,5′-dimethylborazinyl)ethane (56).[284]

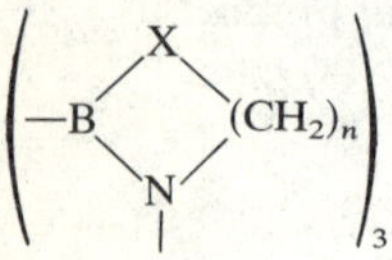

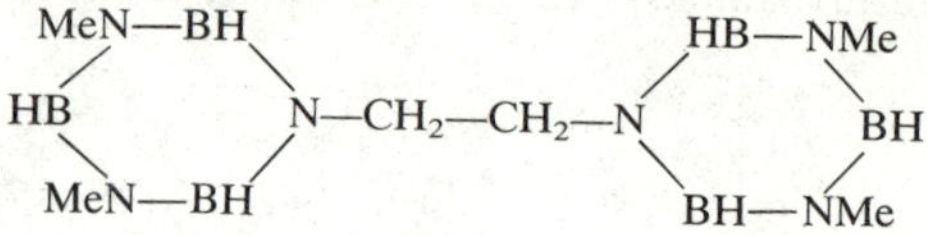

The number of known (borazine)Cr(CO)$_3$ complexes has been increased by the discovery of ligand-exchange reactions of the type (34).[285]

(Et$_3$B$_3$N$_3$Me$_3$)Cr(CO)$_3$ + Et$_3$B$_3$N$_3$Et$_3$ → (Et$_3$B$_3$N$_3$Et$_3$)Cr(CO)$_3$ + Et$_3$B$_3$N$_3$Me$_3$
(34)

Semi-empirical MO calculations on boroxine, B$_3$O$_3$H$_3$, and some substituted derivatives show that the σ-electron drift in the ring system is towards the oxygen, the π-electron drift towards the boron.[286] However, the latter is not as great as from N to B in borazine and hence, boroxine could be said to be 'less aromatic' than borazine.

[279] K. E. Blick, E. B. Broadley, K. Iwatani, K. Niedenzu, T. Takusaka, T. Totani, and H. Watanabe, Z. anorg. Chem., 1975, **417**, 19.
[280] G. A. Anderson and J. J. Lagowski, Inorg. Chem., 1975, 1845.
[281] L. J. Turbini and R. F. Porter, Inorg. Chem., 1975, **14**, 1252.
[282] R. H. Cragg and A. F. Weston, J.C.S. Dalton, 1975, 1761.
[283] L. J. Turbini, G. M. Golenwsky, and R. F. Porter, Inorg. Chem., 1975, **14**, 691.
[284] L. J. Turbini, T. J. Mazanec and R. F. Porter, J. Inorg. Nuclear Chem., 1975, **37**, 1129.
[285] M. Scotti and H. Werner, J. Organometallic Chem., 1974, **81**, C17.
[286] D. T. Haworth and V. M. Scheer, J. Inorg. Nuclear Chem., 1975, **37**, 2010.

Complete mass spectra have been reported for the 1,3-dioxaborinane esters (57; R = Ph, PhO, or Bu$^t$O).[287] A series of 2-iminoxy-4,4,6-trimethyl-1,3,2-dioxaborinanes (58; R$^1$ and R$^2$ = H, Me, Et, Pr$^n$, etc.) has been prepared by a variety of synthetic routes.[288]

Bis-(2-hydroxyethyl)-ether or -thioether undergoes condensation reactions with phenyldihydroxyborane, to give new eight-membered boron-containing heterocycles (59).[289]

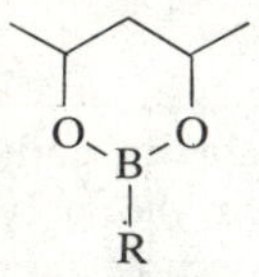

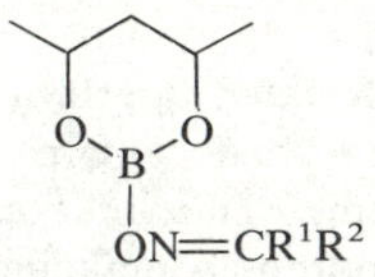

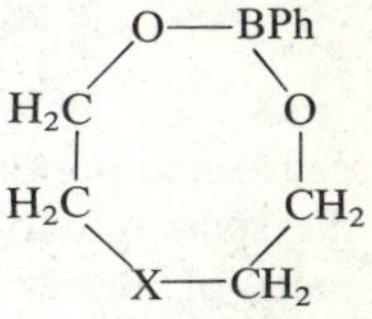

He I photoelectron spectra of boratrane (60) show a large increase in the nitrogen lone-pair binding energy compared with non-cage analogues.[290] This is due to the close approach of B and N, and the presence of bonding interaction of the N lone-pair with a vacant B orbital.

Formation of the 1:1, luminescent, 'boron-2-hydroxy-4-methoxy-4'-chlorobenzophenone' complex in concentrated sulphuric acid takes place in at least two stages, according to recent kinetic and equilibrium studies.[291] An intermediate of the form (61) is involved, which loses $HSO_4^-$ and $H^+$ to form the ring-closed species (62). The experiments also pointed to the possibility of a $B(OH)(HSO_4)_3^-$ species being in equilibrium with the more usual $B(HSO_4)_4^-$ ion in less-concentrated $H_2SO_4$.

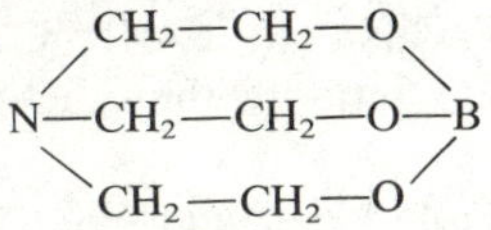

A redetermination of the crystal structure of tribromoborthiin (63) gave the following bond distances and angles: B—S, 1.807 Å; B—Br, 1.895 Å; ∠BSB, 109.2°; ∠SBS, 130.7°. The crystals are monoclinic, belonging to the space group $P2_1/c$.[292]

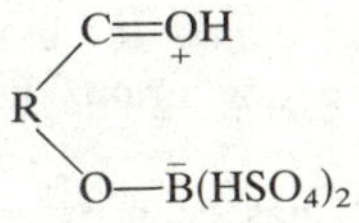

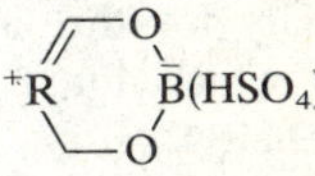

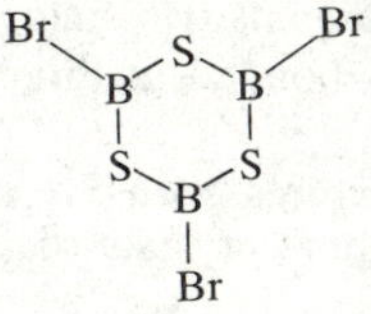

---

[287] P. B. Brindley, R. Davis, B. L. Horner, and D. I. Ritchie, *J. Organometallic Chem.*, 1975, **88**, 321.
[288] A. Singh and R. C. Mehrotra, *Synth. React. Inorg. Metal-Org. Chem.*, 1974, **4**, 557.
[289] U. W. Gerwarth and W. Weber, *Synth. React. Inorg. Metal-Org. Chem.*, 1975, **5**, 175.
[290] S. Cradock, E. A. V. Ebsworth, and I. B. Muiry, *J.C.S. Dalton*, 1975, 25.
[291] M. Marcantonatos and C. Menziger, *Inorg. Chim. Acta*, 1975, **14**, 227.
[292] W. Schwarz, H.-D. Hausen, and H. Hess, *Z. Naturforsch.*, 1974, **29b**, 596.

3,4-Xylenyl-1-di-iodoborane reacts with polymeric iodoboron selenide at 180 °C to form xyleno-1,2,5-selenadiborolen (64).[293]

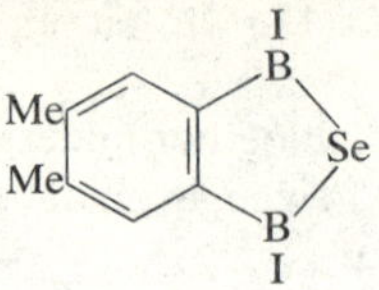

(64)

**Boron Nitride, Metal Borides, and Related Species.**—$Al_{0.06}BeB_{3.05}$, i.e. '$BeB_3$', belongs to the space group $P6/mmm$, and contains $B_{12}$ icosahedra and other polyhedral units of Be and B atoms. The linkages between the polyhedra resemble those in $\beta$-rhombohedral boron. Aluminium atoms occupy interstitial sites.[294]

The mixed boride $\beta$-$(Al,Be)B_{12}$ has been characterized, and a new orthorhombic form identified.[295] It has the same unit cell as the A-phase of $\beta$-$AlB_{12}$.

X-Ray diffraction analysis of a boron carbide containing a very large excess of boron indicates that a rhombohedral structure, probable space group $R\bar{3}m$, related to that of $B_4C$ is maintained. $B_{12}$ icosahedra are present, with about 1/4 of the CBC units replaced by $B_4$ groups. Terminal atoms in the latter are five-co-ordinate, to two other atoms in the group and three of a $B_{12}$ group, as are the bridging B atoms.[296]

Good quality i.r. spectra can be obtained from very finely divided samples of $B_4C$, i.e. $B_{11}C(CBC)$, $B_{12}O$, $B_{12}P_2$, and $B_{12}As_2$.[297] The oxy-species gave the simplest spectrum, and yielded an assignment which could be used to interpret the more complex spectra of the remainder, especially $B_4C$ and $B_{12}P_2$. The greater complexity for the latter is related to disorder within the C—B—C or P—P groups, which for $B_4C$ depends upon the B:C ratio (within the range of homogeneity).

The phase diagram of BN has been extended to 14 kbar and 4000 K.[298] Apparently the wurtzite-like phase is stable at high pressures and low temperatures.

When gaseous $BBr_3$–$PBr_3$ mixtures are reduced by $H_2$, a boron-rich boron phosphide is produced. A crystal with the formula $B_{12}(P_{1.36}B_{0.64})$ possesses a rhombohedral unit cell, space group $R\bar{3}m$. The structure is related to that of $\sigma$-rhombohedral boron, with $B_{12}$ icosahedra at every corner. There is a channel in the centre of each cell, which contains two single atoms (B and/or P) statistically distributed. Each single atom is bonded to three $B_{12}$ units, and the other single atom.[299]

Reduction of $BCl_3$–$TiCl_4$ mixtures with $H_2$ on BN substrates in the range 1050—1250 °C produces tetragonal crystals of $(B_{12})_4B_2Ti_{1.3-2.0}$. X-Ray diffrac-

---

[293] W. Siebert and B. Asgarouladi, Z. Naturforsch., 1975, **30b**, 647.
[294] R. Mattes, K.-F. Tebbe, H. Neidhard, and H. Rethfeld, Z. anorg. Chem., 1975, **413**, 1.
[295] H. J. Becher, H. Rethfeld, and R. Mattes, Z. anorg. Chem., 1975, **414**, 203.
[296] H. L. Yakel, Acta Cryst., 1975, **B31**, 1797.
[297] H. J. Becher and F. Thevenot, Z. anorg. Chem., 1974, **410**, 274.
[298] I. N. Frantsevich, T. R. Balan, A. V. Brokko, S. S. Dzhamarov, G. G. Karyuk, A. V. Kurdyumov, and A. M. Pilyankevich, Doklady Chem., 1974, **218**, 662.
[299] E. Amberger and P. A. Rauh, Acta Cryst., 1974, **B30**, 2549.

tion studies on $(B_{12})_4B_2Ti_{1.37}$ showed that it belonged to the space group $P4_2/nnm$.[300] Forty-eight boron atoms are linked together as four icosahedra, forming a flattened tetrahedron. The Ti atom lies at the centre of this, with an environment constituting a 14-corner polyhedron. The remaining B atoms occupy positions having distorted tetrahedral environments.

$NbB_2$ and $TaB_2$ as crystals oriented along the $c$-axis are formed by the thermal dissociation of $NbB_5$ or $TaB_5$ in the presence of $BBr_3$, at 1400 °C and $2.5 \times 10^{-2}$ Torr.[301]

Heating powdered Cr and B together in Ta foil, contained in a quartz tube to 1400 °C under a vacuum for 2 h produces $Cr_5B_3$.[302] The low-temperature electrical properties of this were studied.

Magnetic measurements have been reported for the solid solutions $Co_{3-x}Fe_xB$, $(Ni, Co)_3B$, and $(Ni, Fe)_3B$, which all have the $Fe_3C$ structure, and for the $\tau$-borides of composition $Ir_{23-x}Fe_xB_6$ ($x = 8-15$).[303]

The $LaB_4$ crystal structure is related to that of $CeB_4$, $ThB_4$, and $UB_4$, and belongs to the tetragonal space group $P4/mbm$. The B—B distances lie between 1.75 and 1.85 Å, and the La—B distances between 2.818 and 3.155 Å.[304]

Uranium-containing ternary borides $UMB_4$ (M = V, Cr, Mo, W, Mn, Re, Fe, or Co) have been prepared.[305] Those for M = Mo, W, or Re have structures isotypic with $ThMoB_4$, while the others crystallize with the $YCrB_4$ structure.

Ferromagnetism has been found at $T_C$ of ca. 31 K for the alloys (U, La)$B_4$, and at ca. 4.5 K for (U,Lu)$B_4$.[306]

Single-crystal structures have been reported for the ternary complex borides $ThMB_4$ (M = Mo or W).[307] The vanadium and rhenium analogues are isotypic. The structures were discussed in relation to the boron layer structures in $AlB_2$, $YCrB_4$, and $Y_2ReB_6$.

## 2 Aluminium

**General.**—Formation cross-sections for the production of $^{26}Al$ from Mg, Al, and Si by $\alpha$-particle bombardment have been measured and tabulated.[308]

LiAl has been shown to possess a crystal structure of the NaTl type.[309] The crystal structure of 'BaAl$_2$' shows that it is actually $Ba_7Al_{13}$, and belongs to the trigonal space group $P\bar{3}m1$.[310] The atomic arrangement is very similar to that in the Laves phase prototype, $MgNi_2$.

All of the compounds $AB_2Al_9$ (A = Ba, B = Fe, Ni, or Co; A = Sr, B = Ni or Co) crystallize in the hexagonal space group $P6/mmm$ $(D_{6h}^1)$.[311]

Investigation of the Sr–Al and Ba–Al phase diagrams revealed the existence of

---

[300] E. Amberger and K. Polborn, *Acta Cryst.*, 1975, **B31**, 949.
[301] B. Armas and F. Trombe, *Compt. rend.*, 1975, **280**, C, 435.
[302] E. Cruceanu, O. Ivanciu, and M. Popa, *J. Less-Common Metals*, 1975, **41**, 339.
[303] R. Sobczak, *Monatsh.*, 1975, **105**, 1071.
[304] K. Kato, I. Kawada, C. Oshima, and S. Kawai, *Acta Cryst.*, 1974, **B30**, 2933.
[305] P. Rogl and H. Novotny, *Monatsh.*, 1975, **106**, 381.
[306] H. H. Hill, A. L. Giorgi, E. G. Sklarz, and J. L. Smith, *J. Less-Common Metals*, 1974, **38**, 239.
[307] P. Rogl and H. Nowotny, *Monatsh.*, 1974, **105**, 1082.
[308] S. Tanaka, K. Sakamoto, K. Komura, and M. Furukawa, *J. Inorg. Nuclear Chem.*, 1975, **37**, 2002.
[309] K. Kuriyama and N. Masaki, *Acta Cryst.*, 1975, **B31**, 1793.
[310] M. L. Fornasini and G. Bruzzone, *J. Less-Common Metals*, 1975, **40**, 335.
[311] K. Turban and H. Schäfer, *J. Less-Common Metals,* 1975, **40**, 91.

the following compounds: $Sr_3Al_2$ (666 °C), $SrAl_2$ (936 °C), $SrAl_4$ (1040 °C), BaAl (730 °C), $BaAl_2$ (914 °C), and $BaAl_4$ (1104 °C).[312] Only the $MAl_4$ species melted congruently.

High-temperature Knudsen cell mass spectrometric measurements on the vapours above Al–Si alloys indicated the presence of $Al_2$, AlSi, and $AlSi_2$. Their atomization energies are $40 \pm 3.6$, $59 \pm 3$, and $150 \pm 5$ kcal mol$^{-1}$, respectively.[313]

The core levels and valence bands of $Fe_3Al$ have been determined by X-ray photoelectron spectroscopy.[314] Shifts in the $Fe2p_{3/2}$ and $Al2p$ levels by comparison with the pure components show that there is a transfer of electrons from the Al to the Fe.

The crystal structures and magnetic properties of $YFe_{2x}Al_{2-2x}$ species have been investigated.[315]

Subjection of trialuminides of the lighter lanthanides, $LnAl_3$, to high pressures gives polymorphic changes in the direction of structures with increased cubic nature. New polymorphs reported were: $GdAl_3$ ($BaPb_3$-type structure), $TbAl_3$ ($HoAl_3$-type structure), and $LuAl_3$ ($Cu_3Au$-type).[316,317]

Phase studies have been made of the aluminium-based rare-earth alloys $RT_2$–$RAl_2$ (R = Pr, Gd, or Er; T = Mn, Fe, Co, Ni, or Cu).[318]

A review on organometallic complexes containing a bond between a Group III metal (Al, Ga, In, or Tl) and a transition metal has been published.[319]

**Aluminium Hydrides.**—PNO-CI (pair natural orbital/configuration interaction) and CEPA-PNO (coupled electron pair approximation with pair natural orbitals) MO calculations on $AlH_3$ give a binding energy (referred to Al + 3H) of ca. 205 kcal mol$^{-1}$.[320]

A direct synthesis of alkali metal aluminium hydrides [reaction (35)] has been carried out in the molten alkali metal as solvent.[321]

$$M(l) + Al(s) + 2H_2(g) \xrightarrow[100-400\text{ atm}]{200-400\,°C} MAlH_4(s) \qquad (35)$$

Reactions (36)—(38) give good yields of tetra-alkylammonium tetrahydro-aluminates (M = Li or Na; $R^1$ = Me, Et, or Bu; $R^2$ = Et or Bu; X = Cl, Br, or I).[322]

$$MH + AlR_3^2 \xrightarrow[30-35\,°C]{Et_2O} (AlHR_3^2) \qquad (36)$$

$$M(AlHR_3^2) + NR_4^1 X \xrightarrow[30-35\,°C]{Et_2O} NR_4^1(AlHR_3^2) + MX \qquad (37)$$

$$NR_4(AlHR_3^2) + LiAlH_4 \xrightarrow[20\,°C]{Et_2O} NR_4^1(AlH_4) + Li(AlHR_3^2) \qquad (38)$$

---

[312] G. Bruzzone and F. Merlo, *J. Less-Common Metals*, 1975, **39**, 1.
[313] C. Chatillon, M. Allibert, and A. Pattoret, *Compt. rend.*, 1975, **280**, C, 1505.
[314] I. N. Shabanova and V. A. Trapeznikov, *J. Electron Spectrocopy*, 1975, **6**, 297.
[315] K. H. J. Buschow, *J. Less-Common Metals*, 1975, **40**, 361.
[316] J. F. Cannon and H. T. Hall, *J. Less-Common Metals*, 1975, **40**, 313.
[317] A. E. Dwight, C. W. Kimball, R. S. Preston, S. P. Taneja, and L. Weber, *J. Less-Common Metals*, 1975, **40**, 285.
[318] H. Oesterreicher, *Inorg. Chem.*, 1974, **13**, 2807.
[319] A. T. T. Hsieh, *Inorg. Chim. Acta*, 1975, **14**, 87.
[320] R. Ahlrichs, F. Keil, H. Lischka, W. Kutzelnigg, and V. Staemmler, *J. Chem. Phys.*, 1975, **63**, 455.
[321] T. N. Dymova, N. G. Elieeva, S. I. Bakum, and Yu. M. Dergashev, *Doklady Chem.*, 1974, **215**, 256.
[322] K. N. Semenenko, E. E. Fokina, and V. A. Kuznetsov, *Russ. J. Inorg. Chem.*, 1974, **19**, 338.

An alternative route to these compounds, which gives yields in the range 85--100%, consists of the double decomposition reaction of $LiAlH_4$ with $R_4NBH_4$ (R = Me, Et, or Bu) in THF at room temperature.[323]

Solutions containing $LiAlH_4$ and $AlH_3$ (solvents $Et_2O$ or THF) gave no evidence (using i.r. spectroscopy) for complex formation between these two species. Likewise, d.t.a.–t.g.a. and $X$-ray powder diffraction experiments on solids obtained by evaporating these solutions showed only these species, and not $LiAl_2H_7$ etc., as had been claimed previously (J. N. Dymova et al., Doklady Akad. Nauk S.S.S.R., 1969, **184**, 1338).[324]

Halogen–hydrogen exchange in the systems $AlH_3$–$MX_2$ (M = Ca, Mg, Zn, Cd, or Cu; X = Cl, Br, or I), depends upon the electronegativity of M. Thus, $CaBr_2$ and $MgBr_2$ do not react, whereas $ZnCl_2$ and $ZnBr_2$ do, giving a new hydride of the type $H_3Zn_2X$ (X = Cl or Br). The interaction of $ZnI_2$ with $AlH_3$ resulted in the formation of $ZnI_2,AlH_3$.[325]

A d.t.a. study of $Na_3AlH_6$ reveals that it decomposes at 245 °C.[326]

A fairly complete vibrational assignment has been proposed for $Al(BH_4)_3$ on the basis of i.r. (gas and crystal) and Raman (liquid) data.[327] $D_{3h}$ symmetry was adequate, and the assignment was more complete than earlier ones based on i.r. spectra and incomplete Raman data.

A reinvestigation of the structure of $Al(BH_4)_3,NH_3$ by single-crystal $X$-ray diffraction confirms the earlier results and the authors concentrate on details of the molecular packing. Intermolecular H $\cdots$ B distances are in the range 2.7—2.8 Å, suggesting that hydrogen bonding is significant.[328]

Phase diagrams have been reported for the systems $Al(BH_4)_3$–$NR_4BH_4$, and $Al(BH_4)_3$–$NR_4I$ (R = Bu$^n$).[329]

Iodide salts of two novel four-co-ordinate aluminium cations have been prepared. They are (65; L–L = TMED or sparteine).[330] They are produced by nucleophilic displacement reactions on $Me_3N,AlH_2I$. The four-co-ordination was confirmed by the values of the $\nu(AlH)$ wavenumbers (ca. 1890 cm$^{-1}$); five-co-ordination would have given lower values. Such a phenomenon was observed for the adduct $TMED,AlHBr_2$ [$\nu(AlH)$ at 1735 cm$^{-1}$], which must therefore be a molecular adduct rather than an ionic compound.

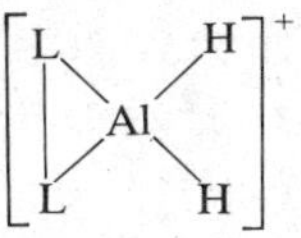

$$\begin{bmatrix} L & H \\ & Al & \\ L & H \end{bmatrix}^+$$

(65)

A number of assignments have been proposed from the i.r. and Raman spectra

---

[323] L. V. Titov and V. D. Sasnovskaya, Russ. J. Inorg. Chem., 1974, **19**, 141.
[324] E. C. Ashby, J. J. Watkins, and H. S. Prasad, Inorg. Chem., 1975, **14**, 583.
[325] E. C. Ashby and H. S. Prasad, Inorg. Chem., 1975, **14**, 1608.
[326] V. A. Kuznetsov, N. D. Golubeva, and K. N. Semenenko, Russ. J. Inorg. Chem., 1974, **19**, 669.
[327] V. B. Polyakova, Yu. Ya. Kharitonov, and K. N. Semenenko, Russ. J. Inorg. Chem., 1974, **19**, 945.
[328] E. B. Lobkovskii, V. B. Polyakova, S. P. Shilkin, and K. N. Semenenko, J. Struct. Chem. 1975, **16**, 66.
[329] K. N. Semenenko, O. V. Kravchenko, and I. I. Korobov, Russ. J. Inorg. Chem., 1974, **19**, 923.
[330] K. R. Skillern and H. C. Kelly, Inorg. Chem., 1974, **13**, 2802.

of $AlH_2Cl,NEt_3$, $AlHCl_2,NEt_3$ and $AlCl_3,NEt_3$, as pure substances, and as solutions in benzene.[331] Thus, in $AlHCl_2,NEt_3$, $\nu(AlN)$ is at 528 cm$^{-1}$, $\nu(AlC)$ at 444 and 516 cm$^{-1}$, and $\nu(AlH)$ at 1873 cm$^{-1}$.

Thermal studies have given some information on the ranges of existence of different solid phases of $MAlH_4$ (M = Na to Cs) and $M_3AlH_6$ (M = Na, K, or Cs).[332]

Molecular weights and conductance studies suggest that ion-pairs and triple-ions are present in THF solutions of $LiAlH_4$, $NaAlH_4$, and $LiBH_4$, and for substituted derivatives containing bulky alkoxide groups. If OMe groups are present, however, linear aggregates may be formed, linked by oxygen-bridge bonds (66). In $Et_2O$ as solvent, $LiAlH_4$ and $LiBH_4$ give larger units than the triple ion.[333]

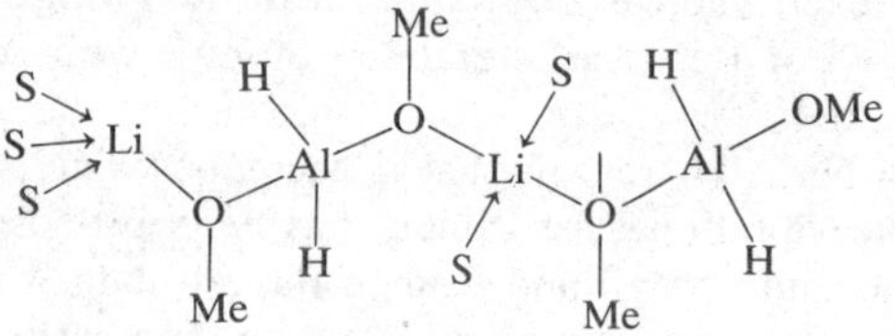

(66)

**Compounds containing Al—C Bonds.**—$^{27}Al$–$^{13}C$ and $^{13}C$–$^{1}H$ coupling constants have been measured in $LiAlMe_4$; they were 71.5 ± 0.5, 106.5 ± 1.0 Hz, respectively.[334]

The crystal structures of $[MoH(C_5H_5)(C_5H_4)]_2Al_3Me_5$ and of $[Mo(C_5H_4)_2Al_2Me_3]_2$ show that in both cases Mo—Al bonds are present, together with $\sigma$-bonding between the Al and the cyclopentadienyl rings. In addition, the former contains an Mo—H—Al bridge linkage.[335]

The bridged system $(\mu\text{-}NPh_2)_2Al_2Me_4$ is cleaved by the bases (B = $NMe_3$, $OEt_2$, and $SMe_2$) forming adducts $Me_2Al(NPh)_2,B$. The relative stabilities of the adducts are in the order $NMe_3 > OEt_2 > SMe_2$. The $SMe_2$ adduct dissociates at −74 °C.[336]

Ab initio MO calculations on $H_2AlCp$ suggest that the equilibrium configuration is of the form $H_2Al(\eta^2\text{-}C_5H_5)$ (67), with the $H_2Al$ unit lying in the symmetry

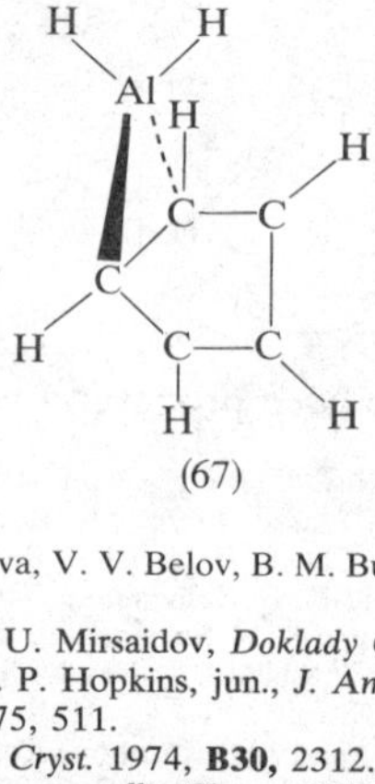

(67)

---

[331] K. N. Semeneenko, V. B. Polyakova, V. V. Belov, B. M. Bulychev, and Yu. Ya. Kharitonov, *Russ. J. Inorg. Chem.*, 1975, **20**, 351.
[332] T. N. Dymova, S. I. Bakum, and U. Mirsaidov, *Doklady Chem.*, 1974, **216**, 285.
[333] E. C. Ahby, F. R. Dobbs, and H. P. Hopkins, jun., *J. Amer Chem. Soc.*, 1975, **97**, 3158.
[334] O. Yamamoto, *Chem. Letters*, 1975, 511.
[335] R. A. Forder and K. Prout, *Acta Cryst.* 1974, **B30**, 2312.
[336] J. E. Rie and J. P. Oliver, *J. Organometallic Chem.*, 1974, **80**, 219.

plane. The bonding between the Al atom and the ring seems to be chiefly due to an interaction of the $a_1$ $\pi$-orbital of the ring with the $3s$ orbital of Al, and of the $e_{1x}$ $\pi$-orbital with the $3p_x$ orbital of the Al.[337]

Hitherto unknown dialkylmetal cyclopentadienyl derivatives of Group III may be prepared by reaction (39) where R = Me or Et; $M^1$ = Al, Ga, or In; $M^2$ = Na

$$R_2M^1Cl + M^2Cp \rightarrow R_2M^1Cp + M^2Cl, \qquad (39)$$

or K. They give a number of $\nu$(CH) bands in the i.r. and Raman spectra, and are therefore probably $h^1$-complexes, although only one signal due to the Cp is seen in the $^1$H n.m.r. spectrum, even at $-100\,°\text{C}$.[338]

**Compounds containing Al—N Bonds.**—Electron-diffraction measurements have yielded the following structural parameters for $AlCl_3,NH_3$; $r(Al—Cl)$ $2.100 \pm 0.005$ Å, $r(Al—N)$ $1.996 \pm 0.019$ Å, $r(Cl \cdots Cl)$ $3.569 \pm 0.011$ Å, $r(Cl \cdots N)$ $3.165 \pm 0.012$ Å, and $\angle ClAlCl$ $116.9°$. A staggered model was assumed, based on the predictions of a CNDO/2 MO calculation.[339] A normal-co-ordinate analysis has also been performed on this adduct.[340]

$^1$H and $^{11}$B n.m.r. spectra (using $^{11}$B, $^{14}$N, and $^{27}$Al double-resonance techniques) of $Al(BH_4)_3,NH_3$ and $Al(BH_4)_3,NMe_3$ show that all the $BH_4$ groups are equivalent, and that the nature of the Al—$BH_4$ binding is the same as in free $Al(BH_4)_3$.[341] Similar data show that $Al(BH_4)_3,6NH_3$ must be formulated as $[Al(NH_3)_6]^{3+}$ $3BH_4^-$.

Dimethyldiazido-aluminates and -gallates are produced by reaction (40),

$$Me_2MN_3 + 3NMe_4N_3 \rightarrow 3NMe_4[Me_2M(N_3)_2] \qquad (40)$$

where M = Al or Ga.[342] They are explosive when heated, or on exposure to mechanical shock. I.r. and Raman spectra of the solids were recorded and partially assigned, as were those of their adducts with $AlMe_3$, $GaMe_3$, and $MgMe_2$.

Gas-phase electron diffraction was used to determine the molecular structure of $[Cl_2AlNMe_2]_2$.[343] the results were consistent with a model of $D_{2h}$ symmetry, in which the $NMe_2$ groups are bridging (Figure 25).

$Cs[AlMe_3(N_3)]$ forms orthorhombic crystals, of space group *Pbcm*. The two N—N distances are significantly different.[344]

The dimer of dimethylaluminium methylphenylamide, $(Me_2AlNMePh)_2$, exists as a mixture of *cis*- and *trans*-isomers (depending on whether or not the two phenyl groups are on the same side of the $Al_2N_2$ ring).[345] Variable-temperature n.m.r. results yield a value of $\Delta H$ for isomerization of $4.47(9)$ kJ mol$^{-1}$, and the

---

[337] O. Gropen and A. Haaland, *J. Organometallic Chem.*, 1975, **92**, 157.
[338] J. Stadelhofer, J. Weidlein, and A. Haaland, *J. Organometallic Chem.*, 1975, **84**, C1.
[339] M. Hargitai, I. Hargitai, V. P. Spiridonov, M. Pelissier, and J.-F. Labarre, *J. Mol. Structure*, 1975, **24**, 27.
[340] S. J. Cyvin, B. N. Cyvin, and I. Hargittai, *J. Mol. Structure*, 1974, **23**, 385.
[341] G. N. Boiko, Yu. I. Malov, K. N. Semenenko, and S. P. Shilkin, *Russ. J. Inorg. Chem.*, 1975, **19**, 810.
[342] K. Dehnicke and N. Röder, *J. Organometallic Chem.*, 1975, **86**, 335.
[343] T. C. Bartke, A. Haaland, and D. P. Novak, *Acta Chem. Scand.* (A), 1975, **29**, 273.
[344] J. L. Atwood and W. R. Newbery, *J. Organometallic Chem.*, 1975, **87**, 1.
[345] K. Wakatsuki and T. Tanaka, *Bull. Chem. Soc., Japan*, 1975, **48**, 1475.

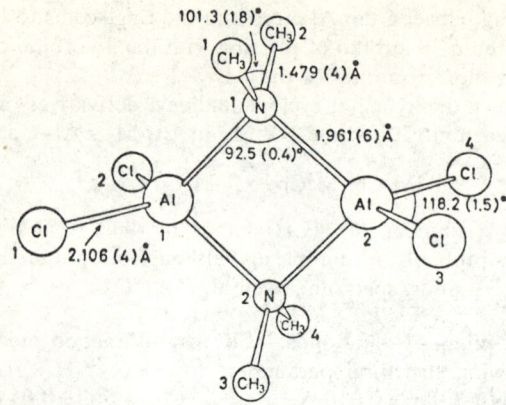

**Figure 25** *Molecular model of* [Cl$_2$AlNMe$_2$]$_2$
[Reproduced by permission from *Acta Chem. Scand.* (*A*), 1975, **29**, 273]

most probable mechanism is one involving the breaking of the Al—N bond, followed by rotation of the non-bridged nitrogen atom.

$NN'$-Dimethyloxamide reacts with Al, Ga, or In trialkyls to give (R$_2$M)$_2$[O$_2$C$_2$(NMe)$_2$] (M = Al, Ga, or In; R = Me or Et). The skeletal M$_2$O$_2$C$_2$N$_2$ unit comprises an almost planar system of two fused five-membered rings (symmetry $S_2$), (68).[346]

(68)

Poly($N$-alkylaminoalanes) may be prepared directly from Al and RNH$_2$. Small amounts of MAlH$_4$ must be present as an activator.[347] Pure, well-defined oligomers result from this process, in contrast to other reactions, which give mixtures or insoluble products. Another good route to these compounds involves the interaction of LiAlH$_4$ or NaAlH$_4$ with primary amines having $\alpha$- or $\beta$-secondary carbon atoms, or with t-butylamine, in hydrocarbon solvents.[348]

The same group of workers have reported several crystallographic studies on these and related substances. Thus, (HAlNPr$^i$)$_6$ possesses a cage structure, built up from two almost planar (AlN)$_3$ six-membered rings, joined by six transverse Al—N bonds.[349,350] The mean bond lengths were: Al—N (in six-membered rings) 1.898(2) Å, (in transverse bonds) 1.956(2) Å; Al—H 1.49(1) Å and N—C

---

[346] H. U. Schwering, J. Weidlein, and P. Fischer, *J. Organometallic Chem.*, 1975, **84**, 17.
[347] S. Cucinella, A. Mazzei, and G. Dozzi, *J. Organometallic Chem.*, 1975, **84**, C19.
[348] S. Cucinella, G. Dozzi, A. Mazzei, and T. Salvatori, *J. Organometallic Chem.*, 1975, **90**, 257.
[349] S. Cucinella, T. Salvatori, C. Busetto, G. Perego, and A. Mazzei, *J. Organometallic Chem.*, 1974, **78**, 185.
[350] M. Cesari, G. Perego, G. Del Piero, S. Cucinella and E. Cernia, *J. Organometallic Chem.*, 1974, **78**, 203.

*Elements of Group III* 119

1.514(2) Å. The crystal structure of the adduct $(HAlNPr^i)_6,AlH_3$ shows that this structure is maintained, with the $AlH_3$ unit co-ordinated *via* two Al—N and two Al—H—Al bonds.[351] $H(HAlNPr^i)_5AlH_2,LiH,Et_2O$ contains a pseudo-hexameric cage with a five-membered Al—N—Al—N—Al unit cross-linked to a six-membered $(AlN)_3$ ring. The H of LiH is indistinguishable from the other hydridic hydrogens, and so the *N*-isopropyliminoalane unit could be regarded formally as an anion (charge $-1$). The $Li^+$ is linked *via* Li—H—Al bridges, its fourth co-ordinate position being occupied by $OEt_2$.[352] $(HAlNPr^i)_2(H_2AlNHPr^i)_3$ possesses a molecular structure made up from a cyclohexane-type ring $(HAlNPr^i)_2(H_2AlNHPr^i)$ in a skew-boat conformation, with an $H_2AlNHPr^i$ bridging unit attached to each side, being bonded to an N and Al atom of the ring.[353]

**Compounds containing Al—O Bonds.**—Laser-induced fluorescence has been used to investigate reaction (41).

$$Al + O_2 \rightarrow AlO + O \qquad (41)$$

A value of $D_0^0(AlO) = 121.5 \pm 1$ kcal mol$^{-1}$ was derived for the ground-state dissociation energy of AlO.[354]

Ab initio MO calculations carried out on $H_3AlOH_2$ and $(H_2AlOH)_2$ show that in each case the configuration about O is intermediate between trigonal and tetrahedral.[355] The energy minima are, however, very shallow, and the planar forms lie only 0.19 and 0.35 kcal mol$^{-1}$, respectively, above the equilibrium conformation. No evidence was found for $(p-d)\pi$-bonding between O and Al.

Similar calculations for $Al(H_2O)_n^{3+}$ ($n = 1$—7) suggest that the binding energies increase steadily to $n = 6$, with the values for $n = 6$ and 7 being about equal. The most likely mechanism for the exchange of $H_2O$ molecules from the first hydration shell of $Al^{3+}$ appears to be dissociative, $S_N1$, if the leaving molecule stays in the second hydration shell.[356]

Separate —OH $^1H$ n.m.r. signals for free and bound MeOH and EtOH can be detected in $AlX_3$ ($X = ClO_4$, $NO_3$, or Cl) and $Ga(ClO_4)_3$ in MeOH solutions and of $AlCl_3$ in EtOH. These data give the following solvation numbers: *ca.* 6 for the perchlorates [except $Ga(ClO_4)_3$ in MeOH at less than 45 °C, for which it is *ca.* 7]; *ca.* 5 for $Al(NO_3)_3$ in MeOH; and *ca.* 4 for $AlCl_3$ in MeOH and EtOH. Ligand-exchange processes are believed to occur by whole ligand exchange, and not by proton exchange.[357]

A number of assignments have been proposed for the vibrational modes of 1:1 adducts of $Et_2O$ with $AlHCl_2$, $AlDCl_2$, and $AlHBr_2$. In each case $\nu(AlO)$ was close to 330 cm$^{-1}$, and shifted very little on changing Cl for Br.[358]

[351] G. Perego, M. Cesari, G. Del Piero, A. Balducci, and E. Cernia, *J. Organometallic Chem.*, 1975, **87**, 33.
[352] M. Cesari, G. Perego, G. Del Piero, M. Corbellini, and A. Immirzi, *J. Organometallic Chem.*, 1975, **87**, 43.
[353] G. Perego, G. Del Piero, M. Cesari, A. Zazzetta, and G. Dozzi, *J. Organometallic Chem.*, 1975, **87**, 53.
[354] P. J. Dagdigian, H. W. Cruse, and R. N. Zare, *J. Chem. Phys.*, 1975, **62**, 1824.
[355] O. Gropen, R. Johansen, A. Haaland, and O. Stokkeland, *J. Organometallic Chem.*, 1975, **92**, 147.
[356] H. Veillard, J. Demnynck, and A. Veillard, *Chem. Phys. Letters*, 1975, **33**, 221.
[357] D. Richardson and T. D. Alger, *J. Phys. Chem.*, 1975, **79**, 1733.
[358] K. N. Semenenko, Yu. Ya. Kharitonov, V. B. Polyakova, and V. N. Fokin, *Russ. J. Inorg. Chem.*, 1974, **19**, 1437.

The crystal structure of $Cs_2Al(NO_3)_5$ shows that the central aluminium atom is in distorted octahedral co-ordination, with the apices occupied by one bidentate and four unidentate nitrate groups. The major molecular parameters associated with the discrete $[Al(NO_3)_5]^{2-}$ anion are: $r(Al-O)_{unidentate}$, 1.89(4) and 1.93(4) Å; $r(Al-O)_{bidentate}$, 1.98(3) Å; the angles at the aluminium ($\angle OAlO$) vary from 65(1)° for the bidentate nitrate group to 105(1)°.[359]

$[Al(urea)_6](ClO_4)_3$ forms crystals at room temperature which are rhombohedral, space group $R\bar{3}c$. The crystal structure is isomorphous with that of the analogous $Ti^{III}$ compound, having similar disorder among the $ClO_4$ groups. Comparison of these data with e.s.r. measurements in connection with the phase transition at ca. 25 °C suggests that above that temperature, the actual structure possesses $R\bar{3}$ symmetry.[360]

$^{27}Al$ N.m.r. data for solutions of $Al(ClO_4)_3$ in nitromethane containing DMF or DMSO showed that all seven species $Al(DMSO)_{6-x}(DMF)_x^{3+}$ ($x = 0-6$) are present at various DMF:DMSO ratios.[361] The equilibrium constants for reaction (42) are: $K_1 = 2.08 \pm 0.09$; $K_2 = 1.57 \pm 0.10$; $K_3 = 1.51 \pm 0.06$; $K_4 = 1.50 \pm 0.06$; $K_5 = 1.77 \pm 0.13$.

$$Al(DMSO)_{7-x}(DMF)_{x-1}^{3+} + Al(DMSO)_{5-x}(DMF)_{x+1}^{3+} \rightleftharpoons 2Al(DMSO)_{6-x}(DMF)_x^{3+}, \quad (42)$$

Compound (69) is monomeric in benzene solution, dimerization being prevented by steric strain.[362] Aluminium ethylene dioxide is prepared by treating aluminium s-butoxide with ethylene glycol, followed by heating to 150 °C. It is tetrameric, with seven ethylene glycol molecules present. Spectroscopic considerations indicate a structure like (70).[363]

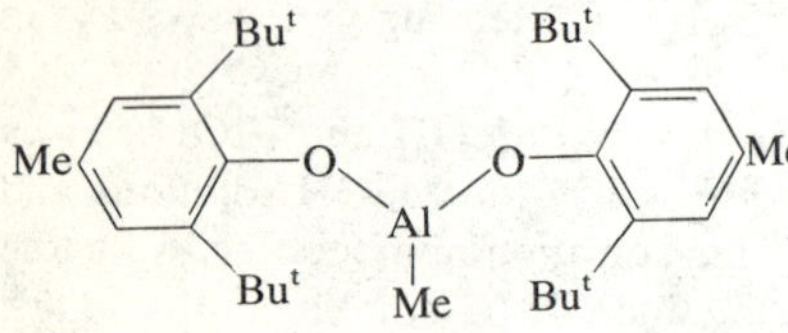

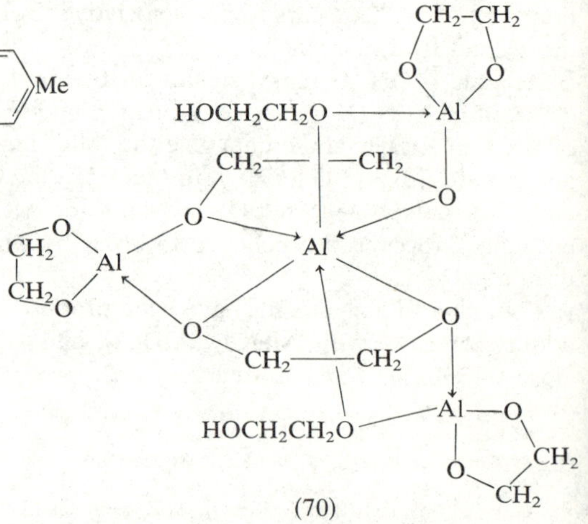

[359] O. A. D'yachenko and L. O. Atovmyam, *J. Struct. Chem.*, 1975, **16**, 73.
[360] J. H. M. Mooy, W. Krieger, D. Heijdenrijk, and C. H. Stam, *Chem. Phys. Letters*, 1974, **29**, 179.
[361] D. Gudlin and H. Schneider, *J. Magn. Resonance*, 1974, **16**, 362.
[362] K. B. Starowieysky, S. Pasinkiewicz, and M. Skowrońska-Ptasińska, *J. Organometallic Chem.*, 1975, **90**, C43.
[363] P. Maleki and M. J. Schwing-Weill, *J. Inorg. Nuclear Chem.*, 1975, **37**, 435.

$Al^{3+}$ forms octahedral solvates with trialkyl phosphates and phosphonates and dialkyl hydrogen phosphites, whereas with hexamethylphosphoramide it gives a tetrahedral species. These conclusions were drawn from $^{27}Al$ and $^{31}P$ n.m.r. measurements.[364] The kinetics of ligand exchange point to dissociation for the six-co-ordinate and association for the four-co-ordinate complexes, with different activation enthalpies and entropies. Thus, for $Al(dimethylmethylphosphonate)_6^{3+}$, $\Delta H^{\ddagger} = 19.8$ kcal mol$^{-1}$; $\Delta S^{\ddagger} = 6.9$ e.u.; for $Al(HMPA)_4^{3+}$ the corresponding values are 7.7 kcal mol$^{-1}$ and $-10.2$ e.u.

I.r. and Raman spectra of the dimethylmetal hypophosphites, $Me_2M(OOPH_2)$, and dimethylmetal dimethylthiophosphinates, $Me_2M(OSPMe_2)$ (M = Al, Ga, In, or Tl) have been assigned. The former contain eight-membered rings, the latter four-membered rings.[365]

**Table 8** Stability constants of aluminium-propionate and -lactate complexes

| Propionate | 25 °C | 35 °C | 45 °C |
|---|---|---|---|
| $\beta_1$ | $60 \pm 2.0$ | $64 \pm 2.0$ | $78 \pm 2.0$ |
| $\beta_2$ | $(2.6 \pm 0.4) \times 10^3$ | $(1.1 \pm 0.1) \times 10^4$ | $(2.4 \pm 0.4) \times 10^4$ |
| Lactate | | | |
| $\beta_1$ | $(1.8 \pm 0.1) \times 10^4$ | $(1.9 \pm 0.2) \times 10^2$ | $2.5 \pm 0.2) \times 10^2$ |
| $\beta_2$ | $(7.5 \pm 1.0) \times 10^4$ | $(8.7 \pm 1.0) \times 10^4$ | $(1.8 \pm 0.4) \times 10^5$ |

Stepwise formation constants and free energy, enthalpy, and entropy changes have been calculated for the 1:1, 1:2, and 1:3 complexes of $Al^{3+}$ with malonate and succinate ions.[366] Table 8 gives the calculated values for the stability constants of aluminium complexes with propionate and lactate ions in aqueous solutions, with $I = 1$.[367] Solution calorimetry was used to determine values for the enthalpies of formation of tris(tropolonato)$Al^{III}$ and tris(4-methyltropolonato)$Al^{III}$ (Table 9); they were used to derive gas-phase enthalpies of formation, and hence the Al—O bond energies.[368]

**Table 9** Thermodynamic characteristics of tris(tropolonato)- and tris-(4-methyltropolonato)-aluminium(III), all figures in kcal mol$^{-1}$

| | $\Delta H_f^{\circ}(c)$ | $\Delta H_{subl}^{\circ}$ | $\Delta H_f^{\circ}(g)$ | E(Al—O) |
|---|---|---|---|---|
| Al(trop)$_3$ | $-316.96 \pm 0.66$ | $30.0 \pm 5.0$ | $-287.0 \pm 5.0$ | $61 \pm 3$ |
| Al(4-Metrop)$_3$ | $-329.8 \pm 2.2$ | $30.0 \pm 5.0$ | $-299.8 \pm 5.5$ | $60 \pm 3$ |

Formation constants of mononuclear hydroxo-complexes of aluminium were determined spectrophotometrically at 15 °C using alizarin-3-sulphonic acid as a competing ligand. The hydrolysis constants of Al ions at ionic strengths in the range 0.1—1.0 were also calculated.[369] The results for $Al(OH)_n^{(3-n)-}$ are shown in Table 10.

The electronic structure of $AlO_4^{5-}$ was calculated by the SCF-Xα scattered-

[364] J.-J. Delpuech, M. R. Khaddar, A. A. Peguy, and P. R. Rubini, *J. Amer. Chem. Soc.*, 1975, **97**, 3373.
[365] B. Schaible, K. Roessel, J. Weidlein, and H.-D. Hausen, *Z. anorg. Chem.*, 1974, **409**, 176.
[366] P. H. Tedesco and J. G. Quintana, *J. Inorg. Nuclear Chem.*, 1974, **36**, 2628.
[367] P. H. Tedesco and V. B. De Rumi, *J. Inorg. Nuclear Chem.*, 1975, **37**, 1833.
[368] R. J. Irving and M. A. V. Ribeiro da Silva, *J.C.S. Dalton*, 1975, 1257.
[369] V. A. Nazarenko and E. A. Biryuk, *Russ. J. Inorg. Chem.*, 1974, **19**, 341.

**Table 10** *Formation and hydrolysis constants of mononuclear hydroxo-complexes of aluminium*(III)

| | Overall formation constants | | | Hydrolysis constants | | |
|---|---|---|---|---|---|---|
| $I$ | $\beta_1$ | $\beta_2$ | $\beta_3$ | $K_1$ | $K_2$ | $K_3$ |
| 0.1 | $1.25 \times 10^9$ | $4.50 \times 10^{17}$ | $5.60 \times 10^{25}$ | $1.25 \times 10^{-5}$ | $3.60 \times 10^{-6}$ | $1.24 \times 10^{-6}$ |
| 0.3 | $2.50 \times 10^9$ | $18.70 \times 10^{17}$ | $56.0 \times 10^{25}$ | $2.50 \times 10^{-5}$ | $7.40 \times 10^{-6}$ | $3.00 \times 10^{-6}$ |
| 0.5 | $5.00 \times 10^9$ | $75.0 \times 10^{17}$ | $416 \times 10^{25}$ | $5.00 \times 10^{-5}$ | $15.0 \times 10^{-6}$ | $5.54 \times 10^{-6}$ |
| 1.0 | $25.0 \times 10^9$ | $2.26 \times 10^{20}$ | $1.45 \times 10^{30}$ | $25.0 \times 10^{-5}$ | $104 \times 10^{-6}$ | $55.3 \times 10^{-6}$ |

wave cluster method.[370] The predicted orbital energies are in good agreement with experimental data on minerals containing this anion.

If solutions of sodium silicate and aluminate, having the same pH and resistance, are mixed, and if the concentrations are so low that gel formation does not occur, then a considerable increase in conductivity is noted. This must be due to formation of soluble aluminosilicate complexes.[371]

Methods for the hydrothermal synthesis of the aluminosilicates of Mg, Ca, and Ba have been reviewed.[372] A number of lithium aluminium double phosphates have been isolated from the reactions of lithium hydroxide dialuminate with $H_3PO_4$ solutions.[373]

I.r. spectra of low- ($\alpha$) and high-temperature ($\beta$) forms of $Li_5AlO_4$ have been reported.[374] Both give bands between 750 and 810 cm$^{-1}$ derived from the $T_{1u}$ $AlO_4$ stretch, while the $A_1$ $AlO_4$ stretch gives a band at ca. 700 cm$^{-1}$ for the $\alpha$-form only.

Single-crystal n.m.r. spectroscopy has been used to determine the temperature dependence of the nuclear quadrupole coupling constants $|e^2qQ/h|$ in the $\beta$-alum $CsAl(SO_4)_2, 12H_2O$.[375]

Thermogravimetric and X-ray powder diffraction methods have revealed[376] the dehydration of sodium aluminium alum to proceed by reactions (43)—(45).

$$NaAl(SO_4)_2, 12H_2O \xrightarrow{100\,°C} NaAl(SO_4)_2, 6H_2O + 6H_2O \quad (43)$$

$$NaAl(SO_4)_2, 6H_2O \xrightarrow{190\,°C} NaAl(SO_4)_2, 2H_2O + 4H_2O \quad (44)$$

$$NaAl(SO_4)_2, 2H_2O \xrightarrow{280\,°C} NaAl(SO_4)_2 + 2H_2O. \quad (45)$$

X-Ray powder data and i.r. spectra have been reported for the kyanite phases $Al_2SiO_5$ and $Al_2GeO_5$, synthesized at 20 kbar and 1000 °C.[377]

Shock compression experiments on the andalusite phase of $Al_2SiO_5$ show that disproportionation to $Al_2O_3$ and $SiO_2$ occurs at pressures in excess of 585 kbar.[378]

---

[370] J. A. Tossell, *J. Amer. Chem. Soc.*, 1975, **97**, 4840.
[371] J.-L. Guth, P. Caullet, and R. Wey, *Bull. Soc. chim. France*, 1974, 2363.
[372] N. A. Ovramenko and F. D. Ovcharenko, *Doklady Chem.*, 1974, **219**, 815.
[373] T. O. Ashchyan, V. P. Danilov, O. F. Kostyleva, and I. N. Lepeshkov, *Russ. J. Inorg. Chem.*, 1975, **20**, 341.
[374] V. A. Kolesova, *Russ. J. Inorg. Chem.*, 1974, **19**, 1585.
[375] N. Weiden and A. Weiss, *Ber. Bunsengesellschaft phys. Chem.*, 1975, **79**, 557.
[376] E. B. Gitis, E. F. Dubrava, and V. F. Annopol'ski, *Russ. J. Inorg. Chem.*, 1974, **19**, 801.
[377] K. Langer and K. K. Sharma, *J. Appl. Cryst.*, 1975, **3**, 329.
[378] H. Schneider and U. Hornemann, *Naturwiss.*, 1975, **62**, 296.

*Elements of Group III* 123

Rapid quenching of a fused $0.8Al_2O_3$–$0.2Nd_2O_3$ mixture produces a glass showing no short-range order; above 900 °C it crystallizes into $\beta$-$Al_2O_3$ and $NdAlO_3$.[379]

The spinel $FeAl_2O_4$ is reduced by $H_2$ in the range 750—1030 °C, with an activation energy of $45 \pm 3$ kcal $mol^{-1}$.[380] No single mechanism appeared to fit the available data. The cationic distribution can be altered by heat treatment. Thus, two samples prepared at 900 and 1200 °C gave formulae $(Fe_{1-\tau}Al_\tau)_A(Fe_\tau Al_{1-\tau})_B O_4$ in which $\tau = 0.08$ and 0.13, respectively.[381]

The reactivities of $\eta$-, $\gamma$-, and $\alpha$-$Al_2O_3$ in the formation of $ZnAl_2O_4$ are in the sequence $\eta > \gamma > \alpha$. This can be explained by assuming that the first two give an imperfect, the last a perfect spinel structure for the $ZnAl_2O_4$.[382]

Sodium aluminium sulphate, $3Na_2SO_4,Al_2(SO_4)_3$, undergoes a reversible polymorphic transformation at 600 °C, and melts incongruently at 700 °C. Among the decomposition products is a new polymorph of $Na_2SO_4,Al_2(SO_4)_3$.[383]

The crystal structure of holmquistite, $Na_{0.03}K_{0.01}Ca_{0.03}Li_{1.79}Mg_{1.76}Fe^{II}_{1.21}Fe^{III}_{0.24}Mn_{0.03}Ti_{0.02}Al^{(6)}_{1.84}Si_{7.89}Al^{(4)}_{0.11}O_{22}$, has been found to contain a chain geometry close to those found in anthophyllite and gedrite, although the two chains are more alike, and the tetrahedral rotations are smaller than in the latter.[384]

Scapolites are solid-solution species based on marialite, $Na_4Al_3Si_9O_{24}Cl$, and meionite, $Ca_4Al_6Si_6O_{24}(CO_3,SO_4)$.[385] The crystal structure of a specimen having the composition $(Ca_{4.17}Na_{3.31}K_{0.46})(Si_{14.99}Al_{8.69})O_{47.8}Cl_{0.73}(SO_4)_{0.37}(CO_3)_{0.87}$ has been determined, and the geometry of the central anion cage elucidated.

A new phase, $SiAl_4N_4O_2$, has been characterized in the Si–Al–O–N system. It is believed to be a thermolysis product of the recently described $\beta'$-$Si_3N_4$.[386]

The following could be isolated from the reaction of $Al(NO_3)_3$ with Na or K orthophosphates in the pH range 3—7 (in aqueous solutions at room temperature): $K_3H_6Al_4(PO_4)_7,16H_2O$, $M_xH_{3-x}Al_4(PO_4)_5,16H_2O$ (M = K, $x = 3$; M = Na, $x = 1, 2,$ or 3), $M_xH_{3-x}Al_3(PO_4)_4,12H_2O$ (M = K, $x = 1.5, 2$ or 3; M = Na, $x = 1, 2,$ or 3), $Na_4Al_4(PO_4)_5(OH),16H_2O$, and $K_2H_4Al_5(PO_4)_7,18H_2O$.[387] Similar series of products result from the analogous reaction with ammonium phosphates.[388]

The zeolites phillipsite ($K_{\sim 2}Ca_{\sim 1.5}Na_{\sim 0.4}Al_{\sim 5}Si_{\sim 10}O_{32},12H_2O$) and harmotome ($Ba_{\sim 2}Ca_{\sim 0.5}Al_{\sim 5}Si_{\sim 11}O_{32},12H_2O$) are isostructural, with the same cationic sites and $H_2O$ molecule distribution within an almost identical framework. There was no evidence for Si,Al ordering.[389]

Calcining a $Li_2CO_3$–$\gamma$-$Al_2O_3$ catalyst mixture at up to 500 °C gives $\alpha$-$LiAlO_2$. Higher temperatures lead to formation of $LiAl_5O_8$. The K and Cs analogues give no compound formation at calcination temperatures up to 900 K. The carbonates

---

[379] J.-P. Coutures, A. Rouanet, G. Benezech, E. Antic, and P. Caro, *Compt. rend.*, 1975, **280,** C, 693.
[380] I. Gaballah, F. Jeannot, and C. Gleitzer, *Compt. rend.*, 1975, **280,** C, 697.
[381] I. Gaballah, A. Courtois, F. Jeannot, and C. Gleitzer, *Compt. rend.*, 1975, **280,** C, 1367.
[382] T. Tsuchida, R. Furuichi, and T. Ishii, *Z. anorg. Chem.*, 1975, **415,** 175.
[383] F. L. Glekel', M. M. Kazakov, and N. A. Parpiev, *Russ. J. Inorg. Chem.*, 1974, **19,** 493.
[384] M. C. Irusteta and E. J. W. Whittaker, *Acta Cryst.*, 1975, **B31,** 145.
[385] S. B. Lin and B. J. Barley, *Acta Cryst.*, 1975, **B31,** 1806.
[386] D. Brachet, P. Goursat, and M. Billy, *Compt. rend.*, 1975, **280,** C, 1207.
[387] A. M. Golub and I. I. Boldog, *Russ. J. Inorg. Chem.*, 1974, **19,** 955.
[388] A. M. Golub and I. I. Boldog, *Russ. J. Inorg. Chem.*, 1974, **19,** 499.
[389] R. Rinaldi, J. J. Pluth, and J. V. Smith, *Acta Cryst.*, 1974, **B30,** 2426.

merely react with surface hydroxyls, and the large size of the K and Cs precludes their ready diffusion into the $\gamma$-$Al_2O_3$ lattice.[390]

AlN and $SiO_2$ react on heating to 1700 °C to form $\alpha$-$Al_2O_3$ and $\beta$-$Si_3N_4$. After 70 h these react to give $2Si_3N_4,3Al_2O_3$, which is, however, only stable in the presence of excess $O_2$.[391] Tricalcium aluminate, $Ca_9Al_6O_{18}$, belongs to the space group $Pa3$ ($a$ = 15.263 Å).[392] The structure is built up from six $AlO_4$ tetrahedra forming rings (eight per unit cell). The $AlO_4$ tetrahedra are only slightly distorted from regular geometry.

$Ca_{8.5}NaAl_6O_{18}$ possesses an orthorhombic unit cell, space group $Pbca$. The mean Al—O bond length is 1.751(1) Å, and the Al atoms are distributed at positions very near the corners of pseudo-cubic sub-cells.[393] The crystal structures of a number of nickel aluminosilicates have been determined.[394—396]

A single crystal of the new zeolite mazzite, $Na_{\sim 0.3}K_{2.5}Ca_{1.4}Mg_{2.1}$-$(Al_{9.9}Si_{26.5}O_{72}),28H_2O$, was dehydrated at 600 °C and ca. $10^{-5}$ Torr. The resulting crystal belonged to the space group $P6_3/mmc$. The aluminosilicate framework is built up of gmelinite-type cages, staggered at $+1/4$ and $-1/4$ in $z$. Cross-linking forms 12-membered rings perpendicular to $c$, and these are so connected as to form isolated channels. The walls of these consist of alternating ladders of four- and five-membered rings.[397]

Alterations in the structural framework of low-soda Y zeolites of varying lanthanum content, due to a variety of thermal treatments, have been monitored by i.r. spectroscopy, chiefly in the $\nu$(OH) region.[398,399] Thus, 'steaming' at 540 and 820 °C results in partial framework dealumination and some structural rearrangement of the framework.

X-Ray photoelectron spectra reveal that the Si:Al ratio of the external surfaces of the zeolites NaA, NaX, NaY, NaZ, HZ, cerium-exchange Y-zeolite, and oxidized CeY is approximately twice that in the bulk.[400] Progressive breakdown of the crystal structure and porous framework of NaY zeolite occurs on grinding. The transformation to the amorphous state is thought to be due to local overheating of the sample.[401] The site selectivity of $Tl^+$ in ion exchange of zeolite Linde 4A is the same as that of $K^+$, and different from that of $Ag^+$.[402] No evidence was found for the replacement of Si or Al of the framework by P during zeolite synthesis. Zeolites A, NaP(gismondine-type), sodalite hydrate, and KF (edingtonite-type) were prepared from $AlPO_4$–silica mixtures, and very little phosphorus was found in the products, the upper limit being 0.00056—0.0117 P atom per Al atom.[403]

---

[390] W. H. J. Stork and G. T. Pott, *J. Phys. Chem.*, 1974, **78**, 2496.
[391] J.-P. Torre and A. Mocellin, *Compt. rend.*, 1974, **279**, C, 943.
[392] P. Mondal and J. W. Jeffrey, *Acta Cryst.*, 1975, **B31**, 689.
[393] F. Niwhi and Y. Takéuchi, *Acta Cryst.*, 1975, **B31**, 1169.
[394] C.-B. Ma, K. Sahl, and E. Tillmanns, *Acta Cryst.* 1975, **B31**, 2137.
[395] C.-B. Ma and E. Tillmanns, *Acta Cryst.*, 1975, **B31**, 2139.
[396] C.-B. Ma and K. Sahl, *Acta Cryst.*, 1975, **B31**, 2142.
[397] R. Rinaldi, J. J. Pluth, and J. V. Smith, *Acta Cryst.*, 1975, **B31**, 1603.
[398] J. Scherzer and J. L. Bass, *J. Phys. Chem.*, 1975, **79**, 1194.
[399] J. Scherzer, J. L. Bass, and F. D. Hunter, *J. Phys. Chem.*, 1975, **79**, 1200.
[400] J.-F. Tempere, D. Delafosse, and J. P. Coutour, *Chem. Phys. Letters*, 1975, **33**, 95.
[401] B. Moraweck, P. Gallezot, A. Renauprez, and B. Imelik, *J. Phys. Chem.*, 1974, **78**, 1959.
[402] K. Ogawa, M. Nitta, and K. Aomura, *J.C.S. Chem. Comm.*, 1975, 88.
[403] R. M. Barrer and M. Liquornik, *J.C.S. Dalton*, 1974, 2126.

# Elements of Group III

A number of studies of phase relationships in Al—O containing systems have been made.[404—420]

**Aluminium Halides.**—Three i.r. bands due to $AlF_3$ are found in a matrix of $AlF_3$ in solid argon: $\nu_2 = 286.2$ cm$^{-1}$, $\nu_3 = 909.4$ cm$^{-1}$, and $\nu_4 = 276.9$ cm$^{-1}$. No multiplet pattern at ca. 950 cm$^{-1}$ (previously reported) could be seen in carefully isolated $AlF_3$. No bands due to $\nu_1$ could be assigned definitely, but non-planarity could not be ruled out conclusively.[421]

A further contribution has been made to the contentious literature on the composition of aluminium fluoride-containing melts.[422,423] Gilbert et al. have examined the Raman spectra of alkali fluoride–$AlF_3$ melts in the temperature range 700—900 °C. Changes in the spectra with melt composition were interpreted in terms of an equilibrium between $AlF_4^-$ and $AlF_6^{3-}$. Wavenumbers due to the latter were seen at 550(pol), 390(depol) and 335(depol) cm$^{-1}$; this left features which must be assigned to $AlF_4^-$ at 750(depol), 622(pol), 322(depol), and 201(depol) cm$^{-1}$. These were assigned to the modes $\nu_3$, $\nu_1$, $\nu_4$, and $\nu_2$ of a regular tetrahedral species. Calculated values of the force constants (all in mdyn Å$^{-1}$) were $f_d = 4.33$, $f_\alpha/d^2 = 0.26$, $f_{\alpha\alpha'}/d^2 = -0.10$, and $(f_{r\alpha} - f_{r\alpha'})/d = 0.53$; these are in agreement with trends observed for other $AlX_4^-$. Measurements of the relative intensities of the $\nu_1$ bands led to an estimate of the stoicheiometric dissociation constant: $K = [AlF_6^{3-}][F^-]^2/[AlF_6^{3-}] = (3 \pm 1) \times 10^{-2}$, where the concentrations are expressed in mole fractions.

New fluoro-derivatives of the Group III elements, i.e. $Me_4NMF_4$ (M = Al, Ga, or In), have been isolated during investigation of the system $MF_3$–$Me_4NF$–HF–$H_2O$.[424] The initial products are hydrated, but anhydrous materials can be obtained on heating. In the Al compound it seems that tetrahedral co-ordination is most probable, which gives added impetus to the claims made in the two previous papers.

X-Ray powder diffraction results for natural and synthetic elpasolite,

---

[404] K. G. Shcherbina, M. V. Mokhosoev, and A. I. Gruba, *Russ. J. Inorg. Chem.*, 1974, **19**, 215.
[405] A. A. Maksimenko and V. G. Shevchuk, *Russ. J. Inorg. Chem.*, 1974, **19**, 291.
[406] M. V. Mokhosoev, K. G. Shcherbina, A. I. Gruba, and V. I. Drivobok, *Russ. J. Inorg. Chem.*, 1974, **19**, 526.
[407] L. P. Ni, O. B. Khalyapina, and T. V. Solenko, *Russ. J. Inorg. Chem.*, 1974, **19**, 532.
[408] L. S. Itkina and N. M. Chaplygina, *Russ. J. Inorg. Chem.*, 1974, **19**, 601,
[409] S. M. Portnova and L. S. Itkina, *Russ. J. Inorg. Chem.*, 1974, **19**, 621.
[410] N. M. Chaplygina, L. S. Itkina, and E. V. Petrova, *Russ. J. Inorg. Chem.*, 1974, **19**, 762.
[411] L. S. Mataitene, A. Yu. Kaminskas, and A. I. Mataitis, *Russ. J. Inorg. Chem.*, 1974, **19**, 766.
[412] V. I. Kuz'menkov, S. V. Plyshevskii, and V. V. Pechkovskii, *Russ. J. Inorg. Chem.*, 1974, **19**, 881.
[413] S. I. Berul' and N. I. Grishina, *Russ. J. Inorg. Chem.*, 1974, **19**, 1702.
[414] N. P. Bobrysheva, B. Ya. Brach, and V. N. Ikorskii, *Russ. J. Inorg. Chem.*, 1974, **19**, 1741.
[415] I. N. Lepshkov, U. P. Danilov, I. S. Zaitesva, E. V. Durinin, and L. T. Kotova, *Russ. J. Inorg. Chem.*, 1974, **19**, 1786.
[416] H. V. Keer, M. G. Bodas, A. Bhaduri, and A. B. Biswas, *J. Inorg. Nuclear Chem.*, 1975, **37**, 1605.
[417] P. Foulatier, M. Lenglet, and J.-C. Tellier, *Compt. rend.*, 1975, **280**, C, 965.
[418] E. H. L. J. Dekker and G. D. Rieck, *Z. anorg. Chem.*, 1975, **415**, 69.
[419] A. Burewicz and U. Politanska, *Russ. J. Inorg. Chem.*, 1975, **20**, 466.
[420] K.-J. Range, G. Engert, and M. Zabel, *Z. Naturforsch.*, 1974, **29b**, 807.
[421] Y. S. Yang and J. S. Shirk, *J. Mol. Spectroscopy*, 1975, **24**, 39.
[422] B. Gilbert, G. Mamontov, and G. M. Begun, *Inorg. Nuclear Chem. Letters*, 1974, **10**, 1123.
[423] B. Gilbert, G. Mamontov, and G. M. Begun, *J. Chem. Phys.*, 1975, **62**, 950.
[424] P. Bukonvec and J. Šiftar, *Monatsh.*, 1975, **106**, 483.

$K_2NaAlF_6$, show that both are face-centred cubic, with the likely space group $Fm3m$, and regular octahedral $AlF_6^{3-}$ anions.[425]

Ab initio MO calculations of the ground-state geometry of $AlCl_3$ yield a value for the Al—Cl bond distance of 2.13 Å, with a preferred planar arrangement. This is in good agreement with experiment (2.06 Å, $\angle ClAlCl = 118 \pm 1.5°$).[426] The emission spectrum of an $AlCl_3$ arc plasma (core temperatures in the range 5600—6150 K) has been analysed.[427]

A new study of the $AlCl_3$–NaCl system (by d.t.a., visual observation, and X-ray powder diffraction of the crystalline phases isolated) shows that $NaAlCl_4$ is formed, melting incongruently at $153 \pm 0.5$ °C.[428]

Pyrex diaphragm and transpiration studies reveal that $NH_4AlCl_4$ vaporizes with simultaneous dissociation into $NH_3AlCl_3$ and HCl. Thermodynamic constants for the various vaporization processes from the liquid and solid phases were derived.[429]

Chloroaluminate melts provide a suitable medium for the production of positive oxidation states of sulphur, selenium, and iodine, by electrochemical oxidation.[430]

No evidence of complex formation involving chloroalanes was found in the reactions of $AlCl_3$ with $LiAlH_4$ in $Et_2O$ at 20 °C.[431]

The crystal structure of $K(MeAlCl_3)$ shows that it forms orthorhombic crystals, space group $Pnma$. Both $K^+$ and the anion lie on the crystallographic mirror plane; the two independent Al—Cl bond lengths are 2.16 and 2.17 Å.[432]

Heating $AlCl_3$ with phosphoric acid in ethanol forms a tetrameric aluminium phosphate, of empirical formula $Al(PO_4)(HCl)(EtOH)_4$, which is water soluble and gives $AlPO_4$ on heating. It forms tetragonal crystals, space group $I\bar{4}$, built up from tetrameric units which are approximately cubic. The Al atoms are six-co-ordinate, being attached to three oxygens from EtOH molecules, and three from $PO_4$ groups.[433]

Adducts of the series $(Me_xH_{3-x}Si)_3P,AlCl_3$ ($x = 1$ or 2) are formed by reactions (46).

$$3Me_xH_{3-x}SiPH_2 + AlCl_3 \rightarrow (Me_xH_{3-x}Si)_3P,AlCl_3 + 2PH_3 \qquad (46)$$

$Me_3SiPEt_2$ and $Me_xH_{3-x}SiPMe_2$ form adducts with $AlCl_3$ which decompose on heating to give, e.g., $(Cl_2AlPMe_2)_3$.[434]

Raman spectra of a number of melts in the $KCl$–$AlCl_3$–$TeCl_4$ system can be interpreted on the basis of a number of different tellurium chloro-complexes, but the only Al-containing species produced appears to be $AlCl_4^-$.[435]

Preliminary single-crystal data confirm the formula $2AlCl_3,3POCl_3$ for the

---

[425] L. R. Morss, J. Inorg. Nuclear Chem., 1974, **36**, 3876.
[426] S. P. So and W. G. Richards, Chem. Phys. Letters, 1975, **32**, 231.
[427] G. Mück, Z. Naturforsch., 1974, **29a**, 1643.
[428] E. M. Levin, J. F. Kinney, R. D. Wells, and J. T. Benedict, J. Res. Nat. Bur. Standards, 1974, **78A**, 505.
[429] W. C. Laughlin and N. W. Gregory, Inorg. Chem., 1975, **14**, 1263.
[430] R. Marassi, G. Mamantov, and J. Q. Chambers, Inorg. Nuclear Chem. Letters, 1975, **11**, 245.
[431] V. M. Mikheeva and N. N. Mal'tseva, Russ. J. Inorg. Chem., 1974, **19**, 140.
[432] J. L. Atwood, D. C. Hrncir, and W. R. Newberry, Cryst. Structure Comm. 1974, **3**, 615.
[433] J. E. Cassidy, J. A. J. Jarvis, and R. N. Rothan, J.C.S. Dalton, 1975, 1497.
[434] G. Fritz and R. Emül, Z. anorg. Chem., 1975, **416**, 19.
[435] F. W. Poulsen, N. J. Bjerrum, and O. F. Nielsen, Inorg. Chem., 1974, **13**, 2693.

adduct, melting at 166 °C, obtained in addition to the well-known 1:1 adduct in the $AlCl_3$–$POCl_3$ system.[436]

N.q.r. ($^{35}Cl$ and $^{127}I$) and n.m.r. ($^{27}Al$) measurements on the species (derived from $ICl$–$AlCl_3$ mixtures) $I_3AlCl_4$, $I_5AlCl_4$, and $I_2ClAlCl_4$, reveal that $AlCl_4^-$ is indeed present in all of them.[437]

Using previously published data on related systems, Semenenko et al. have given assignments for most of the vibrational wavenumbers of $Al(BH_4)_nCl_{3-n}{,}OEt_2$ ($n = 1, 2,$ or $3$) from their i.r. and Raman spectra.[438] $^1H$ and $^{11}B$ n.m.r. data for these compounds, and related adducts of THF have also been reported.[439] The THF complexes underwent slow decomposition in solution, with the formation of butoxy-compounds.

The crystal structure of (71) shows that the $AlCl_3$ group is co-ordinated, via a σ-bond, to a non-planar four-membered ring. The C—C bond lengths confirm this formulation.[440]

```
        Me         Me
          \       /
           +-----+
           |     |-----AlCl_3^-
           | +   |
           +-----+
          /       \
        Me         Me
```

(71)

Phase diagrams have been constructed for the systems $AlCl_3$–$TeCl_4$, $AlCl_3$–$NaCl$–$TeCl_4$;[441] $AlCl_3$–$SbCl_3$, $GaCl_3$–$SbCl_3$;[442] and $HCl$–$AlCl_3$, $Et_2O$–$Et_2O$.[443]

Tensimetric studies on the systems $AlCl_3$–$MCl_n$ (M = Cr, $n = 3$; M = Co, Zn, Cu, Mg, or Mn, $n = 2$), containing 0.02—0.05 mass% of $MCl_n$, show that the volatility of the impurity is in the sequence Cr < Mg < Mn < Co < Cu ≤ Zn.[444]

Chemical transport studies on Cu, Ag, Au, Rh, Ru, Pd, Ir, or Pt in the presence of $Al_2Cl_6$ or $Al_2I_6$ show that stable gaseous complexes are formed, if HCl, $Cl_2$, or $I_2$ are present. The function of the last is to produce an initial amount of metal halide.[445]

Thermodynamic data have been compiled and discussed for a variety of gaseous $MCl_m{,}xAlCl_3$ complexes (for many different M), most of which are very stable below 1200 K.[446]

Spectrophotometric measurements showed the presence of the gaseous complexes $V^{II}Al_3Cl_{11}$, $V^{III}Al_2Cl_9$, and $V^{III}AlCl_6$ in mixtures of $VCl_2$ or $VCl_3$ with $AlCl_3$.[447]

---

[436] P. H. Collons, J. Appl. Cryst., 1975, **8**, 337.
[437] D. J. Merryman, J. D. Corbett, and P. A. Edwards, Inorg. Chem., 1975, **14**, 428.
[438] K. N. Semenenko, Yu. Ya. Kharitonov, V. B. Polyakova, and V. N. Fokin, Russ. J. Inorg. Chem., 1975, **20**, 19.
[439] G. N. Boiko, V. N. Fokin, Yu. I. Malov, and K. N. Semenenko, Russ. J. Inorg. Chem., 1974, **19**, 1644.
[440] C. Krüger, P. J. Roberts, Y.-H. Tsay, and J. B. Koster, J. Organometallic Chem., 1974, **78**, 69.
[441] V. V. Safonov, A. V. Konov, L. N. Myl'nikova, and B. G. Korshunov, Russ. J. Inorg. Chem., 1974, **19**, 446.
[442] L. A. Nisel'son, Z. N. Orshanskaya, and K. V. Tret'yakov, Russ. J. Inorg. Chem., 1974, **19**, 580.
[443] K. N. Semenenko, E. A. Lavut, and A. P. Isaeva, Russ. J. Inorg. Chem., 1975, **20**, 449.
[444] T. N. Naumova, I. G. Bykova, K. N. Gol'toova, A. F. Yakovleva, and B. D. Stepin, Russ. J. Inorg. Chem., 1975, **20**, 22.
[445] H. Schäfer and M. Trenkel, Z. anorg. Chem., 1975, **414**, 137.
[446] H. Schäfer, Z. anorg. Chem., 1975, **414**, 151.
[447] A. Anundskås and H. A. Øye, J. Inorg. Nuclear Chem., 1975, **37**, 1609.

E.s.r. (at room and glass temperatures) and u.v. spectroscopic studies on mixtures of dichlorobis($\eta$-cyclopentadienyl)vanadium and ethylaluminium chloride indicate that a number of species are formed, some of which are active catalysts for ethylene polymerization. The most detailed results were obtained for (the catalytically inactive) species (72).[448]

Electronic absorption spectra suggest that vapour mixtures of $Fe_2Cl_6$ and $Al_2Cl_6$ contain the species $FeAlCl_6$.[449]

The reaction of solid $CoCl_2$ and gaseous $Al_2Cl_6$ may be followed spectrophotometrically: data concerning the range 600—800 K were rationalized in terms of reaction (47).

$$CoCl_2 + Al_2Cl_6 \rightarrow Co(AlCl_4)_2. \qquad (47)$$

This paper indicated that the structure of the product is (73), i.e. the Co is octahedrally co-ordinated, being triply-bridged to Al.[450] A second group of workers, however, on the basis of optical spectroscopy, concluded that the structure is (74), with tetrahedrally co-ordinated Co. This report also mentions a second complex produced in the same reaction, viz. $Co(Al_2Cl_7)_2$, formulated as (75).[451] A third investigation[452] also reported the existence of $CoAl_2Cl_8$. Evidence was presented in this case for a structure in which the Co was co-ordinated by six Cl atoms in a distorted octahedral arrangement.[452]

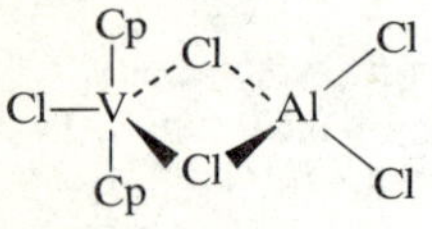

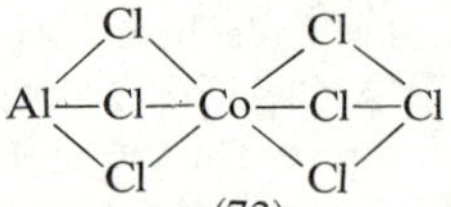

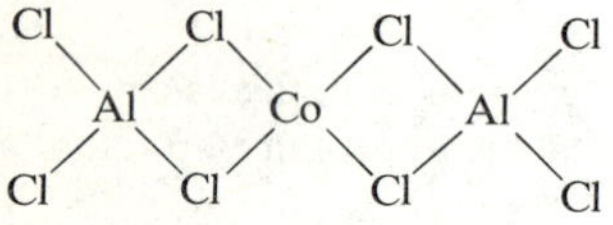

The processes of vaporization and dissociation in the vapour of the adducts $AlBr_3,NH_3$ and $AlI_3,NH_3$ have been studied using a membrane null-manometer;[453] the latter compound undergoes ammonolysis on heating.

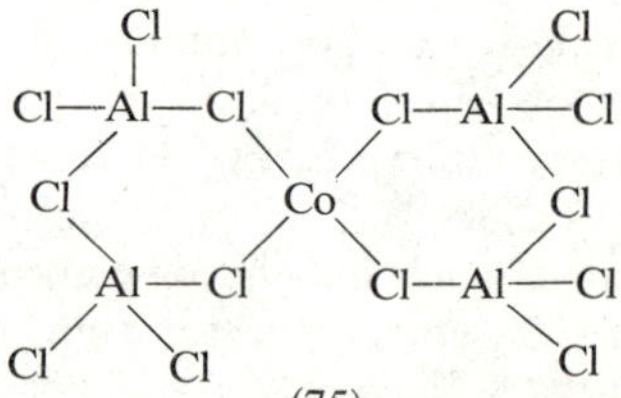

N.q.r. spectra and Zeeman effects of $AlX_3,SbX_3$ (X = Br or I) and $AlBr_3, BiBr_3$ show that the first two are built up of $AlX_4^-$ and $SbX_2^+$ units (with weak additional bonds between the anions and the cations). The data for $AlBr_3, BiBr_3$ did not give such a clear-cut result, however.[454]

---

[448] A. G. Evans, J. C. Evans, and E. H. Monn, J.C.S. Dalton, 1974, 2390.
[449] C.-F. Shieh and N. W. Gregory, J. Phys. Chem., 1975, **79**, 828.
[450] G. N. Papatheodorou. Z. anorg. Chem., 1975, **411**, 153.
[451] A. Dell'Anna and F. F. Emmenegger, Helv. Chim. Acta, 1975, **58**, 1145.
[452] P. J. Thistlethwaite and S. Ciach, Inorg. Chem., 1975, **14**, 1430.
[453] V. I. Trusov and A. V. Suvorov, Russ. J. Chem., 1974, **19**, 1781.
[454] T. Okuda, K. Yamada, H. Ishihara, and H. Negita, Chem. Letters, 1975, 785.

# Elements of Group III

N.q.r. (Br) measurements have been recorded for the sulphone complexes $R_2SO_2,AlBr_3$ and $R_2SO_2,2AlBr_3$, in which co-ordination takes place *via* the oxygen atoms. In all cases there is a low frequency shift (compared with $Al_2Br_6$) which can be related to the enthalpy of formation of the complexes.[455]

$NH_4Al_2Br_7$ forms orthorhombic crystals, space group $Pna2_1$. The anion is made up of two $AlBr_4$ tetrahedra sharing one corner, with a staggered conformation and bent Al—Br—Al bridge (angle = 107.7°). The overall symmetry of the anion is close to $C_{2v}$.[456]

$AlI_3$–ether (ether = THF, tetrahydropyran, oxetan, 2-MeTHF, $Et_2O$, $Bu^n_2O$, anisole, or phenetole), and $AlI_3,SEt_2$ adducts may be prepared by direct reaction of the components. $^1H$ n.m.r. parameters were listed.[457]

## 3 Gallium

**General.**—The surface tension of high-purity liquid gallium has the following values: 723.2, 724.6, 723.9, 723.7, and 725.0 dyn cm$^{-1}$ at 40, 30, 29.6, 25, and 17 °C respectively.[458]

Extraction of micro-amounts of $Ga^{III}$ by a number of reagents (butyl acetate, isobutyl methyl ketone, *etc.*) is suppressed by the addition of macro-amounts of $Sb^{III}$ or $Sb^V$. The extent of the suppression increases with increasing dielectric constant of the organic solvent.[459]

**Gallium Hydrides.**—I.r. and Raman spectra of $MGaH_4$ (M = Na, K, Rb, or Cs) and of $NaGaD_4$, all in diglyme solution, are consistent with $T_d$ symmetry for the anion. The observed wavenumbers are summarized in Table 11. The solid-phase spectra showed many more bands, due to crystal effects. The magnitudes of these were dependent on the nature of the cation.[460]

**Table 11** *Vibrational assignments*/cm$^{-1}$ *for* $NaGaH_4$ *and* $NaGaD_4$

|  | NaGaH$_4$ | | NaGaD$_4$ | |
| --- | --- | --- | --- | --- |
|  | I.r. | Raman | I.r. | Raman |
| $\nu_1(A_1)$ |  | 1752 |  | 1260 |
| $\nu_2(E)$ |  | *a* |  | 566 |
| $\nu_3(T_2)$ | 1700 | 1700 | 1230 | 1230 |
| $\nu_4(T_2)$ | 733 | *a* | 533 | 540 |

*a* Obscured by solvent bands

Crystals of 11$H$, 22$H$-10,10,21,21-tetramethyl-3,5,14,16,-tetra($NN$-dimethyl-2-aminoethyl)-1,12-dioxonia-3,5,7,10,14,16,18,21-octazonia-2,4,6,11,13,15,17,22-octagallanato-nonacyclo[13,7,0,0,$^{2,7}$ 0$^{3,22}$, 0$^{4,12}$, 0$^{7,11}$, 0$^{11,14}$, 0$^{13,18}$, 0$^{18,22}$]-docosane, a novel polycyclic cage compound, $(GaH)_6(GaH_2)_2(\mu_3\text{-O})_2(\mu_3\text{-NCH}_2\text{-} CH_2NMe_2)_4(\mu\text{-NHCH}_2CH_2NMe_2)_2$, are triclinic, space group $P\bar{1}$. The structure

---

[455] E. N. Gur'yanova, V. V. Puchkova, and A. F. Volkov, *J. Struct. Chem.*, 1974, **15**, 374.
[456] E. Rytter, B. E. D. Rytter, H. A. Øye, and J. Krogh-Moe, *Acta Cryst.*, 1975, **B31**, 2177.
[457] P. J. Ogren, L. Steenhoek, K. S. Greve, and W. C. Hutton, *J. Inorg. Nuclear Chem.*, 1975, **37**, 293.
[458] G. J. Abbaschian, *J. Less-Common Metals*, 1975, **40**, 329.
[459] Yu. A. Zolotov, V. S. Vlasov, N. M. Kuz'min, and V. A. Fedorov, *Russ. J. Inorg. Chem.* 1975, **20**, 256.
[460] A. P. Kurbakova, L. A. Leites, V. V. Gavrilenko, Yu. N. Karaksin, and L. I. Zakharkin, *Spectrochim. Acta*, 1975, **31A**, 281.

provides the first example of a four-membered $Ga_2NO$ ring, and six-membered $Ga_3N_2O$ rings. Three of the crystallographically independent Ga atoms are four-co-ordinate, the fourth is five-co-ordinate.[461]

Investigation of hydrometallation reactions of $\alpha$-olefins with alkali-metal tetrahydrogallates have shown that reaction is very slight in ether solvents. In the presence of $GaEt_3$, however, the sodium and potassium salts react readily when the solvent is a hydrocarbon, producing tetra-alkylgallates.[462]

**Compounds containing Ga—C Bonds.**—Okawara has reviewed the 1973 literature on organo-derivatives of Ga and In.[463]

N.m.r. studies on solutions of $GaMe_3$ in benzene can be interpreted as showing that a $\pi$-complex is formed.[464]

A study of the kinetics of ligand exchange in the systems $Et_3Ga + Me_3Ga,NEt_3$ and $Me_3Ga + Me_3Ga,NEt_3$ suggests that the mechanism is primarily dissociative in character.[465]

Tri-t-butylgallium can be prepared in 50—60% yield by the reaction of $Bu^tMgCl$ with $GaBr_3$ in ethereal solution. It is photosensitive, and reacts with $GaX_3$ to give $Bu^t_2GaX$ (X = Cl, Br, etc.). I.r. bands due to $\nu(GaC_2)$ were reported at 543 and 512 $cm^{-1}$.[466]

Reactions of $Me_3P=CH_2$ with $Me_2MCl$ (M = Ga, In, or Tl) form the new dimethylmetaldimethylphosphonium-bis(methylides), $[Me_2MCH_2PMe_2CH_2]_n$. Dimers ($n = 2$) may be isolated, and these possess eight-membered ring structures (76). This structure stabilizes the inherently labile thallium(III) derivative.[467]

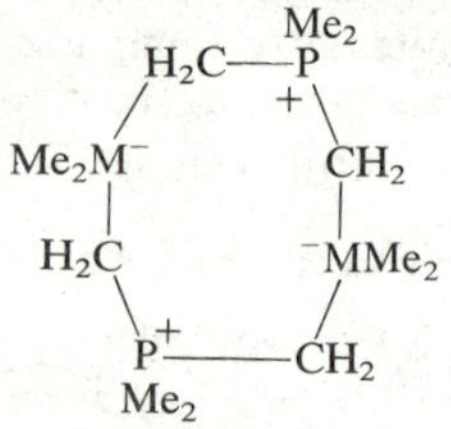

(76)

**Compounds containing Ga—N, Ga—As, or Ga—Sb Bonds.**—The enthalpies of dissociation of $GaX_3,NH_3$ (X = Cl or Br) have been calculated from thermodynamic studies on the vapours of these adducts. They are $(32.1 \pm 0.2)$ kcal $mol^{-1}$ (Cl) and $(32.8 \pm 0.2)$ kcal $mol^{-1}$ (Br). The entropies of dissociation ($\Delta S°_{t,diss}$) are $33.8 \pm 0.6$ and $37.7 \pm 0.6$ e.u., respectively.[468]

$(Me_4N)[Ga(NCS)_4]$ and $(R_4N)_3[Ga(NCS)_6]$ (R = Me, Et, or Bu) may be isolated

---

[461] S. J. Rettig, A. Storr, and J. Trotter, Canad. J. Chem., 1975, **53**, 753.
[462] V. V. Gavrilenko, V. S. Kolesov, and L. I. Zakharkin, J. Gen. Chem. (U.S.S.R.), 1974, **44**, 1867.
[463] R. Okawara, J. Organometallic Chem., 1974, **83**, 55.
[464] G. M. Gusakov, B. I. Kozyrkin, B. G. Gribov, and E. N. Zorina, Doklady Chem., 1974, **215**, 168.
[465] G. M. Gusakov, B. G. Gribov, B. P. Kozyrkin, N. E. Kulagin, and G. K. Chirkiu, Doklady Chem., 1975, **220**, 55.
[466] H.-U. Schwering, E. Jungk, and J. Weidlein, J. Organometallic Chem., 1975, **91**, C4.
[467] H. Schmidbaur and H.-J. Hoffmann, Chem. Ber., 1974, **107**, 3674.
[468] V. I. Trusov, A. V. Suvorov, and R. N. Abdkumova, Russ. J. Inorg. Chem., 1975, **20**, 278.

*Elements of Group III* 131

from mixtures of $(R_4N)NCS$ and $Ga(NCS)_3$ in aqueous solutions at pH 2.[469] The same species (R = Me only) have been isolated from solutions in absolute methanol. Their i.r. spectra are consistent with Ga—N co-ordination, and $\nu(GaN)$ was assigned to a band at 350 cm$^{-1}$ [Ga(NCS)$_4^-$] or 250 cm$^{-1}$ [Ga(NCS)$_6^{3-}$].[470]

$Ga(NMe_2)_3$, $ClGa(NMe_2)_2$, and $Ga[N(SiMe_3)_2]_3$ are formed by reactions of $GaCl_3$ with $LiNMe_2$ or $NaN(SiMe_3)_2$, respectively. An excess of $LiNMe_2$ leads to the formation of $LiGa(NMe_2)_4$. this reacts with further $GaCl_3$ to reform $Ga(NMe_2)_3$. The latter is apparently dimerized *via* $NMe_2$ bridges, whereas the $N(SiMe_3)_2$ analogue remains monomeric, because of steric effects.[471]

The instability constant of $[Ga(NCS)]^{2+}$ has been determined from the decrease in the polarographic catalytic current of the thiocyanato-complexes of Ti$^{IV}$ and In$^{III}$ in the presence of Ga$^{III}$: $K_1^\circ = (4.8 \pm 0.6) \times 10^{-3}$ at 25 °C.[472]

Complexes $Ga(NCS)_3$,Am,MeOH, $Ga(NCS)_3$,1.5Am, and $Ga(NCS)_3$,2Am have been isolated from MeOH solutions and $Ga(NCS)_3$,Am,H$_2$O and AmH[Ga(NCS)$_4$Am] from 1:1 H$_2$O:EtOH solutions (Am = bipy or phen). $Ga(NCS)_3$,Am can be produced by keeping the MeOH or H$_2$O adducts *in vacuo* at 120 °C.[473] Tabulations of i.r. data for all of these species[474] suggest that Ga—NCS bonds are present, with the bidentate Am also co-ordinated to the metal.

Crystals of the *NN*-dimethylethanolaminogallane dimer (77) are monoclinic, space group $P2_1/c$, whereas the compound with R = Me is orthorhombic, space group *Pccn*. However, both contain well-separated centrosymmetric dimers, having $Ga_2O_2$ rings.[475]

Dimethylbis(pyrazol-1-yl)- and bis(3,5-dimethylpyrazol-1-yl)-dimethylgallate ions act as bidentate chelating agents towards transition-metal ions.[476] The $[Me_2B(pz)_2]^-$ ion acts similarly, but the Al and In analogues do not appear to form transition metal complexes. The crystal structure of (78), has been determined. The space group is $P2_1/c$. The whole molecule is in a pseudo-chair conformation, with two six-membered Ga—(N—N)$_2$—Ni rings in boat conformations. The co-ordination at the Ni atom is planar.[477]

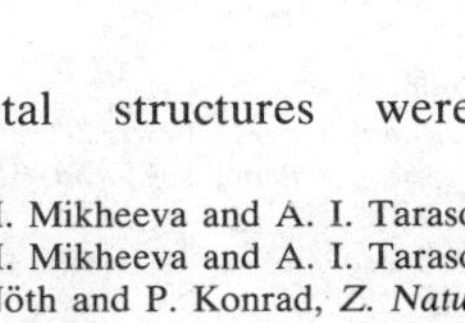

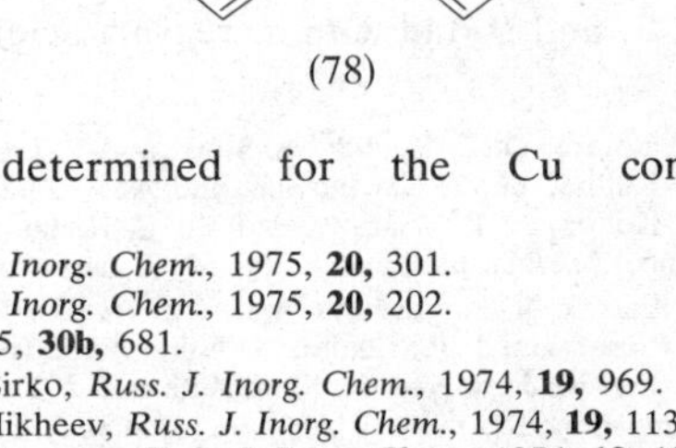

Crystal structures were also determined for the Cu complexes

[469] L. M. Mikheeva and A. I. Tarasova, *Russ. J. Inorg. Chem.*, 1975, **20**, 301.
[470] L. M. Mikheeva and A. I. Tarasova, *Russ. J. Inorg. Chem.*, 1975, **20**, 202.
[471] H. Nöth and P. Konrad, *Z. Naturforsch.*, 1975, **30b**, 681.
[472] Ya. I. Tur'yan, L. M. Makarova, and V. N. Sirko, *Russ. J. Inorg. Chem.*, 1974, **19**, 969.
[473] L. M. Mikheeva, A. I. Tarasova, and N. V. Mikheev, *Russ. J. Inorg. Chem.*, 1974, **19**, 1131.
[474] L. M. Mikheeva, A. I. Grigor'ev, and A. I. Tarasova, *Russ. J. Inorg. Chem.*, 1974, **19**, 1277.
[475] S. J. Rettig, A. Storr, and J. Trotter, *Canad. J. Chem.*, 1975, **53**, 58.
[476] K. R. Breakell, D. J. Patmore, and A. Storr, *J.C.S. Dalton*, 1975, 749.
[477] D. F. Rendle, A. Storr, and J. Trotter, *J.C.S. Dalton*, 1975, 176.

{[Me$_2$Ga(pz)$_2$]$_2$Cu} and {[Me$_2$Ga(dmpz$_2$)$_2$]$_2$Cu} (pz = pyrazolyl and dmpz = 3,5-dimethylpyrazolyl).[478] These show that for the former, the Cu has approximately planar co-ordination, with the Ga—(N—N)$_2$—Cu rings in the boat conformation. Steric requirements for the latter, however, force a pseudo-tetrahedral geometry on the Cu, with the chelate rings being almost planar.

The formation constant, $\beta_1$, for Ga$^{III}$–bipy complex is almost independent of [H$^+$], but for the analogous phen species, $\beta_1$, and $\beta_2$ increase at lower [H$^+$], probably owing to the formation of mixed hydroxoamine complexes.[479] Mixed complexes of Ga$^{III}$ with bipy or phen have been detected for a variety of acido-ligands (anions of oxalic, tartaric, or malonic acids *etc.*).[480]

Complexing of Ga$^{3+}$ by glycine or isoleucine has been studied by i.r. spectrometry and potentiometry. With both ligands, complexes are formed by the ligand in the zwitterionic form (GaL$^{3+}_{+/-}$) and as an anion (GaL$^{2+}_{-}$).[481]

A complex (with co-ordination of the piperidine N and the neighbouring carbonyl O) of GaBr$_3$ with 6-piperidino-*N*-methylanthrapyridone has been prepared and characterized (electronic spectra).[482]

Gallium Schiff-base complexes Ga(OPr$^i$) (SB) and Ga$_2$(SB)$_3$ (SB$^{2-}$ is anion of a bifunctional, terdentate Schiff base containing the donor system O—N—O) have been prepared. Some spectroscopic data were recorded for characterization.[483]

Two new phases, Ca$_5$Ga$_2$As$_6$ and Ca$_4$Ga$_3$As$_5$, have been produced by heating together appropriate mixtures of the binary chalcogenides at 1000 °C.[484]

Excess molar enthalpies of liquid Ga–Sb alloys ware found at 995 K. The enthalpy of formation of GaSb at 978 K is $-(7300 \pm 300)$ cal$_{th}$ mol$^{-1}$.[485]

Phase relationships were studied in the InSb–GaSb system.[486] Evidence was found for the formation of In$_2$GaSb$_3$ and InGa$_2$Sb$_3$.

**Compounds containing Ga—O, Ga—S, or Ga—Se Bonds.**—The X-ray photoelectron and X-ray spectra of gallium oxides, sulphides, and phosphides can only be interpreted adequately if allowance is made for the participation of Ga 3$d$ orbitals in the bonding.[487]

The dialkylmetal alkoxides R$^1_2$MOR$^2$, (R$^1$ = CH$_3$, CD$_3$, or Et; R$^2$ = CH$_3$ or CD$_3$; M = Ga or In) have been prepared by standard methods, and their i.r. and Raman spectra recorded. The methyl derivatives are trimeric, and their vibrational spectra are consistent with the presence of puckered M$_3$O$_3$ ring systems. Some i.r. and Raman data were also recorded for MeClGa(OMe).[488]

Ga$^{III}$ oxalate complexes (1:1, 1:2, or 1:3) are formed over a wide range of pH, but $\beta_1$ and $\beta_2$ fall with increasing acidity of the solution owing, it is believed,

---

[478] D. J. Patmore, D. F. Rendle, A. Storr, and J. Trotter, *J.C.S. Dalton*, 1975, 718.
[479] F. Ya. Kul'ba, Yu. S. Kananovich, and A. P. Zharkov, *Russ. J. Inorg. Chem.*, 1975, **20**, 345.
[480] F. Ya. Kul'ba, A. P. Zharkov, and Yu. S. Kananovich, *Russ. J. Inorg. Chem.*, 1975, **20**, 357.
[481] P. Bianco, J. Haladjian, and R. Pilard, *J. Less-Common Metals*, 1975, **42**, 127.
[482] B. E. Zaitsev, N. P. Vasil'eva, and B. N. Ivanov-Emin, *Russ. J. Inorg. Chem.*, 1974, **19**, 1584.
[483] R. N. Prasad and J. P. Tandon, *J. Inorg. Nuclear Chem.*, 1975, **37**, 35.
[484] P. Verdier, M. Maunaye, R. Marchand, and J. Lang, *Compt. rend.*, 1975, **281**, C, 457.
[485] M. Gambiro and J.-P. Bros, *J. Chem. Thermodynamics*, 1975, **7**, 443.
[486] V. M. Glazov, A. S. Timoshin, and V. B. Ufintsev, *Doklady Chem.*, 1974, **218**, 696.
[487] V. I. Nefedov, Ya. V. Salin, E. P. Domashevskaya, Ya. A. Ugai, and V. A. Terekhov, *J. Electron Spectroscopy*, 1975, **6**, 231.
[488] G. Mann, H. Olapinski, R. Ott, and J. Weidlein, *Z. anorg. Chem.*, 1974, **410**, 195.

to the formation of hydroxo-oxalato complexes of composition $Ga(OH)_i(C_2O_4)^{(1-i)+}$.[489]

Ligand exchange with $Ga(acac)_3$ is thought to take place *via* a dissociative mechanism involving a unidentate ligand. This conclusion followed from a kinetic study of the rate of exchange between the complex and $^{14}C$-labelled acac in THF. The reaction is catalysed by trichloroacetic acid. The rate of exchange for the series of Group III acac complexes is in the sequence: $In > Ga > Al$.[490]

Spectrophotometric studies of the interactions in solutions containing $Ga^{3+}$ and acetylacetonate ions show that at acetylacetone:Ga ratios less than 3, hydrolysis equilibra leading to the formation of hydroxo-complexes are superimposed on the complex-formation equilibria of $Ga^{3+}$.[491]

$^1H$ and $^{71}Ga$ n.m.r. studies, together with equilibrium dialysis of aqueous gallium citrate solutions, show that at low pH, low-molecular weight complexes (Ga:citrate = 1:1) are present. At pH 2—6 polymerization occurred, although in alkaline solutions $Ga(OH)_4^-$ was the main species present.[492]

$\nu(GaO)$ bands have been assigned from the i.r. spectrum of $LaGaO_3$. $\nu_1(T_{1u})$ is at $610 \text{ cm}^{-1}$, with the other stretches at $400 \text{ cm}^{-1}$, and a deformation at $183 \text{ cm}^{-1}$.[493]

Non-empirical MO calculations (neglect of diatomic differential overlap) have been carried out on $GaO_4^{5-}$ and $GaO_3^{3-}$.[494]

Heating $Na_2O + Ga_2O_3$ for 1—2 days at 600—700 °C forms the new species $Na_5GaO_4$, which has orthorhombic crystals, belonging to the space group *Pbca*. The structure is built up from isolated $GaO_4^{5-}$ tetrahedra.[495]

$Ga_2TiO_5$ also forms orthorhombic crystals, the space group here being *Bbmm*. It is isotypic with pseudobrookite, with a statistical distribution of metal positions. The $Ga^{3+}$ is surrounded by a distorted octahedral arrangement of O atoms.[496]

$Ca_2FeGaO_5$ and $Ca_2Fe_{0.5}Ga_{1.5}O_5$ crystallize in the orthorhombic space group *Pcmna*. They are isotypic with $Ca_2Fe_2O_5$, rather than $Ca_2FeAlO_5$ (brownmillerite).[497]

Neutron-diffraction studies on copper gallate, $CuGa_2O_4$, suggest that the cation distribution is as follows: $Cu_{0.21}Ga_{0.79}(Cu_{0.79}Ga_{1.21})O_4$.[498]

The crystal structures of the synthetic feldspars $SrGa_2Si_2O_8$ and $BaGa_2Si_2O_8$ are both similar to that of celsian. There is complete Ga–Si ordering in the former, and probably in the latter also.[499]

Instability constants have been measured for molybdate and tungstate heteropolycomplexes of gallium. These complexes are very stable, with, however, decreasing stability in the series: $GaM_{12} > GaM_9 > GaM_6 > GaM_3$.[500]

---

[489] F. Ya. Kul'ba, N. A. Babkina, and A. P. Zharkov, *Russ. J. Inorg. Chem.*, 1974, **19**, 365.
[490] C. Chattergee, K. Matsuzawa, H. Kido, and K. Saito, *Bull. Chem. Soc. Japan*, 1974, **47**, 2809.
[491] J. Haladjian, P. Bianco, and R. Pilard, *J. chim. Phys.*, 1974, **71**, 1251.
[492] J. D. Glickson, T. P. Pitner, J. Webb, and R. A. Gams, *J. Amer. Chem. Soc.*, 1975, **97**, 1679.
[493] E. J. Baran, *Z. Naturforsch.*, 1975, **30b**, 136.
[494] B. V. Shchegolev and M. E. Dyatkina, *J. Struct. Chem.*, 1974, **15**, 302.
[495] D. Fink and R. Hoppe, *Z. anorg. Chem.*, 1975, **414**, 193.
[496] H. Müller-Buschbaum and H.-R. Freund, *Z. Naturforsch.*, 1974, **29b**, 590.
[497] R. Arpe, R. von Schenk, and H. Müller-Buschbaum, *Z. anorg. Chem.*, 1974, **410**, 97.
[498] J. Lopitaux and M. Lenglet, *J. Inorg. Nuclear Chem.*, 1975, **37**, 332.
[499] M. Calleri and G. Gazzoni, *Acta Cryst.*, 1975, **B31**, 560.
[500] L. P. Tsyganok and T. V. Kleinerman, *Russ. J. Inorg. Chem.*, 1974, **19**, 1225.

GaO reacts at 850 °C with a stream of $H_2S$ gas to give colourless crystals of $CsGaS_2$. These are isostructural with $RbFeS_2$, being monoclinic, space group $C2/c$.[501]

Phase studies have been reported for the systems $Ga_2S_3$–$GeS_2$[502] and Ga–Ge–Se.[503]

$MMo_2S_8$ (M = Al or Ga) and $GaMo_4Se_8$ from cubic crystals, space group $F\bar{4}3m$. The $GaMo_4S_8$ crystal structure shows that the S atoms form a f.c.c. array, with the Ga atoms at tetrahedral, the Mo atoms at octahedral sites.[504]

**Gallium Halides.**—A mixture of $Ga(ClO_4)_3$ and HF, in the range $0.1 < [H^+] < 0.5$, forms complex ions $GaF^{2+}$ and $GaHF^{3+}$. The negative logarithms of their instability constants are (at $I = 1.0$, $T = 25\ °C$) $3.33 \pm 0.05$ and $1.79 \pm 0.05$, respectively.[505]

$^{19}F$, $^{69}Ga$, and $^{71}Ga$ n.m.r. spectra of $Ga^{III}$ in aqueous HF or $NH_4F$ showed the formation of $GaF^{2+}_{aq}$, $GaF^{+}_{2(aq)}$, $GaF_{3(aq)}$, and $GaF^{-}_{4(aq)}$. As it was not possible to detect free $F^-$, only relative formation constants could be evaluated. There was evidence for six-co-ordination of the Ga. Results of ligand competition reactions were also presented.[506]

$LiMnGaF_6$ crystallizes with the $Na_2SiF_6$ structure, being trigonal, and belonging to the space group $P312$.[507]

Chloride-exchange processes in $Ga^{III}$ species in concentrated aqueous chloride solutions were studied using $^{35}Cl$ and $^{71}Ga$ n.m.r. The results may be rationalized in terms of an associative exchange of $Cl^-$ at $GaCl_4^-$. Values were obtained for the enthalpy and entropy of activation.[508]

A systematic study of the formation of complex chlorides in binary systems containing $MCl_3$ (M = Ga or In) or InCl has been made.[509]

$^{69}Ga$ and $^{71}Ga$ n.m.r. spectra of acetonitrile solutions of mixtures of gallium halides have been analysed to determine the disproportionation constants of $GaX_3Y^-$, $GaX_2Y_2^-$, and $GaX_3Y^-$ (X, Y = Cl, Br, I).[510]

The Raman spectra of $GaCl_3Br^-$ and $GaBr_3Cl^-$ have been obtained. They can be analysed assuming $C_{3v}$ symmetry for the ions (Table 12). They were produced by mixing methanolic solutions of $GaCl_3$ or $GaBr_3$ with the appropriate $R_4E^+X^-$ salt.[511]

Solutions of the gallium halides (Cl, Br, or I) in acetonitrile, benzene, or nitromethane, together with some binary solvent mixtures, have been investigated by $^{69}Ga$, $^{71}Ga$, and $^1H$ n.m.r. spectroscopy. A major conclusion is that the solvation numbers of the $Ga^{3+}$ ion increase on passing from chloride to iodide; in

---

[501] D. Schmitz and W. Bronger, *Z. Naturforsch.*, 1975, **30b**, 491.
[502] A.-M. Loireau-Lozac'h and M. Guittard, *Ann. Chim. (France)*, 1975, **10**, 101.
[503] C. Thiebault, L. Guen, R. Eholie, and J. Flahaut, *Bull. Soc. chim. France*, 1975, 967.
[504] C. Perrin, R. Chevrel, and M. Sergent, *Compt. rend.*, 1975, **280**, C, 949.
[505] Yu. I. Mikhailyuk and V. I. Gordienko, *Russ. J. Inorg. Chem.*, 1974, **19**, 1114.
[506] Yu. A. Buslaev, S. P. Petrosyants, and V. P. Tarasov, *J. Struct. Chem.*, 1974, **15**, 187.
[507] W. Viebahn, *Z. anorg. Chem.*, 1975, **413**, 77.
[508] S. F. Lincoln, A. C. Sandercock, and D. R. Stranks. *J.C.S. Dalton*, 1975, 669.
[509] P. I. Fedorov and P. P. Fedorov, *Russ. J. Inorg. Chem.*, 1974, **19**, 117.
[510] Yu. A. Buslaev, V. P. Tarasov, N. N. Mel'nikov, and S. P. Petrosyants, *Doklady Chem.*, 1974, **216**, 370.
[511] R. Rafaeloff and A. Silberstein-Hirsch, *Spectrochim. Acta*, 1975, **31A**, 183.

**Table 12** Vibrational assignments/cm$^{-1}$ for the ions GaCl$_3$Br$^-$ and GaBr$_3$Cl$^-$

|  | (C$_4$H$_9$)$_4$P$^+$GaCl$_3$Br$^-$ | (C$_6$H$_5$)$_4$As$^+$GaBr$_3$Cl$^-$ |
|---|---|---|
| $\nu_6(E)$ | 143 | 117 |
| $\nu_3(A_1)$ | 152 | 136 |
| $\nu_2(A_1)$ | 246 | 212 |
| $\nu_5(E)$ | 260 | 240 |
| $\nu_1(A_1)$ | 348,355 | 230,234 |
| $\nu_4(E)$ | 380,384 | 275,282 |

addition there is evidence for the formation of free halide ion, particularly in the bromide and iodide systems.[512]

The solvating abilities of a number of polar organic solvents towards GaCl$_3$ were deduced from a cryoscopic study, to be in the order dioxan < acetone < THF < DMF.[513]

$^{69}$Ga and $^{35}$Cl n.q.r. measurements for MGa$_2$Cl$_7$ (M = Na, K, Rb, or Ga$^I$), $^{35}$Cl n.q.r. data for MGaCl$_4$ (M as before, plus Cs), and $^{81}$Br and $^{79}$Br n.q.r. data for KAl$_2$Br$_7$ have been reported.[514] Assignments of all the resonances were made, except for KAl$_2$Br$_7$, where no assignment was possible, probably owing to the small degree of $(p-d)\pi$-character in the terminal Al—Br bonds. KGaCl$_4$ and Ga$^I$Ga$^{III}$Cl$_4$ are polymorphic, the high-temperature phase being associated with lower symmetry of the GaCl$_4$ units.

The 1:1 adducts of GaBr$_3$ with esters or nitrobenzene have been shown to ionize by a halogenolytic mechanism, giving the ions GaBr$_2$·2L$^+$, Ga$_2$Br$_7$·L$^-$, and GaBr$_4^-$; acylium or carbonium ions were not formed in appreciable amounts.[515]

Previous work on the analysis of the photoelectron spectrum of GaI$_3$ (J. L. Dehner et al., J. Chem. Phys., 1974, **61**, 594) had suggested that the molecule was non-planar, and of $C_{3v}$ symmetry. It is now thought that the anomalous features in the spectrum which necessitated this suggestion could be explained equally well by a planar model ($D_{3h}$ symmetry), with higher-order spin–orbit interactions taken into account.[516]

The compounds RGaI$_2$ (R = Me, Et, Pr$^n$, or Bu$^n$) can all be prepared by the reaction of Ga + I$_2$ + RI, or by the reaction of 'gallium moniodide' with RI at room temperature. Some vibrational and mass spectroscopic data were recorded for these, e.g. $\nu$(GaC) for MeGaI$_2$ is at 580 cm$^{-1}$ whereas in EtGaI$_2$ it is at 549 cm$^{-1}$.[517]

The KI–GaI$_3$ system gave evidence for the formation of only one compound, KGaI$_4$. For MI–GaI$_3$ (M = Rb or Cs) the analogous compound was detected, but here MGa$_2$I$_7$ are also formed. Thus, Ga$_2$I$_7^-$ is only stabilized by the presence of large cations.[518]

**Other Gallium Compounds.**—Phase relationships in the V–Fe–Ga and V–Mn–Ga systems have been elucidated.[519]

---

[512] Yu. A. Buslaev, V. P. Tarasov, S. P. Petrosyants, and N. N. Mel'nikov, J. Struct. Chem., 1974, **15**, 525.
[513] A. M. Golub, F. V. Cha, and V. N. Samoilenko, Russ. J. Inorg. Chem., 1975, **20**, 41.
[514] T. Degg and A. Weiss, Ber. Bunsengesellschaft phys. Chem., 1975, **79**, 497.
[515] Yu. A. Lysenko, V. V. Pinchuk, and A. A. Kuropatova, J. Gen. Chem. (U.S.S.R), 1974, **44**, 954.
[516] K. Wittel and R. Manne, J. Chem. Phys., 1975, **63**, 1322.
[517] M. Wilkinson and I. J. Worrall, J. Organometallic Chem., 1975, **93**, 39.
[518] D. Mashcerpa-Corral and A. Potier, Bull. Soc. chim. France, 1975, 993.
[519] K. Girgis, Naturwiss., 1974, **61**, 682, 683.

$V_8Ga_{41}$ belongs to the rhombohedral space group $R\bar{3}$, and the unit cell contains one $Ga(VGa_5)_8$ unit. One Ga is in the centre of a cuboctahedron of Ga atoms, whose triangular faces are each shared by the face of a $VGa_{10}$ polyhedron. This latter is formed by 10 triangular and three nearly square faces, and corresponds to half an icosahedron and half a cube.[520]

Three intermediate phases are found in the Nb-rich region of the Nb–Ga system, $Nb_3Ga$, $Nb_5Ga_3$, and $Nb_5Ga_4$. No $Nb_3Ga_2$ phase was detected, but a hexagonal, oxygen-stabilized phase, $Nb_5Ga_3O_x$, was indicated.[521]

$Cr_3Ga_4$ and $Fe_3Ga_4$ are isomorphous, crystallizing in the space group $C2/m$.[522] $Ni_2GaGe$ possesses an orthorhombic crystal structure, related to CoSn.[523] $Pt_3Ga$ has a crystal structure which is a variant of the $Cu_3Au$ structure.[524]

$\gamma_1$-$Cu_9Ga_4$ possesses a typical $\gamma$-brass structure, with a primitive lattice. The Ga—Ga distances are 3.08 Å.[525] The $Au_2Ga$ phase crystallizes with the $Ni_2Si$ structure (orthorhombic). Half of the Ga atoms are eight-co-ordinate by Au, the others being nine-co-ordinate.[526] $Au_{0.05}Cu_{0.45}Ga_{0.05}$ forms orthorhombic crystals, space group *Pbam*, which are isotypic with $CuAu_2Zn$.[527]

## 4 Indium

**Compounds containing Bonds between Indium and the Elements of Group VI.**—Simple $In^I$ compounds with bidentate monoprotic organic ligands have been reported for the first time, *i.e.* InX [X = 4,4,4-trifluoro-1-(thien-2'-yl)butane-1,3-dionate, 6, 6, 7, 7, 8, 8, 8-heptafluoro-2, 2-dimethyloctane-3, 5-dionate, quinolin-8-olate, and the anions of 2-mercapto-pentan-3-one and -cyclohexanone]. The $In^I$ quinolin-8-olate may be oxidized under mild conditions to $In^{III}$ species, suggesting a useful synthetic route to unusual $In^{III}$ compounds.[528]

The hydrolysis constant of $In^{3+}$ in aqueous DMSO decreases as the DMSO concentration rises. In aqueous DMF solutions, however, DMF has very little effect.[529]

$Me_2In(OAc)$, polymeric in the crystalline form, can react with neutral donor ligands to give three types of adduct: (i) unidentate donors form $Me_2In(OAc)L$ (L = DMSO or py), a five-co-ordinate monomer in solution; (ii) bidentate donors produce monomeric six-co-ordinate $Me_2In(OAc)L^1$ [$L^1$ = en or 1,2-bis(diphenylphosphino)ethane] or $[Me_2In(OAc)]_2L^2$ ($L^2$ = bipy or phen).[530]

$^{19}F$ N.m.r. was used to monitor ligand exchange between $Me_2InL$ and HL (L = $CF_3COCHCOR^-$, R = Me or $Bu^t$). The process is first-order in both $Me_2InL$ and HL, and the rate-determining step is rotation of one unidentate diketonate ligand about a partial double bond, before intramolecular proton transfer to a second unidentate ligand.[531]

---

[520] K. Girgis, W. Petter, and G. Pupp, *Acta Cryst.*, 1975, **B31**, 113.
[521] P. W. Brown and F. J. Worzala, *J. Less-Common Metals*, 1975, **41**, 77.
[522] M. J. Philipps, B. Malaman, B. Roques, A. Courtois, and J. Protas, *Acta Cryst.*, 1975, **B31**, 477.
[523] M. K. Bhargava and K. Schubert, *J. Less-Common Metals*, 1974, **38**, 177.
[524] T. Chattopadhyay and K. Schubert, *J. Less-Common Metals*, 1975, **41**, 19.
[525] R. Stokhuyzen, J. K. Brandon, P. C. Chieh, and W. B. Pearson, *Acta Cryst.*, 1974, **B30**, 2910.
[526] M. Pušelj and K. Schubert, *J. Less-Common Metals*, 1974, **38**, 83.
[527] M. Dirand, A. Courtois, J. Hertz, and J. Protas, *Compt. rend.*, 1975, **280**, C, 559.
[528] J. J. Habeeb and D. G. Tuck, *J.C.S. Dalton*, 1975, 1815.
[529] F. Ya. Kul'ba, Yu. B. Yakovlev, and D. A. Zenchenko, *Russ. J. Inorg. Chem.*, 1974, **19**, 502.
[530] J. J. Habeeb and D. G. Tuck, *Canad. J. Chem.*, 1974, **52**, 3950.
[531] H. L. Chung and D. G. Tuck, *Canad. J. Chem.*, 1974, **52**, 3944.

Na[In(acac)$_4$] has been isolated from the reaction of Na(acac) with InCl$_3$.[532]
Dichloro(acetylacetonato)-2,2'-bipyridylindium(III) forms monoclinic crystals, space group $P2_1/c$. The Cl atoms are *cis*, with $r$(In—Cl) = 2.443(1) and 2.394(1) Å, $r$(In—O) = 2.124(3) and 2.164(3) Å, and $r$(In—N) = 2.276(4) and 2.299(4) Å.[533]

In$_2$VO$_5$ forms orthorhombic crystals, space group *Pnma*, and In$_2$TiO$_5$ is isostructural with it. The two In atoms are independent, and both are co-ordinated to form a trigonal antiprism flattened along a three-fold axis. The antiprisms share edges to give infinite ribbons along the short (*b*) axis.[534]

Rb$_2$In$_4$O$_7$ has been isolated from 'RbO$_{0.56}$' + In$_2$O$_3$ mixtures. it forms trigonal crystals, space group $P\bar{3}1\,m$, and the structure is derived from cubic close-packing or O$^{2-}$, with In$^{3+}$ in 1/9 of the tetrahedral and 2/9 of the octahedral sites. Rb substitutes for 2/3 of the O$^{2-}$ ions in every third layer.[535]

I.r. and other data for the complexes [In(Y)$_2$X,(H$_2$O)] (Y = pyridine-2- or -3-carboxylate, X = Cl, Br, I, or OH) are consistent with the presence of unidentate carboxylate ligands (79).[536]

Cl\ \ \ \ \ \ \ \ OH$_2$
\ \ \ \ In
RCO\ \ \ \ \ OCR
\ ‖\ \ \ \ \ \ \ \ \ ‖
\ O\ \ \ \ \ \ \ \ \ O

(79)

In(NO$_2$)$_3$ has been isolated, and its i.r. spectrum shows that the NO$_2$ group is *N*-bonded. Adducts with bipy, phen, 8-quinolinol, and 2-aminoethanol were also reported, and these were characterized by i.r. spectroscopy, and X-ray powder diffraction.[537]

Complex formation by In$^{III}$ with orthophosphoric acid, in the [H$^+$] range 0.1—0.51 mol l$^{-1}$, produces InH$_2$PO$_4^{2+}$ ($K_{stab}$ = 2.7 × 10$^2$). Lower [H$^+$] leads to complexes containing less-protonated ligands.[538]

Amorphous indium arsenate, InAsO$_4$,4H$_2$O, is precipitated from solutions containing In$^{3+}$ and AsO$_4^{3-}$ at pH values between 1.9 and 2.7. Heating this at 70 °C in the mother liquor gradually produces the crystalline InAsO$_4$,2H$_2$O. X-Ray powder diffraction and i.r. data were used to characterize this.[539]

Mass-spectral data for tris(dimethyldithiocarbamato)-complexes of In$^{3+}$ and Cr$^{3+}$ have been collected, showing that the parent ion is virtually absent for the In species, but of high abundance in the Cr. This behaviour was attributed to stronger bonding in the latter, owing to the $\pi$-electron acceptor ability of Cr$^{III}$.[540]

---

[532] F. Ya. Kul'ba, N. B. Platunova, N. T'Li Hang, and F. G. Gavryuchenkov, *Russ. J. Inorg. Chem.*, 1974, **19**, 1409.
[533] J. G. Contreras, F. W. B. Einstein, and D. G. Tuck, *Canad. J. Chem.*, 1974, **52**, 3793.
[534] J. Senegas, J.-P. Maraud, and J. Galy, *Acta Cryst.*, 1975, **B31**, 1614.
[535] D. Fink and R. Hoppe, *Z. anorg. Chem.*, 1974, **409**, 97.
[536] J. K. Bhadra and B. Pandyopadhyay, *J. Inorg. Nuclear Chem.*, 1975, **37**, 1298.
[537] A. M. Golub, R. Akmyradov, and S. L. Uskova, *Russ. J. Inorg. Chem.*, 1974, **19**, 958.
[538] L. N. Filatova and T. N. Kurdyumova, *Russ. J. Inorg. Chem.*, 1974, **19**, 1746.
[539] E. N. Deichman, Kh. A. Ezhova, I. V. Tananaev, and Yu. Ya. Kharitonov, *Russ. J. Inorg. Chem.*, 1974, **19**, 19.
[540] P. J. Hauser and A. F. Schreiner, *Inorg. Chim. Acta*, 1974, **9**, 113.

High-pressure treatment of $KInS_2$ and $RbInS_2$, followed by quenching to ambient temperature and pressure, resulted in the formation of new tetragonal modifications, $KInS_2$-II, and $RbInS_2$-II, having the TlSe structure. The K species was metastable at all temperatures and pressures studied. The stable high-pressure form, $KInS_2$-III, has the $\alpha$-$NaFeO_2$ structure. Similar treatment of $CsInS_2$ produced an unidentifiable new species.[541]

Indium monoselenide, InSe, is rhombohedral, space group $R3m$. The hexagonal unit is formed by double layers of Se parallel to the (001) plane, between which occur pairs of In atoms.[542]

$MnIn_2Te_4$ is formed from the elements at 800 °C. It forms tetragonal crystals (space group $I\bar{4}2m$), the structure of which is derived from zinc blende. It is also related to those of $CdGa_2S_4$ and $\beta$-$Cu_2HgI_4$, although in $MnIn_2Te_4$ the different cations are distributed randomly over the occupied tetrahedral sites.[543]

**Indium Halides.**—Two new synthetic routes to $In^I$ halogeno-complexes have been reported.[544] Thus $(InX_3)^{2-}$ and $(InX_2)^-$ systems result either from reactions of In(Cp), e.g. (48),

$$In(Cp) + n(cat)(X) + HX \rightarrow CpH + (cat)_n(InX_{n+1}), \qquad (48)$$

(cat = a cation of the type $MePh_2PC_2H_4PPh_2Me^+$), or from the electrolytic oxidation of indium metal in organic media.

Changes in enthalpy and entropy associated with the formation of $InF_n^{(3-n)+}$ ($n = 1$—4) have been measured and used to calculate the standard thermodynamic characteristics of these species at 298.15 K (Table 13).[545]

**Table 13** *Thermodynamic characteristics of the complex ions* $InF_n^{(3-n)+}$

|         | $\Delta G°_{f,298.15}$/kcal mol$^{-1}$ | $\Delta H°_{f,298.15}$/kcal mol$^{-1}$ | $S°_{298.15}$/e.u. |
|---------|---------------------------------------|---------------------------------------|-------------------|
| $InF^{2+}$ | 96.65 ± 0.15                       | 108.61                                | −33.26            |
| $InF_2^+$  | 168.14 ± 0.25                      | 185.27                                | −10.78            |
| $InF_3$    | 237.84 ± 0.37                      | 263.41                                | 0.73              |
| $InF_4^-$  | 306.34 ± 0.47                      | 341.00                                | 10.07             |

The decomposition of InCl in aqueous solutions is a complex process, involving disproportionation of $In^I$ and the oxidation of $In^+$ by the components of the solution. The former gives $In^{III}$ and metallic indium, the latter $In^{III}$ and $H_2$ gas.[546]

Various $In^{III}$ species have been prepared by electrochemical oxidation of the metal by an appropriate solute in an organic medium. Thus the adducts $InX_3,3DMSO$ (X = Cl or Br) are produced by the action of $X_2$ as solute in a mixed DMSO–benzene solvent.[547]

Methylindium dichloride forms tetragonal crystals, space group $I\bar{4}$. The In atom is five-co-ordinate, in the form of a very distorted trigonal bipyramid, by one carbon and four Cl atoms, the In-Cl distances being very variable.[548]

---

[541] K.-J. Range and G. Mahlberg, *Z. Naturforsch.*, 1975, **30b**, 81.
[542] A. Likforman, D. Carré, J. Etienne, and B. Bachet, *Acta Cryst.*, 1975, **B31**, 1252.
[543] K.-J. Range and H.-J. Hübner, *Z. Naturforsch.*, 1975, **30b**, 145.
[544] J. J. Habeeb and D. G. Tuck, *J.C.S. Chem. Comm.*, 1975, 600.
[545] V. P. Vasil'ev and E. V. Kozlovskii, *Russ. J. Inorg. Chem.*, 1974, **19**, 807.
[546] L. F. Kozin and A. G. Egorova, *Russ. J. Inorg. Chem.*, 1974, **19**, 588.
[547] J. J. Habeeb and D. G. Tuck, *J.C.S. Chem. Comm.*, 1975, 808.
[548] K. Mertz, W. Schwarz, F. Zettler, and H.-D. Hausen, *Z. Naturforsch.*, 1975, **30b**, 159.

Me$_2$InCl forms orthorhombic crystals, space group $Cmcm$, and the In atom is irregularly six-co-ordinate (distorted octahedron), with two carbon and four Cl atoms. As in the previous example, there was a range of In—Cl distances (2.67—2.95 Å). The same paper listed summary vibrational assignments for Me$_2$InX (X = F, Cl, Br, or I), Et$_2$InX(X = F, Cl, Br, or I), MeInX$_2$ (X = Cl, Br, or I), and EtInX$_2$ (X = Br or I).[549]

Me$_2$InCl reacts with SbCl$_5$, Me$_3$SbCl$_2$, and Me$_3$AsCl$_2$ to give ionic products such as (Me$_4$Sb)$^+$(SbCl$_6$)$^-$, (Me$_4$Sb)$^+$(MeInCl$_3$)$^-$, and (Me$_4$As)$^+$(MeInCl$_3$)$^-$.[550] Some vibrational assignments were given for these.

Phase relationships have been studied in the system InCl$_3$–M$_4$P$_2$O$_7$ (M = Rb or Cs).[551]

[InCl$_4$(H$_2$O)$_2$][S$_4$N$_3$] forms orthorhombic crystals, space group $Pnam$. The cation is of course octahedral, and the H$_2$O molecules are $cis$-co-ordinated.[552]

Single-crystal vibrational measurements and $X$-ray space group determinations show that InCl$_5^{2-}$ and TlCl$_5^{2-}$, as their Et$_4$N$^+$ salts, are isomorphous and isostructural. Vibrational assignments and $X$-ray refinements were carried out on the basis of $C_2$ symmetry for the anions, and the space group $P\bar{4}$.[553]

Reactions of InBr with alkyl bromides appear to follow the sequence (49)—(52).[554]

$$8\text{InBr} + 2\text{RBr} \rightarrow \text{In}_7\text{Br}_9 + \text{R}_2\text{InBr} \qquad (49)$$

$$\text{In}_7\text{Br}_9 + \text{R}_2\text{InBr} + 6\text{RBr} \rightarrow 8\text{RInBr}_2 \qquad (50)$$

$$\text{In}_7\text{Br}_9 + 2\text{RBr} \rightarrow \text{In}_5\text{Br}_7 + 2\text{RInBr}_2 \qquad (51)$$

$$\text{In}_5\text{Br}_7 + 3\text{RBr} \rightarrow 3\text{RInBr}_2 + \text{In}_2\text{Br}_4 \qquad (52)$$

Reactions of In metal with alkyl halides RX (X = Br or I; R = Me, Et, Pr$^n$, or Bu$^n$) generally gave pure R$_3$In$_2$X$_3$, which were characterized spectroscopically.[555]

The crystal structure of Me$_2$InBr, tetragonal, space group $I4mm$, reveals the presence, for the first time in the solid state, of a linear dimethylindium ion, Me$_2$In$^+$. The two In—C distances are not exactly equal [2.116(6) and 2.226(7) Å].[556]

Studies of the electrical conductivity of MBr–InBr$_3$ (M = Li or K) mixtures gave the following values for the activation energy of electrical conductivity of M$_3$InBr$_6$: M = Li, 21.53 kJ mol$^{-1}$; M = K, 16.10 kJ mol$^{-1}$.[557]

Activated indium metal reacts with aryl iodides to give a good yield of diphenyl- or ditolyl-indium iodides.[558]

Two compounds have been identified in the InI–SnI$_2$ system: InSn$_2$I$_5$ and In$_4$SnI$_6$; in the InI–CdI$_2$ system only In$_4$CdI$_6$ is formed.[559]

[549] H.-D. Hausen, K. Mertz, E. Veigel, and J. Weidlein, *Z. anorg. Chem.*, 1974, **410**, 156.
[550] H. J. Widler, H.-D. Hausen, and J. Weidlein, *Z. Naturforsch.*, 1975, **30b**, 645.
[551] E. N. Deichman, I. V. Tananaev, and Zh. A. Ezhova, *Russ. J. Inorg. Chem.*, 1974, **19**, 139.
[552] M. L. Ziegler, H. U. Schlumper, B. Nuber, J. Weiss, and G. Ertl, *Z. anorg. Chem.*, 1975, **415**, 193.
[553] G. Joy, A. P. Gaughan, I. Wharf, D. F. Shriver, and J. P. Dougherty, *Inorg. Chem.*, 1975, **14**, 1795.
[554] L. G. Waterworth and I. J. Worrall, *J. Organometallic Chem.*, 1974, **81**, 23.
[555] M. J. S. Gynane and I. J. Worrall, *J. Organometallic Chem.*, 1974, **81**, 329.
[556] H.-D. Hausen, K. Mertz, J. Weidlein, and W. Schwarz, *J. Organometallic Chem.*, 1975, **93**, 291.
[557] A. G. Dudareva, G. A. Lovetskaya, and K. Gladis, *Russ. J. Inorg. Chem.*, 1974, **19**, 435.
[558] L.-C. Chao and R. D. Rieke, *Synth. React. Inorg. Metal-Org. Chem.*, 1975, **5**, 165.
[559] Yu. N. Denisov, N. S. Malova, and P. I. Fedorov, *Russ. J. Inorg. Chem.*, 1974, **19**, 443.

**Other Indium Compounds.**—The indium hexathiocyanates, $R_3[In(NCS)_6]$ (R = $NMe_4$, $NEt_4$, $NBu_4$, or $NMe_3Ph$), have been isolated from aqueous solutions. Their i.r. spectra were consistent with In—N co-ordination.[560]

Vapour-phase electron diffraction measurements on $InMe_3$ gave the following molecular parameters: $r$(In—O) 209.6 pm, $r$(C—H) 114.0 pm, and $\angle$HCIn 112.7°. A slight non-planarity (3.0 ± 2.5°) could be attributed to a shrinkage effect in $r$(C $\cdots$ C) of 0.4 pm.[561]

$InR_3$ (R = alkyl or aryl) may be prepared conveniently by the reaction of diorganomercury species with activated indium metal. The latter is formed by reducing indium salts by alkali metals in hydrocarbon solvents.[562]

$Me_3In$ reacts with $C_5H_6$ to give the stable species $Me_2InCp$. Its i.r. spectrum was very simple, suggesting that $C_{5v}$ local symmetry may apply for the $C_5H_5^-$ ring.[563]

$Re_2(CO)_8[\mu\text{-}InRe(CO)_5]_2$ crystallizes in the monoclinic space group $P2_1/n$. The complex contains a planar $Re_2In_2$ ring (80), the Re—Re distance, 3.232(1) Å, and the $\angle$ReInRe, 71.07(3)°, being indicative of Re—Re bonding. The In—Re distance was 2.766(1) Å.[564]

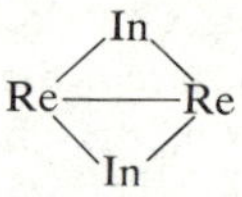

$$Re\underset{In}{\overset{In}{\diamondsuit}}Re$$

(80)

$CoIn_2$ crystallizes with the $CuMg_2$ structure type, space group $Fddd$. The bond lengths agree well with those calculated for a model based upon hard-sphere contact.[565]

$Au_9In_4$(h) crystallizes with the $Cu_9In_4$-type of structure, while $Au_7In_3$ has a hexagonal structure, related to that of $\gamma$-brass.[566]

Phase diagrams for the La–In system have been established. The intermetallic compounds $La_3In$, $La_2In$, and $LaIn_2$ decompose peritectically at 816, 955, and 1153 °C respectively. LaIn, $La_3In_5$, and $LaIn_3$ melt congruently at 1125, 1185, and 1140 °C.[567]

## 5 Thallium

**Thallium(III) Compounds.**—Kurosawa and Okawara have reviewed the literature on organothallium derivatives for 1973.[568]

Thallium(II) ions are generated by flash photolysis of thallium(III) solutions, and rate constants were measured for reactions (53—55).

$$Tl^{2+} + Mn^{2+} \rightarrow Tl^+ + Mn^{3+} \quad (1.9 \pm 0.2) \times 10^6 \text{ l mol}^{-1} \quad (53)$$

$$Tl^{2+} + Fe^{2+} \rightarrow Tl^+ + Fe^{3+} \quad (2.6 \pm 0.1) \times 10^6 \text{ l mol}^{-1} \quad (54)$$

$$Tl^{2+} + Co^{3+} \rightarrow Tl^{3+} + Co^{2+} \quad (9.5 \pm 0.5) \times 10^6 \text{ l mol}^{-1} \quad (55)$$

[560] V. S. Mal'tseva and Yu. G. Eremin, *Russ. J. Inorg. Chem.*, 1974, **19**, 1268.
[561] G. Barbe, J. L. Hencher, Q. Shen, and D. G. Tuck, *Canad. J. Chem.*, 1974, **52**, 3936.
[562] L.-C. Chao and R. D. Rieke, *Synth. React. Inorg. Metal-Org. Chem.*, 1974, **4**, 373.
[563] P. Krommes and J. Lorberth, *J. Organometallic Chem.*, 1975, **88**, 329.
[564] H. Preut and H.-J. Haupt, *Chem. Ber.*, 1975, **108**, 1447.
[565] H. H. Stadelmaier and H. K. Manaktala, *Acta Cryst.*, 1975, **B31**, 374.
[566] M. Pušelj and K. Schubert, *J. Less-Common Metals*, 1975, **41**, 33.
[567] O. D. McMasters and K. A. Gschneidner, *J. Less-Common Metals*, 1974, **38**, 137.
[568] H. Kurosawa and R. Okawara, *J. Organometallic Chem*, 1974, **79**, 1.

These, outer-sphere, one-electron reactions appear to have activation energies of ca. 25 kJ mol$^{-1}$ less than those for related two-electron processes.[569]

A photochemical reaction between Tl$^{III}$ and Ru(bipy)$_3^{2+}$ has been studied. Tl$^{2+}$ ions are produced as intermediates, and these can oxidize Ru(bipy)$_3^{2+}$ and reduce Ru(bipy)$_3^{3+}$.[570]

$^1$H N.m.r. spectra of Me$_2$Tl$^+$ in anionic and cationic detergent nematic phases show that the electrostatic interactions play a significant role in the orientation of the Me$_2$Tl$^+$ ions.[571]

The crystal structures of Me$_2$TlX derivatives have been determined for X = CN, N$_3$, NCS, or NCO;[572] X = acetate, tropolonate, acetylacetonate, or dibenzoylmethide;[573] X = tricyanomethide or dicyanamide.[574] The NCO species[572] is dimorphic, the orthorhombic form containing infinite layers of Me$_2$TlNCO molecules. The trigonal form, however, is built up from dimeric units.

Variable-temperature $^{205}$Tl and $^{203}$Tl n.m.r. spectra (with and without $^1$H decoupling), together with $^1$H–{$^{205}$Tl} INDOR experiments on dimethylthallium ethoxide, (Me$_2$TlOEt)$_2$, provide evidence for chemical exchange.[575]

Bis-cyclopentadienyl- and bis-indenyl-Tl$^{III}$ hydrides can be prepared by reactions (56) and (57).

$$R_2TlBH_4 + Et_3N \xrightarrow{THF} R_2TlH + Et_3N,BH_3 \qquad (56)$$

$$R_2TlCl + NaH \xrightarrow{THF} R_2TlH + NaCl \qquad (57)$$

$\nu$(Tl—H) is at 2105 cm$^{-1}$ (R = Cp), 2200 cm$^{-1}$ (R = indenyl).[576]

I.r. spectra (band positions and intensities) indicate that R$_2$TlNCS (R = Me or Ph) and [Me$_2$Tl(NCS)$_2$]$^-$ have Tl—NCS co-ordination, with some ionic character in the interaction. PhTl(SCN), however, is S-bonded.[577]

The thermal stabilities of the Tl$^{III}$ ammine complexes, Tl(N–N)$_n$X$_3$, are in the following sequences: N–N = en > propylenediamine > trimethylenediamine; X = Cl > Br > I.[578]

The isotropic $^{205}$Tl hyperfine couplings from the e.s.r. spectra of the radical-cations of Tl$^{III}$ meso-tetraphenylporphyrin, octaethylporphyrin and octaethylchlorin can be explained by taking account of both $\sigma$–$\pi$ spin polarization of the Tl—N bonds, and direct $\pi$-interaction of the Tl orbitals with the ligand $\pi$-orbitals.[579]

The 1:1 acid complex Tl$^{III}$H$_5$L (H$_8$L = ethylenediamine-$NNN'N'$-tetramethylphosphonic acid) has been detected spectrophotometrically. Its instability constant (log $K$) is 5.74.[580]

---

[569] B. Falcinella, P. D. Felgate, and G. S. Laurence, *J.C.S. Dalton*, 1975, 1.
[570] G. S. Laurence and V. Balzani, *Inorg. Chem.*, 1974, **13**, 2976.
[571] Y. Lee and L. W. Reeves, *Canad. J. Chem.*, 1975, **53**, 161.
[572] Y. M. Chow and D. Britton, *Acta Cryst.*, 1975, **B31**, 1922.
[573] Y. M. Chow and D. Britton, *Acta Cryst.*, 1975, **B31**, 1929.
[574] Y. M. Chow and D. Britton, *Acta Cryst.*, 1975, **B31**, 1934.
[575] G. M. Sheldrick and J. P. Yesinowski, *J.C.S. Dalton*, 1975, 870.
[576] N. Kumar and R. K. Sharma, *J. Inorg. Nuclear Chem.*, 1974, **36**, 2626.
[577] N. Bertazzi, G. C. Stocco, L. Pellerito, and A. Silvestri, *J. Organometallic Chem.*, 1974, **81**, 27.
[578] A. V. Barsukov and D. M. Markheeva, *Russ. J. Inorg. Chem.*, 1974, **19**, 1126.
[579] C. Mengersen, J. Subramanian, J.-H. Fuhrhop, and K. M. Smith, *Z. Naturforsch.*, 1974, **29a**, 1827.
[580] V. I. Kornev, N. I. Pechurova, and L. I. Martynenko, *Russ. J. Inorg. Chem.*, 1974, **19**, 146.

Paramagnetic complexes of semiquinones with $Tl^{III}$, (81), have been prepared by the reaction of $Et_2TlCl$ with the sodium salt of the o-semiquinones. Their e.s.r. spectra were reported, and hyperfine coupling constants determined.[581]

(81)

Tris(NN-dimethyldithiocarbamato)thallium(III) monohydrate, $Tl(Me_2dtc)_3,H_2O$, forms monoclinc crystals, space group $P2_1/c$. The Tl is six-co-ordinate, with all the Tl—S distances = 2.659 Å. The geometry is intermediate between trigonal-prismatic and trigonal-antiprismatic (twist angle = 33.2°).[582]

$KTlF_4$ forms crystals which are trigonal, belonging to the space group $P3_1$. The structure is that of a superstructure variant of the $CaF_2$ type. The thallium atoms are seven-co-ordinate. The analogous Rb and Cs compounds are made by heating equivalent mixtures of alkaline chlorides with $Tl_2O_3$ in $F_2$ at 450—500 °C. Both give orthorhombic crystals, containing layers of two-dimensional, corner-linked, distorted $TlF_{4/2}F_2$ octahedra, connected by Rb or Cs ions.[583,584]

Kinetic measurements on the reduction of chlorothallium(III) complexes by phosphite give a reactivity sequence for the complexes which agrees with previous work, i.e.:[585] $TlCl_4^- > TlCl_3 > TlCl_2^+ > Tl^{3+} > TlCl^{2+}$.

**Thallium(I) Compounds.**—Isotopic exchange reactions between $^{204}Tl^I$ and $Tl^{III}$ have been studied in a variety of solvents ($H_2O$–EtOH, $H_2O$–glycol, $H_2O$–DMSO).[586]

Pulse-radiolysis of $Tl^+$-containing solutions, with conductivity and spectrophotometry being used as detection probes, provided evidence for the formation of $TlOH^+$, in equilibrium with $Tl^{2+}$.[587]

Two groups of workers have shown that the chemical shift of the $^{205}Tl^+$ ion is very dependent on the nature of the solvent.[588,589] Thus, $Tl^+$ could be a useful probe of the function of univalent ions in biological systems.

$Tl^I$ derivatives of metal carbonyl anions are generally stable when the anion is weakly basic, e.g. $Co(CO)_3[P(OPh)_3]^-$, whereas disproportionation to $Tl^{III}$ derivatives and $Tl^0$ occur when it is more strongly basic, e.g. $Mn(CO)_5^-$.[590]

Hydrated $Tl^+$-exchanged zeolite-X crystallizes in the cubic space group $Fd3m$. The $Tl^+$ ions are placed inside the sodalite cages in front of the hexagonal prisms

---

[581] E. S. Klimov, G. A. Abakumov, E. N. Gladyshev, P. Ya. Bayushkin, V. A. Muraev, and G. A. Razuvaev, Doklady Chem., 1974, **218**, 678.
[582] H. Abrahamson, J. R. Heiman, and L. H. Pignolet, Inorg. Chem., 1975, **14**, 2070.
[583] C. Herbecker, Z. Naturforsch., 1975, **30b**, 305.
[584] C. Herbecker, Z. anorg. Chem., 1975, **412**, 37.
[585] K. S. Gupta and Y. K. Gupta, Inorg. Chem., 1975, **14**, 2000.
[586] I. Cecal and I. A. Schneider, Inorg. Nuclear Chem. Letters, 1974, **10**, 977.
[587] P. O'Neill and D. Schulte-Frohlinde, J.C.S. Chem. Comm., 1975, 387.
[588] J. F. Hinton and R. W. Briggs, J. Magn. Resonance, 1975, **19**, 393.
[589] J. J. Deckler and J. I. Zink, J. Amer. Chem. Soc., 1975, **97**, 2937.
[590] S. E. Pedersen, W. R. Robinson, and D. P. Schussler, J.C.S. Chem. Comm., 1974, 805.

(site I'), and in front of the six-membered ring face of the cage on the supercage side (site II). The shortest $Tl^+$—O distances are 2.64(5) and 2.79(5) Å.[591]

$Tl^I$ forms a 1:1 complex with urea when $TlIO_3$, $TlBrO_3$, or $Tl_2S_2O_3$ is dissolved in 3.0M-$(NH_2)_2CO$ solutions at 20 °C. The stability constant of $[TlOC(NH_2)_2]^+$ is ca. 0.43.[592]

Complex formation between $Tl^I$ and α-alanine, glycine, or serine has been investigated potentiometrically. The stability constants of the 1:1 complexes are 30.4 ± 5.7, 32.3 ± 2.7, and 33.9 ± 2.7, and 33.9 ± 1.5, respectively.[593]

Hexafluoroacetylacetonatothallium(I) forms triclinic crystals, space group $P\bar{1}$. The structure contains infinite chains formed by two different structural units, linked by O atoms of the hfac⁻ anions along the b-axis. $Tl^+$ ions are five-co-ordinate, by three O atoms in a plane and two to one side of this. $r(Tl—O)$ ranges from 2.62 to 3.04 Å.[594]

Thallium(I) carbonate crystallizes in the monoclinic system, space group $C2/m$. The $Tl^+$ $6s^2$ lone pair is stereochemically active in this structure, being situated in channels parallel to [010].[595]

$TlNO_3$(III) belongs to the space group $Pnma$, and the crystal structure is related to that of CsCl. The suggestion was made that the structure phase II is closely related to that of phase III, and that the re-orientations needed to accomplish the transformation are quite small.[596]

$Tl_3Na(SO_3)_2$ is polymorphic. The stable phase above 230 °C at atmospheric pressure is hexagonal, belonging to one of the space groups $P6/mmm$, $P622$, $P6mm$, or $P\bar{6}2m$. A metastable phase is isotypic with $Tl_2SO_3$ itself.[597]

$Tl^IH_5(PO_4)_2$ has been isolated from the $H_2O–H_3PO_4–TlH_2PO_4$ system. The new species forms monoclinic crystals, space group $Cc$ or $C2/c$.[598]

$TlVO_3$ crystallizes in the orthorhombic space group $Pbcm$. In the crystals Tl atom is 10-co-ordinate; this is analogous to the situation for the alkali metals, and thus the $6s^2$ electron pair of the Tl is sterochemically insignificant in this structure.[599]

The thallium(I) polysulphide $Tl_2^IS_5$ has a structure in which the two negative charges in the $S_5^{2-}$ ion are localized at the end two S atoms. The Tl atoms are three-co-ordinate and pyramidal, so here the lone pair is sterochemically active, occupying the fourth tetrahedral position.[600]

Studies of the $Tl_2S–Bi_2S_3$ system have shown that compounds $TlBiS_2$ (rhombohedral, isostructural with $NaFeO_2$) and $Tl_4Bi_2S_5$ (orthorhombic, space group $Pn2_1$ or $Pnam$) are produced.[601]

Diethyldithiophosphinatothallium(I), $Tl(Et_2PS_2)$, forms orthorhombic crystals, space group $Pcca$. The structure is built up from monomeric units, in which each

---

[591] J. J. de Boer and I. E. Maxwell, *J. Phys. Chem.*, 1974, **78**, 2395.
[592] E. A. Gyunner and A. M. Fedorenko, *Russ. J. Inorg. Chem.*, 1974, **19**, 979.
[593] F. Ya. Kul'ba, V. G. Ushakova, and Yu. B. Yakovlev, *Russ. J. Inorg. Chem.*, 1974, **19**, 972.
[594] S. Tachiyashiki, H. Nakayama, R. Kuroda, S. Sato, and G. Saito, *Acta Cryst.*, 1975, **B31**, 1483.
[595] R. Marchand, Y. Piffard, and M. Tournoux, *Canad. J. Chem.*, 1975, **53**, 2454.
[596] W. L. Frazer, S. W. Kennedy, and M. R. Snow, *Acta Cryst.*, 1975, **B31**, 365.
[597] Y. Oddon, G. Coffy, and A. Tranquard, *Bull. Soc. chim. France*, 1975, 1481.
[598] G. Coffy, Y. Oddon, M.-J. Boinon, and A. Tranquard, *Compt. rend.*, 1975, **280**, C, 1301.
[599] M. Ganne, Y. Piffard, and M. Tournoux, *Canad. J. Chem.*, 1974, **52**, 3539.
[600] B. Leclerc and T. S. Kabré, *Acta Cryst.*, 1975, **B31**, 1675.
[601] M. Julien-Pouzol and M. Guittard, *Bull. Soc. chim. France*, 1975, 1037.

Tl atom is co-ordinated by two S atoms, with four further S atoms at a greater distance (one from each of four adjacent molecules).[602]

Thallium(I) diethyldithiocarbamate, Tl($S_2$CNEt$_2$), forms crystals in which the structure is built up from dimeric units, with four-co-ordinate Tl. These are linked into chains by weaker Tl—S bonding.[603] The isobutyl analogue has a very similar molecular structure.[604]

TlSb$_3$F$_{10}$ is produced by dissolving Tl$_2$CO$_3$ and SbF$_3$ in dilute aqueous HF. It is isostructural with the Rb and NH$_4$ salts.[605]

Oxidation of TlF by halogens, BrCl, ICl, or IBr gives the mixed Tl$^{III}$ fluoride halides: TlFI$_2$, TlFBrI, TlFClI, and TlFClBr. All appeared to be true Tl$^{III}$ species, and this is the first report of their existence.[606]

TlCl$_2$ forms tetragonal crystals, space group $I4_1/a$. The structure is that of CaWO$_4$, *i.e.* 'TlCl$_2$' is indeed Tl$^+$TlCl$_4^-$.[607] A series of mixed crystals are formed in the TlCl$_2$–TlBr$_2$ system, *i.e.* TlCl$_{2-x}$Br$_x$ ($0 \leqslant x \leqslant 2$). If $x \leqslant 0.28$, these have the same structure as TlCl$_2$, but when $x > 1.83$, the structures resemble that of orthorhombic TlBr$_2$. The intermediate phases have a new 'TlClBr' type of structure. This can best be described as Tl$^I$ halogenothallate(III) (with a NiAs-like arrangement).[608]

Variations of formation constants of TlBr and TlBr$_2^-$ with the ionic strength of the solution have been noted.[609]

Kinetic measurements on the solid-phase reactions (58)—(60)

$$\text{AgI} + \text{TlI} \rightarrow \text{AgTlI}_2 \quad (58)$$

$$\text{AgI} + \text{AgTl}_2\text{I}_3 \rightarrow 2\text{AgTlI}_2 \quad (59)$$

$$\text{AgTlI}_2 + \text{TlI} \rightarrow \text{AgTl}_2\text{I}_3 \quad (60)$$

show that the governing mechanism is always cation counter-diffusion, with the rate-determining step being diffusion of Tl$^+$.[610]

The Tl$^I$–I$_3$ complex is formed in aqueous HClO$_4$ (0.01 mol l$^{-1}$) with a rate constant of $(1.75 \pm 0.03) \times 10^4$ l mol$^{-1}$ s$^{-1}$ at 25°C. This is rather slow for a reaction involving singly-charged ions, and may indicate ligand dependence for the rate-determining step.[611]

**Other Thallium Compounds.**—A new, low-temperature form of TlS, *i.e.* α-TlS, has been revealed by radiocrystallographic examination of the Tl-S system.[612]

A redetermination of the crystal structure of TlTe shows that the Te sublattice had been incorrectly described previously. There are short Te—Te distances, comparable with covalent bond lengths.[613]

---

[602] S. Esperås and S. Husebye, *Acta Chem. Scand.* (A), 1974, **28**, 1015.
[603] H. Pritzkow and P. Jennische, *Acta Chem. Scand.* (A), 1975, **29**, 60.
[604] H. Anacker-Eickhoff, P. Jennische, and R. Hesse, *Acta Chem. Scand.* (A), 1975, **29**, 51.
[605] B. Ducourant, B. Bonnet, R. Fourcade, and G. Mascherpa, *Bull. Soc. chim. France*, 1975, 1471.
[606] S. S. Batsanov and M. N. Stas', *Russ. J. Inorg. Chem.*, 1975, **20**, 444.
[607] G. Thiele and W. Rink, *Z. anorg. Chem.*, 1975, **414**, 231.
[608] G. Thiele and W. Rink, *Z. anorg. Chem.*, 1975, **414**, 47.
[609] V. A. Fedorov, A. M. Robov, I. D. Isaev, and A. A. Alekseeva, *Russ. J. Inorg. Chem.*, 1974, **19**, 798.
[610] G. Flor, V. Massarotti, and R. Riccardi, *Z. Naturforsch.*, 1975, **30a**, 304.
[611] N. Purdie, M. M. Farrow, M. Steggall, and E. M. Eyring, *J. Amer. Chem. Soc.*, 1975, **97**, 1078.
[612] J. Tudo, B. Dermigny, and B. Jolibois, *Compt. rend.*, 1975, **280**, C, 1375.
[613] J. Weis, H. Schäfer, B. Eisenmann, and G. Schön, *Z. Naturforsch.*, 1974, **29b**, 585.

# 4
# Elements of Group IV

BY P. G. HARRISON AND P. HUBBERSTEY

## 1 Carbon

Papers abstracted for this section of the Report have been restricted to those in which the inorganic chemistry of the carbon allotropes and of the non-catenated molecular carbon species, particularly those containing carbon–non-metal (*i.e.*, hydrogen, nitrogen, phosphorus, oxygen, chalcogen, and halogen) bonds is described. The carbaboranes and carbides are omitted, the former since they are comprehensively reviewed in Chapter 3, the latter since no original data have been published during the period of this Report on Main-Group element carbides.

The current chemistry of the Group IV elements has been reviewed in the second volume of the MTP series.[1] In an analysis of the i.r. band intensities of a number of $MHCl_3$ and $MX_4$ (M = C, Si, or Ge; X = F, Cl, Br, or I) molecules,[2] the electronegativities of the Group IV elements have been shown not to vary smoothly as deduced by Pauling, but to vary erratically, as indicated by the Allred–Rochow and Sanderson Electronegativity scales (Table 1).

Two theoretical analyses of physicochemical parameters of carbon-containing molecules have been effected.[3,4] $C(1s)$ core-electron binding energies, calculated from atomic charges obtained by an electronegativity equilization procedure,[3] and enthalpies of formation, molecular geometries, dipole moments, and first ionization potentials, derived from an improved version (MINDO/3) of the MINDO semi-empirical SCF MO treatment,[4] have all been shown to be in excellent agreeement with experimentally derived values. The results of a theoretical study

**Table 1** *Electronegativities of Group IV elements*

| | | Electronegativity | |
| Element | Pauling | Allred–Rochow | Sanderson |
| --- | --- | --- | --- |
| Carbon | 2.5 | 2.5 | 2.5 |
| Silicon | 1.8 | 1.7 | 1.7 |
| Germanium | 1.8 | 2.0 | 2.3 |
| Tin | 1.7 | 1.7 | 2.0 |

[1] M.T.P. International Reviews of Science, Inorganic Chemistry, Series 2 Vol. 2, 'Main Group Elements, Groups IV and V', ed. D. B. Sowerby, University Park Press, Baltimore, 1975.
[2] J. G. Chambers, T. E. Thomas, A. J. Barnes, and W. J. Orville-Thomas, *Chem. Phys.*, 1975, **9**, 467.
[3] W. L. Jolly and W. B. Perry, *Inorg. Chem.*, 1974, **13**, 2686.
[4] R. C. Bingham, M. J. S. Dewar, and D. H. Lo, *J. Amer. Chem. Soc.*, 1975, **97**, 1285.

of intermolecular hydrogen bonding in dimeric complexes involving small molecular carbon species have also been published.[5]

An electrochemical technique for measuring gas-phase carbon activities, based on the concentration cell:

| gas-phase carbon compounds | membrane electrode | electrolyte with carbon-containing ionic species | carbon reference electrode |
|---|---|---|---|
| e.g. CO | α-Fe | $Li_2CO_3$–$Na_2CO_3$ eutectic (m.pt. 775 K) | carbon-saturated iron rod |

has been developed by Hobdell et al.[6] Meyer and Lynch[7] have perfected a method for analysing the nature of carbon mass loss. The vapour-phase component is determined by reaction with $H_2$ or $O_2$; the extent of mass loss as particulates is then obtained by difference from gravimetric measurement of the total mass loss.

**Carbon Allotropes.**—The chemistry of the allotropic forms of carbon considered here includes their synthetic, structural and chemical properties; those papers in which their catalytic, adsorption, surface, and other similar properties are described have not been abstracted. A group of Russian authors[8] have questioned the existence of a third crystalline form of carbon (i.e. carbyne, chaoite, and carbon VI).[9] They have demonstrated that the diffraction patterns previously attributed to these forms of carbon can be generated by twinned single crystals consisting of a small number of hexagonal carbon (graphite-like) networks. For twinning angles of 27°48′ and 15°15′, the 'interplanar spacings' calculated from double-diffraction reflections agree with those in carbyne and chaoite and with those in carbon VI, respectively.

*Synthetic Studies.* The formation of carbon deposits by hydrocarbon pyrolysis has been extensively studied. Baker et al.[10–12] have described the results of a controlled-atmosphere electron microscopy (CAEM) study of the deposit of filamentous carbon by catalysed decomposition of $C_2H_2$ (on iron doped with tin)[10] and $C_2H_2$,[11] $C_2H_4$,[12] $C_3H_4$,[12] $C_3H_6$,[12] $C_4H_6$,[12] and $C_6H_6$[12] (on cobalt). In the $C_2H_2$–Fe(Sn) system, the filamentous growth was almost entirely in the form of spirals with a common pitch; although spiral filaments have been observed in other systems, they are rare, and growth exclusively as spirals is very unusual.[10] Baker et al.[13] have also used controlled-atmosphere optical microscopy (CAOM) techniques to study the deposits formed when $CH_4$ is pyrolysed on nickel surfaces; the deposits were predominantly nodular clusters of polycrystalline material and graphite flakes. In an independent study of the deposition of carbon

---

[5] P. Kollman, J. McKelvey, A. Johansson, and S. Rothenberg, *J. Amer. Chem. Soc.*, 1975, **97**, 955.
[6] J. R. Gwyther, M. R. Hobdell, and S. P. Tyfield, *Carbon*, 1974, **12**, 698.
[7] R. T. Meyer and A. W. Lynch, *Carbon*, 1974, **12**, 684.
[8] V. G. Nagornyi, A. P. Nabatnikov, V. I. Frolov, A. N. Deev, and V. P. Sosedov, *Russ. J. Phys. Chem.*, 1975, **49**, 840.
[9] P. G. Harrison and P. Hubberstey, 'Elements of Group IV' (Specialist Periodical Reports), The Chemical Society, London, 1973—5, Vols. 1—3.
[10] R. T. K. Baker, P. S. Harris, and S. Terry, *Nature*, 1975, **253**, 37.
[11] R. T. K. Baker, G. R. Gadsby, R. B. Thomas, and R. J. Waite, *Carbon*, 1975, **13**, 211.
[12] R. T. K. Baker, G. R. Gadsby, and S. Terry, *Carbon*, 1975, **13**, 245.
[13] R. T. K. Baker, P. S. Harris, J. Henderson, and R. B. Thomas, *Carbon*, 1975, **13**, 17.

on nickel by pyrolysis of $CH_4$, Derbyshire and Trimm[14] identified both continuous films of laminar graphite and islands of graphite in a uniform graphite matrix. Lieberman et al.[15—18] have published the results of a number of investigations of the pyrolysis of hydrocarbons (including $CH_4$[15—17] and $C_6H_6$[18]). Identification of a wide variety of gas-phase intermediates during the deposition of carbon from $CH_4$ is said to support the contention that this process is dominated by gas-phase reactions.[16] This reaction mechanism has also been proposed as one of two routes for the pyrolysis of aromatic hydrocarbons;[19] the second route involves condensation reactions without loss of aromatic structure. Transmission electron microscopy techniques have also been used to study the microstructures of a number of isotropic carbons[20—22] and of codeposited pyrolytic carbon—silicon carbide mixtures.[23]

The preparation of carbons of high structural order, microcrystalline size, and monocrystalline character has been achieved in the thermal decomposition of magnesium carbide.[24] The carbon was formed via $Mg_2C_3$ as intermediate. The magnesium, necessarily formed in the reaction, appeared to catalyse the rearrangement of the carbon atoms to give a higher degree of order. In view of the relatively low reaction temperature (950—1200 °C), the short reaction time 1—5 hours), and the high yields (86%), it is suggested that the process may have industrial significance.[24]

*Structural Studies.* Structural characteristics of both natural and synthetic diamonds have been studied using electron microscopy[25—27] and i.r.[25,28] and optical absorption spectroscopy.[25,27,29—31] Greatest interest is associated with defect structures of electron-irradiated diamond.[29—31]

Molecular and large-scale structural parameters of vitreous carbon have been elucidated as a function of heat-treatment temperature, using $X$-ray diffraction[32] and electron microscopic[32,33] techniques. Short-range order in the carbon layers is characterized by an interatomic distance of $1.42 \pm 0.01$ Å (*cf.* graphite) but with a

[14] F. J. Derbyshire and D. L. Trimm, *Carbon*, 1975, **13**, 189.
[15] M. L. Lieberman and G. T. Noles, *Carbon*, 1974, **12**, 689.
[16] M. L. Lieberman, *Carbon*, 1975, **13**, 243.
[17] H. O. Pierson and M. L. Lieberman, *Carbon*, 1975, **13**, 159.
[18] R. M. Curlee and M. L. Lieberman, *Carbon*, 1975, **13**, 248.
[19] S. C. Graham, J. B. Homer, and J. L. J. Rosenfeld, *Proc. Roy. Soc.*, 1975, **A344**, 259.
[20] C. S. Yust and H. P. Krautwasser, *Carbon*, 1975, **13**, 125.
[21] J. Pelissier and L. Lombard, *Carbon*, 1975, **13**, 205.
[22] J. L. Kaae, *Carbon*, 1975, **13**, 55.
[23] J. L. Kaae, *Carbon*, 1975, **13**, 51.
[24] D. Osetzky, *Carbon*, 1974, **12**, 517.
[25] Yu. P. Solodova, L. D. Podol'skikh, L. T. Litvin, V. M. Kulakov, V. P. Butuzov, and M. I. Samoilovich, *Soviet Phys. Cryst.*, 1975, **20**, 50.
[26] V. N. Apollonov, N. F. Borovikov, L. F. Vereshchagin, Ya. A. Kalashnikov, L. B. Nepomnyashchii, M. D. Shalimov, and N. N. Shipkov, *Soviet Phys. Cryst.*, 1974, **19**, 406.
[27] T. Evans and P. Rainey, *Proc. Roy. Soc.*, 1975, **A344**, 111.
[28] M. I. Samoilovich, L. D. Podol'skikh, V. P. Butuzov, E. Norullaev, and Sh. A. Vakhidov, *Russ. J. Phys. Chem.*, 1974, **48**, 1422.
[29] J. Walker, L. A. Vermeulen, and C. D. Clark, *Proc. Roy. Soc.*, 1974, **A341**, 253.
[30] G. Davies, *Proc. Roy. Soc.*, 1974, **A336**, 507.
[31] G. Davies and C. M. Penchina, *Proc. Roy. Soc.*, 1974, **A338**, 359.
[32] Yu. S. Lopatto, D. K. Khakimova, V. K. Nikitina, M. A. Audcenko, and G. M. Plavnik, *Doklady Chem.*, 1974, **217**, 471.
[33] L. R. Bunnell, *Carbon*, 1974, **12**, 693.

co-ordination somewhat less than three, *i.e.*, 2.4—2.6.[32] The large-scale data[32,33] complement the understanding of the vitreous carbon structure achieved on the molecular level.[32]

A precise thermodynamic investigation of the equilibrium (1), employing

$$\text{vitreous carbon} \rightleftharpoons \text{graphite} \qquad (1)$$

a solid oxide electrolyte in the temperature range 650—1150 °C, has been undertaken to gain structural information on vitreous carbons.[34] The vitreous carbon samples studied exhibited a wide variation in thermodynamic properties, some of which represent a large disorder relative to graphite. On this basis it is suggested that it is more appropriate to think of vitreous carbon as a broad class of materials rather than as a single material.[34]

An e.s.r. study[35] has also been carried out to assess the influence of both heat-treatment and neutron irradiation on the structure of vitreous carbon.

The transient presence of interstitial carbon atoms in the graphitization of cokes has been deduced from an analysis of radial distribution function curves of variously heat-treated samples.[36] A peak corresponding to an abnormal C—C distance of 1.90 Å is observed for samples heat-treated to 2050 and 2100 °C. This distance is assigned to interstitial carbons located 1.2 Å above or below the hexagonal layers at the centre of hexagonal rings. A theoretical analysis[37] of the empirically based suggestion that the magnetoresistance of graphite is a suitable parameter for the assessment of the degree of graphitization of the sample has been undertaken.

Although considerable research effort has been applied to the chemistry of carbon fibres, little is of relevance to the inorganic chemist. Even those structural investigations which have been carried out are primarily concerned with the morphology of the fibres rather than their detailed molecular structures.[38]

*Chemical Studies.* Chemical reactivity of carbonaceous materials is profoundly influenced by the presence of surface functional groups. For this reason, several investigations of the structure and reactivity of these surface complexes have been undertaken.[39—44] Barton and Harrison[39] have investigated the chemistry of surface oxides on both carbon and graphite. They conclude that cyclic esters are primarily responsible for the surface acidity. Two distinct types of acidic oxide, (1) and (2), were found on the carbon surface; although they both appeared to be lactone groups, only (1) was associated with an active hydrogen. One type of acidic oxide (3) was found on the more conjugated graphite surface; it was, however, associated with an active hydrogen. The presence of oxygen complexes

---

[34] S. K. Das and E. E. Hucke, *Carbon*, 1975, **13**, 33.
[35] S. Orzeszko and K. T. Yang, *Carbon*, 1974, **12**, 493.
[36] J. P. Rouchy and L. Gatineau, *Carbon*, 1975, **13**, 267.
[37] Y. Hishiyama, *Carbon*, 1975, **13**, 244.
[38] D. J. Johnson, I. Tomizuka, and O. Watanabe, *Carbon*, 1975, **13**, 321.
[39] S. S. Barton and B. H. Harrison, *Carbon*, 1975, **13**, 283.
[40] O. P. Mahajan, S. S. Bhardwaj, B. C. Kaistha, S. K. Garg, and B. R. Puri, *Indian J. Chem.*, 1974, **12**, 1183.
[41] P. Cadman, J. D. Scott, and J. M. Thomas, *J.C.S. Chem. Comm.*, 1975, 654.
[42] V. R. Deitz, J. N. Robinson, and E. J. Poziomek, *Carbon*, 1975, **13**, 1831.
[43] E. J. Poziomek, R. A. Mackay, and R. P. Barrett, *Carbon*, 1975, **13**, 259.
[44] E. J. Poziomek, R. A. Mackay, and R. P. Barrett, *Carbon*, 1975, **13**, 347.

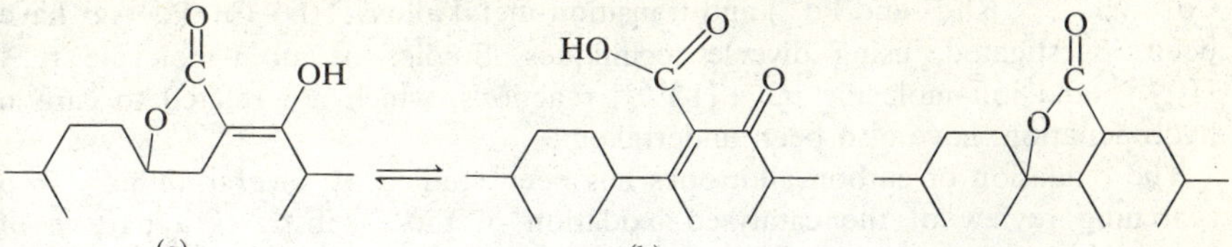

on the surfaces of carbons has been shown[40] to enhance the adsorption of $NH_3$, the effect being more pronounced for acidic $CO_2$-complexes than for acidic CO-complexes. That oxygen which is fixed at unsaturated sites and which does not impart acidity to the surface has no effect.

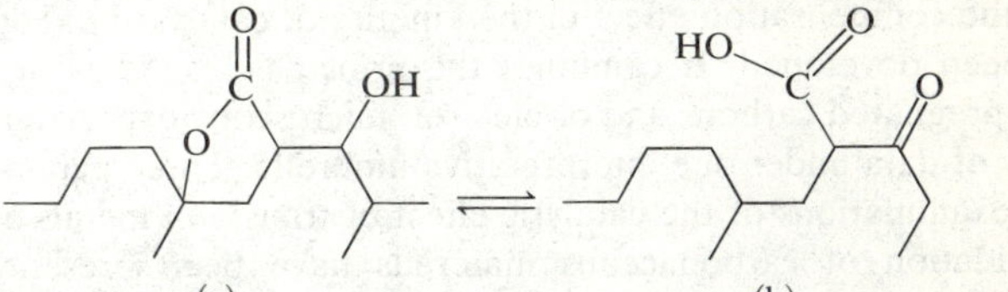

The first unambiguous identification of functional groups formed by chemisorption of heteroatoms on a single-crystal elemental carbon surface has been claimed.[41] $CF_3$, $CF_2$, and CF surface groups that are formed when diamond is exposed to microwave-discharged $SF_6$ have been characterized and quantitatively estimated by XPS (Mg $K_\alpha$). The C (1s) spectrum of the sample contains three satellites, at higher binding energy (ca. 3.1, 5.3, and 7.8 eV) from the main C (1s) peak, corresponding to CF, $CF_2$, and $CF_3$ groups, respectively.

Three papers describing the chemistry of the reactive surface species in whetlerite (a gas-adsorbent charcoal prepared by impregnation of charcoal with ammoniacal solutions of Cu, Cr, and Ag salts) have been published.[42—44] Electron micrographs of simplified whetlerites showed that separate impregnation with ammonium chromate and with ammoniacal basic copper carbonate gave single-crystal electron diffraction patterns having planar separations corresponding to $Cr_2O_3$ and $Cu_2O$, respectively.[42] An ammoniacal solution of the crystals formed in the slow evaporation of a whetlerizing solution gave planar separations which could be correlated with those of $(NH_4)_2CrO_4$, $Cu_2O$, and $Cr_2O_3$; the impregnated copper salts thus appear to protect the chromate ion to a considerable extent.[42] An e.s.r. study[43] has shown that whetlerite reacts with both CNCl and HCN. E.s.r. signals were not obtained after treatment with HCl, t-$C_4H_9NH_2$, and CO. The effect of amine treatments on the reactivity of whetlerites, with respect to CNCl, has also been studied.[44]

The hydrogenation of carbons and graphites has been the subject of a number of communications.[45—49] The catalytic activities of several transition metals (Fe,[47]

[45] R. K. Gould, J. Chem. Phys., 1975, 63, 1825.
[46] H. Imai, S. Nomura, and Y. Sasaki, Carbon, 1975, 13, 333.
[47] A. Tomita and Y. Tamai, J. Phys. Chem., 1974, 78, 2254.
[48] J. Weber and M. Bastick, Compt. rend., 1975, 280, C, 1177.
[49] S. D. Robertson, N. Mulder, and R. Prins, Carbon, 1975, 13, 348.

Co,[47] Ni,[47—49] Rh,[47] and Pd[49]) and transition-metal alloys[49] (Ni–Cu, Pd–Ag) have been investigated, using diverse techniques. Studies of atom–molecule (C + $H_2$)[50,51] and ion–molecule ($C^+ + H_2$)[52,53] reactions, which are related to carbon hydrogenation, have also been undertaken.

The oxidation of carbon allotropes has been studied by several authors. In a searching review of the catalysed oxidation of these substrates a pattern of behaviour amongst catalysts has been identified and a link between rates of catalysed and uncatalysed reactions established.[54] A phenomenological model, based on the compensation effect in the kinetics of catalysed oxidation of carbon, has also been developed. It combines the principal results of bulk oxidation of heavily impregnated carbons and of electron microscopy experiments and unites a large body of data under one interpretative umbrella. Three papers describing the results of examinations of the catalytic effect of transition metals and their oxides in the oxidation of carbonaceous materials have been presented.[55—57] In an extension of their CAEM studies of carbon chemistry, Baker et al.[55] have concluded that the same reactive species is operative in the catalytic oxidation of graphite using vanadium and $V_2O_5$; their pronounced activity was attributed to their ability to wet and spread over the graphite surface at moderately low temperatures. Addition of copper (as $CuSO_4$) to charcoal has been shown to lead to a marked reduction in the charcoal ignition temperature and an increase in the $CO_2$:CO ratio in the gas-phase products.[56] The influence of Cr, Fe, Co, Ni, Cu, and Ag on the gasification of carbon by NO–$O_2$ mixtures has also been examined.[57] The results of further oxidation studies of diamond,[58] graphite,[58] vitreous carbon,[59] and carbon fibres,[60,61] in which the kinetics and mechanism of the reaction are elucidated, have been reported during the period of this Report. Charge-transfer processes in atom–molecule reactions between C and $O_2$ have been studied; products found include $C^+$, $C^-$, $O_2^+$, $O_2^-$, and free electrons.[62] The oxidation of carbon allotropes, by reaction with either $CO_2$[61,63—65] or $H_2O$,[66,67] has been the subject of a number of recent communications. The reaction of OH radicals with graphite has been examined in a flow system, using mass-spectroscopic techniques;[68] it is concluded that OH radicals, which react rapidly at 298 K to produce almost equal amounts of CO and $CO_2$, are much more

---

[50] G. I. Bebeshko and E. N. Eremin, *Russ. J. Phys. Chem.*, 1974, **48**, 52.
[51] R. J. Blint and M. D. Newton, *Chem. Phys. Letters*, 1975, **32**, 178.
[52] D. H. Liskow, C. F. Bender, and H. F. Schaefer, *J. Chem. Phys.*, 1974, **61**, 2507.
[53] J. Appell, D. Brandt, and C. Ottinger, *Chem. Phys. Letters*, 1975, **33**, 131.
[54] F. S. Feates, P. S. Harris, and B. G. Reuben, *J.C.S. Faraday I*, 1974, **70**, 2011.
[55] R. T. K. Baker, R. B. Thomas, and M. Wells, *Carbon*, 1975, **13**, 141.
[56] J. W. Patrick and A. Walker, *Carbon*, 1974, **12**, 507.
[57] H. Marsh and R. Adair, *Carbon*, 1975, **13**, 327.
[58] D. V. Fedoseev and K. S. Uspenskaya, *Russ. J. Phys. Chem.*, 1974, **48**, 897.
[59] F. Rodriguez-Reinoso and P. L. Walker, *Carbon*, 1975, **13**, 7.
[60] W. Watt and W. Johnson, *Nature*, 1975, **257**, 210.
[61] F. Molleyre and M. Bastick, *Compt. rend.*, 1975, **280**, C, 1121.
[62] G. P. Konnen, A. Haring, and A. E. De Vries, *Chem. Phys. Letters*, 1975, **30**, 11.
[63] L. A. Rusakova and E. N. Eremin, *Russ. J. Phys. Chem.*, 1975, **49**, 439.
[64] L. A. Rusakova and E. N. Eremin, *Russ. J. Phys. Chem.*, 1975, **49**, 604.
[65] G. S. Rellick, F. Rodriguez-Reinoso, P. A. Thrower, and P. L. Walker, *Carbon*, 1975, **13**, 81.
[66] I. I. Kulakova and A. P. Rudenko, *Russ. J. Phys. Chem.*, 1974, **48**, 506.
[67] J. Chenion and F.-M. Lang, *Bull. Soc. chim. France*, 1975, 62.
[68] M. F. R. Mulcahy and B. C. Young, *Carbon*, 1975, **13**, 115.

reactive than oxygen atoms under identical conditions. Finally, the electrochemical oxidation of graphites in acids has been investigated by cyclic voltammetry techniques;[69] it was shown that in 70% $HClO_4$, for example, a graphite oxide is formed with almost 100% efficiency.

**Graphite Intercalation Compounds.**—Interest in the inorganic chemistry of these compounds has been maintained during the period of this Report. An improved method[70] for the preparation and isolation of the non-conducting intercalation compound graphite oxide has been reported by a group of Russian authors. The oxide is prepared by oxidation of graphite with a mixture of $H_2SO_4$ and $KMnO_4$ (the method of Hummers and Offeman[71]) and isolated in the following two stages: (i) washed with 5% HCl solution (to dissolve oxidant reduction products) and (ii) treated with dry acetone (to wash it free of HCl). This isolation procedure obviates the difficulties associated with (a) the use of large amounts of water to wash the product, and (b) the consequent peptization of the oxide, and it makes it possible to obtain a product with a higher oxygen and lower water content.[70] Graphite oxide has also been produced,[69] with almost 100% efficiency, during the electrochemical oxidation of graphite in acids such as 70% $HClO_4$.

The structure of graphite fluoride has been examined in two independent investigations using X-ray diffraction[72] and n.m.r. second-moment studies.[73] The interpretation of the results, however, differs considerably; whereas the X-ray data are interpreted in terms of the previously accepted infinite array of *trans*-linked cyclohexane chairs (Figure 1), those of the n.m.r. analysis can only be understood in terms of an infinite array of *cis-trans*-linked cyclohexane boats (Figure 2). The newly described structure is said to be compatible with other

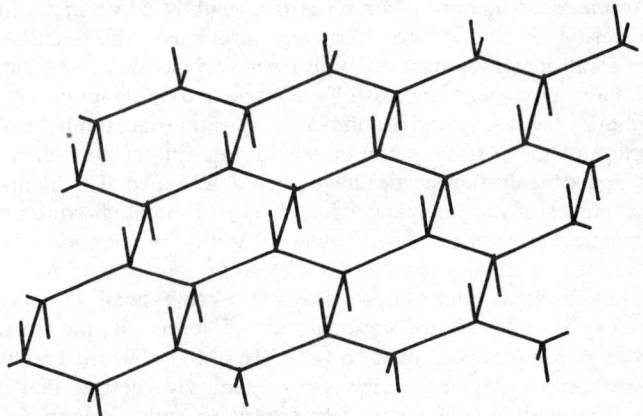

**Figure 1** *Structure of poly(carbon monofluoride) composed of an infinite array of trans-linked cyclohexane chairs.*

[69] J. O. Besenhard and H. P. Fritz, *Z. anorg. Chem.*, 1975, **416**, 106.
[70] A. V. Nikolaev, A. S. Nazarov, and V. V. Lisitsa, *Russ. J. Inorg. Chem.*, 1974, **19**, 1862.
[71] W. S. Hummers and R. E. Offeman, *J. Amer. Chem. Soc.*, 1958, **80**, 1339.
[72] V. K. Mahajan, R. B. Badachhape, and J. L. Margrave, *Inorg. Nuclear Chem. Letters*, 1974, **10**, 1103.
[73] L. B. Ebert, J. I. Brauman, and R. A. Huggins, *J. Amer. Chem. Soc.*, 1974, **96**, 7841.

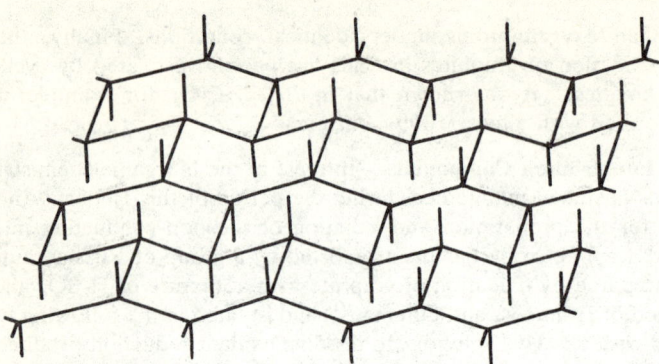

**Figure 2** Structure of poly(carbon monofluoride) composed of an infinite array of cis–trans-linked cyclohexane boats.

experimental measurements, and it is suggested that if $(CF)_n$ does exist in the boat rather than the chair modification, thorough investigation of the $X$-ray diffraction pattern should reveal the existence of an orthorhombic unit cell with approximate cell constants $a = 2.51$, $b = 5.13$, and $c = 6.16$ Å[73] [based on $r(C{-}C) = 1.54$ Å, $r(C{-}F) = 1.4$ Å, tetrahedral co-ordination at carbon, and AAA stacking of layers].

XPS studies[74] of carbon and graphite anodes used in the molten KF–2HF system at 100 °C have indicated that $(CF)_n$ film forms on the electrode surface. In a related study,[75] good agreement has been found between theoretically computed and experimentally determined XPS spectra of $(CF)_n$.

*Metals.* A tremendous amount of data has been published on metal intercalates during the period of this Report. Although previously this section has been restricted to alkali-metal intercalates, both transition metals[76] and alkaline-earth-metals[77—79] have now been successfully intercalated in graphite. A group of Russian authors[76] have prepared a number of transition-metal intercalates (Table 2) by reduction of the corresponding metal chloride intercalates; their magnetic, Mössbauer, and structural properties have been determined. The identity periods ($I_c$) and thicknesses of the intercalated layers ($I'_c$) of metal chloride and derived metal intercalates are collected in Table 2. With the exception of the Mo derivative, $I'_c$ for the metal intercalates varies from 5.80 to 6.15 Å. (*cf.* those of the heavier alkali-metal intercalates). The remarkably small $I'_c$ values for Mo intercalates (3.65—3.95 Å) are rationalized in terms of the formation of sandwich-type $\pi$-complexes, similar to $(h^6\text{-}C_6H_6)_2Mo$, in which graphite acts as an aromatic $\pi$-ligand. Arguments are also presented suggesting that in intercalates of Fe, Co, and Ni the metal forms weak complexes with the graphite

---

[74] H. Imoto, T. Nakajima, and N. Watanabe, *Bull. Chem. Soc. Japan*, 1975, **48**, 1633.
[75] D. E. Parry, J. M. Thomas, B. Bach, and E. L. Evans, *Chem. Phys. Letters*, 1974, **29**, 128.
[76] M. E. Vol'pin, Yu. N. Novikov, N. D. Lapkina, V. I. Kasatochkin, Yu. T. Struchkov, M. E. Kazakov, R. A. Stukan, V. A. Povitskij, Yu. S. Karimov, and A. V. Zvarikina, *J. Amer. Chem. Soc.*, 1975, **97**, 3366.
[77] D. Guerard and A. Hérold, *Compt. rend.*, 1975, **280**, C, 729.
[78] D. Guerard and A. Hérold, *Compt. rend.*, 1974, **279**, C, 455.
[79] D. Billaud and A. Hérold, *Compt. rend.*, 1975, **281**, C, 305.

# Elements of Group IV

**Table 2** Identity periods ($I_c$) and thicknesses of the intercalated layers ($I'_c$) of metal chloride ($C_nMCl_x$) and metal ($C_nM$) intercalates

| Metal chloride | n | Stage | $I_c$/Å | $I'_c$/Å | Metal | Stage | $I_c$/Å | $I'_c$/Å |
|---|---|---|---|---|---|---|---|---|
| FeCl$_3$ | 9 | 1st | 9.42 | 9.42 | Fe | 1st | 5.9 | 5.9 |
| FeCl$_3$ | 18 | 2nd | 12.66 | 9.31 | Fe | 2nd | 9.5 | 6.15 |
| CoCl$_2$ | 7 | 1st | 9.45 | 9.45 | Co | 1st | 5.8 | 5.8 |
| CoCl$_2$ | 13 | 2nd | 12.74 | 9.39 | Co | 2nd | 9.3 | 5.95 |
| MnCl$_2$ | 9 | 1st | 9.50 | 9.50 | Mn | 1st | 5.8 | 5.8 |
| NiCl$_2$ | 16 | 2nd | 12.70 | 9.35 | Ni | 2nd | 9.3 | 5.95 |
| CuCl$_2$ | 6 | 1st | 9.42 | 9.42 | Cu | 1st | 5.8 | 5.8 |
| CuCl$_2$ | 12 | 2nd | 12.75 | 9.40 | Cu | 2nd | 9.3 | 5.95 |
| MoCl$_5$ | 21 | 2nd | 12.54 | 9.19 | Mo | 2nd | 7.3 | 3.95 |
| MoCl$_5$ | 34 | 3rd | 16.02 | 9.32 | Mo | 3rd | 10.4 | 3.7 |
| MoCl$_5$ | 42 | 4th | 19.37 | 9.32 | Mo | 4th | 13.7 | 3.65 |

lattice.[76] Contrasting results have been obtained in independent studies of the reduction of FeCl$_3$[80] and AuCl$_3$[81] intercalates, using potassium and hydrogen, respectively. Tricker et al.[80] suggest that the product from reduction of FeCl$_3$ is α-Fe; Vangelisti and Hérold[81] believe that AuCl$_3$ is reduced to Au, which gathers in aggregates that are several thousand angströms in diameter.

Hérold[77,78] has reported the direct synthesis of binary intercalates of Sr[77] and Ba.[78] Although first- to sixth-stage compounds have been prepared in both systems, only the first-stage compounds, C$_6$Sr and C$_6$Ba, have been obtained in a pure state. The variation in $I_c$ for these compounds (Table 3) shows that the graphite layers conserve their separation throughout all stages, the $I'_c$ values being 4.94 and 5.255 Å for C$_6$Sr and C$_6$Ba, respectively. Intercalation of Ca and Sm has also been reported, but no details are given.[77] Hérold[79] has extended this work by the preparation of new ternary intercalation compounds containing both Ba and Na. The compound richest in metal, C$_{7.5}$(Na, Ba), belongs to the second stage, with an $I_c$ value of 10.73 Å. The value of $I'_c$ (7.38 Å) and the intensity of the 00*l* reflections are in good agreement with the existence of a triple metallic layer formed of a central Ba layer surrounded by two Na layers (4).

Hérold[82—86] has made a further substantial contribution to the chemistry of alkali-metal intercalates. Lithium intercalates of the first, second, third, and fourth stages have been prepared[82] by (i) heating graphite and lithium at 400 °C in a copper or stainless steel tube, and (ii) compressing lithium powder with crushed graphite in an argon atmosphere. The unit cells of first-stage C$_6$Li(5) and

**Table 3** Identity periods ($I_c$/Å) of strontium and barium intercalates

| Stage: | 1st | 2nd | 3rd | 4th | 5th | 6th |
|---|---|---|---|---|---|---|
| $I_c$ of strontium intercalates | 4.94 | 8.34 | 11.70 | 15.01 | 18.25 | 21.53 |
| $I_c$ of barium intercalates | 5.255 | 8.63 | 11.97 | 15.21 | 18.56 | 21.88 |

[80] M. J. Trickèr, E. L. Evans, P. Cadman, N. C. Davies, and B. Bach, Carbon, 1974, **12**, 499.
[81] R. Vangelisti and A. Hérold, Compt. rend., 1975, **280**, C, 571.
[82] D. Guerard and A. Hérold, Carbon, 1975, **13**, 337.
[83] D. Billaud and A. Hérold, Bull. Soc. chim. France, 1974, 2715.
[84] D. Billaud, D. Balesdent, and A. Hérold, Bull. Soc. chim. France, 1974, 2402.
[85] D. Billaud and A. Hérold, Bull. Soc. chim. France, 1974, 2407.
[86] G. Furdin, B. Carton, A. Hérold, and C. Zeller, Compt. rend., 1975, **280**, B, 653.

second-stage $C_{12}Li$ compounds (6) belong to $P6/mmm$ space group, with identity periods of 3.706(1) and 7.065(2) Å, respectively. Lithium also intercalates into soft and hard carbons, including fibres, like the heavier alkali metals.[82]

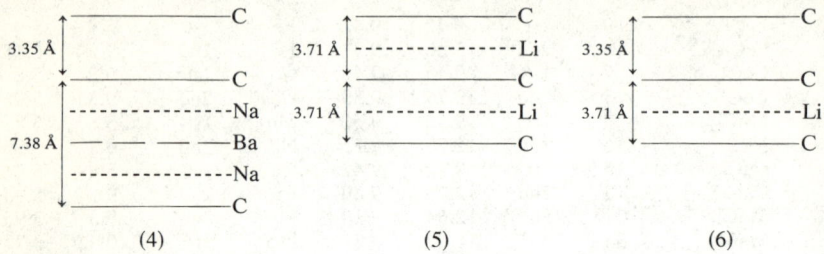

(4)   (5)   (6)

In an intriguing study of the action of binary solutions of two heavy alkali metals (Na+K,[83] Na+Cs,[83] K+Rb,[84] K+Cs,[84] Rb+Cs[84]) on graphite, Hérold has prepared new ternary intercalates, of which those richest in metal have the formula $C_8(M_xM'_{1-x})$, $0 \leq x \leq 1$ [cf. $C_{7.5}$(Na, Ba)]. Although these compounds are generally stable, those containing sodium are metastable; on prolonged contact with the liquid metal or its vapour, equilibrium is established, with loss of Na, leading to the compounds $C_8K$ and $C_8Na_xCs_{1-x}$ ($x \leq 0.02$).[83] The inserted mixture invariably contains more of the heavier alkali metal than the solution with which it is in contact, the extent of enrichment being given by the coefficient $K_e$, as defined by equation (2),

$$K_e = \frac{(1-x)}{x} \frac{X}{(1-X)} \qquad (2)$$

where $X$ is the mole fraction of the lighter element in the metal solution. X-Ray studies of all five ternary products indicate that the metals are inserted as solid solutions; the interplanar distance, however, does not vary with the composition in an ideal manner. The magnetic susceptibilities[86] of those intercalates formed from K, Rb, and Cs have been determined as a function of composition. They are small, positive, independent of temperature, and virtually isotropic. Their composition dependence, however, includes anomalies at $x = 0.35$ (for $C_8K_xRb_{1-x}$), 0.5 (for $C_8K_xCs_{1-x}$), and 0.8 (for $C_8Rb_xCs_{1-x}$). These anomalies, which do not seem to have been observed previously for paramagnetic phases, are situated in the zones of rapid variation of the $c$ parameter with composition.[86] Finally, it has been reiterated that in the equilibrium between Rb and $C_8Rb$, the inserted metal is richer in the heavier isotope than is the free metal. An enrichment coefficient of 1.003 has been obtained.[86] A thermodynamic interpretation of these systems leads to a correlation between $K_e$ and the standard enthalpies and entropies of formation of the pure $C_8M$ phases and to a discussion of the ideality of the $C_8M_xM'_{1-x}$ phases.[84]

Three papers describing the reaction chemistry of alkali-metal intercalates have been published.[87–89] The reaction of $C_8K$ with metal carbonyls provides an easy

---

[87] C. Ungurenasu and M. Palie, J.C.S. Chem. Comm., 1975, 388.
[88] F. Beguin and R. Setton, Carbon, 1975, **13**, 293.
[89] P. Lagrange and A. Hérold, Compt. rend., 1975, **281**, C, 381.

## Elements of Group IV

synthetic route to dinuclear anionic carbonyls of Cr, Mo, W, and Fe[87] [equations (3) and (4)].

$$2M(CO)_6 + 2C_8K \xrightarrow[\text{THF}]{25\,°C} K_2M_2(CO)_{10} + CO + 17C \quad (M = Cr, Mo, \text{or } W) \quad (3)$$

$$2Fe(CO)_5 + 2C_8K \xrightarrow[\text{THF}]{25\,°C} K_2Fe_2(CO)_8 + 2CO + 16C \quad (4)$$

The reaction of $C_8K$ with $Co_2(CO)_8$ proceeds smoothly, according to equation (5).

$$Co_2(CO)_8 + 2C_8K \xrightarrow[\text{THF}]{25\,°C} 2KCo(CO)_4 + 16C \quad (5)$$

The action of $C_8K$, $C_{24}K$, $C_{36}K$ or $C_{48}K$, on solutions of conjugated aromatic hydrocarbons (phenanthrene or anthracene in THF) yields well-defined ternary compounds containing graphite, potassium, and THF.[88] Depending on the parent binary compound and the hydrocarbon, two different first-stage [$C_{24}K(THF)$ and $C_{36}K(THF)$], one second-stage [$C_{48}K(THF)$], and one third-stage [$C_{72}K(THF)$] compound can be produced. Physisorption of $H_2$ by intercalation compounds of potassium with hard and soft carbons has been examined at 77 K by Hérold et al.[89] Although hard carbon derivatives do not generally physisorb $H_2$, soft carbon derivatives will absorb $H_2$, the amount depending on the initial intercalate. $C_8K$ will not absorb $H_2$, but $C_{24}K$, $C_{36}K$, etc. will physisorb up to a maximum $H_2:K$ ratio of 1.9:1, i.e. each cavity contains 4 molecules of $H_2$. In increasing the $C:K$ ratio from 8 to 24 (i.e. for $C_8K$–$C_{24}K$ mixtures) the amount of $H_2$ physisorbed increases from zero to a $H_2:K$ ratio of 1.9:1.[89]

*Halides and Acids.* Intercalation of fluorides of Main-Group elements in graphite has been the subject of four papers.[90—93] Although intercalation of $AsF_5$ (to form $C_{10.0\pm0.3}AsF_5$) has been established, no compounds were formed with $PF_5$, $POF_3$, $BF_3$, $NF_3$, or $AsF_3$, under identical conditions.[90] The reaction of pure liquid, or gaseous, $SbF_5$ with graphite[91] has led to first-stage ($C_{6.5}SbF_5$) and second-/third-stage ($C_{13.2}SbF_5$) products, with $I_c$ values of 8.46 and 11.76/15.11 Å, respectively. Evidence for liquid-type behaviour of $SbF_5$, intercalated in graphite, at temperatures well below the freezing point of pure $SbF_5$ has been obtained from a separate wide-line $^{19}F$ n.m.r. study.[92] Intercalation of $IF_5$ and of $XeOF_4$ has been shown to lead[93] to $C_{8.5}IF_5$ and $C_{8.7}XeOF_4$; the similarity in stoicheiometries is explained on the basis of analogous molecular shapes and dimensions. $C_{8.5}IF_5$ starts to decompose at ca. 80 °C with liberation of $IF_5$. The compound $C_{8.7}XeOF_4$ is stable at 0 °C, but decomposes slowly above room temperature, liberating mainly $XeOF_4$, with minor amounts of xenon.[93]

Considerably more attention has been paid to transition-metal halide intercalates than to those of halides of Main-Group elements, compounds of $TiF_4$,[94]

---

[90] L. Chun-Hsu, H. Selig, M. Rabinowitz, I. Agranat, and S. Sarig, *Inorg. Nuclear Chem. Letters*, 1975, **11**, 601.
[91] J. Melin and A. Hérold, *Compt. rend.*, 1975, **280**, C, 641.
[92] L. B. Ebert, R. A. Huggins, and J. I. Brauman, *J.C.S. Chem. Comm.*, 1974, 924.
[93] H. Selig and O. Gani, *Inorg. Nuclear Chem. Letters*, 1975, **11**, 75.
[94] E. Buscarlet, P. Touzain, M. Armand, and L. Bonnetain, *Compt. rend.*, 1975, **280**, C, 1313.

**Table 4** Identity periods ($I_c$/Å) of transition-metal intercalates

| Intercalate | Stage | $I_c$ | Ref. |
|---|---|---|---|
| $C_{21}TiF_4$ | 3rd | 15.10 | 94 |
| $C_{16.6}NbF_5$ | 2nd | 11.78 | 91 |
| $C_{17.6}TaF_5$ | 2nd | 11.76 | 91 |
| $C_{22}TaF_5$ | 3rd | 15.14 | 91 |

$NbF_5$,[91] $TaF_5$,[91] $MoF_6$,[95] $FeCl_3$,[96—100] $CuCl_2$,[99] and $AuCl_3$[99] being studied. Intercalation of $TiF_4$,[94] $NbF_5$,[91] and $TaF_5$[91] will only occur in the presence of $Cl_2$ or $Br_2$, and gives second- or third-stage compounds (Table 4). $MoF_6$ reacts with graphite with the formation of the first-stage product $C_{7.64}MoF_6$.[95] Metz and Hohlwein[96,97] have described the preparation and characterization of a complete series of $FeCl_3$ intercalates; the value of $I'_c$ (9.38 Å) is larger than those of $TiF_4$ (8.40 Å), $NbF_5$ (8.43 Å), and $TaF_5$ (8.41 Å). The reaction of $C_7FeCl_3$ with potassium has been monitored, using Mössbauer spectroscopy;[98] the reduction results in the initial production of $FeCl_2$ intercalates and eventually the total conversion of ferric ions into bulk α-Fe. Although reduction of $FeCl_3$ and $CuCl_2$ intercalated in graphite is not completely achieved with $H_2$, $AuCl_3$ is reduced to metallic Au at 150 °C.[99] The metal gathers in aggregates of diameter several thousand angström, the smallest being located at the heart of the carbon particles. The ease of reduction of intercalated $AuCl_3$ is explained by the facile dissociation of $AuCl_3$ [equation (6)] and the tendency of AuCl to disproportionate [equation (7)].[99] The stability of $C_{7.5}FeCl_3$ in conc. $H_3PO_4$ (at 160 °C) and 1M-$H_2SO_4$ (at 23 °C) has also been examined by electro-chemical methods.[100]

$$AuCl_3 \rightarrow AuCl + Cl_2 \quad (6)$$

$$3AuCl \rightarrow 2Au + AuCl_3 \quad (7)$$

The use of graphite bisulphate, $C_{24}^+,HSO_4^-,2H_2SO_4$, as an esterification reagent in organic synthesis has been considered.[101] It is very efficient for the production of formates and acetates, reaction times being less than one hour; esterification of other acids is slower, but in most cases the yields are very high.

**Methane and its Substituted Derivatives.**—A large proportion of the recent literature in which the chemistry of these compounds is described is associated with their molecular properties. Following the pattern adopted in the previous Report, therefore, the abstracted data will be considered in the four sections dealing with theoretical, structural, spectroscopic, and chemical studies.

---

[95] A. A. Opalovskii, Z. M. Kuznetsova, Yu. V. Chichagov, A. S. Nazarov, and A. A. Uminskii, *Russ. J. Inorg. Chem.*, 1974, **19**, 1134.
[96] W. Metz and D. Hohlwein, *Carbon*, 1975, **13**, 84.
[97] W. Metz and D. Hohlwein, *Carbon*, 1975, **13**, 87.
[98] M. J. Tricker, E. L. Evans, P. Cadman, N. C. Davies, and B. Bach, *Carbon*, 1974, **12**, 499.
[99] R. Vangelisti and A. Hérold, *Compt. rend.*, 1975, **280**, C, 571.
[100] K. Kinoshita, *Carbon*, 1974, **12**, 686.
[101] J. Bertin, H. B. Kagan, J.-L. Luche, and R. Setton, *J. Amer. Chem. Soc.*, 1974, **96**, 8113.

# Elements of Group IV

*Theoretical Studies.* Five theoretical investigations of $CH_4$ have been undertaken.[102—106] The data have been compared with those for the hydrides of the heavier Group IV elements,[102] for the isoelectronic species $NH_3$,[103,104] $H_2O$,[103,104] $HF$,[103] and $Ne$,[103,105] and for the carbanion $CH_3^-$.[106] Independent investigations of the electronic structures of $CH_3^-$[107] and of the $CH_3^.$ radical[108,109] have also been effected.

Independent *ab initio* calculations[110,111] on $CH_3OH$, $CH_3SH$, and the related anions $^-CH_2OH$, $^-CH_2SH$, $CH_3O^-$, and $CH_3S^-$ have confirmed the relative unimportance of $(p-d)\pi$-bonding between the carbanion lone pair and sulphur $3d$ orbitals in interpreting the enhanced acidity of C—H bonds in $CH_3SH$ relative to those in $CH_3OH$. Whereas Bernadi *et al.*[110] noted two reasons for the stabilization of the sulphur-containing carbanions, *viz.*, the greater polarizability of sulphur and the longer C—S bond length, Streitwieser and Williams[111] conclude that they are stabilized by polarization alone.

*Structural Studies.* Polymorphism in solid $CCl_4$ has been studied by independent groups of authors.[112,113] Although the results establish unambiguously the existence of two $CCl_4$ phases, fcc Ia and orthorhombic Ib, at temperatures immediately below its m. pt., the authors' interpretations of the phase relationships differ. Koga and Morrison[112] consider Ia, which they invariably found on freezing $CCl_4$, to be the stable high-temperature modification, and Ib to be stable at lower temperatures, the transition temperature lying between 240 and 249 K. Abassalti and Mechaud,[113] however, found two crystallization points and consider Ia (crystallization temperature 245.9 K) to be the metastable phase, and Ib (crystallization temperature 250.2 K) to be the stable phase. Whereas Koga and Morrison[112] were unsure of the exact nature of the lower-temperature phase I–phase II transition (225.5 K), Abassalti and Michaud[113] claim that only the orthorhombic phase transforms reversibly into the monoclinic II phase. The phase transitions can be summarized as in equation (8).

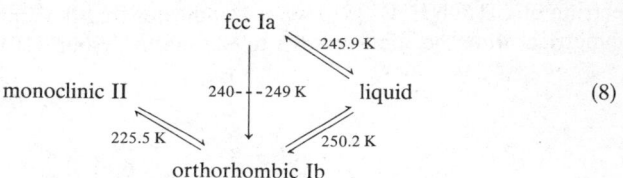

(8)

---

[102] J. P. Desclaux and P. Pyykko, *Chem. Phys. Letters*, 1974, **29**, 534.
[103] T. P. Debies and J. W. Rabalais, *J. Amer. Chem. Soc.*, 1975, **97**, 487.
[104] T. Ahlenius and P. Lindner, *Chem. Phys. Letters*, 1975, **34**, 123.
[105] P. W. Deutsch and A. B. Kunz, *J. Chem. Phys.*, 1975, **62**, 4069.
[106] R. Ahlrichs, F. Driessler, H. Lischka, V. Staemmler, and W. Kutzelnigh, *J. Chem. Phys.*, 1975, **62**, 1235.
[107] J. E. Williams and A. Streitwieser, *J. Amer. Chem. Soc.*, 1975, **97**, 2634.
[108] Y. Ellinger, A. Rassat, R. Subra, and G. Berthier, *J. Chem. Phys.*, 1975, **62**, 1.
[109] Y. Ellinger, R. Subra, B. Levy, P. Millie, and G. Berthier, *J. Chem. Phys.*, 1975, **62**, 10.
[110] F. Bernardi, I. G. Csizmadia, A. Mangini, H. B. Schlegel, M.-H. Whangbo, and S. Wolfe, *J. Amer. Chem. Soc.*, 1975, **97**, 2209.
[111] A. Streitwieser and J. E. Williams, *J. Amer. Chem. Soc.*, 1975, **97**, 191.
[112] Y. Koga and J. A. Morrison, *J. Chem. Phys.*, 1975, **62**, 3359.
[113] M. Abassalti and M. Michaud, *Rev. Chim. minérale*, 1975, **12**, 134.

The transition between phases I and II of $CF_4$ has been studied, using far-i.r. spectroscopic techniques.[114] As a result, the published X-ray diffraction data for phase II have been reinterpreted in terms of a unit cell with space group $C2/c$ ($Z=4$) to give a structure which is consistent with the far-i.r. spectra. A theoretical consideration[115] of the crystal growth and orientational disorder in $CHBr_3$ has also been undertaken.

Structural parameters for $CH_2ClF$,[116] $CH_2DNH_2$,[117] and $CH_3ONH_nD_{2-n}$ ($0 \leq n \leq 2$)[118] have been derived from microwave data. Refined bond lengths and angles have been determined for $CH_2ClF$ (7).[116] The spectra of $CH_2DNH_2$ shows

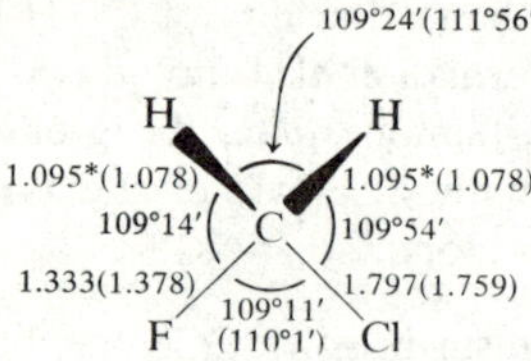

109°24′(111°56′)

H       H
1.095*(1.078)    1.095*(1.078)*
109°14′   C   109°54′
1.333(1.378)      1.797(1.759)
F   109°11′   Cl
   (110°1′)

The angle ($\theta$) between the C—Cl bond and the projection of the C—H bonds on the F—C—Cl plane is 126° 6′ (125° 41′). All values given in parentheses are earlier data. (N. Muller, *J. Amer. Chem. Soc.*, 1953, **75**, 860.) * means that values are assumed ones.

All distances/Å
(7)

it to exist as two distinguishable [*trans* (8) and *gauche* (9)] conformers;[117] this has been confirmed by theoretical[119] and gas-phase i.r. studies.[120] The microwave

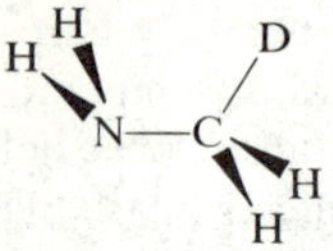

(8) 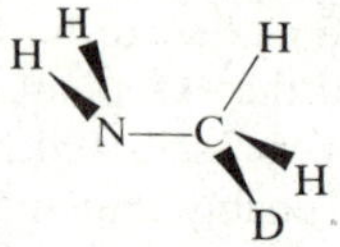          (9)

spectrum of $CH_3ONH_nD_{2-n}$ ($0 \leq n \leq 2$) indicates that the amino-protons occupy a symmetrical *trans*-position relative to the methyl group (10);[118] this is in accord

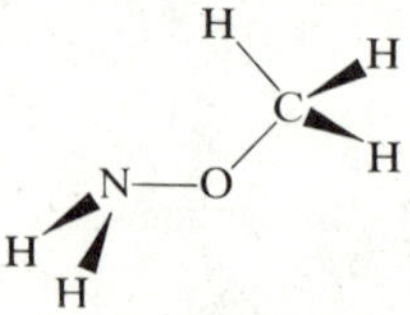

(10)

with the results of *ab initio* calculations.[121] In contrast, a theoretical conformational analysis of the analogous $CH_3OPF_2$[122] has revealed the existence of two

---

[114] Y. A. Sataty, A. Ron, and F. H. Herbstein, *J. Chem. Phys.*, 1975, **62**, 1094.
[115] C. A. Coulson and D. Emerson, *Proc. Roy. Soc.*, 1974, **A337**, 151.
[116] R. N. Nandi and A. Chatterji, *Spectrochim. Acta*, 1975, **31A**, 603.
[117] K. Tamagake and M. Tsuboi, *J. Mol. Spectroscopy*, 1974, **53**, 204.
[118] M. Y. Fong, L. J. Johnson, and M. D. Harmony, *J. Mol. Spectroscopy*, 1974, **53**, 45.
[119] K. Tamagake and M. Tsuboi, *J. Mol. Spectroscopy*, 1974, **53**, 189.
[120] M. Tsuboi, K. Tamagake, and A. Y. Hirakawa, *Spectrochim. Acta*, 1975, **31A**, 495.
[121] L. Radom, W. J. Hehre, and J. A. Pople, *J. Amer. Chem. Soc.*, 1972, **94**, 2371.
[122] G. Robinet, J.-F. Labarre, and C. Leibovici, *Chem. Phys. Letters*, 1974, **29**, 449.

*gauche* (11) equivalent preferred conformations, in which one F atom would be virtually in the POC plane, in disagreement with an earlier [*cis* (12)] interpretation of microwave data.[123]

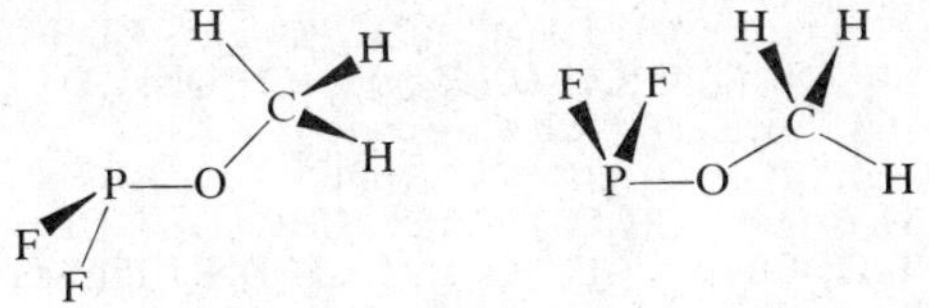

(11)    (12)

*Ab initio* calculations of the equilibrium geometry of the radical CH$_2$OH (13) have shown it to be non-planar, with a CH$_2$ out-of-plane angle ($\alpha$) of 25°.[124] The barriers to internal rotation of the OH group are calculated to be 7.5 (*trans*) and 14.6 (*cis*) kJ mol$^{-1}$.[124]

(a)    (b)
(13)

The vibrational spectra of solutions of [C(N$_3$)$_3$]MCl$_6$ (M = Sb or U)[125] and solid CF$_3$OCs[126] have been recorded and assigned, assuming the C(N$_3$)$_3^+$ ion to possess $C_3$ point symmetry and the CF$_3$O$^-$ ion a tetrahedral structure with $C_{3v}$ symmetry. The bonding in CF$_3$O$^-$ is similar to that in the isoelectronic species NF$_3$O; the C—O bond possesses almost double character and the three C—F bonds are highly polar, implying very strong contributions from resonance structures such as (14).[126]

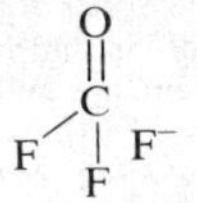

(14)

*Spectroscopic Studies.* The spectroscopic properties of methane and its substituted derivatives have been described in a large number of papers; the molecules examined, using techniques such as microwave, i.r., Raman, u.v., *etc.*, are listed in Table 5. Structural information has been derived from both microwave[116—118] and i.r.[114,120,125,126] spectra; the data are considered in detail in the preceding section.

*Chemical Studies.* Increasing concern at the presence of commerical halogenomethanes (*e.g.* CF$_2$Cl$_2$, CFCl$_3$) in the atmosphere has been manifest in an increase in the number of communications in which their chemistry is described. Fluorocarbons are thought to be inert in the troposphere and to undergo vertical transport into the stratosphere, with subsequent dissociation there. Furthermore,

---

[123] E. G. Codding, C. E. Jones, and R. H. Schwendeman, *Inorg. Chem.*, 1974, **13**, 178.
[124] T.-K. Ha, *Chem. Phys. Letters*, 1975, **30**, 379.
[125] U. Muller and W. Kolitsch, *Spectrochim. Acta*, 1975, **31A**, 1455.
[126] K. O. Christie, E. C. Curtis, and C. J. Schack, *Spectrochim. Acta*, 1975, **31A**, 1035.

**Table 5** *Spectroscopic studies of methane and its substituted derivatives*

| Spectroscopic technique | Molecules examined |
|---|---|
| Microwave | $CH_3NO_2(g)*$;[a] $CH_2DI(g)$;[b] $CH_2FCl(g)$;[c] $CD_2F_2(g)$;[d] $CH_2(CN)_2(g)$;[e] $CH_2DNH_2(g)$;[f] $CH_3ONH_nD_{2-n}$ ($o \leq n \leq 2$)(g);[g] |
| Far-i.r. | $CH_4(g)$;[h] $CF_4(s)$;[i] $CF_3Cl(g$ and 1);[j] |
| I.r. | $CH_4(g)$;[k] $CH_2D_2(g)$;[l] $CHD_3(g)$;[m] $CH_3OH(g)$;[n] $CH_3SCl(g)$;[o] $CH_3Br(g)$;[p] $CH_2Cl_2$(soln.);[q] $CHCl_3(g)$;[r] $CDF_3(g)$;[s] $CF_3Cl(g)$;[t] $CF_3OOF(m)$;[u] $CF_3OCs(s)$;[v] $CH_2DNH_2(g)$;[w] $[C(N_3)_3]MCl_6$ (M = Sb or U)(soln.);[x] |
| Raman | $CH_4(g)$;[y] $CH_3Br(g)$;[z] $CD_3Br(g)$;[aa] $CH_3I(g)$;[ab] $CHF_3(1)$;[ac] $CXCl_3$ (X = H or D)(1);[ad] $CHBr_3(1)$;[ad] $CHI_3(s)$;[ae] $CDX_3$ (X = Cl or Br)(1);[af] $CF_4$(soln.);[ag] $CCl_4(g)$;[ah] $CXCl_3$ (X = F or Br) (g);[ai] $CH_3SCl(g)$;[o] $CF_3OCs(s)$;[v] |
| U.v. | $CH_nBr_{4-n}$ ($n = 0, 1,$ or 3)(g);[aj] $CF_2BrX$ (X = F, Cl, or Br)(g);[ak] |
| N.m.r. | $[^1H]CH_nD_{4-n}$ ($1 \leq n \leq 4$)(g);[al] $[^2H]CDX_3$ (X = Cl or Br)(1);[af] $[^2H]CD_3F(1$ and s);[am] $^{19}F]CD_3F(1$ and s);[am] |
| U.P.S. (He I) | $CF_2BrX$ (X = F, Cl, or Br)(g);[ak] $CFX_3$ (X = Cl or Br)(g);[an] |
| X.P.S. (Y $M_\zeta$) | $CH_nF_{4-n}$ ($0 \leq n \leq 4$)(g);[ao] |
| X.P.S. (Al $K_\alpha$) | $CH_nF_{4-n}$ ($0 \leq n \leq 4$)(g);[ap] $CH_3NO_2(g)$;[aq] $CH_3CN(g)$;[aq] |
| Electron-impact ionization | $CH_4(g^{ar}$ and $s^{as}$); $CX_4$ (X = H or D)(g);[at] $CH_3Br(g)$;[au] $CX_3CN$ (X = H or D)(g);[av] $CF_4(g)$;[aw] $CCl_4(g)$;[ax] |
| Photo-ionization | $CH_3Cl(g)$;[ay] $CH_nBr_{4-n}$ ($1 \leq n \leq 3$)(g);[az] $CH_nI_{4-n}$ ($1 \leq n \leq 3$)(g).[az] |

* g, gas phase; l, liquid phase; m, matrix-isolated; s, solid phase; soln., solution.

[a] F. Rohart, *J. Mol. Spectroscopy*, 1975, **57**, 301; [b] P. D. Mallinson, *J. Mol. Spectroscopy*, 1975, **55**, 94; [c] Ref. 116; [d] E. Hirota and M. Sahara, *J. Mol. Spectroscopy*, 1975, **56**, 21; [e] R. L. Cook, R. T. Walden, and G. E. Jones, *J. Mol. Spectroscopy*, 1974, **53**, 370; [f] Ref. 117; [g] Ref. 118; [h] A. R. H. Cole and F. R. Honey, *J. Mol. Spectroscopy*, 1975, **55**, 492; A. Rosenberg and I. Ozier, *ibid.*, 1975, **56**, 124; [i] Ref. 114; [j] G. J. Davies and M. Evans. *J.C.S. Faraday II*, 1975, **71**, 1275; [k] G. Tarrago, M. Dang-Nhu, G. Poussigue, G. Guelachvili, and C. Amiot, *J. Mol. Spectroscopy*, 1975, **57**, 246; A. S. Pine, *ibid.*, 1975, **54**, 132; [l] J. C. Deroche and G. Guelachvili, *J. Mol. Spectroscopy*, 1975, **56**, 76; J. C. Deroche, G. Graner and A. Cabina, *ibid.*, 1975, **57**, 331; [m] D. E. Jennings and W. E. Blass, *J. Mol. Spectroscopy*, 1975, **55**, 445; [n] R. G. Lee, R. H. Hunt, and E. K. Plyler, *J. Mol. Spectroscopy*, 1975, **57**, 138; [o] F. Winther, A. Guarnieri, and O. F. Nielsen, *Spectrochim. Acta*, 1975, **31A**, 689.; [p] N. Bensari-zizi *Compt. rend.*, 1974, **279**, B, 427; M. Betrencourt, M. Morillonchapey, C. Amiot, and G. Guelachvili, *J. Mol. Spectroscopy*, 1975, **57**, 402; [q] K. Tanabe, *Spectrochim. Acta*, 1974, **30A**, 1891; [r] K. H. Schmidt, W. Hauswirth, and A. Muller, *J. Mol. Spectroscopy*, 1975, **57**, 316; [s] A. Ruorff, H. Burger, S. Biedermann, and R. Antilla, *Spectrochim. Acta*, 1975, **31A**, 1099; [t] S. Biedermann, H. Burger, A. Ruoff, C. Alamichael, and M. Morillon, *J. Mol. Spectroscopy*, 1975, **56**, 367; [u] R. R. Smarzdzewski, D. A. De Marco, and W. B. Fox, *J. Chem. Phys.*, 1975, **63**, 1083; [v] Ref. 126; [w] Ref. 120; [x] Ref. 125; [y] J. P. Champion and H. Berger, *J. Mol. Spectroscopy*, 1975, **55**, 15; [z] T. H. Edwards and S. Brodersen, *J. Mol. Spectroscopy*, 1975, **56**, 376; [aa] T. H. Edwards and S. Brodersen, *J. Mol. Spectroscopy*, 1975, **54**, 121; [ab] P. A. Freedman and W. J. Jones, *J.C.S. Faraday II*, 1975, **71**, 650; [ac] J. De Zwaan, D. W. Hess, and C. S. Johnson, *J. Chem. Phys.*, 1975, **63**, 422; [ad] A. Ruoff, I. Rossi-Sonnichsen, C. Brodbeck, and Nguyen Van Thanh, *Compt. rend*, 1974, **279**, C, 997; [ae] P. Dawson and B. J. Berenblut, *Spectrochim. Acata*, 1975, **31A**, 1049; [af] D. A. Wright and M. T. Rogers, *J. Chem. Phys*, 1975, **63**, 909; [ag] F. P. Daly, A. G. Hopkins, and C. W. Brown, *Spectrochim. Acta*, 1974, **30A**, 2159; [ah] R. J. H. Clark and P. D. Mitchell, *J.C.S. Faraday II*, 1975, **71**, 515; [ai] R. J. H. Clark and O. H. Ellastad, *J. Mol. Spectroscopy*, 1975, **56**, 386; [aj] G. C. Causley and B. R. Russell, *J. Chem. Phys.*, 1975, **62**, 848; [ak] J. Douchet, R. Gilbert, P. Sauvageau, and C. Sandorfy, *J. Chem. Phys.*, 1975, **62**, 366; [al] G. C. Marconi and G. Orlando, *Chem. Physics*, 1975, **10**, 439; [am] P. K. Battacharyya and B. P. Dailey, *J. Chem. Phys.*, 1975, **63**, 1336; [an] F. T. Chau and C. A. McDowell, *J. Electron Spectroscopy*, 1975, **6**, 357; [ao] M. S. Banna and D. A. Shirley, *Chem. Phys. Letters*, 1975, **33**, 441; [ap] M. S. Banna, B. E. Mills, D. W. Davis, and D. A. Shirley, *J. Chem. Phys.*, 1974, **61**, 4780; [aq] T. Fujikawa, T. Ohta, and H. Kuroda, *Chem. Phys. Letters*, 1974, **28**, 433; [ar] J. Milhaud, *Internat. J. Mass Spectrometry Ion Phys.*, 1975, **16**, 327; A. M. Peers and J. Milhaud, *ibid.*, 1974, **15**, 145; [as] T. Huang and W. H. Hamill, *J. Phys. Chem.*, 1974, **78**, 2077; [at] T. G. Finn, B. L. Carnahan, W. C. Wells, and E. C.

it is postulated that the Cl and ClO radicals so formed lead to destruction of the ozone layer *via* a Cl–$O_3$ chain reaction analogous to the NO–$O_3$ chain reaction. Measurements of the concentrations of halogenomethanes in the atmosphere, however, have revealed $CH_3Cl$ as the dominant halogenomethane.[127] Since $CH_3Cl$ is a product of microbial fermentation and of smouldering combustion of vegetation, it has been suggested that the biosynthesis of these compounds might respond to some function of stratospheric $O_3$ density, thus acting as an ozone regulator.[127]

Decomposition of halogenomethanes ($CF_2Cl_2$,[128–131] $CFCl_3$,[129,131] $CF_2Br_2$,[129] $CClF_3$,[131] and $CBrClF_2$[132]) on metals and/or metal oxides has been investigated by a number of French authors. Whereas the reaction[128] of $CF_2Cl_2$ with Zn, Cd, and Hg yielded perfluorinated alkanes ($C_1$ to $C_7$), alkenes ($C_5$ and $C_6$), and mixed fluoro-chloro-derivatives of $C_2H_6$, $C_3H_8$, and $C_5H_{12}$, that[129–132] of halogenomethanes with metal oxides invariably yielded $CO_2$, metal halides and/or metal oxyhalides. It is suggested from an analysis of the results that decomposition of $CF_2Cl_2$ in the presence of $Al_2O_3/CaCO_3$,[130] $Gd_2O_3$,[131] and $CuO$[131] may be the basis of an accurate technique for analysis of $CF_2Cl_2$. A study of the decomposition of chemically activated $CHF_3$ and $CHF_2Cl$ has also been undertaken.[133]

I.r. measurements show that the fluorocarbons containing higher halogens (*e.g.* $CF_2Br,CCl_3F$) can break hydrogen bonds.[134] The order of increasing hydrogen-bond-breaking potency is F < Cl < Br < I for fluorocarbons containing no hydrogen. The presence of hydrogen in chloro- or bromo-fluorocarbons substantially increases the potency. This phenomenon is thought to be due to a competitive mechanism of association, consisting in the formation of donor–acceptor complexes.

Several papers have been published describing donor–acceptor interactions in systems involving $CH_4$[135] and its substituted derivatives.[136–144] Interaction

[127] J. E. Lovelock, *Nature*, 1975, **256**, 193.
[128] D. Barthes and M. Chaigneau, *Compt. rend.*, 1974, **279**, C, 1017.
[129] R. Badre, G. Fevrier, P. Mignon, and J.-L. Vernet, *Compt. rend.*, 1975, **280**, C, 735.
[130] G. Fevrier, P. Mignon, and J.-L. Vernet, *Bull. Soc. chim. France*, 1975, 1587.
[131] M. Chaigneau, D. Barthes, and M. Santarromana, *Compt. rend.*, 1975, **281**, C, 99.
[132] M. Chaigneau and M. Chastagnier, *Bull. Soc. chim. France*, 1974, 2357.
[133] V. I. Vedeneev, A. I. Voronin, and M. A. Teitelboum, *Chem. Phys. Letters*, 1975, **34**, 73.
[134] T. DiPaolo and C. Sandorfy, *Canad. J. Chem.*, 1974, **52**, 3612.
[135] S. R. Ungemach and H. F. Schaefer, *J. Amer. Chem. Soc.*, 1974, **96**, 7898.
[136] V. D. Simonov, T. M. Shamsvidinov, V. E. Pogulyai, and L. N. Popova, *Russ. J. Phys. Chem.*, 1974, **48**, 2659.
[137] D. Guillen, S. Otin, M. Gracia, and C. Gutierrez-Losa, *J. Chim. phys.*, 1975, **72**, 425.
[138] J.-M. Dumas and M. Gomez, *J. Chim. phys.*, 1975, **72**, 953.
[139] H. Peurichard, J.-M. Dumas, and M. Gomez, *Compt. rend.*, 1975, **281**, C, 147.
[140] H. Peurichard, J.-M. Dumas, and M. Gomez, *Compt. rend.*, 1975, **281**, C, 205.
[141] K. F. Wong and S. Ng, *J.C.S. Faraday II*, 1975, **71**, 622.
[142] A. C. Legon, D. J. Millen, and S. C. Rogers, *J.C.S. Chem. Comm.*, 1975, 580.
[143] C. N. R. Rao, *J.C.S. Faraday I*, 1975, **71**, 980.
[144] M. Abassalti, G. Papin, and M. Michaud, *Compt. rend.*, 1975, **280**, C, 365.

---

Zipf, *J. Chem. Phys.*, 1975, **63**, 1596; [au] S. Ikuta, K. Yoshihara, and T. Shiokawa, *Bull. Chem. Soc. Japan*, 1975, **48**, 2134; [av] I. Tokue, I. Nishiyama, and K. Kuchitsu, *Chem. Phys. Letters*, 1975, **35**, 69; [aw] G. R. Wight and C. E. Brion, *J. Electron Spectroscopy*, 1974, **4**, 327; [ax] J. S. Lee, T. C. Wong, and R. A. Bonham, *J. Chem. Phys.*, 1975, **63**, 1609; [ay] T. Baer, A. S. Werner, B. P. Tsai, and S. F. Lin *J. Chem. Phys.*, 1974, **61**, 5468; [az] B. P. Tsai, T. Baer, A. S. Werner, and S. F. Lin, *J. Phys. Chem*, 1975, **79**, 570.

between both $CH_4$[135] and $CCl_4$[136] and $H_2O$ has been substantiated. The theoretically derived equilibrium geometry of $H_2O$—$CH_4$ (15) can be compared to the recently calculated conformations[145] of the ion–non-polar complexes $H_3O^+$—$CH_4$ (16) and $NH_4^+$—$CH_4$ (17); the O—C distance in the weak donor–acceptor complex is, as expected, markedly larger than that in $H_3O^+$—$CH_4$.

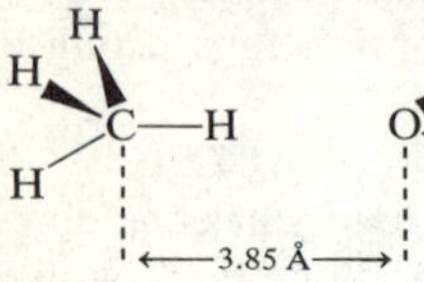

(15) 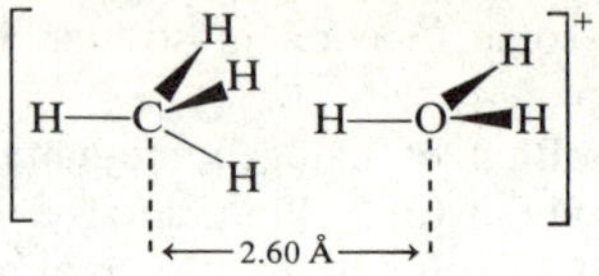  (16)

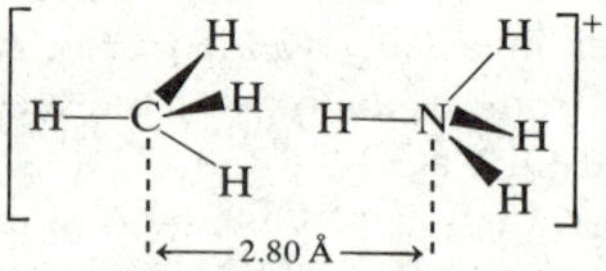

(17)

Specific liquid-phase interactions between perhalogenated methanes and organic bases have been inferred from thermodynamic,[137] dielectric,[137–139] and spectroscopic[138] data. Complex formation between $CHF_3$ and tertiary amines[141] and ethers[141] has been established, using $^1H$ n.m.r. techniques. Microwave data[142] show that the gas-phase complex between $ClF_3$ and $Me_3N$ has the conformation (18), with an N—I distance of 2.93 Å. The interaction of $CH_3OH$ with electron-donor molecules[143] has been studied as a function of deuteriation of the hydroxyl hydrogen.

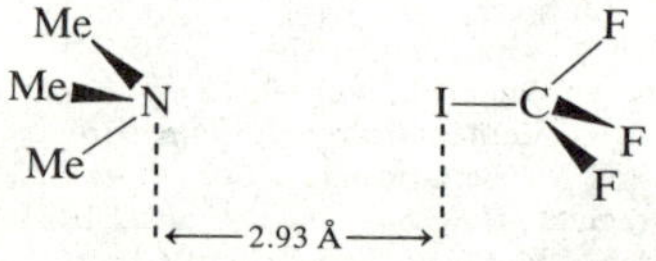

(18)

Finally, in an analysis[144] of solid–liquid equilibrium in the $CHCl_3$–$Bu_4^nN^+$ $Br^-$ system, the presence of three complexes $Bu_4^nN^+$ $Br^-,xCHCl_3$ ($1 \leqslant x \leqslant 3$) in the phase diagram has been established.

As in previous years, a considerable proportion of the chemistry of the substituted methanes involves their reaction with small atomic and molecular species. The pertinent atom–molecule and radical–molecule systems that have been studied are summarized in Table 6. A notable increase in the number of ion cyclotron resonance (I.C.R.) studies of ion–molecule reactions in mixtures including carbon-containing species has occurred; the molecules and mixtures in which these reactions have been investigated are summarized in Table 7. In an extension of the I.C.R technique, which is usually handicapped since the identity of only the

[145] K. Tanaka, T. Yamabe, H. Kato, and K. Fukui, *Bull. Chem. Soc. Japan*, 1975, **48**, 1740.

## Elements of Group IV

**Table 6** *Recently studied reactions of methane and its substituted derivatives with atoms and radicals*

| | | | |
|---|---|---|---|
| Li, Na | $+CF_nCl_{3-n}(n=1$ or $2)^a$ | H, D | $+CH_3OH^l$ |
| Li, Na | $+CF_2ClBr^a$ | H, D | $+CX_3I(X=H$ or $D)^m$ |
| Li, Na | $+CHFCl_2{}^a$ | H | $+CH_3X(X=F, Cl,$ or $Br)^n$ |
| Li, Na | $+CCl_nBr_{4-n}(0 \le n \le 3)^b$ | O | $+CH_3X(X=Cl$ or $Br)^o$ |
| Li, Na | $+CI_4{}^b$ | O | $+CF_3Br^p$ |
| Li, Na, K | $+CCl_4{}^c$ | O | $+CH_3NO_2{}^q$ |
| Li, Na, K, Cs | $+CCl_nBr_{4-n}(2 \le n \le 4)^d$ | Br | $+CH_3X(X=F, Cl, Br,$ or $I)^r$ |
| K | $+CH_3I^{e-g}$ | OH | $+CF_nCl_{4-n}(n=1$ or $2)^s$ |
| K | $+CH_3X(X=Br, CN,$ or $NC)^h$ | OH | $+CHF_2Cl^s$ |
| K | $+(CH_3I)_n{}^i$ | NH | $+CH_nF_{4-n}(0 \le n \le 3)^t$ |
| Rb | $+CH_3I^{j,k}$ | | |
| Rb | $+(CH_3I)_n{}^i$ | | |

[a] D. E. Tevault and L. Andrews, *J. Mol. Spectroscopy*, 1975, **54**, 54; [b] D. E. Tevault and L. Andrews, *J. Amer. Chem. Soc.*, 1975, **97**, 1707; [c] D. E. Tevault and L. Andrews, *J. Mol. Spectroscopy*, 1975, **54**, 110; [d] D. A. Hatzenbuhler, L. Andrews, and F. A. Carey, *J. Amer. Chem. Soc.*, 1975, **97**, 187; [e] H. K. Shin, *Chem. Phys. Letters*, 1975, **34**, 546; [f] G. Rotzoll, R. Viard, and K. Schugerl, *Chem. Phys. Letters*, 1975, **35**, 353; [g] G. Marcelin and P. R. Brooks, *J. Amer. Chem. Soc.*, 1975, **97**, 1710; [h] R. H. Goldbaum and L. R. Martin, *J. Chem. Phys.*, 1975, **62**, 1181; [i] A. Gonzalez-Urena, R. B. Bernstein, and G. R. Phillips, *J. Chem. Phys.*, 1975, **62**, 1818; [j] H. E. Litvak, A. Gonzalez-Urena, and R. B. Bernstein, *J. Chem. Phys.*, 1974, **61**, 4091; [k] A. Cabello and A. Gonzalez-Urena, *Chem. Phys. Letters*, 1975, **35**, 255; [l] J. F. Meacher, P. Kim, J. H. Lee, and R. B. Timmons, *J. Phys. Chem.*, 1974, **78**, 2650; [m] M. R. Levy and J. P. Simons, *J.C.S. Faraday II*, 1975, **71**, 561; [n] A. A. Westenberg and N. DeHaas, *J. Chem. Phys.*, 1975, **62**, 3321; [o] A. A. Westenberg and N. DeHaas, *J. Chem. Phys.*, 1975, **62**, 4477; [p] R. G. Gann, T. C. Frankiewicz, and F. W. Williams, *J. Chem. Phys.*, 1974, **61**, 3488; [q] I. M. Campbell and K. Goodman, *Chem. Phys. Letters*, 1975, **34**, 105; [r] M. E. Berg, W. M. Graver, R. W. Helton, and E. P. Rack, *J. Phys. Chem.*, 1975, **79**, 1327; [s] R. Atkinson, D. A. Hansen, and J. N. Pitts, *J. Chem. Phys.*, 1975, **63**, 1703; [t] P. R. Poole and G. C. Pimental, *J. Chem. Phys.*, 1975, **63**, 1950.

**Table 7** *Systems, including carbon-containing molecules, in which ion–molecule reactions have been studied recently*

| | | |
|---|---|---|
| $CH_4{}^a$ | $CH_4-CO_2{}^e$ | $CH_4-CF_4{}^i$ |
| $CH_3X$ $(X=SCN$ or $NCS)^b$ | $CH_4-COS^f$ | $CH_4-C_2H_6{}^j$ |
| $CH_4-Ar^c$ | $CH_4-CD_4{}^g$ | $CH_4-HCO_2H^k$ |
| $CH_4-H_2{}^d$ | $CX_4-SiH_4(X=H$ or $D)^h$ | $CH_3SH$–small |
| $CH_4-N_2O^e$ | | organic molecules$^l$ |

[a] M. D. Sefcik, J. M. S. Henis, and P. P. Gasper, *J. Chem. Phys.*, 1974, **61**, 4321; M. Riggin and R. C. Dunbar, *Chem. Phys. Letters*, 1975, **31**, 539; [b] T. McAllister, *Internat. J. Mass Spectrometry Ion Phys.*, 1974, **15**, 303; [c] J. R. Wyatt, L. W. Strattan, S. C. Snyder, and P. M. Hierl, *J. Chem. Phys.*, 1975, **62**, 2555; [d] J. K. Kim, L. P. Theard, and W. T. Huntress, *J. Chem. Phys.*, 1975, **62**, 45; J. K. Kim, L. P. Theard, and W. T. Huntress, *Internat. J. Mass Spectrometry Ion Phys.*, 1974/75, **15**, 223; V. G. Anicich, J. H. Futrell, W. T. Huntress, and J. K. Kim, *Internat. J. Mass Spectrometry Ion Phys.*, 1975, **18**, 63; W. T. Huntress, J. K. Kim, and L. P. Theard, *Chem. Phys. Letters*, 1974, **29**, 189; C. J. Cook, N. P. A. Smyth, and O. Heintz, *J. Chem. Phys.*, 1975, **63**, 1218; [e] K. R. Ryan and P. W. Harland, *Internat. J. Mass Spectrometry Ion Phys.*, 1974, **15**, 197; [f] A. Matsumoto. S. Okada, T. Misaki, S. Taniguchi, and T. Hayakawa, *Bull. Chem. Soc. Japan*, 1975, **48**, 794; [g] J. M. S. Henis and M. K. Tripodi, *J. Chem. Phys.*, 1974, **61**, 4863; [h] J. M. S. Henis, G. W. Stewart, and P. P. Gasper, *J. Chem. Phys.*, 1974, **61**, 4860; [i] E. Heckel and R. J. Hanrahan, *J. Chem. Phys.*, 1975, **62**, 1027; [j] K. Hiraoka and P. Kebarle, *Canad. J. Chem.*, 1975, **53**, 970; K. Hiraoka and P. Kebarle, *J. Chem. Phys.*, 1975, **63**, 394; [k] R. H. Staley and J. L. Beauchamp, *J. Chem. Phys.*, 1975, **62**, 1998; [l] B. H. Solka and A. G. Harrison, *Internat. J. Mass Spectrometry Ion Phys.*, 1975, **17**, 379.

charged products can be ascertained, Lieder and Brauman[146] have reported the development of a new technique for the detection of neutral products, e.g. $CH_3Cl, CH_3Br$. Positive-ion reactions of $CH_3Cl$, in the gas mixtures $CH_3Cl-H_2$ and $CH_3Cl-N_2$, have been examined in a high-pressure ion source mass spectrometer;[147] the formation of $CH_2Cl^+(CH_3Cl)_n$, $CH_3ClH^+(CH_3Cl)_n$, and $(H_3CClCH_3)^+$ $(CH_3Cl)_n$ clusters has been observed.

In a thermodynamic analysis[148] of reaction (9), structures of the gas-phase $CH_5^+(CH_4)_n$ clusters have been predicted. The enthalpy and free-energy data are compatible with a three-centre-bond structure for $CH_5^+$ with $C_s$ symmetry (19);

$$\begin{bmatrix} H & & H \\ H \blacktriangleright C \cdots \cdots & \\ H & & H \end{bmatrix}^+$$

(19)

this is in agreement with the results of a recent theoretical (CNDO) study of $CH_5^+$.[149] In the clusters, the first two $CH_4$ molecules interact with the two hydrogen atoms with lowest electron densities, forming bridged three-centre bonds (20). Similar structures are also proposed for the higher clusters.[148] In the

$$\begin{bmatrix} H & & & & H \\ H \blacktriangleright C \cdots H \cdots C \blacktriangleleft H \\ H & H & H & H \end{bmatrix}^+$$

(20)

theoretical analysis,[149] the geometrical structures of isolated and hydrated $CH_5^+$ and $CH_5^-$ were studied. The cation $CH_5^+$ was found to have a conformation with $C_{4v}$ symmetry in solution (cf. $C_s$ symmetry for the isolated cation), whereas the anion $CH_5^-$ was found to have a conformation of $D_{3h}$ symmetry both in solution and in the gaseous phase.

$$CH_5^+(CH_4)_{n-1} + CH_4 \rightleftharpoons CH_5^+(CH_4)_n \qquad (9)$$

The interactions of low-energy electrons with carbon-containing molecules have been examined, using electron cyclotron resonance (E.C.R.) techniques; molecules studied included $CH_3Cl$,[150,151] $CH_2Cl_2$,[151,152] $CHCl_3$,[151,152] $CCl_4$,[151] $CF_4$,[153] $COCl_2$,[151] and $CNCl$.[151]

The photochemistry of $CH_4$,[154,155] $CD_4$,[154] and halogenomethanes including

[146] C. A. Lieder and J. I. Brauman, *Internat. J. Mass Spectrometry Ion Phys.*, 1975, **16**, 307.
[147] Z. Luczynski, W. Malicki, and H. Wincel, *Internat. J. Mass Spectrometry Ion Phys.*, 1974, **15**, 321.
[148] K. Hiraoka and P. Kebarle, *J. Amer. Chem. Soc.*, 1975, **97**, 4179.
[149] P. Cremaschi and M. Simonetta, *Theor. Chim. Acta*, 1975, **37**, 341.
[150] A. A. Christodoulides, R. Schumacher, and R. N. Schindler, *J. Phys. Chem.*, 1975, **79**, 1904.
[151] E. Schultes, A. A. Christodoulides, and R. N. Schindler, *Chem. Physics*, 1975, **8**, 354.
[152] A. A. Christodoulides, R. Schumacher, and R. N. Schindler, *Z. Naturforsch.*, 1975, **30a**, 811.
[153] J. E. Ahnell and W. S. Koski, *J. Chem. Phys.*, 1975, **62**, 4474.
[154] J. Masanet and C. Vermeil, *J. Chim. phys.*, 1975, **72**, 820.
[155] J. P. Ferris and C. T. Chen, *J. Amer. Chem. Soc.*, 1975, **97**, 2962.

$CH_3I$,[156—158] $CH_2I_2$,[159] $CHI_3$,[159] $CF_2Cl_2$,[160] $CFCl_3$,[161] and $CCl_4$[162] has been investigated. In particular, the photo-oxidation of $CH_3I$ in solid $Ar^{158}$ at 10 K has yielded $H_2CO$, $H_2O$, HI, CO, $CO_2$, $HO_2$, and hydrogen hypoiodide, hydrogen-bonded to formaldehyde (21). Other photochemical studies are collated in Table 5, in the section on photoionization spectroscopy.

$$\begin{matrix} H & & I \\ & \diagdown & \diagup \\ & C=O\cdots H-O & \\ & \diagup & \\ H & & \end{matrix}$$
(21)

Radiolysis studies of $CH_4$,[163] $CH_3OH$,[164] $CH_3CN$,[165] $CX_2Cl_2$ (X = H or D),[166] $CH_2X_2$ (X = Cl or Br),[167] $CHX_3$ (X = Cl or Br),[168] and $CCl_4$,[169] in varying conditions have also been undertaken.

The results of several investigations of the oxidation of $CH_4$,[170—174] $CH_3OH$,[175—181] $CH_2Cl_2$,[182] and $CH_3Cl^{182}$ have been reported; kinetic, thermodynamic, mechanistic, and catalytic effects have been examined. In the study of the oxidation of $CH_3OH$ by lead(IV) acetate[181] [equations (10), (11)], it has been shown that the product, $H_2CO$, reportedly stable in the presence of lead(IV) acetate, is, in fact, further oxidized via the methanol hemi-acetal to methyl formate [equations (12)—(14)].

$$CH_3OH + Pb(OAc)_4 \rightleftharpoons CH_3OPb(OAc)_3 + HOAc \qquad (10)$$

$$CH_3OPb(OAc)_3 \rightarrow H_2CO + HOAc + Pb(OAc)_2 \qquad (11)$$

$$H_2CO + CH_3OH \rightleftharpoons CH_3OCH_2OH \qquad (12)$$

$$CH_3OCH_2OH + Pb(OAc)_4 \rightleftharpoons CH_3OCH_2OPb(OAc)_3 + HOAc \qquad (13)$$

$$CH_3OCH_2OPb(OAc)_3 \rightarrow HCO_2CH_3 + HOAc + Pb(OAc)_2 \qquad (14)$$

[156] M. R. Levy and J. P. Simons, *J.C.S. Faraday II*, 1975, **71**, 561.
[157] A. J. Barnes, H. E. Hallam, and J. D. R. Howells, *J.C.S. Faraday II*, 1974, **70**, 1682.
[158] J. F. Ogilvie, V. R. Salares, and M. J. Newlands, *Canad. J. Chem.*, 1975, **53**, 269.
[159] M. Kawasaki S. J. Lee, and R. Bersohn, *J. Chem. Phys.*, 1975, **63**, 809.
[160] R. Milstein and F. S. Rowland, *J. Phys. Chem.*, 1975, **79**, 669.
[161] D. E. Tevault and L. Andrews, *J. Mol. Spectroscopy*, 1975, **54**, 54.
[162] D. D. Davis, J. F. Schmidt, C. M. Neeley, and R. J. Hanrahan, *J. Phys. Chem.*, 1975, **79**, 11.
[163] Y. Siderer and S. Sato, *Bull. Chem. Soc. Japan*, 1975, **48**, 2383.
[164] D. W. Johnson and G. A. Salmon, *J.C.S. Faraday I*, 1975, **71**, 583.
[165] J. L. Baptista and H. D. Burrows, *J.C.S. Faraday I*, 1974, **70**, 2066.
[166] A. Lund, T. Gillbro, D.-F. Feng, and L. Kevan, *Chem. Physics*, 1975, **7**, 414.
[167] H. Ogura, T. Fujimura, T. Unnai, and M. Kondo, *Bull. Chem. Soc. Japan*, 1975, **48**, 1631.
[168] B. S. Ault and L. Andrews, *J. Chem. Phys.*, 1975, **63**, 1411.
[169] L. Andrews, J. M. Grzybowski, and R. O. Allen, *J. Phys. Chem.*, 1975, **79**, 904.
[170] M. Cathonnet and H. James, *J. Chim. phys.*, 1975, **72**, 247.
[171] M. Cathonnet and H. James, *J. Chim. phys.*, 1975, **72**, 253.
[172] N. Pelini and S. Antonik, *Bull. Soc. chim. France*, 1974, 2735.
[173] A. Perche and M. Lucquin, *Compt. rend.*, 1975, **281**, C, 475.
[174] A. B. Nalbandyan, *Doklady Chem.*, 1975, **220**, 100.
[175] H. Baussart, R. Delobel, G. Francois, and J.-M. Leroy, *Compt. rend.*, 1975, **281**, C, 375.
[176] Yu. G. Medvedevskikh, A. Ya. Bulakh, and S. K. Chuchmarev, *Russ. J. Phys. Chem.*, 1974, **48**, 1259.
[177] E. Momot and G. Bronoel, *Compt. rend.*, 1974, **279**, C, 619.
[178] P. Sideswaren, *Indian J. Chem.*, 1974, **12**, 1077.
[179] L. N. Kurina and N. V. Vorontsova, *Russ. J. Phys. Chem.*, 1975, **49**, 675.
[180] L. N. Kurina, L. I. Shakirova, and L. I. Durofeeva, *Russ. J. Phys. Chem.*, 1975, **49**, 695.
[181] Y. Pocker and B. C. Davis, *J.C.S. Chem. Comm.*, 1974, 803.
[182] E. Sanueza and J. Heicklen, *J. Phys. Chem.*, 1975, **79**, 7.

The kinetics of the reaction of $CH_3OH$ with $NH_3$ [equation (15)], on an industrial $SiO_2/Al_2O_3$ catalyst, have been determined as a function of temperature (318—418 °C);[183] they are interpreted in terms of the surface properties of the catalyst.

$$CH_3OH + NH_3 \rightarrow CH_3NH_2 + H_2O \qquad (15)$$

Thermodynamic calculations have been undertaken for gas-phase chlorination/fluorination[184] and disproportionation[185] reactions in the $CH_4$–$Cl_2$–HF system; the formation of Freon 11 and 12 in the gas-phase chlorofluorination of $CH_4$ is included.[184]

A number of investigations of the chemistry of bistrifluoromethyl peroxide and related compounds have been undertaken.[186—191] $CF_3OOCF_3$ has been established as a major product of the photolysis of $CF_3OCl$ in Ar matrices at 8 K;[186] $COF_2$ was also formed, together with smaller quantities of $CF_3OF$, $ClF$, and $COFCl$. Photolysis of $CF_3OF$ yielded, under identical conditions, $COF_2$ only.[186] The kinetics of the gas-phase pyrolysis of $CF_3OOCF_3$, in the presence of $SO_3F$ dimers, have been investigated as a function of temperature (197—244 °C) and pressure (5—100 Torr);[187] $CF_3OOSO_2F$ is the sole product of the reaction. In an analysis of the kinetic data, a value (193.3 kJ mol$^{-1}$) for the peroxide bond dissociation energy $D(CF_3O\text{—}OCF_3)$ has been calculated (cf. earlier values of $169.8 \pm 20.9$[188] and $193.3 \pm 1.4$ kJ mol$^{-1}$ [189]).

The results of a study[190] of the behaviour of $CF_3OOF$ towards selected inorganic substrates indicate that a primary step in the reaction sequence involves formation of OF radicals, which lead to a complex series of oxidation and fluorination reactions; the reaction of $CF_3OOF$ with $SO_2$ [equations (16)—(21)] is quoted as an example. Finally, $CF_3OOCl$[191] undergoes addition reactions with

$$CF_3OOF + SO_2 \rightarrow CF_3OSO_2^{\cdot} + OF^{\cdot} \qquad (16)$$

$$OF^{\cdot} \rightarrow F^{\cdot} + O^{\cdot} (\text{and } O_2) \qquad (17)$$

$$CF_3OSO_2^{\cdot} + F^{\cdot} \rightarrow CF_3OSO_2F \qquad (18)$$

$$SO_2^{\cdot} + F^{\cdot} \rightarrow SO_2F^{\cdot} \qquad (19)$$

$$CF_3OSO_2^{\cdot} + O^{\cdot} \rightarrow CF_3OSO_2O^{\cdot} \qquad (20)$$

$$CF_3OSO_2O^{\cdot} + FSO_2^{\cdot} \rightarrow CF_3OSO_2OSO_2F \qquad (21)$$

alkenes at low temperatures (<0 °C), giving trifluoromethylperoxy derivatives in high yield; the reaction is thought to proceed via an electrophilic mechanism in which the positive Cl of $CF_3OOCl$ adds to the more negative alkene carbon [equation (22)].

$$CF_3OOCl + CR^1R^2\!=\!CR^3R^4 \rightarrow CF_3OOCR^1R^2CR^3R^4Cl \qquad (22)$$

---

[183] G. Schmitz, J. Chim. phys., 1975, **72**, 579.
[184] G. P. Chernyuk, Russ. J. Phys. Chem., 1974, **48**, 1549.
[185] G. P. Chernyuk, Russ. J. Phys. Chem., 1975, **49**, 466.
[186] R. R. Smardzewski and W. B. Fox, J. Phys. Chem., 1975, **79**, 219.
[187] B. Descamps and W. Forst, Canad. J. Chem., 1975, **53**, 1442.
[188] J. B. Levy and R. C. Kennedy, J. Amer. Chem. Soc., 1972, **94**, 3302.
[189] R. C. Kennedy and J. B. Levy, J. Phys. Chem., 1972, **76**, 3480.
[190] R. A. DeMarco and W. B. Fox, Inorg. Nuclear Chem. Letters, 1974, **10**, 965.
[191] N. Walker and D. D. DesMarteau, J. Amer. Chem. Soc., 1975, **97**, 13.

## Elements of Group IV

**Formaldehyde and its Substituted Derivatives.**—Interest in the inorganic chemistry of formaldehyde and its substituted derivatives appears to be declining. Indeed, very few papers in which their properties are described have been published during the period of this Report. The majority of those abstracted are considered in the following sub-sections; studies in which spectroscopic data are reported, however, have been collated in Table 8.

**Table 8** *Spectroscopic studies of formaldehyde and its substituted derivatives*

| Spectroscopic technique | Molecules examined |
|---|---|
| Microwave | FClCS(g)*;[a] HCOSH(g);[b] |
| Far-i.r. | (HCOOX)$_2$ (X = H or D)(g);[c] |
| I.r. | H$_2$CO(g[d] and soln.[e,f]); HCO$_2$H(g);[g,h] DCO$_2$H(g);[g] XCO$_2$K (X = H or D)(s);[i] |
| U.v. | H$_2$CO(g);[j] FClCS(g);[k] HCO$_2$H(g);[l] |
| X.P.S. | H$_2$CO(g);[m,n] |
| Electron-impact ionization | H$_2$CO(g).[o] |

* g, gas phase; s, solid phase; soln., solution.

[a] Ref. 209; [b] Ref. 215; [c] W. G. Rothschild, *J. Chem. Phys.*, 1974, **61**, 3422; [d] K. Tanaka, K. Yamada, T. Nakagawa, J. Overend, and K. Kuchitsu, *J. Mol. Spectroscopy*, 1975, **54**, 243; [e] O. A. Lavrova, Zh. A. Mateeva, T. M. Lesteva, and B. I. Pantukh, *Russ. J. Phys. Chem.*, 1975, **49**, 373; [f] O. A. Lavrova, Zh. A. Mateeva, T. M. Lesteva, and B. I. Pantukh, *Russ. J. Phys. Chem.*, 1975, **49**, 389; [g] J. Bournay and Y. Marechal, *Spectrochim. Acta*, 1975, **31A**, 1351; [h] E. Willemot, D. Dangoisse, and J. Bellet, *Compt. rend.*, 1974, **279**, B, 247; [i] E. Spinner, *Spectrochim. Acta*, 1975, **31A**, 1545; [j] C. R. Lessard, D. C. Molule, and S. Bell, *Chem. Phys. Letters*, 1974, **29**, 603; [k] C. R. Subramaniam and D. C. Moule, *J. Mol. Spectroscopy*, 1974, **53**, 443; [l] S. Bell, T. L. Ng, and A. D. Walsh, *J.C.S. Faraday II*, 1975, **71**, 393; [m] Ref. 200; [n] Ref. 201; [o] A. Chutjian, *J. Chem. Phys.*, 1974, **61**, 4279.

*Formaldehyde, Carbonyl Halides*, etc. The production of thioformaldehyde monomer, by direct photolysis of thietan vapour, has been reported.[192] Also studied is a method whereby the monomer may be trapped and analysed quantitatively *via* its reaction with cyclopentadiene.[192]

Theoretical anaylses of the electronic structures of H$_2$CO,[193–201] Br$_2$CO,[202] and X$_2$CY (X = F or Cl; Y = O or S)[203] have been undertaken. Furthermore, the geometries of H$_2$CO,[204] H$_2$CS,[204] and their protonated derivatives H$_2$$\overset{+}{C}$OH and H$_2$$\overset{+}{C}$SH, and of the radical H$\dot{C}$O[205] have been predicted from theoretical calculations; that of H$_2$$\overset{+}{C}$OH has also been elucidated in a mass spectrometric analysis[206] of reaction (23). SCF MO calculations[204] show that whereas the methylene group

---

[192] D. R. Dice and R. P. Steer, *Canad. J. Chem.*, 1974, **52**, 3518.
[193] L. S. Cederbaum, W. Domcke, and W. von Niessen, *Chem. Phys. Letters*, 1975, **34**, 60.
[194] W. C. Johnson, *J. Chem. Phys.*, 1975, **63**, 2144.
[195] B. J. Garrison, W. A. Lester, and H. F. Schaefer, *J. Chem. Phys.*, 1975, **63**, 1449.
[196] L. P. Batra and O. Robaux, *Chem. Phys. Letters*, 1975, **28**, 529.
[197] S. R. Langhoff, S. T. Elbert, C. F. Jackels, and E. R. Davidson, *Chem. Phys. Letters*, 1974, **29**, 247.
[198] B. J. Garrison and H. F. Schaefer, *J. Chem. Phys.*, 1974, **61**, 3039.
[199] T. D. Davis, G. M. Maggiora, and R. E. Christoffersen, *J. Amer. Chem. Soc.*, 1974, **96**, 7878.
[200] I. H. Hillier and J. Kendrick, *J. Electron Spectroscopy*, 1975, **6**, 325.
[201] H. Basch, *Chem. Physics*, 1975, **10**, 157.
[202] A. B. M. S. Bassi and R. E. Bruns, *J. Phys. Chem.*, 1975, **79**, 1880.
[203] A. B. M. S. Bassi and R. E. Bruns, *J. Chem. Phys.*, 1975, **62**, 3235.
[204] F. Bernardi, I. G. Csizmadia, I. H. B. Schlegel, and S. Wolfe, *Canad. J. Chem.*, 1975, **53**, 1144.
[205] P. Botschwina, *Chem. Phys. Letters*, 1974, **29**, 98.
[206] K. Hiraoka and P. Kebarle, *J. Chem. Phys.*, 1975, **63**, 1688.

of $H_2CO$ is strongly positive, that of $H_2CS$ is slightly negative; *i.e.*, whereas oxygen behaves towards carbon as a $\pi$-donor and $\sigma$-acceptor, sulphur behaves both as a $\pi$- and $\sigma$-donor. The results of these calculations[204] also suggest that the stable conformations of $H_2COH^+$ and $H_2CSH^+$ correspond to a structure (22) in which all atoms lie in the same plane. In contrast, those of the thermodynamic analysis[206]

$$HCO^+ + H_2 \rightarrow H_2C^+OH \qquad (23)$$

indicate that the structure is not that of protonated formaldehyde, $H_2COH^+$, but is that of $H_3CO^+$ (23), which contains a non-symmetric three-centre bond between three hydrogens; this is in agreement with an earlier suggestion by Fehsenfeld.[207]

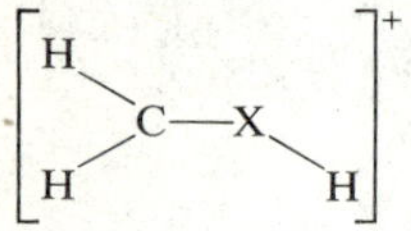

(22) 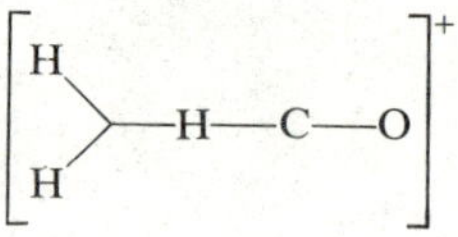            (23)

The theoretically derived[205] geometry of $HCO^·$ has been compared to that measured experimentally.[208] The OCH bond angles (128.0° and 127.4°, respectively) agree closely; whereas $r(C-H)$ is calculated (1.100 Å) to be shorter than experimentally determined (1.110 Å), the calculated $r(C-O)$ value (1.180 Å) is larger than the experimental value (1.171 Å). The theoretician[205] finds this latter observation astonishing, and, based on previous correlations between theoretical and experimental bond distances, would predict a real bond length of 1.195 Å, *ca.* 0.015 Å larger than the calculated value.

The structure of FClCS has been determined by Ziel *et al.*, using both microwave[209] and electron-diffraction[210] techniques. The two sets of data (Table 9) are in excellent agreement, confirming the planar structure.

**Table 9** *Geometrical parameters of* FClCS

|  | Ref. | $r(S-C)$/Å | $r(C-F)$/Å | $r(C-Cl)$/Å | $\angle SCF$/° | $\angle SCCl$/° |
|---|---|---|---|---|---|---|
| Microwave data | 209 | 1.595 | 1.326 | 1.715 | 123.82 | 127.12 |
| Electron-diffraction data | 210 | 1.5931 | 1.3387 | 1.7178 | 123.58 | 127.28 |

The oxidation of $H_2CO$ has been the subject of a number of recent communications;[211,212] both kinetic and thermodynamic factors are assessed. Ion–molecule chemistry in $H_2CO$[213] and in $H_2CO-NH_3$ mixtures[214] has also been investigated, using I.C.R. techniques.

[207] F. C. Fehsenfeld, D. B. Dunkin, and E. E. Ferguson, *Astrophys. J.*, 1974, **188**, 43.
[208] J. A. Austin, D. H. Levy, C. A. Gottlieb, and H. E. Radford, *J. Chem. Phys.*, 1974, **60**, 207.
[209] H. J. Kohrmann and W. Zeil, *Z. Naturforsch.*, 1975, **30a**, 183.
[210] F. Gleisberg, A. Haberl, and W. Zeil, *Z. Naturforsch.*, 1975, **30a**, 549.
[211] P. Sideswaren, *Indian J. Chem.*, 1974, **12**, 1077.
[212] Yu. G. Medvedevskikh, A. Ya. Bulakh, and S. K. Chuchmarev, *Russ. J. Phys. Chem.*, 1974, **48**, 1259.
[213] Z. Karpas and F. S. Klein, *Internat. J. Mass Spectrometry Ion Phys.*, 1975, **16**, 289.
[214] Z. Karpas and F. S. Klein, *Internat. J. Mass Spectrometry Ion Phys.*, 1975, **18**, 65.

*Formic Acid, Formates*, etc. The structure of monothioformic acid has been determined in an analysis of the microwave spectra of isotopically substituted species of HCOSH.[215] Two distinct but similar spectra were observed which could be assigned to a mixture of *trans* (24) and *cis* (25) isomers of HCOSH. No evidence for either isomer of HCSOH was obtained. Preliminary values of $r(C-S)$ and $\angle OCS$ were calculated for both isomers.

(24)        (25)

\* assumed values; all distances/Å

Accurate X-ray and neutron-diffraction data have been combined to study the electron distribution in non-centrosymmetric $HCO_2Li,H_2O$.[216] *Ab initio* calculations of difference electron density have also been effected. Both techniques indicate a significant dissimilarity in electron density distributions associated with the two OH bonds of the $H_2O$ molecule.[216] The i.r. spectra of $XCO_2K$ (X = H or D) in KBr discs have been found to change far more extensively on pulverizing and repressing than is normal.[217] The various spectra were all superpositions, in varying proportions, of two independent spectra; these were assigned to the original solid and to a transformation product, formed presumably under the action of pressure.[217] The results of a $^{13}C$ n.m.r. study[218] of the two crystallographically non-equivalent ions in $(HCO_2)_2Ca$ indicate a strong perturbation of the ions by the crystal field; this is considered to be evidence for the non-planarity of the ions.[218]

The catalytic decomposition[219] of $HCO_2H$ vapour over $MnMoO_4$ and its component oxides, MnO and $MoO_3$, has been found to lead predominantly to dehydrogenation. The thermal decomposition of alkali-metal formates has also been examined;[220] the principal reaction products were alkali-metal carbonates and oxalates.

**Derivatives of Group VI Elements.**—*Oxides, Sulphides, and Related Species.* Following the pattern adopted in the previous Report, and as a result of the paucity of reported data, the chemistry of these species will be considered in a single section. M.O. calculations of the electronic structures of the molecules $CO$,[221,222] $CS$,[223] $CO_2$,[224] and the carbon monoxide dimer (ethylenedione) $C_2O_2$,[225] of the radical ions $CO^+$,[222] $CO^-$,[226] $CO_2^-$,[224,226] and $CO_3^-$,[226] and of the

---

[215] W. H. Hocking and G. Winnewisser, *J.C.S. Chem. Comm.*, 1975, 63.
[216] J. O. Thomas, R. Tellgren, and J. Almof, *Acta Cryst.*, 1975, **B31**, 1946.
[217] E. Spinner, *Spectrochim. Acta*, 1975, **31A**, 1545.
[218] J. L. Ackerman, J. Tegenfeldt, and J. S. Waugh, *J. Amer. Chem. Soc.*, 1974, **96**, 6843.
[219] P. Rajaram, B. Viswanathan, M. V. C. Sastri, and V. Srinivasan, *Indian J. Chem.*, 1974, **12**, 1267.
[220] T. Meisel, Z. Halmos, K. Seybold, and E. Pungor, *J. Therm. Analysis*, 1975, **7**, 73.
[221] G. DeWith and D. Feil, *Chem. Phys. Letters*, 1975, **30**, 279.
[222] J. Cambray, J. Gasteiger, A. Streitwieser, and P. S. Bagus, *J. Amer. Chem. Soc.*, 1974, **96**, 5978.
[223] P. J. Bruna, W. E. Kammer, and K. Vasudevan, *Chem. Physics*, 1975, **9**, 91.
[224] J. Pacansky, U. Wahlgren, and P. S. Bagus, *J. Chem. Phys.*, 1975, **62**, 2740.
[225] R. C. Haddon, D. Poppinger, and L. Radom, *J. Amer. Chem. Soc.*, 1975, **97**, 1645.
[226] I. Matousek, A. Fojtik, and R. Zahradnik, *Coll. Czech. Chem. Comm.*, 1975, **40**, 1679.

complexes $CO_2,H_2O^{227}$ and $CO^-,H_2O^{226}$ have been undertaken. Although the principal reason for these studies is the elucidation of the spectroscopic parameters of these species, both thermodynamic and geometrical data have been derived. The conclusion obtained from an examination of the dissociative pathways available to $C_2O_2,^{225}$ *i.e.* that $C_2O_2$ is both kinetically (singlet) and thermodynamically (singlet and triplet) unstable with respect to two CO molecules, is consistent with the experimental evidence but contrasts with the findings of earlier theoretical studies. The most stable conformation of $CO_2,H_2O$ (26) has been found to be planar, the carbon atom being bound to the water oxygen atom $[r(C-O) = 2.63 \text{ Å}].^{227}$

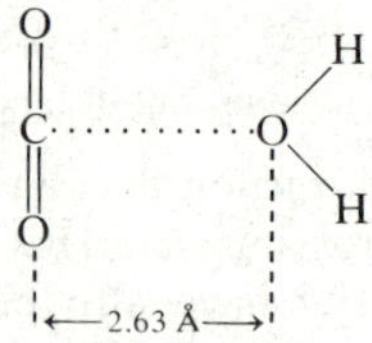

$$\underset{\underset{O}{\overset{\|}{C}}}{\overset{O}{\|}}\cdots\cdots O\overset{H}{\underset{H}{\diagdown}}$$

$\longleftarrow$ 2.63 Å $\longrightarrow$

(26)

Although many spectroscopic studies of these molecules have been undertaken (Table 10), very few are of more than peripheral interest to the inorganic chemist. The most interesting is a $^{13}C$ n.m.r. study of $C_3O_2$ in $CDCl_3$ at $-40\,°C.^{228}$ The data obtained are consistent with considerable oxonium ion character at C-1 and a high electron density at C-2, as represented by resonance structure (27).$^{228}$

$$\overset{+}{O}\equiv C(2)-\overset{-}{C}(1)=C(2)=O$$

(27)

A major part of the research effort in this field is associated with the oxidation of CO.$^{229-240}$ The catalytic activity of Pt,$^{229}$ Pd—SnO$_2$,$^{230}$ Fe$_2$O$_3$,$^{231}$ and ZnO$^{232}$ has been ascertained; furthermore, several spectroscopic (i.r.$^{235}$ and p.e.s.$^{236-239}$) and theoretical$^{240}$ studies of the species formed when CO is adsorbed on transition metals have been undertaken. The conversion of CO into $CO_2$ by dinitrosyl complexes of Ir and Rh has been studied;$^{233}$ evidence supporting a mechanism involving the formation of dinitrogen dioxide intermediates in these reactions is given. The related CS–O$_2$ reaction chemistry has been examined;$^{241}$ the mixture is

---

[227] B. Jonsson, G. Karlstrom, and H. Wennerstrom, *Chem. Phys. Letters*, 1975, **30**, 58.
[228] E. A. Williams, J. D. Cargioli, and A. Ewo, *J.C.S. Chem. Comm.*, 1975, 366.
[229] S. Kishimoto, *Bull. Chem. Soc. Japan*, 1975, **48**, 1937.
[230] G. C. Bond, L. R. Molloy, and M. J. Fuller, *J.C.S. Chem. Comm.*, 1975, 796.
[231] Y. Sugi, N. Todo, and T. Sato, *Bull. Chem. Soc. Japan*, 1975, **48**, 337.
[232] K. Tanaka and K. Miyahara, *J. Phys. Chem.*, 1974, **78**, 2303.
[233] S. Bhaduri, B. F. G. Johnson, C. J. Savory, J. A. Segal, and R. H. Walter, *J.C.S. Chem. Comm.*, 1974, 809.
[234] Yu. P. Andreev, Yu. M. Voronkov, and I. A. Semiokhin, *Russ. J. Phys. Chem.*, 1975, **49**, 540.
[235] N. P. Sokolova, *Russ. J. Phys. Chem.*, 1974, **48**, 744.
[236] R. W. Joyner and M. W. Roberts, *J.C.S. Faraday I*, 1974, **70**, 1819.
[237] A. M. Bradshaw, D. Menzel, and M. Steinkilberg, *Chem. Phys. Letters*, 1974, **28**, 516.
[238] R. W. Joyner and M. W. Roberts, *Chem. Phys. Letters*, 1974, **29**, 447.
[239] K. Kishi and M. W. Roberts, *J.C.S. Faraday I*, 1975, **71**, 1715.
[240] G. Blyholder, *J. Phys. Chem.*, 1975, **79**, 756.
[241] R. J. Richardson, *J. Phys. Chem.*, 1975, **79**, 1153.

**Table 10** *Recent spectroscopic studies of carbon oxides, sulphides, and related species*

| Spectroscopic technique | Molecules examined. |
|---|---|
| Far-i.r. | $CO_2(s)$;[a] |
| I.r. | $C_3O_2(g)$;[b]  $CO(s)$;[c]  $CO(m)$;[c]  $CO_2(g)$;[d—f]  $CO_2(m)$;[g]  $COSe(g)$;[h] |
| U.v. | $CO(g)$;[i]  $CO_2(s)$;[j]  $COSe(g)$;[h,k,l]  $CS_2(g)$;[m]  $CSe_2(g)$;[n] |
| N.m.r. | $[^{13}C]C_3O_2$(soln.);[o] |
| Photoelectron | $C_3O_2$ (XPS);[p]  CO (XPS);[p]  CO (UPS);[q]  COSe (UPS);[r]  CSSe (UPS);[r] |
| Electron-impact ionization | $CO(g)$;[s—v]  $CO_2(g)$;[w—y]  $COS(g)$;[z]  $CS_2(g)$;[w,z] |
| Photoionization | $CO(g)$;[q]  $CO_2(g)$;[aa]  $CO_2(s)$;[aa]  $COS(g)$.[ab] |

\* g, gas phase; m, matrix-isolated; s, solid phase; soln., solution.

[a] Y. A. Sataty and A. Ron, *J. Chem. Phys.*, 1974, **61**, 5471; [b] A. W. Mantz, P. Connes, G. Guelachvili, and C. Amiot, *J. Mol. Spectroscopy*, 1975, **54**, 43; [c] G. J. Jiang, W. B. Person, and K. G. Brown, *J. Chem. Phys.*, 1975, **62**, 1201; [d] R. A. Toth, *J. Mol. Spectroscopy*, 1974, **53**, 1; [e] J. P. Aldridge, R. F. Holland, H. Flicker, K. Nill, and T. C. Harman, *J. Mol. Spectroscopy*, 1975, **54**, 328; [f] W. G. Planet, J. R. Aronson, and J. F. Butler, *J. Mol. Spectroscopy*, 1975, **54**, 331; [g] L. Fredin, B. Nelander, and G. Ribbegard, *J. Mol. Spectroscopy*, 1975, **53**, 410; [h] E. J. Finn and G. W. King, *J. Mol. Spectroscopy*, 1975, **56**, 39; [i] S. G. Tilford and J. D. Simmons, *J. Mol. Spectroscopy*, 1974, **53**, 436; [j] K. M. Monahan and W. C. Walker, *J. Chem. Phys.*, 1975, **63**, 1676; [k] S. Cradock, R. J. Donavan, W. Duncan, and H. M. Gillespie, *Chem. Phys. Letters*, 1975, **31**, 344; [l] E. J. Finn and G. W. King, *J. Mol. Spectroscopy*, 1975, **56**, 52; [m] K. E. J. Halin, D. N. Malm, and A. J. Merer, *J. Mol. Spectroscopy*, 1975, **54**, 318; [n] S. Cradock, R. J. Donovan, W. Duncan, and H. M. Gillespie, *J.C.S. Faraday II*, 1975, **71**, 156; [o] Ref. 228; [p] I. H. Hillier and J. Kendrick, *J.C.S. Faraday II*, 1975, **71**, 1369; [q] J. L. Gardner and J. A. R. Samson, *J. Chem. Phys.*, 1975, **62**, 1447; [r] S. Cradock and W. Duncan, *J.C.S. Faraday II*, 1975, **71**, 1262; [s] K. Kollman, *Internat. J. Mass Spectrometry Ion Phys.*, 1975, **17**, 261; [t] R. G. Hirsch, R. J. Van Brunt, and W. D. Whitehead, *Internat. J. Mass Spectrometry Ion Phys.*, 1975, **17**, 335; [u] J. Mazeau, C. Schermann, and G. Joyez, *J. Electron Spectroscopy*, 1975, **7**, 269; [v] N. Swanson, J. A. Celotta, C. E. Kuyatt, and J. W. Cooper, *J. Chem. Phys.*, 1975, **62**, 4880; [w] M. J. Hubin-Franskin and J. E. Collin, *J. Electron Spectroscopy*, 1975, **7**, 139; [x] J. A. D. Stockdale, B. P. Pullen, and A. E. Carter, *Internat. J. Mass Spectrometry Ion Phys.*, 1975, **17**, 241; [y] M. Misakian, M. J. Mumaia, and J. F. Faris, *J. Chem. Phys.*, 1975, **62**, 3442; [z] G. R. Wight and C. E. Brion, *J. Electron Spectroscopy*, 1974, **4**, 335; [aa] K. M. Monahan and W. C. Walker, *J. Chem. Phys.*, 1974, **61**, 3886; [ab] D. L. Judge and L. C. Lee, *Internat. J. Mass Spectrometry Ion Phys.*, 1975, **17**, 329.

hypergolic at room temperature and exhibits the characteristics of a branched-chain reaction, the dominant reaction path producing O and OCS.

Conditions for the formation of $CH_3OH$ in the catalytic hydrogenation of CO have been defined; it is obtained in small quantities at high temperature, under low pressure, and with low $CO:H_2$ ratio.[242] The interaction between CO and $NH_3$ or $N_2H_4$ on transition-metal (V or Fe) surfaces at 25 °C has been investigated using i.r. spectroscopic techniques.[243] The high-pressure reaction of $PF_5$ with $CO_2$, $CS_2$, or COS has also been investigated.[244]

Aspects of the photochemistry of $CO_2$[245—248] and COS[249—251] have been studied

[242] C. Aharoni and H. Starer, *Canad. J. Chem.*, 1974, **52**, 4044.
[243] R. W. Sheets and G. Blyholder, *J. Phys. Chem.*, 1975, **79**, 1572.
[244] A. P. Hagan and B. W. Calloway, *Inorg. Chem.*, 1975, **14**, 1622.
[245] T. G. Slanger, R. L. Sharpless, G. Black, and S. V. Filseth, *J. Chem. Phys.*, 1974, **61**, 5022.
[246] R. F. Phillips and D. S. Sethi, *J. Chem. Phys.*, 1974, **61**, 5473.
[247] I. Koyano, T. S. Wauchop, and K. H. Welge, *J. Chem. Phys.*, 1975, **63**, 110.
[248] L. F. Loucks and R. C. Michaelson, *J. Chem. Phys.*, 1975, **63**, 404.
[249] M. C. Lin, *Chem. Physics*, 1975, **7**, 433.
[250] M. C. Lin, *Chem. Physics*, 1975, **7**, 438.
[251] R. B. Klemm, S. Glicker, and L. J. Stief, *Chem. Phys. Letters*, 1975, **33**, 512.

in a number of flash photolysis experiments; other related data are collected in Table 10 in the section on photoionization spectroscopy. The heterogeneous thermal decomposition of $C_3O_2$ on a hot filament occurs according to equation (24) at temperatures up to 2000 °C;[252] the production of free radicals is not observed. The reaction rates are low, presumably due to the high thermal stability of $C_3O_2$.

$$C_3O_2 \rightarrow 2CO + C$$

Several ion–molecule and ion-clustering reactions involving these molecules have been investigated. Negative-ion reactions involving $CO_3^-$ (with O,[253] H,[254] $NO_2$,[255] $CO_2$,[255] $H_2O$,[255] and $SO_2^{255}$) and $CO_4^-$ (with H[254] and $O_3^{255}$) have led to, *inter alia*, the complex $CO_3^-,H_2O$.[255] Association reactions of $CO^+$ (with $H_2^{256}$ and $CO^{257,258}$) and of $HCO^+$ (with $CO$)[258] have yielded, *inter alia*, the complexes $HCO^+,CO$ and $CO^+,(CO)_n$; in fact, $CO^+$ was found to add up to three molecules of CO;[257] the first and second appear to be attached irreversibly, probably by true chemical bonds. In view of the readiness with which $CO^+,CO$ is formed, it is suggested that it may be possible to achieve synthesis of $C_2O_2$ by indirect means; *i.e.*, by electron neutralization of clusters of $CO^+(CO)_n$.[257] Formation of negative ions from collisions of CS with $CO_2$, $CS_2$, and COS has been observed;[259] the parent ions $CO_2^-$, $CS_2^-$, and $COS^-$ are obtained, together with other fragment ions. Ion–molecule reactions involving carbon monoxide and carbon dioxide as molecular substrates are summarized in Table 11.

**Table 11** *Recently studied reactions of CO and of $CO_2$ with ionic species*

| | |
|---|---|
| $CO+X^+$ (X = He, Ne, Ar, Kr, or Xe)[a] | $CO_2+O^{-e}$ |
| $CO+O_2^{+b,c}$ | $CO_2+OH^{-e}$ |
| $CO+H_3^{+d}$ | $CO_2+O_2^{-e}$ |
| $CO+OH^-(H_2O)_n$ ($2 \leq n \leq 4$)[e] | $CO_2+O_3^{-e}$ |
| $CO_2+X^+$ (X = He, Ne, Ar, Kr, or Xe)[a] | $CO_2+O_3^-(H_2O)_n$ ($n = 1$ or $2$)[e] |
| $CO_2+O_2^{+b}$ | $CO_2+OH^-(H_2O)_n$ ($2 \leq n \leq 4$)[e] |

[a] J. B. Laudenslager, W. T. Huntress, and M. T. Bowers, *J. Chem. Phys.*, 1974, **61**, 4600; [b] W. Lindinger, D. L. Albritton, M. McFarland, F. C. Fehsenfeld, A. L. Schmeltekopf, and E. G. Ferguson, *J. Chem. Phys.*, 1975, **62**, 4101; [c] J. M. Ajello, *J. Chem. Phys.*, 1975, **63**, 1863; [d] J. K. Kim, L. P. Theard, and W. T. Huntress, *Chem. Phys. Letters*, 1975, **32**, 610; [e] Ref. 255.

Atom–molecule reactions of hydrogen with COS,[260] of carbon with CO,[261] $CS$,[262] COS,[263—265] and $CS_2$,[263] of argon with COS,[266] and of barium with $CO_2^{267}$ have

[252] A Wehrer, X. Duval, and P. Wehrer, *Bull. Soc. chim. France*, 1975, 1099.
[253] J. Fletcher and J. L. Moruzzi, *Internat. J. Mass Spectrometry Ion Phys.*, 1975, **18**, 57.
[254] F. C. Fehsenfeld, *J. Chem. Phys.*, 1975, **63**, 1686.
[255] F. C. Fehsenfeld and E. E. Ferguson, *J. Chem. Phys.*, 1974, **61**, 3181.
[256] J. K. Kim, L. P. Theard, and W. T. Huntress, *J. Chem. Phys.*, 1975, **62**, 45.
[257] R. L. Horton, J. L. Franklin, and B. Mazzeo, *J. Chem. Phys.*, 1975, **62**, 1739.
[258] M. Meot-Ner and F. H. Field, *J. Chem. Phys.*, 1974, **61**, 3742.
[259] S. Y. Tang, E. W. Rothe, and G. P. Reck, *J. Chem. Phys.*, 1974, **61**, 2592.
[260] S. Tsunashima, T. Yokata, I. Safarik, H. E. Gunning, and O. P. Strausz, *J. Phys. Chem.*, 1975, **79**, 775.
[261] G. R. Smith and W. Weltner, *J. Chem. Phys.*, 1975, **62**, 4592.
[262] I. R. Slagle, R. E. Graham, J. R. Gilbert, and D. Gutman, *Chem. Phys. Letters*, 1975, **32**, 184.
[263] C. N. Wei and R. B. Timmons, *J. Chem. Phys.*, 1975, **62**, 3240.
[264] R. B. Klemm and L. J. Stief, *J. Chem. Phys.*, 1974, **61**, 4900.
[265] R. G. Shortridge and M. C. Lin, *Chem. Phys. Letters*, 1975, **37**, 146.
[266] S. J. Harris, K. C. Janda, S. E. Novick, and W. Klemperer, *J. Chem. Phys.*, 1975, **63**, 881.
[267] P. J. Dagdigian, H. W. Cruse, A. Schultz, and R. N. Zare, *J. Chem. Phys.*, 1974, **61**, 4450.

*Elements of Group IV*

also been investigated. CCO, produced in the C—CO reaction was trapped in various matrices at 4 K; e.s.r. spectra were determined and isotope and matrix effects were estimated.[261] The microwave spectrum of the van der Waals molecule ArOCS, produced in the expansion of a 1 atm. gas mixture of 99% Ar and 1% COS through a supersonic nozzle at −80 °C, has been measured.[266] ArOCS is a prolate, slightly asymmetric, top with a non-linear T-shaped structure (28). It is suggested that this structure provides evidence for the proposal[268] that van der Waals molecules mimic the structures of their isoelectronic and isovalent chemical analogues (*i.e.* $OCS_2^{2-}$ or $CO_3^{2-}$).

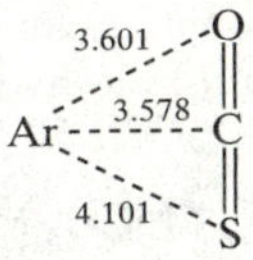

(28)

All distances/Å

*Carbonates, Thiocarbonates, and Related Anions.* Data for this section of the Report have been collected only for the simple compounds of the Main-Group elements; those concerning the chemistry of for example, rare-earth-metal carbonates or transition-metal carbonato-complexes have been excluded. The charge distribution in the $CO_3^{2-}$ anion has been the subject of further discussion;[269—271] Ladd[269] has refuted the challenge, made by Jenkins and Waddington,[270] on earlier results[271] by showing that the value for the charge on the $CO_3^{2-}$ oxygen is closer to $-0.80^{269}$ than $-0.20$, as suggested by Jenkins and Waddington.[270] Yamamoto *et al.*[272] have calculated effective charges for the atoms in $CaCO_3$ and $MgCO_3$ from an analysis of the optically active vibrations of calcite and magnesite using a rigid-ion model: those for oxygen, $-0.47$ (29) and $-0.50$ (30), respectively, lie between those discussed earlier.[269—271] Yamamoto *et al.*[273] have also analysed the corresponding vibrations of dolomite, $CaMg(CO_3)_2$, with reference to those of calcite and magnesite. The electronic structure of the $CO_3^{2-}$ anion has also been examined in an XPS study of the valence region of a number of carbonates.[274]

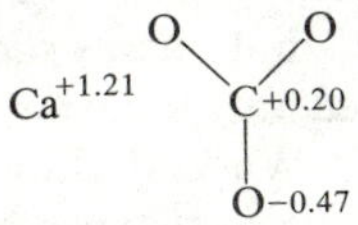

(29) 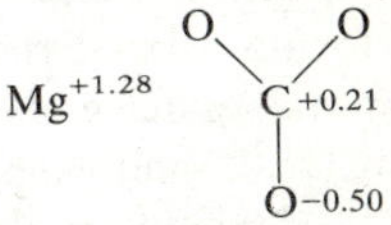                                      (30)

MO calculations show that $CO_4^{4-}$ and $CS_4^{4-}$, so far experimentally unknown, are capable of existence, perhaps in either molten-salt or non-aqueous media.[275]

---

[268] S. J. Harris, S. E. Novick, W. Klemperer, and W. E. Falconer, *J. Chem. Phys.*, 1974, **61**, 193.
[269] M. F. C. Ladd, *J. Inorg. Nuclear Chem.*, 1975, **37**, 1529.
[270] H. D. B. Jenkins and T. C. Waddington, *Nature Phys. Sci.*, 1972, **238**, 126.
[271] M. F. C. Ladd, *Nature Phys. Sci.*, 1972, **238**, 125.
[272] A. Yamamoto, Y. Shiro, and H. Murato, *Bull. Chem. Soc. Japan*, 1975, **48**, 1102.
[273] A. Yamamoto, T. Utida, H. Murato, and Y. Shiro, *Spectrochim. Acta*, 1975, **31A**, 1265.
[274] A. Calabrese and R. G. Hayes, *J. Electron Spectroscopy*, 1975, **6**, 1.
[275] D. K. Johnson and J. R. Wasson, *Inorg. Nuclear Chem. Letters*, 1974, **10**, 891.

The Raman spectra of the highly symmetrical squarate, $C_4O_4^{2-}$ ($D_{4h}$), and croconate, $C_5O_5^{2-}$ ($D_{5h}$), anions have been measured at various excitation wavelengths in aqueous solutions of $K_2C_4O_4$ and $Na_2C_5O_5$.[276] The observed vibrational (i.r. and Raman) spectra of anyhdrous $K_2CO_3$ and $Rb_2CO_3$[277] are in excellent agreement with a $C_{2h}$ unit-cell analysis based on isomorphous crystal structures with space group $P2_1/c$.

X-Ray studies of the crystal chemistry of $4MgCO_3,Mg(OH)_2,4H_2O$,[278] $CaCO_3$,[279] $SrCO_3$,[280] $BaCO_3$,[280] and $Tl_2CO_3$[281] have been undertaken. $MgCO_3,Mg(OH)_2,4H_2O$,[278] $CaCO_3(II)$[279] (a high-pressure metastable phase), and $Tl_2CO_3$[281] all crystallize with monoclinic symmetry; their unit-cell parameters are collected in Table 12. The structure of hydromagnesite, which has a strong

**Table 12** Unit-cell parameters of $MgCO_3,Mg(OH)_2,4H_2O$, $CaCO_3$ (II), and $Tl_2CO_3$

| Compound | a/Å | b/Å | c/Å | β/° | Space group | Ref. |
|---|---|---|---|---|---|---|
| $MgCO_3,Mg(OH)_2,4H_2O$ | 10.11(1) | 8.97(1) | 8.38(1) | 114.6(9) | $P2_1/c$ | 278 |
| $CaCO_3$ (II) | 6.334(20) | 4.948(15) | 8.033(25) | 107.9 | $P2_1/c$ | 279 |
| $Tl_2CO_3$ | 12.486 | 5.382 | 7.530 | 122.35 | $C2/m$ | 281 |

pseudo-orthorhombic character (space group $Bbcm$), is based on a three-dimensional framework of $MgO_6$ octahedra and triangular $CO_3$ ions;[278] both independent Mg atoms have octahedral co-ordinations, with average Mg—O distances of 2.10 and 2.04 Å, respectively. The transformation of $CaCO_3$ from hexagonal calcite to monoclinic $CaCO_3(II)$, under a pressure of 15 kbar, is displacive;[279] it involves both the rotation of $CO_3^{2-}$ groups by 11° and a small displacement of Ca atoms. The related topochemical conversion of aragonite into calcite has been studied, using dilatometric[282] and mechanicochemical[283,284] techniques. In a discussion of the interrelationships of room-temperature [orthorhombic $MCO_3(III)$] and high-temperature [rhombohedral $MCO_3(II)$ and face-centred-cubic $MCO_3(I)$] phases of $SrCO_3$ and $BaCO_3$, it is suggested that the structure of the rhombohedral phases is probably more complex than the model previously proposed for disordered rhombohedral phases such as $NaNO_3$ (I).[280]

Neutron-diffraction studies have been used to locate the hydrogen atoms in $Na_2CO_3,H_2O$,[285] $KHCO_3$,[286] and $KDCO_3$.[286] Wu and Brown have shown that those in $Na_2CO_3,H_2O$[285] lie between the positions predicted by Baur[287] and those

---

[276] M. I. Ijima, Y. Udagawa, K. Kaya, and M. Ito, *Chem. Physics*, 1975, **9**, 229.
[277] M. H. Brooker and J. B. Bates, *Spectrochim. Acta*, 1974, **30A**, 2211.
[278] M. Akao, F. Marumo, and S. Iwai, *Acta Cryst.* 1974, **B30**, 2670.
[279] L. Merrill and W. A. Bassett, *Acta Cryst.*, 1975, **B31**, 343.
[280] K. O. Stromme, *Acta Chem. Scand.*(A), 1975, **29**, 105.
[281] R. Marchand, Y. Piffard, and M. Tournoux, *Canad. J. Chem.*, 1975, **53**, 2454.
[282] L. I. Tolokonnikova, N. D. Topor, B. M. Kadenatsi, and T. V. Solov'eva, *Russ. J. Phys. Chem.*, 1975, **48**, 1552.
[283] J. M. Craido and J. M. Trillo, *J.C.S. Faraday I*, 1975, **71**, 961.
[284] N. Kawashima and K. Meguro, *Bull. Chem. Soc. Japan*, 1975, **48**, 1857.
[285] K. K. Wu, and I. D. Brown, *Acta Cryst.*, 1975, **B31**, 890.
[286] J. O. Thomas, R. Tellgren, and I. Olovsson, *Acta Cryst.*, 1974, **B30**, 2540.
[287] W. H. Baur, *Acta Cryst.*, 1972, **B28**, 1456.

postulated by Dickens et al.[288] Bond valence calculations show that the Na—O and H···O bonds are equivalent as regards their strength, and that all three oxygen atoms of the $CO_3^{2-}$ ions have identical trigonal-bipyramidal environments.[285] In an extension of their comprehensive hydrogen-bond studies, Thomas et al.[286] have collected neutron-diffraction data for $KHCO_3$ and $KDCO_3$ at 298 K. The data are complementary to those obtained in the single-crystal X-ray diffraction study described in the previous Report.[289] Contrary to the X-ray findings, the H (and D) atoms of the centrosymmetric $(HCO_3)_2^{2-}$ dimers occupy two possible sites in the hydrogen bonds in the rough proportions 4:1; the dimers must also be subject to the same degree of disordering. The extent to which this effect is dependent on the specimen could not be ascertained from this work alone.[282] The geometries of the disordered dimers as determined in the neutron-diffraction study are compared with the X-ray values in Figure 3. The data

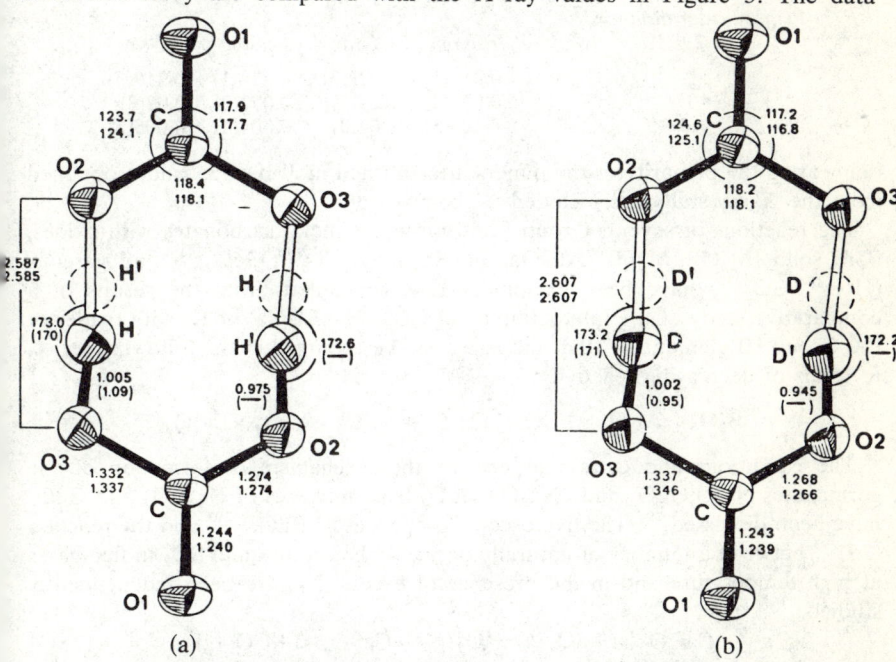

**Figure 3** *Distances/Å and angles/°, derived from neutron-diffraction data, within the disordered dimeric anions in* (a) $KHCO_3$ *and* (b) $KDCO_3$. *The values derived from X-ray data are included for comparison below the neutron values. The figures have been simplified by drawing the centrosymmetrically related H and D atoms as dashed circles. The thermal vibration ellipsoids for the atoms are otherwise drawn to include 50% probability.*

(Reproduced by permission from *Acta Cryst.*, 1974, **B30**, 2540)

---

[288] B. Dickens, F. Mauer, and W. E. Brown, *J. Res. Nat. Bur. Stand, Sect. A*, 1970, **74**, 319.
[289] P. G. Harrison and P. Hubberstey, in 'Inorganic Chemistry of the Main-Group Elements', ed. C. C. Addison, (Specialist Periodical Reports), The Chemical Society, London, 1975, Vol. 3, p. 190.

**Table 13** *The planarity of the $(HCO_3)_2^{2-}$ dimer: values obtained from the X-ray study are included for comparison. The parallel planes through the oxygen atoms of each $HCO_3^-$ ion are taken as references; the notation is defined as follows*

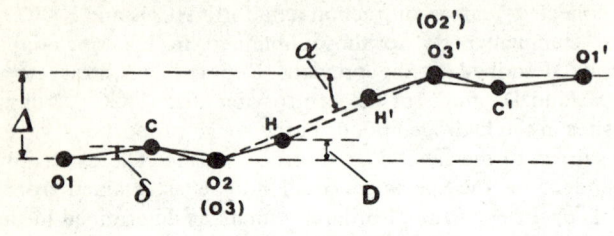

| Perpendicular distances/Å at 298 K: | KHCO$_3$ | | KDCO$_3$ | |
|---|---|---|---|---|
| | neutron | X-ray | neutron | X-ray |
| Δ | 0.214(1) | 0.222(3) | 0.214(1) | 0.219(3) |
| D | 0.077(2) | 0.09(3) | 0.072(2) | 0.03(3) |
| δ | 0.003(1) | 0.004(3) | 0.004(1) | 0.005(3) |

delineating the planarity of the dimers are collected in Table 13; values obtained from the X-ray study are included for comparison.

The reactions of several Group I and Group II metal carbonates with oxides, both solid [$M_2O_3$ (M = B, Al, Ga, In, Sc, La or Y;[290] $GeO_2$[291]] and gaseous ($H_2O$[292] $SO_2$[293]) have been examined. It is concluded, from the results of a comparative study of the interaction of $M_2CO_3$ (M = Li, Na, or K) with oxides of the Group III elements,[290] that the latter can be arranged in the following series, in terms of decreasing reactivity:

$$B_2O_3 > Al_2O_3 > In_2O_3 > Ga_2O_3 > Sc_2O_3 > Y_2O_3 > La_2O_3$$

The conditions and certain aspects of the mechanism of formation of the germanates $Na_2O,GeO_2$ and $2Na_2O,9GeO_2$ from mixtures of $Na_2CO_3$ and $GeO_2$ have been discussed.[291] The hygroscopic properties of $BaCO_3$[292] and the reaction (25)[293] between a number of naturally occurring limestones and $SO_2$ in flue gases at high temperature and in the presence of excess $O_2$ have been independently studied.

$$CaCO_3(s) + SO_2(g) + \tfrac{1}{2}O_2(g) \rightarrow CaSO_4(s) + CO_2(g) \quad (25)$$

The decomposition of $CaCO_3$[294,295] and of solid solutions of $CaCO_3$ and $SrCO_3$[295] has been investigated, using vapour-pressure techniques. The results show that the solid solutions exhibit positive deviation from ideality and adopt the aragonite structure.

In a continuing study of the chemistry of the $CO_3^-$ radical, Hoffman et al. have measured rate constants for the reaction of $CO_3^-$ and $HCO_3$ radicals with

---

[290] V. F. Annopol'skii, E. K. Belyaev, and I. P. Knigavko, *Russ. J. Inorg. Chem.*, 1975, **20**, 164.
[291] E. K. Belyaev and V. F. Annopol'skii, *Russ. J. Inorg. Chem.*, 1974, **19**, 30.
[292] E. V. Polyakov, V. V. Davituliani, and T. G. Akhmetov, *Russ. J. Phys. Chem.*, 1975, **49**, 846.
[293] M. Hartman, *Coll. Czech. Chem. Comm.*, 1975, **40**, 1466.
[294] D. Beruto and A. W. Searcy, *J.C.S. Faraday I*, 1974, **70**, 2145.
[295] M. M. Evstigneeva, N. A. Iofis, and A. A. Bundel', *Russ. J. Phys. Chem.*, 1974, **48**, 800.

tryptophan,[296] its derivatives,[296] and a series of substituted benzene derivatives;[297] it is concluded that $CO_3^-$ is a selective electrophilic reagent.[297] A mass spectrometric study of the photochemistry of mixtures of $CO_3^-$ and $H_2O$ has been effected;[298] $CO_3^-$ photodissociates into $CO_2$ and $O^-$, but $CO_3^-,H_2O$ into $CO_3^-$ and $H_2O$. A semi-empirical M.O. Study of, *inter alia*, $CO_3^-$ and $HCO_3$ has also been undertaken.[299]

Several papers describing the chemistry of perthiocarbonates have been published by a number of French authors, principally Robineau and Zins. The preparation and characterization of $K_2CS_4$,[300] its addition compounds $K_2CS_4$, R (R = $\frac{1}{2}H_2O$, MeOH, or $Me_2NH$),[300] $(Me_4N)_2CS_4$,[301] and the addition compound $(Me_4N)_2CS_4MeOH$[301] have been reported; the i.r. spectra of the $CS_4^{2-}$ anion in these compounds[300,301] and $(Ph_4As)_2CS_4$[302] and of the $CS_3^{2-}$ anion in $(Ph_4As)_2CS_3$[302] have been recorded and analysed. The crystal structure of $K_2CS_4,MeOH$ has been described.[303,304] It is triclinic (space group $P\bar{1}$), with unit-cell parameters $a = 8.00$, $b = 9.75$, $c = 13.41$ Å, $\alpha = 69.02$, $\beta = 73.78$, $\gamma = 82.36°$.[304] The stacking of the $CS_4^{2-}$ and $K^+$ ions is such that a channel is available for the location of the MeOH molecules. Two forms of the planar $S_2CSS^{2-}$ anion occur in the crystal lattice; the first with a localized C=S double bond (31), the other with a delocalized $\pi$-system (32).[303,304] The different structures are rationalized in terms of the environment of sulphur atoms in the crystal.[304]

(31)

(a) bond lengths: 1.745(31), 1.728(31), 1.649(31), 2.021(12)
(b) bond angles: 109.8(3.2)°, 124.9(4.0)°, 125.3(4.1)°, 110.9(2.2)°

(32)

(a) bond lengths: 1.708(29), 1.728(29), 1.702(29), 2.032(11)
(b) bond angles: 111.9(3.1)°, 124.1(3.7)°, 124.0(3.7)°, 112.1(2.0)°

The structure of a new perthiocarbonate anion $C_2S_6^{2-}$ (33) in solid $(Me_4N)_2C_2S_6,\frac{1}{2}CS_2$ has been examined.[305] It consists of two planar $CS_3$ groups bonded by two sulphur atoms. The dihedral angle (87.12°) between the planes

[296] S. Chen and M. Z. Hoffman, *J. Phys. Chem.*, 1974, **78**, 2099.
[297] S. Chen, M. Z. Hoffman, and G. H. Parsons, *J. Phys. Chem.*, 1975, **79**, 1911.
[298] J. T. Moseley, P. C. Cosby, R. A. Bennett, and J. R. Peterson, *J. Chem. Phys.*, 1975, **62**, 4826.
[299] I. Matousek, A. Fojtik, and R. Zahradnik *Coll. Czech. Chem. Comm.*, 1975, **40**, 1679.
[300] M. Abrouk, *Rev. Chem. minérale*, 1974, **11**, 726.
[301] M. Robineau and D. Zins, *Compt. rend.*, 1975, **280**, C, 759.
[302] J.-N. Pons, J. Roger, and M. Stern, *Compt. rend.*, 1975, **280**, C, 763.
[303] P. Silber, D. Zins, and M. Robineau, *Rev. Chim. minérale*, 1975, **12**, 347.
[304] D. Zins, M. Robineau, and M.-C. Brianso-Perucaud, *Compt. rend.*, 1975, **280**, C, 875.
[305] P. Silber, M. Robineau, D. Zins, and M.-C. Brianso-Perucaud, *Compt. rend.*, 1975, **280**, C, 1517.

and the S—S bridging distance (2.034 Å) are consistent with the sulphur stereochemistry. The geometries of the $CS_3$ and $S_2CSS$ groups are related to those of the $CS_3^{2-}$ and $CS_4^{2-}$ anions, respectively.[305]

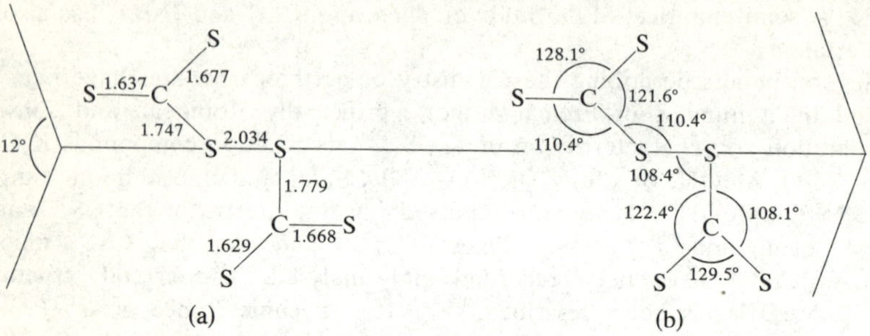

(a) (b)

(33)

All distances/Å

**Derivatives of Group V Elements.**—*Cyanogen, Cyanides, Cyanates, and Related Species.* As for previous Reports, the data abstracted for this section fall into categories associated with the molecular and chemical properties of these species. Particular interest has been shown in the structure of HCP, and in the reactions of $CN^-$ in MeCN. Several detailed analyses of the spectroscopic properties of these molecules have been undertaken; they are summarized in Table 14. *Ab initio* M.O. calculations have been carried out to facilitate the analysis of the vibrational

**Table 14** *Recent spectroscopic studies of cyanides, cyanates, and related species*

| Spectroscopic technique | Molecules examined |
|---|---|
| Microwave | $HCN(g)$;[a] $HNC(g)$;[a] $TCN(g)$;[b] $DCNO(g)$[c] |
| I.r. | $MCN(M=Na$ or $K)(m)$;[d] $DCNO(g)$;[e] $CsNCO(s)$;[f,g] $CsNCS(s)$;[f] $NCO^-$ (in $CsI)$[f—h] $NCO^-$ (in $CsX$; $X=Cl$ or $Br)$;[i] $NCS^-$ (in $MI$; $M=Na$, $K$, or $Rb)$;[g] $NCS^-$ (in $CsI)$;[f] |
| Raman | $XCN$ ($X=Cl$, $Br$, or $I)(s)$;[j] |
| Photoelectron | $HCN$ [He(I), He(II)](g);[k] $DCN$ [He(I)](g);[k] $CN.N_3$[He(I), Ne(I)](g);[l] |
| Electron-impact ionization | $XCN$ ($X=H$ or $D)(g)$;[k] |
| N.m.r. | $[^{13}C]HCP(soln.)$;[m] |
| E.s.r. | $HNCO(s)$.[n] |

* g, gas phase; m, matrix-isolated; s, solid phase; soln., solution.

[a] G. L. Blackman, R. D. Brown, P. D. Godfrey, and H. I. Gunn, *Chem. Phys. Letters*, 1975, **34**, 241; [b] F. C. DeLucia, *J. Mol. Spectroscopy*, 1975, **55**, 271; [c] B. P. Winnewisser and M. Winnewisser, *J. Mol. Spectroscopy*, 1975, **56**, 471; [d] Ref. 320; [e] E. L. Feretti and K. N. Rao, *J. Mol. Spectroscopy*, 1975, **56**, 494; [f] D. J. Gordon and D. F. Smith, *Spectrochim. Acta*, 1974, **30A**, 1953; [g] D. J. Gordon and D. F. Smith, *Spectrochim. Acta*, 1974, **30A**, 2047; [h] D. F. Smith and R. J. York, *J. Chem. Phys.*, 1974, **61**, 5028; [i] D. F. Smith, *J. Mol. Spectroscopy*, 1975, **57**, 447; [j] T. S. Sun and A. Anderson, *J. Raman Spectroscopy*, 1974, **2**, 573; [k] C. Fridh and L. Asbrink, *J. Electron Spectroscopy*, 1975, **7**, 119; [l] Ref. 310; [m] Ref. 319; [n] Ref. 321.

spectra of HCN,[306] HCNO,[307,308] and HNCO[308] and of the photoelectron spectra of HCN[309] and CN,N$_3$.[310] Theoretical calculations have also been successfully completed for HCN,[311-314] HNC,[313,315] and FCN;[314] in particular, the photodissociation of HCN,[312] the HNC–HCN isomerization,[313] and the dipole moment of isolated HCN molecules[311] has been studied.

The equilibrium geometry of HCP has been derived in two independent *ab initio* theoretical analyses;[316,317] the results of both calculations are in reasonable agreement with the experimental data[318] (Table 15). The nature of the C—P bond

**Table 15** *Geometrical parameters/Å in HCP.*

| Analysis | r(H—C) | r(C—P) | Ref. |
|---|---|---|---|
| Theoretical | 1.0579 | 1.5441 | 316 |
| Theoretical | 1.061 | 1.51 | 317 |
| Experimental | 1.0692 | 1.5398 | 318 |

has been estimated from n.m.r. parameters determined for HCP solutions.[319] The carbon atom is *sp* hybridized (the C—H bond has *ca.* 43% *s*-character), leading to a triple-bond description, presumably involving $2p_\pi - 3p_\pi$ overlap, of the C—P bond.[319]

The i.r. spectra of matrix-isolated MCN (M = Na or K) have been examined;[320] they were interpreted assuming monomers, dimers, and higher polymers to be present. The dimers are thought to have the form of a rhombic tetra-atomic ring, composed of metal and carbon atoms (34).[320]

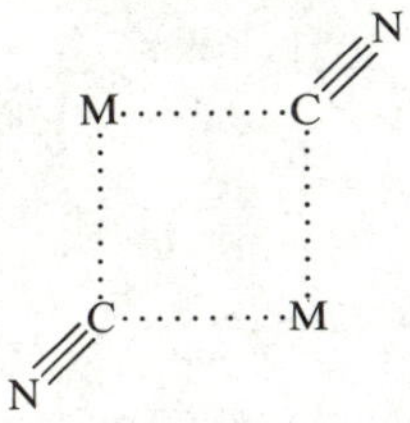

(34)

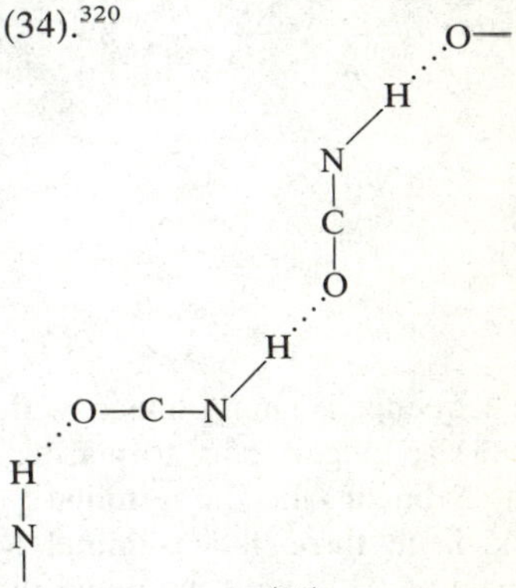

(35)

[306] T. Miyazaki, M. Ikeda, and M. Shibata, *Bull. Chem. Soc. Japan*, 1975, **48**, 1138.
[307] K. Yamada, B. P. Winnewisser, and M. Winnewisser, *J. Mol. Spectroscopy*, 1975, **56**, 449.
[308] J. M. R. Stone, *J. Mol. Spectroscopy*, 1975, **54**, 1.
[309] S. P. So and W. G. Richards, *J.C.S. Faraday II*, 1975, **71**, 62.
[310] B. Bak, P. Jansen, and H. Stafast, *Chem. Phys. Letters*, 1975, **35**, 247.
[311] D. G. Bounds, A. Hinchcliffe, R. W. Munn, and R. J. Newham, *Chem. Phys. Letters*, 1974, **29**, 600.
[312] Y. B. Band and K. F. Freed, *Chem. Phys. Letters*, 1975, **28**, 328.
[313] P. K. Pearson, H. F. Schaefer, and U. Wahlgren, *J. Chem. Phys.*, 1975, **62**, 350.
[314] M. Dixon, G. Doggett, and G. Howat, *J.C.S. Faraday II*, 1975, **71**, 452.
[315] G. M. Schwenzer, H. F. Schaefer, and C. F. Bender, *J. Chem. Phys.*, 1975, **63**, 569.
[316] P. Botschwina, K. Pecul, and H. Preuss, *Z. Naturforsch.*, 1975, **30a**, 1015.
[317] C. Thomson, *Theor. Chim. Acta*, 1974, **35**, 237.
[318] G. Strey and I. M. Mills, *Mol. Phys.*, 1973, **26**, 129.
[319] S. Panderson, H. Goldwithe, D. Ko, and A. Letsou, *J.C.S. Chem. Comm.*, 1975, 744.
[320] Z. K. Ismail, R. H. Hauge, and J. L. Margrave, *J. Mol. Spectroscopy*, 1975, **54**, 402.

The alignment of the isocyanic acid molecule in the oriented solid (35) has been determined from an analysis of the e.s.r. spectrum of the irradiated solid;[321] the NCO axis of the molecules is perpendicular to that of the adjacent acid molecule, through a hydrogen bond.

The crystal structures of $Pb(NCS)_2$,[322] $Cs(SeCN)_3$,[323] $KSe(SCN)_3.\frac{1}{2}H_2O$,[324] and $KSe(SeCN)_3.\frac{1}{2}H_2O$[325] have been determined by X-ray methods; the unit-cell parameters for all four compounds are summarized in Table 16. The NCS⁻ ion in

**Table 16** Unit-cell parameters of $Pb(NCS)_2$,[322] $Cs(SeCN)_3$,[323] $KSe(SCN)_3.\frac{1}{2}H_2O$,[324] and $KSe(SeCN)_3.\frac{1}{2}H_2O$.[325]

| Compound | Symmetry | Space group | a/Å | b/Å | c/Å | α/° | β/° | γ/° |
|---|---|---|---|---|---|---|---|---|
| $Pb(NCS)_2$ | monoclinic | C2/c | 9.661 | 6.544 | 8.253 | — | 92.37 | — |
| $Cs(SeCN)_3$ | monoclinic | C2/c | 7.969 | 21.156 | 5.593 | — | 98.84 | — |
| $KSe(SCN)_3.\frac{1}{2}H_2O$ | triclinic | $P\bar{1}$ | 8.775 | 15.067 | 8.956 | 119.64 | 101.03 | 99.62 |
| $KSe(SeCN)_3.\frac{1}{2}H_2O$ | triclinic | $P\bar{1}$ | 9.170 | 13.377 | 9.057 | 106.22 | 100.64 | 99.07 |

the lead salt has the geometry (36) and makes contact with four lead atoms *via* two Pb—N and two Pb—S bonds; the Pb atoms lie on a two-fold axis and are surrounded by four N and four S atoms.[322] The triselenocyanate ion (37)[323] possesses a two-fold axis of symmetry on which the central SeCN group is located.

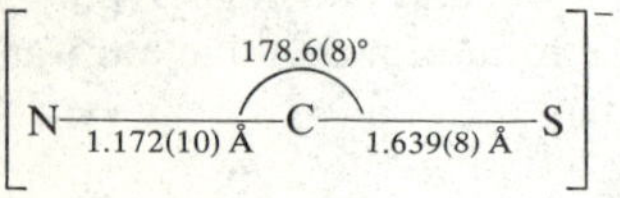

$$\left[ N \underset{1.172(10)\,\text{Å}}{\overset{178.6(8)°}{\text{---C---}}} \underset{1.639(8)\,\text{Å}}{\text{---S}} \right]$$

(36)

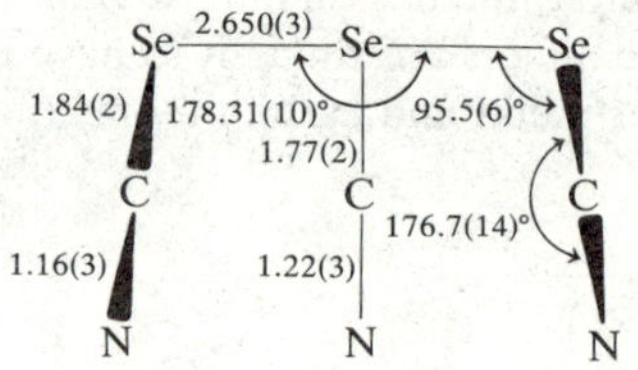

(37) — Se—Se 2.650(3), Se—C 1.84(2), 1.77(2); C—N 1.16(3), 1.22(3); angles 178.31(10)°, 95.5(6)°, 176.7(14)°

All distances/Å

The SeSeSe sequence is nearly linear, with Se—Se bond distances of 2.650(3) Å, which are 0.31 Å longer than normal single Se—Se bonds. The central SeCN group is exactly linear, and the terminal SeCN groups are linear, within error. A least-squares plane through a terminal SeCN group and the central Se atom makes an angle of 43.9° with the plane through the middle SeCN group and the terminal Se atoms.

The gas-phase oxidation of $(CN)_2$,[326] the thermal dissociation of CNBr,[327] and the fluorination of CNCl[328] and of $ClN=CCl_2$[328] have been investigated.

Bond dissociation energies of the H—CN[329] and M—CN[330] bonds have been

[321] H. Hirai and M. Fujiwara, *Bull. Chem. Soc. Japan*, 1975, **48**, 669.
[322] J. A. A. Mokoulu and J. C. Speakman, *Acta Cryst.*, 1975, **B31**, 172.
[323] S. Hauge, *Acta Chem. Scand.(A)*, 1975, **29**, 163.
[324] S. Hauge and P. A. Hendriksen, *Acta Chem. Scand.(A)*, 1975, **29**, 778.
[325] S. Hauge, *Acta Chem. Scand.(A)*, 1975, **29**, 771.
[326] J. Jeanjean, F. Gaillard, and H. James, *Bull. Soc. chim. France*, 1975, 432.
[327] K. Tabayashi, O. Kajimoto, and T. Fueno, *J. Phys. Chem.*, 1975, **79**, 204.
[328] J. D. Cameron and B. W. Tattershall, *Angew. Chem. Internat. Edn.*, 1975, **14**, 166.
[329] D. Betowski, G. MacKay, J. Payzant, and D. Bohme, *Canad. J. Chem.*, 1975, **53**, 2365.
[330] J. N. Mulvihill and L. E. Philips, *Chem. Phys. Letters*, 1975, **33**, 608.

# Elements of Group IV

calculated from thermodynamic analyses of equilibria (26) and (27), respectively; they are collected in Table 17.

$$SH^- + HCN \rightleftharpoons CN^- + H_2S \qquad (26)$$

$$M + HCN \rightleftharpoons MCN + H \ (M = Li, Na, K, Rb, or Cs) \qquad (27)$$

**Table 17** Bond dissociation energies/kJ mol$^{-1}$ for H—CN[329] and M—CN[330] bonds

| $D$(H—CN) | $D$(M—CN) | | | | |
|---|---|---|---|---|---|
| | Li | Na | K | Rb | Cs |
| 519±8 | 497±22 | 474±22 | 497±22 | 487±22 | 501±22 |

Pulse radiolysis of $N_2O$-saturated solutions of HCN and $CN^-$ has been studied, in the pH range 3.7—14.0, using spectroscopic techniques;[331] $CN^-$ anions were found to react with OH radicals in an addition reaction, and not by electron transfer.

In a comprehensive series of papers, Austad has described the reactions of ionic $CN^-$, in MeCN, with $S(CN)_2$,[332] $Se(CN)_2$,[332] $S_5O_6^{2-}$,[333] $SeS_4O_6^{2-}$,[333] $TeS_4O_6^{2-}$,[334] $SeS_2O_6^{2-}$,[335] $Se_2S_4O_6^{2-}$,[335] and aromatic sulphinates of bivalent sulphur and selenium.[336] $S(CN)_2$ and $Se(CN)_2$ react quantitatively with $CN^-$ in the mole ratio 1:3, according to reaction (28).[332] The reaction between $TeS_4O_6^{2-}$ and $CN^-$ [reaction (29)][334] is completely different to that between $S_5O_6^{2-}$ or $SeS_4O_6^{2-}$ and $CN^-$ [reaction (30)].[333] The overall reaction, of both $SeS_2O_6^{2-}$ and $Se_2S_2O_6^{2-}$ with $CN^-$ [reactions (31) and (32)] has also been ascertained.[335] Complete reaction mechanisms have been proposed for all seven reactions.[332—335]

$$X(CN)_2 + 3CN^- \rightarrow XCN^- + (CN)_4^{2-} \ (X = S \text{ or } Se) \qquad (28)$$

$$TeS_4O_6^{2-} + 6CN^- \rightarrow TeCN^- + SCN^- + S_2O_3^{2-} + SO_3^{2-} + (CN)_4^{2-} \qquad (29)$$

$$XS_4O_6^{2-} + 3CN^- \rightarrow O_3SCN^- + XCN^- + SCN^- + S_2O_3^{2-} \ (X = S \text{ or } Se) \qquad (30)$$

$$SeS_2O_6^{2-} + 5CN^- \rightarrow 2SO_3^{2-} + SeCN^- + (CN)_4^{2-} \qquad (31)$$

$$Se_2S_2O_6^{2-} + 6CN^- \rightarrow 2SO_3^{2-} + 2SeCN^- + (CN)_4^{2-} \qquad (32)$$

The formation of the pseudohalogen $(OCN)_2$ as an unstable intermediate in the electrochemical oxidation of $OCN^-$ on Pt in MeCN has been postulated.[337] The proposed reaction mechanism involves the dimer as degradation product of the OCN radical. Finally, the kinetics of oxidation of $SCN^-$ by chromic acid in aqueous $H_2SO_4$ media have been described.[338] A complete reaction mechanism,

---

[331] D. Behar, *J. Phys. Chem.*, 1974, **78**, 2660.
[332] T. Austad and S. Esperas, *Acta Chem. Scand.*(A), 1974, **28**, 892.
[333] T. Austad, *Acta Chem. Scand.*(A), 1974, **28**, 935.
[334] T. Austad, *Acta Chem. Scand.*(A), 1974, **28**, 927.
[335] T. Austad, *Acta Chem. Scand.*(A), 1975, **29**, 71.
[336] T. Austad, *Acta Chem. Scand.*(A), 1975, **29**, 241.
[337] G. Cauquis and G. Pierre, *Bull. Soc. chim. France*, 1975, 997.
[338] U. Muralikrishna and K. V. Bapanaiah, *Indian J. Chem.*, 1974, **12**, 880.

leading to the formation of HCN and $H_2SO_4$, which is consistent with the observed rate law, has been proposed.

## 2 Silicon, Germanium, Tin, and Lead

**Hydrides of Silicon, Germanium, and Tin.**—Fixed-frequency laser photoelectron spectrometry has been utilized to study the ions $Si^-$, $SiH^-$, and $SiH_2^-$, and values for the electron affinities of the species Si, SiH, and $SiH_2$ of 1.385, 1.277, and 1.124 eV, respectively, has been deduced. The internuclear distance in the ground state of $SiH^-$ was determined to be 1.474 Å.[339] The two solid phases of monosilane are body-centred tetragonal, and orientation ordering of molecules appears to be extensive in both.[340] Ion–molecule reactions in both monosilane–acetylene[341] and monosilane–ethylene[342] mixtures have been studied by tandem and high-pressure mass spectrometry. Above 0.05 Torr, the principal reaction of acetylene with ions derived from monosilane appears to be that of simple association, leading to ionic polymerization of acetylene.[341] A persistent or 'sticky' complex $SiC_2H_7^+$ was observed in the collisions of $SiH_3^+$ with ethylene, which at high pressures successively adds two more molecules of ethylene. The products were suggested to be the alkylsiliconium ions $Et_nSiH_{3-n}^+$ ($n = 1—3$).[342] The same technique has been used to study ion–molecule reactions in methylsilane.[343,344] The dominant mode of reaction proceeds *via* a direct stripping-type process, yielding $MeSiH(D)_2^+$ ions with very little kinetic energy, and by a complex mechanism that is particularly prominent at low collision energies. No scrambling of hydrogen atoms between carbon and silicon was observed, and the hydride abstraction occurs solely at silicon. The rates of reaction of hydrogen atoms with both mono-silane and -germane at 20 Torr and of $H_2$ and $D_2$ with monosilane and the methylsilanes $Me_{4-n}SiH_n$ ($n = 1—3$) at 3 Torr have been studied in a discharge flow system.[345,346] The thermal decomposition of monosilane and disilane in the presence of acetylene has been examined. The product of the reaction of silylene (generated from disilane) with acetylene was found to be $H_3SiC\equiv CH$. The absence of this product near zero reaction time in the $SiH_4$–$C_2H_2$ pyrolysis system suggests that silylene is not the dominant radical species in monosilane pyrolysis. The products of the latter system are, however, consistent with silyl radicals as the dominant silicon radical species. The absence of silylene species in the monosilane pyrolysis can be accounted for by orbital symmetry considerations.[347] The kinetics of the gas-phase thermal decomposition of $H_2MeSiSiH_3$ and $Si_3H_8$ have been examined in the temperature ranges 539.6—569.6 K and 529.6—560.5 K, respectively. The relatively low activation energies observed argue for 1,2-hydrogen migration *via* a hydrogen-atom-bridged transition state in the initial reactions, which produce, in the case of methyldisilane,

---

[339] A. Kasdan, E. Herbs, and W. C. Lineberger, *J. Chem. Phys.*, 1975, **62**, 541.
[340] W. M. Sears and J. A. Morrison, *J. Chem. Phys.*, 1975, **62**, 2736.
[341] T. M. Mayer and F. W. Lampe, *J. Phys. Chem.*, 1974, **78**, 2644.
[342] T. M. Mayer and F. W. Lampe, *J. Phys. Chem.*, 1974, **78**, 2433.
[343] T. M. Mayer and F. W. Lampe, *J. Phys. Chem.*, 1974, **78**, 2422.
[344] T. M. Mayer and F. W. Lampe, *J. Phys. Chem.*, 1974, **78**, 2428.
[345] J. A. Cowfer, K. P. Lynch, and J. V. Michael, *J. Chem. Phys.*, 1975, **79**, 1139.
[346] K. Y. Choo, P. P. Gaspar, and A. P. Wolf, *J. Phys. Chem.*, 1975, **79**, 1752.
[347] C. H. Haas and M. A. Ring, *Inorg. Chem.*, 1975, **14**, 2253.

methylsilylene and monosilane, and, in the case of trisilane, silylene and disilane and SiH$_3$SiH and monosilane. Silylene insertion into Si—C and C—H bonds was not observed.[348] The i.r. spectra of silaethylene H$_2$Si=CH$_2$ and its deuteriated analogue H$_2$Si=CD$_2$ have been predicted by theoretical calculations, giving an optimized structure of $C_{2v}$ symmetry. The dipole moment of 1.1 D indicates a highly polarized Si=C bond.[349]

Monogermane reacts with the alkali metals in dimethoxyethane to give the alkali-metal germyl salts M$^I$GeH$_3$ (M$^I$ = Li, Na, K, Rb, or Cs). The potassium and rubidium salts are isostructural with the corresponding alkali-metal silyl compounds (NaCl structure). CsGeH$_3$, however, has the TlI-type structure, and the co-ordination polyhedron at germanium is shown in Figure 4. Wide-line $^1$H n.m.r. studies show that the GeH$_3$ group rotations are frozen at low temperatures, and a value of 92.5 ± 4° was deduced for the HGeH valence angle of the fixed germyl groups.[350] Varma and his co-workers have investigated the mercury-photosensitized reactions of monogermane with nitric oxide.[351] Digermane,

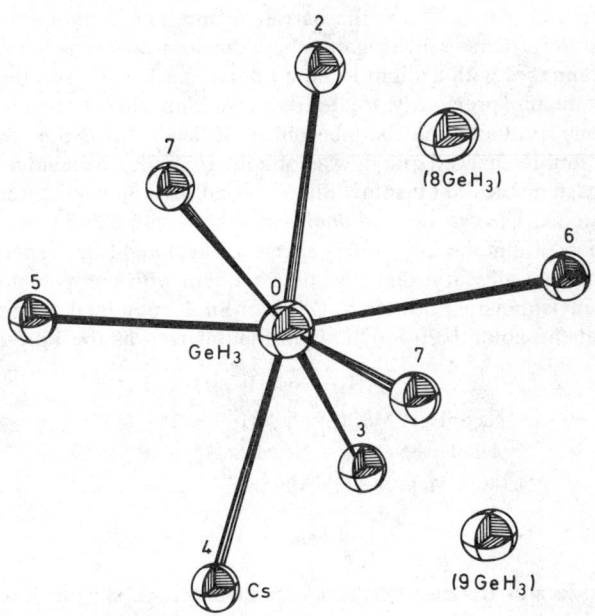

**Figure 4** *The co-ordination polyhedron of the GeH$_3$ group in CsGeH$_3$*
Reproduced by permission from *Z. anorg. Chem.*, 1975, **417,** 221)

[48] A. J. Vanderwielen, M. A. Ring, and H. E. O'Neal, *J. Amer. Chem. Soc.*, 1975, **97,** 993.
[49] H. B. Schlegel, S. Wolfe, and K. Mislow, *J.C.S. Chem. Comm.*, 1975, 246.
[50] G. Thirase, E. Weiss, H. J. Hennig, and H. Lechert, *Z. anorg. Chem.*, 1975, **417,** 221.
[51] R. Varma, K. R. Ramaprasad, A. J. Signorelli, and B. K. Sahay, *J. Inorg. Nuclear Chem.*, 1975, **37,** 563.

digermyl ether, nitrogen, and hydrogen were observed as products, and Scheme 1 was proposed to account for them.

$$GeH_4 + Hg^* \rightarrow GeH_3 + H + Hg$$
$$GeH_4 + H \rightarrow GeH_3 + H_2$$
$$GeH_3 + NO \rightarrow GeH_3ON$$
$$2GeH_3ON \rightarrow GeH_3ONNOGeH_3$$
$$(H_3Ge)_2O \xleftarrow{GeH_3} 2GeH_3O + N_2$$

**Scheme 1**

The reactions of bromogermane with lead(II) carboxylates provide convenient routes to germyl formate, acetate, and trifluoroacetate. Silyl formate may also be obtained similarly. Lead(II) cyanate and thiocyanate react to yield the corresponding germyl isocyanate and isothiocyanate, respectively. The latter compound reacts with protic reagents, with cleavage of the Ge—N bond.[352] The microwave spectra of five isotopic species of $ClCH_2GeH_3$ have been measured, leading to a value of $1.733 \pm 30$ kcal mol$^{-1}$ for the barrier to internal rotation.[353]

Birchall and Pereira have investigated the behaviour of stannane and the alkyl- and phenyl-stannanes with sodium in liquid ammonia.[354] The addition of methyl chloride to stannane previously treated with sodium gave both dimethyl- and methyl-stannane, probably *via* the mechanism of Scheme 2. No evidence for the formation of doubly charged anions was obtained. Similar behaviour is observed by the methylstannanes, but some Sn—C bond fission was observed in the benzyl–tin systems. The reactions of phenylstannanes with alkali metals also result in the formation of anionic $Ph_{3-n}SnH_n^-$ species, which yield the expected methyl-stannanes and other alkylarylstannanes on treatment with methyl chloride. In the case of diphenylstannane, however, $(Ph_2Me)_2Sn$ is obtained as one product, suggesting that the anion $[(Ph_2Sn)_2]^{2-}$ is formed as well as the $Ph_2SnH^-$ anion.

$$Na + NH_3 \rightarrow NaNH_2 + \tfrac{1}{2}H_2$$
$$Na[SnH_3] + MeCl \rightarrow MeSnH_3 + NaCl$$
$$MeSnH_3 + NaNH_2 \rightarrow Na[MeSnH_2] + NH_3$$
$$Na[MeSnH_2] + MeCl \rightarrow Me_2SnH_2 + NaCl$$

**Scheme 2**

**Silicon(IV) Oxide and Related Silicates.**—In this section, silicon dioxide and the silicates will be described separately; emphasis will be laid on the inorganic chemistry of these compounds, and papers describing solely their catalytic, adsorption, diffusion, and other similar properties will not be discussed. The chemistry of the aluminosilicates and zeolites will not be discussed here; recent publications in which their chemistry is described are considered in detail in

---

[352] P. C. Angus and S. R. Stobart, *J.C.S. Dalton*, 1975, 2342.
[353] J. Nakagawa and M. Hayashi, *Chem. Letters*, 1974, 1379.
[354] T. Birchall and A. R. Pereira, *J.C.S. Dalton*, 1975, 1087.

Chapter 3. Surprisingly few papers have been published, during the period of this Report, which describe aspects of the chemistry of silicon dioxide; indeed, the majority of the reported data deal with surface properties and the chemistry of the species adsorbed thereon. Interest in the silicates has been maintained, with a marked increase in the number of communications relating to the chemistry of the interlamellar complexes formed by the layered silicates, particularly montmorillonite.

A method has been described by independent authors[355,356] for evaluating bond strengths and valence bond distributions in known ionic structures according to the Pauling electronegativity principle. Results for simple silicates (*e.g.* olivines and pyroxenes) and related compounds have been discussed.[355,356] Bond-angle (O—Si—O) distortions in silicate tetrahedra, calculated from electrostatic-potential models, have been compared with experimental values;[357] it is concluded that repulsion between M and Si atoms is proportional to the ideal ionic charge (*i.e.* valence) of the M atoms rather than to the effective ionic charge calculated from electronegativity data. A statistical thermodynamic approach to the distribution of cations in silicate minerals has also been reported.[358]

The electronic structure of the $SiO_4^{4-}$ ion has been the subject of both experimental (an XPS study of $Li_4SiO_4$[359]) and theoretical (M.O. calculations[359,360]) analyses; the calculated data are in good agreement with available XPS, X-ray emission, and optical data for quartz and other silicates.[360] The XPS data for $Li_4SiO_4$ are compared with those for $Li_3PO_4$, $Li_2SO_4$, and $LiClO_4$;[359] it is concluded that the contribution of the oxygen $2s$ orbitals to the bonding and the extent of $\pi$-back-bonding between oxygen $2p$ and central atom $3d$ orbitals in these oxyanions both increase from Si to Cl.

A normal-co-ordinate analysis of the vibrations of $Mg_2SiO_4$ has been carried out to investigate the group behaviour of the $SiO_4^{4-}$ ion.[361] An i.r. and Raman study[362] of a number of silicates with $Si_3O_9^{6-}$ ring structures has been undertaken to show that the essential spectroscopic features of these rings are modified under the influence of structure change with or without modification of the local symmetry of the ring.

Thermodynamic properties of fused silicates have been calculated[363] on the basis of a polymeric model[364] assuming ideal ionic solutions; satisfactory agreement with experimental data was achieved.

*Silicon Dioxide.* The nature of the $\alpha-\beta$ transition in quartz has been considered;[365] it occurs in a temperature interval of $\not> 0.05$ K and is characterized by temperature hysteresis ($1.4 \pm 0.3$ K). The nature of the defects produced in non-crystalline $SiO_2$ and $\alpha$-quartz single crystals by fast neutron irradiation has

---
[355] R. B. Ferguson, *Acta Cryst.*, 1974, **B30**, 2527.
[356] R. Allmann, *Monatsh.*, 1975, **106**, 779.
[357] M. E. Fleet, *Acta Cryst.*, 1975, **B31**, 1095.
[358] W. J. Mortier, *J. Phys. Chem.*, 1975, **79**, 1447.
[359] R. Prins, *J. Chem. Phys.*, 1974, **61**, 2580.
[360] J. A. Tosselil, *J. Amer. Chem. Soc.*, 1975, **97**, 4840.
[361] V. Devarajan and F. Funck, *J. Chem. Phys.*, 1975, **62**, 3406.
[362] J. Choisnet, A. Deschanvres, and P. Tarte, *Spectrochim. Acta*, 1975, **31A**, 1023.
[363] O. A. Esin, *Russ. J. Phys. Chem.*, 1974, **48**, 1247.
[364] O. A. Esin, *Russ. J. Phys. Chem.*, 1974, **48**, 1249.
[365] O. N. Brevsov and V. F. Tatsii, *Russ. J. Inorg. Chem.*, 1975, **20**, 482.

been studied by i.r. and Raman techniques as a function of neutron flux.[366] At $2 \times 10^{20}$ neutrons cm$^{-2}$, the quartz sample was transformed into a non-crystalline material. At $9 \times 10^{19}$ neutrons cm$^{-2}$, however, remnant crystallites of the original $\alpha$-quartz remained. On annealing ($\not> 1000\,°C$), the structures of the irradiated samples transformed into that of unirradiated $SiO_2$.[366] The $Fe^{3+}$ e.p.r. spectrum in irradiated quartz has also been investigated.[367]

The effect of trace quantities of metals (Li, Na, or K) on the sintering of silica gel has been studied.[368] The observed acceleration in sintering and crystallization has been attributed to a decrease in the covalency of the Si—O—Si bonds caused by the electronic effect of the alkali metals. An i.r. investigation[369] of $SiO_2$-MgO (<20%MgO) gels has revealed that some $Mg^{2+}$ ions are four-co-ordinate. Crystallization of the gels by thermal treatment (>800 °C) leads to monoclinic clinoenstatite, in which all $Mg^{2+}$ ions are six-co-ordinate. The thermal characteristics of coprecipitated gels in the $SiO_2$-$Al_2O_3$-$In_2O_3$-$H_2O$ system have also been studied[370] in order to determine the role of the $M^{3+}$ ions in the gel structure.

The hydrothermal chemistry of the $SiO_2$-BaO-$K_2O$-$H_2O$ (A),[371] $SiO_2$-BaO-$B_2O_3$-$H_2O$ (B),[372] $SiO_2$-CaO-$B_2O_3$-$H_2O$ (C),[373] and $SiO_2$-$Al_2O_3$-$Fe_2O_3$-$Na_2O$-$H_2O$ (D)[374] systems has been examined. In system (A),[371] five barium silicates, together with the mixed silcate phase $(K, Ba)_2Si_4O_9,1.3H_2O$, were distinguished. In the baric[372] and calcic[373] systems, (B and C, respectively), silicates, borates, and borosilicates of barium and calcium (as appropriate) were established as reaction products; cristobalite and quartz were also formed in the calcic system. The previously unknown sodium hydrosilicoferrite has been obtained in system (D);[374] conditions for its formation have been defined and its optical, thermal, and structural parameters determined.

Mechanicochemical reactions of various gases (e.g. $H_2O$, $NH_3$, NO, $NO_2$, and $Cl_2$) with quartz have been carried out in a glass vibromill at 80 and 300 K.[375] The quartz, normally accepted as being an inert mineral, was surprisingly reactive; this was attributed to the formation of short-lived centres on freshly formed surfaces.

The reaction of $SiO_2$ ($\alpha$-cristobalite) with powdered AlN at 1700 °C for 70 hours in a current of $N_2$ yields $\alpha$-$Al_2O_3$ and $\beta$-$Si_3N_4$ in the interfacial regions [reaction (33)].[376] These products react further, to form a compound with

$$8AlN + 6SiO_2 \rightarrow 4Al_2O_3 + 2Si_3N_4 \qquad (33)$$

stoicheiometry close to $2Si_3N_4,3Al_2O_3$ or a solid solution of $Al_2O_3$ in $\beta$-$Si_3N_4$. $2Si_3N_4,3Al_2O_3$ can only exist in the presence of excess $O_2$, produced in reactions

---

[366] J. B. Bates, R. W. Hendricks, and L. B. Shaffer, *J. Chem. Phys.*, 1974, **61**, 4163.
[367] M. M. Zaitov, M. M. Zaripov, M. I. Samoilovich, V. E. Khadzhi, and L. I. Tsinober, *Soviet Phys. Cryst.*, 1975, **19**, 674.
[368] V. V. Strelko, T. N. Burushkina, and V. N. Belyakov, *Doklady Chem.*, 1974, **215**, 195.
[369] M. Briend-Faure, M. Kermarec, D. Delafosse, and O. Cornu, *Bull. Soc. chim. France*, 1974, 2393.
[370] M. Kasai, Y. Kudo, and S. Hamada, *Bull. Chem. Soc. Japan*, 1975, **48**, 1608.
[371] Yu. A. Malinovskii, E. A. Pobedimskaya, V. A. Kuznetsov, and N. V. Belov, *Soviet Phys. Cryst.*, 1975, **19**, 561.
[372] R. M. Barrer and E. F. Freund, *J.C.S. Dalton*, 1974, 2054.
[373] R. M. Barrer and E. F. Freund, *J.C.S. Dalton*, 1974, 2060.
[374] L. P. Ni, O. B. Khalyapina, and T. V. Solenko, *Russ. J. Inorg. Chem.*, 1974, **19**, 532.
[375] I. V. Kolbanev and P. Yu. Butyagin, *Russ. J. Phys. Chem.*, 1974, **48**, 670.
[376] J.-P. Torre and A. Mocellin, *Compt. rend.*, 1974, **279**, C, 943.

(34) and (35); in excess $N_2$ it decomposes with the formation of solid solutions

$$3N_2 + 6SiO_2 \rightarrow 3Si_2N_2O + 9/2 O_2 \qquad (34)$$

$$N_2 + 3Si_2N_2O \rightarrow 2Si_3N_4 + 3/2 O_2 \qquad (35)$$

of $Al_2O_3$ in $\beta$-$Si_3N_4$ and of $Si_3N_4$ in $\alpha$-$Al_2O_3$.[376] The thermal decomposition of $Si_2N_2O$, studied in the temperature range 1300—1700 °C, in vacuum and in the presence of $N_2$, Ar, or He, always occurs according to reaction (36).[377]

$$3Si_2N_2O \rightarrow Si_3N_4 + 3SiO + N_2 \qquad (36)$$

The first-stage products appear to be $Si_3N_4$ and $SiO_2$; the $SiO_2$ then reacts with Si that has been formed in the dissociation of $Si_3N_4$ to form SiO, which condenses out, thus continuously displacing the equilibrium. A new phase, denoted $\delta$-phase, and characterized by the empirical formula $SiAl_4N_4O_2$, has been prepared[378] from mixtures of $Si_2N_2O$ and $Al_2O_3$ by heating (1650—1700 °C) in an Ar atmosphere; it must be considered as a thermolysis product of the recently discovered solid solution $\beta'$-$Si_3N_4$.[379]

The chemistry of the functional groups present on silica surfaces continues to arouse interest;[380—387] the relationships between surface chemistry and catalytic activity of oxides of varying acidity have been reviewed.[381] Several theoretically based analyses of i.r. data, in which the nature of the isolated —OH group present on the surface is considered in detail, have been completed.[382—384] The acid–base properties of $SiO_2$–$MgO$[385] and $SiO_2$–$Al_2O_3$ (0—50% $Al_2O_3$)[386] gels have also been examined by i.r. study of the adsorption of weak hydrogen-bond-accepting molecules onto the gels. The surface $H_2O$ content of $SiO_2$, $Al_2O_3$, and $SiO_2$–$Al_2O_3$ gels has been determined by both a successive ignition-loss method and chemical reaction between active hydrogen and MeMgI;[387] the results obtained by the two methods were in good agreement.

The interaction between $H_2O$ and both hydroxylated[388] and dehydroxylated[389] $SiO_2$ has been investigated using physisorption techniques. The results for the hydroxylated surface were compared to similar data for hydroxylated $Al_2O_3$ and $SiO_2$–$Al_2O_3$ gel surfaces.[388] Three different kinds of adsorbed $H_2O$ occur in these systems; sites of type (38) occur on $SiO_2$ and $SiO_2$–$Al_2O_3$, whereas sites (39) are found on $Al_2O_3$ and $SiO_2$–$Al_2O_3$ surfaces. The strong $H_2O$ physisorption sites of type (40), with protonic acid character, which are peculiar to $SiO_2$–$Al_2O_3$ surfaces, are thought to appear when the two component oxides come into contact.[388—390]

---

[377] P. Lortholary and M. Billy, *Bull. Soc. chim. France*, 1975, 1057.
[378] D. Brachet, P. Goursat, and M. Billy, *Compt. rend.*, 1975, **280**, C, 1207.
[379] Y. Oyama and O. Kamigaito, *Japan J. Appl. Phys.*, 1971, **10**, 1637.
[380] H.-J. Tiller, W. Kuhn, and K. Meyer, *Z. Chem.*, 1974, **14**, 450.
[381] W. K. Hall, *Accounts Chem. Res.*, 1975, **8**, 257.
[382] P. R. Ryason and B. G. Russell, *J. Phys. Chem.*, 1975, **79**, 1276.
[383] V. I. Lygin and V. V. Smolikov, *Russ. J. Phys. Chem.*, 1974, **48**, 581.
[384] A. V. Kiselev, V. I. Lygin, and V. V. Smolikov, *Russ. J. Phys. Chem.*, 1975, **49**, 600.
[385] M. Kermarec, M. Briend-Faure, and D. Delafosse, *J.C.S. Faraday I*, 1974, **70**, 2180.
[386] P. G. Rouxhet and R. E. Sempels, *J.C.S. Faraday I*, 1974, **70**, 2021.
[387] H. Naono, T. Kadota, and T. Morimoto, *Bull. Chem. Soc. Japan*, 1975, **48**, 1123.
[388] T. Morimoto, M. Nagao, and J. Imai, *Bull. Chem. Soc. Japan*, 1974, **47**, 2994.
[389] B. A. Morrow and I. A. Cody, *J. Phys. Chem.*, 1975, **79**, 761.
[390] T. Morimoto, M. Nagao, and J. Imai, *Bull. Chem. Soc. Japan*, 1971, **44**, 1282.

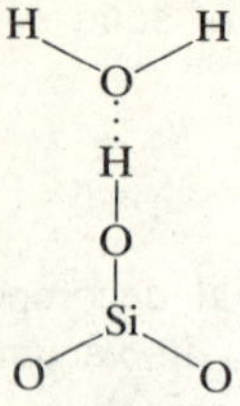

                (38) 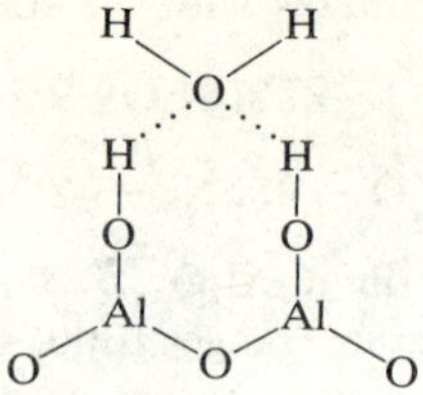                    (39) 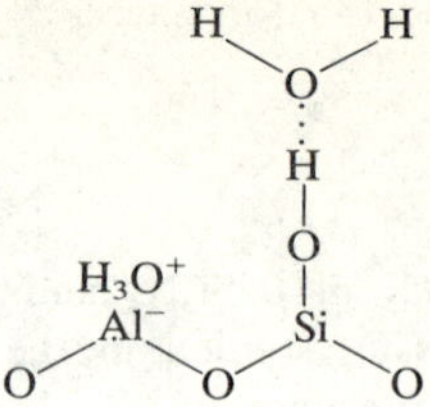                    (40)

The study of the chemisorption of $H_2O$ on dehydroxylated $SiO_2$ is part of a wider investigation, including $NH_3$ and MeOH as adsorbates.[389] The results provide further evidence that the reactive site previously recognized in reactions of $BF_3$ and $BCl_3$ with dehydroxylated $SiO_2$[391,392] is a strained siloxane-bridge type site. Assuming this to be so, the stoicheiometry of the reactions can be represented by equations (37) and (38).

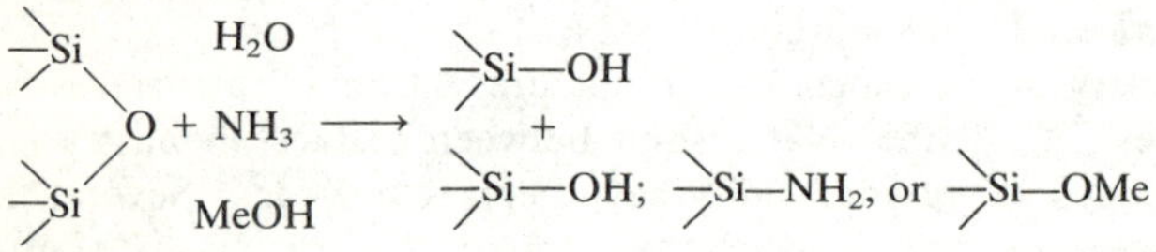

$$\begin{array}{c}\ce{>Si}\\ \ce{>Si}\end{array}\ce{O} + \begin{array}{c}H_2O\\ NH_3\\ MeOH\end{array} \longrightarrow \begin{array}{c}\ce{>Si-OH}\\ +\\ \ce{>Si-OH;\ >Si-NH_2,\ or\ >Si-OMe}\end{array} \quad (37)$$

$$\begin{array}{c}\ce{>Si}\\ \ce{>Si}\end{array}\ce{O} + BX_3 \longrightarrow \begin{array}{c}\ce{>Si-X}\\ +\\ \ce{>Si-BX_2}\end{array} \quad (X = F\ or\ Cl) \quad (38)$$

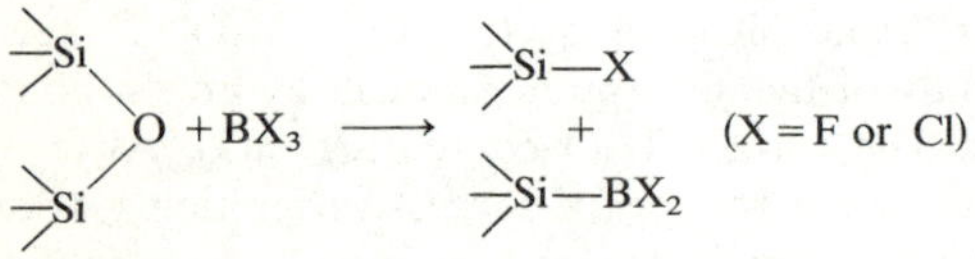

Several papers, principally by Russian authors, have been published recently which describe the reactions leading to molecular layering on $SiO_2$ surfaces.[393—397] The reaction of $BBr_3$ with hydroxylated $SiO_2$ (41) involves the replacement of two —OH groups by a —BBr functional group; subsequent hydrolysis yields a layer of —BOH groups on the $SiO_2$ surface (42).[393] Interaction of $TiCl_4$ with boron-containing $SiO_2$ gel (42), followed by hydrolysis, leads to the formation of —Ti(OH)$_2$ groups on the surface (43).[394] These groups can react further with $TiCl_4$ to produce, after hydrolysis, new —Ti(OH)$_2$ surface groups (44). Repeated alternation of these reactions leads to the stepwise growth of molecular oxide layers on $SiO_2$ surfaces.[394] Similar $Ti^{IV}$ (45)[395] and $V^V$ (46)[396] surfaces have been formed directly on $SiO_2$ surfaces by the molecular layering technique. In an i.r. and vis.—u.v. spectral study of the hydrolysis products,[396] it has been shown that

[391] B. A. Morrow and A. Devi, J.C.S. Chem. Comm., 1971, 1237.
[392] B. A. Morrow and A. Devi, J.C.S. Faraday I, 1972, 68, 403.
[393] T. V. Ukhova, A. A. Malygin, A. N. Volkova, S. I. Kol'tsov, and V. B. Aleskovskii, Russ. J. Phys. Chem., 1974, 48, 921.
[394] S. I. Kol'tsov, A. N. Volkova, T. V. Ukhova, and V. B. Aleskovskii, Russ. J. Inorg. Chem., 1975, 20, 203.
[395] V. N. Pak, Russ. J. Phys. Chem., 1974, 48, 1383.
[396] V. N. Pak, A. M. Boldyreva, A. A. Malygin, S. I. Kol'tsov, and V. B. Aleskovskii, Russ. J. Phys. Chem., 1975, 49, 299.
[397] W. Hanke, R. Bienert, and H.-G. Jerschkewitz, Z. anorg. Chem., 1975, 414, 109.

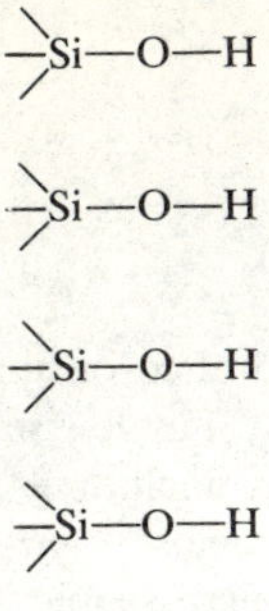

(41) 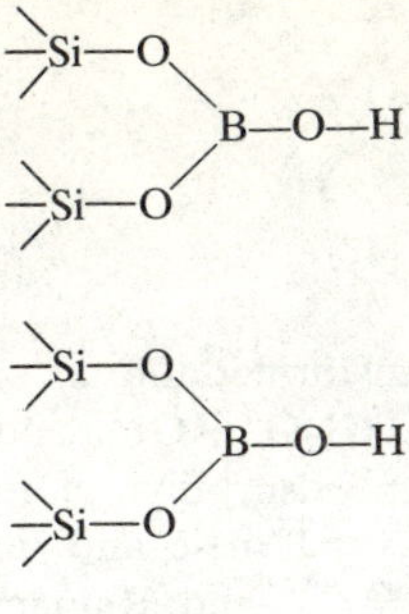 (42) 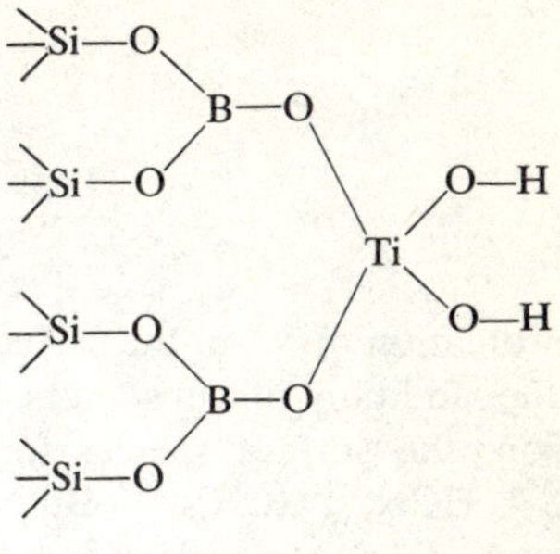 (43)

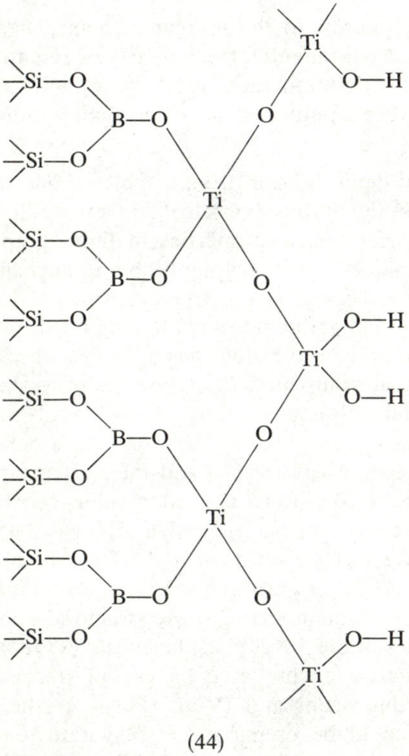

(44)

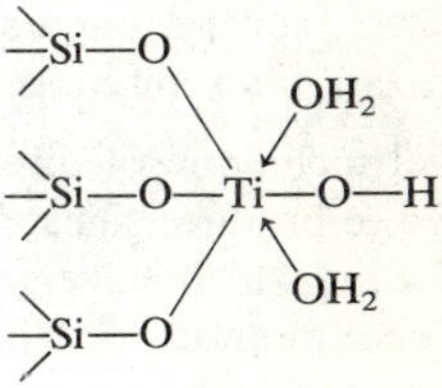

(45)

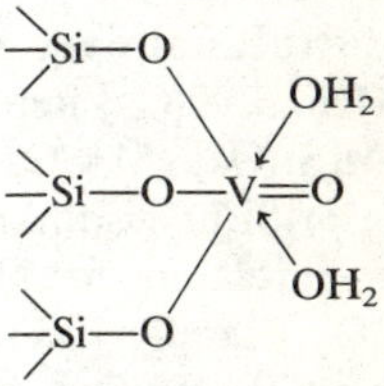

(46)

both transition metals are octahedrally co-ordinated. The $H_2O$ adsorption that occurs to provide these octahedral co-ordination spheres occurs at highly active centres. These include functional groups of the original $SiO_2$ gel whose Si atom is attached through an oxygen bridge to a Ti atom on the surface. This observation supports the view that such centres may be produced on the surfaces of mixed oxide catalysts.[395]

The product (47) of an independent study[397] of the reaction of $VOCl_3$ with $SiO_2$ surfaces has been reported. As observed by the Russian authors, the tetrahedral

$$\text{\textbackslash Si—O—V(Cl)(Cl)=O}$$

$$\begin{array}{c} \phantom{xx}\diagdown\phantom{xxxx}Cl \\ \phantom{xx}\phantom{xx}\diagup \\ \diagdown Si-O-V=O \\ \diagup\phantom{xxxxx}\diagdown \\ \phantom{xxxxxxxxx}Cl \end{array}$$

(47)

co-ordination of $V^V$ in the surface is transformed into an octahedral co-ordination by the addition of substances such as $H_2O$, ROH, and $NH_3$. Finally, in this section, the surface species formed on adsorption of moieties including $I_2$,[398] ICl,[398] HCN,[399] $(CN)_2$,[399] $Bu^nNCO$,[400] and oleic and linoleic acids[401] on $SiO_2$ surfaces have been identified by i.r.[399—401] and Raman[398] spectroscopic techniques.

*Silicates.* Silicates are of particular interest to the inorganic chemist because of their complex molecular structures. Consequently, the majority of research effort in this field is associated with the determination of their crystal structures. Nevertheless, several papers have been published in which their synthesis and characterization are discussed.

A novel technique of directional devitrification from a molten zone has been developed to produce silicate ($\beta$-Ca[SiO$_3$]) rods, crystallized in a truly aligned and fibrous habit, which have far superior tensile properties to those prepared by conventional methods.[402] The technique is probably applicable to any silicate (or other system) which crystallizes in a sheet or chain structure.

The synthesis of ternary silicates (and germanates) containing alkali-metal (Na or K) and alkaline-earth-metal (Ca, Sr, or Ba) cations has been described;[403] they are prepared either by direct reaction of mixtures of carbonates or oxalates with $SiO_2$ (or $GeO_2$) or by fusion and subsequent recrystallization of the glass. Preliminary structural data have also been obtained.[403] Thus, representatives of the cubic $M_4^+M^{2+}[X_3O_9]$ family include $Na_4Ca[Si_3O_9]$ and the isostructural compounds $K_4M[Si_3O_9]$ (M = Ca or Sr); $K_4Ba[Si_3O_9]$ is pseudo-cubic, but the symmetry of $Na_4Sr[Si_3O_9]$ is unknown. the orthopyroxene polymorph $Na_2Ba[Si_2O_6]$ is structurally unrelated to $Na_2M[Ge_2O_6]$ (M = Ca or Sr). The rhombohedral $M_8^+M^{2+}[X_{10}O_{25}]$ family includes $K_8M[Si_{10}O_{25}]$ (M = Ca, Sr, or Ba). Finally, the pyrosilicate $Na_2Ba_2[Si_2O_7]$ and its germanium analogue are structurally similar.[403] The preparation and some properties of the novel phosphosilicate $(VO)^{2+}[P_2SiO_8]$ have been reported.[404] Small crystals were prepared by vapour transport, in a quartz tube, using $I_2$ as transporting agent and $(VO)^{2+}(PO_3)_2$ as the starting material; polycrystalline material could be prepared by solid-state reaction of $(VO)^{2+}(PO_3)_2$ and $SiO_2$ for 48 hours at 1040 °C in a sealed Pt tube. The crystal symmetry is tetragonal, with $a = 8.697(8)$, $c = 8.119(8)$ Å. The stoicheiometry of the condensed tetrahedra $[X_3O_8]$, is unusual; it is suggested that the compound,

---

[398] T. Nagasao and H. Yamada, *J. Raman Spectroscopy*, 1975, **3**, 153.
[399] B. A. Morrow and I. A. Cody, *J.C.S. Faraday I*, 1975, **71**, 1021.
[400] A. Guillet, M. Coudurier, and J. B. Donnet, *Bull. Soc. chim. France*, 1975, 1563.
[401] K. Marshall and C. H. Rochester, *J.C.S. Faraday I*, 1975, **71**, 1754.
[402] A. Maries and P. S. Rogers, *Nature*, 1975, **256**, 401.
[403] R. P. Gunawardane and F. P. Dent Glasser, *Z. anorg. Chem.*, 1975, **411**, 163.
[404] B. C. Tofield, G. R. Crane, P. M. Bridenbarugh, and R. C. Sherwood, *Nature*, 1975, **253**, 722.

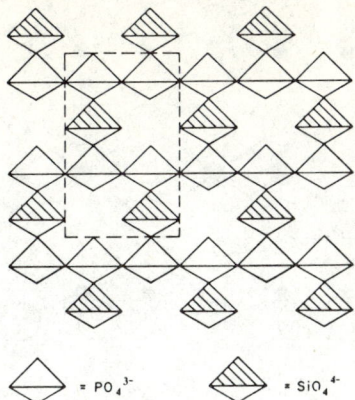

**Figure 5** *One possible schematic representation of tetrahedra in the* $[P_2SiO_8]^{2-}$ *framework which may occur in* $VO[P_2SiO_8]$; *the dotted line encloses a* $[P_4Si_2O_{16}]^{4-}$ *unit*

(Reproduced by permission from *Nature*, 1975, **253**, 722)

thought to be the first condensed stoicheiometric phosphosilicate to be reported, may adopt a novel structural arrangement. One possible way of linking the tetrahedra in such a situation, which would also provide a basis for ordering of P and Si, is shown schematically in Figure 5.[404] $(VO)^{2+}[P_2SiO_8]$ may be heated ($\not> 1000\,°C$) in air without oxidation or decomposition, and it is soluble in most non-oxidizing solvents.[404]

A recent renascence of interest in intercalation of various moieties between individual sheets of layered silicates, particularly montmorillonite, has resulted in a large number of published papers on these materials. High-resolution electron-microscopic studies[405] of structural faults in kaolinite, zussmanite, and stilpnomelane, and γ-resonance spectroscopy studies[406] of thermally initiated structural changes of montmorillonite and hydromica, have been undertaken. The nature of the exchange acidity[407] and the thermal stability[408] of the surface —OH groups of, *inter alia*, montmorillonite, montronite, vermiculite, and kaolinite, have also been examined.

In an extensive series of papers, Adams, Thomas, and co-workers[409—414] have described the initial results of a systematic study of organic intercalates of montmorillonite. Various novel reactions (relevant in inorganic, bio-organic, and

---

[405] D. A. Jefferson and J. M. Thomas, *J.C.S. Faraday II*, 1974, **70**, 1691.
[406] A. S. Plachinda, Yu. I. Tarasevich, V. I. Gol'danskii, F. D. Ovcharenko, E. F. Madarov, I. P. Suzdalev, and Z. E. Suyunova, *Soviet Phys. Cryst.*, 1975, **19**, 477.
[407] N. G. Vasilev, F. D. Ovcharenko, and M. A. Buntova, *Doklady Chem.*, 1974, **216**, 407.
[408] N. G. Vasilev, F. D. Ovcharenko, and L. V. Golovko, *Doklady Chem.*, 1974, **217**, 536.
[409] D. T. B. Tenakoon, J. M. Thomas, M. J. Tricker, and J. O. Williams, *J.C.S. Dalton*, 1974, 2207.
[410] D. T. B. Tennakoon and M. J. Tricker, *J.C.S. Dalton*, 1975, 1802.
[411] D. T. B. Tennakoon, J. M. Thomas, and M. J. Tricker, *J.C.S. Dalton*, 1974, 2211.
[412] J. M. Adams, *J.C.S. Dalton*, 1974, 2286.
[413] J. M. Adams, J. M. Thomas, and M. J. Walters, *J.C.S. Dalton*, 1975, 1459.
[414] M. J. Tricker, D. T. B. Tennakoon, J. M. Thomas, and S. H. Graham, *Nature*, 1975, **253**, 110.

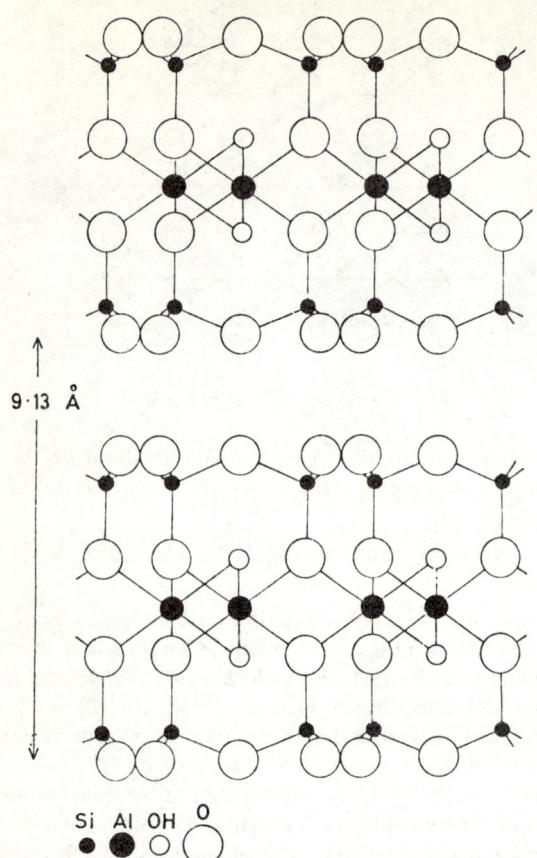

**Figure 6** *Schematic representation of a sheet silicate of general composition* $Al_4[Si_8O_{20}](OH)_2$. *In montmorillonite, substitution of Al or Si by other cations (e.g.* $Mg^{2+}$ *or* $Fe^{2+}$ *for Al, and* $Al^{3+}$ *for Si) takes place*

(Reproduced from *J.C.S. Dalton*, 1974, 2207).

prebiological contexts), which can be effected between the layers, or at the surface of montmorillonite, of idealized formula $Al_4[Si_8O_{20}](OH)_4$ (Figure 6), have been summarized.[409] The colour reactions displayed by selected nitrogen-containing aromatic molecules [*e.g.* benzidine (bzn), *NNN'N'*-tetramethylbenzidine (tbzn), and *trans*-4,4'-diaminostilbene (dsn)] when complexed with montmorillonite have been briefly reviewed, together with the results of a quantitative adsorptive study of bzn.[409] As an example, bzn reacts with montmorillonite in the presence of $H_2O$ to produce a blue material, which reversibly yields a yellow-green material on dehydration.[409] The origin of the different colours associated with montmorillonite–bzn, –tbzn, and –dsn complexes has been elucidated by visible —u.v. reflectance and i.r. spectroscopy.[410] The role of lattice-substituted iron(III)

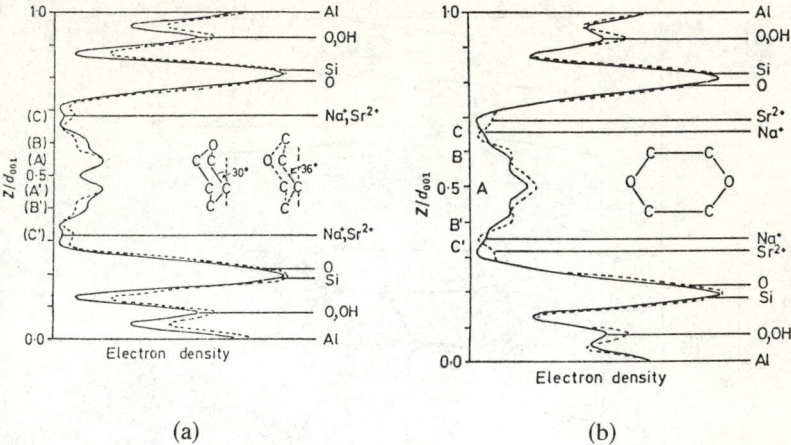

**Figure 7** *Electron-density maps for (a) tetrahydropyran and (b) 1,4-dioxan intercalates with $Na^+$-exchanged (------) and $Sr^{2+}$-exchanged (----) montmorillonite*

(Reproduced from *J.C.S. Dalton*, 1974, 2286)

in the colour reactions of montmorillonite–bzn complexes has also been examined in an $^{57}Fe$ Mössbauer study;[411] it simply acts as an electron-accepting site, being reduced to iron(II) in the production of bzn radical cations.

Intercalation of tetrahydropyran (THP),[412] 1,4-dioxan,[412] pyridine,[413] ethanol,[415] acrylonitrile (anl),[416] and other aromatic molecules[414] in cation-exchanged montmorillonite samples has been examined, using a number of physicochemical techniques. In a structural investigation[412] of THP and 1,4-dioxan intercalates of $Na^+$- and $Sr^{2+}$-exchanged montmorillonites, Adams has prepared electron-density maps (Figures 7a and 7b) which show that the interlamellar molecules are oriented essentially perpendicular to the silicate layers. Adams also notes that these complexes provide a good illustration of the necessity of obtaining and interpreting electron-density maps, whenever possible, since changes in basal spacing, used alone, may yield misleading information concerning the orientation adopted by the organic guest species.[412]

Adams and Thomas[413] have also undertaken crystallographic, XPS, and kinetic studies on the $Na^+$–montmorillonite–pyridine system. Neutron-diffraction studies of the stable pyridine intercalate (48) (of basal spacing 14.8 Å) show that the plane of the guest molecule is perpendicular to, and the long C—N axis is inclined at 60° to, the silicate sheets (Figure 8). The kinetics of the formation of this stable intercalate from the initially produced, less stable, intercalate (49) (basal spacing 23.3 Å) [reaction (39)] have been followed by X-ray diffraction techniques. XPS shows that the pyridine is retained by (48) even *in vacuo* ($10^{-6} N\,m^{-2}$) at room

---

[415] M. S. Stul and J. B. Uytterhoeven, *J.C.S. Faraday I*, 1975, **71**, 1396.
[416] S. Yamanaka, F. Kanamaru, and M. Koizumi, *J. Phys. Chem.*, 1975, **79**, 1285.

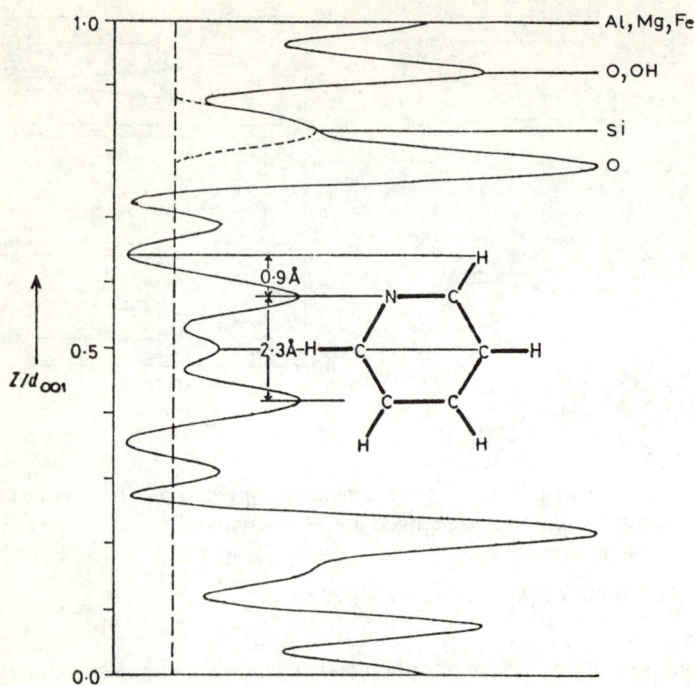

**Figure 8** One-dimensional projection of nuclear scattering density perpendicular to (001) for pyridine intercalates of $Na^+$-exchanged montmorillonite

(Reproduced from *J.C.S. Dalton*, 1975, 1459)

temperature and that the nitrogen atoms are situated within essentially one type of environment.[413]

$$[M(py)_4],2H_2O \xrightarrow[+H_2O]{-py} [M(py)_2],4H_2O \quad \{M = 2Na^+_{0.5}(Al_{3.5}Mg_{0.5}[Si_8O_{20}](OH)_4\} \quad (39)$$

(49)                 (48)

The arrangement of the adsorbed molecules in alkali-metal- and alkaline-earth-metal–montmorillonite–acrylonitrile complexes has been determined by a group of Japanese authors[416] from the results of studies of i.r. pleochroism, deuterium exchange, and basal spacing. The complexes fall into three different groups, depending on the interlayer cation. The basal spacing (13.0 Å, *cf.* 9.13 Å for montmorillonite) of the first group ($Li^+$, $Na^+$, $K^+$, and $Ba^{2+}$) indicates that the sorbed acrylonitrile molecules are oriented with their planes parallel to the silicate sheets. The second group ($Ca^{2+}$, $Mg^{2+}$, $Co^{2+}$, and $Ni^{2+}$) adopted two different basal spacings, corresponding to the amount of $H_2O$ retained; complexes with spacing of 13.0 Å (*cf.* the first group of complexes) are dehydration forms of those with spacing of 17.8 Å. A possible arrangement of the adsorbed molecules in complexes with 17.8 Å spacing is shown in Figure 9a. The third group contained

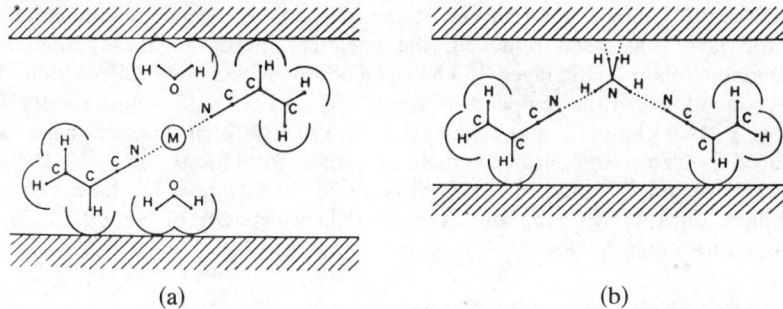

**Figure 9** *Proposed orientation of interlayer moieties in (a) acrylonitrile intercalates of $Mg^{2+}$-, $Ca^{2+}$-, $Co^{2+}$-, or $Ni^{2+}$-exchanged montmorillonite with a basal spacing of 17.8 Å (M indicates one of the exchanged cations), and (b) acrylonitrile intercalates of $NH_4^+$-exchanged montmorillonite*
(Reproduced by permission from *J. Phys. Chem.*, 1975, **79**, 1285)

complexes formed only by montmorillonite containing $NH_4^+$ cations; the basal spacing (16.3 Å), smaller than that of the chemically similar $K^+$–montmorillonite complex (17.8 Å), was explained by the formation of hydrogen bonds between the $NH_4^+$ ions and the nitrogen of the intercalated acrylonitrile. The proposed orientation of acrylonitrile molecules and the $NH_4^+$ ion in the complex is shown in Figure 9b.[416]

Thomas *et al.*[414] have reviewed the formation of complexes between transition-metal-ion-exchanged montmorillonites and such aromatic molecules as benzene, toluene, xylene, and anisole and other species such as *trans*-stilbene and indene. Two types of complex are formed with, for example, interlayer $Cu^{II}$ ions: (i) an edge $\pi$-bonded copper arene moiety, and (ii) a bond-type complex which causes associated distortion of the aromatic ring and some delocalization of the circumferential double bonds. These complexes display a rich range of unusual reactions. For example, on heating in a mass spectrometer, products of high molecular weight (trimers and dimers) appeared, specifically for toluene and all three xylenes [which form type (i) complexes], but only in negligible proportions for benzene [which forms type (ii) complexes].[414]

Thermodynamic parameters for acid-activated montmorillonite–benzene complexes have been determined by a group of Russian authors.[417] It is concluded that a phase transition resembling fusion of the guest species occurs in the intercalated complex. The phase-transition temperature increases to that of the melting point of bulk-phase benzene (5.5 °C) on attainment of the limiting degree of adsorption.

The catalytic hydrogenation of hex-1-ene by $Rh^{2+}$-exchanged hectorite {parent compound $Na_{0.66}$ ($Li_{0.66}$, $Mg_{5.34}$) [$Si_8O_{20}$] $(OH, F)_4$} has also been investigated.[418]

The main characteristics of the acidic attack on chrysotile have been established;[419] it occurs *via* progressive elimination of the brucitic octahedral layer

---

[417] S. Z. Muminov, A. A. Rakhimov, and E. A. Aripov, *Russ. J. Phys. Chem.*, 1975, **49**, 160.
[418] T. J. Pinnavaia and P. K. Welty, *J. Amer. Chem. Soc.*, 1975, **97**, 3819.
[419] C. Bleiman and J. P. Mercier, *Bull. Soc. chim. France*, 1975, 529.

[reaction (40)]. Since chlorination of the silicic layer occurs readily after the brucitic layer has been removed, the chemical inertia of these minerals is attributed to the brucitic layer.[419] The reduction of $NiSiO_3$ by CO [reaction (41)] has been studied in the temperature range 1073—1273 K;[420] values for the free energy ($-44.6$ kJ mol$^{-1}$) and enthalpy ($-47.3$ kJ mol$^{-1}$) of reaction have been deduced. The standard enthalpies of formation of $Na_2ZrSi_2O_7$ ($-3622.9$ kJ mol$^{-1}$)[421] and of $Na_4Zr_2Si_3O_{12}$ ($-6356.8$ kJ mol$^{-1}$)[421] have been determined, together with the entropy of and heat capacity of $Na_2ZrSiO_5$ in the temperature range 55—300 K.[422]

$$\begin{array}{c}\ge\!\!Si\!-\!O\!-\!Mg\!-\!OH\\ \\HO\!-\!Mg\!-\!OH + 6H^+ \longrightarrow \begin{array}{c}\ge\!\!Si\!-\!OH\\ +\\ \ge\!\!Si\!-\!OH\end{array} + 3Mg^{2+} + 4H_2O\\ \\ \ge\!\!Si\!-\!O\!-\!Mg\!-\!OH\end{array} \quad (40)$$

$$NiSiO_3 + CO \rightarrow Ni + SiO_2 + CO_2 \quad (41)$$

Thermodynamic and structural aspects of the isopolymorphism in the binary systems $Ca_2SiO_4$–$Sr_2GeO_4$,[423] $SrAl_2Si_2O_8$–$SrGa_2Si_2O_8$,[424] $SrGa_2Si_2O_8$–$BaGa_2Si_2O_8$,[424] $Y_2Si_2O_7$–$Sc_2Si_2O_7$,[425] and the quaternary reciprocal system $Ba_2SiO_4$, $Ba_2GeO_4 \parallel MSiO_3$, $MGeO_3$ (M = Mg or Zn)[426] have been examined.

The crystal structures of many silicates have been studied during the period of this Report; following the pattern previously adopted, the data will be considered in order of decreasing O:Si ratio (*i.e.* increasing structural complexity). Refinement of the structure of the high-pressure modification of $Ni_2[SiO_4]$ spinel ($a = 8.0424$ Å) has been achieved.[427] The crystal structures of the orthosilicates $\gamma$-$Li_2Be[SiO_4]$,[428] $NaCa[HSiO_4]$,[429] $Na_2UO_2[SiO_4]$,[430] $Na_2Zn_3[SiO_4]_2$ (phase F),[431] and the sulphide–orthosilicate $Ag_8S_2[SiO_4]$[432] have been determined using single-crystal techniques; pertinent unit-cell parameters are collected in Table 18. The crystal structure of $\gamma$-$Li_2Be[SiO_4]$ is directly related to the high-temperature $\gamma$-polymorphs of $LiAlO_2$, $NaAlO_2$, and $NaFeO_2$.[428] The crystal structure of $NaCa[HSiO_4]$ has been compared to those of $Na_2Ca[SiO_4]$ and $Na_2SO_4$.[429] The U—O bond in $Na_2UO_2[SiO_4]$ is insufficiently clearly defined for uranyl sodium silicate;[430] indeed, this compound has character intermediate between primary and secondary silicates of uranium. The structure of $Na_2Zn_3[SiO_4]_2$ is characterized[431] by alternation of layers of networks of Na polyhedra and parallel 'layers' of

---

[420] A. S. Kurnygin, I. N. Piskunov, and F. T. Bumazhnov, *Russ. J. Phys. Chem.*, 1974, **48**, 601.
[421] E. V. Shibanov and V. G. Chukhlantsev, *Russ. J. Phys. Chem.*, 1974, **48**, 133.
[422] E. V. Shibanov and V. G. Chukhlantsev, *Russ. J. Phys. Chem.*, 1974, **48**, 299.
[423] R. G. Grebenshchikov, V. I. Shitova, and L. Yu. Dmitrieva, *Russ. J. Inorg. Chem.*, 1975, **20**, 271.
[424] M. Calleri and G. Gazzoni, *Z. Krist.*, 1975, **141**, 293.
[425] A. N. Sokolov, I. F. Andreev, and T. V. Ostapenko, *Doklady Chem.*, 1974, **215**, 244.
[426] N. A. Sirazhiddinov, N. N. Mirababaeva, R. G. Grebenshchikov, and E. V. Stroganov, *Russ. J. Inorg. Chem.*, 1974, **19**, 817.
[427] C.-B. Ma, *Z. Krist.*, 1975, **141**, 126.
[428] R. A. Howie and A. R. West, *Acta Cryst.*, 1974, **B30**, 2434.
[429] V. F. Kazak, V. I. Lyutin, V. V. Ilyukhin, and N. V. Belov, *Soviet Phys. Cryst.*, 1975, **19**, 592.
[430] D. P. Shashkin, E. A. Lur'e, and N. V. Belov, *Soviet Phys. Cryst.*, 1975, **19**, 595.
[431] G. F. Plakhov, M. A. Simonov, and N. V. Belov, *Soviet Phys. Cryst.*, 1975, **20**, 24.
[432] E. Schultze-Rhonhof, *Acta Cryst.*, 1974, **B30**, 2553.

bands, with the $[Zn_9O_{24}]$ motif extended along the $a$-axis and fixed by individual $[SiO_4]$ tetrahedra into a unified three-dimensional framework. The silver sublattice of $Ag_8S_2[SiO_4]$ may be described[432] as a distorted and expanded superlattice of the structure of metallic silver; the octahedral holes of this structure are partially occupied in an ordered way, by S atoms or $[SiO_4]$ groups.

Two Russian papers describing the physicochemical properties of calcium–rare-earth oxysilicate apatities have been published.[433,434] Single crystals of $CaY_4[SiO_4]_3O$[433] and $Ca_2Ln_3[SiO_4]_2(PO_4)O$ (Ln = Y, La, Nd, Sm, Gd, Dy, Ho, Er, or Yb)[434] have been grown and their structures determined; the unit-cell parameters of these hexagonal compounds are summarized in Table 19. Compounds of both types can be regarded as derivatives of the phosphate apatite $M_5(PO_4)_3(F, OH)$, since the valence balance does not change and the structure of the apatite is retained.

The rare-earth silicate $Er_4Pb[Si_5O_{17}]$,[435] prepared by crystallization from PbO–$SiO_2$–$Er_2O_3$ melts, is unusual since its structure (Table 20) possesses a combination of $[Si_2O_7]^{6-}$ and $[Si_3O_{10}]^{2-}$ anions; a space-group requirement is that these anions have centro- and mirror-symmetric symmetry, respectively.

The crystal structures of the monoclinic and orthorhombic pyroxene polymorphs of $Zn[SiO_3]$ and of the orthorhombic pyroxene $ZnMg[Si_2O_6]$ have been determined[436] (Table 20); that of $ZnMg[Si_2O_6]$ is intermediate between that of enstatite $\{Mg[SiO_3]\}$ and of orthorhombic $Zn[SiO_3]$. Single-crystal photographs[437] have shown that carpholite, $MnAl_2[Si_2O_6](OH)_4$, has an orthorhombic structure (Table 20); it is based on zig-zag strips of cation octahedra and metasilicate chains parallel to the $c$-axis. Data comparing carpholite with pyroxenes and micas have also been given.

Two new types of chain silicate have been discovered.[438,439] The recently determined crystal structure of inesite, $Ca_2Mn_7[Si_{10}O_{28}](OH)_2,5H_2O$ (Table 20), provides the first known example of double silicate chains with periodicity of 5 tetrahedra.[438] In fact, the inesite structure comprises two components: (i) silicate double chains with alternate six- and eight-membered rings, and (ii) a polyhedral band of which the building blocks are a distorted pentagonal bipyramid around Ca and octahedra around four crystallographically different Mg atoms. The structure model proposed for the new fibrous phase $NaMg_4[Si_6O_{15}(OH)](OH)_2$ (Table 20), synthesized in the $Na_2O$–MgO–$SiO_2$–$H_2O$ system at 1000 atm, is based on strips of three pyroxene chains.[439] In the projection on the (010) plane the structure resembles those of monoclinic pyroxenes and amphiboles.

The dehydration of cavansite, $Ca(VO)[Si_4O_{10}],4H_2O$, has been studied *in situ* at 220 °C and $10^{-5}$ atm, using $X$-ray diffraction techniques.[440] The orthorhombic cell dimensions (space group *Pcmn*) decrease from $a = 9.792$, $b = 13.644$, $c = 9.624$, to $a = 9.368$, $b = 12.808$, $c = 9.550$ Å, with an 11% reduction in volume.

[433] I. A. Bondar', L. P. Mezentseva, and A. I. Domanskii, *Russ. J. Inorg. Chem.*, 1974, **19**, 1650.
[434] N. F. Fedorov, I. F. Andreev, and N. S. Meliksetyan, *Soviet Phys. Cryst.*, 1975, **20**, 280.
[435] G. B. Ansell and B. Wanklyn, *J.C.S. Chem Comm.*, 1975, 794.
[436] N. Morimoto, Y. Nakajima, Y. Syano, S. Akimoto, and Y. Matsui, *Acta Cryst.*, 1975, **B31**, 1041.
[437] I. S. Naumova, E. A. Pobedimskaya, and N. V. Belov, *Soviet Phys. Cryst.*, 1975, **19**, 718.
[438] C. Wan and S. Ghose, *Naturwiss.*, 1975, **62**, 96.
[439] V. A. Drits, Yu. I. Goncharov, V. A. Alexandrova, V. E. Khadzhi, and A. L. Dmitrik, *Soviet Phys. Cryst.*, 1975, **19**, 737.
[440] R. Rinaldi, J. J. Pluth, and J. V. Smith, *Acta Cryst.*, 1975, **B31**, 1598.

**Table 18** *Unit-cell parameters for a number of orthosilicates*

| *Silicate* | *Symmetry* | *Space group* | *a*/Å | *b*/Å | *c*/Å | *α*/° | *β*/° | *γ*/° | *Ref.* |
|---|---|---|---|---|---|---|---|---|---|
| γ-$Li_2Be[SiO_4]$ | orthorhombic | *C*222$_1$ | 6.853 | 6.927 | 6.125 | — | — | — | 428 |
| $NaCa[HSiO_4]$ | monoclinic | 2/*mP*2$_1$ | 5.71 | 5.45 | 7.03 | — | — | 122 | 429 |
| $Na_2UO_2[SiO_4]$ | tetragonal | *I*4$_1$/*acd* | 12.718 | — | 13.376 | — | — | — | 430 |
| $Na_2Zn_3[SiO_4]_2$ | triclinic | $P\bar{1}$ | 5.124 | 8.830 | 13.504 | 72.59 | 101.70 | 89.11 | 431 |
| $Ag_8S_2[SiO_4]$ | tetragonal | *I*4$_1$/*amd* | 7.005 | — | 17.750 | — | — | — | 432 |

**Table 19** *Unit-cell parameters/Å of calcium–rare-earth oxysilicate apatites*

| $CaY_4[SiO_4]_3O$[433] | | $Ca_2Ln_3[SiO_4]_2(PO_4)O$[434] | | | | | | | | |
|---|---|---|---|---|---|---|---|---|---|---|
| | | Y | La | Nd | Sm | Gd | Dy | Ho | Er | Yb |
| *a* | 9.51 | 9.36 | 9.63 | 9.51 | 9.41 | 9.38 | 9.36 | 9.31 | 9.25 | 9.19 |
| *c* | 7.03 | 6.82 | 7.10 | 6.97 | 6.91 | 6.88 | 6.82 | 6.80 | 6.76 | 6.72 |
| *c*/*a* | 0.739 | 0.728 | 0.737 | 0.732 | 0.734 | 0.733 | 0.728 | 0.730 | 0.731 | 0.731 |

**Table 20** *Unit-cell parameters for a number of silicates*

| *Silicate* | *Symmetry* | *Space group* | *a*/Å | *b*/Å | *c*/Å | *α*/° | *β*/° | *γ*/° | *Ref.* |
|---|---|---|---|---|---|---|---|---|---|
| $Er_4Pb[Si_5O_{17}]$ | monoclinic | *P*2$_1$/*m* | 5.534 | 10.58 | 6.960 | — | 107.2 | — | 435 |
| $Zn[SiO_3]$ | monoclinic | *C*2/*c* | 9.787 | 9.161 | 5.296 | — | 111.42 | — | 436 |
| $Zn[SiO_3]$ | orthorhombic | *Pbca* | 18.204 | 9.087 | 5.278 | — | — | — | 436 |
| $ZnMg[Si_2O_6]$ | orthorhombic | *Pbca* | 18.201 | 8.916 | 5.209 | — | — | — | 436 |
| $MnAl_2[Si_2O_6](OH)_4$ | orthorhombic | *Ccca* | 13.381 | 20.296 | 5.121 | — | — | — | 437 |
| $Ca_2Mn_7[Si_{10}O_{28}](OH)_2,5H_2O$ | triclinic | $P\bar{1}$ | 8.849 | 9.277 | 11.980 | 88.11 | 131.88 | 96.76 | 438 |
| $NaMg_4[Si_6O_{15}(OH)](OH)_2$ | monoclinic | *C*2/*c* | 10.132 | 27.12 | 5.257 | — | 106.9 | — | 439 |

# Elements of Group IV

The reduction is due mainly to removal of three $H_2O$ molecules (per Ca atom) from between silicate layers. Loss of the last $H_2O$ molecule probably corresponds to the breakdown of the structure, which occurs at ca. 400 °C.[440] Finally, X-ray studies of milarite, armenite, and sogdianite have shown their structures to be similar, with hexagonal unit cells of the $P6/mcc$ space group;[441] their formulae and relevant crystal parameters are collated in Table 21.

**Table 21** Unit-cell parameters of some isostructural, naturally occurring, silicate minerals[441]

| Silicate | Formula | a/Å | c/Å |
|---|---|---|---|
| milarite | $KCa_2(Be_{2.3}Al_{0.7})[Si_{12}O_{29.7}(OH)_{0.3}],(H_2O)_{0.7}$ | 10.40 | 13.80 |
| armenite | $BaCa_2Al_3[Al_3Si_9O_{30}],(H_2O)_2$ | 10.69 | 13.90 |
| sogdianite | $(K_{1.05}Na_{0.95})(Zr_{0.8}Fe^{III}_{0.6}Ti_{0.4}Fe^{II}_{0.2})(Li_{2.55}Al_{0.15}\square_{0.30})[SiO_{30}]$ | 10.09 | 13.98 |

**Germanium(IV), Tin(IV), and Lead(IV) Oxide Phases.**—Phase equilibria in several metal germanate systems have been studied by a variety of techniques. The $Na_2O-GeO_2$ and $Li_2O-GeO_2$ systems have been investigated by dilatometry, t.d.a., and radiocrystallography. $Li_2Ge_7O_{15}$ is the only stable phase in the latter system; other phases previously described undergo irreversible decomposition of heating, and the powder data attributed to $Li_6Ge_8O_{19}$ correspond to a binary mixture. In the $Na_2O-GeO_2$ system, the known phases $Na_2GeO_3$, $Na_2Ge_4O_9$, and $Na_4Ge_9O_{20}$ all take part in equilibria. The digermanate $Na_2Ge_2O_5$ does not appear, and the tetragermanate $Na_2Ge_4O_9$ is completely metastable. The phases $Na_4GeO_2$, $Na_6Ge_2O_7$, $Li_4GeO_4$, $Li_6Ge_2O_7$, and $Li_8GeO_6$ have all been characterized for the first time.[442] X-Ray diffraction and derivatographic studies have confirmed that the two compounds $Na_2O,GeO_2$ and $2Na_2O,9GeO_2$ are formed in mixtures of $Na_2CO_3$ and $GeO_2$, the latter being the primary compound of the system.[443] The ternary germanate phases $K_4MGe_3O_9$ (M = Ca or Sr), $Na_2MGe_2O_6$ (M = Ca or Sr), and $Na_2Ba_2Ge_2O_7$ have been obtained from the solid-state reactions of mixtures of the metal carbonates and oxalates and $GeO_2$.[444] Two types of solid solutions exist in the low-temperature region of the $Ca_2GeO_4-Sr_2GeO_4$ system. One has the olivine lattice and contains ca. 8 mol % $Sr_2GeO_4$ in $\gamma$-$Ca_2GeO_4$, whilst the other has compositions ranging from equimolecular to pure $SrGeO_4$.[445] The formation of a continuous series of solid solutions with a tetragonal lattice of the hardystonite type has been established in the $Ba_2ZnGe_2O_7-Sr_2ZnGe_2O_7$ system by X-ray diffraction and crystal optical analysis.[446] Those in the quaternary reciprocal '2BaO,MgO', '2BaO,ZnO' ∥ $2SiO_2$, $2GeO_2$ or $Ba_2SiO_4$, $Ba_2GeO_4$ ∥ $MSiO_3$, $MGeO_3$ (M = Mg or Zn) system possess the okermanite structure.[447] The structures of two uranyl germanates have been

---

[441] V. V. Bakakin, V. P. Balko, and L. P. Solov'eva, *Soviet Phys. Cryst.*, 1975, **19**, 460.
[442] B. Monnaye, *Rev. Chim. minérale*, 1975, **12**, 268.
[443] E. K. Belyaev and V. F. Annopol'skii, *Russ. J. Inorg. Chem.*, 1974, **19**, 30.
[444] R. P. Gunawardane and F. P. Glasser, *Z. anorg. Chem.*, 1975, **411**, 163.
[445] R. G. Grebenshchikov and V. I. Shitova, *Doklady Chem.*, 1974, **218**, 596.
[446] N. A. Sirazhiddinov, N. N. Mirbabaeva, and R. G. Grebenshchikov, *Russ. J. Inorg. Chem.*, 1974, **19**, 576.
[447] N. A. Sirazhiddinov, N. N. Mirababaeva, R. G. Grebenshchikov, and E. V. Stroganov, *Russ. J. Inorg. Chem.*, 1974, **19**, 817.

determined by single-crystal $X$-ray diffraction. That of copper uranyl germanate $[Cu(H_2O)_4][UO_2HGeO_4]_2,2H_2O$ is made up of infinite sheets of $(UO_2HGeO_4)_n^{n-}$ anions between which are water molecules and $[Cu(H_2O)_4]$ groups. Chains of uranium atoms, which are bonded together in sheets by double asymmetric oxygen bridges, are linked together by $GeO_4$ tetrahedra, which share an edge with one chain and a corner with another.[448] Uranyl germanate $(UO_2)_2GeO_4(H_2O)_2$ possesses a structure in which each uranium atom enjoys pentagonal-bipyramidal co-ordination by seven oxygen atoms. Four of the five equatorial oxygen atoms belong to $GeO_4$ tetrahedra, the fifth being that of a water molecule. Infinite chains formed by the sharing of an edge between the pentagonal bipyramids are linked together by the $GeO_4$ tetrahedra, thus forming a three-dimensional network (Figure 10).[449] The crystal structure of potassium orthostannate $K_2SnO_4$ consists of discrete tetrahedral $SnO_4$ groups bonded by potassium ions.[450]

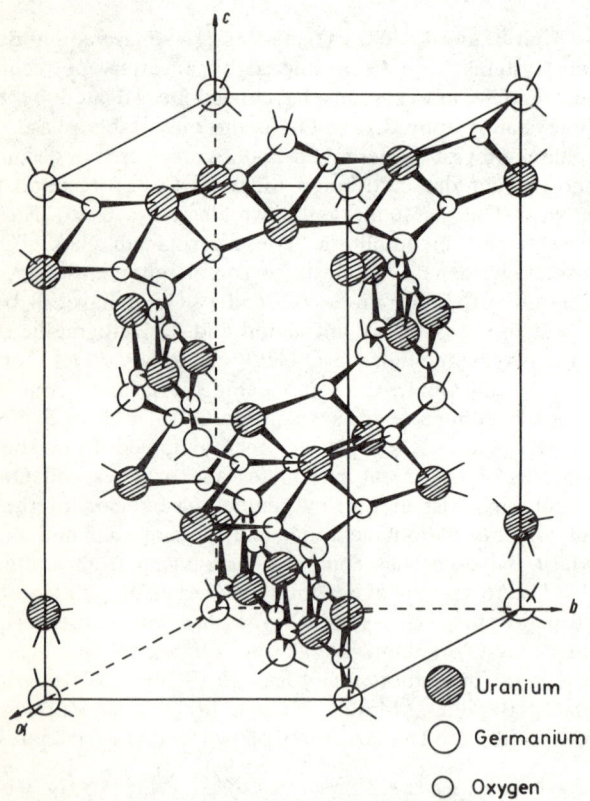

**Figure 10** *The structure of uranyl germanate dihydrate*

(Reproduced by permission from *Acta Cryst.*, 1975, **B31**, 1140)

---

[448] J. P. Legros and Y. Jeannin, *Acta Cryst.*, 1975, **B31**, 1133.
[449] J. P. Legros and Y. Jeannin, *Acta Cryst.*, 1975, **B31**, 1140.
[450] R. Marchand, Y. Piffard, and M. Tournoux, *Acta Cryst.*, 1975, **B31**, 511.

# Elements of Group IV

The majority of papers on tin(IV) oxide systems are concerned with adsorption and heterogeneous catalysis. The surface of tin(IV) oxide gel has been studied extensively, using the transmission i.r. technique in the temperature range $293 \leq T/K \leq 773$. Molecular water is largely removed by evacuation at 320 K and fully removed at 473 K. Hydrogen-bonded OH groups are present after evacuation at 773 K. Treatment of the oxide surface thermally pretreated in the temperature range $320 < T/K < 618$ with carbon dioxide results in the formation of surface carbonate and bicarbonate species. Treatment with carbon monoxide results in severe reduction, to bivalent tin.[451] The adsorption of ammonia and pyridine on the oxide surface shows that it exhibits Lewis acidity, but not Brönsted acidity, towards these bases even in the presence of water. Ammonium cations formed on a surface previously exposed to HCl are not free to rotate. When HCl is added to a damp surface, the molecular water present is protonated to give hydroxonium ions.[452] The tin(IV) oxide surface is strongly oxidizing towards small organic molecules. Methanol and dimethyl carbonate are chemisorbed to give surface methoxy-groups, but these are readily oxidized to a surface formate at temperatures $\geq 320$ K. Acetone and acetaldehyde are adsorbed predominantly as acetates, but some evidence was found for an enol form which may be responsible for the rapid deuterium exchange between $(CD_3)_2CO$ and surface hydroxy-groups. Formic and acetic acids also give rise to surface carboxylate species, but undissociated acid molecules are also present on the oxide surface at low temperatures. As expected, the acetate is thermally more stable than the surface formate.[453] A Russian paper reports the i.r. spectra of molecular oxygen adsorbed onto tin(IV) oxide.[454] Fuller and Warwick[455] have investigated the kinetics of the reaction $CO + N_2O \rightarrow CO_2 + N_2$ at 190—220 °C on tin(IV) oxide gel that had been activated at 450 °C, using a differential flow reactor. The steady-state kinetics are adequately represented by the empirical relationship (42). The activation energy

$$\text{rate} = A \, \exp[-E/RT] \rho_{CO}^{0.2} \rho_{N_2O}^{0.5} \tag{42}$$

was deduced to be 24.3 kcal mol$^{-1}$, and no product inhibition by $CO_2$ or $N_2$ was observed. The results indicate that the reaction occurs principally by catalyst redox involving CO chemisorption, $CO_2$ desorption *via* abstraction of lattice oxygen, and re-oxidation of the catalyst by $N_2O$, the last process being the rate-determining step. Pd–SnO$_2$ catalysts are more active for CO oxidation than either Pd–SiO$_2$ or tin(IV) oxide itself, but are relatively inactive for $C_2H_4$ hydrogenation. The catalysts may be made either by impregnating tin(IV) oxide gel with $H_2PdCl_4$ solution, by cation exchange with $[Pd(NH_3)_3(OH)_2]$, or by coprecipitation of $Pd(OH)_2$ and hydrated $SnO_2$ from chloride solution with potassium hydroxide, followed by washing, drying, and re-washing. The grinding together of Pd–SiO$_2$ and –SnO$_2$ mixtures enhances CO oxidation activity, indicating that the palladium induces activity of the SnO$_2$ surface.[456]

---

[451] E. W. Thornton and P. G. Harrison, *J.C.S. Faraday I*, 1975, **71**, 461.
[452] P. G. Harrison and E. W. Thornton, *J.C.S. Faraday I*, 1975, **71**, 1013.
[453] E. W. Thornton and P. G. Harrison, *J.C.S. Faraday I*, 1975, **71**, 2468.
[454] T. A. Gundrizer and A. A. Davydov, *Reaction Kinetics Catalysis Letters*, 1975, **3**, 63.
[455] M. J. Fuller and M. E. Warwick, *J. Catalysis*, 1975, **39**, 412.
[456] G. C. Bond, L. R. Molloy, and M. J. Fuller, *J.C.S. Chem. Comm.*, 1975, 796.

Ai has investigated the amounts of both acidic and basic sites of series of $SnO_2-V_2O_5$,[457] $SnO_2-MoO_3$,[458] and $SnO_2-P_2O_5$[458] catalysts by the adsorption of basic and acidic molecules from the gas phase. All three catalyst systems are basic when $SnO_2$-rich. The $V_2O_5$-rich catalysts are acidic, and the activities and selectivities of the vapour-phase oxidations of but-1-ene and butadiene, the isomerization of but-1-ene, and the dehydration and dehydrogenation of isopropyl alcohol may be relatively well explained by the acid–base properties between the catalyst and reactant.[457] The acidities of $SnO_2-MoO_3$ catalysts are dramatically high at a molybdenum content of 30—60 atom%, but those of $MoO_3$-rich (Mo>80%) catalysts are fairly low. The introduction of $P_2O_5$ into $SnO_2$ increases the acidity and decreases the basicity only to a small extent, and the $SnO_2-P_2O_5$ catalysts are rather basic.[458] A variety of mixed tin(IV) oxide catalyst systems have been used for the catalytic ammoxidation of hydrocarbons. Tin(IV) oxide itself is highly selective for the formation of acrylonitrile in the ammoxidation of propene.[459] $SnO_2-Sb_2O_5$ and $SnO_2-MoO_3$ catalysts also give good selectivity.[460,461] The latter catalyst system gives high initial selectivity for the formation of benzonitrile from toluene and, successively, p-toluonitrile and terephthalonitrile from p-xylene.[462] Kinetic studies of the ammoxidation of propene and isobutene over a $SnO_2-V_2O_5-P_2O_5$ catalyst are consistent with a mechanism in which the abstraction of an allylic hydrogen from adsorbed alkene by adsorbed peroxy-species is rate-determining. The reaction was suggested to occur at $Sn^{4+}$ sites from which electron transfer to neighbouring vanadium ion is possible.[463] The presence of steam in propene-ammoxidation over a Sn–Sb–Fe–O catalyst at 450 °C does not improve the yields of acrylonitrile as in the Bi–Mo–O system.[464]

The combination of α-stannic acid with aqueous solutions of saturated monocyclic amines yields solid products, the composition of which varies according to the base used. The thermal decomposition of these materials in an inert atmosphere occurs in several steps corresponding to loss of water, of base, and finally of $CO_2$, the latter due to oxidation by $SnO_2$ of carbon resulting from the pyrolysis of traces of amine.[465] Modifications of the lattices of the substances accompany some steps of the thermal degradation. d.t.a. and t.g.a. studies in air allow the compounds to be assigned the general formula $xSnO_2,(x-1)H_2O$,base.[466]

XPS studies of the bulk oxides of lead reveal no clear evidence for the two expected oxidation states of lead in $Pb_3O_4$. The 4f binding energies of lead in the rhombic and tetragonal modifications of PbO are the same within experimental error, but are greater than those for β-$PbO_2$.[467] Carbon monoxide undergoes

---

[457] M. Ai, *J. Catalysis*, 1975, **40**, 318.
[458] M. Ai, *J. Catalysis*, 1975, **40**, 327.
[459] J. E. Germain and R. Perez, *Bull. Soc. chim. France*, 1975, 735.
[460] J. E. Germain and R. Perez, *Bull. Soc. chim. France*, 1975, 739.
[461] J. E. Germain and R. Perez, *Bull. Soc. chim. France*, 1975, 1216.
[462] G. Simon and J. E. Germain, *Bull. Soc. chim. France*, 1975, 2617.
[463] Z. Onsan and D. L. Trimm, *J. Catalysis*, 1975, **38**, 257.
[464] E. Ghenassia and J. E. Germain, *Bull. Soc. chim. France*, 1975, 731.
[465] S. Durand and E. Masdupuy *Bull. Soc. chim. France*, 1975, 453.
[466] S. Durand and E. Masdupuy, *Bull. Soc. chim. France*, 1975, 456.
[467] J. M. Thomas and M. J. Tricker, *J.C.S. Faraday II*, 1975, **97**, 329.

catalytic oxidation at the $Pb_3O_4$ surface, the reaction between CO and adsorbed oxygen being the slowest step in the overall reaction.[468]

**Molecular Silicon(IV)–, Germanium(IV)–, Tin(IV)–, and Lead(IV)–Oxygen Compounds.**—*Oxides.* The structures of $(F_3Si)_2O$[469] and $(Cl_3Si)_2O$[470] have been investigated by vibrational spectroscopy in the gaseous, liquid, and solid phases. The spectra in all three phases for $(F_3Si)_2O$ and in the gaseous phase for the chloro-compound indicate $C_{2v}$ symmetry for the molecules in these phases. However, structural changes occur upon condensation of $(Cl_3Si)_2O$, the Si—O—Si angle approaching linearity in the solid, probably due to crystal-packing factors. Tricyclohexyltin hydroxide has been assigned a hydroxy-bridged structure analogous to that deduced for the methyltin homologue on the basis of its Mössbauer and i.r. spectroscopic properties.[471] The Sn—O bond energy terms $D(Ph_3Sn—O)$ and standard enthalpies of formation of $Ph_3SnOH$ and $(Ph_3Sn)_2$ have been derived by calorimetric methods. The bond energies were deduced to be $91 \pm 5$ kcal mol$^{-1}$ for $Ph_3SnOH$ and $81 \pm 5$ kcal mol$^{-1}$ for $(Ph_3Sn)_2$.[472] Bis(tributyltin) oxide reacts exothermically with arylsulphonyl isocyanates to give an equilibrium mixture of 1:1 adduct tautomers, as shown in Scheme 3. The adducts are

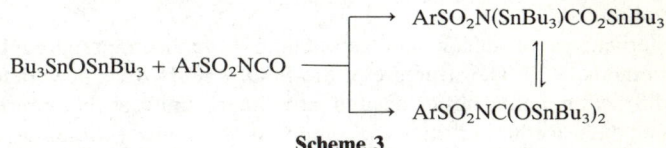

**Scheme 3**

thermally unstable and are readily protolysed. With two moles of sulphonyl isocyanate a 1:2 adduct is formed which decarboxylates to give 1,3-bisarylsulphonyl-1,3-bistributylstannylureas.[473] Harada has investigated the structures of the compounds formed between $Me_2SnO$ and $Me_2SnBr_2$ under various conditions.[474]

*Alkoxides and Related Derivatives.* The structure of $(MeO)_4Si$ has been studied by electron diffraction. The molecule possesses $S_4$ symmetry, being slightly flattened along the axis. The Si—O bonds are shorter (1.613 Å) than in $MeOSiH_3$, demonstrating the effect of electronegative substituents on silicon.[475] $(MeO)_4Si$ undergoes three different types of reaction with $CT^+$ ions (from $CT_4$ exposed to β-radiation): by hydride abstraction [to give $(MeO)_3SiOĊH_2$], abstraction of methoxy-groups [to give $MeOCT_3$ and $^+Si(OMe)_3$], or $Me^+$ displacement [to give $(MeO)_3SiOCT_3$].[476] The thermodynamic and spectral characteristics of complexes of iodine with $Si(OR)_4$ (R = Me, Et, or Pr$^n$) have been studied spectrophotometrically in n-heptane solution. The donor strengths of the tetraalkoxysilanes are

---

[468] M. Kobayashi and H. Kobayashi, *J. Catalysis*, 1975, **36**, 74.
[469] J. R. Durig, V. F. Kalsinsky, and M. J. Flanagan, *Inorg. Chem.*, 1975, **14**, 2839.
[470] J. Durig, M. J. Flanagan, and V. F. Kalsinsky, *J. Mol. Structure*, 1975, **27**, 241.
[471] B. Y. K. Ho and J. J. Zuckerman, *J. Organometallic Chem.*, 1975, **96**, 41.
[472] A. S. Carson, J. Franklin, P. G. Laye, and H. Morris, *J. Chem. Thermodynamics*, 1975, **7**, 763.
[473] N. I. Mysin and Yu. I. Dergunov, *J. Gen. Chem.* (U.S.S.R), 1974, **44**, 1491.
[474] T. Harada, *J. Inorg. Nuclear Chem.*, 1975, **37**, 288.
[475] L. H. Boonstra, F. C. Mijlhoff, G. Renes, A. Spelbos, and I. Hargittai, *J. Mol. Structure*, 1975, **28**, 129.
[476] V. D. Nefedov, N. P. Kharitonov, E. N. Sinotova, T. A. Kochina, and M. V. Korsakov, *J. Gen. Chem.* (U.S.S.R), 1975, **45**, 477.

similar, and all are weaker donors than diethyl ether.[477] Aryloxytrimethylsilanes are conveniently synthesized by the reaction of $(Me_3Si)_2S$ with the appropriate phenol.[478]

The dissolution of $GeO_2$ in aqueous solutions or pyrocatechol initially proceeds more slowly than in pure water, and is accompanied by a fall in the pH and the formation of tripyrocatechol-germanic acid. As $GeO_2$ accumulates in solution, it is converted into bis(pyrocatecholato)germanium, which then precipitates.[479] The interaction of $Ge^{IV}$ with 4-chloro- and 4-bromo-catechol has been studied spectrophotometrically and potentiometrically at 25 °C. Only one soluble complex is formed in solution, which has a 1:3 Ge:L ratio and the properties of a dibasic acid.[480] Co-ordination compounds of germanium(IV) derivatives of polyhydric phenols (pyrocatechol, pyrogallol, gallic acid, and methyl gallate) and DMSO, of composition $GeL_2R_2$ (L = residue of the corresponding phenol and L = DMSO), have been obtained by treating the appropriate $[GeL_2(H_2O)_2]$ compounds with DMSO.[481] Tin(IV) derivatives of 2,2'-dihydroxyphenol, 2,2'-dihydroxybinaphthyl, and pyrocatechol have been synthesized by the reaction of tetraphenyltin and the phenol at 150—160 °C. When the reactions are carried out in a 3:1 molar ratio, compounds of composition $L_2SnLH$ are formed. With amines, $L_2Sn,2base$ complexes are obtained.[482]

Oxime derivatives of silicon and germanium have been prepared by fairly standard methods.[483,484] The structure of $Me_3SnON{=}C_6H_{10}$ has been determined by X-ray diffraction, and is not associated into dimeric units, as first inferred from spectroscopic data. Rather, crystals are composed of infinite $-(Me_3SnO)-_\infty$ chains with $-N{=}C_6H_{10}$ residues pendant from each oxygen atom (Figure 11).[485] Similar structures have been proposed for the triorganotin derivatives of 2- and 3-hydroxypyridine from Mössbauer quadrupole splitting and i.r. data, the pyridyl nitrogen taking no part in co-ordination.[486]

Compounds containing another donor site further remote from the metal in the alkoxide residue have been studied extensively. Triorganotin derivatives of β-diketones have been synthesized from the triorganotin chloride and the sodium salt of the β-diketone in benzene. From the Mössbuer quadrupole splitting data, it was inferred that the trimethyltin derivatives possessed the *meridional* configuration (50), whilst the corresponding triphenyltin derivatives had the all-*cis*-structure (51). The latter structure was confirmed by X-ray diffraction studies for $[Ph_3Sn(dbm)]$ (dbm = dibenzoylmethanate).[487] The dynamics of several β-keto-enolato-silicon, -germanium, and -tin compounds have been studied. Several silicon(IV) complexes of β-keto-enolates have been resolved into their optical

---

[477] Z. Kokot and M. Tamres, *Inorg. Chem.*, 1975, **14**, 2441.
[478] E. P. Lebedev, V. A. Baburina, and V. O. Reikhsfel'd, *J. Gen. Chem. (U.S.S.R)*, 1975, **45**, 337.
[479] D. Ya Evdokimov and E. A. Kogan, *J. Gen. Chem. (U.S.S.R)*, 1974, **44**, 1498.
[480] V. A. Nazarenko, L. I. Vinarova, and N. V. Lebedeva, *Russ. J. Inorg. Chem.*, 1974, **19**, 1295.
[481] G. I. Kurnevich, V. B. Vishnevskii, and E. M. Luiko, *Russ. J. Inorg. Chem.*, 1974, **19**, 375.
[482] K. Andrä and H. R. Hopper, *Z. anorg. Chem.*, 1975, **413**, 97.
[483] A. B. Goel and V. D. Gupta, *J. Organometallic Chem.*, 1975, **85**, 327.
[484] A. Singh and R. C. Mehrotra, *Synth. React. Inorg. Metal-Org. Chem.*, 1974, **4**, 549.
[485] P. F. R. Ewings, P. G. Harrison, T. J. King, R. C. Phillips, and J. A. Richards, *J.C.S. Dalton*, 1975, 1950.
[486] P. G. Harrison and R. C. Phillips, *J. Organometallic Chem.*, 1975, **99**, 79.
[487] G. M. Bancroft, B. W. Davies, N. C. Payne, and T. K. Sham, *J.C.S. Dalton*, 1975, 973.

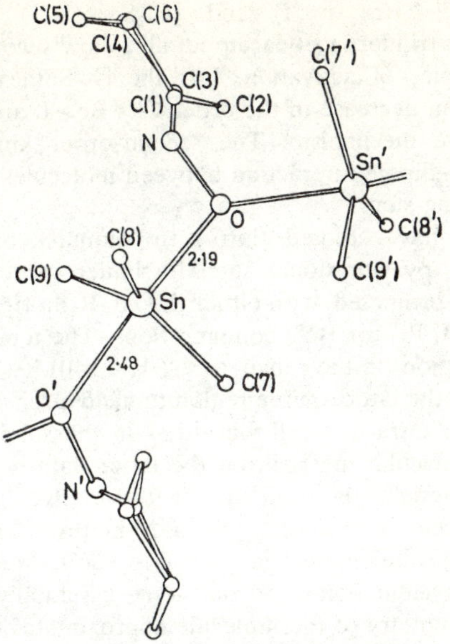

**Figure 11** *Part of the chain structure of* $Me_3SnON=C_6H_{10}$
(Reproduced from *J.C.S. Dalton*, 1975, 1950)

(50)      (51)

isomers, and their racemization has been studied kinetically in 1,1,2,2-tetrachloroethane and acetonitrile in the range 30—80 °C. U.v. data and isotopic labelling verified that the process occurred by an intramolecular mechanism.[488] The phenylchlorometal complexes [PhClM(acac)] (M = Si, Ge, or Sn) adopt predominantly a *cis*-(Ph, Cl) structure in solution, with a small amount ($\leqslant 5\%$) of the *trans*-(Ph, Cl) form. The methyltin analogue has a similar constitution in solution, but the methylsilicon and methylgermanium analogues appear to be totally *cis*. The solution structures of [$Ph_2Ge(acac)_2$] and [$Me_2Sn(acac)_2$] could not be determined. The environmental averaging of the acac ligands is believed to proceed *via* a twist mechanism in which a metal–oxygen bond is ruptured to give a five-co-ordinated intermediate. The rate-determining sequence is thought to involve the proper orientation of two five-co-ordinated species prior to the formation of a bis(acac)-bridged dimer.[489—492] The hydrolyses of the complexes

[488] T. Inoue, J. Fujita, and K. Saito, *Bull. Chem. Soc. Japan*, 1975, **48**, 1228.
[489] N. Serpone and K. A. Hersh, *Inorg. Chem.*, 1974, **13**, 2901.
[490] D. G. Bickley and N. Serpone, *Inorg. Chem.*, 1974, **13**, 2908.
[491] N. Serpone and K. A. Hersh, *J. Organometallic Chem.*, 1975, **84**, 177.
[492] K. A. Hersh and N. Serpone, *Canad. J. Chem.*, 1975, **53**, 448.

[X$_2$Sn(acac)$_2$] (X = Cl, Br, or I) and [Cl$_2$Sn(ket)$_2$] (ket = benzoylacetone, dibenzoylmethane, or trifluoroacetylacetone) all follow second-order kinetics and have negative entropies of activation. For the [X$_2$Sn(acac)$_2$] compounds, the enthalpies of activation decrease in the order Cl > Br > I, and order which argues against an associative mechanism. The second-order kinetics observed were attributed to a dipole–dipole interaction between molecules of the complex prior to the rate-determining step.[493]

Rein and Herber have studied tin(IV) tropolonates of the types R$_2$SnT$_2$, PhSnT$_3$, and Cl$_2$SnT$_2$ by vibrational and Mössbauer spectroscopy. The R$_2$SnT$_2$ derivatives are six-co-ordinated, with either the cis-R$_2$Sn (R = Me, Ph, or PhCH$_2$) or trans-R$_2$Sn (R = Et, Pr$^n$, or Bu$^n$) configurations. The temperature dependence of the recoil-free fraction (in the range $78 \leq T/K \leq 140$) has been correlated with the Raman spectra in the lattice-mode region to yield a self-consistent assignment on the intermolecular intra-unit-cell vibrations in the solids and a value of the effective vibrating molecular mass. From the latter data it is inferred that these molecules are monomeric in the solid state, with (relatively) weak intermolecular forces between adjacent molecules.[494] Both N-methyl-N-acetylhydroxylamino-ligands function as anisobidentate ligands in Me$_2$Sn(ONMeCOMe)$_2$, each ligand forming one short covalent bond and one long, essentially co-ordinate, Sn—O bond. The overall symmetry of the molecule approximates to $C_{2v}$ (Figure 12).[495] The tin atoms are pentaco-ordinated in the dimethyltin derivative of N-(2-hydroxyphenyl)salicylaldimine, with the Schiff base functioning as a terdentate ligand via the two axial sites and one of the equatorial sites of a distorted trigonal bipyramid (Figure 13).[496] The dimethyltin derivatives of similar ONO and SNO

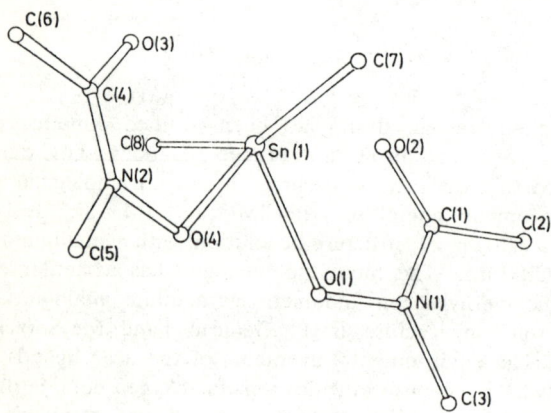

**Figure 12** *The structure of* Me$_2$Sn(ONMeCOMe)$_2$
(Reproduced from *J.C.S. Dalton*, 1975, 826)

---

[493] M. J. Frazer and L. I. B. Haines, *J.C.S. Dalton*, 1975, 1471.
[494] A. J. Rein and R. H. Herber, *J. Chem. Phys.*, 1975, **63**, 1021.
[495] P. G. Harrison, T. J. King, and J. A. Richards, *J.C.S. Dalton*, 1975, 826.
[496] D. L. Evans and B. R. Penfold, *J. Cryst. Mol. Structure*, 1975, **5**, 93.

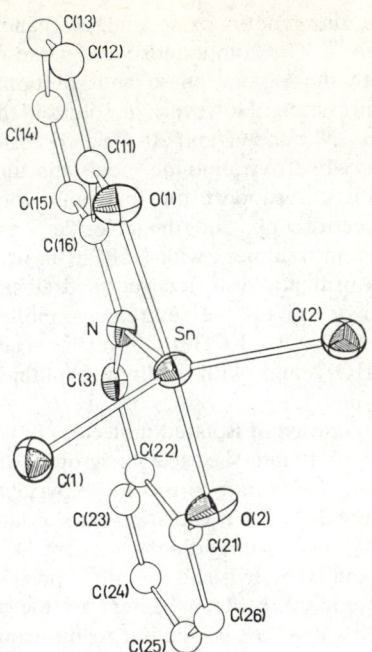

**Figure 13** *The structure of the dimethyltin derivative of N-(2-hydroxyphenyl)-salicylaldimine*

(Reproduced by permission from *J. Cryst. Mol. Structure*, 1975, **5**, 93)

terdentate Schiff-base ligands are readily converted into the corresponding dihalogenotin derivatives by reaction with tin(IV) halide. Similar trigonal-bipyramidal monomeric structures were proposed for these compounds on the basis of spectroscopic data, although bridged dimeric or polymeric structures could not be totally excluded.[497]

The crystal structure of the second modification of 1-phenylsilatrane, the γ-form, has been determined, and is (not surprisingly) very similar to that of the β-form reported previously. In both modifications, the Si—N bond distance is indicative of a bonding interaction.[498] The same conclusion has resulted from the He (I) photoelectron spectra of the silatranes [N(CH$_2$CH$_2$O)$_3$]SiR (R = Me, H, or OEt) compared with non-cage triethoxysilanes and triethanolamine itself. The silatranes exhibit large increases in the binding energy of the nitrogen lone pair, showing that the close approach of the nitrogen atom to silicon is associated with a bonding interaction with vacant orbitals on the metalloid.[499] Organostannatranes [N(CH$_2$CH$_2$O)$_3$]SnR have been obtained by two synthetic routes: from the appropriate organotin tris(ethoxide) and triethanolamine,[500] or from the

---

[497] R. Barbieri, F. DiBianca, G. Alonzo, A. Silvestri, L. Pellerito, N. Bertazzi, and G. C. Stocco, *Z. anorg. Chem.*, 1975, **411**, 173.
[498] L. Parkanyi, J. Nagy, and K. Simon, *J. Organometallic Chem.*, 1975, **101**, 11.
[499] S. Cradock, E. A. V. Ebsworth, and I. B. Murray, *J.C.S. Dalton*, 1975, 25.
[500] M. Zeldin and J. Ochs, *J. Organometallic Chem.*, 1975, **86**, 369.

organostannonic acid or diorganotin oxide and triethanolamine, using a KOH catalyst in the latter case.[501] The compounds were white crystalline solids which were monomeric in both the vapour phase and solution, and again a N → Sn bonding interaction is proposed. However, in solution there appears to be an equilibrium between Sn—N bonded and dissociated forms.[500] The reaction of silicon(IV) chloride with 8-hydroxyquinoline yields the unusual ionic compound [Si(ox)$_2$]$^{2+}$ 2Cl$^-$. Unequivocal support for this ionic formulation comes from X-ray photoelectron spectroscopy, and the chloride ions may be replaced by tetraphenylborate anions on treatment with NaBPh$_4$ in water. The cation probably possesses a four-co-ordinate distorted tetrahedral structure. Treatment of SiCl$_4$ with salicylaldehyde gives the extremely moisture-sensitive hexaco-ordinated compound SiCl$_2$(OC$_6$H$_4$CHO)$_2$, which reacts with ethanol to give Si(OEt)$_2$(OC$_6$H$_4$CHO)$_2$ and with sodium naphthalide to afford Na$_2$Si(C$_6$H$_4$OCHO)$_2$.[502]

Crystals of Si(O$_2$CMe)$_4$ consist of isolated molecules in which each silicon atom is tetrahedrally co-ordinated and the acetate groups are unidentate.[503] The carboxylate groups are also unidentate in the 2,2'-bipyridyl complex of divinyltin bis(trifluoroacetate) (Figure 14),[504] but chelating carboxylate groups are indicated for diaryltin bis(carboxylates) from dipole-moment and i.r. data.[505] Both trimethyltin glycinate[506] and trimethyl-lead acetate[507] possess bridging carboxylate groups, although the mode of bridging differs in the two compounds. The structure of the latter derivative is very similar to the trimethyltin analogue, i.e. planar trimethylmetal moieties bridged by carboxy-groups resulting in linear chain

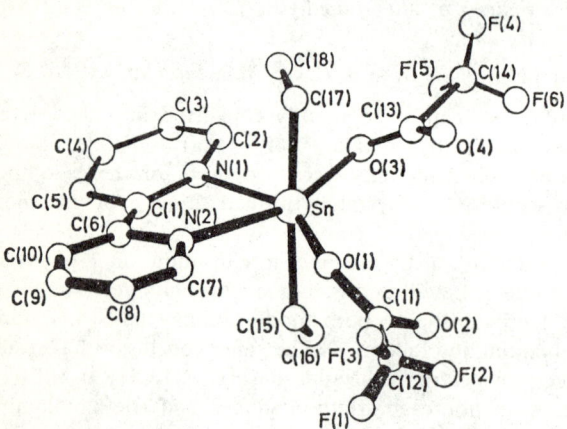

**Figure 14** *The structure of divinyltin bis(trifluoroacetate) 2,2'-bipyridyl* (Reproduced from *J.C.S. Dalton*, 1975, 562)

[501] A. Tzschach and K. Pönicke, *Z. anorg. Chem.*, 1975, **413**, 136.
[502] M. M. Millard and G. Urry, *Inorg. Chem.*, 1975, **14**, 1982.
[503] B. Kamenar and M. Bruvo, *Z. Krist.*, 1975, **141**, 97.
[504] C. D. Garner, B. Syutkina, and T. J. King, *J.C.S. Dalton*, 1975, 562.
[505] E. N. Gur'yanova, O. P. Syutkina, E. M. Panov, and K. A. Kocheskov, *J. Gen. Chem.* (*U.S.S.R*), 1974, **44**, 1937.
[506] B. Y. K. Ho, J. A. Zubieta, and J. J. Zuckerman, *J.C.S. Chem. Comm.*, 1975, 88.
[507] G. M. Sheldrick and R. Taylor, *Acta Cryst.*, 1975, **B31**, 2740.

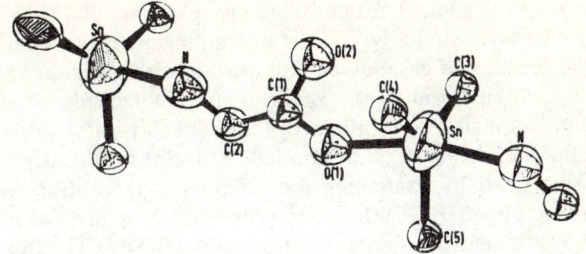

**Figure 15** *The chain structure of trimethyltin glycinate*
(Reproduced from *J.C.S. Chem. Comm.*, 1975, 88)

polymers. Chain polymers are also present in trimethyltin glycinate, but in this case bridging takes place *via* the amino-nitrogen atom (Figure 15), and the Me₃Sn units are no longer planar. Bridging carboxylate groups are also thought to be present in the dibromobis(carboxylato)tin(IV) complexes (obtained from SnBr₄ and silver or sodium carboxylate).[508]

Equilibrium constants of complexes formed between Ge$^{IV}$ and glycollic, lactic, malic, and 2,3,4-trihydroxyglutaric acids have been determined spectrophotometrically. The compositions of the compounds formed correspond to either 1:2 or 1:3 Ge:acid, depending on the concentration of the acid.[509,510] The properties of the crystalline complex $H_2[Ge(C_2O_4)_3]$,6H₂O have been investigated by chemical, X-ray diffraction, crystal-optical, and t.g.a. methods. Ir. data show that the oxalate groups are bidentate in nature. Thermolysis is accompanied by three endothermic effects, which were attributed to fusion of the hydrate (58 °C), removal of water (170 °C), and decomposition of anhydrous trisoxalatogermanic acid (240 °C).[511] The products of the interaction of *trans*-cyclohexane-1,2-diaminetetra-acetic acid (cdta) and SnCl₄ depend on the particular solvent used. In benzene the product has the composition $Sn(cdta)_{1.5}Cl_{1.5}$, but in CH₂Cl₂ solution $Sn(cdta)_{1.3}Cl_2$ is obtained. In methanol with added amine base the ionic complex $(RNH_3[Sn_2(hcdta)_3])$ results.[512] Tetrakis(carboxylato) tin(IV) and dichlorobis(carboxylato) tin(IV) compounds take part in redistribution reactions with cyclopentadienyltin(IV) compounds, giving cyclopentadienyltin(IV) carboxylates. The same compounds may also be obtained by the protolysis of cyclopentadienyl groups from tetracyclopentadienyltin(IV) by the appropriate carboxylic acid.[513] Lead(IV) trifluoroacetate, $Pb(O_2CCF_3)_4$, dissolves exothermically in fluorosulphuric acid to give a bright yellow solution from which yellow crystals of lead(IV) fluorosulphate, $Pb(O_3SF)_4$, may be obtained quantitatively. The compound is stable at room temperature for several weeks, but is readily hydrolysed to PbO₂, and it thermolyses at 150—210 °C to lead(II) fluorosulphate, $Pb(O_3SF)_2$,

---

[508] E. B. Jessen and K. Taugbol, *J. Inorg. Nuclear Chem.*, 1975, **37**, 1057.
[509] A. F. Pozharitskii, I. I. Seifullina, and E. M. Belousova, *J. Gen. Chem. (U.S.S.R)*, 1975, **45**, 1026.
[510] A. F. Pozharitskii, I. I. Seifullina, and E. M. Belousova, *J. Gen. Chem. (U.S.S.R)*, 1975, **45**, 1285.
[511] T. B. Shkodina, E. I. Krylov, and V. A. Sharov, *Russ. J. Inorg. Chem.*, 1974, **19**, 1595.
[512] S. K. Dhar and W. E. Kurcz, *J. Inorg. Nuclear Chem.*, 1975, **37**, 2003.
[513] N. D. Kolosova, N. N. Zemlyanskii, A. A. Azizov, P. I. Zhakharov, Yu. A. Ustynyuk, N. P. Barminova, and K. A. Kocheskov, *Doklady Chem.*, 1974, **218**, 641.

$S_2O_5F_2$, and oxygen. Lead(IV) trifluoromethylsulphonate, $Pb(O_3SCF_3)_4$, may be prepared, and it behaves similarly.[514] Bistriorganotin sulphites $(R_3Sn)_2SO_3$ may be prepared by the reaction of gaseous sulphur dioxide with bis(triorganotin) oxide at 20 °C. At higher temperatures, disproportionation takes place, and organotin sulphinates and sulphates are obtained.[515] The solid-state structures of bis-(tributyltin) sulphate, selenate, and chromate $(Bu_3Sn)_2OEO_3$ (E = S, Se, or Cr) have been investigated by examining the Mössbauer recoil-free fraction in the temperature range $80 \leq T/K \leq 260$. A non-polymeric structure was inferred for all three compounds.[516] Bis(triorganotin) carbonates $(R_3Sn)_2CO_3$ (R = Me or cyclohexyl), for which the polymeric structure (52) has been proposed, may be prepared by passing carbon dioxide through suspensions of the corresponding triorganotin hydroxide in benzene.[471] Carbamoyloxytrimethylsilanes $Me_3SiO_2$-CNHAr result from the reaction of carbon dioxide, hexamethyldisilazane, and an aromatic amine $ArNH_2$.[517] The thermal decomposition of the complex iodatostannic acids $H_2[Sn(IO_3)_6],2H_2O$ and $H_2[Sn(IO_3)_4(OH)_4],2H_2O$ as well as the salts $M_2[Sn(IO_3)_6]$ (M = K, Rb, or Cs) proceeds stepwise, with the liberation of iodine and oxygen. The final thermolysis product of the acids is $SnO_2$ and of the salts a mixture of $SnO_2$ and MI.[518]

Trialkylsilyl phosphites $R'_3SiOP(OR^2)_2$ are readily obtained by the cleavage of the Si—N bonds of hexaorganodisilazanes or triorganosilylamines by dialkyl phosphites $(R^2O)P(O)H$.[519,520] $Me_3SiOP(OEt)_2$ reacts with succinimide and N-butylsuccinimide, affording the adducts (53),[521] and the compounds $Me_3SiOP(OR)_2$ (R = Me or Et) react with methyl diazoacetate to afford the adducts (54), which rearrange completely at room temperature to the Si—N bonded isomers (55).[522] Readily hydrolysable trimethylsilylphosphoramides $Me_3SiOP(NR_2)_2$ may be prepared from $Me_3SiCl$ and the Grignard reagent $(R_2N)_2POMgBr$,[523] and they react with substituted benzaldehydes to form the adducts (56).[524]. The potentially amibidentate nucleophile $(Me_2CHO)_2POS^-$ reacts with trimethyl- and triphenyl-tin chloride to form $R_3SnOP(S)(OHCMe_2)_2$ derivatives.[525] The cubic tin and lead pyrophosphates $M^{4+}P_2O_7^{4-}$ (M = Sn or Pb) have been examined by X-ray powder, i.r., Raman, and Mössbauer spectroscopies. The tin compound appeared to be of Chaunac's type I (with $P_2O_7$ groups oriented at random), but could be converted into type II (with ordered $P_2O_7$ groups) by heating to high temperatures. The lead compound is of Chaunac's type II.[526]

---

[514] H. A. Carter, C. A. Milne, and F. Aubke, *J. Inorg. Nuclear Chem.*, 1975, **37**, 282.
[515] U. Kunze and H. P. Völke, *Chem. Ber.*, 1974, **107**, 3818.
[516] H. Sano and Y. Mekata, *Chem. Letters*, 1975, 155.
[517] V. D. Sheludyakov, A. D. Kirilin, and V. F. Mironov, *J. Gen. Chem.* (*U.S.S.R*), 1975, **45**, 471.
[518] L. A. Cheburina and T. G. Balicheva, *Russ. J. Inorg. Chem.*, 1974, **19**, 143.
[519] E. A. Chernyshev, E. F. Bugerenko, A. S. Akat'eva, and A. D. Naumov, *J. Gen. Chem.* (*U.S.S.R*), 1975, **45**, 231.
[520] M. A. Pudovik, M. D. Medvedeva, and A. N. Pudovik, *J. Gen. Chem.* (*U.S.S.R*), 1975, **45**, 682.
[521] A. N. Pudovik, E. S. Batyeva, and G. U. Zamaletdinova, *J. Gen. Chem.* (*U.S.S.R*), 1975, **45**, 922.
[522] A. N. Pudovik and R. D. Gareev, *J. Gen. Chem.* (*U.S.S.R*), 1975, **45**, 220.
[523] A. N. Pudovik, E. S. Batyeva, and V. A. Al'fonsov, *J. Gen. Chem.* (*U.S.S.R*), 1975, **45**, 240.
[524] A. N. Pudovik, E. S. Batyeva, and V. A. Al'fonsov, *J. Gen. Chem.* (*U.S.S.R*), 1975, **45**, 921.
[525] I. A. Duncan and C. Glidewell, *J. Organometallic Chem.*, 1975, **97**, 51.
[526] C. H. Huang, O. Knop, D. A. Othen, F. W. D. Woodhams, and R. A. Home, *Canad. J. Chem.*, 1975, **53**, 79.

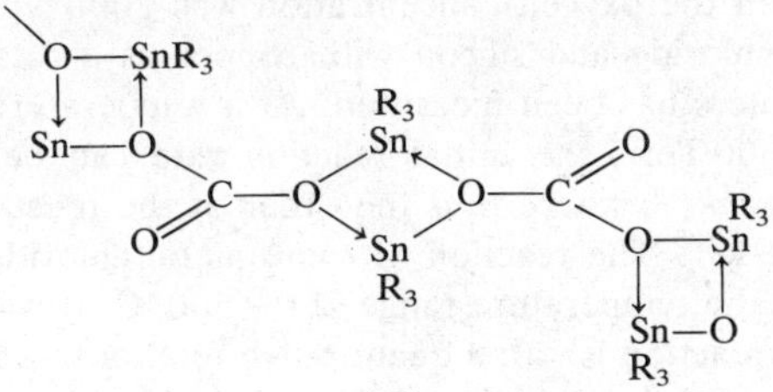

(52)

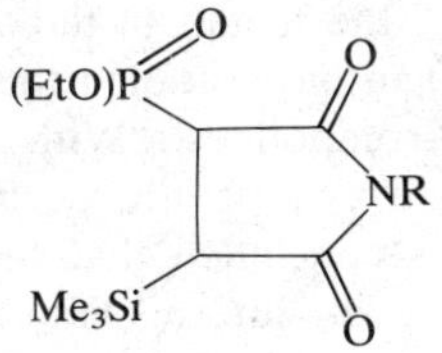

(53)

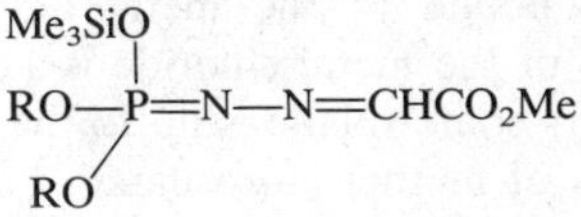

(54) Me₃SiO\\RO—P=N—N=CHCO₂Me\\RO/

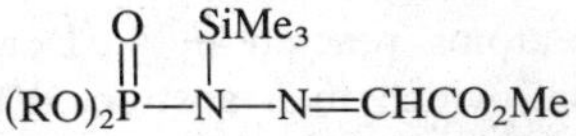

(55) (RO)₂P(=O)—N(SiMe₃)—N=CHCO₂Me

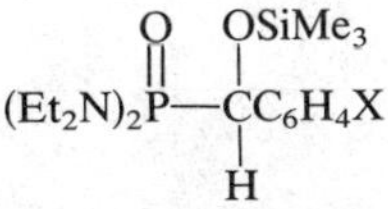

(56) (Et₂N)₂P(=O)—C(OSiMe₃)(H)C₆H₄X    X = H or NO₂

**Halides of Silicon, Germanium, Tin, and Lead.**—The 'direct' synthesis of methylchlorosilanes has been further investigated. By subjecting a partially chlorinated silicon surface (thus containing surface >SiCl, >SiCl₂, and —SiCl₃ groups) with methylmagnesium chloride the then further chlorination, the surface species are converted into methylchlorosilanes, which desorb into the gas phase. The proportions of the —SiCl$_n$ groups on the surface depend greatly on the conditions of the prechlorination treatment.[527] The reaction of surface —SiCl$_n$ species with dimethyl ether has also been investigated. At temperatures of 350 °C or higher, chloromethyl- and chloromethoxy-silanes such as Me$_n$SiCl$_{4-n}$ ($n = 1$—3) and (MeO)$_n$SiCl$_{4-n}$ ($n = 1$ or 2) are formed. Without pretreatment of the surface with either chlorine or HCl, none of the observed products are formed.[528] The influence of oxygen on the direct synthesis using a copper catalyst has been studied at 320 °C and 1 atm pressure. It was found that oxygen sharply reduces the rate of reaction, although the product composition was hardly influenced. The maximum degree of silicon conversion fell from ~70% when no oxygen was

---

[527] S. A. Golubtsov, K. A. Andrianov, N. T. Ivanova, R. A. Turetskaya, N. S. Fel'dshtein, E. A. Chernyshev, and V. G. Dzvonar, *J. Gen. Chem.* (*U.S.S.R*), 1974, **45**, 795.
[528] R. A. Turetskaya, N. T. Ivanova, N. S. Fel'dshtein, and E. A. Chernyshev, *J. Gen. Chem.* (*U.S.S.R*), 1974, **44**, 2738.

present to 50% when the oxygen concentration was 2000 p.p.m.[529] The rate of reaction of methyl chloride and silicon with copper as a catalyst and zinc and aluminium as promoters has been measured. At a temperature of 333 °C and at pressures of 300—800 Torr, the initial reaction rate can be expressed as the power function: rate = $kP^n$, where $n$, is the order of the reaction with respect to methyl chloride = 0.94.[530] The reaction of some metal chlorides with silicon has been investigated in the temperature range 200—500 °C. It was observed that, if the free enthalpy of reaction is <0, a quantitative relationship exists between the threshold temperature of the reaction of the metal chloride with silicon and the product of the average standard free enthalpy of formation per gram atom of chlorine bound to the metal times the temperature/K at which the vapour pressure of the metal chloride is 1 Torr. The threshold temperature for reaction also bears some relationship to the catalytic activity of the metals in the 'direct' synthesis of methylchlorosilanes.[531] McGlinchey and Tan have studied the reactions of germanium atoms with various chloro-compounds. With $CCl_4$, $CCl_3GeCl_3$, and some hexachloroethane are formed, whilst $SiCl_4$ afforded $Cl_3GeSiCl_3$ and chloroform gave $CHCl_2GeCl_3$. It was concluded that the germanium atoms reacted in a $^3P$ ground state to give an intermediate triplet germylene, which then abstracted halogen atoms in a stepwise manner characteristic of radicals,[532] as shown in Scheme 4. α-Iodothiophen reacts readily with tin metal, with an amine hydrochloride as catalyst, to give bis(α-thienyl)tin di-iodide in good yield.[533]

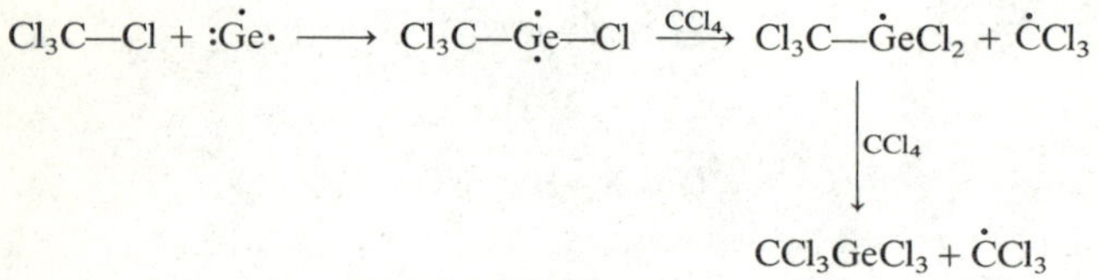

$$Cl_3C\text{—}Cl + :\dot{G}e\cdot \longrightarrow Cl_3C\text{—}\dot{G}e\text{—}Cl \xrightarrow{CCl_4} Cl_3C\text{—}\dot{G}eCl_2 + \dot{C}Cl_3$$

$$\downarrow CCl_4$$

$$CCl_3GeCl_3 + \dot{C}Cl_3$$

**Scheme 4**

Germanium(IV) bromide has been reinvestigated by electron diffraction. Constraining the molecule symmetry to $T_d$ afforded a value of 2.272(1) Å.[534] The electronic radial distribution functions for germanium(IV) and tin(IV) chlorides in the liquid state at 23 °C have been calculated from X-ray diffraction intensity distributions obtained by use of theta–theta reflection diffractometry. Both liquids show intermolecular effects at distances equivalent to the Cl—Cl intermolecular distance.[535] Values of 0.9 D (Si—Cl), 1.5 D (Ge—Cl), and 2.7 D (Sn—Cl) have been derived for the bond dipole moments of these bonds in the metal(IV)

---

[529] M. G. R. T. de Cooker, R. P. A. Van den Hof, and P. J. Van den Berg, *J. Organometallic Chem.*, 1975, **84**, 305.
[530] M. G. R. T. de Cooker and P. J. Van den Berg, *Rec. Trav. chim.*, 1975, **94**, 192.
[531] M. G. R. T. de Cooker, J. W. de Jong, and P. J. Van den Berg, *J. Organometallic Chem.*, 1975, **86**, 175.
[532] M. L. McGlinchey and T. S. Tan, *Inorg. Chem.*, 1975, **14**, 1209.
[533] C. Gopinathan, S. K. Pandit, S. Gopinathan, A. Y. Sonsale, and P. A. Awasarkar, *Indian J. Chem.*, 1975, **13**, 576.
[534] G. G. B. Souza and J. D. Wieser, *J. Mol. Structure*, 1975, **25**, 442.
[535] C. T. Rutledge and G. T. Clayton, *J. Chem. Phys.*, 1975, **63**, 2211.

# Elements of Group IV

chlorides from the intensities of the $\nu_3$ and $\nu_4$ vibrations in the gas phase.[536] The number of external modes observed in the Raman spectrum of $SnI_4$ is considerably less than the number expected on the basis of group-theoretical predictions. In addition, some crystal-field splitting is seen in the asymmetric stretching mode.[537] The enthalpy change of the reaction of Sn(c) and $Br_2(l)$ in $CS_2$ as solvent, giving $SnBr_4(s)$, has been determined by calorimetry to be $-374.2 \pm 1.4$ kJ mol$^{-1}$, affording a value of the heat of solution of $SnBr_4(c)$ in $CS_2$ of $11.9 \pm 0.3$ kJ mol$^{-1}$ and a value of the standard heat of formation of $SnBr_4(c)$ or $-386 \pm 1.5$ kJ mol$^{-1}$. By substitution of this latter value in the thermochemical cycle, the value of $-124.3$ kcal mol$^{-1}$ is obtained for the standard heat of formation of $SnCl_4(l)$,[538] close to the value of $-123.16 \pm 0.25$ kcal mol$^{-1}$ obtained by calorimetric studies of the dissolution of $SnCl_4(l)$ in HCl solutions containing hydrogen peroxide.[539] The interactions of the components in the $FeCl_3$–$GeCl_4$–$SnCl_4$,[540] $FeCl_3$–$GeCl_4$–$TeCl_4$,[540] $GeCl_4$–$SiCl_4$–$TeCl_4$,[541] and $GeCl_4$–$SnCl_4$–$TeCl_4$[541] have been studied by d.t.a. and visual polythermal analysis. Extensive regions of liquid immiscibility occur in all four systems.

The passage of a d.c. pulse discharge through a mixture of $SiCl_4$ and $BCl_3$ yields pentachloroboroslinae, $Cl_3SiBCl_2$, which rapidly decomposes at 80 °C to $SiCl_4$, $BCl_3$, and $(BSiCl_3)_n$. Treatment with $SbF_3$ gave $SiF_4$ as the sole volatile product. Attempts to prepare analogous germanium–, tin–, or titanium–boron compounds failed.[542] Hass has studied halogen exchange between silicon(IV) halides and other metal halides. Exchange between $SiCl_4$ and $BBr_3$, $SbBr_3$, $MgBr_2$, $AsBr_3$, and $AlBr_3$ results in the formation of mixtures of silicon(IV) chloride bromides.[543] Mixed silicon(IV) chloride bromides also result from the reaction of $SiCl_4$ and $Br_2$ in the gas phase at 400 °C.[544] Cl/I Exchange between AgCl and $SiI_4$ proceeds stepwise, with the intermediate formation of the mixed compounds $SiCl_nI_{4-n}$ ($n = 1$—3). With a vast excess of AgCl (1 : 12), the $SiI_4$ is soon totally removed from the reaction mixture, and the concentrations of $SiI_3Cl$, $SiI_2Cl_2$, and $SiICl_3$ reach a maximum after 18, 70, and 255 minutes, respectively. $SiCl_4$ is formed completely after 680 minutes.[545] The mixed silicon(IV) chloride iodides may be separated by gas chromatography.[546] $SiF_4$ shows no reaction with nitric oxide, but $Cl_3SiH$ and $Me_3SiCl$ do react under thermal and photolytic conditions. The thermal reactions yield only a trace of NOCl in the case of $Cl_3SiH$, but NOCl, $N_2O$, and disiloxane from $Me_3SiCl$. Photolysis produces a much wider range of products, and $N_2O$, NOCl, $N_2$, $H_2$, $Si_2Cl_6$, disiloxane, and $SiCl_4$ are found in the $Cl_3SiH$ reaction mixture, and $N_2$, $N_2O$, NOCl, disiloxane, and $(Me_2SiO)_4$ in the

---

[536] H. Stoekli-Evans, A. J. Barnes, and W. J. Orville-Thomas, *J. Mol. Structure*, 1975, **24**, 73.
[537] P. Dawson, *Spectrochim. Acta*, 1975, **31A**, 1101.
[538] J. Mikler and A. Janitsch, *Monatsh.*, 1975, **106**, 1307.
[539] V. P. Vasil'ev, V. N. Vasil'eva, N. I. Kokurin, and P. N. Vorob'ev, *Russ. J. Inorg. Chem.*, 1974, **19**, 1620.
[540] Zh. K. Fes'kova, V. V. Safonov, B. G. Korshunov, and V. I. Ksenzenko, *Russ. J. Inorg. Chem.*, 1974, **19**, 281.
[541] Zh. K. Fes'kova, V. V. Safonov, N. M. Grigor'eva, B. G. Korshunov, and V. I. Ksenzenko, *Russ. J. Inorg. Chem.*, 1974, **19**, 759.
[542] M. Zeldin, D. Solan, and B. Dickman, *J. Inorg. Nuclear Chem.*, 1975, **37**, 25.
[543] D. Hass, S. Goldstein, and M. Nimz, *Z. Chem.*, 1975, **15**, 156.
[544] D. Hass, S. Goldstein, and M. Nimz, *Z. Chem.*, 1975, **15**, 240.
[545] D. Hass and H. Schwarz, *Z. Chem.*, 1975, **15**, 320.
[546] D. Hass and H. Schwarz, *Z. Chem.*, 1975, **15**, 455.

case of $Me_3SiCl$.[547] Treatment of $Si_2Cl_6$ with isopropyl-lithium in variable proportions leads to stepwise replacement of chlorine by isopropyl groups, giving, progressively, 1-, 1,1-, 1,1,2-, and 1,1,2,2-substituted products. Further reaction with $Pr^iLi$ at low temperatures affords 1- and 1,2-hydrosilanes.[548] $Si_2Cl_6$ reacts with diazomethane to yield $ClCH_2Si_2Cl_5$, which may then be reduced by $LiAlH_4$ to $ClCH_2Si_2H_5$. The bromo-analogue $BrCH_2Si_2H_5$ may be obtained from the reaction of $CH_2N_2$ and $BrSi_2H_5$.[549] The Hg-sensitized photolysis of $Cl_3SiH$ at 55 °C at pressures of 50—450 Torr gives dendritic crystals of $neo$-$Si_5Cl_{12}$. $H_2$, $SiCl_4$, $Si_2Cl_6$, $Si_3Cl_8$, $Si_4Cl_{10}$, and a viscous yellow oil are also produced.[550] Treatment of $(Cl_3SiCCl_2)_2SiCl_2$ with excess methyl-lithium results in the formation of a multitude of different methylsilanes.[551] Germanium(IV) halides are reduced in high yields (75—90%) to $HGeX_3,2Et_2O$ complexes (X = Cl, Br, or I) by treatment with $(HMe_2Si)_2O$ in ether at 20—40 °C. $GeCl_4$ reacts much more slowly in dioxan or $Bu_2O$, giving much lower yields of $GeCl_2$,dioxan and $HGeCl_3,2Bu_2O$, respectively. The iodo-compound $HGeI_3,2Et_2O$, a new compound, is a heavy black liquid which is insoluble in ether. It reacts with $CCl_4$ to give $CHCl_3$, with MeI to give $MeGeI_3$, with acrylic acid to give $I_3GeCH_2CH_2CO_2H$, and $in$ $vacuo$ loses HI and ether, affording $GeI_2$. In solution it slowly decomposes to $EtGeI_3$, EtOH, and $Et_2O$.[552]

The role of copper metal in the synthesis of mono-organogermanium trihalides from $GeX_4$ and alkyl halide has been investigated. The formation of an organocopper intermediate is unlikely because of the high temperatures usually involved. Rather it appears that copper first reduces the germanium(IV) halide to germanium, which then reacts with the alkyl halide to give the observed product.[553] Tin(IV) halides have been used in several cases as halogenating reagents. An improved high-yield synthesis of $Me_3GeCl$ by the $AlCl_3$-catalysed redistribution of $Me_4Ge$ with $SnCl_4$ has been described.[554] With $SiH_4$, $Si_2H_6$, $SiH_3F$, $SiH_3Mn(CO)_5$, and $(SiH_3)_3N$, reaction with $SnCl_4$ leads to monochlorination at silicon, producing the compounds $SiH_2FCl$, $SiH_2ClMn(CO)_5$, $ClSiH_2N(SiH_3)_2$, $(ClSiH_2)_2NSiH_3$, and $(ClSiH_2)_3N$. Reactions with other functionally substituted silyl compounds generally lead to the formation of $SiH_3Cl$.[555] $SnCl_4$ chlorinates $(C_5H_5)_2Sn$, giving $(C_5H_5)_2SnCl_2$. Cyclopentadienyltin(IV) halides can more generally be prepared by substitution of tin(IV) halides using sodium cyclopentadienide or the corresponding Grignard reagent. They participate one with another and with $SnCl_4$ in surprisingly fast conproportionation reactions.[556] The reaction of germanium(IV) and tin(IV) chlorides with the $\alpha,\omega$-dithiols $X(CH_2CH_2SH)_2$ (X = O or S) leads to the formation of 2,2-dichloro-1,3,6,2-trithiametallocans (X = S) and 5,5-dichloro-1,4,6,5-oxathiametallocans (X = O) (57). Solutions of these compounds contain two chiral boat–chair conformers with are equilibrated by passing

[547] R. Varma, P. Orlander, and A. K. Ray, *J. Inorg. Nuclear Chem.*, 1975, **37**, 1797.
[548] M. Weidenbruch and W. Peter, *J. Organometallic Chem.*, 1975, **84**, 151.
[549] J. A. Morrison and J. M. Bellama, *Inorg. Chem.*, 1975, **14**, 1614.
[550] K. G. Sharp, P. A. Sutor, T. C. Farrar, and K. Ishibitsu, *J. Amer. Chem. Soc.*, 1975, **97**, 5612.
[551] G. Fritz, J. W. Chang, and N. Braunagel, *Z. anorg. Chem.*, 1975, **416**, 211.
[552] V. F. Mironov and T. K. Gar, *J. Gen. Chem. (U.S.S.R)*, 1975, **45**, 94.
[553] V. V. Pozdeev and V. Ya. Dvdarev, *J. Gen. Chem. (U.S.S.R)*, 1975, **45**, 946.
[554] J. Grobe and J. Hendriock, *Synth. React. Inorg. Metal.-Org. Chem.*, 1975, **5**, 393.
[555] S. Cradock, E. A. V. Ebsworth, and N. Hosmane, *J.C.S. Dalton*, 1975, 1624.
[556] U. Schröer, H. J. Albert, and W. P. Neumann, *J. Organometallic Chem.*, 1975, **102**, 291.

(57)

through an achiral chair–chair conformation.[557] The structures of three of the compounds [(57; M = Ge, X = S),[558] (57; M = Sn, X = O),[559] and (57; M = Sn, X = S)[560]] have been determined. All three compounds have similar structures, with distorted trigonal-bipyramidal co-ordination at the metal, a transannular X···M interaction completing the co-ordination. The dithiolato ligand is bonded *via* two equatorial sites, the remaining equatorial and axial sites being occupied by the two chlorine atoms. The two thiametallocans differ from the oxastannocan in the conformation of the nine-membered heterocyclic rings. The thiametallocans exist in enantiomeric boat–chair forms (Figure 16), but the oxastannocan has the achiral chair–chair form only in the crystal.

The structures of PhSiF$_3$ (by electron diffraction),[561] bis(1,8-naphthylene)-dichloro-disilane and -disiloxane (by X-ray difffaction),[562] and Me$_3$GeCl (from microwave rotational spectra)[563] have been determined. PhSiF$_3$ and Me$_3$GeCl are tetrahedral molecules. The structures of the two chlorosilanes are shown in Figure

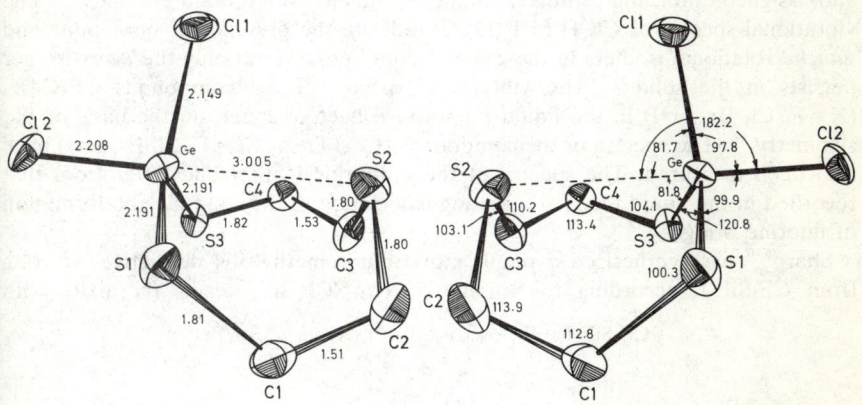

**Figure 16** *The enantiomeric pair of molecules in crystals of* Cl$_2$Ge(SCH$_2$CH$_2$)$_2$S
(Reproduced by permission from *Chem. Ber.*, 1975, **108**, 1723)

---

[557] M. Dräger and L. Ross, *Chem. Ber.*, 1975, **108**, 1712.
[558] M. Dräger, *Chem. Ber.*, 1975, **108**, 1723.
[559] M. Dräger and R. Engler, *Z. anorg. Chem.*, 1975, **413**, 229.
[560] M. Dräger and R. Engler, *Chem. Ber.*, 1975, **108**, 17.
[561] T. M. Il'enko, N. N. Veniamov, and N. V. Alekseev, *J. Struct. Chem.*, 1975, **16**, 270.
[562] O. A. D'yachenko, L. O. Atovmyan, V. I. Ponomarev, V. I. Andianov, Yu. V. Nekrasov, L. A. Muradyan, N. G. Komalenkova, and E. A. Chernyshev, *J. Struct. Chem.*, 1975, **16**, 144.
[563] J. R. Durig and K. L. Hellams, *J. Mol. Structure*, 1975, **29**, 349.

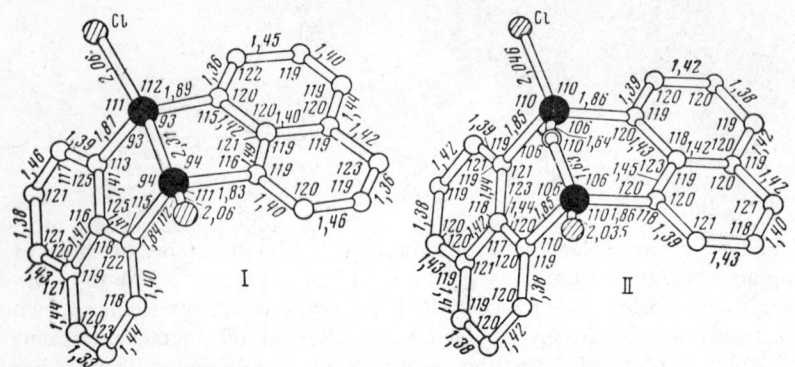

**Figure 17** *The structures of bis(1,8-naphthylene)dichloro-disilane and -disiloxane* (Reproduced by permission from *J. Struct. Chem.*, 1975, **16**, 144)

17; in each, the silicon has distorted tetrahedral geometry. The electric dipole moments of H$_3$SiF[564] and H$_3$GeF[565] have been measured. A CNDO/2 localized M.O. study of the former compound shows that the major contribution arises from the SiF bond moment, whilst that of H$_3$GeF is substantially larger than the moment of H$_3$GeCl, excluding 4$d$ orbital interactions. The i.r. and Raman spectra of several triorgano-silicon, -germanium, and -tin fluorides have been recorded and assigned, affording estimates of the M—C and M—F bond energies.[566] The vibrational spectra of ClCH$_2$SiH$_2$(D$_2$)Cl indicate the presence of both *trans* and *gauche* rotational isomers in the gas and liquid phases, but only the *trans*-isomer persists in the solid.[567] The vibrational spectra of the compounds CF$_3$GeX$_3$ (X = F, Cl, Br, or I) in the liquid phase have been analysed on the basis of C$_{3v}$ symmetry.[568] The spectra of the compounds (CF$_3$)$_2$GeX$_2$ (X = F, Cl, Br, or I) have also been measured. The spectra of the difluoride in the solid differ from that recorded in the liquid phase, suggesting association in the solid *via* the formation of fluorine bridges.[569]

Sharp[570] has synthesized a number of trifluoromethylsilyl derivatives starting from CF$_3$SiF$_2$I, according to Scheme 5. HF$_2$CCF$_2$SiF$_3$ reacts thermally with

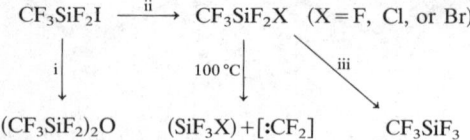

Reagents: i, HgO; ii, SbX$_3$; iii, H$_2$O vapour

**Scheme 5**

[564] P. M. Kuznesof, *Inorg. Chim. Acta*, 1975, **15**, L13.
[565] J. M. Bellama and J. A. Morrison, *Inorg. Nuclear Chem. Letters*, 1975, **11**, 127.
[566] K. Licht, P. Koehler, and H. Kriegsman, *Z. anorg. Chem.*, 1975, **415**, 31.
[567] K. Sera, *Bull. Chem. Soc. Japan*, 1975, 649.
[568] H. Bürger and R. Eiyen, *Spectrochim. Acta*, 1975, **31A**, 1645.
[569] H. Bürger and R. Eiyen, *Spectrochim. Acta*, 1975, **31A**, 1655.
[570] K. G. Sharp, *Inorg. Chem.*, 1975, **14**, 1231.

Me$_3$SiOMe, (Me$_3$Si)$_2$NH, or Me$_3$SiOH, possibly *via* co-ordination of silicon of the SiF$_3$ group with the oxygen or nitrogen atom to give Me$_3$SiF and HCF$_2$CF$_2$SiOMe, tar, or polymeric material [possibly the silsequioxane (CHF$_2$CF$_2$SiO$_{1.5}$)$_n$], respectively. SiF$_4$ reacts under the same conditions with the disilazane or the methoxosilane to afford Me$_3$SiF and F$_3$SiNHSiMe$_3$ and FSi(OMe)$_3$, respectively.[571] The gas-phase thermal decomposition of CCl$_3$SiF$_3$ at 100—160 °C and 10—100 Torr initial pressure in the presence of alkenes is first-order in the silane. The rate constant is independent of the alkene used and the surface to volume ratio, providing strong evidence for the participation of free dichlorocarbene in the reactions.[572] Tri-t-butylsilyl halides have been obtained starting from HSiCl$_3$ according to Scheme 6.[573]

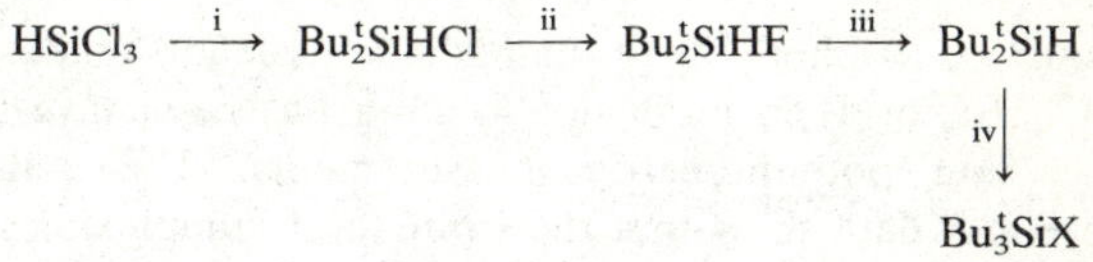

$$HSiCl_3 \xrightarrow{i} Bu^t_2SiHCl \xrightarrow{ii} Bu^t_2SiHF \xrightarrow{iii} Bu^t_2SiH \xrightarrow{iv} Bu^t_3SiX$$

Reagents: i, 2Bu$^t$Li; ii, ZnF$_2$; iii, Bu$^t$Li; iv, X$_2$ (X = Cl, Br, or I)

**Scheme 6**

Trichlorogermane adds to allyl chloride to give Cl$_3$Ge(CH$_2$)$_3$Cl, which eliminates HCl at 180 °C to afford Cl$_3$GeCH$_2$CH=CH$_2$ together with compounds of high molecular weight. At a pressure of 45 Torr, the thermolysis produces (CH$_2$=CHCH$_2$)(Cl)$_2$Ge(CH$_2$)$_3$Cl.[574] The treatment of Cl$_2$Si(OMe)$_2$ with magnesium in the presence of isoprene in THF gives the silylene addition product (58), the spiro-compound (59), and Si(OEt)$_4$. The formation of the two latter

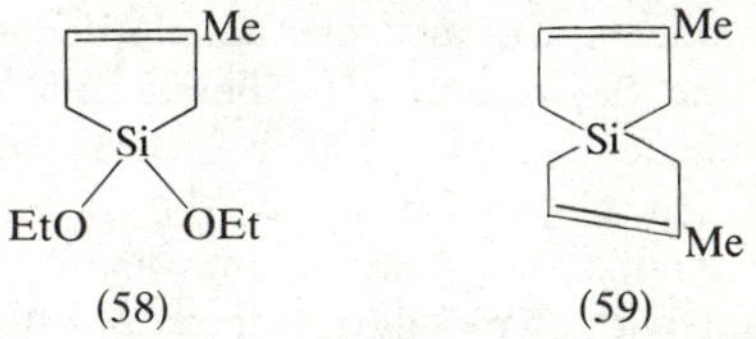

(58) — Si ring with Me, EtO, OEt
(59) — spiro Si with Me, Me

products was presumed to occur *via* the reaction of (58) with further magnesium and isoprene, giving Mg(OEt)$_2$, which then reacts with SiCl$_2$(OEt)$_2$, producing Si(OEt)$_4$.[575]

The methylfluorogermanes Me$_n$GeF$_{4-n}$ ($n = 1$—3) are conveniently prepared by treating (Me$_3$Ge)$_2$O, (Me$_2$GeO)$_n$ ($n = 3$ or 4), and [(MeGeO)$_2$O]$_n$, respectively, with AsF$_3$.[576] The extent of $^{18}$F exchange between Me$_3$SiF and the tungsten(vi) fluorides WF$_{6-n}$X$_n$ (X = OMe, $n = 1, 2$ or 4; X = OPh or OC$_6$F$_5$, $n = 1$; X = NEt$_2$, $n = 1$ or 4) depends on the identity and number of the substituents of tungsten.[577]

[571] R. N. Haszeldine, A. E. Tipping, and R. O'B. Watts, *J.C.S. Dalton*, 1975, 1431.
[572] F. Anderson, J. M. Birchall, R. N. Haszeldine, and B. J. Tyler, *J.C.S. Perkin II*, 1975, 1051.
[573] M. Weidenbruch and W. Peter, *Angew. Chem. Internat. Edn.*, 1975, **14**, 642.
[574] P. Boudjouk, *Inorg. Nuclear Chem. Letters*, 1975, **11**, 679.
[575] D. Terunuma, S. Hatta, and T. Araki, *Chem. Letters*, 1974, 1321.
[576] S. C. Pace, *Synth. react. Inorg. Metal.-Org. Chem.*, 1975, **5**, 373.
[577] C. J. W. Fraser, A. Majid, G. Oates, and J. M. Winfield, *J. Inorg. Nucl. Chem.*, 1975, **37**, 1535.

The kinetics of halogen exchange in binary mixtures of trimethyltin halides in solution have been investigated by total lineshape analysis. By measurements in solvents of different permittivities it was concluded that ionization does play a part in the kinetics, and that exchange proceeds by a five-co-ordinated intermediate[578] according to Scheme 7.

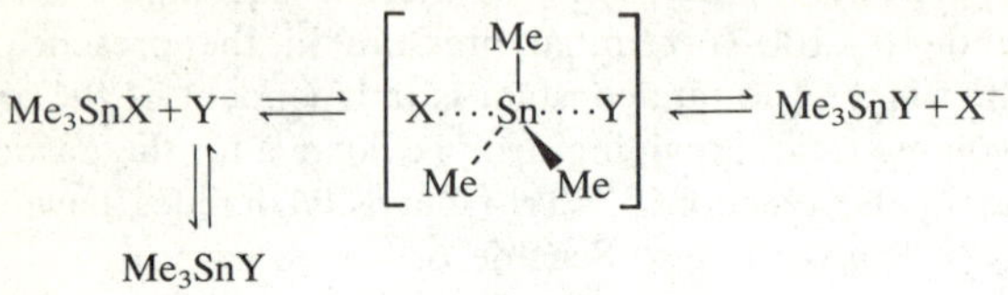

$$Me_3SnX + Y^- \rightleftharpoons \begin{bmatrix} & Me & \\ X\cdots & Sn & \cdots Y \\ & Me\;\;\;Me & \end{bmatrix} \rightleftharpoons Me_3SnY + X^-$$

$$\Updownarrow$$

$$Me_3SnY$$

**Scheme 7**

The behaviour of triphenyl-lead and diphenyl-lead halides $Ph_3PbX$ and $Ph_2SnX_2$ (X = Cl, Br, or I) in methanol solution has been investigated by u.v., conductiometric, and potentiometric measurements. The stability constants evaluated from e.m.f. data show that the strength of complexing increases in the order $Cl^- < Br^- < I^-$ for both the triphenyl-lead and diphenyl-lead cases. E.m.f. measurements of the $Ph_2Pb^{2+}-I^-$ system indicated the formation of dinuclear complexes such as $Ph_4Pb_2I_2^{2+}$, $Ph_4Pb_2I_3^+$, or $Ph_4Pb_2I_4$ in addition to the usual mononuclear compounds.[579]

As usual, there have been a large number of reports of complexes between silicon, germanium, and tin tetrahalides and various donors. However, the majority are trivial, and will be mentioned only briefly. 1:1 or 1:2 adducts form between $SiF_4$ and $o$-, $m$-, and $p$-phenylenediamines,[580] ethylamines,[581] hexane-1,6-diamine and tetraethylenepentamine,[582] and ethanolamines.[583] In the latter case, fluorosilicates are also formed, depending on the ratio of reactants and the solvent used. The 1:2 adducts with aromatic amines decompose on heating, with heterolytic rupture of the Si—N bond.[584] The thermal decompositions of the bivalent metal fluorosilicate ammines $M[SiF_6],nNH_3$ ($n = 6$, M = Mg or Mn; $n = 12$, M = Fe, Co, Ni, Zn, Cu, or Cd; $n = 2$, M = Ca or Sr) have been studied. The dodecammines decompose *via* pentammines and/or tetrammines. The final $NH_3$ loss is concomitant with fluorosilicate decomposition.[585] Xenon difluoride forms a 2:1 adduct with $SnF_4$, which Mössbauer and i.r. spectra show to have the ionic constitution $(XeF^+)_2(SnF_6^-)$.[586] The hexafluorostannate systems $(NH_4)_2SnF_6$–$HF$–$H_2O$ and $(en)SnF_6$–$HF$–$H_2O$ (en = ethylenediamine) have been studied by the isothermal solubility method at 0 °C, and the following solid phases isolated $[(NH_4)_2SnF_6],nHF$ ($n = 0, 2,$ or $4$), $[(en)SnF_6],2H_2O$, and $[(en)SnF_6],nHF$ ($n = 0$, 1 or 3).[587]

---

[578] J. A. Ladd and B. R. Glasberg, *J.C.S. Dalton*, 1975, 2378.
[579] S. Stafford, H. J. Haupt, and F. Huber *Inorg. Chim. Acta*, 1974, **11**, 207.
[580] A. A. Ennan, B. M. Kats, and L. V. Ostapchuk, *Russ. J. Inorg. Chem.*, 1974, **19**, 1302.
[581] A. A. Ennan and L. A. Gavrilova, *Russ. J. Inorg. Chem.*, 1974, **19**, 1652.
[582] A. A. Ennan and L. A. Gavrilova, *J. Gen. Chem. (U.S.S.R)*, 1975, **45**, 1373.
[583] A. A. Ennan, L. A. Gavrilova, and I. N. Kirichenko, *Russ. J. Inorg. Chem.*, 1974, **19**, 1784.
[584] A. A. Ennan and B. M. Kats, *Russ. J. Inorg. Chem.*, 1974, **19**, 26.
[585] K. C. Patil and E. A. Secco, *Canad. J. Chem.*, 1975, **53**, 2426.
[586] V. N. Zarubin and A. S. Marinin, *Russ. J. Inorg. Chem.*, 1974, **19**, 1599.
[587] I. I. Tychinskaya, N. F. Yedunov, Z. A. Grankina, and R. S. Ivchev, *Russ. J. Inorg. Chem.*, 1973, **18**, 1659.

Oxidation of [SiCl$_2$(bipy)$_2$] (bipy = 2,2'-bipyridyl) with chlorine yields ionic [SiCl$_2$(bipy)$_2$]Cl$_2$, which is unusually stable to solvolysis. Methanol solutions are stable for weeks at 25 °C, the [Si(OMe)$_2$(bipy)$_2$]$^{2+}$ cation being formed only after several months, but in aqueous solution [Si(OH)$_2$(bipy)$_2$]$^{2+}$ can be observed after only a few hours, and complete conversion is attained after several days. The corresponding perchlorate [SiCl$_2$(bipy)$_2$](ClO$_4$)$_2$ may be isolated, following ion exchange in methanol, as sparingly soluble colourless crystals.[588] The reactions of the mixed chloromethoxysilanes SiCl$_n$(OMe)$_{4-n}$ ($n$ = 0—4) with amides have been studied. With $NN$-dimethylformamide, 1:2 adducts are formed in all cases, but with acetanilide, formamide, and benzamide, ($n$ + 2)amide adducts result.[589] Farona[590] has examined the relative positions of the Sn—X stretching vibrations for the available data for $cis$- and $trans$-SnX$_4$L$_2$ (X = halogen) in terms of the theory of hard and soft acids and bases. For chloride complexes, the Sn—Cl stretching frequency decreases in the order hard > borderline > soft; for the bromide complexes the order for the Sn—Br frequency is hard ≈ borderline > soft; but the order for the corresponding iodides is the reverse of that for the chloride, i.e. hard < borderline ≈ soft. Solid SnCl$_4$,2Me$_2$O has the $cis$-octahedral geometry, whereas the analogous selenide SnCl$_4$,2Me$_2$Se has the $trans$-structure. Both complexes, however, participate in a fast $cis$-$trans$ equilibrium in either CH$_2$Cl$_2$ or CH$_2$Br$_2$ solution.[591] The Raman spectrum of K$_2$SnBr$_6$ indicates tetragonal symmetry at room temperature and cubic symmetry at 140 °C.[592] The vibrational spectra of the mixed phases A$_2$[(Sn, Te)X$_6$] (A = NH$_4$, K, Rb, or Cs; X = Cl or Br) have been measured and compared with those of the parent hexahalogenostannates(IV) and -tellurates(IV). Population of delocalized valence-shell bands by Te$^{IV}$ non-bonding electrons was indicated.[593] Other similar complexes which have been examined are, with special points of interest in parentheses: GeX$_4$ (X = Br or I) with α-, $m$-, and $p$-aminobenzoic acids (composition GeX$_4$,4L; stability in the orders $p > m > o$ and Br > i)[594] and with benzocaine and procaine (composition GeX$_4$,4L; stability in the order Cl > Br > I);[595] GeCl$_4$ and SnX$_4$ (X = Cl, Br, or I) with different substituted pyridines (composition MX$_4$,2L; ligands L $trans$);[596] GeCl$_4$, SnCl$_4$, and SnBr$_4$ with haematoxylin (composition SnX$_4$,2L and GeCl$_4$,L; stability constants and free energies of formation;[597] GeBr$_4$ and GeI$_4$ with $o$-, $m$-, and $p$-phenylenediamines (composition GeX$_4$,$n$L, $n$ = 2,4, or 6);[598] SnCl$_4$ with EtRP(O)Cl,[599] dicarboxylic esters,[600] Schiff

---

[588] D. Kummer and T. Seshadri, *Angew. Chem. Internat. Edn.*, 1975, **14**, 1974, 699.
[589] N. Yoshino and T. Yoshino, *Synth. React. Inorg. Metal.-Org. Chem.*, 1974, **4**, 263.
[590] M. F. Farona, *Inorg. Chem.*, 1975, **14**, 2020.
[591] S. J. Ruzicka and A. E. Merbach, *Helv. Chim. Acta*, 1975, **58**, 584.
[592] J. W. Anthonson, *Acta Chem. Scand.(A)*, 1974, **28**, 974.
[593] J. D. Donaldson, S. D. Ross, J. Silver, and P. J. Watkiss, *J.C.S. Dalton*, 1975, 1980.
[594] E. M. Belousova, V. N. Reznichenko, R. G. Yankelevich, and A. A. Dzhambek, *Russ. J. Inorg. Chem.*, 1974, **19**, 24.
[595] E. M. Belousova, V. N. Reznichenko, R. G. Yankelevich, and O. I. Navaly, *J. Gen. Chem. (U.S.S.R)*, 1975, **45**, 99.
[596] S. A. A. Zaidi and K. S. Siddiqi, *Indian J. Chem.*, 1975, **13**, 182.
[597] S. A. A. Zaidi and K. S. Siddiqi, *Indian J. Chem.*, 1975, **13**, 624.
[598] E. M. Belousova, V. N. Reznichenko, and K. I. Varchuk, *J. Gen. Chem. (U.S.S.R)*, 1974, **44**, 1031.
[599] A. N. Pudovik, I. Ya. Kuramshii, A. B. Remizov, A. A. Muratova, and R. A. Manapov, *J. Gen. Chem. (U.S.S.R)*, 1974, **44**, 41.
[600] K. C. Malhotra and S. M. Sehgal, *Indian J. Chem.*, 1975, **13**, 599.

bases derived from pentane-2,4-dione and 2-hydroxy-1-naphthaldehyde,[601] disilazanes (heats of formation),[602] $p$-halogenoanilines,[603] substituted benzylideneanilines (heats of mixing of $SnCl_4$ with bases),[604] $(R_3SiO)_3PO$ (R = Me, Et, or Ph) (heats of complexation),[605] tetra-alkoxysilanes (dipole moments and heats of formation),[606] dialkyl sulphoxides (heats of formation),[607] $Me_3PO$, $Me_2P(O)OMe$, and $Me_2P(O)SMe$ (activation energies for *cis–trans* interconversion in solution)[608] (all with composition $SnCl_4;2L$); $SnCl_4$ with tetraphenylphosphinimines,[609] diaminosilanes (heats of formation),[602] dibenzoyl sulphide,[610] nickel(II) aldoximates[611] (all with composition 1:1); and 2:1 complexes with the spiro-substituted methanes $C(CH_2PPh_2Cl_2)_4$,[612] and $C[CH_2P(O)Ph_2]_4$;[613] $SnX_4$ (X = F, Cl, Br, or I) with pyruvonitrile (composition $SnX_4,L$);[614] $SnX_4$ (X = Cl or Br) with 1,2-dithiole-3-thiones (Mössbauer spectra)[615] and phosphorus acid amides;[616] $SnX_4$ (X = Br or I) with $Me_3SiOCHMe_2$ or acetic anhydride (equilibrium constants by Mössbauer spectroscopy);[617] $[SnCl_4(N_3)_2]^{2-}$ and $[Sn_2Cl_8(N_3)_2]^{2-}$;[618] complexes of the types $X_2SnL_2$ (L = Multidentate Schiff bases).[619—621]

The structure of the carbonyl-stabilized phosphorus ylide complex of $Me_3SnCl$, *viz* ·$Ph_3PCHCOMe·Me_3SnCl$, has been reported, and shows unequivocally that the donor site of the ylide is the carbonyl oxygen atom only. The geometry at tin is that of a trigonal bipyramid with a planar $Me_3Sn$ moiety (Figure 18).[622] The 1:1 adduct between $Me_3SnCl$ and ethylenediamine adopts a similar geometry.[623] Five-co-ordination and somewhat distorted trigonal-bipyramidal stereochemistry are also enjoyed by the tin atoms in quinolinium dimethyltrichlorostannate(IV)[624] and

---

[601] O. P. Singh and J. P. Tandon, *Monatsh.*, 1975, **106**, 1525.
[602] Yu. V. Kolodyazhnyi, M. G. Gruntfest, L. S. Baturina, M. M. Morgunova, N. I. Sizova, G. S. Gol'din, and O. A. Osipov, *J. Gen. Chem. (U.S.S.R.)*, 1975, **45**, 1033.
[603] I. Ya. Shternberg and Ya. F. Freimanis, *J. Gen. Chem. (U.S.S.R.)*, 1975, **45**, 877.
[604] V. A. Kogan, A. S. Egorov, and O. A. Osipov, *Russ. J. Inorg. Chem.*, 1974, **19**, 980.
[605] Yu. V. Kolodyazhnyi, V. G. Tkalenko, O. A. Osipov, N. A. Kardanov, and N. N. Tsapkova, *J. Gen. Chem. (U.S.S.R.)*, 1975, **45**, 743.
[606] K. S. Puskareva, Yu. V. Kolodyazhnyi, N. E. Drapkina, and O. A. Osipov, *J. Gen. Chem. (U.S.S.R.)*, 1975, **45**, 747.
[607] V. V. Puchkova and E. N. Gur'yanova, *J. Gen. Chem. (U.S.S.R.)*, 1975, **45**, 949.
[608] A. V. Aganov, A. A. Musina, I. Ya. Kuramshin, E. G. Yarkova, Yu. Yu. Samitov, and A. A. Muratova, *J. Struct. Chem.*, 1974, **15**, 686.
[609] N. A. Vvanova, V. A. Kogan, O. A. Osipov, and N. I. Dorokhova, *J. Gen. Chem. (U.S.S.R.)*, 1974, **44**, 1974.
[610] K. C. Malhotra and J. K. Puri, *Indian J. Chem.*, 1975, **13**, 184.
[611] N. S. Biradar, B. R. Patil, and V. H. Kulkarni, *J. Inorg. Nuclear Chem.*, 1975, **37**, 1901.
[612] J. Ellerman and M. Thierling, *Z. anorg. Chem.*, 1975, **411**, 15.
[613] J. Ellerman and M. Thierling, *Z. anorg. Chem.*, 1975, **411**, 28.
[614] S. C. Jain and B. P. Hajela, *Indian J. Chem.*, 1974, **12**, 843.
[615] F. Petillon and J. E. Guerchais, *J. Inorg. Nuclear Chem.*, 1975, **37**, 1863.
[616] I. Ya. Kuramshin, S. S. Bashkirov, A. A. Muratova, R. A. Manapov, A. S. Khramov, and A. N. Pudovik, *J. Gen. Chem. (U.S.S.R.)*, 1975, **45**, 684.
[617] A. Vertes, S. Nagy, I. Czako-Nagy, and E. Csakvary, *J. Phys. Chem.*, 1975, **79**, 149.
[618] B. Busch, J. Pebler, and K. Dehnicke, *Z. anorg. Chem.*, 1975, **416**, 203.
[619] S. N. Podder and N. S. Das, *Indian J. Chem.*, 1974, **12**, 1105.
[620] L. Pellerito, N. Bertazzi, G. C. Stocco, A. Silvestri, and R. Barbieri, *Spectrochim. Acta*, 1975, **31A**, 303.
[621] J. N. R. Ruddick and J. R. Sams, *J. Inorg. Nuclear Chem.*, 1975, **37**, 565.
[622] J. Buckle, P. G. Harrison, T. J. King, and J. A. Richards, *J.C.S. Dalton*, 1975, 1552.
[623] G. Eng and J. H. Terry, *Inorg. Chim. Acta*, 1975, **14**, L19.
[624] A. J. Buttenshaw, M. Duchen, and M. Webster, *J.C.S. Dalton*, 1975, 2231.

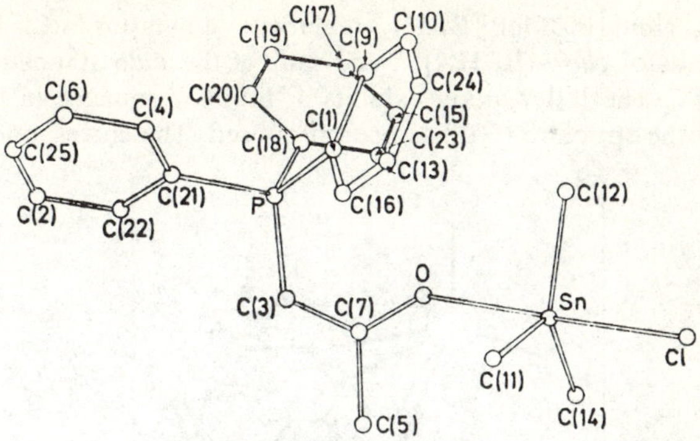

**Figure 18** *The structure of* Ph$_3$PCHCOMe,Me$_3$SnC.
(Reproduced from *J.C.S Dalton*, 1975, 1552)

the 1:1 adduct of Me$_2$SnCl$_2$ with salicylaldehyde.[625] In the latter compound, the salicylaldehyde group is bonded to the metal *via* the carbonyl oxygen at an axial site (Figure 19), and in both compounds the methyl groups occupy equatorial sites. Chlorine atoms occupy the remaining co-ordination positions in both species. Greenwood and Youll have reported the reaction of some tin(II) and tin(IV) compounds with the dodecahydro-*nido*-decaborate ion [B$_{10}$H$_{12}$]$^{2-}$.

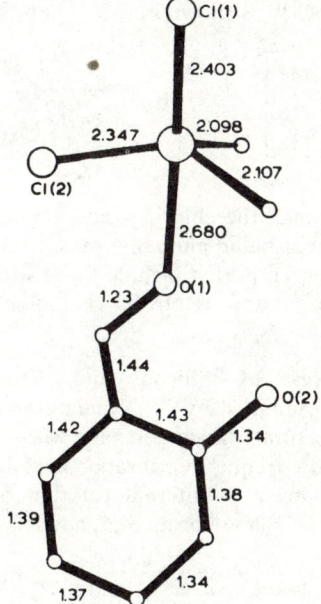

**Figure 19** *The structure of the 1:1 adduct of* Me$_2$SnCl$_2$ *with salicylaldehyde*
(Reproduced by permission from *J. Organometallic Chem.*, 1975, **90**, C23)

[625] D. Cunningham, I. Douek, M. J. Fraser, M. MacPartlin, and J. D. Matthews, *J. Organometallic Chem.*, 1975, **90**, C23.

[Ph$_4$As][B$_{10}$H$_{12}$] and [Ph$_3$MeP][B$_{10}$H$_{12}$] react with anhydrous SnCl$_2$ to give metallic tin, HCl, salts of *closo*-[B$_{10}$H$_{12}$]$^{2-}$, and salts of the *nido*-stannadecaborate ion [B$_{10}$H$_{12}$SnCl$_2$]$^{2-}$. The latter anion exhibits a 'tin(II)' resonance in its Mössbauer spectrum, and the structure (60) has been proposed. The corresponding reactions

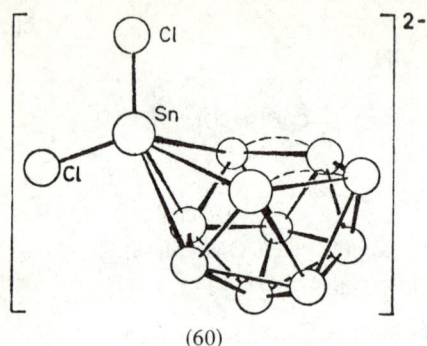

(60)

of Me$_2$SnCl$_2$ yield [B$_{10}$H$_{10}$]$^{2-}$ salts and also the tin(II) ion [Me$_2$SnCl$_2$]$^{2-}$ as well as [B$_{10}$H$_{12}$Me$_2$SnCl$_2$]$^{2-}$, which exhibits a 'tin(IV)' Mössbauer resonance. The use of Et$_2$SnCl$_2$ leads to the formation of [Ph$_4$As][B$_{10}$H$_{12}$Et$_2$SnCl$_2$], but Ph$_2$SnCl$_2$ did not afford any isolable stannaborate products. The [Me$_2$SnCl$_2$]$^{2-}$ anion is isoelectronic with Me$_2$TeCl$_2$, and probably has the structure (61). The proposed structure of the [B$_{10}$H$_{12}$Me$_2$SnCl]$^{2-}$ is shown in (62).[626] Ph$_2$SnCl$_2$,2DMSO possesses an

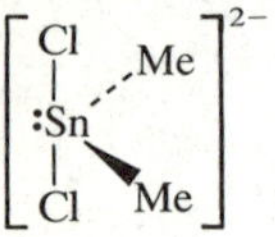

(61) 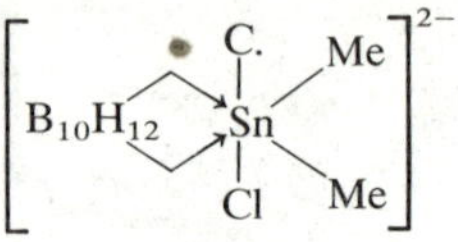    (62)

octahedral structure in which the chlorines and DMSO ligands occupy *cis* positions, the two phenyl groups being mutually *trans*.[627] Other complexes of organotin halides which have been reported include those with isoquinoline,[628] diphenyl sulphoxide,[629] hydrazine,[630] and (RO$_2$P(O)H (equilibrium constants in benzene).[631]

**Pseudohalide Derivatives of Silicon and Tin.**—Dimethyl cyanosilane Me$_2$Si(CN)H, has been examined by i.r., Raman, and microwave spectroscopy from which precise structural parameters and the dipole moment have been determined. From the low-frequency vibrational assignment, an upper limit of 2.8 kcal mol$^{-1}$ for the barrier to internal rotation of the methyl groups was deduced.[632] Trimethylcyanosilane forms 1:1 adducts of composition (63) with

[626] N. N. Greenwood and B. Youll, *J.C.S. Dalton*, 1975, 158.
[627] L. Coghi, C. Pelizzi, and G. Pelizzi, *Gazzetta*, 1974, **104**, 873.
[628] T. N. Srivastava, P. C. Srivastava, and K. Srivastava, *J. Inorg. Nuclear Chem.*, 1975, **37**, 1803.
[629] K. L. Jaura, R. K. Chadha, and K. K. Sharma, *Indian J. Chem.*, 1974, **12**, 766.
[630] K. L. Jaura, B. Singh, and R. K. Chadha, *Indian J. Chem.*, 1974, **12**, 1304.
[631] A. N. Pudovik, A. N. Muratova, N. R. Safiullina, E. G. Yarkova, and V. P. Plekhov, *J. Gen. Chem.* (*U.S.S.R.*), 1975, **45**, 515.
[632] J. R. Durig, P. J. Cooper, and Y. S. Li, *Inorg. Chem.*, 1975, **14**, 2845.

ketens $R^1R^2C=C=O$,[633] and, with carbodi-imides in the presence of $AlCl_3$, of structure (64).[634] Kinetic studies show that the reaction between tri-

$$\begin{array}{cc} R^1 \quad OSiMe_3 & CN \\ \diagdown \quad \diagup & | \\ C=C & RN-C=NR \\ \diagup \quad \diagdown & | \\ R^2 \quad CN & SiMe_3 \\ (63) & (64) \end{array}$$

alkylcyanosilanes and sulphur in 1-chloronaphthalenes is second-order. The ease of C—X bond cleavage increases sharply for the heavier chalcogens in the order X = S < Se < Te. Organoisotellurocyanatosilanes are not stable at room temperature.[635] Crystals of $Ph_3SiNCS$ consist of isolated tetrahedral molecules in which the NCS group is approximately linear.[636] Triaryl-tin and -lead isothiocyanates are readily obtained by the reaction of dithiocyanogen with either tetra-aryl-lead or hexa-aryl-ditin or -dilead. Tetra-arylstannanes react only very slowly.[637] Trimethylsilyl isocyanate reacts with acyl chlorides in the presence of catalytic amounts of $SnCl_4$ at temperatures of 90—150 °C, giving acyl isocyanates and $Me_3SiCl$. The reaction probably proceeds *via* the formation of an intermediate tin isocyanate which undergoes exchange with the acyl chloride.[638] The basic tin pseudohalide derivatives $Me_3SnOH,Me_3SnNCO$, $Me_3SnOH,Me_3SnNCS$, $Me_3SnOH,Me_3PbN_3$, and $Ph_3SnOH,Ph_3SnN_3$ are prepared by mixing equimolecular amounts of the components in a hot solvent and allowing it to cool. Their i.r. and Mössbauer spectra indicate a polymeric structure in which OH and pseudohalide groups alternately bridge planar $R_3M$ moieties. In addition, the trimethyl-metal derivatives show evidence of interchain hydrogen-bonding.[639] The tin pseudohalide complex anions $[Ph_3Sn(N_3)_2]^-$, $[Ph_3Sn(N_3)(NCS)]^-$, $[Ph_2Sn(N_3)_2(NCS)_2]^{2-}$, and $[Me_2Sn(N_3)_4]^{2-}$ have been obtained similarly from the appropriate organotin pseudohalide and either the tetra-alkylammonium or tetraphenylarsonium pseudohalide.[640] The reactions of bis(trialkyl-silyl, -germyl, and -stannyl)carbodi-imides with phenyl isocyanate have been examined. With the silylcarbodi-imide, 1,3-diphenyl-5,$N^6$-bis(trimethylsilyl)isoammelide (65) is produced, but bis(triethylgermyl)carbodi-imide yields the germyliminodiphenyl-uretidinone (66).[641] The reaction of bis(tributylstannyl)carbodi-imide in a 1:1 molar ratio affords the equilibrium mixture (67) as a green oil, but in a 1:2 molar ratio the products (68) and (69) are formed. A 1:3 adduct is also formed, though excess phenyl isocyanate results in the formation of the isocyanurate.[642]

---

[633] U. Hertenstein and S. Hünig, *Angew. Chem. Internat Edn.*, 1975, **14**, 179.
[634] I. Ojima, S. I. Inaba, and Y. Nagai, *J. Organometallic Chem.*, 1975, **99**, C5.
[635] J. A. Secker and J. S. Thayer, *Inorg. Chem.*, 1975, **14**, 573.
[636] G. M. Sheldrick and R. Taylor, *J. Organometallic Chem.*, 1975, **87**, 145.
[637] S. N. Bhattacharya, P. Raj, and R. C. Srivastava, *J. Organometallic Chem.*, 1975, **87**, 279.
[638] V. Mironov, V. P. Kozyukov, A. S. Tkacher, and E. K. Dobrovinskaya, *J. Gen. Chem. (U.S.S.R.)*, 1975, **45**, 467.
[639] N. Bertazzi, G. Alonzo, F. Di Bianca, and G. C. Stocco, *Inorg. Chim. Acta*, 1975, **12**, 123.
[640] R. Barbieri, N. Bertazzi, C. Tomarchio, and R. H. Herber, *J. Organometallic Chem.*, 1975, **84**, 39.
[641] Yu. I. Dergunov, A. S. Gordetsov, I. A. Vostokov, and V. F. Gerega, *J. Gen. Chem. (U.S.S.R.)*, 1974, **44**, 1494.
[642] Yu. I. Dergunov, A. S. Gordetsov, I. A. Vostokov, and V. F. Gergega, *J. Gen. Chem. (U.S.S.R.)*, 1974, **44**, 2127.

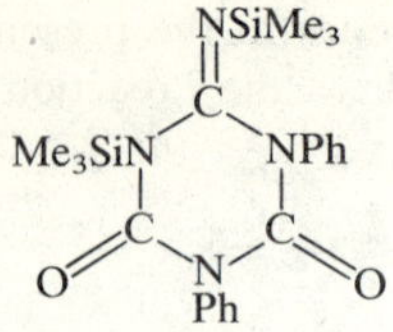

(65) 

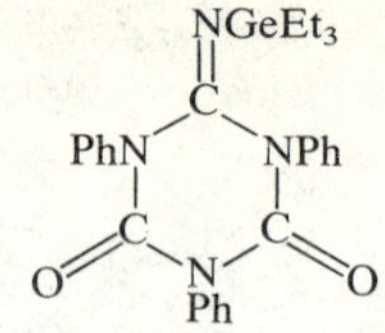

Structure: Central C ring with NSiMe3 (double bond up), Me3SiN and NPh on sides, two C=O groups at bottom, N-Ph in center.

(66)

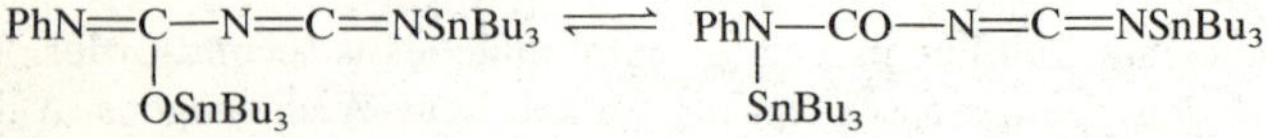

Structure: Central C ring with NGeEt3 (double bond up), PhN and NPh on sides, two C=O groups at bottom, N-Ph in center.

$PhN{=}C{-}N{=}C{=}NSnBu_3 \rightleftharpoons PhN{-}CO{-}N{=}C{=}NSnBu_3$
      |                                                |
    OSnBu₃                                     SnBu₃

(67)

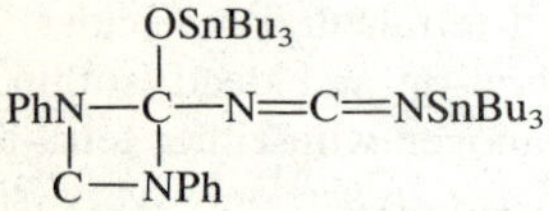

(68) $PhN{-}C{-}N{=}C{=}NSnBu_3$ with OSnBu₃ above C and C—NPh below

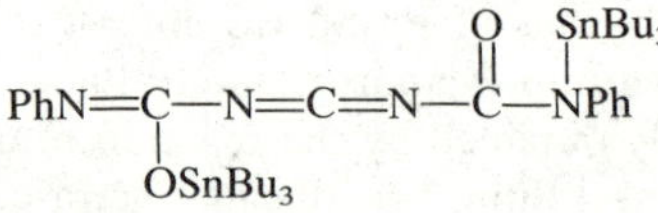

(69) $PhN{=}C{-}N{=}C{=}N{-}C{-}NPh$ with OSnBu₃ below left C and O=/SnBu₃ on right C

Bis(triphenylstannyl)carbodi-imide reacts with acid chlorides, alkylchlorocarbonates, and benzenesulphonyl chloride in THF at 26 °C to give high yields of $Ph_3SnCl$ and $Ph_3SnN(CN)R$ (R = EtCO, PhCO, PhCH₂CO, $p$-NO₂C₆H₄, $p$-COC₆H₄CO, CF₃CO, MeOCO, EtOCO, or PhSO₂) derivatives. The acetyl and trifluoroacetyl derivatives may also be obtained by reaction with the corresponding acid anhydride.[643] Thioamides react to afford organic nitriles, whilst N-substituted thioamides give N'-substituted-N-cyanoamidines.[644] Dergunov has investigated the physical characteristics (density, viscosity, surface tension, dipole moment, *etc.*) of $Et_3SnNCO$ and $Et_3SnNCNSnEt_3$.[645]

The Edinburgh group has characterized the unstable compound silylsulphinylamine, $H_3SiNSO$. The compound may be obtained by the exchange of groups between either $Me_3GeNSO$ or $Bu_3SnNSO$ and silyl bromide. The reaction of trisilylamine and $SOCl_2$ gave an inseparable mixture of silylamine and silyl chloride. The structure of silylsulphinylamine, from electron-diffraction data, shows that the molecular skeleton is bent at both the nitrogen and sulphur atoms and has an unusually long Si—N bond [1.726(6) Å].[646]

**Sulphur and Selenium Derivatives of Silicon, Germanium, Tin, and Lead.**—Crystals of the high-temperature monoclinic form of $GeS_2$ consist of chains of $GeS_4$ tetrahedra parallel to the $a$-axis, which are connected to layers perpendicular to the $c$-axis by double tetrahedra, which share a common edge.[647] Those of $Pr_6Ge_{2.5}S_{14}$ have germanium atoms in two different types of co-ordination; one type is tetrahedrally co-ordinated, with Ge—S distances of 2.16 and 2.23 Å, and the second type is in almost perfect octahedral co-ordination, with much longer Ge—S bond distances (2.63, 2.67 Å).[648] $K_2SnS_3,2H_2O$ contains $SnS_4$ tetrahedra

---

[643] E. J. Kupchik and J. A. Feiccabrino, *J. Organometallic Chem.*, 1975, **93**, 325.
[644] E. J. Kupchik and H. E. Hanke, *J. Organometallic Chem.*, 1975, **97**, 39.
[645] E. Z. Zhuravlev, V. F. Gerega, V. D. Selivanov, P. V. Mulyanov, and Yu. T. Dergunov, *J. Gen. Chem. (U.S.S.R.)*, 1975, **45**, 1030.
[646] S. Cradock, E. A. V. Ebsworth, G. D. Meikle, and D. W. H. Rankin, *J.C.S. Dalton*, 1975, 805.
[647] G. Dittmar and H. Schäfer, *Acta Cryst.*, 1975, **B31**, 2060.
[648] V. V. Bakakin, E. N. Ipatova, and L. P. Solov'eva, *J. Struct. Chem.*, 1974, **15**, 393.

which are connected by sharing corners to form infinite anionic $(SnS_3)_n^{2n-}$ chains.[649] Mössbauer spectra of $M^I$ and $M^{II}$ thiostannates, selenostannates, and selenothiostannates indicate that discrete dinuclear $Sn_2S_6^{4-}$, tetrahedral $SnS_4^{4-}$, dinuclear $Sn_2S_7^{6-}$, and the related seleno-ions exist in the solids.[650] The vibrational spectra of $M_2^{II}GeS_4$ ($M^{II}$ = Ba, Sr, or Pb), $Ba_2SnS_4$, and $Na_4SnS_4$ have been recorded, and normal-co-ordinate analyses performed for the $MS_4^{4-}$ anions in the $T_d$ point group, except for $Na_4SnS_4$, which was treated in the $D_{2d}$ group.[651] A new polythiostannate anion, $Sn_{10}O_4S_{20}^{8-}$, has been synthesized as its caesium salt. The anion has idealized $T_d$ symmetry, and has ten corner-linked $SnS_4$ tetrahedra (Figure 20). The octahedral sites lying between the tetrahedra are occupied by the oxygen atoms.[652]

Mixed silylgermyl sulphides have been obtained by exchange between a germyl halide and either bis(silyl) or bis(trimethylsilyl) sulphide.[653] Organosilthianes are conveniently prepared by the reactions of ammonium hydrosulphide with $Me_3SiCl$ or $R^1R^2SiCl_2$.[654] $(Me_3Si)_2S$ reacts with salts of mineral acids to form the trimethylsilyl derivative of the mineral acid.[655] Cyclic dithiastanna-alkanes may be prepared from the diorganotin dichloride and the corresponding lead(II) dithiolate.[656] Tetra(methylsilicon) hexasulphide has the adamantane-type structure, as do the corresponding germanium and tin compounds.[657] $(Me_2SnS)_3$ has the twisted boat conformation in the crystal.[658] The same structure was also inferred for $(Me_2SnSe)_3$ from its vibrational spectra.[659] The crystal structures of $Ph_3SnSC_6H_4$-Me-2 and $Ph_3SnSC_6H_2$-F-4-$Br_2$-2,6 have also been determined, and contain

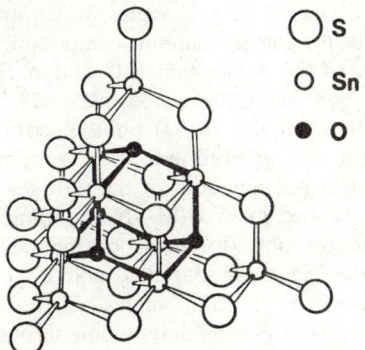

**Figure 20** The structure of the $[Sn_{10}O_4S_{20}]^{8-}$ anion
(Reproduced by permission from *Angew. Chem. Internat. Edn.*, 1975, **14**, 436)

[649] W. Schiwy, C. Blutau, D. Gäthje, and B. Krebs, *Z. anorg. Chem.*, 1975, **412**, 1.
[650] S. Ichiba, M. Katada, and H. Negita, *J. Inorg. Nuclear Chem.*, 1975, **37**, 2249.
[651] M. Neyrand, M. Ribes, E. Philippot, and M. Maurin, *Rev. Chim. minérale*, 1975, **12**, 406.
[652] W. Schiwy and B. Krebs, *Angew. Chem. Internat Edn.*, 1975, **14**, 436.
[653] M. A. Finch and C. H. Van Dyke, *Inorg. Chem.*, 1975, **14**, 134.
[654] E. P. Lebedev, D. V. Fridland, and V. O. Reikhsfel'd, *J. Gen. Chem. (U.S.S.R.)*, 1974, **44**, 2737.
[655] E. P. Lebedev, V. A. Baburina, and V. O. Reikhsfel'd, *J. Gen. Chem. (U.S.S.R.)*, 1974, **44**, 1736.
[656] R. H. Cragg and A. Taylor, *J. Organometallic Chem.*, 1975, **99**, 391.
[657] J. C. J. Bart and J. J. Daly, *J.C.S. Dalton*, 1975, 2063.
[658] B. Menzebach and P. Bleckmann, *J. Organometallic Chem.*, 1975, **91**, 291.
[659] B. Mathiasch and A. Blecher, *Bull. Soc. chim. belges*, 1975, **84**, 1045.

tetrahedrally co-ordinated tin.[660] Germanium and tin dithiocarbamates [Cl$_2$Ge(dtcEt)$_2$], [X$_2$Sn(dtcEt)$_2$], and [Sn(dtcEt)$_4$] (dtcEt = diethyldithiocarbamate; X = Cl, Br, or I) have been prepared by substitution of the metal(IV) chloride.[661] Organolead phosphorodithioates have been obtained either by substitution of the organolead chloride, by azeotropic dehydration of the organolead oxide and HS(S)P(OR)$_2$, or by reaction of Ph$_4$Pb and HS(S)P(OR)$_2$.[662]

**Nitrogen and Phosphorus Derivatives of Silicon, Germanium, and Tin.**—The structures of several silicon–nitrogen-bonded compounds have been determined, mostly by $X$-ray diffraction. (Ph$_2$MeSi)$_3$N possesses crystallographic $C_{3v}$ symmetry and is pyramidal at nitrogen.[663] (Me$_3$Si)$_2$N—N=N—N(SiMe$_3$)$_2$[664] and Me$_3$Si—N=N—SiMe$_3$[665] both have the *trans* configuration at the N=N double bond and unusually long Si—N bonds. Crystals of (Ph$_3$Si)$_2$N$_2$C contain three crystallographically independent molecules, which have very similar structures, with linear Si—N=C=N—Si chains.[666] (Me$_2$SiNH)$_3$ has been studied by electron diffraction in the gas phase. The six-membered (SiN)$_3$ ring is puckered, but the deviation from planarity is small.[667] Slightly puckered five-membered rings are also present in Ti(—NMe—SiMe$_2$SiMe$_2$—MeN—)$_2$, which is obtained as yellow crystals from the reaction between TiBr$_4$ and LiNMeSiMe$_2$SiMe$_2$N-MeLi.[668] Alcock has determined the structures of tetrakis(diphenylketimino)-silicon(IV), -germanium(IV), and -tin(IV), prepared by substitution from the metal(IV) chloride and LiNCPh$_2$.[669,670] All the crystals contain discrete molecules which, like the other molecules reported above, contain tetrahedrally co-ordinated metal. The principal differences lie in the M—N=C angles, which are 134.7, 139.5° for the silicon compound, 127.0° for the germanium compound, and 121.3° for the tin derivative. The deviation of this angle from 120° and its change was attributed to the presence of $p_\pi \rightarrow d_\pi$ bonding in these molecules, which decreases in the order Si > Ge > Sn. The first band in the He (I) photoelectron spectra of M(NMe$_2$)$_4$ (M = C, Si, or Ge) has been assigned to ionization from molecular orbitals which are regarded as being linear combinations of the nitrogen lone-pair orbitals. The next band corresponds to the main $\sigma$(M—N) bonding orbital.[671] The dipole moments and ionization constants (for those compounds with N—H bonds) of cyclic and acyclic diaminosilanes, disilazanes, and cyclotrisilazanes have been determined. The lowering of the p$K_a$ values of the disilazanes and cyclotrisilazanes was attributed to higher $\pi$ interaction through Si—N bonds of the Si$_2$N groups.[672]

---

[660] N. G. Bokii, Yu. T. Struchkov, D. V. Kravstsov, and E. M. Rokhlina, *J. Struct. Chem.*, 1974, **15**, 424.
[661] S. A. A. Zaidi, K. S. Siddiqi, and N. Islam, *Indian J. Chem.*, 1974, **12**, 1197.
[662] I. Haiduc, F. Martinas, D. Ruse, and M. Curtui, *Synth. React. Inorg. Metal.-Org. Chem.*, 1975, **5**, 103.
[663] J. J. Daly and F. Sanz, *Acta Cryst.*, 1974, **B30**, 2766.
[664] M. Veith, *Acta Cryst.*, 1975, **B31**, 678.
[665] M. Veith and H. Bärnighausen, *Acta Cryst.*, 1974, **B30**, 1806.
[666] G. M. Sheldrick and R. Taylor, *J. Organometallic Chem.*, 1975, **101**, 19.
[667] B. Rozsondai, I. Hargittai, A. V. Golubinskii, L. V. Vilkov, and V. S. Mastryukov, *J. Mol. Structure*, 1975, **28**, 339.
[668] H. Bürger, K. Wiegel, U. Thewalt, and D. Schomburg, *J. Organometallic Chem.*, 1975, **87**, 301.
[669] N. W. Alcock, M. Pierce-Butler, G. R. Willey, and K. Wade, *J.C.S. Chem. Comm.*, 1975, 183.
[670] N. W. Alcock and M. Pierce-Butler, *J.C.S. Dalton*, 1975, 2469.
[671] S. G. Gibbins, M. F. Lappert, J. B. Pedley, and G. J. Sharp, *J.C.S. Dalton*, 1975, 72.
[672] Yu. V. Kolodyazhnyi, M. G. Gruntfest, V. A. Bren, L. S. Baturina, M. I. Morgunova, D. Ya. Zhinkin, G. S. Gol'din, and O. A. Osipov, *J. Gen. Chem. (U.S.S.R.)*, 1975, **45**, 1069.

Several new types of metal(IV)–nitrogen compounds have been synthesized. Fluoroaminosilanes have attracted considerable interest. At relatively low temperatures, trisilylamine reacts with $PF_5$ by two pathways: the predominant pathway involves Si—N bond cleavage, giving silyl fluoride, but the second is an exchange reaction that results in the formation of fluorosilyl derivatives such as $(H_3Si)_2NSiH_2F$, $H_3SiN(SiH_2F)_2$ and $(FH_2Si)_3N$.[673] The principal method of synthesis is, however, by substitution of $SiF_4$ or organofluorosilanes by lithium or sodium alkylsilylamines or disilylazanes, and compounds of the types $R^1R^2SiF(NR^3SiMe_3)$, $R^1SiF(NR^3SiMe_3)_2$, $F_2Si(NR^3SiMe_3)_2$, and $FSi(NR^3SiMe_3)_3$, and $RSiF_2N(SiR_3)_2$, $R_2SiFN(SiR_3)_2$, and $R^1R^2SiFN(SiMe_3)_2$ are formed.[674—676] $(Me_3Si)_2NSiF_2R$ derivatives may be converted into $(Me_3Si)_2NSiFR^1(OR^2)$ compounds by treatment with sodium alkoxides.[677] The reaction of $LiNRSiMe_3$ ($R = SiMe_3$ or $Bu^t$) with $AsF_3$ yields the silylaminoarsenic fluorides $AsF_n(NRSiMe_3)_{3-n}$ ($n = 1$ or 2). Pyrolysis of the compound when $n = 1$ at 150 °C results in the elimination of $Me_3SiF$ and the formation of (70). The other compound ($n = 2$) also eliminates $Me_3SiF$ at high temperature (220 °C), but the decomposition is rather complex.[678] Diphenylsilatriazole trimer (71) are formed in

$$
\begin{array}{cc}
\text{NSiMe}_3 & \\
| & \\
\text{As} & \\
\diagup \quad \diagdown & \\
\text{Me}_3\text{SiN} \qquad \text{NSiMe}_3 & \\
\diagdown \quad \diagup & \\
\text{As} & \\
| & \\
\text{NSiMe}_3 & \\
(70) &
\end{array}
\qquad
\begin{array}{c}
\text{Ph}_2\text{Si}\text{—NR} \\
\text{...(71)}
\end{array}
$$

40—60% yield by the reaction of $Ph_2SiCl_2$ with the sodium salts of 2,4,6-trihydrazino-s-triazines.[679] Veith has prepared four-membered stannasilylamines, according to Scheme 8. Compound (72) is a red monomeric liquid, whilst (73) is a colourless crystalline compound which preliminary $X$-ray data show to possess the spiro-structure.[680] The primary germylamine $Pr^i_3GeNH_2$ has been obtained by the reaction of $Pr^i_3GeCl$ with $KNH_2$ in liquid $NH_3$ at 5 °C.[681] The treatment of $Bu^tSnCl_2$ with $Bu^tNHLi$ yields $Bu^t_2Sn(Bu^tNH)_2$, which may be converted into the dilithium derivative $Bu^t_2Sn(Bu^tNLi)_2$, by reaction with BuLi in benzene. Reaction of the latter compound with $Me_2SnCl_2$ yields the four-membered-ring stannazane (74), whilst with $Ph_3SnCl$ (75) is obtained. cyclo-$(Bu^t_2SnNH)_3$ is formed by the reaction of $Bu^t_2SnCl_2$ with $KNH_2$ in liquid ammonia, and reacts with $Me_3SiCl$ to afford $Bu^t_2ClSn$—NH—$SiMe_3$.[682] Hexamethyldisilazane can convert diorganodichlorosilanes $R_2SiCl_2$ (R = Me, Et, or Ph) into the corresponding cyclosilazanes $(R_2SiNH)_n$ ($n = 3$ or 4).[683] N-Silylphosphoramides react with benzaldehyde to

---

[673] L. H. Marcus and C. H. Van Dyke, *Inorg. Chem.*, 1975, **14**, 3124.
[674] U. Klingebiel, D. Enterling, and A. Meller, *J. Organometallic Chem.*, 1975, **101**, 45.
[675] U. Klingebiel and A. Meller, *Chem. Ber.*, 1975, **108**, 155.
[676] U. Klingebiel and A. Meller, *J. Organometallic Chem.*, 1975, **88**, 149.
[677] U. Klingebiel, D. Fischer, and A. Meller, *J. Organometallic Chem.*, 1975, **85**, 141.
[678] E. Niecke and W. Bitter, *Synth. React. Inorg. Metal.-Org. Chem.*, 1975, **5**, 231.
[679] S. V. Sunthanker and S. T. Mahadik, *J.C.S. Chem. Comm.*, 1975, 281.
[680] M. Veith, *Angew. Chem. Internat. Edn.*, 1975, **14**, 263.
[681] H. J. Götze, *Chem. Ber.*, 1975, **108**, 988.
[582] D. Hänssgen, J. Kuna, and B. Ross, *J. Organometallic Chem.*, 1975, **92**, C49.
[683] E. P. Lebedev, R. G. Valimukhametova, E. N. Korol, and V. O. Reikhsfel'd, *J. Gen. Chem. (U.S.S.R.)*, 1974, **44**, 1906.

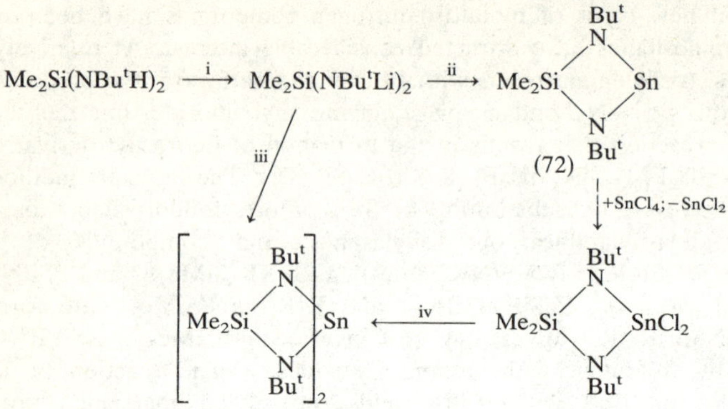

Me$_2$Si(NBu$^t$H)$_2$ $\xrightarrow{i}$ Me$_2$Si(NBu$^t$Li)$_2$ $\xrightarrow{ii}$ Me$_2$Si(μ-NBu$^t$)$_2$Sn

(72)

↓ +SnCl$_4$; −SnCl$_2$

[Me$_2$Si(μ-NBu$^t$)$_2$Sn]$_2$ $\xleftarrow{iv}$ Me$_2$Si(μ-NBu$^t$)$_2$SnCl$_2$

(73)

Reagents: i, 2BuLi; ii, SnCl$_2$; iii, SnCl$_4$; iv, Me$_2$Si(NBu$^t$Li)$_2$

**Scheme 8**

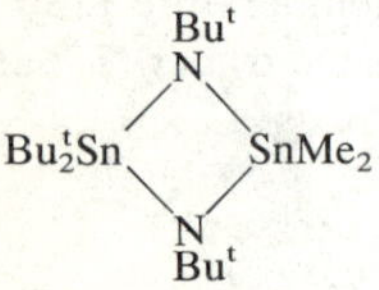

Bu$^t_2$Sn(μ-NBu$^t$)$_2$SnMe$_2$ (74)

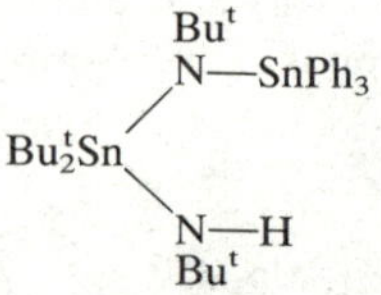

Bu$^t_2$Sn(NBu$^t$—SnPh$_3$)(NBu$^t$—H) (75)

give the benzaldehyde imine.[684] Transamination occurs when sulphonylamines are added to triorganostannylamines, and R$^1_3$SnNR$^2$S(O)R$^3$ derivatives result.[685]

The crystal structures of (pc)Si(OSiMe$_3$)$_2$ (pc = phthalocyanine)[686] and a germànium hemiporphyrazine[687] have been determined. The central metal atom in each case enjoys octahedral co-ordination, whilst the other two silicon atoms in the former compound have usual tetrahedral geometry. The two Si$_{pc}$—O—Si bond angles are quite large, 157.8 and 156.6°. The kinetic stabilities of pcGeX$_2$ (X = Cl or OH) derivatives have been examined, the hydroxy-compound being less stable than the chloride.[688] Dichlorotetraphenyltin(IV) complexes give resolvable quadrupole splittings in their Mössbauer spectra.[689]

The thermal decomposition of Si$_2$N$_2$O in the temperature range 1300—1700 °C under vacuum or in atmospheres of N$_2$, Ar, or He always occurs according to:

$$3Si_2N_2O \to Si_3N_4 + 3SiO + N_2$$

The first stage of the decomposition appears to be the formation of Si$_3$N$_4$ and silica, which then reacts with silicon from the dissociation of Si$_3$N$_4$ to form SiO.[690] The preparation of the two isotypic monoclinic phases Ca$_{10+x}$Si$_{12-2x}$M$_{16}$ (M = P or As) has been described. The materials may only be obtained in a pure state when

---

[684] M. A. Pudovik, M. D. Medvedeva, L. K. Kibardina, and A. N. Pudovik, *J. Gen. Chem. (U.S.S.R.)*, 1975, **45**, 924.
[685] E. Wenschuh, W. D. Riedmann, L. Korecz, and K. Burger, *Z. anorg. Chem.*, 1975, **413**, 143.
[686] J. R. Mooney, C. K. Choy, K. Knox, and M. E. Kenney, *J. Amer. Chem. Soc.*, 1975, **97**, 3033.
[687] H. J. Hecht and P. Luger, *Acta Cryst.*, 1974, **B30**, 2843.
[688] B. D. Berezin and A. S. Akopov, *J. Gen. Chem. (U.S.S.R.)*, 1974, **44**, 1047.
[689] N. W. G. Debye and A. D. Adler, *Inorg. Chem.*, 1974, **13**, 1037.
[690] P. Lortholary and M. Billy, *Bull. Soc. chim. France*, 1975, 1057.

0.66 < $x$ < 2.5, and they have structures which are related to the NaCl type, with slight distortion.[691]

A microwave study of silylphosphine, $H_3SiPH_2$, has yielded accurate structural parameters, a value for the barrier to hindered internal rotation (1.535 ± 0.040 kcal mol$^{-1}$), and the dipole moment (0.59 ± 0.50 D). Partial double bonding of the $(p-d)\pi$ type was discounted.[692] The gas-phase i.r. and liquid Raman spectra of $F_3SiPH(D)_2$ have been assigned on the basis of a normal-co-ordinate analysis.[693] Maya and Burg have investigated the reactions of $H_3SiP(CF_3)_2$ with a variety of reagents. These are summarized in Table 22.[694] Schumann and Dumont

**Table 22** Reactions of $(CF_3)_2PSiH_3$

| Reactant (mmol) | Conditions | Products (mmol, where known) |
|---|---|---|
| $BCl_3$ (0.75) | 20 h at 25 °C | 0.4 $BF_3$, 0.02 $SiH_3Cl$, 0.73 $H_2SiCl_2$, 0.13 $(CF_3)_2PH$, 0.125 $P_2(CF_3)_4$, trace $(CF_3)_2PH$; $(H_2PBF_2)_x$ empirical formula of residue by difference |
| $B(OMe)_3$ (0.15 used of 0.50) | 24 h at 25 °C nil, 24 h at 50 °C | |
| $MeOBCl_2$ (0.45) | Fast at <25 °C | 0.2 $SiH_4$, 0.24 $(CF_3)_2PH$, 0.26 $(CF_3)_2PMe$, non-volatiles |
| $Me_2PCl$ (0.46) | Fast at <25 °C | 0.36 $(CF_3)_2PMe$, 0.3 $SiH_3Cl$, 0.2 $H_2SiCl_2$, <0.05 $(CF_3)_2PH$ |
| $(CF_3)_2PCN$ (0.5) | 28 h at 25 °C | Yellow ppt, 0.12 $Me_2P-P(CF_3)_2$, nil $SiH_3Cl$, 0.45 $H_2SiCl_2$, 0.07 $(CF_3)_2PH$, 0.02 $P_2(CF_3)_4$ |
| $PF_3$ | 1 h at 105 °C | 0.22 $P_3(CF_3)_4$, 0.34 $SiH_3CN$, 0.17 $(CF_3)_2PH$ |
| $PF_5$ | Fast at <25 °C | No reaction |
| MeCN (0.5) | 3 h at 65 °C | Obsd. $(CF_3)_2PH$, $(CF_3)_2PF$, $H_2SiF_2$; minor unidentified products |
| MeNC (0.45) | Fast at <25 °C | Obsd. $P_2(CF_3)_4$, $SiH_4$ |
| $BrSiF_3$ (0.07 used of 0.5) | 16 h at 50 °C nil, 24 h at 120 °C | Brown solids, 0.1 $SiH_4$, 0.03 $(CF_3)_2PH$, 0.06 $P_2(CF_3)_4$ Total of $BrSiH_3$, $(CF_3)_2PH$, and $SiF_4$ amounts to 0.13 |

have used exchange reactions between $Me_3SiPBu_2^i$ and (methyl)chloro-silanes[695] or -germanes[696] to obtain the mixed chlorophosphines $Bu_2^iPMMe_{3-n}Cl_n$ ($n$ = 1—3). $Ph_2PGeMe_2Cl$ and $(Me_2ClGe)_3P$ are similarly prepared from $Me_2GeCl_2$ and $Me_3SiPPh_2$ or $(Me_3Si)_3P$.[696] A new high-yield preparation of $(Me_3Si)_3P$ has been devised. White phosphorus is added to Na–K alloy in monoglyme, forming 'Na$_3$P–K$_3$P' *in situ*, to which is added $Me_3SiCl$.[697] The chemistry of germanium phosphines has been extensively studied. $R_2Ge(PH_2)_2$ (R = Me or Et) derivatives condense at 120—140 °C to give high yields of $(R_2Ge)_6P_4$ compounds. An X-ray crystal structural examination of the methyl compound shows it to possess tetrahedral symmetry, with the four phosphorus atoms at the apices of a tetrahedron and the $Me_2Ge$ situated along the six edges. The compounds $(Me_2GePH_2)_2PH$ and $(Me_2GePH_2)_3P$ were also characterized as intermediates in the condensation.[698] The new compounds $MeGe(PH_2)_2H$ and $MeGe(PH_2)_3$ have been prepared

---

[691] M. Hamon, J. Guyader, and J. Lang, *Rev. Chim. minérale*, 1975, **12**, 1.
[692] R. Varma, K. R. Ramaprasad, and J. F. Nelson, *J. Chem. Phys.*, 1975, **63**, 915.
[693] R. Demuth, *Spectrochim. Acta*, 1975, **37A**, 238.
[694] L. Maya and A. B. Burg, *Inorg. Chem.*, 1975, **14**, 698.
[695] H. Schumann and W. W. Dumont, *Z. anorg. Chem.*, 1975, **418**, 259.
[696] H. Schumann and W. W. Dumont, *Chem. Ber.*, 1975, **108**, 2261.
[697] G. Becker and W. Hölderich, *Chem. Ber.*, 1975, **108**, 2484.
[698] A. R. Dahl, A. D. Norman, H. Shenav, and R. Schaeffer, *J. Amer. Chem. Soc.*, 1975, **97**, 6364.

from MeGeCl$_3$ and LiAl(PH$_2$)$_4$ in glyme solvents. Phosphino- and hydrido-groups on germanium undergo rapid redistribution. Thus, solutions of H$_3$GePH$_2$ contain GeH$_2$(PH$_2$)$_2$, GeH(PH$_2$)$_3$, and GeH$_4$, and the Me$_2$Ge(PH$_2$)$_x$H$_{2-x}$ systems, at equilibrium, contain their respective redistribution species in a ratio close to that for statistical sorting of PH$_2$ and hydrido ligands.[699] Me$_3$GePH$_2$ and Me$_2$Ge(PH$_2$)$_2$ react rapidly with gaseous oxygen to give a mixture of oxidation products from which the new phosphonoxygermoxanes (Me$_3$GeO)$_2$P(O)H and [Me$_2$GeOP(O)HO]$_2$ could be isolated.[700] At ambient temperatures, Me$_2$Ge(PH$_2$)$_2$, Me$_2$Ge(PH$_2$)H and Me$_3$GePH$_2$ are hydrolysed to phosphine and germoxanes. The hydrolysis of Me$_2$Ge(PH$_2$)$_2$ is complex, owing to the lability of the Ge—H groups. With excess water and long reaction times (>2 hours) phosphine, Me$_2$GeH$_2$, and (Me$_2$GeO)$_{3,4}$ are formed, but with a deficiency of water and short reaction times (20—100 minutes) (Me$_2$GeH)$_2$O is formed in addition to the other products.[701] Me$_3$SiPEt$_2$ and Me$_n$H$_{3-n}$SiPMe$_2$ form adducts with AlCl$_3$ which eliminate Me$_3$SiCl on heating. The former compound reacts with H$_2$AlCl and HAlMe$_2$ to give (HClAl—PMe$_2$)$_3$ and (Me$_2$Al—PMe$_2$)$_4$, respectively. Redistribution accompanies complex formation of Me$_n$H$_{3-n}$SiPHMe and Me$_n$H$_{3-n}$SiPH$_2$ with AlCl$_3$, and the complexes (Me$_n$ H$_{3-n}$Si)$_2$PMe=AlCl$_3$ and (Me$_n$H$_{3-n}$Si)$_3$P=AlCl$_3$ result.[702] (Me$_3$Sn)$_3$P displaces bicyclo[2,2,1]hepta-2,5-diene from (C$_7$H$_8$)M(CO)$_4$ (M = Cr, Mo, or W) complexes to give [(Me$_3$Sn)$_3$P]$_2$M(CO)$_4$ complexes. Both cis- and trans-isomers are formed.[703]

**Derivatives of Silicon, Germanium, Tin, and Lead containing Bonds to Main-group Elements.**—An electron-diffraction study of Si$_2$F$_6$ has shown that the two SiF$_3$ residues are twisted 34.6° with respect to each other.[704] The thermolysis of Me$_3$SiSiMe$_3$ has been studied in a stirred-flow system in the temperature range 770—872 °C. The initial process in the decomposition is the homolytic fission of the Si—Si bond, and a value of 337 kJ mol$^{-1}$ for $D$(Me$_3$Si—SiMe$_3$) was deduced.[705] The stability of the Si—Si bond towards oxidation by molecular oxygen or Me$_3$SiOOSiMe$_3$ is greatly reduced when the linkage is part of a strained ring system or if the silicon atoms are substituted by fluorine atoms. The main product is usually the corresponding disiloxane.[706] Halogenodisilane derivatives are obtained by cleavage of phenyl groups from methylphenyldisilanes, using HCl or HBr.[707] Cleavage of methyl groups from silicon in di-, tri-, and tetra-silane may be accomplished using SbCl$_5$. However, the process was only a useful synthetic method for the preparation of Me$_5$Si$_2$Cl from Me$_5$Si$_2$, reaction in all other cases yielding inseparable mixtures.[708] The methoxy-substituted disilane [Me(MeO)$_2$Si]$_2$ is conveniently obtained by photolysis of Me(MeO)$_2$SiH in the gas phase.[709]

---

[699] A. R. Dahl, C. A. Heil, and A. D. Norman, Inorg. Chem., 1975, **14**, 1095.
[700] A. R. Dahl and A. D. Norman, Inorg. Chem., 1975, **14**, 1093.
[701] A. R. Dahl, C. A. Heil, and A. D. Norman, Inorg. Chem., 1975, **14**, 2562.
[702] G. Fritz and R. Emül, Z. anorg. Chem., 1975, **416**, 19.
[703] H. Schumann and J. Opitz, J. Organometallic Chem., 1975, **85**, 357.
[704] D. W. H. Rankin and A. Robertson, J. Mol. Structure, 1975, **27**, 438.
[705] I. M. T. Davidson and A. V. Howard, J.C.S. Faraday I, 1975, **71**, 69.
[706] K. Tamao, M. Kumada, and T. Tokahashi, J. Organometallic Chem., 1975, **94**, 367.
[707] E. Hengge, G. Bauer, E. Brandstätter, and G. Kollmann, Monatsh., 1975, **106**, 887.
[708] E. Carberry, T. Keene, and J. Johnson, J. Inorg. Nuclear Chem., 1975, **37**, 839.
[709] M. E. Childs and W. P. Weber, J. Organometallic Chem., 1975, **86**, 169.

ClSi[Si(OMe)$_3$]$_3$, from the action of chlorine on HSi[Si(OMe)$_3$]$_3$, may be transformed into the perhalogenated tetrasilanes ClSi(SiX$_3$)$_x$ (X = F or Cl), using the corresponding boron trihalide.[710] Treatment of Ph$_6$Ge$_2$ with either HCl or HBr produces the 1,1,2,2-tetrahalogeno-1,2-diphenyldigermane in almost quantitative yield, but under the same reaction conditions the Si—Ge bond of Ph$_3$SiGePh$_3$ undergoes scission.[711] Cocondensation of thermally evaporated germanium vapour with trimethylsilane yields Me$_3$SiGeH$_2$SiMe$_3$ in addition to Me$_6$Si$_2$ and unchanged Me$_3$SiH. The reaction presumably proceeds *via* insertion of Ge into the Si—H bonding, forming an intermediate germylene HGeSiMe$_3$, which then inserts into a second Si—H bond.[712]

In the trialkyltin-halide-catalysed decomposition of Me$_6$Sn$_2$ to Me$_4$Sn and 'Me$_2$Sn', the halide itself is the electrophilic reagent, not the dissociated cation, as previously suggested. The decomposition was thought to occur *via* an $S_E$i mechanism involving single-stage cleavage of both the Sn—C and Sn—Sn bonds[713] (Scheme 9). Me$_6$Sn$_2$ reacts rapidly with HgCl$_2$ in methanol to yield Hg

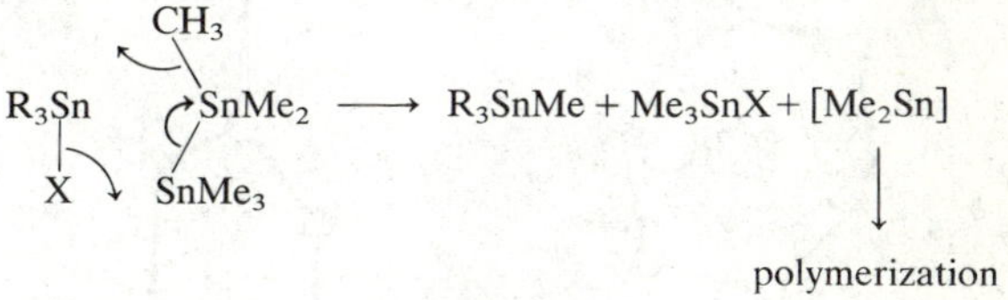

**Scheme 9**

and Me$_3$SnCl. With alkylmercury salts the reaction is complex, giving tetraalkyltin and/or dialkylmercury, depending on the reactivity of the alkylmercury salt. An electrophilic mechanism for these reactions has also been suggested which involves the intermediacy of stannylmercurial species.[714] The thermal decomposition of Me$_6$Pb$_2$ is first-order in benzene and toluene solutions, and dissociation to afford Me$_2$Pb (as a transient intermediate) and Me$_4$Pb is the proposed first step.[715] The reactions of Me$_6$Pb$_2$ with 7,7,8,8-tetracyanoquinodimethane (tcnq) and tetracyanoethylene (tcne) afford the divalent lead derivatives Pb(tcnq)$_2$ and Pb(tcne)$_2$ respectively.[716]

Recrystallization of the initial product from the reaction of triphenylarsine with Ph$_2$Sn(DMSO)$_3$(NO$_3$)$_2$ yields crystals of Sn(NO$_3$)(SnPh$_3$)$_3$, the structure of which is illustrated in Figure 21. The central tin atom is five-co-ordinated by three Ph$_3$Sn groups and a symmetrically bidentate nitrato-ligand.[717]

[Me$_3$SiCH$_2$]$_3$SnLi has been prepared by Hg–Li exchange from {[Me$_3$SiCH$_2$]$_3$Sn}$_2$Hg and lithium metal in THF. If the reaction is performed in benzene, only the corresponding hexa-alkylditin compound is obtained. The hexa-alkylditin compound is also formed when the lithium reagent is treated with

---

[710] F. Höfler and R. Jannach, *Inorg. Nuclear Chem. Letters*, 1975, **11**, 743.
[711] F. Höfler and E. Brandstätter, *Monatsh.*, 1975, **106**, 893.
[712] R. T. Conlin, S. H. Lockhart, and P. P. Gaspar, *J.C.S. Chem. Comm.*, 1975, 825.
[713] D. C. McWilliam and P. R. Wells, *J. Organometallic Chem.*, 1975, **85**, 165.
[714] D. C. McWilliam and P. R. Wells, *J. Organometallic Chem.*, 1975, **85**, 335.
[715] D. P. Arnold and P. R. Wells, *J.C.S. Chem. Comm.*, 1975, 642.
[716] A. W. S. Dick, A. K. Holliday, and R. J. Puddephatt, *J. Organometallic Chem.*, 1975, **96**, C41.
[717] G. Pelizzi, *J. Organometallic Chem.*, 1975, **87**, C1.

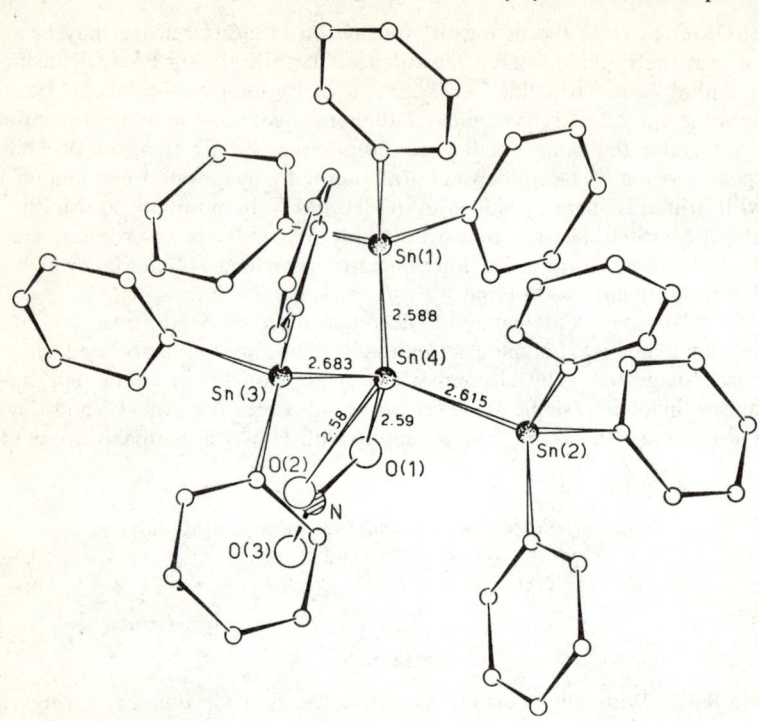

**Figure 21** *The structure of* [(Ph₃Sn)₃Sn(NO₃)]
(Reproduced by permission from *J. Organometallic Chem.*, 1975, **87**, C1)

Bu$^t$Br, C$_6$H$_{11}$Br, ClCH$_2$CO$_2$Et, or C$_6$F$_5$Cl, but reaction with Bu$^t$Cl results in the formation of the corresponding tin hydride, isobutene, and lithium chloride. 'Normal' substitution products are obtained with EtBr and Me$_3$SiCl.[718] The reactions with peroxy-compounds have also been investigated.[719] The reaction of silyl bromide with (Me$_3$Si)$_2$Hg affords solutions of Me$_3$SiHgSiH$_3$, from which Me$_3$SiSiH$_3$ is slowly evolved over a period of a few days. Me$_3$SiHgGeH$_3$ may be formed similarly,[720] whilst (ClMe$_2$Si)$_2$Hg is obtained as white crystals from the reaction of Me$_2$ClSiH and Bu$_2^t$Hg.[721] Hexa-alkylditins are obtained from the reaction of triorganotin compounds with (Me$_3$Si)$_2$Hg whereas diorganotin compounds afford the corresponding (R$_2$Sn)$_n$ polymers.[722] With hydroxy-compounds, (Ph$_3$Si)$_2$Hg reacts to yield triphenylsilyl ethers, hydrogen, and mercury as the

---

[718] O. A. Kruglaya, T. A. Basalgina, G. S. Kalinina and N. S. Vyazankin, *Zhur. obshchei Khim.*, 1974, **44**, 1068.
[719] G. S. Kalinina, T. A. Basalgina, N. S. Vyazankin, G. A. Razuvaev, V. A. Yablokov, and N. V. Yablokova, *J. Organometallic Chem.*, 1975, **96**, 213.
[720] S. Cradock E. A. V. Ebsworth, N. S. Hosmane, and K. M. Mackay, *Angew. Chem. Internat. Edn.*, 1975, **14**, 167.
[721] T. F. Schaaf, R. R. Kao, and J. P. Oliver, *Inorg. Chem.*, 1975, **14**, 2288.
[722] T. N. Mitchell, *J. Organometallic Chem.*, 1975, **92**, 311.

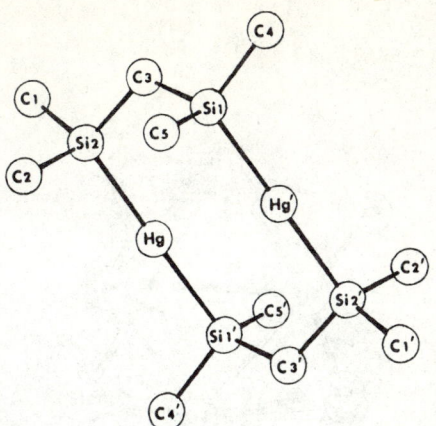

**Figure 22** The structure of 2,2,4,4,6,6,8,8-octamethyl-2,4,6,8-tetrasila-1,5,-dimercuracyclo-octane
(Reproduced by permission from *J. Amer. Chem. Soc.*, 1975, **97**, 6261)

major products.[723] Photolysis of (Me$_3$M)$_2$Hg (M = Si or Ge) results in the formation of Me$_3$M' radicals, which combine to give disilane or digermane. Me$_3$MHg· compounds [(C$_6$F$_5$)$_3$Sn]$_2$Hg and [(C$_6$F$_5$)$_3$SnHgGe(C$_6$F$_5$)$_3$]. Both are unusually thermally stable and demercurate slowly at temperatures >80—90 °C.[725] The cyclic compound 2,2,4,4,6,6,8,8-octamethyl-2,4,6,8-tetrasila-1,5-mercuracyclo-octane is a centrosymmetric molecule containing two Si—Hg—Si moieties joined by methylene bridges (Figure 22). Lithium tetrakis(dimethylphenylsilyl)mercurate(II), Li$_2$(Me$_2$PhSi)$_4$Hg, may be described as an isolated contact ion pair in which the anion comprises a mercury atom tetrahedrally co-ordinated by four Me$_2$PhSi groups, with the two lithium cations entrapped in symmetry-related cages comprised of the mercury, three silicon atoms, and five carbon atoms (Figure 23).[726] (Et$_3$Ge)$_2$Hg reacts in benzene at 20 °C with (C$_5$H$_5$)$_2$TiCl$_2$ to give a quantitative yield of mercury and Et$_3$GeCl and (C$_5$H$_5$)$_2$TiCl. The corresponding reaction with (Et$_3$Ge)$_2$Cd yields the complex [(C$_5$H$_5$)$_2$TiCl$_2$,Cd(GeEt$_3$)$_2$], which decomposes in toluene at 20 °C to yield cadmium metal, Et$_3$GeCl, and [(C$_5$H$_5$)$_2$TiCl(GeEt$_3$)].[727] Mössbauer isomer-shift values for triphenyltin derivatives of zinc and cadmium suggest a high *s* character for the metal–metal bonds.[728]

The results of an electron-diffraction study of 1- and 2-silylpentaborane are consistent with extended Hückel calculations regarding the stabilities of the compounds. Thus, the predicted more-stable isomer 1-SiH$_3$B$_5$H$_8$ has a Si—B

---

[723] C. Eaborn, R. A. Jackson, and M. T. Rahman, *J. Organometallic Chem.*, 1975, **84**, 15.
[724] M. Lehnig, F. Werner, and W. P. Neumann, *J. Organometallic Chem.*, 1975, **97**, 375.
[725] M. N. Bochkarev, S. P. Korneva, L. P. Maiorova, V. A. Kuznetsov, and N. S. Vyazankin, *J. Gen. Chem. (U.S.S.R.)*, 1974, **44**, 293.
[726] M. J. Albright, T. F. Schaaf, W. M. Butler, A. K. Horland, M. D. Glick, and J. P. Oliver, *J. Amer. Chem. Soc.*, 1975, **97**, 6261.
[727] G. A. Razuvaev, V. N. Latyaeva, L. I. Vishinskaya, V. T. Bytchkov, and G. A. Vasilyeva, *J. Organometallic Chem.*, 1975, **87**, 93.
[728] R. Barbieri, L. Pellerito, N. Bertazzi, G. Alonzo, and J. G. Noltes, *Inorg. Chim. Acta*, 1975, **15**, 201.

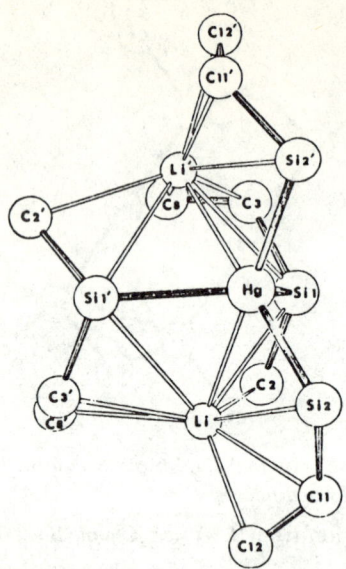

**Figure 23** *The structure of lithium tetrakis(dimethylphenylsilyl)mercurate*(II)
(Reproduced by permission from *J. Amer. Chem. Soc.*, 1975, **97**, 6261)

bond distance some 0.025 Å shorter than the 2-isomer; this is rationalized in terms of increased $\pi$ interactions with the boron cage for the apical isomer.[729] Anions derived from 1- and 2-trimethyl-silyl- and -germyl-pentaboranes by radicals have a very short, if any, lifetime.[724] Photolysis of mixtures of $(C_6F_5)_3SnBr$ with either $(Et_3Ge)_2Hg$ or $EtGeHg(C_6F_5)_3$ yields the perfluorinated proton abstraction react with $Me_2BCl$ to produce 1- and 2-[$Me_3M$]-$\mu$-[$Me_2B$]$B_5H_7$ derivatives.[730] Lithium pentaborane reacts with dichloro-silanes and -germanes to afford $(B_5H_8)MRMe$ (M = Si or Ge; R = H or Me) compounds as mixtures of isomers.[731] Flash thermolysis of 1,2-bis(trimethylsilyl)pentaborane(9) gives a number of *C*-silyl derivatives, including 1-$H_3Si$-1,5-$C_2B_3H_3$, 1-$MeH_2Si$-1,5-$C_2B_3H_4$, 2-Me-1-($H_3Si$)-1,5-$C_2B_3H_3$, and an equilibrium mixture of 2- and 4-methyl derivatives of $CB_5H_7$.[732]

Tris(trimethylstannyl)bismuth has been prepared by the hydrostannolysis of $Et_3Bi$ by $Me_3SnH$. The reaction of $Me_3SnCl$ with $Na_3Bi$ in liquid ammonia yields only $Me_6Sn_2$. With metal carbonyl derivatives such as $M(CO)_5THF$ (M = Cr, Mo, or W) and $Ni(CO)_4$, the stannylbismuthine affords the complexes $M(CO)_5,Bi(SnMe_3)_3$ and $Ni(CO)_3,Bi(SnMe_3)_3$.[733]

**Derivatives of Silicon, Germanium, Tin, and Lead containing Bonds to Transition Metals.**—The substitution of organo-silicon, -germanium, -tin, and -lead halides

---

[729] J. D. Wieser, D. C. Moody, J. C. Huffman, R. L. Hilderbrandt, and R. Schaeffer, *J. Amer. Chem. Soc.*, 1975, **97**, 1074.
[730] D. F. Gaines and J. Ullman, *J. Organometallic Chem.*, 1975, **93**, 281.
[731] D. F. Gaines and J. Ullman, *Inorg. Chem.*, 1974, **13**, 2792.
[732] J. B. Leach, G. Oates, S. Tang, and T. Onak, *J.C.S. Dalton*, 1975, 1018.
[733] H. Schumann and H. J. Breunig, *J. Organometallic Chem.*, 1975, **87**, 83.

by transition-metal carbonyl anions has been employed to synthesize [($\eta$-$C_7H_7$)Mo(CO)$_2$(MPh$_3$)] (M = Ge or Sn),[734] [($C_5H_5$)Mo(CO)$_2$(CNMe)CR$_3$] (MR$_3$ = SnMe$_3$, GeMe$_3$, or PbPh$_3$),[735] [($C_5H_5$)(CO)$_2$Fe(SiMe$_n$Cl$_{2-n}$CH$_2$X)] (X = Cl, $n = 0$—1; X = Br, $n = 2$);[736] [($C_5H_5$)(CO)$_n$M-SiR$_2$CH=CH$_2$] (M = Fe, $n = 2$; M = Mo or W, $n = 3$; R = Me or Cl),[737] R$_{4-n}$Pb[Mn(CO)$_4$(PPh$_3$)]$_n$, and Ph$_2$Sn[Mn(CO)$_4$(PPh$_3$)],[738] [MeH$_2$GeMn(CO)$_5$],[739] and [R$_3$SnMn(CO)$_4$(EPh$_3$)] (E = P, As, or Sb).[740] In addition, the nucleophilic substitution reactions of Sn—Co, Sn—Mo, Sn—Mn, and Sn—Re bonded compounds by other transition-metal carbonyl anions have been investigated.[741] Other methods by which the synthesis of Group IV metal–transition metal bonds has been accomplished include the reactions of SnCl$_2$, HGeCl$_3$, or HSiCl$_3$ with [($\eta$-$C_7H_7$)Mo(CO)$_2$Cl], giving [($\eta$-$C_7H_7$)Mo(CO)$_2$MCl$_3$] (M = Si, Ge, or Sn) complexes,[734] the photolysis of [PhMeSnFe(CO)$_4$]$_2$, yielding [(MePhSn)$_2$Fe$_2$(CO)$_7$],[742] the reaction of [($C_5H_5$)$_2$V] with [(R$_3$Ge)$_2$Cd], giving [($C_5H_5$)$_2$VGeR$_3$] (R = Me or Ph),[743] the reactions of organotin acetates and chlorides with Co$_2$(CO)$_8$, giving tetracarbonyl-cobalt–tin derivatives,[744] and of Me$_2$MH$_2$ (M = Ge or Sn) with Co$_2$(CO)$_8$, giving [(OC)$_3$Co($\mu$-MMe$_2$)Co(CO)$_3$] compounds,[745] the reaction of ($C_5H_5$)$_2$Sn with [HMn(CO)$_5$], giving {H[Mn(CO)$_5$]$_2$Sn}$_2$,[746] the reaction of GeCl$_4$ with [(diars)Mn(CO)$_3$Cl] to give fac-[(diars)Mn(CO)$_3$GeCl$_3$], and with [(diphos)Mn(CO)$_3$Br] to give mer-[(diphos)Mn(CO)$_3$GeCl$_3$],[747] and of SnCl$_4$ with [(L$_2$)M(CO)$_4$] (M = Mo or W; L$_2$ = diphos or diars), giving [(L$_2$)M(CO)$_4$SnCl$_3$]-[SnCl$_5$OH$_2$],[748] the addition of Ph$_3$MCl (M = Ge, Sn, or Pb) to Cs$_2$M'(CO)$_5$ (M' = Cr, Mo, or W), producing [Ph$_3$MM'(CO)$_5$]$^-$ anions,[749] the reactions of R$_3$MLi (M = Si, Ge, Sn, or Pb) with Et$_4$N[ClM'(CO)$_5$] salts, giving again [R$_3$MM'(CO)$_5$]$^-$ anions,[750] or with Et$_4$N[HFe$_3$(CO)$_{11}$], giving Et$_4$N[Ph$_3$MFe(CO)$_4$] (M = Si, Ge, or Sn) salts,[751] the reaction of Ru$_3$(CO)$_{12}$ with Me$_3$SnCH$_2$NMe$_2$, giving [(Me$_3$Sn)$_2$Ru(CO)$_4$],[752] the reaction of R$_3$SiH and R$_3$SnH with [Pt(CO$_3$)PMe$_2$Ph)$_2$], [Pt(PMe$_2$Ph)$_4$], or [PtHCl(PMe$_2$Ph)$_2$], giving platinum–silicon and

---

[734] E. E. Isaacs and W. A. G. Graham, *Canad. J. Chem.*, 1975, **53**, 975.
[735] R. D. Adams, *J. Organometallic Chem.*, 1975, **88**, C38.
[736] C. Windus, S. Sujishi, and W. P. Giering, *J. Organometallic Chem.*, 1975, **101**, 279.
[737] W. Malisch and P. Panster, *Chem. Ber.*, 1975, **108**, 2554.
[738] W. Schubert, H. J. Haupt, and F. Huber, *Z. anorg. Chem.*, 1975, **412**, 77.
[739] B. W. L. Graham, K. M. Mackay, and S. R. Stobart, *J.C.S. Dalton*, 1975, 475.
[740] S. Onaka and H. Sano, *Bull. Chem. Soc. Japan*, 1975, 258.
[741] A. N. Nesmeyanov, N. E. Kolobova, V. N. Khandozhko, and K. N. Anisimov, *J. Gen. Chem. (U.S.S.R.)*, 1974, **44**, 298.
[742] T. J. Marks and G. W. Grynkewich, *J. Organometallic Chem.*, 1975, **91**, C9.
[743] G. A. Razuvaev, V. T. Bychkov, L. I. Vyshinskaya, V. N. Latyaeva, and N. N. Spiridonova, *Doklady Akad. Nauk S.S.S.R.*, 1975, **220**, 854.
[744] A. N. Nesmeyanov, K. N. Anisimov, N. E. Kolobova, and V. N. Khandozhko, *J. Gen. Chem. (U.S.S.R.)*, 1974, **44**, 1265.
[745] R. D. Adams, F. A. Cotton, W. R. Cullen, D. L. Hunter, and L. Mihichuk, *Inorg. Chem.*, 1975, **14**, 1395.
[746] K. D. Bos, E. J. Bulten, J. G. Noltes, and A. L. Spek, *J. Organometallic Chem.*, 1975, **92**, 33.
[747] W. R. Cullen, F. W. B. Einstein, R. K. Pomeroy, and P. L. Vogel, *Inorg. Chem.*, 1975, **14**, 3017.
[748] W. R. Cullen and R. K. Pomeroy, *Inorg. Chem.*, 1975, **14**, 939.
[749] J. E. Ellis, S. G. Hentges, D. G. Kalina, and G. P. Hagen, *J. Organometallic Chem.*, 1975, **97**, 79.
[750] E. E. Isaacs and W. A. G. Graham, *Canad. J. Chem.*, 1975, **53**, 467.
[751] E. E. Isaacs and W. A. G. Graham, *J. Organometallic Chem.*, 1975, **85**, 237.
[752] M. R. Churchill, B. G. De Boer, F. J. Rotella, E. W. Abel, and R. J. Rowley, *J. Amer. Chem. Soc.*, 1975, **97**, 7158.

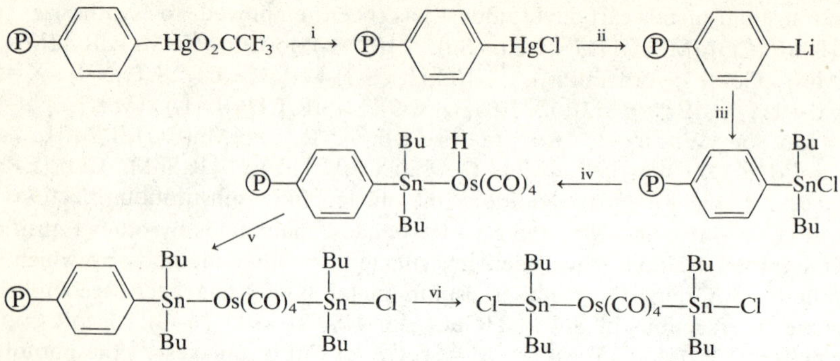

## Scheme 10

Ⓟ represents the macroreticular styrene–divinylbenzene copolymer template
Reagents: i, $Cl^-$; ii, $Bu^nLi$; iii, $Bu_2SnCl_2$; iv, $H_2Os(CO)_4$–$Et_2NH$; v, $Bu_2SnCl_2$–$Et_2NH$; vi, HCl

**Scheme 10**

platinum–tin complexes,[753,754] and the reactions of $Me_3MH$ (M = Si, Ge, Sn, or Pb) with $[Rh(C_5H_5)(CO)_2]$, both thermally and photochemically, to afford Rh—M compounds such as $[Rh(C_5H_5)(CO)(MMe_3)_2]$ and $[Rh(CO)_3(SnMe_3)_3]$.[755] Platinum–tin compounds may also be obtained from the reaction of $Me_3SnH$ with alkylplatinum compounds. $Me_3SnH$ displaces alkane from $[PtR_2(phos)]$ {phos = $(Ph_2P)_2CH_2$; R = Me or Et} at room temperature or at 50 °C when R = Ph. The $d^0$ product $[PtH(SnMe_3)_3(phos)]$ reversibly dissociates in solution to $[Pt(SnMe_3)_2(phos)]$. $Me_3SiH$ or $Me_3GeH$ displace only 1 mole equivalent of alkane, forming $[Pt(MMe_3)(Me)(phos)]$ (M = Si or Ge) complexes. $Me_3SnH$ displaces $Ph_6Pb_2$ from the lead complex $[Pt(PbPh_3)_2(phos)]$.[756] Burlitch has devised an elegant method for the synthesis of $[(ClBu_2Sn)_2Os(CO)_4]$ according to Scheme 10.[757]

The structures of several transition-metal derivatives have been determined either by gas-phase electron diffraction or by X-ray crystallography. The structures of the pentacarbonylmanganese derivatives $[H_3SiMn(CO)_5]$,[758] $[H_3GeMN(CO)_5]$,[758] $[F_3SiMn(CO)_5]$,[759] and $[Cl_3SnMn(CO)_5]$[760] are similar, with distorted tetrahedral co-ordination for the Group IV metal and distorted octahedral geometry for the manganese. He (I) photoelectron data for the methyl, silyl, and germyl pentacarbonylmanganese compounds indicate that the main effect responsible for changes in the metal–metal bond in these compounds is the change in σ-acceptor character of the Group IV metal residue.[758] The trichlorogermyl group is present in the complexes $[(\eta\text{-}C_6H_6)Ru(CO)(GeCl_3)_2]$,[761] and $fac$-$[(diphos)Mn(CO)_3GeCl_3]$,[747] whilst the trichlorostannyl group also occurs in the ionic complex $[(diphos)Mo(CO)_4SnCl_4]^+[SnCl_5OH_2]^-, C_6H_6$.[748,762] The anion

---

[753] C. Eaborn, A. Pidcock, and B. R. Steele, *J.C.S. Dalton*, 1975, 809.
[754] C. Eaborn, T. N. Metham, and A. Pidcock, *J.C.S. Dalton*, 1975, 2212.
[755] R. Hill and S. A. R. Knox, *J.C.S. Dalton*, 1975, 2622.
[756] F. Glockling and R. J. Pollock, *J.C.S. Dalton*, 1975, 497.
[757] J. M. Burlitch and R. C. Winterton *J. Amer. Chem. Soc.*, 1975, **97**, 5605.
[758] D. W. H. Rankin and A. Robertson, *J. Organometallic Chem.*, 1975, **85**, 225.
[759] D. W. H. Rankin, A. Robertson, and R. Serp, *J. Organometallic Chem.*, 1975, **88**, 191.
[760] S. Onaka, *Bull. Chem. Soc. Japan*, 1975, **48**, 319.
[761] L. Y. Y. Chan and W. A. G. Graham, *Inorg. Chem.*, 1975, **14**, 1778.
[762] F. W. B. Einstein and J. S. Field, *J.C.S. Dalton*, 1975, 1628.

in the latter complex has the expected distorted octahedral geometry. Woodward and his co-workers have determined the structures of the ruthenium complexes [(μ-Me₃Si-cyclo-C₇H₆)(CO)₅(Me₃Si)Ru₂],[763] [Ru₂(CO)₅{Me₂SiCH₂CH₂SiMe₂(C₈H₈)},[764] [Ru₂(SiMe₃)(CO)₄(C₈H₈SiMe₃)],[764] and [(Me₃Sn)₂(CO)₈Ru₂],[765] which all possess a linear M—Ru—Ru—M (M = Si or Sn) backbone and again a distorted tetrahedrally co-ordinated RuMR₃ group. The structures of these complexes are illustrated in Figures 24—27, and show other features of the molecules. The dimethylgermyl group functions as a bridging group in the ring compound [(Me₂Ge)Mn₂(CO)₉] (Figure 28),[766] whilst the geometry at tin in the complexes

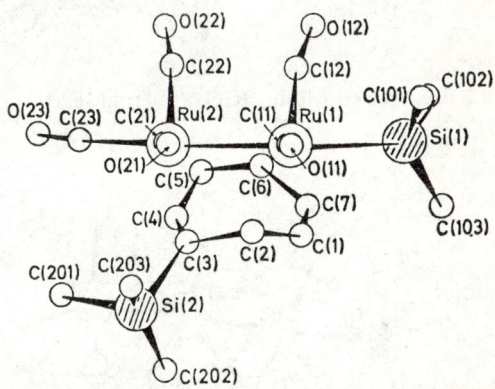

**Figure 24** *The structure of* [(μ-Me₃Si-cyclo-C₇H₆)(CO)₅(Me₃Si)Ru₂] (Reproduced from *J.C.S. Dalton*, 1975, 59)

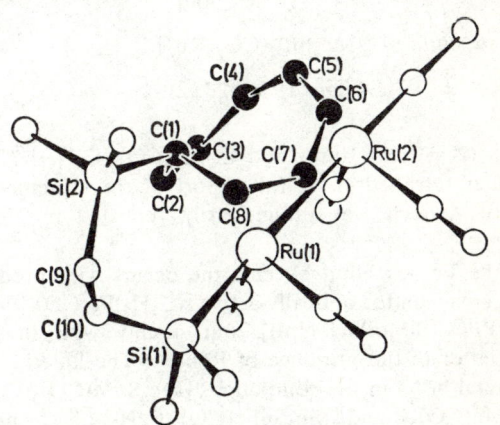

**Figure 25** *The structure of* [Ru₂(CO)₅{Me₂SiCH₂CH₂SiMe₂(C₈H₈)}] (Reproduced from *J.C.S. Chem. Comm.*, 1975, 828)

---

[763] J. Howard and P. Woodward, *J.C.S. Dalton*, 1975, 59.
[764] J. D. Edwards, R. Goddard, S. A. R. Knox, R. J. McKinney, F. G. A. Stone, and P. Woodward, *J.C.S. Chem. Comm.*, 1975, 828.
[765] J. A. K. Howard, S. C. Kellett, and P. Woodward, *J.C.S. Dalton*, 1975, 2332.
[766] K. Triplett and M. D. Curtis, *J. Amer. Chem. Soc.*, 1975, **97**, 5747.

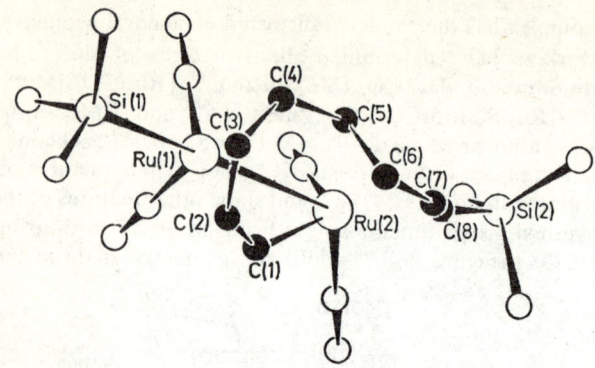

**Figure 26** The structure of $[Ru_2(SiMe_3)(CO)_4(C_8H_8SiMe_3)]$
(Reproduced from *J.C.S. Chem. Comm.*, 1975, 828)

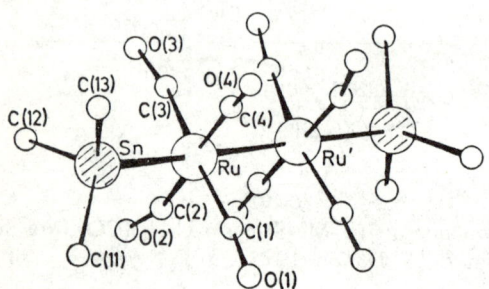

**Figure 27** The structure of $[(Me_3Sn)_2(CO)_8Ru_2]$
(Reproduced from *J.C.S. Dalton*, 1975, 2332)

$X_2Sn[Mn(CO)_5]_2$ (X = Cl or Br)[767] and $Cl_2[Cr(CO)_3(C_5H_5)]_2$[768] deviates substantially from regular tetrahedral. A full report of the structure determination of $\{H[Mn(CO)_5]_2Sn\}_2$, which was preliminarily reported in Volume 3, has appeared.[746]

Cleavage of the Fe—Si bond by chlorine occurs with predominant retention of configuration at silicon for optically active $[(C_5H_5)Fe(CO)_2\{SiMePh(1-Np)\}]$ and $[(C_5H_5)(CO)Fe(PPh_3)\{SiMePh(1-Np)\}]$, but mainly with inversions of configuration for the latter in the presence of $PPh_3$.[769] The kinetics of cleavage of the tin–transition metal bond in the compounds $[Me_3SnM(CO)_3(C_5H_5)]$ (M = Cr, Mo, or W), $[Me_3SnMn(CO)_5]$, and $[Me_3SnFe(CO)_2(C_5H_5)]$,[770,771] and of the Group IV metal–transition metal bond in the compounds $[Me_3MMn(CO)_5]$ (M = Si, Ge, Sn, or Pb) and $[Me_3MFe(CO)_2(C_5H_5)]$ (M = Si, Ge, or Sn)[772] by iodine in $CCl_4$ have

---

[767] H. Preut, W. Wolfes, and H. J. Haupt, *Z. anorg. Chem.*, 1975, **412**, 121.
[768] F. S. Stephens, *J.C.S. Dalton*, 1975, 230.
[769] G. Caerveau, E. Colomer, R. Corriu, and W. E. Douglas, *J.C.S. Chem. Comm.*, 1975, 40.
[770] J. R. Chipperfield, J. Ford, and D. E. Webster, *J.C.S. Dalton*, 1975, 2042.
[771] J. R. Chipperfield, A. C. Hayter, and D. E. Webster, *J.C.S. Dalton*, 1975, 2048.
[772] J. R. Chipperfield, A. C. Hayter, and D. E. Webster, *J.C.S. Chem. Comm.*, 1975, 625.

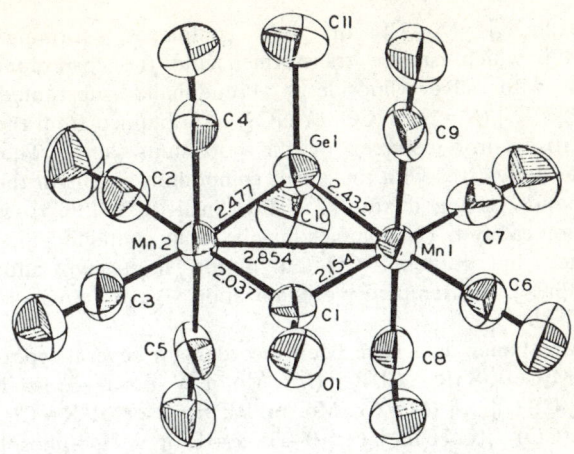

**Figure 28** The structure of [(Me$_2$Ge)Mn$_2$(CO)$_9$]
(Reproduced by permission from *J. Amer. Chem. Soc.*, 1975, **97**, 5747)

been investigated. A similar mechanism was proposed for all the systems, in which an intermediate formed by the electrophilic attack of iodine on the transition metal is stabilized by delocalization of the induced positive charge over the other ligands. The products of the reactions of the compounds [Ph$_3$MMn(CO)$_5$] (M = Si or Sn) with chlorine, bromine, and iodine in CCl$_4$ have been determined. The silicon compound does not react with iodine, whereas chlorine and bromine cleave the Si—Mn bond. [Ph$_3$SnMn(CO)$_5$] reacts in a much more complex manner, Ph—Sn bond cleavage occurring with all three halogens, giving mono-, di-, and (with the exception of iodine) tri-halogenated derivatives. Bromine in high concentration and iodine also cleave the Sn—Mn bond.[773] NO$^+$ PF$_6^-$ reacts with the complexes [NEt$_4$][R$_3$MM'(CO)$_5$] in dichloromethane at −78° C to afford the complexes [R$_3$MM'(CO)$_4$NO] (R = Me or Ph; M = Si, Ge, Sn, or Pb; M' = Mo or W). The corresponding chromium complexes could not be obtained. The reaction of the complexes [Et$_4$N][X$_3$SnW(CO)$_5$] (X = Cl or Br) resulted in reduction of the tin, and SnX$_2$ and XW(CO)$_4$NO were isolated.[774] Cycloheptatriene reacts with [(Me$_3$M)$_2$Ru(CO)$_4$] (M = Si or Ge), with migration of a MMe$_3$ group, giving the complexes [Ru(MMe$_3$)(CO)$_2$(1—5-$\eta$-C$_7$H$_8$-MMe$_3$-6)]. The reaction with [Ru(MMe$_3$)(CO)$_4$]$_2$ is more complex.[775] Arenes react with [Ru(CO)$_4$(GeCl$_3$)$_2$] at 150 °C or higher to form the very stable complexes [($\eta$-arene)Ru(CO)(GeCl$_3$)$_2$].[776] The protonation of Et$_4$N[Ph$_3$GeFe(CO)$_4$], using HCl in ether—THF, yields the air-sensitive, thermally unstable *cis*-[Ph$_3$GeFeH(CO)$_4$].[751] Triphenyl-phosphine, -arsine, and -stibine displace CO from tetracarbonylcobalt–tin complexes,[777] as does P(OPh)$_3$ with

---

[773] J. R. Chipperfield, J. Ford, and D. E. Webster, *J. Organometallic Chem.*, 1975, **102**, 417.
[774] E. E. Isaacs and W. A. G. Graham, *J. Organometallic Chem.*, 1975, **99**, 110.
[775] A. Brookes, S. A. R. Knox, V. Riera, B. A. Sosinsky, and F. G. A. Stone, *J.C.S. Dalton*, 1975, 1641.
[776] R. K. Pomeroy and W. A. G. Graham, *Canad. J. Chem.*, 1975, **53**, 2985.
[777] A. N. Nesmeyanov, K. N. Anisimov, N. E. Kolobova, and V. N. Khandozhko, *J. Gen. Chem. (U.S.S.R.)*, 1974, **44**, 1038.

[$(C_5H_5)Fe(CO)_2SnX_3$] (X = Cl or Br) complexes, forming [$(C_5H_5)Fe$-$[P(OPh)_3]_2SnX_3$], which can be transformed into the corresponding fluoride by treatment with silver fluoride.[778] Pseudohalide-substituted derivatives [$(C_5H_5)Fe(CO)_2SiX_3$] (X = $N_3$, NCO, or NCS) are obtained from the substitution of the trichlorosilyl–iron complex by the potassium salt.[779] The reaction of ethynylbenzene derivatives with tin–cobalt compounds results in the cleavage of the Sn—Co bonds, giving mixtures.[780] [$PtH(SnCl_3)(CO)(PPh_3)_2$] is a selective hydroformylation catalyst. The complex may be ionic, containing $SnCl_3^-$ anions, or five-co-ordinate, and it loses CO on drying *in vacuo*, affording *trans*-[$PtH(SnCl_3)(PPh_3)_2$]. Attempted crystallization gave $SnCl_2$ and *trans*-[$PtHCl(PPh_3)_2$].[781]

Tin-119$m$ Mössbauer data have been recorded for several types of complex, including [$(L_2)M(CO)_3(SnR_{3-n}Cl_n)Cl$] (M = Mo or W; $n$ = 1—3; R = Me or Ph),[782] [$SnX_n\{M(CO)_3(C_5H_5)\}_{4-n}$] (M = Cr, Mo, or W; $n$ = 2 or 3; X = Cl, Br, or I),[783] [$Ph_{3-n}Cl_nSnFe(CO)_{2-x}(C_5H_5)L_x$] ($n$ = 0—3; $x$ = 1 or 2; L = phosphine),[784] and chlorostannyl–nickel complexes.[785] The molecular electric dipole moments for the complexes $R_nX_{m-n}Sn[Co(CO)_4]_{4-m}$ ($m$ = 1—3; $n \leq m$; R = alkyl or phenyl; X = halogen) show that the charge transfer in the cobalt–tin bond is mainly determined by the inductive properties of the ligands attached to tin.[786] The bond dissociation energies for the complexes [$Me_3MM'(CO)_x$] (M = Si, Ge, or Sn; M' = Co, $x$ = 4; M' = Re, $n$ = 5) have been determined from appearance-potential data.[787]

**Bivalent Derivatives of Silicon, Germanium, Tin, and Lead.**—*Silylenes.* The pyrolysis of disilane provides a clean source of singlet silylene $SiH_2$, which readily inserts into Si—H bonds. Thus copyrolysis of disilane with $Me_nSiH_{4-n}$ ($n$ = 1—3) gives only the alkyldisilane as product.[788] The thermolysis of hexamethylsilacyclopropane at 60—80 °C affords a method for generation of dimethylsilylene $Me_2Si$ under mild conditions. As with silylene itself, $Me_2Si$ inserts into the Si—H bond of silanes to give disilanes.[789] 3-Silacyclopentanes are obtained from the reaction of butadiene and $SiH_2$, $ClSiH$, and $MeSiH$ (generated from disilane pyrolysis). 1-Silyl-3-silacyclopentene and 1,2-disilacyclohex-4-ene are obtained in nearly equal quantities from the reaction of butadiene with $Si_2H_4$ ($H_3SiSiH$ or $H_2Si{=}SiH_2$) generated from the pyrolysis of $Si_3H_8$. Competitive reactions of $SiH_2$ with butadiene and $Si_2H_6$, and of $GeH_2$ with butadiene and $Ge_2H_6$, show that,

---

[778] B. Herber and H. Werner, *Synth. React. Inorg. Metal.-Org. Chem.*, 1975, **5**, 381.
[779] M. Höfler, J. Scheuren, and D. Spilker, *J. Organometallic Chem.*, 1975, **102**, 205.
[780] A. N. Nesmeyanov, K. N. Anisimov, N. E. Kolobova, and V. N. Khandozhko, *J. Gen. Chem.* (U.S.S.R.), 1974, **44**, 302.
[781] C.-Y. Hsu and M. Orchin, *J. Amer. Chem. Soc.*, 1975, **97**, 3553.
[782] W. R. Cullen, R. K. Pomeroy, J. R. Sams, and T. B. Tsin, *J.C.S. Dalton*, 1975, 1216.
[783] R. J. Dickinson, R. V. Parish, P. J. Rowbotham, A. R. Manning, and P. Hackett, *J.C.S. Dalton*, 1975, 424.
[784] G. M. Bancroft and A. T. Rake, *Inorg. Chim. Acta*, 1975, **13**, 175.
[785] G. M. Bancroft and K. D. Butler, *Canad. J. Chem.*, 1975, **53**, 307.
[786] L. F. Wuyts and G. P. Van der Kelen, *J. Organometallic Chem.*, 1975, **97**, 453.
[787] R. A. Burnham and S. R. Stobart, *J. Organometallic Chem.*, 1975, **86**, C45.
[788] B. Cox and H. Purnell, *J.C.S. Faraday I*, 1975, **71**, 859.
[789] D. Seyferth and D. C. Annarelli, *J. Amer. Chem. Soc.*, 1975, **97**, 7162.

## Elements of Group IV

while addition of $SiH_2$ to butadiene can compete favourably with insertion into $Si_2H_6$, addition of $GeH_2$ to butadiene cannot compete with insertion into $Ge_2H_6$.[790] The additions of $SiH_2$, $SiCl_2$, and $SiMe_2$ to cyclopentadiene and of $SiMe_2$ to cyclohexa-1,3-diene have been carried out by copyrolysis of the appropriate disilane and the diene. All the observed products are believed to arise *via* the 1,2-addition of the silylenes, forming bicyclic vinylsilacyclopropane derivatives which undergo subsequent non-concerted rearrangements.[791] A mixture of singlet and triplet $^{31}SiF_2$ is formed by the reaction of recoil $^{31}Si$ atoms with $PF_3$. The triplet species interacts with paramagnetic molecules, forming $^{31}SiF_2$–donor complexes towards butadiene, whereas the singlet species reacts with butadiene to give difluorosilacyclohexene.[792] Cocondensation of $SiF_2$ with phosphine yields the new compounds $SiF_2HPH_2$ and $SiF_3PH_2$ in addition to $Si_2F_5H$ and thermally unstable compounds of high molecular weight.[793] $SiF_2BrSiF_2H$ is obtained as the primary product of the low-temperature reaction of $SiF_2$ and HBr. The compound shows a marked tendency towards decomposition at 25 °C involving redistribution of fluorine and bromine (but not hydrogen) atoms by both intra- and intermolecular processes.[794]

*Germylenes.* New procedures for the preparation of $GeBr_2$[795] and $GeI_2$[796] have been described. Functionally substituted germylenes PhGeY may be synthesized by the action of a protonic reagent HY (Y = Halogen, OH, OR, SR, $NR_2$, $PR_2'$, $PhCl_2Ge$, or $O_2CMe$) on $PhGeNMe_2,HCl$.[797] Germylene species have been shown to undergo many reactions characteristic of carbenes. Thus, $GeF_2$ exhibits strong carbene reactivity towards insertion into $\sigma$-bonds. For example, the reaction with $Bu_3SnH$ is violent, $Bu_3SnF$, $GeH_2$, and $GeF_2$ being formed by the decomposition of the intermediate insertion product $Bu_3SnGeF_2H$, whilst $Ph_3SnGeF_2SnMe_3$ may be isolated from the reaction of $GeF_2$ with $Ph_3SnSnMe_3$ at 130 °C. PhGeF exhibits a generally lower reactivity, but both species insert into the $M^{IV}$—X bonds of organometal(IV) compounds. The insertion reactions into the intracyclic Ge—O bond of diasteroisomeric oxagermacycloalkanes are generally stereospecific.[798] Mono- or di-halogenated germylenes insert into the Ge—P bond of germylphosphines to form phosphorylated digermanes. The phenylchlorogermylene adducts, which are stable at room temperature, thermolyse by an $\alpha$-elimination to phosphorylated germylenes, which can be trapped by dimethylbuta-2,3-diene.[799] du Mont and Schumann have isolated and characterized the stable phosphorylated germylene $Bu_2^tPGeCl$ by the reaction of $Bu^tP$-$SiMe_3$ and $Ph_3PGeCl_2$ in benzene. The compound is a yellow, thermally stable,

---

[790] R. L. Jenkins, R. A. Kedrowski, L. E. Elliot, D. C. Tappen, D. J. Schlyer, and M. A. Ring, *J. Organometallic Chem.*, 1975, **86**, 347.
[791] R. J. Huang, R. T. Conlin, and P. P. Gaspar, *J. Organometallic Chem.*, 1975, **94**, C38.
[792] O. F. Zeck, Y. Y. Su, and Y. N. Tang, *J.C.S. Chem. Comm.*, 1975, 156.
[793] G. R. Langford, D. C. Moody, and J. D. Odom, *Inorg. Chem.*, 1975, **14**, 134.
[794] K. G. Sharp and J. F. Bald, *Inorg. Chem.*, 1975, **14**, 2553.
[795] O. V. Zakolodyazhnaya, R. L. Magunov, and I. P. Kovalevskaya, *Russ. J. Inorg. Chem.*, 1974, **19**, 1240.
[796] O. V. Zakolodyaznaya, R. L. Magunov, and Yu. V. Belyuga, *Russ. J. Inorg. Chem.*, 1974, **19**, 1861.
[797] P. Riviere, M. Riviere-Baudet, and J. Satge, *J. Organometallic Chem.*, 1975, **96**, C7.
[798] P. Riviere, J. Satge, and A. Boy, *J. Organometallic Chem.*, 1975, **96**, 25.
[799] C. Couret, J. Escudie, P. Riviere, J. Satge, and G. Redoules, *J. Organometallic Chem.*, 1975, **84**, 191.

but easily oxidized solid.[800,801] The intermediate $R_3PGeCl_2$ complexes are obtained by the reaction of the phosphine with $Bu_2^tPGeCl_3$, which itself is prepared either by the redistribution of $GeCl_4$ and $(Bu_2^tP)_2GeCl_2$ or by the insertion of $GeCl_2$ into $Bu_2^tPCl$.[800] The reaction of $GeI_2$ with $CF_3I$ to produce $(CF_3)GeI_3$ has been optimized.[802] The electrophilic addition of the functional germylenes PhGeY (Y = OR or Sr) to the carbonyl group of aldehydes leads to the formation of germanium(IV) oligomers of the type $+\!\!-\!GePhY\!-\!O\!-\!CHR\!-\!\!+_n$. For the analogous amino- and phosphino-germylenes PhGeY (Y = $NR_2$ or $PR_2$), addition is followed by an insertion reaction of a carbonyl group into the Ge—Y bond.[803]

*Tin*(II) *and Lead*(II) *Halide Systems.* The crystal structure of $SnBr_2$ has been determined. Each tin atom is surrounded by eight bromines, six of which lie at the apices of a trigonal prism, with the remaining two located outside the prism faces.[804] The structures of four more modifications of $PbI_2$ have also been determined,[805,806] and phase transformations in basic and long-period polytypes investigated.[807] Laurionite, PbOHCl, is not isotypic with $PbCl_2$, but rather each lead atom is surrounded by five chlorine atoms and three OH groups in a polyhedron which may be described as a strongly distorted square antiprism.[808] The heat of formation of $SnI_2$ has been deduced by a calorimetric investigation of the reaction of $SnI_2$ with $I_2$ in $CS_2$.[809]

Bulten has investigated the synthesis of alkyltin(IV) trihalides from tin(II) halides and alkyl halides. The reactions were carried out in the absence of solvent with 200% excess of alkyl halide and a trialkylantimony compound as catalyst. Yields in excess of 90% were obtained.[810] The reaction of an alcohol or oxime, ROH, and $SnCl_2$ with a chlorine compound containing positively polarizable chlorine (*e.g.* N-chlorosuccinimide) results in the formation of $Cl_3SnOR$ compounds or their disproportionation products, $Cl_2Sn(OR)_2$ and $SnCl_4$, sometimes as ROH adducts.[811] With N-acylhydroxylamines, and $SnCl_2$, six-co-ordinate $Cl_2Sn(ONArCOPH)_2$ derivatives are formed.[812] Tetramethyl- and tetraethyl-dithio-oxamide form 1:1 complexes with $SnCl_2$ and $SnBr_2$ which are non-electrolytes in DMF.[813] Thiourea forms complexes of stoicheiometry $PbX_2,nL$ (X = Cl, br, or I; $n = 1$ or 2; X = I, $n = 3$).[814] Dissolution of $BaCl_2,2H_2O$ in a methanolic solution of $SnCl_2$ followed by removal of the solvent affords $BaSnCl_4$, which contains the $SnCl_4^{2-}$ anion.[815] The ideal perovskite structure of $CsSnBr_3$ at room temperature has been confirmed by single-crystal X-ray diffraction. The

---

[800] W. W. Du Mont and H. Schumann, *J. Organometallic Chem.*, 1975, **85**, C45.
[801] W. W. Du Mont and H. Schumann, *Angew. Chem. Internat. Edn.*, 1975, **14**, 368.
[802] R. Eujen and H. Burger, *J. Organometallic Chem.*, 1975, **88**, 165.
[803] P. Riviere, M. Riviere-Baudet, and J. Satge, *J. Organometallic Chem.*, 1975, **97**, C37.
[804] J. Anderson, *Acta Chem. Scand.(A)*, 1975, **29**, 956.
[805] M. Chand and G. C. Trigunayat, *Acta Cryst.*, 1975, **B31**, 1222.
[806] M. Chand and G. C. Trigunayat, *Z. Krist.*, 1975, **141**, 59.
[807] R. Prasad and O. N. Srivastava, *Acta Cryst.*, 1974, **B30**, 1748.
[808] C. C. Venetopoulos and P. J. Rentzeperis, *Z. Krist.*, 1975, **141**, 246.
[809] J. Mikler and A. Janitsch, *Monatsh.*, 1975, **106**, 399.
[810] E. J. Bulten, *J. Organometallic Chem.*, 1975, **97**, 167.
[811] M. Masaki, K. Fukui, I. Uchida, and H. Yasuno, *Bull. Chem. Soc. Japan*, 1975, **48**, 2311.
[812] G. C. Pellacani and G. Peyronel, *Spectrochim. Acta*, 1975, **31A**, 1641.
[813] M. K. Das and M. R. Ghosh, *Indian J. Chem.*, 1975, **13**, 515.
[814] Ya. D. Fridman, S. D. Gorokhov, and T. V. Fokina, *Russ. J. Inorg. Chem.*, 1974, **19**, 1140.
[815] M. Goldstein and P. Tiwari, *J. Inorg. Nuclear Chem.*, 1975, **37**, 1550.

*Elements of Group IV* 243

high-symmetry environment for the tin(II) atom in this compound is proposed to arise because the distorting effect of the non-bonding electrons is reduced by their population of an empty low-energy band in the solid, thus giving rise to the black colour and metallic conducting properties. The high-temperature phase of $CsSn_2Br_5$, $Cs_4SnBr_6$, and of compositions from the $CsSn_2Br_5$–$CsSn_2Cl_5$ system show similar properties.[816] The i.r. spectra of the high-temperature forms of $CsSnCl_3$ and $CsSnBr_3$ are consistent with the perovskite structure. The spectra of phases in the $CsSnCl_3$–$CsSnBr_3$ system have also been recorded and the data interpreted in terms of a valence force field.[817] The complexes $[R_4N][PbX_3]$ and $[R_4N][PbX_2Y]$ (R = Et or Bu$^n$; X or Y = Cl, Br, or I) have been shown to be distinct phases, and appear to be halogen-bridged polymeric structures in the solid state.[818] The $SnF_3^-$ anion displaces CO from chromium, molybdenum, and tungsten hexacarbonyls to form the $[M(CO)_5SnF_3]^-$ anions (M = Cr, Mo, or W). The $[Cr(CO)_5SnI_3]^-$ anion is formed similarly. The trifluorostannite complexes $[NEt_4][Mo(CO)_5SnF_3]$ and $[(C_5H_5)Fe(CO)_2SnF_3]$ may be obtained by treatment of the chloro-analogues with $AgBF_4$. Mixed anions of the types $[M(CO)_5SnX_2Cl]^-$ (M = Cr, Mo, or W, X = F; M = Cr, X = I) have been synthesized by insertion of $SnX_2$ into the chlorocarbonyl anions $[M(CO)_5Cl]^-$.[819,820] $MCl_3^-$ (M = Ge or Sn) anions displace hexamethylborazine from $[B_3N_3Me_6Cr(CO)_3]$ to afford *mer*-$[Cr(CO)_3(MCl_3)_3]^{3-}$ anions.[821] Three compounds have been isolated from mixtures of $K_3IrCl_6$ and $SnCl_2$ in formic acid as solvent, depending on the conditions, and on addition of $NEt_4^+$ cations: $KIrCl_4,SnCl_2$, for which the structure (76)

```
        Cl        ··
         \       Sn
    Cl—Ir—Cl  /    \
         /      Cl   Cl
        Cl     |
               Cl
```

(76)

is proposed from Mössbauer data, $(NEt_4)$ $[IrCl_3(CO)(SnCl_3)_2]$, and $(NEt_4)$ $IrCl_3$-$(SnCl_3)_3]$, which contain Sn—Ir bonds.[822] $SnCl_2$ and $SnBr_2$ react with $Fe_2(CO)_9$ to give complexes of composition $[X_2SnFe(CO)_4]$, which may exist as dimers with $Fe_2Sn_2$ rings or oligomers.[823]

Several mixed halide systems have been studied, including the Pb,Rb,Tl$^I$–Cl,[824] K,Pb,Rb–Cl,[825] Na,Rb,Pb–Br,[826] Li–$PbCl_2$–$UCl_4$,[827] and $BiCl_3$–$SnCl_2$[828] systems.

---

[816] J. D. Donaldson, J. Silver, S. Hadjiminolis, and S. D. Ross, *J.C.S. Dalton*, 1975, 1500.
[817] J. D. Donaldson, S. D. Ross, and J. Silver, *Spectrochim. Acta*, 1975, **31A**, 239.
[818] M. Goldstein and G. C. Tok, *Spectrochim. Acta*, 1975, **31A**, 1993.
[819] B. Herber and H. Werner, *Synth. React. Inorg. Metal.-Org. Chem.*, 1975, **5**, 189.
[820] T. Kruck, K. Ehlert, W. Molls, and M. Schless, *Z. anorg. Chem.*, 1975, **414**, 277.
[821] B. Herber, M. Scotti, and H. Werner, *Helv. Chim. Acta*, 1975, **58**, 1225.
[822] G. Elizarova, E. N. Yurchenko, V. A. Varnek, V. R. Sokolova, and L. G. Matvienko, *Russ. J. Inorg. Chem.*, 1974, **19**, 246.
[823] A. B. Cornwell and L. G. Harrison, *J.C.S. Dalton*, 1975, 2017.
[824] Yu. G. Litvinov and I. I. Il'yasov, *Russ. J. Inorg. Chem.*, 1974, **19**, 441.
[825] M. Davranov, I. I. Il'yasov, and M. Ashurova, *Russ. J. Inorg. Chem.*, 1974, **19**, 885.
[826] M. Davranov and I. I. Il'yasov, *Russ. J. Inorg. Chem.*, 1974, **19**, 1390.
[827] V. N. Desyatnik, V. A. Korol'kov, N. N. Kurbatov, and S. P. Raspopin, *Russ. J. Inorg. Chem.*, 1974, **19**, 1417.
[828] N. V. Karpenko, *Russ. J. Inorg. Chem.*, 1974, **19**, 876.

The interaction of isomolar mixtures of $PbCl_2$ and $MH_2PO_4$ (M = K or Na) is accompanied by the precipitation of $[Pb_2Cl(PO_4)]$ irrespective of the ratio of reactants.[829]

*Oxides and Molecular Oxygen Derivatives of Bivalent Tin and Lead.* The chemistry of molecular tin(II)–oxygen bonded compounds has attracted a great deal of attention during the past year. Gsell and Zeldin have prepared the tin(II) alkoxides $Sn(OR)_2$ (R = Me, Et, or Bu) either from $SnCl_2$ and the alcohol in the presence of triethylamine or by transesterification using $Sn(OMe)_2$.[830,831] Transesterification has also been used to obtain tin(II) derivatives of triethanolamines, $Sn(OCH_2CH_2)_2NR$ (R = H, alkyl, or aryl).[831] Tin(II) phenoxides are available from the protolysis of $Sn(C_5H_4Me)_2$ by phenols.[832] Redistribution of alkoxy-group and halide occurs when THF solutions of $Sn(OMe)_2$ and tin(II) halides are mixed, and tin(II) halide methoxides $Sn(OMe)X$ (X = Cl, Br, or I) result.[833] Harrison and his co-workers have synthesized a wide range of bis($\beta$-keto-enolato)tin(II) derivatives by either the protolysis of $Sn(OMe)_2$ or $Sn(C_5H_4Me)_2$ by the parent diketone or by substitution of tin(II) halide by the corresponding sodium salt. The compounds were crystalline solids except for the derivatives of alkylacetoacetates, which were undistillable oils.[834,835] Except for the derivative of 1,3-cyclohexanedione, where intramolecular co-ordination was impossible, all the compounds were proposed to have the pseudo–trigonal-bipyramidal geometry deduced by X-ray diffraction studies for $Sn\{OPhC:CHC(O)Me\}_2$ (Figure 29), in which both $\beta$-keto-enolato-groups chelate the tin atom and the third equatorial site is occupied by a

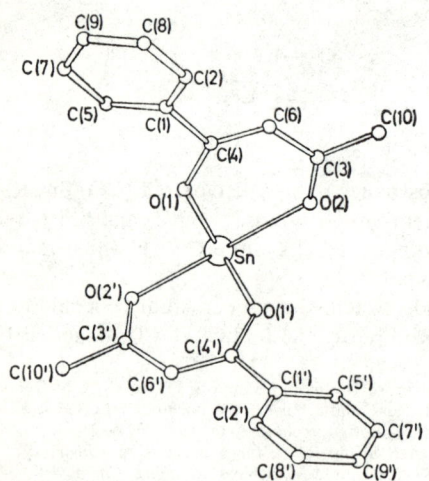

**Figure 29** *The structure of* $[Sn\{OCPh:CHC(O)Me\}_2]$
(Reproduced from *J.C.S. Dalton*, 1975, 1455)

[829] I. S. Vel'mozhnyi, E. A. Gyunner, and L. M. Mel'nichenko, *Russ. J. Inorg. Chem.*, 1974, **19**, 821
[830] R. Gsell and M. Zeldin, *J. Inorg. Nuclear Chem.*, 1975, **37**, 1133.
[831] M. Zeldin and R. Gsell, *Synth. React. Inorg. Metal.-Org. Chem.*, 1976, **6**, 11.
[832] P. F. R. Ewings and P. G. Harrison, *J.C.S. Dalton*, 1975, 2015.
[833] P. F. R. Ewings and P. G. Harrison, *J.C.S. Dalton*, 1975, 1717.
[834] P. F. R. Ewings, D. E. Fenton, and P. G. Harrison, *J.C.S. Dalton*, 1975, 821.
[835] A. B. Cornwell and P. G. Harrison, *J.C.S. Dalton*, 1975, 1722.

stereochemically active lone pair.[836] The stereochemical activity of the lone pair in these compounds has been demonstrated by the formation of [(OC)$_5$MSnX$_2$] (M = Cr, Mo, or W; X = β-keto-enolate) complexes.[837] Reaction with Fe$_2$(CO)$_9$ afforded complexes of composition [X$_2$SnFe(CO)$_4$], which existed as dimers with a Sn$_2$Fe$_2$ central ring if the groups on tin were small or as monomers if the groups on tin were bulky. Dissolution in pyridine caused fission of the Sn$_2$Fe$_2$ ring in the monomers.[823] Tin(II) bis(propane-1,3-dionate) forms adducts with ethyl and phenyl isocyanates,[838] and tin(II) alkoxides, bis(β-keto-enolates), and aryl oxides will catalyse the trimerization of phenyl isocyanate.[839] Tin(II) arylcarboxylates and sulphonates have also been prepared by the protolysis of Sn(OMe)$_2$ and Sn(C$_5$H$_4$Me)$_2$.[833]

A kinetic study of the reaction between stannate(II) ions and alkyl halides in basic solution (Pfeiffer's reaction) has been carried out. Mono- and di-organotin species are formed.[840] Two reports have appeared of the structure of the basic tin(II) sulphate [Sn$_3$(O)(OH)$_2$](SO$_4$),[841,842] although one is significantly more detailed. The material contains discrete [Sn$_3$(O)(OH)$_2$]$^{2+}$ cations, which have tin atoms in two different environments. One tin is bonded to all three oxygen atoms, which form a ring, to give trigonal-pyramidal co-ordination, with short tin–oxygen distances. The remaining tin atoms are bonded to two ring oxygen atoms with short tin–oxygen distances. Longer tin–oxygen contacts from two separate sulphate groups complete distorted square-pyramidal geometry for these tin atoms, forming a layer structure. The principal features of the structure of the basic lead sulphate 4PbO,PbSO$_4$ have been determined.[843]

Lead hafnate, PbHfO$_3$, has been prepared by fusion of HfO$_2$ and PbCO$_3$ at 750—900 °C.[844] The mixed oxide phases B$_2$O$_3$–PbO–V$_2$O$_5$,[845] KF–PbZrO$_3$–NdTaO$_4$,[846] 'Pb(BO$_2$)$_2$'–PbZrO$_3$–NdTaO$_4$,[846] La$_2$(WO$_4$)$_3$–PbWO$_4$,[847] PbSO$_4$(PO$_4$)–MTiO$_3$ (M = Ca, Sr, Ba, Zn, or Cd),[848] PbSO$_4$–MZrO$_3$ (M = Ca, Sr, or Ba),[848] B$_2$O$_3$–MoO$_3$–PbO,[849] and PbO–ZnO–Nb$_2$O$_5$[850] have all been the subject of study. The vibrational spectra and crystal structure of Pb$_2$V$_2$O$_7$ have been determined.[851] The tin-119m Mössbauer data for α- and β-SnWO$_4$ and Sn(SO$_3$X)$_2$ (X = F or Cl) have been recorded.[852] Lead acetate reacts with the phosphates MH$_2$(PO$_4$) (M = Na or K) to precipitate PbHPO$_4$ and Pb$_3$(PO$_4$)$_2$; lead nitrate reacts to give PbHPO$_4$ and Pb$_2$(NO$_3$)PO$_4$.[853]

---

[836] P. F. R. Ewings, P. G. Harrison, and T. J. King, *J.C.S. Dalton*, 1975, 1455.
[837] A. B. Cornwell and P. G. Harrison, *J.C.S. Dalton*, 1975, 1486.
[838] I. Wakeshima and I. Kijima, *Bull. Chem. Soc. Japan*, 1975, 953.
[839] I. Wakeshima, H. Suzuki, and I. Kijima, *Bull. Chem. Soc. Japan*, 1975, **48**, 1069.
[840] M. Devaud and M. C. Madec, *J. Organometallic Chem.*, 1975, **93**, 85.
[841] S. Grimvall, *Acta Chem. Scand.(A)*, 1975, **29**, 590.
[842] C. G. Davies, J. D. Donaldson, D. R. Laughlin, R. A. Howie, and R. Beddoes, *J.C.S. Dalton*, 1975, 2241.
[843] K. Sahl, *Z. Krist.*, 1975, **141**, 145.
[844] I. V. Vinarov, A. N. Grinberg, and L. Ya. Filatov, *Russ. J. Inorg. Chem.*, 1973, **19**, 612.
[845] V. T. Mal'tsev, P. M. Chobanyan, and V. L. Volkov, *Russ. J. Inorg. Chem.*, 1974, **19**, 879.
[846] I. N. Belyaev and E. N. Efshfeev, *Russ. J. Inorg. Chem.*, 1974, **19**, 1704.
[847] A. A. Evdokimo and V. K. Trunov, *Russ. J. Inorg. Chem.*, 1974, **19**, 127.
[848] I. N. Belyaev, I. I. Belyaeva, and L. N. Aver'yanova, *Russ. J. Inorg. Chem.*, 1974, **19**, 123.
[849] V. T. Mal'tsev, P. M. Chobanyan, and V. L. Volkov, *Russ. J. Inorg. Chem.*, 1974, **19**, 271.
[850] J. Bachelier, Fr. Mathieu, and E. Quemeneur, *Bull. Soc. chim. France*, 1975, 1989.
[851] E. J. Baran, J. C. Pedregosa, and P. J. Aymonino, *Monatsh.*, 1975, **106**, 1085.
[852] J. G. Ballard and T. Birchall, *Canad. J. Chem.*, 1975, **53**, 3371.
[853] E. A. Gyunner, I. S. Vel'mozhnyi, and L. M. Mel'nichenko, *Russ. J. Inorg. Chem.*, 1974, **19**, 326.

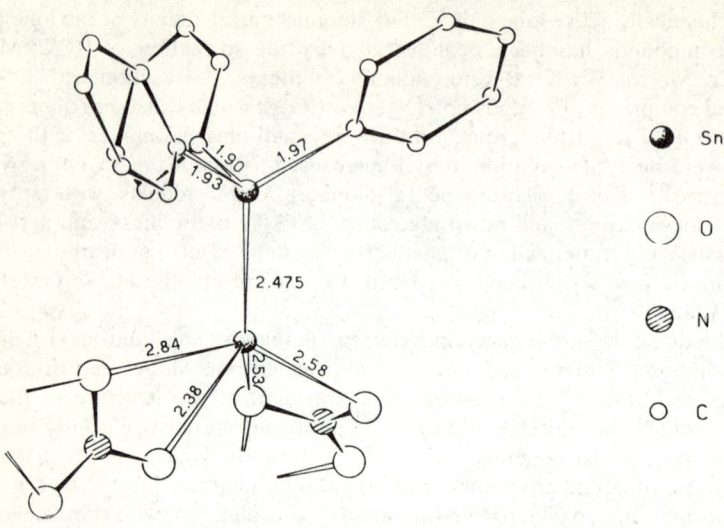

**Figure 30** *The structure of* [Ph$_3$Sn$^{IV}$Sn$^{II}$(NO$_3$)]
(Reproduced by permission from *J. Organometallic Chem.*, 1975, **85**, C43)

Four remarkable X-ray structure determinations have been reported. Two are of nitratotin(II) derivatives and two are of complex lead(II) perchlorates. In Ph$_3$Sn$^{IV}$Sn$^{II}$NO$_3$ the tin(II) atom is co-ordinated in a highly irregular manner by the Ph$_3$Sn$^{IV}$ group, one almost symmetrically bidentate, and one unsymmetrically bidentate nitrate group (Figure 30). All nitrate groups are bridging with all oxygen atoms participating in co-ordination, resulting in a polymeric structure.[854] (2-Aminobenzothiazolato)nitratotin(II) also has a polymeric structure, for the same reason. The immediate environment of the tin atom is shown in Figure 31, where it can be seen that, besides the co-ordination of the two nitrate groups and the 2-aminobenzothiazole ligand, there is an unusual short contact between the tin atom and the phenyl ring of an adjacent benzothiazole molecule.[855] Only one of the perchlorate groups in Pb(ClO$_4$),4phen (phen = 1,10-phenanthroline) is ionic, and crystals are made up of [Pb(phen)$_4$OClO$_3$]$^+$ cations and ClO$_4^-$ anions. The cations enjoy nine-fold co-ordination, with all four phenanthroline ligands bidentate, two being almost symmetrically and two somewhat unsymmetrically so. The ninth co-ordination position is taken up by a unidentate perchlorate group (Figure 32).[856] The lead atoms in bis(10-methylisoalloxazine).lead(II) perchlorate tetrahydrate are disordered, in two sites 0.75 Å above and below a plane of four oxygen atoms, two of which are the centrosymmetrically related carbonyl oxygens

---

[854] M. Nardelli, C. Pelizzi, and G. Pelizzi, *J. Organometallic Chem.*, 1975, **85**, C43.
[855] M. Nardelli, C. Pelizzi, and G. Pelizzi, *J.C.S. Dalton*, 1975, 1595.
[856] A. V. Ablov, A. Yu. Kon, I. F. Burshtein, T. I. Maliovskii, and Z. G. Levitskaya, *Doklady Chem.*, 1974, **217**, 569.

Elements of Group IV

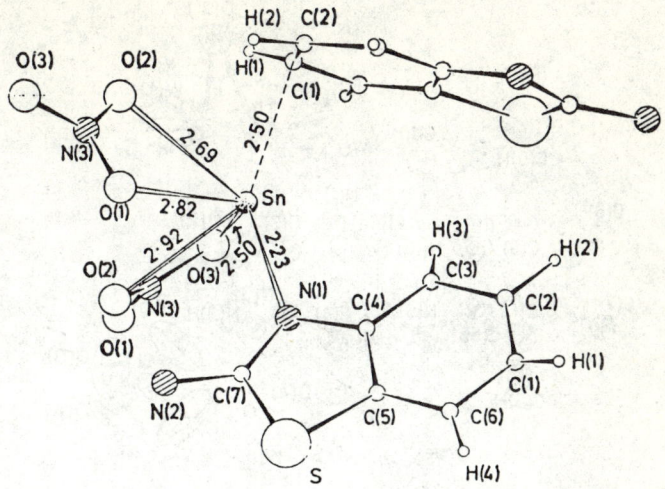

**Figure 31** *The structure of 2-aminobenzothiazolato)nitratotin(II)*
Reproduced from *J.C.S. Dalton*, 1975, 1595)

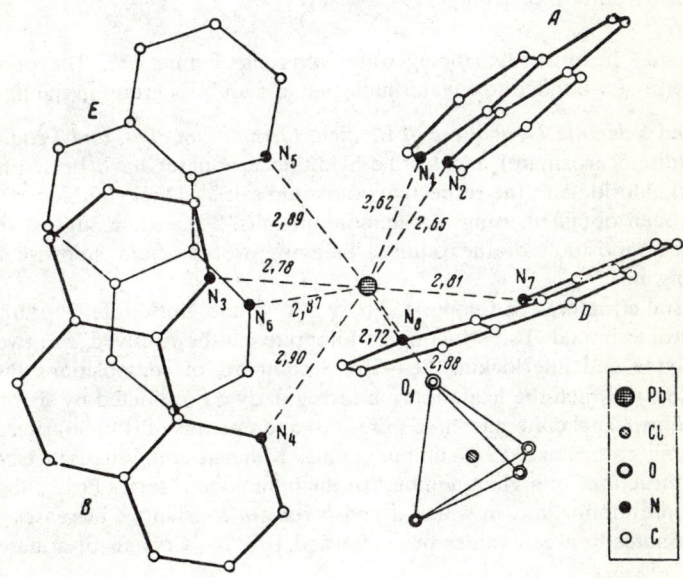

**Figure 32** *The structure of* [Pb(phen)₄OClO₃](ClO₄)
Reproduced by permission from *Doklady Chem.*, 1974, **217,** 569)

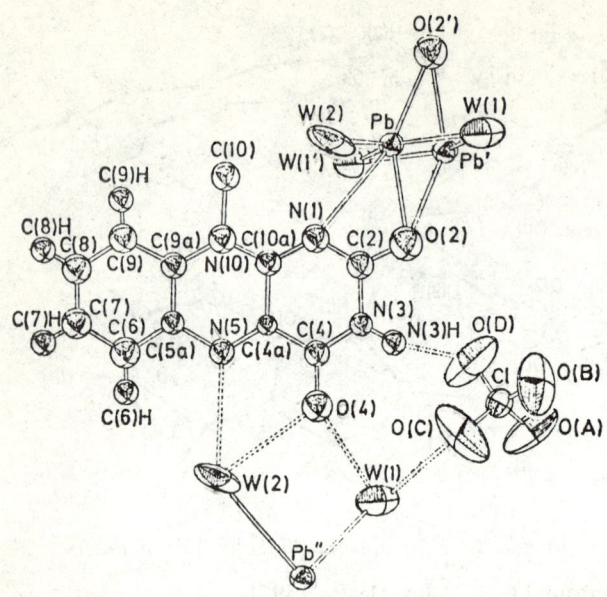

**Figure 33** *The structure of bis(10-methylisoalloxazine)lead(II) perchlorate tetrahydrate*

(Reproduced from *J.C.S. Dalton*, 1975, 377)

of the ligand, the other two being water oxygens (Figure 33). The perchlorate group is hydrogen-bonded to a water molecule and an $NH_2$ group in the ligand.[857]

*Sulphur and Selenium Derivatives of Bivalent Germanium, Tin, and Lead.* Tin(II) bis(diethyldithiocarbamate) and bis(methyldithiocarbonate) have been prepared from tin(II) chloride and the respective potassium salt.[832] Dialkyldithiocarbamates have also been obtained using the ammonium salts. T.g.a. data suggest that the thermal decomposition of the dithiocarbamates proceeds *via* stepwise loss of complete ligands.[858]

The crystal structures of fülöppite, $Pb_3Sb_8S_{15}$,[859] and zinckenite, $Pb_6Sb_{14}S_{27}$,[860] have been determined. The structure of fülöppite can be resolved into two types of interleaving and interlocking Pb—Sb—S complex, of compositions $Pb_2Sb_4S_6$ and $PbSb_4S_9$, in which the lead atoms are irregularly co-ordinated by six or seven sulphur atoms. Zinckenite has three pure (Pb) and one mixed (Pb, Sb) sites, either eight or nine co-ordinated by sulphur atoms. Kohatsu and Wuensch have predicted the structures of higher members of the homologous series $Pb_{3+2n}Sb_8S_{15+2n}$. The series are monoclinic, in which *a* and *b* remain constant, *c* increases with *n*, and *β* alternates between values of *ca.* 94 and 107°.[861] Crystals of a material of

---

[857] M. W. Yu and C. J. Fritchie, *J.C.S. Dalton*, 1975, 377.
[858] D. Perry and R. A. Geanangel, *Inorg. Chim. Acta*, 1975, **13**, 185.
[859] E. W. Nuffield, *Acta Cryst.*, 1975, **B31**, 151.
[860] J. C. Portheine and W. Nowacki, *Z. Krist.*, 1975, **141**, 79.
[861] J. J. Kohatsu and B. J. Wuensch, *Acta Cryst.*, 1974, **B30**, 2935.

*Elements of Group IV* 249

composition PbS,2Bi$_2$S$_3$ (stable between 675 and 736 °C) have been studied. Three different crystal types were identified which appeared to differ in the order of stacking of layers.[862] The phase diagram of the ternary reciprocal system, Sn$_3$S$_3$ + Sb$_2$Se$_3$ ⇌ Sn$_3$Se$_3$ + Sb$_2$S$_3$ has been determined.[863] Glass formation in the PbSe–GeSe–GeSe$_2$ system has been studied.[864] The sublimation kinetics of (001)-orientated GeSe single-crystal platelets have been studied by means of high-temperature mass spectroscopy, microbalance techniques, and hot-stage optical microscopy. The activation enthalpy and entropy for the sublimation were deduced.[865] The orthorhombic to cubic phase change of GeSe has been studied by X-ray diffraction. The transformation takes place at 651 ± 5 °C, and the material remains cubic up to its melting point.[866]

*Nitrogen and Phosphorus Derivatives of Bivalent Tin and Lead.* Following the first syntheses of tin(II)– and lead(II)–nitrogen bonds in 1974, Zeldin has reported the preparation and properties of bis(dimethylamino)tin(II).[867] The compound is a white crystalline solid which is dimeric in cyclohexane solution, but monomeric in the vapour. Variable-temperature n.m.r. studies in chlorobenzene solution indicate exchange of dimethylamino-groups between bridge and terminal positions of the dimer structure (77). With pyridine, a 1:1 adduct is formed; alcohols and

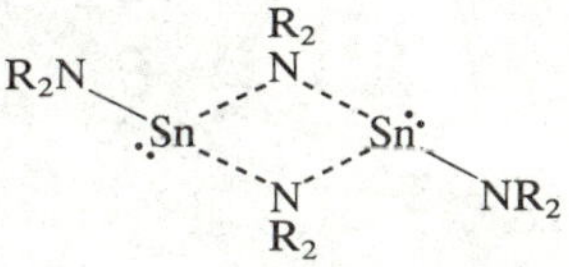

(77)

N-methyldiethanolamine protolyse the Sn—N bond, forming the corresponding alkoxide. The insertion of SnCl$_2$ into ArSO$_2$NCl$_2$ compounds affords the isolable derivatives ArSO$_2$N(Cl)SnCl$_3$. Further insertion of SnCl$_2$ is followed by rapid disproportionation of the intermediate ArSO$_2$N(SnCl$_3$)$_2$, and SnCl$_4$ and ArSO$_2$N=SnCl$_2$ derivatives are obtained.[868] The crystal structure of Pb(NCS)$_2$ has been determined. The lead atom lies on a two-fold axis and is co-ordinated by four nitrogen and four sulphur atoms.[869] The stable tin(II) phosphine Bu$^t_2$PSnCl has been prepared by the reaction of SnCl$_2$ and Bu$^t_2$PSiMe$_3$.[801]

*Bivalent Organo-tin and -lead Derivatives.* Photolysis of dialkyltin(IV) oligomers at room temperature produces, *via* intermediate oligomeric diradicals, dialkylstannylenes, which insert into Sn—H, Sn—C, and Sn—Sn bonds and react with dienes and carbonyl compounds.[870] Dialkylstannylenes are also produced by the thermolysis of XR$_2$SnSnR$_2$X (X = Cl or H) at 120—130 °C. The stannylenes so

---

[862] Y. Takeuchi, J. Takagi, and T. Yamanka, *Z. Krist.*, 1974, **140**, 249.
[863] G. G. Gospodinov, I. N. Odin, and A. V. Novoselova, *Russ. J. Inorg. Chem.*, 1974, **19**, 895.
[864] A. Feltz and L. Senf, *Z. Chem.*, 1975, **15**, 119.
[865] E. A. Irene and H. Wiedemeier, *Z. anorg. Chem.*, 1975, **411**, 182.
[866] H. Wiedemeier and P. A. Siemers, *Z. anorg. Chem.*, 1975, **411**, 90.
[867] P. Foley and M. Zeldin, *Inorg. Chem.*, 1975, **14**, 2264.
[868] A. M. Pinchuk and A. M. Khmaruk, *J. Gen. Chem. (U.S.S.R.)*, 1974, **44**, 1620.
[869] J. A. A. Mokuolu and J. C. Speakman, *Acta Cryst.*, 1975, **B31**, 172.
[870] W. P. Neumann and A. Schwarz, *Angew. Chem. Internat. Edn.*, 1975, **14**, 812.

produced may be trapped by alkyl halides.[871] The Utrecht group has investigated in detail the reaction of $(C_5H_5)_2Sn$ with alkyl halides. The reaction with MeI needs light if it is to proceed, and $Me(C_5H_5)_2SnI$ is formed, but by a much more complex reaction path than the stoicheiometry suggests. With benzyl halides, cleavage of a cyclopentadienyl group from tin occurs, and $(C_5H_5)SnX$ and a mixture of 1- and 2-benzylcyclopentadienes results. Trityl bromide reacts similarly, whilst allyl bromide gives both modes of reaction.[872] The crystal structure of the product of the reaction of $(C_5H_5)Sn$ and $Fe_2(CO)_9$ has been determined, and is shown in Figure 34. The product of stoicheiometry $[(C_5H_5)_2SnFe(CO)_4]_2$ is characterized by a lozenge-shaped $Fe_2Sn_2$ ring and *monohapto*-cyclopentadienyl rings.[873] The cyclopentadienyl ring in $(C_5H_5)SnCl$ is essentially *pentahapto*, but is somewhat tilted such that the tin atom is preferentially associated with two of the ring carbon atoms. The crystals consist of monomeric molecules with only one short Sn—Cl bond.[874] The Jahn–Teller effect has been invoked to rationalize the structures of $(C_5H_5)_2Sn$ and $(C_5H_5)_2Pb$.[875] The full details of the crystal-structure determination of $[(\eta^6-C_6H_6)Sn(AlCl_4)_2],C_6H_6$ have been published.[876]

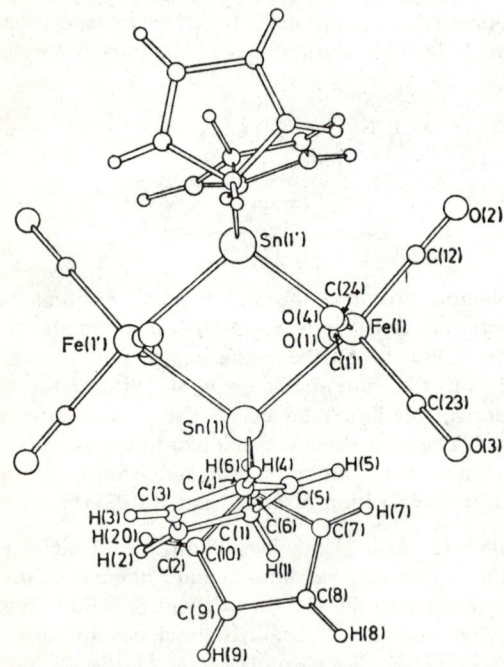

**Figure 34** *The structure of* $[(C_5H_5)_2SnFe(CO)_4]_2$

(Reproduced from *J.C.S. Dalton*, 1975, 2097)

[871] U. Schröer and W. P. Neumann, *Angew. Chem. Internat. Edn.*, 1975, **14**, 246.
[872] K. D. Bos, E. J. Bulten, and J. G. Noltes, *J. Organometallic Chem.*, 1975, **99**, 397.
[873] P. G. Harrison, T. J. King, and J. A. Richards, *J.C.S. Dalton*, 1975, 2097.
[874] K. D. Bos, E. J. Bulten, J. G. Noltes, and A. L. Spek, *J. Organometallic Chem.*, 1975, **99**, 71.
[875] C. Glidewell, *J. Organometallic Chem.*, 1975, **102**, 339.
[876] P. F. Rodesiler, Th. Auel, and E. L. Amma, *J. Amer. Chem. Soc.*, 1975, **97**, 7405.

# Elements of Group IV

**Miscellaneous Studies.** Kinetic studies of the oxidation of $Sn^{II}$ by $Co^{VI}$ in perchloric acid[877] and by $Mo^{VI}$ in hydrochloric acid[878] have been performed, and complexation between $Sn^{2+}$ and acetate ions[879] and between $Pb^{2+}$ and adenine,[880] thiourea,[881] and maleate ions[882] has been studied. Stability constants were usually determined. A model has been proposed for the interpretation of tin-119$m$ Mössbauer data for tin(II) compounds in terms of the distribution of tin valence electrons. The model accounts for the positive signs of the quadrupole coupling constants observed for tin(II) compounds, and for the variation in the Mössbauer parameters of tris(halogenato)stannate(II) anions as the cation is changed.[883]

## 3 Intermetallic Phases

**Binary Systems.**—Very few of the published data on intermetallic phases are of direct interest to the inorganic chemist; those abstracted, however, describe, in general, structural and thermodynamic properties of these systems. In an XPS study,[884] the core levels and valence bands of, *inter alia*, $Fe_3C$ and $Fe_3Si$ and their components have been determined (20—1000 °C). Shifts in the Fe ($3p_{3/2}$), C ($1s$), and Si ($2p$) core levels (Table 23) confirm that migration of electrons occurs from Fe to C and from Si to Fe, rendering C negative but Si positive.

**Table 23** *Binding energies/eV of core levels in* $Fe_3C$, $Fe_3Si$, *and their constituent elements.*[884]

|  | $Fe_3C$ | $Fe_3Si$ | Fe | C | Si |
|---|---|---|---|---|---|
| Fe($3p_{3/2}$) | 708.3 | 707.7 | 708.1 | — | — |
| C($1s$) | 284.1 | — | — | 284.4 | — |
| Si($2p$) | — | 99.7 | — | — | 99.3 |

The crystal structures of several binary intermetallics containing Group IV elements have been reported.[885—895] Independent X-ray studies[885,886] have shown that the true symmetry of $Ru_2Ge_3$ is not tetragonal but orthorhombic; they differ, however, in the assignment of the space group, and the description of the

---

[877] N. A. Daugherty and J. K. Erbacher, *Inorg. Chem.*, 1975, **14**, 683.
[878] B. C. Choi and N. Jespersen, *J. Inorg. Nuclear Chem.*, 1975, **37**, 1945.
[879] S. Gobom, *Acta Chem. Scand. (A)*, 1974, **28**, 1180.
[880] S. J. Beveridge and W. R. Walker, *Austral. J. Chem.*, 1974, **27**, 2563.
[881] V. A. Fedorov, A. V. Fedorova, G. G. Nifant'eva, and L. I. Grubev, *Russ. J. Inorg. Chem.*, 1974, **19**, 538.
[882] A. Olin and P. Svanström, *Acta Chem. Scand. (A)*, 1975, **29**, 849.
[883] J. D. Donaldson, D. C. Puxley, and M. J. Tricker, *J. Inorg. Nuclear Chem.*, 1975, **37**, 655.
[884] I. N. Shabanova and V. A. Trapeznikov, *J. Electron Spectroscopy*, 1975, **6**, 297.
[885] D. J. Poutcharovsky and E. Parthe, *Acta Cryst.*, 1975, **B30**, 2692.
[886] H. Vollenkle, *Monatsh.*, 1974, **105**, 1217.
[887] P. Israiloff and H. Vollenkle, *Monatsh.*, 1974, **105**, 1313.
[888] D. J. Poutcharovsky, K. Yvon, and E. Parthe, *J. Less-Common Metals*, 1975, **40**, 139.
[889] P. Israiloff, H. Vollenkle, and A. Wittman, *Monatsh.*, 1974, **105**, 1387.
[890] W. Wopersnow and K. Schubert, *J. Less-Common Metals*, 1975, **41**, 97.
[891] U. Frank, W. Muller, and H. Schafer, *Z. Naturforsch.*, 1975, **30b**, 10.
[892] U. Frank, W. Muller, and H. Schafer, *Z. Naturforsch.*, 1975, **30b**, 6.
[893] U. Frank, W. Muller, and H. Schafer, *Z. Naturforsch.*, 1975, **30b**, 1.
[894] J. K. Brandon, W. B. Pearson, and D. J. N. Tozer, *Acta Cryst.*, 1975, **B31**, 774.
[895] P. I. Kripyakevich, *Soviet Phys. Cryst.*, 1975, **20**, 168.

structure. Whereas Poutcharovsky and Parthe[885] relate the structure (space group *Pbcn*) to the tetragonal $Ru_2Sn_3$ structure, Vollenkle[886] considers it (space group *Pnca*) to be a member of the $Mn_{11}Si_{19}$ structure family. Both groups of authors conclude that $Ru_2Ge_3$, $Ru_2Si_3$, $Os_2Ge_3$, and $Os_2Si_3$ are isostructural;[885—887] their pertinent unit-cell parameters are collected in Table 24. Poutcharovsky *et al.* have

**Table 24** *Unit-cell parameters/Å of the orthorhombic intermetallics* $M_2X_3$ (M = Ru *or* Os, X = Si *or* Ge) *and* $M_{11}Ge_8$ (M = V, Cr, *or* Mn)

| Compound | Ref. | Space group | a | b | c |
|---|---|---|---|---|---|
| $Ru_2Si_3$ | 885 | *Pbcn* | 11.057 | 8.934 | 5.533 |
| $Ru_2Si_3$ | 887 | *Pnca* | 5.530 | 11.060 | 8.952 |
| $Ru_2Ge_3$ | 885 | *Pbcn* | 11.436 | 9.238 | 5.716 |
| $Ru_2Ge_3$ | 886 | *Pnca* | 5.718 | 11.436 | 9.240 |
| $Os_2Si_3$ | 885 | *Pbcn* | 11.124 | 8.932 | 5.570 |
| $Os_2Ge_3$ | 885 | *Pbcn* | 11.544 | 9.281 | 5.783 |
| $V_{11}Ge_8$ | 889 | *Pnam* | 13.398 | 16.135 | 5.017 |
| $Cr_{11}Ge_8$ | 889 | *Pnam* | 13.171 | 15.775 | 4.939 |
| $Mn_{11}Ge_8$ | 889 | *Pnam* | 13.201 | 15.878 | 5.087 |

also effected a temperature-dependent X-ray examination (−170 to 1000 °C)[888] of $Ru_2X_3$ (X = Si, Ge, or Sn). All three intermetallic phases undergo a transformation from a low-temperature, orthorhombic, centrosymmetric, to a high-temperature, tetragonal, non-centrosymmetric crystal structure. Whereas the transformation temperatures for $Ru_2Si_3$ and $Ru_2Ge_3$ are above, that for $Ru_2Sn_3$ is below room temperature. These transformations are reversible, result from the displacement of Si, Ge, and Sn atoms, and occur gradually over a wide temperature range.[888]

The crystal structures of the isotypic compounds $M_{11}Ge_8$ (M = V, Cr, or Mn) have been determined by direct methods from Weissenberg data of a $Cr_{11}Ge_8$ single crystal;[889] their unit-cell parameters are included in Table 24. The structural relationship between $Cr_{11}Ge_8$, $Cr_5Ge_3$ ($W_5Si_3$-type), and $V_6Si_5$ has been discussed. $Pd_{25}Ge_9$ has been shown to crystallize in a hexagonal structure, of space group $P\bar{3}$, $a$ = 7.35, $c$ = 10.60 Å.[890]

The structural chemistry of $Li_{13}Si_4$,[891] $Li_7Sn_2$,[892] and $Li_5Sn_2$[893] has been elucidated. The crystal structure of $Li_{13}Si_4$, earlier reported as $Li_7Si_2$, has been redetermined;[891] structural relationships between $Li_{13}Si_4$, $Li_7Pb_2$, and $Li_3Pb$ are discussed. $Li_7Sn_2$ is shown to be isotypic with $Li_7Ge_2$.[892] Structural relationships between $Li_5Sn_2$, $Li_2Si$, and $Li_9Ge_4$ are also considered.[893] The redetermined structural parameters are collected in Table 25.

**Table 25** *Unit-cell parameters/Å for* $Li_{13}Si_4$, $Li_7Sn_2$, *and* $Li_5Sn_2$

| Compound | Ref. | Symmetry | Space group | a | b | c |
|---|---|---|---|---|---|---|
| $Li_{13}Si_4$ | 891 | orthorhombic | *Pbam* | 7.99 | 15.21 | 4.43 |
| $Li_7Sn_2$ | 892 | orthorhombic | *Cmmm* | 9.80 | 13.80 | 4.75 |
| $Li_5Sn_2$ | 893 | hexagonal | $R\bar{3}m$ | 4.74 | — | 19.85 |

# Elements of Group IV

A new structure for the $\zeta$-bronze $Cu_{20}Sn_6$ has been reported in a single-crystal study.[894] Previous powder-diffraction data indicated the structure to be a $\gamma$-brass defect superstructure with trigonal space group $P\bar{3}1m$. The single-crystal work shows it to be hexagonal, $P6_3$, with cell constants $a = 7.330$, $c = 7.864$ Å. Rather than being related to $\gamma$-brass, the new $Sn_{20}Cu_6$ model can be visualized as a superstructure based on $\zeta$-Ag–Zn.[894] The structure of $AuSn_2$ has been described as a hybrid of the pyrite and marcasite structure types;[895] fragments of the component structures alternate in the direction of the $z$-axis. The geometries of the $AuSn_2$, pyrite, and marcasite units are compared.

In an analysis of an electron-diffraction study of short-range order in amorphous films of $Ni_5Ge_3$,[896] the relative arrangement of the atoms in the films has been shown to differ from the co-ordination that is characteristic of the intermetallic compound.

The enthalpies of formation of binary manganese–silicon,[897] rare earth–tin,[898,899] and rare earth–lead[900,901] intermetallics have been measured by independent workers. Muradov[897] has calculated the enthalpies of formation of manganese silicides (Table 26), on the basis of reaction (43), from the results of a

$$(1-x)Mn\,(s) + xSi\,(s) \rightarrow Mn_{1-x}Si_x\,(s) \qquad (43)$$

**Table 26** *Enthalpies of formation/kJ mol$^{-1}$ of several intermetallic phases containing Group IV elements*

|  | $\Delta H_f$ |  | $\Delta H_f$ |  | $\Delta H_f$ |  | $\Delta H_f$ |
|---|---|---|---|---|---|---|---|
| $MnSi_{1.7}$ | −89.2 | $NdSn_3$ | −259.8±3.4 | $Y_5Pb_3$ | −535.2±2.1 | $La_5Pb_3$ | −575.7±2.1 |
| $MnSi$ | −77.8 | $LuSn_3$ | −156.5±3.4 | — | — | $La_4Pb_3$ | −489.1±2.1 |
| $Mn_5Si_3$ | −317.9 |  |  | $Y_5Pb_4$ | −542.3±2.1 | $La_5Pb_4$ | −621.4±2.1 |
| $Mn_3Si$ | −147.3 |  |  | — | — | $La_3Pb_4$ | −448.1±2.1 |
| $Mn_{4.4}Si$ | −146.8 |  |  | $YPb_2$ | −111.7±2.1 | $LaPb_2$ | −180.8±2.1 |
| $Mn_{7.3}Si$ | −145.8 |  |  | $YPb_3$ | −112.1±2.1 | $LaPb_3$ | −229.3±2.1 |

thermodynamic analysis of Mn–Si solid solutions in the concentration range 9.3—79.9 mol% Si and temperature range 680—1000 °C. Russian authors[898,899] have undertaken e.m.f. studies of the thermodynamic properties of $NdSn_3$[898] and $LuSn_3$[899] (Table 26); collation of the enthalpies of formation of $LnSn_3$ intermetallics has shown that they diminish from −253.6 (±3.4) kJ mol$^{-1}$ for $LaSn_3$[902] to −156.5 (±3.4) kJ mol$^{-1}$ for $LuSn_3$.[899] Using a differential direct isoperibol calorimeter, some Italian authors[900,901] have measured the enthalpies of formation of a number of $Y_xPb_y$[900] and $La_xPb_y$[901] phases (Table 26). The results are discussed and compared with those calculated by the method suggested by Miedema.[903]

---

[896] E. S. Levin, P. V. Gel'd, I. A. Pavars, and V. P. Yakubchik, *Soviet Phys. Cryst.*, 1975, **20**, 67.
[897] V. G. Muradov, *Russ. J. Phys. Chem.*, 1974, **48**, 1274.
[898] N. G. Kulagina and A. P. Bayanov, *Russ. J. Phys. Chem.*, 1974, **48**, 273.
[899] A. P. Bayanov, E. N. Ganchenko, and N. G. Kulagina, *Russ. J. Phys. Chem.*, 1974, **48**, 1258.
[900] A. Borsese, R. Ferro, R. Capelli, and S. Delfino, *J. Less-Common Metals;* 1975 **42**, 179.
[901] R. Ferro, A. Borsese, R. Capelli, and S. Delfino, *Z. anorg. Chem.*, 1975, **413**, 279.
[902] J. R. Guadagno, M. J. Pool, S. S. Shen, and P. J. Spencer, *Trans. Met. Soc. A.I.M.E.* 1968, **242**, 2018.
[903] A. R. Miedema, *J. Less-Common Metals*, 1973, **32**, 117.

Heat capacities of $Cr_3Si$, $Cr_3Ge$, $V_3Si$, and $V_3Ge$ (13—300 K)[904] and of $Pd_3Sn$, $Pd_2Sn$, and $PdSn$(150—298 K)[905] have also been determined.

Thermodynamic parameters of Mn–Si,[906] Ag–Ge,[907] Au–Sn,[908] and Lu–Pb[909] liquid metal solutions have been examined. Atomic ordering in Zn–Sn melts, assessed from the results of X-ray diffraction[910] and adiabatic compressibility[911] studies, has been reported; it is concluded that interactions between like elements are stronger than between unlike elements, leading to positive deviations from ideality.[911]

Phase equilibria in the Sn–As,[912] Pb–In,[913] and Pb–Sb[914] binary systems have been re-evaluated. Two intermetallic compounds, $Sn_4As_3$ (m.pt. 605 °C) and SnAs (m.pt. 610 °C), were found in the Sn–SnAs section of the Sn–As phase diagram.[912] The significant discrepancies between experimental data for the Pb–In solidus (at high Pb concentrations) have been rationalized in an analysis of the freezing characteristics of these solutions.[913] The temperature and composition of the Pb–Sb eutectic (251.2 °C, 17.7 mol% Sb) have been observed to be pressure-dependent;[914] an increase in pressure leads to a higher eutectic temperature and to an increase in the Sb concentration of the eutectic mixture. Above 23.9 kbar there are indications of the appearance of a new, denser intermetallic compound, possibly corresponding to one of the phases already known in the Sn–Bi and Pb–Bi systems.[914]

$TaSi_2$ has been reported to exhibit only weak oxidation resistance at high temperatures when compared with $MSi_2$ (M = Cr, Mo, or W).[915] The addition of Si to $TaSi_2$ up to a total composition of $Ta_{32}Si_{68}$, however, causes a passivation of the specimen, and this alloy was unaffected by annealing for 1500 hours at 1200 °C. Further improvement of the oxidation resistance (1500 hours at 1500 °C) was obtained by adding small amounts of Co and Fe to the $Ta_{32}Si_{68}$ alloy or by $SiO_2$ additions in sintered powder compacts.[915]

**Ternary Systems.**—As part of a comprehensive investigation of ternary systems containing both Li and a Group IV element (Si, Ge, Sn, or Pb), Schuster et al. have published the results of X-ray studies of LiGaGe,[916] $Li_5Fe_7Ge_8$,[917] $Li_2CeGe$,[918] $Li_2MX$,[919] and $LiM_2X$[919] (M = Pd, Pt, or Ir; X = Ge, Sn, or Pb) together with ternary phases crystallizing from Li–M–X (M = Al, Ga, or In; X = Si, Ge or Sn) systems.[920] All relevant unit-cell parameters have been collected in Tables 27 and 28. In LiGaGe, the Ga and Ge atoms form a wurtzite lattice, the Li atoms occupying the octahedral holes.[916] Two modifications (monoclinic and

---

[904] V. I. Surikov, G. I. Kalishevich, and P. V. Gel'd, *Russ. J. Phys. Chem.*, 1975, **49**, 326.
[905] A. W. Bryant, J. M. Bird, and J. N. Pratt, *J. Less-Common Metals*, 1975, **42**, 249.
[906] Yu. V. Gorbunov, Yu. O. Esin, and P. V. Gel'd, *Russ. J. Phys. Chem.*, 1974, **48**, 1244.
[907] L. Martin-Garin, M. Gomez, P. Bedon, and P. Desre, *J. Less-Common Metals*, 1975, **41**, 65.
[908] C. Chatillon-Colinet, J.-L. Deneuville, J.-C. Mathieu, and E. Bonnier, *J. Chim. phys.*, 1975, **72**, 878.
[909] V. R. Roshchina and A. P. Bayanov, *Russ. J. Phys. Chem.*, 1975, **49**, 162.
[910] Ya. I. Dutchak, V. S. Frenchko, V. M. Zalkin, L. S. Kuznetsova, and O. M. Voznyak, *Russ. J. Phys. Chem.*, 1975, **49**, 569.
[911] L. D. Kruglov, *Russ. J. Phys. Chem.*, 1975, **49**, 767.
[912] T. Z. Vdovina and Z. S. Medeva, *Russ. J. Inorg. Chem.*, 1974, **19**, 1234.
[913] E. S. Kucherenko, *Russ. J. Phys. Chem.*, 1975, **49**, 753.
[914] J. B. Clark and C. W. F. T. Pistorius, *J. Less-Common Metals*, 1975, **42**, 59.
[915] E. Fromm, K. Fischer, R. Kirchheim, and E. Gebhardt, *J. Less-Common Metals*, 1974, **38**, 201.
[916] W. Bockelmann and H.-U. Schuster, *Z. anorg. Chem.*, 1974, **410**, 233.

## Elements of Group IV

**Table 27** Unit-cell parameters of LiGaGe, $Li_5Fe_7Ge_8$, $Li_2CeGe$, $Ca_{10-x}Si_{12-2x}As_{16}$, and $Ca_{10-x}Si_{12-2x}P_{16}$

| Intermetallic phase | Symmetry | a/Å | b/Å | c/Å | β/Å | Ref. |
|---|---|---|---|---|---|---|
| LiGaGe | hexagonal | 4.17 | — | 6.78 | — | 916 |
| $Li_5Fe_7Ge_8$ | monoclinic | 8.74 | 5.04 | 14.84 | 101.42 | 917 |
| $Li_5Fe_7Ge_8$ | hexagonal | 8.74 | — | 8.03 | — | 917 |
| $Li_2CeGe$ | orthorhombic | 18.73 | 6.92 | 4.51 | — | 918 |
| $Ca_{10-x}Si_{12-2x}As_{16}$ | monoclinic | 7.134 | 17.650 | 7.267 | 111.74 | 921 |
| $Ca_{10-x}Si_{12-2x}P_{16}$ | monoclinic | 6.924 | 17.251 | 7.093 | 111.81 | 921 |

**Table 28** Unit-cell parameter, a/Å, of the cubic intermetallic phases $Li_2MX$ (M = Pd, Pt, or Ir; X = Ge, Sn, or Pb), $LiPd_2X$ (X = Ge, Sn, or Pb), LiInX (X = Ge or Sn), LiAlGe, $Li_2AlGe$, $Li_2InSi$, $Li_8In_5Si_3$, and $Li_8Ga_5Ge_3$

| Intermetallic phase | Ref. | a | Intermetallic phase | Ref. | a | Intermetallic phase | Ref. | a |
|---|---|---|---|---|---|---|---|---|
| $Li_2PdGe$ | 919 | 6.03 | $LiPd_2Ge$ | 919 | 6.01 | LiAlGe | 920 | 5.98 |
| $Li_2PdSn$ | 919 | 6.31 | $LiPd_2Sn$ | 919 | 6.26 | $Li_2AlGe$ | 920 | 6.16 |
| $Li_2PdPb$ | 919 | 6.44 | $LiPd_2Pb$ | 919 | 6.38 | $Li_2InSi$ | 920 | 6.77 |
| $Li_2PtSn$ | 919 | 6.27 | LiInGe | 920 | 6.304 | $Li_8In_5Si_3$ | 920 | 6.78 |
| $Li_2IrSn$ | 919 | 6.24 | LiInSn | 920 | 6.676 | $Li_8Ga_5Ge_3$ | 920 | 6.13 |

hexagonal) of $Li_5Fe_7Ge_8$[917] and a single modification (orthorhombic) of $Li_2CeGe$[918] have been described. Lithium forms f.c.c. ternary compounds with the transition metals [Pd, Pt, and Ir (M)] and Group IV elements [Ge, Sn, and Pb (X)] of formulae $Li_2MX$ and/or $LiM_2X$ (Table 28).[919] Single-crystal studies show that $LiPd_2Ge$ crystallizes in a Heusler lattice ($L2_1A$-type), whereas $Li_2PdSn$ has a modified $Li_3BC$ structure ($L2_1C$-type).[919] New ternary phases in the Li–M–X (M = Al, Ga, or In; X = Si, Ge, or Sn) systems have been prepared by heating elemental mixtures in Ta crucibles;[920] the resulting phases have been characterized by X-ray investigation and some chemical studies.

The preparation and characterization of the two phases $Ca_{10-x}Si_{12-2x}As_{16}$ and $Ca_{10-x}Si_{12-x}P_{16}$ ($0.66 \leq x \leq 2.50$) have been described.[921] Their structures are isotypic and crystallize with monoclinic symmetry, of space group $P2_1/m$;[921,922] they may be related to a slightly distorted NaCl-type. Their unit-cell parameters are included in Table 27. A comparison of the different methods of crystal growth of $ZnSiP_2$, $ZnSiAs_2$, and $CdGeAs_2$ (chalcopyrite structures) has also been undertaken.[923]

Johnson[924] has shown that stoicheiometric MnCoGe and MnNiGe are orthorhombic, ordered $Co_2P$-type at room temperature, and not hexagonal $Ni_2In$-type as previously reported. These germanides and the corresponding silicides

---

[917] H.-U. Schuster and E. Welk, *Z. Naturforsch.*, 1974, **29b**, 698.
[918] H.-U. Schuster and A. Czybulka, *Z. Naturforsch.*, 1974, **29b**, 697.
[919] C.-J. Kistrup and H.-U. Schuster, *Z. anorg. Chem.*, 1974, **410**, 113.
[920] W. Bockelmann and H.-U. Schuster, *Z. anorg. Chem.*, 1974, **410**, 241.
[921] M. Hamon, J. Guyader, and J. Lang, *Rev. Chim. minérale*, 1975, **12**, 1.
[922] M. Hamon, J. Guyader, P. L'Haridon, and Y. Laurent, *Acta Cryst.*, 1975, **B31**, 445.
[923] B. B. Mercey and A. Deschanvres, *Compt. rend.*, 1975, **280**, C, 1239.
[924] V. Johnson, *Inorg. Chem.*, 1975, **14**, 1117.

**Table 29** Structural chemistry of MnCoSi, MnCoGe, MnNiSi, and MnNiGe[924]

| Intermetallic phase | Temperature /°C | Structure type | a/Å | b/Å | c/Å |
|---|---|---|---|---|---|
| MnCoSi | 25 | O—$Co_2P$ | 5.8643(5) | 3.6872(4) | 6.8548(5) |
| MnCoSi | 1000 | H—$Ni_2In$ | 4.03 | — | 5.29 |
| MnNiSi | 25 | O—$Co_2P$ | 5.8967(6) | 3.6124(5) | 6.9162(6) |
| MnNiSi | 1000 | H—$Ni_2In$ | 4.04 | — | 5.38 |
| MnCoGe | 25 | O—$Co_2P$ | 5.9572(8) | 3.8168(8) | 7.0542(11) |
| MnCoGe | 135 | H—$Ni_2In$ | 4.10 | — | 5.36 |
| $Mn_{0.975}$CoGe | 25 | H—$Ni_2In$ | 4.0835(4) | — | 5.3097(5) |
| $Mn_{0.95}$CoGe | 25 | H—$Ni_2In$ | 4.067 | — | 5.300 |
| MnNiGe | 25 | O—$Co_2P$ | 6.0421(6) | 3.7550(4) | 7.0860(6) |
| $Mn_{0.975}$NiGe | 25 | O—$Co_2P$ | 6.0268(5) | 3.7581(4) | 7.0714(5) |

(Table 29) transform diffusionlessly at elevated temperatures into the $Ni_2In$-type structure. For the germanides, the orthorhombic → hexagonal transition temperature decreases rapidly with small compositional changes; indeed, $Mn_{0.975}$CoGe adopts the $Ni_2In$-type structure at room temperature (Table 29).[924] These results have been substantially corroborated by Jeischko[925] in an independent study of $Mn_{0.98}$CoGe, its phase-transformation temperature (42—67 °C) lying between those determined by Johnson for stoicheiometric MnCoGe (125—185 °C) and $Mn_{0.975}$CoGe (room temperature). The structural analyses differ, however, in the detailed assignment of the low-temperature phase,; Jeischko suggests it to be an ordered $PbCl_2$ (TiNiSi)-type[925] (cf. ordered $Co_2P$-type[924]). Examination of the ternary phases FeCrGe and FeCoGe, formed by dissolution of Cr or Co in the cubic form of FeGe, has been undertaken;[926] they possess metallic character, and substitution of Fe leads to a diminishing Curie temperature.

A new family of ternary silicides of general formula $MM'Si_3$ (M = Ti Nb, or Ta; M' = Fe, Co, or Ni) have been prepared by direct reaction of the elements;[927] they crystallize with hexagonal symmetry, with a and c parameters close to 17 and 8 Å, respectively. It is suggested that the ternary compounds $Ti_{12}Fe_{54}Si_{36}$, $TiCo_3Si_2$, and $Ti_{14}Ni_9Si_{37}$, obtained earlier by Russian authors,[928,929] might be these members of the $MM'Si_3$ family. The structure of $Ni_2GaGe$, the last unknown structure in the NiGa–NiGe pseudo-binary system, has been analysed as orthorhombic, of space group Cmcm, a = 3.92, b = 12.08, c = 11.03 Å.[930] It belongs to the CoSn structural family. The structural relations between NiGa, $Ni_4Ga_3Ge$, $Ni_2GaGe$, $Ni_4GaGe_3$, and NiGe may be described as a structural expansion with increasing valence-electron concentration.[930]

Interest in $Pb_{1-x}Sn_xTe$ solid solutions and related systems has been maintained, presumably because of their application in semiconductor technology. Three

---

[925] W. Jeitschko, Acta Cryst., 1975, **B31**, 1187.
[926] C. LeCaer and B. Roaves, Compt. rend., 1975, **281**, C, 527.
[927] J. Steinmetz, J.-M. Albrecht, and B. Malaman, Compt. rend., 1974, **279**, C, 1119.
[928] V. Ya. Markiv, L. A. Lysenko, and E. I. Gladyshevsky, Izvest. Akad. Nauk. S.S.S.R., Neorg. Materialy, 1966, **11**, 1980.
[929] V. Ya. Markiv, E. I. Gladyshevsky, P. I. Kripyakevich, and T. I. Fedoruk, Izvest. Akad. Nauk. S.S.S.R., Neorg. Materialy, 1966, **7**, 1317.
[930] M. K. Bhargava and K. Schubert, J. Less-Common Metals, 1974, **58**, 177.

papers,[931—933] in which phase relationships in the Pb–Sn–Te system have been elucidated, have been published during the period of this Report. Phase equilibria (crystallization fields) in the related Pb–Sn–Se,[934] Ga–Ge–Se,[935] As–Ge–Te,[936] and Cu–Sn–Te[937] systems have also been established. The existence of ternary intermetallic compounds in the Ga–Ge–Se and Cu–Sn–Te systems has been demonstrated; both $Ga_2GeSe_5$ (d. 625 °C) and $Cu_2SnTe_3$ (d. 412 °C) are considered to have a superstructure of a zinc-blende type.

Diverse investigations of the miscellaneous ternary systems Sn–Pb–Cd,[938] Sn–Ga–In,[939] Sn–Sb–Bi,[940] Pb–In–Sb,[941] and Pb–Bi–Hg[942] have been undertaken by a number of Russian authors. It was concluded from the results of an ultrasonic study of Sn–Pb–Cd liquid solutions that intermetallic compounds are not formed in this system.[938] Thermodynamic analysis of the Sn–Sb–Bi system shows both positive and negative deviations from ideality in the liquid state.[940] Finally, interpretation of the atomic distribution functions of Pb–In–Sb solutions has led to the conclusion that the melts have a microheterogeneous structure in the fusion region.[941]

---

[931] V. P. Zlomanov, L. K. Fu, A. M. Gas'kov, and A. V. Novoselova, *Russ. J. Inorg. Chem.*, 1974, **19**, 1058.
[932] V. P. Zlomanov, L. K. Fu, A. M. Gas'kov, and A. V. Novoselova, *Russ. J. Inorg. Chem.*, 1974, **19**, 1385.
[933] A. M. Gas'kov, L. K. Fu, V. P. Zlomanov, and A. V. Novoselova, *Russ. J. Inorg. Chem.*, 1975, **20**, 479.
[934] V. I. Shtanov, A. K. Gapeev, and A. V. Novoselova, *Doklady Chem.*, 1974, **218**, 637.
[935] C. Thiebault, L. Guen, R. Eholie, and J. Flahaut, *Bull. Soc. chim. France*, 1975, 967.
[936] G. Z. Vinogradova, S. A. Dembovskii, A. N. Kopeikina, and N. P. Luzhnaya, *Russ. J. Inorg. Chem.*, 1975, **20**, 769.
[937] C. Carcaly, J. Rivet, and J. Flahaut, *J. Less-Common Metals*, 1975, **41**, 1.
[938] A. S. Rubtsov and V. I. Stremovsov, *Russ. J. Phys. Chem.*, 1974, **48**, 29.
[939] P. E. Shurai, V. N. Danilin, and I. T. Srylalin, *Russ. J. Phys. Chem.*, 1974, **48**, 780.
[940] A. A. Vecher, E. I. Voronova, L. A. Mechkovskii, and A. S. Skoropanov, *Russ. J. Phys. Chem.*, 1974, **48**, 584.
[941] Ya. I. Dutchak and V. S. Frenchko, *Russ. J. Phys. Chem.*, 1975, **49**, 353.
[942] G. V. Malyutin and M. V. Nosek, *Russ. J. Phys. Chem.*, 1974, **48**, 616.

# 5
# Elements of Group V

BY N. LOGAN AND D. B. SOWERBY

Several reviews encompassing a number of aspects of the inorganic chemistry of Group V elements have appeared. These include surveys of X-ray diffraction studies of a number of N, P, As, Sb, and Bi compounds,[1] the stereochemistry of compounds containing Si—N, P—N, S—N, and Cl—N bonds,[2] and ESCA-derived group shifts for a substantial number of nitrogen, phosphorus, and arsenic compounds. The latter describe the experimental shifts to within 0.5 eV, and facilitate the predictive and analytical use of ESCA.[3]

## 1 Nitrogen

Reviews dealing specifically with nitrogen compounds provide coverage of the mechanisms of reactions in solutions involving simple inorganic nitrogen species;[4] the structures, spectra, ionization potentials, electron affinities, reactions, and complexes of nitrogen oxides and oxyacids;[5] and chemical and biological aspects of the fixation and reduction of molecular nitrogen.[6] The latter includes consideration of methods of preparation of metal–dinitrogen complexes, bonding in this type of complex, reactivity of complexed $N_2$, models for nitrogenase, and routes for the reduction of complexed nitrogen.

**Elementary Nitrogen.**—Semi-empirical LCAO calculations of electronic and dynamical properties of $\alpha$- and $\gamma$-nitrogen crystals and nitrogen aggregates have been reported.[7] Complete valence-shell configuration interaction calculations have been carried out for a large number of non-Rydberg states of $N_2$, $N_2^-$, $N_2^+$, and $N_2^{2+}$. Complete potential curves for these states were used for the calculation of the spectroscopic constants. Comparison has been made with the available experimental data and several new states and their properties have been predicted.[8] The ionization potentials of molecular nitrogen have been calculated by a

---

[1] M. B. Hursthouse, in 'Molecular Structure by Diffraction Methods', ed. G. A. Sim and L. E. Sutton (Specialist Periodical Reports), The Chemical Society, London, 1974, Vol. **2**, p. 459.
[2] L. V. Volkov and L. S. Kjaikin, *Topics Current Chem.*, 1975, **53**, 25.
[3] B. J. Lindberg and J. Hedman, *Chemica Scripta*, 1975, **7**, 155.
[4] M. N. Hughes, 'M.T.P. International Review of Science; Inorganic Chemistry, Series Two', ed. M. L. Tobe, Butterworth, London, 1974, Vol. **9**, p. 21.
[5] A. J. Vosper, 'M.T.P. International Review of Science; Inorganic Chemistry, Series Two', ed. D. B. Sowerby, Butterworth, London, 1975, Vol. **2**, p. 123.
[6] I. Fischler and E. Körner von Gustorf, *Naturwiss.*, 1975, **62**, 63.
[7] A. Zunger, *Mol. Phys.*, 1974, **28**, 713.
[8] E. W. Thulstrup and A. Andersen, *J. Phys. (B)*, 1975, **8**, 965.

Green's function approach.[9] A simple model of the low-temperature phases of diatomic molecular solids has been examined. The calculations[10] show that ortho-$H_2$ and $N_2$ have the optimum quadrupole structure, $Pa3$. The self-consistent phonon approximation of anharmonic lattice dynamics has been applied to solid $\beta$-$N_2$. The phonon spectrum and thermal expansion as a function of temperature at zero pressure were calculated and the $\alpha$-$\beta$ f.c.c.–h.c.p. transition temperature has been estimated.[11]

The refractive index of gaseous nitrogen has been measured interferometrically at several wavelengths, temperatures in the range $-30$ to $+90\,°C$, and pressures up to $ca.$ 60 bar. The data were used to calculate the density and mean electronic polarizability.[12] The boiling points, freezing points, vapour pressures, and triple points (from 56 K to the normal boiling point) for pure $N_2$ (99.999%) and various $N_2$ samples doped with 100 vol. p.p.m. of $O_2$, Ar, Kr, and CO impurities have been determined. The heats of sublimation, vaporization, and fusion at the triple point of pure $N_2$ are 6.7738, 6.0496, and 0.7243 kJ mol$^{-1}$, respectively, and values of these quantities for the doped samples have also been given.[13] The time constant for the collisional deactivation of the $v=1$ vibrational level of $N_2$ has been found to be $1.5\pm0.5$ s, in liquid $N_2$ of 99.9995% purity at 78 K. This result is consistent with a simple binary collision theory of vibrational relaxation for liquids.[14] A new theory of active nitrogen has been reported.[15]

The natural-abundance $^{15}$N n.m.r. (18.25 MHz) line position of $^{15}$N$^{14}$N in liquid $N_2$ at its boiling point is 65.6 p.p.m. upfield from $^{15}$NO$_3^-$ or 288.8 p.p.m. downfield from the $^{15}$NH$_4^+$ resonance. The $^{15}$N resonance is sharp (linewidth $<0.5$ p.p.m.) and, because of rapid quadrupolar relaxation of the $^{14}$N nucleus, it displays no evidence of $^{15}$N–$^{14}$N scalar coupling.[16] The scattering of 4880 Å light from liquid nitrogen has been observed along the liquid–vapour coexistence line at 69.4—125 K. The light-scattering spectrum of gaseous nitrogen at 127.8 K and 32 atm. has also been observed, and the spectra were interpreted in terms of molecular reorientations, correlations of molecular reorientational motions, and intermolecular short-range interactions.[17] Theoretical investigations of the temperature dependence of the n.q.r. frequency for crystals of $\alpha$-nitrogen in terms of the anharmonic libration of a molecule in a crystal, obtain better results when conventional harmonic approximation treatments are used, and this leads to good agreement between calculated and experimental results.[18] The rotational and rotation–vibrational Raman spectra of $^{14}$N$_2$, $^{14}$N$^{15}$N, and $^{15}$N$_2$ have been recorded, and the internuclear distances $r_e$ were calculated independently for each species from an analysis of the bands. Within the experimental accuracy, the distances are

---

[9] L. S. Cederbaum and W. Von Niessen, *J. Chem. Phys.*, 1975, **62**, 3824.
[10] C. A. English and J. A. Venables, *Proc. Roy. Soc.*, 1974, **A340**, 57.
[11] J. C. Raich, N. S. Gillis, and T. R. Koehler, *J. Chem. Phys.*, 1974, **61**, 1411.
[12] N. J. Timoshenko, E. P. Kholoclov, and A. L. Yamnov, *Zhur. fiz. Khim.*, 1974, **48**, 2139.
[13] J. Ancsin, *Canad. J. Phys.*, 1974, **52**, 1521.
[14] W. F. Calaway and G. E. Ewing, *Chem. Phys. Letters*, 1975, **30**, 485.
[15] R. A. Young, *J. Chem. Phys.*, 1974, **60**, 5050.
[16] C. H. Bradley, G. E. Hawkes, E. W. Randall, and J. D. Roberts, *J. Amer. Chem. Soc.*, 1975, **97**, 1958.
[17] P. E. Schoen, P. S. Y. Cheung, D. A. Jackson, and J. G. Powles, *Mol. Phys.*, 1975, **29**, 1197.
[18] K. Fujita, M. Suhara, and A. Sado, *J. Magn. Resonance*, 1975, **17**, 314.

identical, as required by the Born–Oppenheimer approximation. The mean value for $r_e$ obtained is $1.097701 \pm 0.000004$ Å.[19]

Neutron-diffraction studies[20] show that the internuclear separation of 1.10 Å in gaseous nitrogen is unchanged in liquid nitrogen. A polemic on the crystal structure of $\alpha$-$N_2$ has been published.[21]

The thermal reaction of Ba with $N_2$ yields $Ba_3N_2$ and $Ba_2N_2$. The latter compound is formed by pernitridation of $Ba_3N_2$ and contains the $N_2^{4-}$ ion. After the reaction has proceeded for 50 h at 380—400°C, the product evolves *ca.* 31% of its nitrogen content as $N_2H_4$ when it is hydrolysed.[22]

The distribution of ion types in a cylindrical hollow-cathode discharge in $N_2$ containing traces of $H_2O$ has been explored,[23] using a small radially movable probe incorporating an entrance aperture to a mass spectrometer. $N_2^+$ and $N^+$ at the edge of the glow give way to $N_2H^+$ towards the axis. Quantitative rate-constant measurements were obtained for formation and destruction of $N_2H^+$ and $N^+$. The rate constant for proton transfer from $H_3^+$ to $N_2$ has been measured as a function of the partial pressure of the $H_2$ buffer gas, and it shows a very large pressure dependence.[24]

$N_2$ *Complexing.* LCAO–MO calculations have been made on the electronic structures of the complexes $M,N_2$ ($M = H^+$, $Li^+$, Be, or $B^+$), $N_2,BX_3$ ($X = F$ or H), and $M,N_2M$ ($M = H^+$, $Li^+$, $B^+$, or $BH_3$). The calculations were used to discuss models of nitrogen fixation by enzymes in biological systems.[25] Quantum mechanical calculations by the CNDO/2 method confirm the possibility of $N_2$ complex formation with $Li^+$ and $F^-$ ions. The contribution of covalent bonds to the energy of the complexes has been analysed and their lability discussed.[26]

The products of the co-condensation reactions of Ni, Pd, and Pt atoms with mixtures of $O_2$, $N_2$, and Ar at 6—10 K, investigated by matrix-isolation i.r. spectroscopy, have been established to be $(O_2)M(N_2)$ and $(O_2)M(N_2)_2$, containing side-on-bonded dioxygen and end-on-bonded dinitrogen. Bond stretching force-constants enable a unique insight to be gained into the bonding of $N_2$ and $O_2$ to a transition metal in a situation where both ligands are competing for the same bonding electrons.[27]

Protonation and reduction of dinitrogen in the mononuclear species *trans*-$[M(N_2)_2(diphos)_2]$ ($M = Mo$ or W; diphos = $Ph_2PCH_2CH_2PPh_2$ or $Et_2PCH_2$-$CH_2PEt_2$) or *cis*-$[W(N_2)_2(PMe_2Ph)_4]$ occurs on reaction with excess HCl or HBr, to yield mononuclear diazene-$N$-(HN=NH) or hydrazido(2−)$N$-(H_2N—N^{2-}) complexes. This work provides the first chemical evidence that dinitrogen is as likely to be reduced at a mono- as at a bi-metal site in nitrogenase.[28] *trans*-$[ReCl(N_2)(PMe_2Ph)_4]$ reacts with $Na[S_2CNR_2]$ ($R = Me$ or Et) to give *mer*-

---

[19] J. Bendtsen, *J. Raman Spectroscopy*, 1974, **2**, 133.
[20] D. I. Page and J. G. Powles, *Mol. Phys.*, 1975, **29**, 1287.
[21] W. N. Lipscomb, *J. Chem. Phys.*, 1974, **60**, 5138.
[22] K. H. Linke and K. Schroedter, *Z. anorg. Chem.*, 1975, **413**, 165.
[23] F. Howarka, W. Lindinges, and R. N. Varney, *J. Chem. Phys.*, 1974, **61**, 1180.
[24] J. K. Kim, L. P. Theard, and W. T. Huntress jun., *Chem. Phys. Letters* 1975, **32**, 610.
[25] I. B. Golovanov and V. M. Sobolev, *Teor. i eksp. Khim.*, 1974, **10**, 327.
[26] S. G. Agarin and I. A. Lygina, *Zhur. strukt. Khim.*, 1975, **16**, 112.
[27] G. A. Ozin and W. E. Klotzbuecher, *J. Amer. Chem. Soc.*, 1975, **97**, 3965.
[28] J. Chatt, G. A. Heath, and R. L. Richards, *J.C.S. Dalton*, 1974, 2074.

[Re(S$_2$CNR$_2$)(N$_2$)(PMe$_2$Ph)$_3$]. It was hoped that the high electron-donor power of the dialkyldithiocarbamate ligand would induce protonation of dinitrogen by hydrogen halides, but the dithiocarbamate derivatives, although much more reactive towards HCl and HBr than the parent complexes, undergo loss of N$_2$ to give a new series of rhenium hydrido-complexes.[29] However, two reports of the reduction of co-ordinated molecular nitrogen to ammonia in a protic environment have now appeared. $^{15}$NH$_3$ is obtained on exposure of trans-[Mo($^{15}$N$_2$)$_2$(diphos)$_2$] (diphos = Ph$_2$PCH$_2$CH$_2$PPh$_2$) to HCl or HBr at room temperature. For highest yields, a HBr:Mo ratio of 100:1 and N-methylpyrrolidone as solvent are employed, and intermediates for the overall process have been suggested.[30] Likewise, the reaction of [M(N$_2$)$_2$(PMe$_n$Ph$_{3-n}$)$_4$] (M = Mo or W; n = 1 or 2) with H$_2$SO$_4$ in MeOH or THF gives NH$_3$ yields of up to 90% and provides a model for the feed of electrons from the metal atom to N$_2$ along the atom chain M—N—N.[31] trans-[IrCl(N$_2$)-PPh$_3$)$_2$] undergoes an oxidative addition reaction[32] with MeO$_3$SCF$_3$, with retention of the dinitrogen ligand, to give [IrCl(N$_2$)(O$_3$SCF$_3$)Me(PPh$_3$)$_2$], and a nitrogen-carrying metalloporphine, dinitrogen(octaethylporphinato)tetrahydrofuranosmium(II), (1; X = N$_2$, X' = THF), has been reported.[33]

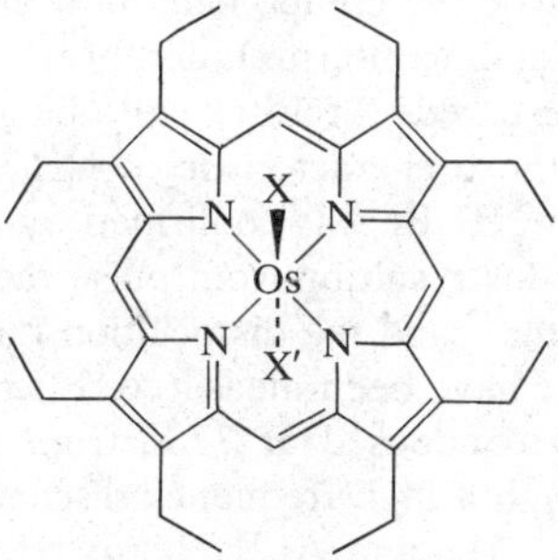

(1)

**Bonds to Hydrogen.**—NH$_3$. The ionization potentials of O$_3$ and NH$_3$ have been calculated by the LCAO-X$_\alpha$ method. For NH$_3$, the grouping and ordering of energy levels and the ordering within groups were determined. The N—H bond length, the HNH angle, and the inversion barrier were also found.[34] The contribution of the correlation energy to this barrier has been shown to be small (<10% of the total).[35] The adequacy of approximate SCF–MO calculations has been studied for selected molecular properties of NH$_3$ and H$_2$O. The analysis involved four ab initio treatments; the basic SCF–MO scheme, the same scheme with a small basis set, and two different mixed-basis schemes.[36] The electric field gradient in ammonia has been calculated from ab initio wavefunctions and a new

---

[29] J. Chatt, R. H. Crabtree, J. R. Dilworth, and R. L. Richards, J.C.S. Dalton, 1974, 2358.
[30] C. R. Brulet and E. E. Van Tamelen, J. Amer. Chem. Soc., 1975, **97**, 911.
[31] J. Chatt, A. J. Pearman, and R. L. Richards, Nature, 1975, **253**, 39.
[32] D. M. Blake, J.C.S. Chem. Comm., 1974, 815.
[33] J. W. Buchler and P. D. Smith, Angew. Chem. Internat. Edn., 1974, **13**, 745.
[34] H. Sambe and R. H. Felton, J. Chem. Phys., 1974, **61**, 3862.
[35] N. C. Dutta and M. Karplus, Chem. Phys. Letters, 1975, **31**, 455.
[36] J. Mrozek and R. Nalewajeski, Acta Phys. Polon., 1974, **A46**, 199.

value of the quadrupole moment for $^{14}$N suggested.[37] As a contribution towards the simulation of hydrogen bonding in biological systems, *ab initio* calculations have been performed for $(NH_3)_2$ and $NH_3(NH_4^+)$ to study the effects of geometric distortion on the N—H···N—H bonds. The optimum separations between the N atoms in linear hydrogen bonds are 3.37 and 2.79 Å for $(NH_3)_2$ and $NH_3(NH_4^+)$ respectively.[38]

The activity of Na in liquid $NH_3$ has been obtained from measurements of vapour pressure together with previously published electrochemical data.[39] For the reaction (1) in the hypothetical 1 molal standard state, $\Delta H° = 6.10$ kJ mol$^{-1}$. The Tait equation of state has been used to calculate the molar volumes of $NH_3$ at 100—10 000 bar and 50—200 °C. The calculations showed that the equation is valid at ≥100 bar and 50 °C and at ≥200 bar and 100—200 °C. Temperature dependences of the constants $B$ and $C$ of the Tait equation and the equation accuracy are discussed.[40]

$$Na(s) \longrightarrow Na^+_{NH_3} + e^-_{NH_3} \qquad (1)$$

The low-temperature i.r. spectra of $NH_3$ 'complexes' with $Cl_2$ and $Br_2$, obtained by simultaneous admission of the components into an optical cryostat at about 72 K, have been recorded, and an interpretation of the main bands of ammonia in these species has been presented. At 150 K, the bands assignable to the 'complexes' were replaced by those characteristic of $NH_4^+$, and the low-temperature reaction yielding $NH_4Cl$ or $NH_4Br$ was confirmed by a calorimetric method.[41]

Yields of solvated electrons resulting from pulse radiolysis of liquid ammonia, methylamine, and ethylamine[42] and the dissociation rate of ammonia in glow and high-frequency discharges[43] have been measured. E.s.r. measurements have been made on $NH_3$ (and $ND_3$) condensed at 77 K from the vapour of a fast flow system, after passage through a high-frequency discharge and after reaction with discharge products from $N_2$, $H_2$, and Ar. $NH_2$ (or $ND_2$) radicals were formed in each case, although the e.s.r. spectra differed somewhat for the different experimental conditions employed.[44] From electrochemical measurements, the limiting stage in the reaction of sulphur with liquid ammonia at −40 °C was found to be the physical dissolution of the sulphur.[45] The principal reaction occurring was (2), in accordance with which, the number of moles of solid sulphur dissolving equalled the number of $NH_4^+SCN^-$ ion pairs formed in the presence of $K^+SCN^-$. I.r. and X-ray photoelectron spectroscopy have been used[46] to identify the products of the reactions of $NH_3$, $N_2H_4$, and $NH_2OH$ with $Na_2[Fe(CN)_5(NO)]$. Reactions of $H^+$ with $CH_4$, $NH_3$, $H_2O$, and $O_2$ have been examined by the ion

---

[37] C. T. O'Konski and J. W. Jost, *U.S.N.T.I.S., AD Report*, 1974, No. 787 837/4GA.
[38] N. C. Baird, *Internat. J. Quantum Chem., Quantum Biol. Symp.*, 1974, **1**, 49.
[39] G. Lepoutre, M. De Backer, and A. Demortier, *J. Chim. phys.*, 1974, **71**, 113.
[40] S. S. Tsimmerman, *Zhur. fiz. Khim.*, 1974, **48**, 1859.
[41] E. V. Belousova, Y. M. Kimel'fel'd, and A. P. Shvedchikov, *Zhur. fiz. Khim.*, 1975, **49**, 1075.
[42] W. A. Seddon, J. W. Fletcher, F. C. Sopchyshyn, and J. Jevcak, *Canad. J. Chem.*, 1974, **52**, 3269.
[43] I. D. Zimina, A. I. Maksimov, and V. I. Svettsov, *Izvest. V.U.Z. Khim. i khim. Tekhnol.*, 1974, **17**, 1644.
[44] F. W. Froben, *J. Phys. Chem.*, 1974, **78**, 2047.
[45] R. Guiraud, M. Aubry, and B. Gilot, *Bull. Soc. chim. France*, 1975, 490.
[46] K. B. Yatsimirskii, V. V. Nemoshkalenko, Yu. P. Nazarenko, V. G. Aleshin, V. V. Zhilinskaya, and Yu. D. Taldenko, *Teor. i eksp. Khim.*, 1974, **10**, 653.

cyclotron resonance trapped-ion and ion-ejection methods, and rate constants for these reactions were measured.[47] The reaction with $NH_3$ proceeds exclusively by charge transfer to produce $NH_3^+$. Rate constants have also been obtained for proton-transfer reactions of the type (3) (X = $H_2$, $NH_2$, $CH_4$, $H_2O$, CO, $N_2$, $C_2H_4$, $C_2H_6$, $C_3H_6$, $N_2O$, or $C_4H_8$) at 297 K by using the flowing afterglow technique, and comparisons were made with classical theories and the exothermicity of the reactions.[48] The kinetics and mechanism of $NH_3$ oxidation on Pt catalysts have been studied at the relatively low temperature of 220—235 °C, where the reaction products are $N_2$ and $N_2O$.[49] The experimental activation energy is 125.4 kJ mol$^{-1}$.

$$3S + 5NH_3 \rightarrow N_2 + 3NH_4^+ + 3SH^- \qquad (2)$$

$$XH^+ + NH_3 \rightarrow NH_4^+ + X \qquad (3)$$

$NH_4^+$. A Roothan LCAO SCF procedure with a Gaussian basis set has been used to investigate the preferences of the ammonium ion for binding ammonia or water and to determine the type of binding involved. In the first co-ordination shell, the affinity for $NH_3$ is larger, and a configuration involving fixation along the line of a NH bond is strongly preferred to a bisecting position. In the second solvation shell, binding of water is favoured.[50]

Precise values of the activity coefficients of aqueous ammonium chloride solutions at 25 °C, determined from e.m.f. measurements of cells with transference, have been reported for the concentration range 0—0.2 mol l$^{-1}$. The results show no anomalous behaviour with respect to the Debye–Hückel limiting law. An interpretation of excess thermodynamic functions of potassium and ammonium chloride solutions has been made in terms of ionic influences on solvent structure.[51]

The laser Raman spectra of $NH_4NO_3$ and $ND_4NO_3$ have been measured between 210 and 320 K and used to study phase transitions,[52] and an e.s.r. study of γ- and u.v.-irradiated $NH_4NO_3$ at 77 K gave resonances assigned to the axially symmetric species $NO_3^{2-}$ and $NO_3$ tumbling freely in the crystal lattice, and $O_2^-$ at various surface sites.[53]

The temperature dependence of the internal vibrations of sulphate ions in $(NH_4)_2SO_4$ has been studied by i.r. and Raman spectroscopy. The phase transition in this compound is correlated with disordering connected mainly with the reorientational motion of $SO_4^{2-}$ ions.[54]

The rotational motion of $NH_4^+$ in $NH_4ClO_4$ has been studied by quasi-elastic neutron scattering by a polycrystalline sample at 66—150 K. The shapes and widths of the quasi-elastic-scattering peaks as a function of momentum transfer

---

[47] W. T. Huntress jun., J. K. Kim, and L. P. Theard, *Chem. Phys. Letters*, 1974, **29**, 189.
[48] R. S. Hemsworth, J. D. Payzant, H. I. Schiff, and D. K. Bohme, *Chem. Phys. Letters*, 1974, **26**, 417.
[49] N. I. Ul'chenko, G. I. Golodets, and I. M. Avilova, *Teor. i eksp. Khim.*, 1975, **11**, 56.
[50] A. Pullman and A. M. Armbruster, *Internat. J. Quantum Chem., Symp.*, 1974, **8**, 169.
[51] R. E. Verall, *J. Solution Chem.*, 1975, **4**, 319.
[52] D. W. James, M. T. Carrick, and W. H. Leong, *Chem. Phys. Letters*, 1974, **28**, 117.
[53] F. H. Jarke and N. A. Ashford, *J. Chem. Phys.*, 1975, **62**, 2923.
[54] M. V. Belousov, V. A. Kamyshev, and A. A. Shultin, *Izvest. Akad. Nauk S.S.S.R., Ser. fiz.*, 1975, **39**, 744.

are consistent with instantaneous reorientation about the four $C_3$ axes.[55] $NH_3$ does not significantly affect the thermal decomposition of orthorhombic $NH_4ClO_4$ (at 239 °C) but completely inhibits the process in the case of the cubic form (at 252.5 °C). This may be related to differences in the defect structures of the two polymorphs.[56] The crystal structure of $NH_4ClO_4$ has been studied at 278, 78, and 10 K by neutron diffraction. The unit cell is orthorhombic, space group *Pnma*, with $Z = 4$ at all three temperatures. The $ClO_4^-$ group and $NH_4^+$ group each have essentially ideal tetrahedral structures and are linked together by N—H···O-type hydrogen bonds, one for each hydrogen, to form a three-dimensional network.[57] The neutron-diffraction results suggest that the rotational motions of the ammonium ions are quite complex, even at 10 K.

The reaction of $CF_3CO_2NH_4$ with LiD and of $CF_3CO_2ND_4$ with LiH has been found to give hydrogen and ammonia, with extensive isotopic exchange. The detailed results of these experiments lead to the very interesting conclusion[58] that this reaction provides the first example of an $S_N2$-like nucleophilic displacement reaction on quaternary nitrogen, proceeding through pentaco-ordinated ammonium hydride, *i.e.* $NH_5$.

$NH_2OH$ *and Derivatives.* The kinetics and mechanism of the reaction between $NH_2OH$ and $O_2$ have been investigated in aqueous buffered media. The reaction proceeds to give $N_2$ as the final product, $H_2O_2$ being detected as an intermediate which is later reduced to $H_2O$. The reactions between $O_2$ and $H_2O_2$ with $NH_2OH$ are first-order with respect to each of the reagents, and only unprotonated $NH_2OH$ is the reactive species. The rate constant and activation parameters for the $O_2$–$NH_2OH$ reaction at 25 °C have been given.[59] Further work has been reported on the kinetics of the alkaline hydrolysis of hydroxylamine-*O*-sulphonic acid, and the kinetics of reaction of this compound with $N_2H_4$ and $NH_2OH$ have also been studied in alkaline solution. It was concluded that nucleophilic attack at the $sp^3$ nitrogen of $H_2NOSO_3^-$ proceeds in the order $H_2NO^- > N_2H_4 \gg OH^-$, which supports an earlier proposal that polarizability and ease of oxidation of the nucleophile are more important than proton basicity for attack at this nitrogen centre.[60] The kinetics and mechanism of the alkaline hydrolysis of hydroxylamine-*N*-monosulphonate at 80 °C, giving sulphite, $N_2O$, and amine sulphonate, have been studied. The rate law has been explained by a reversible heterolytic decomposition of the $ONH·SO_3^{2-}$ ion into HNO and $SO_3^{2-}$ and a simultaneous irreversible homolysis of the O—N bond of the ion.[61] The same authors have also reported that molecular oxygen reacts with dilute alkaline solutions of hydroxylamine-*N*-sulphonate to yield peroxynitrite quantitatively. In more concentrated solutions, peroxynitrite combines with unreacted hydroxylamine-*N*-sulphonate to form *N*-nitrosohydroxylamine-*N*-sulphonate.[62]

---

[55] H. J. Prask, S. F. Trevino, and J. J. Rush, *J. Chem. Phys.*, 1975, **62**, 4156.
[56] B. I. Kaidymov and V. S. Gavazova, *J. Inorg. Nuclear Chem.*, 1974, **36**, 3848.
[57] C. S. Choi, H. J. Prask, and E. Prince, *J. Chem. Phys.*, 1974, **61**, 3523.
[58] G. A. Olah, D. J. Donovan, J. Shen, and G. Klopman, *J. Amer. Chem. Soc.*, 1975, **97**, 3559.
[59] R. Tomat, A. Rigo, and R. Salmaso, *J. Electroanalyt. Chem. Interfacial Electrochem.*, 1975, **59**, 255.
[60] W. E. Steinmetz, D. H. Robison, and M. N. Ackermann, *Inorg. Chem.*, 1975, **14**, 421.
[61] F. Seel and J. Kaschuba, *Z. phys. Chem.* (Frankfurt), 1974, **92**, 235.
[62] F. Seel and J. Kaschuba, *Z. anorg. Chem.*, 1975, **414**, 56.

**Bonds to Nitrogen.**—$N_2H_2$. The i.r. spectra (600—2000 and 2100—3500 cm$^{-1}$) of solid diazene, $N_2H_2$, at 5 K and of $N_2H_2$, $N_2HD$, and $N_2D_2$ in a nitrogen matrix have been recorded and assigned in terms of a *trans* structure of point group $C_{2h}$. The compounds were prepared by thermolysis of the corresponding tosyl hydrazides.[63] The photoelectron spectra of $N_2H_2$ and its deuterio-analogue have also been measured. The simplicity of the spectrum of $N_2H_2$ indicates that it is in a singlet ground state.[64a] A mass spectroscopic investigation of $N_2H_2$ and its deuteriated derivatives has also been carried out.[64b]

$N_2H_4$. The polarized i.r. and Raman spectra of crystalline $N_2H_5Cl$ and $N_2D_5Cl$ have been investigated over the range 50—3370 cm$^{-1}$ at 77 K. The vibrational assignment agrees with the Teller–Redlich isotopic product rule, and the results are explained by assuming that the structure of the molecular ion departs from the staggered conformation.[65]

The thermal decomposition of hydrazine occurs mainly by reaction (4), as studied[66] in the transition range of the unimolecular $N_2H_4$ decomposition by incident and reflected shock waves at 1000—400 K and total density *ca.* $10^{-4}$ mol cm$^{-3}$ using 0.3—2.0% $N_2H_4$ mixtures in Ar containing NO ($N_2H_4$:NO ratio 1:1—50). The rate constant of the $N_2H_4$ decomposition is not influenced by NO. The catalytic decomposition of $N_2H_4$ and $N_2D_4$ on Pt has been studied in $H_2SO_4$ solution at 323—363 K. With the assumption that the kinetics of the catalytic decomposition are limited by the splitting of an N—H bond, a maximum kinetic isotope effect, in good agreement with the experimental value, was estimated.[67] The deuterium isotope effect of the $NH_3$ synthesis was newly discussed, and it was assumed that $N_2H_4$ is the adsorbed species which is in equilibrium with the formed $NH_3$.

$$N_2H_4 \rightarrow NH_3 + NH \qquad (4)$$

The kinetics of the oxidation of $N_2H_4,H_2SO_4$ by $Cr^{VI}$ have been investigated both in the presence and absence of aqueous $H_2SO_4$. At low concentrations of $N_2H_4,H_2SO_4$ (<0.02 mol l$^{-1}$) the reaction is overall of second order (first-order with respect to each reactant), and the results point to the formation of an equilibrium complex between the reactants, prior to the main reaction. A stoicheiometry of $[Cr^{VI}]:[N_2H_4] = 4:3$ and the formation of $N_2$ as the only nitrogen-containing product indicate that $Cr^{VI}$ behaves as a two-electron oxidant.[68] Kinetics have been studied and activation parameters determined for the protonation of $N_2H_4$, $NH_2OH$, $NH_3$, and $[Pt(NH_3)_4(NH_2)Cl]^{2+}$ in aqueous solution,[69] and the reaction of active nitrogen with $N_2H_4$ has been investigated.[70]

---

[63] R. Minkwitz, *Z. anorg. Chem.*, 1975, **411**, 1.
[64] (a) D. C. Frost, S. T. Lee, C. A. McDowell, and N. P. C. Westwood, *Chem. Phys. Letters*, 1975, **30**, 26; (b) G. Holtzmann and R. Minkwitz, *Z. anorg. Chem.*, 1975, **413**, 72.
[65] V. Schettino and R. E. Solomon, *Spectrochim. Acta*, 1974, **30A**, 1445.
[66] E. Meyer and H. G. Wagner, *Z. phys. Chem. (Frankfurt)*, 1974, **89**, 329.
[67] G. Shulz-Ekloff, K. Appell, D. Baresel, W. Gellert, and W. Sarholz, *Ber. Bunsengesellschaft phys. Chem.*, 1975, **79**, 263.
[68] V. M. S. Ramanujam, S. Sundaram, and N. Venkatasubramanian, *Inorg. Chim. Acta*, 1975, **13**, 133.
[69] M. N. Bargaftik, L. A. Katsman, and Ya. K. Surkin, *Izvest. Akad. Nauk S.S.S.R., Ser. khim.*, 1974, **12**, 2697.
[70] Beng-Tiong Yo, *Diss. Abs. Internat. (B)*, 1975, **35**, 4862.

$N_3H$. Second- and third-row Group VIII metals catalyse the decomposition of $N_3H$ at 323—363 K to give $N_2$ and $NH_4N_3$. Metal–N–H surface sites appear to be intermediates in the catalysed decomposition. Fe, Co, and Ni react exothermically at *ca.* 373 K to give nitrides, which do not catalyse the decomposition, but this reaction provides the basis for a unique, low-temperature synthesis of these nitrides.[71] Electronic absorption spectra indicate the formation of NH and $NF_2$ intermediates in the pulse photolysis of $N_3H$–$N_2F_4$ mixtures under adiabatic conditions. A reaction mechanism for the formation of these intermediates has been discussed.[72]

$N_4H_4$. An exciting development in the chemistry of nitrogen is the preparation and isolation of the new compound tetrazene, $N_4H_4$, which is only the fifth example of an unambiguously characterized neutral hydrogen compound of nitrogen (*cf.* $NH_3$, $N_2H_2$, $N_2H_4$, and $N_3H$). The compound is synthesized by treatment of tetrakis(trimethylsilyl)tetrazene with trifluoroacetic acid in methylene dichloride at $-78\,°C$ and is precipitated as a colourless solid which can be purified by sublimation. The compound is structurally *trans*-2-tetrazene and is unexpectedly thermostable; thermolysis of the solid begins only at *ca.* 0 °C, and gaseous tetrazene is metastable even at room temperature.[73] The photoelectron spectrum of tetrazene has been measured; in the energy range up to 21.21 eV it shows eight bands, which have been assigned in terms of qualitative MO considerations.[74]

$$(Me_3Si)_2N-N=N-N(SiMe_3)_2 \xrightarrow[-4F_3CCO_2SiMe]{+4F_3CCO_2H} H_2N-N=N-NH_2 \quad (5)$$

**Bonds to Oxygen.**—*General.* A compilation of physical data for $N_2O$, NO, and $NO_2$ has been published.[75]

Electronic absorptions determined for $N_2O_2$ (the *cis*-dimer, present in condensed phases of NO), $N_2O_3$, and $N_2O_4$ agree reasonably well with transitions predicted by the CNDO/S method.[76]

$HNO_3$ has been pyrolysed in shock waves in the presence of $NO_2$, and the formation of $HO_2$ in reaction (6) was monitored by the u.v. absorption. From the yields and lifetimes of $HO_2$, a rate constant, $(4.5\pm1)\times10^{12}\,cm^3\,mol^{-1}\,s^{-1}$ for reaction (7) was derived[77] in the temperature range 1350—1700 K. The interaction of $H_2O$ vapour under electrical discharge with aqueous $NO_3^-$ and $NO_2^-$ leads to a steady-state ratio of the two ions, irrespective of whether the starting solution contains only $NO_3^-$, $NO_2^-$, or their mixture. This ratio varies with the pH of the solution. The steady states attained due to electrical discharge and to radiolysis are closely similar, and the same sequences of reactions with virtually identical mechanisms are probably involved in the two methods.[78]

$$HO + NO_2 \rightarrow HO_2 + NO \quad (6)$$
$$HO_2 + NO \rightarrow HO + NO_2 \quad (7)$$

[71] E. L. Muetterties, W. J. Evans, and J. C. Sauer, *J.C.S. Chem. Comm.*, 1974, 939.
[72] E. N. Moskvitina, V. M. Zamanskii, and Yu. Ya. Kuzyakov, *Vestnik. Moskov. Univ., Khim.*, 1974, **15**, 486.
[73] N. Wiberg, H. Bayer, and H. Bachhuber, *Angew. Chem. Internat. Edn.*, 1975, **14**, 177.
[74] J. Kroner, N. Wiberg, and H. Bayer, *Angew. Chem. Internat. Edn.*, 1975, **14**, 178.
[75] C. L. Yaws and J. R. Hopper, *Chem. Eng.*, 1974, **81**, 99.
[76] J. Mason, *J.C.S. Dalton*, 1975, 19.
[77] K. Glänzer and J. Troe, *Ber. Bunsengesellschaft phys. Chem.*, 1975, **79**, 465.
[78] T. S. Rao and A. N. Mehetre, *Z. phys. Chem. (Frankfurt)*, 1975, **95**, 227.

A laser magnetic resonance spectrometer has been used in combination with a discharge–flow system to measure the gas-phase reaction rates of the OH radical with NO, $NO_2$ (and CO) at 296 K and over a pressure range 0.4—5 Torr.[79] A study of the behaviour of a number of oxo-nitrogen species in molten $NaCl$–$AlCl_3$ mixtures as solvents has been made. The $NO^+$ ion undergoes reversible one-electron reduction–oxidation at vitreous carbon electrodes. $NO_2^-$ reacts to produce high yields of $NO^+$ ion, which is separable from the solvent phase as NOCl. $NO_2$, $NO_2^+$, and $NO_3^-$ are reduced to varying extents, whereas $N_2O$ and NO remain unaffected for 24 hours. Possible mechanisms for these processes have been discussed.[80]

The effects of the gaseous phase composition (NO, $NO_2$) and of the gas–liquid contact time on the kinetics of the $^{15}N$–$^{14}N$ isotopic exchange between NO–$NO_2$ mixtures and $HNO_3$ has been determined. The influence of $HNO_2$ on the isotopic exchange was also determined, and $N_2O_4$, $N_2O_3$, and $HNO_2$ are intermediates in the proposed mechanism.[81]

$N_2O$ *and Other Nitrogen*(I) *Species.* The results of accurate configuration interaction calculations on the excited states of $N_2O$ have been used to provide accurate assignments of the observed excitation features. Current assignments of the $N_2O$ spectra have been compared, and results for the higher-lying valence states presented. This study also provides a comparison of the Hartree–Fock, INDO, and configuration interaction methods.[82]

Vacuum-u.v. photolysis of $N_2O$ isolated in Ar matrixes at 4 K gives direct luminescent evidence for the photodissociative production of both $O(^1S)$ and $N(^2D)$ atoms. The matrix results have been compared to relative atomic quantum yields measured in the gas phase.[83]

The net probabilities of reaction of hot H atoms with $N_2O$ to give either $N_2$ and OH or NO and NH have been determined as a function of the initial kinetic energy of H, by analysis of the products of photolysis of HI–$N_2O$ mixtures. The reaction giving $N_2$ and OH predominates at all energies.[84] In the slow reaction of $H_2$ and $N_2O$ at 540 and 600 °C the order of reaction was zero in $H_2$ and 1.2 in $N_2O$. Termination occurred by the reaction of H with $N_2O$ to give NH and NO, which competed with the propagation reaction giving OH and $N_2$, followed by the reaction of NH with $N_2O$ to give HNO and $N_2$. Of the possible reactions of HNO discussed, reaction (8) was predominant.[85]

$$HNO + HNO \rightarrow H_2O + N_2O \qquad (8)$$

SCF calculations have been carried out on the molecule HNO ('nitrosyl hydride') and its isomer HON ('nitrogen hydroxide').[86] The system is found to have a low-lying triplet state, which is expected to aid in the dimerization to hyponitrous acid, $H_2O_2N_2$.

---

[79] C. J. Howard and K. M. Evenson, *J. Chem. Phys.*, 1974, **61**, 1943.
[80] R. J. Gale and R. A. Osteryoung, *Inorg. Chem.*, 1975, **14**, 1232.
[81] D. Axente, G. Lacoste, and J. Mahenc, *J. Inorg. Nuclear Chem.*, 1974, **36**, 2057.
[82] N. W. Winter, *Chem. Phys. Letters*, 1975, **33**, 300.
[83] J. M. Brown jun., and H. P. Broida, *Chem. Phys. Letters*, 1975, **33**, 384.
[84] G. A. Oldershaw and D. A. Porter, *J.C.S. Faraday I*, 1974, **70**, 1240.
[85] R. R. Baldwin, A. Gethin, J. Plaistowe, and R. W. Walker, *J.C.S. Faraday I*, 1975, **71**, 1265.
[86] G. A. Gallup, *Inorg. Chem.*, 1975, **14**, 563.

NO *and Other Nitrogen*(II) *Species.* Visible and near-i.r. spectra have been determined for liquid nitric oxide. Pure, condensed phases of NO, which contain the *cis*-dimer, are found to be colourless, the colours commonly reported being due to contamination, notably with $N_2O_3$.[76,87] An i.r. study of nitric oxide adsorbed on silica-supported chromia indicates the presence of both monomeric NO and dimeric $N_2O_2$ as surface species.[88] These assignments are reinforced by isotopic substitution experiments.[89] The experimentally observed bands attributed to $^{15}N_2O_2$ and $^{14}N^{15}NO_2$ agree with those calculated from the assumption that the observed bands for $^{14}N$ species arise from the *cis*-dimer of $N_2O_2$. Slow isotopic mixing in the dimer species was observed at 195 K and rapid mixing at 298 K.

The steady-state concentration of NO obtained from air in a glow discharge at 1550—2000 K varies only with the gas temperature and is independent of gas pressure and density.[90] The kinetics of NO decomposition in a barrier (silent) discharge have been studied. Under the experimental conditions employed, NO decomposes by a fast first-order reaction with an activation energy of 7.5 kJ mol$^{-1}$. The mechanism of the decomposition has been discussed.[91]

The reaction of nitric oxide with hydrogen at 673 K is first-order in $H_2$ and independent of NO pressure. The reaction proceeds by means of a free-radical chain mechanism:

$$H_2 + NO(\text{surface}) \longrightarrow HNO + H \quad (9)$$

$$H + NO \rightleftharpoons HNO \quad (10)$$

$$2HNO \longrightarrow 2OH + N_2 \quad (11)$$

$$OH + H_2 \longrightarrow H_2O + H \quad (12)$$

$$2HNO \longrightarrow H_2O + N_2O \quad (13)$$

This reaction scheme resembles that proposed for the HCHO–NO pyrolysis at 773 K except for the initiating step (9), which takes place on the wall of the vessel in the $H_2$–NO pyrolysis.[92] The temperature dependence of rate constants for the reaction of atomic oxygen ($^3P$) with NO (M = $N_2O$) has been determined over the range 300—392 K.[93] The rate of oxidation of NO in gaseous and liquid phases is described by a third-order kinetic equation and the rate constant of the oxidation in the liquid phase is considerably higher than in the gaseous phase. The oxidation rate increases with temperature with an activation energy of 11.7 ± 0.8 kJ mol$^{-1}$ in the temperature region 288—303 K.[94] The rate constant, $k$/cm$^3$ mol$^{-1}$ s$^{-1}$, of reaction (14), measured[95] in an isothermal flow reactor at $298 \leq T/K \leq 670$ by e.s.r. measurements, is $k = (2 \pm 1) \times 10^{13} \exp(-1430 \text{ K}/T)$. The reaction of NO with $KO_2$ gives $KNO_2$ as the major and $KNO_3$ as the minor product.[96] The catalytic reaction (15), carried out over $NH_4$–Y zeolite calcined at various

---

[87] J. Mason, *J. Chem. Educ.*, 1975, **52**, 445.
[88] E. L. Kugler, R. J. Kokes, and J. W. Gryder, *J. Catalysis*, 1975, **36**, 142.
[89] E. L. Kugler and J. W. Gryder, *J. Catalysis*, 1975, **36**, 152.
[90] V. F. Gvozd, M. M. Bogorodskii, and Yu. M. Emel'yanov, *Zhur. fiz. Khim.*, 1974, **48**, 2250.
[91] E. N. Eremin and E. A. Rubtsova, *Zhur. fiz. Khim.*, 1974, **48**, 1190.
[92] K. Tadasa, N. Imai, and T. Inaba, *Bull. Chem. Soc. Japan*, 1974, **47**, 2979.
[93] R. Atkinson and J. N. Pitts jun., *Chem. Phys. Letters*, 1974, **27**, 467.
[94] V. L. Pogrebnaya, A. P. Usov, A. I. Baranov, A. I. Nesterenko, and P. I. Bez'yazychnyi, *Zhur. priklad. Khim.*, 1975, **48**, 954.
[95] W. Hack, K. Hoyermann, and H. G. Wagner, *Z. Naturforsch.*, 1974, **29a**, 1236.
[96] T. P. Firsova and E. Ya. Filatov, *Izvest. Akad. Nauk S.S.S.R, Ser. khim.*, 1974, **12**, 1416.

*Elements of Group V* 269

temperatures, has been described, and the origin of the catalytic activity of the reaction discussed. No products other than NOCl were detected by gas chromatography. The effects of the temperature of calcination and of the initial pressure of $Cl_2$ on the initial reaction rate were plotted.[97]

$$NO + HO_2 \rightarrow NO_2 + OH \quad (14)$$

$$2NO + Cl_2 \rightarrow 2NOCl \quad (15)$$

A detailed study of the thermal decomposition of sodium trioxodinitrate(II), $Na_2N_2O_3$, in aqueous solution has been made over a wide range of pH values by means of u.v. spectroscopy and tracer experiments, using the labelled form $Na_2(O^{15}NNO_2)$. Angeli's hypothesis of HNO (or NOH) as intermediate in the decomposition and direct precursor of $N_2O$ has been confirmed, and it is postulated that N=N bond breakage to produce HNO and $NO_2^-$ is the primary controlling process in acidic as well as in basic solutions.[98]

*Nitrogen*(III) *Species.* The relative permittivity and electrolytic conductance of liquid $N_2O_3$, $N_2O_4$, and their mixtures have been determined, together with values for the refractive index and dipole moment of $N_2O_3$. The extent of the self-ionization (16) has been estimated, and its influence on chemical reactions in this solvent discussed. The influence of $N_2O_4$ and organic solvents on the reactions of $N_2O_3$ has also been considered.[99]

$$N_2O_3 \rightleftharpoons NO^+ + NO_2^- \quad (16)$$

The gas-phase photodetachment of electrons from $NO_2^-$ ions has been investigated at wavelengths 280—740 nm, and these experiments[100] suggested the existence of a peroxy isomer of $NO_2^-$. To aid in the interpretation of these photodetachment studies, *ab initio* electronic structure calculations were carried out on 3 isomers of $NO_2^-$; (*a*) The normal or expected form, (*b*) the peroxy-form, and (*c*) the ring form, analogous to the isomer in the form of an equilateral triangle theoretically predicted for ozone. These three isomers are predicted to lie at relative energies 0.00, 3.20, and 4.35 eV. The geometrical structures and electronic charge distributions of these isomers have been discussed briefly.[101]

The clustering of $NO^+$ to $O_3$ without reaction, *i.e.* reaction (17), where M is a third body (usually He), has been investigated.[102] The studies were conducted to explain the slowness of the exothermic reaction (18).

$$NO^+ + O_3 + M \rightarrow NO^+,O_3 + M \quad (17)$$
$$NO^+ + O_3 \rightarrow NO_2^+ + O_2 \quad (18)$$

The photolysis at 330—380 nm of mixtures containing p.p.m. concentrations of gaseous $HNO_2$ in pure nitrogen and oxygen has been studied in a flow system. The rates of formation of the products NO and $NO_2$ were measured at low conversion of $HNO_2$, and from these data it was observed that the predominant

---

[97] I. Suzuki, K. Sasaki, and Y. Kaneko, *J. Catalysis*, 1975, **37**, 555.
[98] F. T. Bonner and B. Ravid, *Inorg. Chem.*, 1975, **14**, 558.
[99] A. W. Shaw, A. J. Vosper, and M. Pritchard, *J.C.S. Dalton*, 1974, 2172.
[100] J. H. Richardson, L. M. Stephenson, and J. I. Brauman, *Chem. Phys. Letters*, 1974, **25**, 318.
[101] P. K. Pearson, H. F. Schaefer, J. H. Richardson, L. M. Stephenson, and J. I. Brauman, *J. Amer. Chem. Soc.*, 1974, **96**, 6778.
[102] F. C. Fehsenfield, *J. Chem. Phys.*, 1974, **61**, 1588.

process following absorption by $HNO_2$ in this wavelength region is the dissociation (19). There is some evidence for the occurrence of the alternative process (20), which may account for ca. 10% of the overall photodissociation. An estimate of the rate constant for the reaction (21) was also obtained.[103] A kinetic study of the reaction (22) at 298 K has been reported.[104]

$$HNO_2 \xrightarrow{h\nu} OH + NO \quad (19)$$

$$HNO_2 \xrightarrow{h\nu(\lambda < 366 nm)} H + NO_2 \quad (20)$$

$$OH + HNO_2 \rightarrow H_2O + NO_2 \quad (21)$$

$$2HNO_2 + 2I^- + 2H^+ \rightarrow I_2 + 2NO + 2H_2O \quad (22)$$

$NO_2$–$N_2O_4$. The perplexing geometric parameters of the $N_2O_4$ molecule, viz. the long N—N bond, the preference for the eclipsed $D_{2h}$ structure, and the relatively high rotational barrier, have attracted further attention in ab initio computations by two independent groups of workers.[105,106] It is concluded that the N—N bonding electron pair is of $\sigma$ character, and that the long N—N bond is due to (a) the delocalization of this bonding pair over the whole molecule and (b) a large repulsion between the doubly occupied M.O.'s of the $NO_2$ fragments. The coplanarity of $N_2O_4$ results from a delicate balance of the repulsive forces, which favour the skew structure, and of the effects of bonding, which favour the planar structure.[105] The rotational barrier of $N_2O_4$ has been calculated and analysed in terms of lone-pair interactions between non-adjacent oxygen atoms, along with lone-pair donation into the central $\sigma^*$ bond. Previously suggested electronic configurations, where the $NO_2$ units are bonded only by $\pi$-interactions, have been considered for the ground state of $N_2O_4$ but were found to be dissociative, and thus of little importance.[106]

Laser spectroscopy of a supersonic molecular beam of 5% $NO_2$ in Ar permitted a high-resolution study of free molecules, unperturbed even by gas-phase collisions. The technique also resulted in a substantial reduction in rotational structure, so that the highly complicated and largely unassigned absorption spectrum of $NO_2$ was simplified to a point where individual vibronic bands were well separated.[107] The e.s.r. spectrum of gaseous $NO_2$ has been studied at 0.05—10 Torr at X band in polarizing magnetic fields of 1.5—8 kG. The spectra were interpreted by a theory which includes the effects of anisotropic electronic and molecular magnetism, intermediate coupling of angular momenta, and centrifugal distortion. Well isolated spectral lines at high magnetic field have been employed to give quantitative information on the electronic g tensor.[108]

The reaction (23) may be important in stratospheric photochemistry, and has attracted much attention on this account. Rate constants for this reaction have

---

[103] R. A. Cox, J. Photochem., 1974, **3**, 175.
[104] L. Dozsa, I. Szilassy, and M. T. Beck, Magyar Kém. Folyóirat, 1974, **80**, 267.
[105] R. Ahlrichs and F. Keil, J. Amer. Chem. Soc., 1974, **96**, 7615.
[106] J. M. Howell and J. R. Van Wazer, J. Amer. Chem. Soc., 1974, **96**, 7902.
[107] R. E. Smalley, B. L. Ramakrishna, D. H. Levy, and L. Wharton, J. Chem. Phys., 1974, **61**, 4363.
[108] D. S. Burch, W. H. Tanttila, and M. Mizushima, J. Chem. Phys., 1974, **61**, 1607.

*Elements of Group V*

now been determined from further kinetic investigations involving the stopped-flow technique in conjunction with beam sampling[109] (259—362 K) or time-of-flight[110] (260—343 K) mass spectrometry. A kinetic study of the gas-phase reaction of water vapour and $NO_2$ has also been reported.[111]

$$NO_2 + O_3 \longrightarrow NO_3 + O_2 \qquad (23)$$

*Nitric Acid.* Evidence has been obtained from vapour pressure data for the stability of the species $H_3O^+(H_2O)_2$ and $NO_3^-(HNO_3)_2$ in the $HNO_3,H_2O$ solvent system.[112] A possible ionic dissociation of the acid in the region of the monohydrate is thus (24). It is therefore of related interest to note that an X-ray crystallographic investigation of $HNO_3,3H_2O$ shows the structure to contain $H_3O^+$ ions, each of which is bonded to two $H_2O$ molecules by short hydrogen bonds [2.482(2) and 2.576(2) Å] to form $H_7O_3^+$ ions. A longer H-bond [2.800(2) Å] connects the $H_7O_3^+$ groups with one another to form spirals. These spirals are, in turn, H-bonded to $NO_3^-$ ions, thus forming a three-dimensional network.[113] The electrical conductivity, viscosity, density, vapour pressure, and the i.r. and Raman spectra of the system $HNO_3$–$CCl_3CO_2H$ reveal that the components are involved in a mutual acid–base interaction in which the $HNO_3$ behaves like a base.[114] Boiling points and liquid–vapour equilibria in the $HNO_3$–$NaNO_3$–$H_2O$ system have been determined at 130, 400, and 760 Torr.[115]

$$3HNO_3,H_2O \rightleftharpoons H_3O^+(H_2O)_2 + NO_3^-(HNO_3)_2 \qquad (24)$$

The i.r. spectra of four isotopic species of matrix-isolated $HNO_3$ have been obtained. The vibrational assignments are in general agreement with previous gas- and condensed-phase observations, and the results of diffusion experiments indicate that $HNO_3$ may form cyclic dimers, similar to the carboxylic acids.[116] The vacuum-u.v. absorption spectrum of anhydrous nitric acid has also been reported.[117] I.r. spectra of the $Ac_2O$–$HNO_3$–$H_2O$ system have been examined as a function of composition. Strong absorption bands of $AcONO_2$ and $AcOH$ were observed in the spectra of mixtures containing *ca.* 50 mol% of $HNO_3$ and $Ac_2O$. For a mixture containing 77 mol% $HNO_3$, weak absorption bands assigned to $NO_3^-$ and $NO_2^+$ appeared, and with a further increase of $HNO_3$ content in the mixture, the spectrum coincided with that of $AcOH$, but some additional bands attributed to $N_2O_5$ and $HNO_3$ appeared.[118]

In addition to their X-ray work on $HNO_3,3H_2O$, referred to above,[113] the same group of workers, in continuation of their studies on hydrogen-bonding, has reported the crystal structure of $HNO_3,H_2O$ at both 85 K and 225 K. At each

---

[109] R. E. Huie and J. T. Herron, *Chem. Phys. Letters*, 1974, **27**, 411.
[110] D. D. Davis, J. Prusazcyk, M. Dwyer, and P. Kim, *J. Phys. Chem.*, 1974, **78**, 1775.
[111] C. England and W. H. Corcoran, *Ind. and Eng. Chem. (Fundamentals)*, 1974, **13**, 373.
[112] J. G. Dawber, *J. Inorg. Nuclear Chem.*, 1975, **37**, 1043.
[113] I. Taesler, R. G. Delaplane, and I. Olovsson, *Acta Cryst.*, 1975, **B31**, 1489.
[114] M. I. Usanovich, E. B. Mel'nichenko, and T. N. Sumarokova, *Izvest. Akad. Nauk Kazakh. S.S.R., Ser. khim.*, 1974, **24**, 18.
[115] A. N. Efimov, M. I. Zhikharev, and Yu. P. Zhirnov, *Zhur. priklad. Khim.*, 1975, **47**, 1652.
[116] W. A. Guillory and M. L. Bernstein, *J. Chem. Phys.*, 1975, **62**, 1058.
[117] G. S. Beddard, D. J. Giachardi, and R. P. Wayne, *J. Photochem.*, 1974, **3**, 321.
[118] A. A. Stotskii, N. V. Bukhvalova, S. L. Gorbunova, and N. M. Stotskaya, *Zhur. priklad. Khim.*, 1974, **47**, 2606.

temperature, the structure consists of $H_3O^+$ and $NO_3^-$ ions, each of which has pseudo-three-fold symmetry. The oxonium ion is H-bonded to three different nitrate ions to form infinite layers.[119]

The formation of $NO_2$ and $NO_3$ radicals during the γ-radiolysis of 6—15M-$HNO_3$ solutions at 77 K has been studied by e.s.r. spectroscopy.[120]

The kinetics have been determined for the oxidation of $I_2$ by hyperazeotropic $HNO_3$. The data were analysed assuming that the two reactions (25) and (26) occur. Numerical values were determined for the rate constants of the forward reactions at 298 K in 16—21M-$HNO_3$.[121] Reactions of atomic oxygen and hydrogen with $HNO_3$ were studied in a discharge flow system, the kinetics being followed by using resonance fluorescence for measurement of the atom concentrations. The observed reaction in the H–$HNO_3$ system was partially due to an autocatalytic chain removal of both reactants. Diagnostic tests suggested that OH, $NO_2$, and $NO_3$ are the chain carriers.[122] The tributyl phosphate–nitric acid and tributyl phosphate–nitric acid–water systems have been studied by $^{31}P$ spin-echo and high-resolution $^1H$ n.m.r. methods. The proton chemical shifts of $H_2O$ and $HNO_3$ and the $^{31}P$ relaxation data were used to describe complex formation and to calculate kinetic parameters of exchange reactions.[123] Complex formation in the systems dibutylphosphoric acid–nitric acid and monobutylphosphoric acid–nitric acid has also been examined.[124]

$$I_2 + 3HNO_3 \rightleftharpoons 2I^+ + 2NO_3^- + HNO_2 + H_2O \quad (25)$$

$$I^+ + 6HNO_3 \rightleftharpoons I^{5+} + 4NO_3^- + 2HNO_2 + 2H_2O \quad (26)$$

*Nitrates.* Energies for various configurations of $H(NO_3)_2^-$ have been calculated using a Slater-type orbital basis set. A double-minimum, hydrogen-bond potential function was found for the configuration with the lowest energy, but the energy differences between this and the other possible configurations were not large.[125] The stability constant of $H(NO_3)_2^-$ and the dissociation constant of $HNO_3$ have been determined by e.m.f. measurements and electrical conductivity changes during the neutralization of a number of organic bases by $HNO_3$.[126]

The origin of the low-frequency asymmetry on the symmetric stretching mode of the nitrate ion in crystalline nitrates of Na, Ag, and Cs has been studied. The asymmetry arises from a combination of correlation field splitting and the presence of hot bands. Some factors influencing the position and intensity of the hot bands have been considered.[127] Comparison of vibrational spectra within the series $NO_2$, $FNO_2$, $ClNO_2$, $FONO_2$, and $ClONO_2$ has allowed unambiguous assignments for the halogen nitrate molecules. In this study,[128] Raman polarization measurements indicated that in halogen nitrates the halogen atom is perpendicular to the $ONO_2$

---

[119] R. G. Delaplane, I. Taesler, and I. Olovsson, *Acta Cryst.*, 1975, **B31**, 1486.
[120] V. N. Belevskii and L. T. Bugaenko, *Khim. vysok. Energii*, 1975, **9**, 247.
[121] J. C. Mailen and T. O. Tiffany, *J. Inorg. Nuclear Chem.*, 1975, **37**, 127.
[122] C. J. Chapman and R. P. Wayne, *Internat. J. Chem. Kinetics*, 1974, **6**, 617.
[123] A. A. Vashman and I. S. Pronin, *Zhur. neorg. Khim.*, 1975, **20**, 1043.
[124] E. Kozlowska-Milner, *Bull. Acad. polon. Sci., Sér. Sci. chim.*, 1974, **22**, 793.
[125] R. Gunde, T. Solmajer, A. Azman, and D. Hadzi, *J. Mol. Structure*, 1975, **24**, 405.
[126] Z. Pawlak, T. Jasinski, and C. Dobrogowska, *Roczniki Chem.*, 1974, **48**, 1609.
[127] D. W. James, M. T. Carrick, and H. F. Shurvell, *Austral. J. Chem.*, 1975, **28**, 1129.
[128] K. O. Christe, C. J. Schack, and R. D. Wilson, *Inorg. Chem.*, 1974, **13**, 2811.

plane, whereas a similar investigation, specifically on chlorine nitrate,[129] was taken to imply a planar structure. The vibrational spectrum of $I(NO_3)_3$ was found to be consistent with predominantly covalent nitrato-ligands, and experimental evidence has been obtained for the formation of the new and thermally unstable compound $CF_3I(NO_3)_2$ in the $CF_3I–ClONO_2$ system. Attempts to convert this compound into $CF_3ONO_2$ were, however, unsuccessful.[128]

Crystal data for $Ca(NO_3)_2,2H_2O$[130] and $Cd(NO_3)_2,2H_2O$[130,131] have been reported.

The complexes $(R_4N)_2[Be(NO_3)_4]$ (R = Me or Et) have been prepared by the reaction of $BeCl_2$ and $R_4NCl$, taken in a 1:2 molar ratio, with a $N_2O_4$–NOCl mixture at −4 to −6 °C for 1.5—2 h. On the basis of i.r. spectroscopy, the nitrate ligands are considered to be unidentate, and it has also been proposed that the previously reported compound $Be(NO_3)_2,2N_2O_4$ should be formulated as $(NO)_2[Be(NO_3)_4]$.[132] The synthesis of the tetranitratogallates(III) $M[Ga(NO_3)_4]$ (M = K or Rb), by condensation of excess $N_2O_5$ on to an equimolar MBr–$GaBr_3$ mixture, has also been reported. The tetranitratogallates were identified by $X$-ray diffraction, chemical analyses, and i.r. spectra.[133]

$$O_3 + ClONO_2 \rightarrow NO_2^+ClO_4^- \qquad (27)$$

The reaction (27) produces nitronium perchlorate exclusively, constituting a new synthesis of this powerful oxidizer.[134] $NH_3$, $CO_3^{2-}$, and $NO_2^-$ were major products of the oxidation of $CH_3CO_2^-$ by molten $NaNO_3$–$KNO_3$. The $O_2N$-$CH_2CO_2^-$ ion has been suggested as an intermediate.[135]

Nitrate ions of aqueous 0.01M-$KNO_3$ saturated with hydrogen or oxygen are reduced to $NO_2^-$ by the action of ultrasonic waves, and $H_2O_2$ is also formed. The yield of $NO_2^-$ and $H_2O_2$ depends on the pH of solutions and the duration of the treatment with ultrasound. A redox mechanism involving the radical and molecular products of cleavage of $H_2O$ by ultrasound has been proposed, and the rate equations for the formation of $NO_2^-$ and $H_2O_2$ derived by using a steady-state approximation.[136] γ-Radiolysis of neutral $NaNO_3$ solutions also yields $NO_2^-$, and the radiolysis mechanism has been discussed.[137]

Conductometric titration of $SOCl_2$ with $AgNO_3$ in acetone solution indicates the formation of $SO(NO_3)_2$.[138]

A potentiometric method using a bismuth amalgam electrode has been used to determine formation constants of mixed chloride–nitrate complexes of $Bi^{3+}$ at 298 K. Complex composition was determined by the Sillèn method, and the stability constants of $BiCl_n(NO_3)_m^{3-n-m}$ compounds are given at zero ionic strength.[139]

[129] D. W. Amos and G. W. Flewett, *Spectrochim. Acta*, 1975, **31A**, 213.
[130] B. Ribar, N. Milinski, and S. Djuric, *Z. Krist.*, 1974, **140**, 417.
[131] N. Milniski, B. Ribar, R. Herak, and S. Djuric, *Cryst. Struct. Comm.*, 1974, **3**, 757.
[132] L. B. Serezhkina, N. S. Tamm, and A. I. Grigor'ev, *Zhur. neorg. Khim.*, 1975, **20**, 1826.
[133] B. N. Ivanov-Emin, Z. K. Odinets, S. F. Yushchenko, B. E. Zaitsev, and A. I. Ezhov, *Izvest. V.U.Z., Khim. i khim. Tekhnol.*, 1975, **18**, 1351.
[134] C. J. Schack and K. O. Christe, *Inorg. Chem.*, 1974, **13**, 2378.
[135] J. D. Burke and D. H. Kerridge, *J. Inorg. Nuclear Chem.*, 1975, **37**, 751.
[136] M. A. Margulis, *Zhur. fiz. Khim.*, 1974, **48**, 2968.
[137] I. A. Kulikov and M. V. Vladimirova, *Khim. vysok. Energii*, 1975, **9**, 228.
[138] S. N. Nabi, A. Hussain, and N. N. Ahmed, *J.C.S. Dalton*, 1974, 1199.
[139] V. A. Fedorov, T. N. Kalosh, and L. I. Shmyd'ko, *Zhur. neorg. Khim.*, 1974, **19**, 1820.

Nitrate is the major source of nitrogen for most green plants and fungi, and a highly significant development in the understanding of the mode of action of the nitrate reductases (enzymes responsible for the first step in nitrate assimilation and for nitrate respiration) has resulted from the study of possible model reactions.[140] The nitrate reductases convert nitrate into nitrite and require molybdenum, probably as Mo$^V$, for their activity; the ability of simple Mo$^V$ complexes to reduce nitrate to nitrite has therefore been investigated. The reaction between an excess ($\geqslant 6:1$) of Et$_4$NNO$_3$ and [MoOCl$_3$(OPPh$_3$)$_2$] in CH$_2$Cl$_2$ solution at room temperature has been investigated, using stopped-flow kinetic techniques, and the results obtained were consistent with the mechanism shown in Scheme 1. The

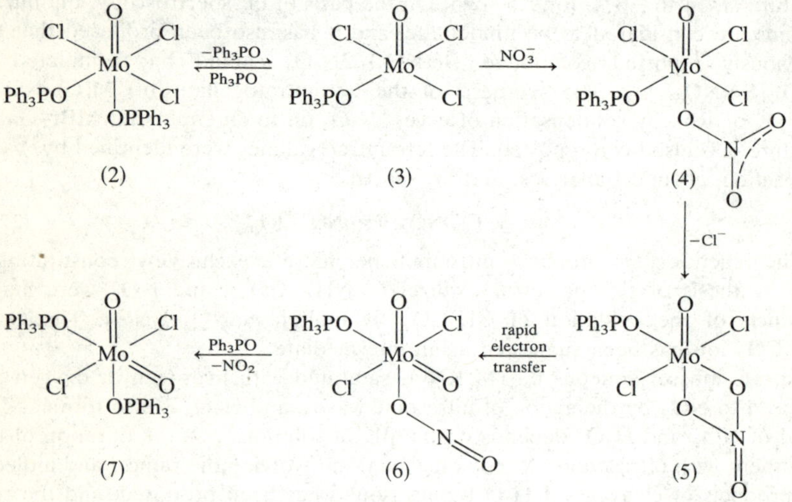

**Scheme 1**

geometry of (5) is ideal for a rapid intramolecular electron transfer from the Mo$^V$ centre to the nitrato-group, and in the presence of water the NO$_2$ produced would hydrolyse to give nitrite and nitrate. In the enzyme a non-aqueous environment at the metal site will favour nitrate co-ordination, and it is suggested that the molybdenum centre of nitrate reductases is in a hydrophobic region of the protein. Co-ordination of nitrate to the metal centre could be important in lowering the activation energy for electron transfer to this group. Moreover, it has been proposed that facile electron transfer from an oxo-molybdenum(v) centre to a nitrate group will only occur if the latter is able to co-ordinate by way of one oxygen atom at a site *cis* to the oxo-group, with the plane of the NO$_3$ moiety containing the Mo=O axis.[140]

*Other Nitrogen*(v) *Species.* Extended Hückel M.O. calculations indicate that the anions NS$_3^-$, NS$_4^{3-}$, and NO$_4^{3-}$ are capable of existence, perhaps in molten salt or non-aqueous media and stabilized with large counter-ions, although for NS$_3^-$ and NS$_4^{3-}$ only a transient existence seems likely.[141]

---

[140] C. D. Garner, M. R. Hyde, F. E. Mabbs, and V. I. Routledge, *Nature*, 1974, **252**, 579.
[141] D. K. Johnson and J. R. Wasson, *Inorg. Nuclear Chem. Letters*, 1974, **10**, 891.

# Elements of Group V

A doubly filled graphite–iron(III) chloride–dinitrogen pentoxide intercalation complex of empirical formula $C_{31},FeCl_3,(N_2O_5)_{1.7}$ has been prepared in which a fourth stage graphite–iron(III) chloride complex is independently interleaved by a second stage graphite–dinitrogen pentoxide complex, after treatment with $N_2O_5$ at 273 K. The most ordered possible structure based upon the observed formulation is shown in (8). The repeat distance of 31.9 Å along the c-axis is based on

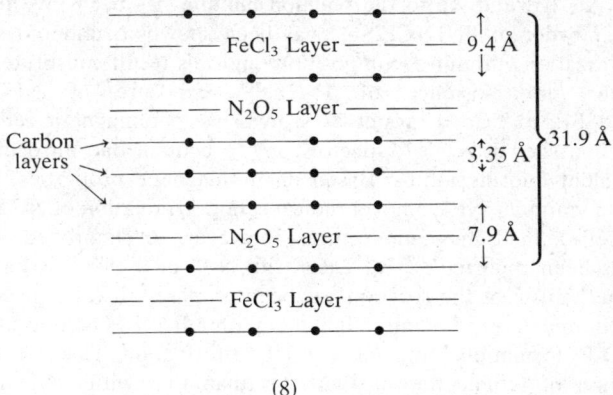

(8)

observed inter-carbon layer separation distances of 7.8 Å for the compound $C_9,N_2O_5$ and 9.4 Å for $C_8,FeCl_3$.[142] I.r. spectroscopy has been used to follow the rates of the chemical changes in gaseous $N_2O_5–SO_2$ and $N_2O_5–SO_2–O_3$ mixtures. $SO_3$ was not a detectable product of these reaction systems, and no significant $SO_2$ removal occurred. The near-u.v. absorption spectrum of pure $N_2O_5$ has also been determined.[143]

**Bonds to Fluorine.**—Calculation of e.s.r. coupling constants of the $\pi$-radicals $NF_2$ and $PF_2$ by the UHF method gives poor agreement with experimental data, whereas C.I. calculations yield coupling constants in good agreement with experiment.[144]

New data obtained from a Raman spectroscopic study of gaseous $N_2F_4$ at 760 Torr necessitate a reassignment of some of the fundamentals for both the gauche $(C_2)$ and trans $(C_{2h})$ conformers. The Raman spectrum of the solid at 18 K has also been recorded. Both conformers were found in the solid, even after annealing the sample for 9 h at 101.6 K and for 3 h at 106.8 K; however, bands attributable to the gauche-isomer became more intense relative to those for the trans-isomer after several hours of annealing.[145]

The geometries of a number of molecules, including $NF_2$ and $NF_3$, have been rationalized using a hard-atom model similar to that used in describing the structures of simple ionic compounds in the solid state. The calculated F—N—F angles/° (observed angles in parentheses) are $NF_2$ 104.8 (102.5) and $NF_3$ 104.6 (102.4). The results indicate that factors other than electron-pair repulsions

---

[142] A. G. Freeman, *J.C.S. Chem. Comm.*, 1974, 746.
[143] R. L. Daubendick and J. G. Calvert, *Environ. Letters*, 1975, **8**, 103.
[144] J. Kendrick, I. H. Hillier, and M. F. Guest, *Chem. Phys. Letters*, 1975, **33**, 173.
[145] J. R. Durig and R. W. MacNamee, *J. Raman Spectroscopy*, 1974, **2**, 635.

may be important in determining the structures of simple inorganic derivatives of first-row elements.[146]

$N_2F_2$ and $NF_3$ have been synthesized by irradiation of a nitrogen–fluorine mixture with uranium fission fragments and with mixed n,γ-radiation from a nuclear reactor. The kinetics of the process have been studied over a wide interval of the initial composition, pressure, temperature, and dose rate, and the effect of adding He, Ar, Kr, and Xe to the reaction mixture has been investigated.[147] A compressed powder of $FNH_3^+CF_3SO_2^-$ has been shown to undergo simple two-species vaporization without decomposition, and this result constitutes additional evidence for the existence of $H_2NF$.[148] Exposure of $NF_4^+AsF_6^-$ and $NF_4^+SbF_6^-$,$0.8SbF_5$ to $^{60}Co$ γ-rays at 77 K gives two paramagnetic centres, one of which has been shown by e.s.r. spectroscopy to contain one nitrogen atom and three equivalent fluorine atoms. Based on its magnetic properties, it has been identified as pyramidal $\cdot NF_3^+$, a novel radical cation. Irradiation of $NF_4^+BF_4^-$ did not result in well-defined magnetic centres; however, an improved synthesis of $NF_4^+BF_4^-$ has been reported.[149] $F_2$ reacts with $N_2O$ in a glow discharge to form $NF_3O$. A mechanism of the reaction has been proposed, involving the formation of NO and atomic fluorine as intermediates. With a $F_2:N_2:O_2$ concentration ratio of 3:1:1, $O_2F_2$ formation competes with $NF_3O$ formation. This is attributed to a faster reaction of atomic fluorine with $O_2$ than that with NO, and to $O_2F_2$ stabilization on cold reaction-vessel walls.[150]

**Bonds to Chlorine and Iodine.**—The hard-atom model, referred to above in connection with the molecular geometries of $NF_2$ and $NF_3$, has also been applied to $NCl_3$. The calculated and observed values quoted for the Cl—N—Cl angle are 108.9 and 107.1°, respectively.[146] The crystal structure of $NCl_3$ at 148 K has been determined by X-ray analysis, using direct methods, and was refined to $R = 9.1\%$. Besides dipole–dipole forces there are weak N···Cl and Cl···Cl interactions determining the crystal structure built up by $NCl_3$ molecules. The average N—Cl bond length is $1.75 \pm 0.01$ Å and the average Cl—N—Cl angle is $106.8 \pm 2°$. The crystal structure of $NCl_3$ strongly resembles that of $HCCl_3$.[151] Force-constant calculations[152] on $NCl_3$ and $PCl_3$ and the kinetics of the thermal decomposition of $NCl_3$ in $CCl_4$ containing electron-donor compounds such as benzene, toluene, cumene, p-xylene, and mesitylene (with which $NCl_3$ forms π-complexes)[153] have also been reported. In the presence of PhMe the activation energy of $NCl_3$ thermal decomposition is 95.8 kJ mol$^{-1}$, i.e. approximately 40 kJ mol$^{-1}$ lower than that of thermal decomposition in pure $CCl_4$.[153]

NOCl is separated into Cl and NO radicals when photolysed by 3000—4000 Å radiation in the vapour phase.[154] The reaction of excited iodine atoms, produced

---

[146] C. Glidewell, *Inorg. Nuclear Chem. Letters*, 1975, **11**, 353.
[147] V. A. Dmitrievskii, E. K. Il'in, and A. I. Migachev, *Khim. vysok. Energii*, 1974, **8**, 433.
[148] R. T. V. Kung and R. Roberts, *J. Phys. Chem.*, 1974, **78**, 1433.
[149] S. P. Mishra, M. C. R. Symons, K. O. Christe, R. D. Wilson, and R. I. Wagner, *Inorg. Chem.*, 1975, **14**, 1103.
[150] I. V. Nikitin and V. Ya Rosolovskii, *Zhur. fiz. Khim.*, 1974, **48**, 369.
[151] H. Hartl, J. Schoener, J. Jander, and H. Schulz, *Z. anorg. Chem.*, 1975, **413**, 61.
[152] G. Cazzoli, *J. Mol. Spectroscopy*, 1974, **53**, 37.
[153] B. A. Radbil, *Kinetika i Kataliz*, 1975, **16**, 360.
[154] G. Lucas, *Nuova cimento*, 1974, **50**, 39.

in the photolysis of $CF_3I$, with NOCl, with the formation of NO and ICl, has been investigated. The quantum yield of NO was measured as a function of the photolysis time and partial pressures of NO, NOCl, $CF_3I$, and $CO_2$. Kinetic measurements indicated that $CF_3$ radical is captured by NO, and only excited iodine atoms are reactive with NOCl.[155] The i.r. chemiluminescence technique has been used[156] to investigate the energy distribution among the reaction products for reaction (28), and the kinetics of oxidation of NOCl by $N_2O_5$ have been studied in the presence and absence of $HNO_3$.[157]

$$H + NOCl \rightarrow HCl + NO \qquad (28)$$
$$NO + NO_2Cl \rightarrow NOCl + NO_2 \qquad (29)$$

The kinetics of reaction (29) have been investigated by rapid-scan i.r. spectroscopy. The experimental rate constant is in reasonable agreement with an earlier calorimetric value, but the use of isotopic NO has now established the reaction mechanism to be a Cl atom transfer. Rate constants for the nitrogen isotope exchange reactions (30) and (31) have also been determined.[158]

$$^{15}NO + NOCl \rightarrow NO + {}^{15}NOCl \qquad (30)$$
$$^{15}NO_2 + NO_2Cl \rightarrow NO_2 + {}^{15}NO_2Cl \qquad (31)$$

The crystal structure of the 1:1 adduct of nitrogen tri-iodide and pyridine, $NI_3,C_5H_5N$, like 'nitrogen tri-iodide', $NI_3,NH_3$, contains $NI_4$ tetrahedra as essential structure elements. The tetrahedra are connected by common corners, forming indefinite chains. The pyridine molecule is bonded by its lone electron pair to one of the two iodine atoms that do not participate in the connection of the tetrahedra. In contrast to $NI_3,NH_3$, there are very weak intermolecular interactions between iodine atoms of neighbouring chains.[159] The compound of 1:1:1 stoicheiometry between nitrogen tri-iodide, iodine, and hexamethylenetetramine, $NI_3,I_2,C_6H_{12}N_4$, has also been the subject of an X-ray crystal structure determination. Each $NI_3$ and $C_6H_{12}N_4$ molecule is bonded to four molecules of the other kind. The connections are formed by three intermolecular N—I bonds and an $I_2$ bridge between the nitrogen atom of the $NI_3$ and one of the nitrogen atoms of the hexamethylenetetramine. Structure (9) shows a projection of the structure on the $a-b$ plane.[160]

## 2 Phosphorus

**Phosphorus and Phosphines.**—Red phosphorus is transported, using a low-pressure hydrogen discharge, to give a product purer than the starting material, in amorphous layers.[161] There is evidence for the suggestion that the transport mechanism involves the formation and subsequent decomposition of unstable volatile hydrides. Applications of a low-pressure silent electric discharge in

---

[155] L. G. Karpov, A. M. Pravilov, and F. I. Vilesov, *Khim. vysok. Energii*, 1974, **8**, 483.
[156] M. A. Nazar, J. C. Polanyi, and W. J. Skralac, *Chem. Phys. Letters*, 1974, **29**, 473.
[157] R. A. Wilkins, *Diss. Abs. Internat.* (*B*), 1974, **35**, 174.
[158] R. A. Wilkins jun., M. C. Dodge, and I. C. Hisatsune, *J. Phys. Chem.*, 1974, **78**, 2073.
[159] H. Hartl and D. Ullrich, *Z. anorg. Chem.*, 1974, **409**, 228.
[160] H. Pritzkow, *Z. anorg. Chem.*, 1974, **409**, 237.
[161] S. Veprěk and H. R. Oswald, *Z. anorg. Chem.*, 1975, **415**, 190.

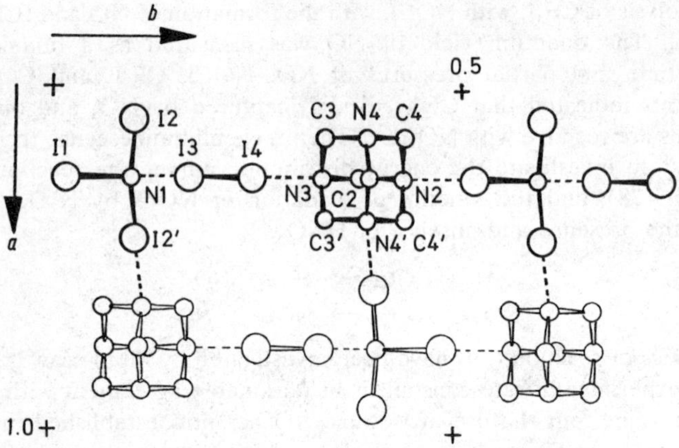

phosphorus chemistry have been reported.[162] With phosphine, yields of up to 50% of diphosphine can be achieved, using 5000 V, while with methylphosphine the products include MePHPH$_2$ and MePHPHMe. The $PP'$-dimethyldiphosphine, from n.m.r. data, is the expected mixture of (±)- and *meso*-forms. Experiments with CF$_3$PH$_2$ and (CF$_3$)$_2$PH have also been carried out, and mixtures of PH$_3$ and acetylene led to HC⫶CPH$_2$, which is unstable above −20 °C.

M.O. calculations, both with and without $d$-orbital participation, have been reported for PH$_3$,[163,164] PH$_2$F,[164] PH$_3$F$_2$, PH$_5$, PH$_4$F, and PH$_2$F$_3$. PH$_4$ radicals result when PH$_3$ is irradiated with $^{60}$Co, and e.s.r. parameters have been tabulated for PH$_4$, PH$_3$D, and PD$_4$.[165] There is, however, some evidence from recent calculations[166] that the radical identified as PH$_4$ is, in fact, the P$_2$H$_6^+$ species.

Problems associated with $(p-d)\pi$-bonding in four-co-ordinated phosphorus compounds, *i.e.* R$_3$PX (R = H, Me, or Cl; X = O, NH, BH$_3$, S or CH$_2$), have been approached by a comparison of the $X$-ray p.e. spectra of those species with those of the corresponding nitrogen compounds, and correlation with results from M.O. calculations.[167] The overall conclusion is that there is little influence of $d$-orbitals on the bonding in these compounds.

**Phosphides.**—An arc melting technique, which offers advantages in purity of product and higher attainable temperatures over the conventional approach to phosphide preparation, has been used to re-examine the V–P system (the V–As and Cr–As systems were also investigated).[168] The phases in the V–P system were confirmed, but in the corresponding arsenic system two new orthorhombic phases have been identified.

A manganese phosphide, MnP$_4$, has been prepared from the elements in a tetrahedral high-pressure device at 30–55 kbar and shown to contain condensed

---

[162] J. P. Albrand, S. P. Anderson, H. Goldwhite, and L. Huff, *Inorg. Chem.*, 1975, **14**, 570.
[163] J. G. Norman, *J. Chem. Phys.*, 1974, **61**, 4630.
[164] F. Keil and W. Kutzelnigg, *J. Amer. Chem. Soc.*, 1975, **97**, 3623.
[165] A. J. Colussi, J. R. Morton, and K. F. Preston, *J. Chem. Phys.*, 1975, **62**, 2004.
[166] T. A. Claxton, B. W. Fullam, E. Platt, and M. C. R. Symons, *J.C.S. Dalton*, 1975, 1395.
[167] W. B. Perry, T. F. Schaaf, and W. L. Jolly, *J. Amer. Chem. Soc.*, 1975, **97**, 4899.
[168] R. Berger, *Acta Chem. Scand.* (A), 1975, **29**, 641.

## Elements of Group V

MnP$_6$ octahedral units and phosphorus atoms surrounded tetrahedrally by manganese and phosphorus atoms.[169] New ternary phases, KMnP and KMnAs,[170] in which the manganese atoms are tetrahedrally co-ordinated by the Group V atom, and Mg$_{1.75}$Zn$_{1.25}$P$_2$ and Mg$_2$ZnAs$_2$,[171] probably with a modified anti-La$_2$O$_3$ structure, have been prepared from the elements.

Lanthanum metal and phosphorus react at 1080 °C in the presence of iodine to give LaP$_2$[172] and at 540 °C in the NaCl–KI eutectic with iodine, giving LaP$_7$.[173] The latter can be described as a Zintl phase, and has a structure based on seven-membered phosphorus rings fused in pairs to give P$_{12}$ units. The monoclinic cell of the diphosphide contains 16 formula units, which make up P$_3^{5-}$ and P$_5^{7-}$ chains, justifying formulation as La$_4$P$_3$P$_5$. P—P Bond distances/Å are shown in (10) and (11). Experiments involving the preparation of LaP and CeP crystals by

$$
\begin{array}{cc}
\text{P} \diagdown 2.207 & \text{P} \diagdown 2.217 \\
107.5° \text{P} & 113.6° \text{P} \\
\diagup 2.233 & 2.198 \diagdown 102.6° \quad 110.3° \diagup 2.216 \\
\text{P} & \text{P} \underline{\quad 2.315 \quad} \text{P}
\end{array}
$$

(10)                      (11)

chemical transport in the presence of iodine have been reported,[174] but polyphosphides of La, Pr, or Nd could not be obtained similarly. Three new ternary phases Li$_2$MP$_2$, with M = La, Ce, or Pr, are the products of reactions between Li$_3$P and the required monophosphide in the presence of phosphorus vapour at 900 °C.[175] Powder data for Pd$_6$P indicate a structure closely related to the Re$_3$B type.[176]

The phosphide chlorides Ca$_2$PCl and Sr$_2$PCl have been identified as the products of heating mixtures of the metal, metal chloride, and red phosphorus to 1000 °C,[177] and the general formation of alkaline-earth iodo-pnictides has been investigated by an examination of the M–Y–I systems, where M = Ca, Sr, or Ba and Y = P, As, or Sb.[178] The results point to the existence of two series of phases, i.e. M$_2$YI and M$_{3+x}$ Y$_{1+x}$I$_{3-x}$, which have structures closely related to the NaCl structure.

**Compounds containing P—P Bonds.**—Electron-diffraction measurements on P$_2$F$_4$ and P$_2$(CF$_3$)$_4$ can be interpreted in each case in terms of the *trans*-conformer only; thus at room temperature the concentration of the *gauche*-form probably does not exceed 10%.[179] Important parameters for these two molecules are given in Table 1. Ionization potentials for P$_2$F$_4$(9.3 eV) and a number of other compounds,

---

[169] W. Jeitschko and P. C. Donohue, *Acta Cryst.*, 1975, **B31**, 574.
[170] L. Linowsky and W. Bronger, *Z. anorg. Chem.*, 1974, **409**, 221.
[171] A. Mewis, P. Klüfers, and H.-U. Schuster, *Z. Naturforsch.*, 1975, **30b**, 132.
[172] H. G. von Schnering, W. Wichelhaus, and M. Schulze-Nahrup, *Z. anorg. Chem.*, 1975, **412**, 193.
[173] W. Wichelhaus and H. G. von Schnering, *Naturwiss.*, 1975, **62**, 180.
[174] V. I. Torbov, V. I. Chukalin, V. N. Doronin, L. G. Nikolaeva, and Z. S. Medvedeva, *Russ. J. Inorg. Chem.*, 1974, **19**, 21.
[175] A. El Maslout, J.-P. Motte, A. Courtois, and C. Gleitzer, *Compt. rend.*, 1975, **280**, C, 21.
[176] Y. Andersson, V. Kaewchansilp, M. R. C. Soto, and S. Rundqvist, *Acta Chem. Scand. (A)*, 1974, **28**, 797.
[177] C. Hadenfeldt, *Z. Naturforsch.*, 1975, **30b**, 165.
[178] C. Hamon, R. Marchand, and J. Lang, *Rev. Chim. minérale*, 1974, **11**, 287.
[179] H. L. Hodges, L. S. Su, and L. S. Bartell, *Inorg. Chem.*, 1975, **14**, 599.

**Table 1** Structural parameters for the molecules $P_2F_4$ and $P_2(CF_3)_4$

| $P_2F_4$ | | $P_2(CF_3)_4$ | |
|---|---|---|---|
| $r_g(P-P)/Å$ | 2.281(6) | $r_g(P-P)/Å$ | 2.182(16) |
| $r_g(P-F)/Å$ | 1.587(3) | $r_g(P-C)/Å$ | 1.914(4) |
| ∠PPF/° | 95.4(3) | ∠PPC/° | 107.6(7) |
| ∠FPF/° | 99.1(4) | ∠CPC/° | 103.8(3) |

including $PF_3$ (11.6 eV) and $PF_2I$ (10.8 eV), have been determined by the electron-impact method, and from appearance-potential data the P—P bond dissociation energy in $P_2F_4$ has been estimated as $238 \pm 40$ kJ mol$^{-1}$.[180]

Photolysis of mixtures of tetrafluorodiphosphine and olefins provides a ready route to bis(difluorophosphino) species *via* addition of $PF_2$ radicals across the double bond.[181] Among the olefins used were $C_2H_4$, $C_3H_6$, but-2-ene, cyclohexene, $C_2F_4$, and $C_3F_6$, giving as products $F_2PCH_2CH_2PF_2$, $MeCHPF_2CH_2PF_2$, $MeCHPF_2CHPF_2CH_3$, *trans*-bis(difluorophosphino)cyclohexane, $F_2PCF_2CF_2PF_2$, and $CF_3CF(PF_2)CF_2PF_2$, respectively. A thermally stable diphosphine, $(CF_3)$-HPPH$(CF_3)$, can be obtained simply from $CF_3PI_2$ and mercury in the presence of *ca.* 1 mol of HI, and the method can be extended to the arsenic analogue.[182] The direct synthesis of a compound containing a P—P bond has been reported from an unusual reaction between equimolar quantities of $MeP(OMe)_2$ and $Ph_2PCl$.[183] The product is $Ph_2PP(O)(Me)(OMe)$, but if diphenylchlorophosphine is in excess the reaction gives tetraphenyldiphosphine. These products are in direct contrast to those from the $MeP(OMe)_2$-$MePCl_2$ system.

Cleavage of the P—P bond in tetra-alkoxy-diphosphines occurs on acetylation,[184] the products with acetyl chloride being phosphorochloridites $(RO)_2PCl$ and the previously unknown dialkylacetylphosphonites $(RO)_2PC(O)OMe$. The latter have been isolated in small yield, the bulk of the product being in the form of a dimer to which structure (12) has been assigned. The P—P bond in alkoxy-

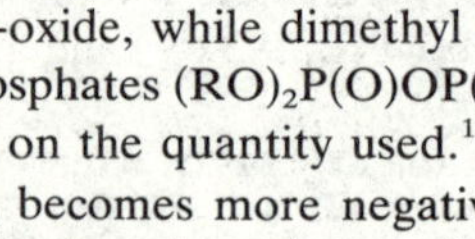

$$(RO)_2P-\underset{\underset{C(O)Me}{|}}{\overset{\overset{Me}{|}}{C}}-OP(OR)_2$$

(12)

diphosphines (tetra-alkylhypodiphosphites) is also labile with respect to oxidation by nitric oxide or pyridine 1-oxide, while dimethyl sulphoxide gives diphosphites $(RO)_2POP(OR)_2$, isohypophosphates $(RO)_2P(O)OP(OR)_2$, or diphosphates $(RO)_2$-$P(O)OP(O)(OR)_2$ depending on the quantity used.[185] The coupling constant $^1J_{PP'}$ in organodiphosphines $R_4P_2$ becomes more negative as R is varied from Me to Bu$^t$, a trend that has been attributed to changes in hybridization at phosphorus leading to less *s*-character in the orbitals used in the P—P bond.[186]

---

[180] C. R. S. Dean, A. Finch, P. J. Gardner, and D. W. Payling, *J.C.S. Faraday I*, 1974, **70**, 1921.
[181] J. G. Morse and K. W. Morse, *Inorg. Chem.*, 1975, **14**, 565.
[182] R. C. Dobbie and P. D. Gosling, *J.C.S. Chem. Comm.*, 1975, 585.
[183] K. M. Abraham and J. R. van Wazer, *Inorg. Chem.*, 1975, **14**, 1099.
[184] M. V. Proskurnina, A. L. Chekhun, and I. F. Lutsenko, *J. Gen. Chem. (U.S.S.R.)*, 1974, **44**, 1216.
[185] M. V. Proskurnina, A. L. Chekhun, and I. F. Lutsenko, *J. Gen. Chem. (U.S.S.R.)*, 1974, **44**, 2080.
[186] H. C. E. McFarlane and W. McFarlane, *J.C.S. Chem. Comm.*, 1975, 582.

## Elements of Group V

The formation and detection of phosphanediyl (phosphinidine) radicals RṖ:, which result from thermolysis of cyclophosphanes $(RP)_n$, has been reviewed critically.[187] Data from mass spectrometry and chemical trapping experiments are considered, as is the production of species thought to be PhPO and PhPS. The tetrameric and pentameric pentafluoroethylcyclopolyphosphines $(C_2F_5P)_{4\,and\,5}$ have been prepared, and evidence has been presented for conversion of the tetramer into pentamer on heating.[188] A trinuclear complex of triphenylcyclotriphosphine (13) has been confirmed by X-ray data as the product from demetallation of the

```
              Ph\    ,-Mn(CO)₂Cp
                 P
           Ph\  / \
              P----P---Ph
    Cp(CO)₂Mn'     \
                    Mn(CO)₂Cp
```

(13)

lithium derivative $(C_5H_5)Mn(CO)_2(PPhLi_2)$ with N,N-dichlorocyclohexylamine, in direct contrast to behaviour in the corresponding arsenic system.[189]

Support for the suggestion that $(p-d)\pi$-bonding in cyclopolyphosphines is relatively unimportant comes from a study of their electrochemical reduction, where the reduction potential appears to be solely dependent on the inductive effect of the attached R group, with no effect of differing ring size.[190] Problems associated with the determination of ring size in polyphosphines by conventional molecular weight methods have often led to confusion, but it has now been shown[191] that tetra- and penta-phosphines can be distinguished in solution by observing the proton-decoupled $^{31}$P n.m.r. spectrum, when the former collapse to a singlet between +50 and +70 p.p.m. and the latter show a complex multiplet between −20 and +10 p.p.m. from $H_3PO_4$. Further investigations in this area point to $(PhP)_6$, giving a spectrum completely different from that of the pentamer and undergoing ring-size redistribution only after months in solution at 30 °C.[192] The pentamer, on the other hand, on melting, shows the presence of small amounts of the hexamer and a second species.

**Bonds to Boron.**—With the exception noted below, the papers in this section are concerned with aspects of donor–acceptor behaviour. The exception deals with the formation of P—B σ-bonds by the stepwise replacement of chlorine in $BCl_3,OEt_2$ by $LiPEt_2$ at low temperature to give the dimeric products (14), (15), and (16).[193]

I.r. and Raman spectra of $H_3P,BX_3$ and $D_3P,BX_3$, where X = F, Br, or I, have been assigned on the basis of $C_{3v}$ symmetry, but no trend can be discerned in the

---

[87] U. Schmidt, *Angew. Chem. Internat. Edn.*, 1975, **14**, 523.
[88] A. N. Lavrent'ev, I. G. Maslennikov, V. A. Efanov, and E. G. Sochilin, *J. Gen. Chem. (U.S.S.R.)*, 1974, **44**, 2550.
[89] G. Huttner, H.-D. Müller, A. Frank, and H. Lorenz, *Angew. Chem. Internat. Edn.*, 1975, **14**, 572.
[90] T. J. DuPont, L. R. Smith, and J. L. Mills, *J.C.S. Chem. Comm.*, 1974, 1001.
[91] L. R. Smith and J. L. Mills, *J.C.S. Chem. Comm.*, 1974, 808.
[92] P. R. Hoffman and K. G. Caulton, *Inorg. Chem.*, 1975, **14**, 1997.
[93] G. Fritz and E. Sattler, *Z. anorg. Chem.*, 1975, **411**, 193.

$$\begin{array}{ccc}
\text{Cl}_2\text{B}-\text{PEt}_2 & \text{Cl}-\overset{\overset{\displaystyle\text{PEt}_2}{|}}{\underset{\underset{\displaystyle\text{PEt}_2}{|}}{\text{B}}}-\text{PEt}_2 & (\text{Et}_2\text{P})_2\text{B}-\text{PEt}_2 \\
\text{Et}_2\text{P}-\text{BCl}_2 & \text{Et}_2\text{P}-\text{B}-\text{Cl} & \text{Et}_2\text{P}-\text{B}(\text{PEt}_2)_2 \\
(14) & (15) & (16)
\end{array}$$

force constants for the boron–phosphorus bond obtained after a normal co-ordinate analysis.[194] Spectra for the 1:1 adducts of $PH_3$ and $PD_3$ with $B_2Cl_4$ have been interpreted on the basis of a *trans*, $C_{2h}$, conformation in the solid state.[195]

Data are now available on the n.m.r. behaviour of compounds in the series $PH_3, BCl_nBr_{3-n}$, $PH_3, BCl_nI_{3-n}$, $PH_3, BBr_nI_{3-n}$ (where $n = 0—3$), and the analogous adducts where $MePH_2$ is the donor.[196] The value of $J_{PB}$ is considered less reliable than $J_{PH}$ as an indicator of donor–acceptor strength because, at the temperatures used, exchange phenomena, causing inaccuracies, are likely. Trends in the dipole moments of $Me_nH_{3-n}P$ and the corresponding borane adducts obtained by CNDO methods give results in general agreement with empirically derived moments.[197] The analogous methylamine systems have also been investigated, giving results which point to both the B—P and B—N bonds being highly polar. In the $HPF_2, BH_3$ adduct, *ab initio* calculations show that stabilization of the molecule by 8.4 kJ mol$^{-1}$ is achieved by tilting the $BH_3$ group.[198] The calculations reproduce satisfactorily the rotation barrier about the P—B bond, and indicate that phosphorus $d$-orbitals contribute to bonding in both the adduct and free $HPF_2$, but to a lesser extent than in the phosphine oxide $HF_2PO$.

Further work[199] using the base-displacement method toward $BH_3$ gives the order of base strengths as $PF_2Bu^t \gtrsim PF_2Et > PF_2C\vdotsCMe > PF_2Me > PF_2NMe_2 > PF_2OPr^i > PF_2OEt > PF_2OMe > PF_2OCH_2CF_3 \gtrsim PF_2SMe > PF_3 > PF_2Cl > PF_2Br$. As considered above, this order is not mirrored by the values of $J_{PB}$,[196] nor by values for $\nu(BH)$.

Diborane(4) complexes with $PF_2X$ (X = F, Cl, or Br) can be obtained from $B_3H_7,Me_2O$ or $B_4H_{10}$,[200] and structures with a B—P bond similar to that found by electron diffraction for $(F_3P)_2B_2H_4$ are most probable in view of similarities in vibrational and n.m.r. data. Data for heat of reaction have been obtained by gas-phase calorimetry for 11 of the 15 possible adducts in the $Me_3P$, $Me_3As$, or $Me_3Sb$ systems with $BF_3$, $BCl_3$, $BBr_3$, $B_2H_6$, or $Me_3B$.[201] In general, the order of base strength is $Me_3P > Me_3As > Me_3Sb$, with the acid strengths decreasing in the order $BBr_3 > BCl_3 \approx BH_3 > BF_3 > BMe_3$, but trimethylborane did not react with the arsine or stibine, and diborane and $BF_3$ reacted anomalously with trimethylstibine. $^1$H N.m.r. data for these complexes did not seem to provide useful

---

[194] J. R. Durig, S. Riethmiller, V. F. Kalasinsky, and J. D. Odom, *Inorg. Chem.*, 1974, **13**, 2729.
[195] J. D. Odom, V. F. Kalasinsky, and J. R. Durig, *Inorg. Chem.*, 1975, **14**, 434.
[196] J. E. Drake and B. Rapp, *J. Inorg. Nuclear Chem.*, 1974, **36**, 2613.
[197] P. M. Kuznesof, F. B. T. Pessine, R. E. Bruns, and D. F. Shriver, *Inorg. Chim. Acta*, 1975, **14**, 271.
[198] D. R. Armstrong, *Inorg. Chim. Acta*, 1975, **13**, 121.
[199] E. L. Lines and L. F. Centofanti, *Inorg. Chem.*, 1974, **13**, 2796.
[200] R. T. Pain and R. W. Parry, *Inorg. Chem.*, 1975, **14**, 689.
[201] D. C. Mente, J. L. Mills, and R. E. Mitchell, *Inorg. Chem.*, 1975, **14**, 123.

information, but the vibration spectra and results of gas-phase displacement reactions[202] were in accord with the calorimetry described above.

The trimethylphosphine adducts with $BCl_3$, $BBr_3$, and $BI_3$ are isomorphous, and have a staggered conformation with effective $C_{3v}$ symmetry, according to X-ray diffraction results.[203] The P—B bond distance decreases from 1.957 Å in the chloride to 1.918 Å in the iodide, as expected from the acid strengths, but the differences are within the uncertainty ranges.

$^1$H N.m.r. spectroscopy, used to follow halogen redistribution reactions in the boron trihalide adducts with $Me_3P$, $Me_3PO$, and $Me_3PS$, gives results that are complicated by the low solubility and possible reaction with the dichloromethane solvent.[204] In agreement with previous results, only small amounts of fluorine-containing boron mixed halides are present with $Me_3PS$ as donor. Complex formation between boron halides and substituted aryl chlorophosphines $(XC_6H_4)_{3-n}PCl_n$ has also been examined by n.m.r. spectroscopy.[205]

**Bonds to Carbon.**—*Phosphorus*(III) *Compounds*. Although $(p-p)\pi$-bonds between carbon and the heavier elements in the third, fourth, and fifth Groups of the Periodic Table are generally considered to be unstable, a remarkable number of such species, particularly those containing P, As, or Sb, have in fact been stabilized. This whole area of chemistry has been reviewed in depth,[206] discussing preparative methods, characteristic reactions, and criteria for the existence of such compounds.

A near-Hartree–Fock wavefunction for HCP confirms the triple C≡P bond and yields values for a number of electron properties which agree quite well with the limited available experimental data.[207] N.m.r. parameters for this compound, obtained by passing $PH_3$ through a carbon arc, have been obtained, the value of 211 Hz for $^1J(^{13}C-H)$ being rationalized on the basis of $sp$ hybridization at carbon.[208] The appearance potential of HCP$^+$ was determined as 11.4 eV, a value considerably higher than that (9.5 eV) calculated by *ab initio* M.O. methods.

The results of a microwave determination of the molecular structure of phosphabenzene are summarized in (17),[209] and in 2-phenyl-1-phosphanaphthalene (18) the condensed rings are almost planar.[210]

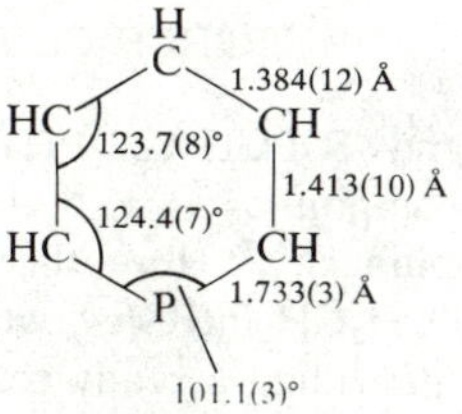

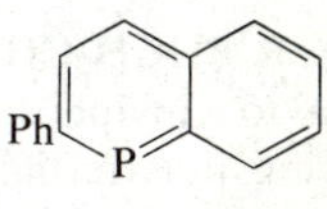

[202] D. C. Mente and J. L. Mills, *Inorg. Chem.*, 1975, **14**, 1862.
[203] D. L. Black and R. C. Taylor, *Acta Cryst.*, 1975, **B31**, 1116.
[204] M. J. Bula, J. S. Hartman, and C. V. Raman, *Canad. J. Chem.*, 1975, **53**, 326.
[205] E. Muylle, G. P. van der Kelen, and Z. Eeckhaut, *Spectrochim. Acta*, 1975, **31A**, 1039; E. Muylle and G. P. van der Kelen, *ibid.*, p. 1045.
[206] P. Jutzi, *Angew. Chem. Internat. Edn.*, 1975, **14**, 232.
[207] C. Thomson, *Theor. Chim. Acta*, 1974, **35**, 237.
[208] S. P. Anderson, H. Goldwhite, D. Ko, A. Letsou, and F. Esparza, *J.C.S. Chem. Comm.*, 1975, 744.
[209] T. C. Wong and L. S. Bartell, *J. Chem. Phys.*, 1974, **61**, 2840.
[210] J. J. Daly and F. Sanz, *J.C.S. Dalton*, 1974, 2388.

Coupling constants for MePH$_2$ have been calculated by a simple LCAO—MO method, and agree qualitatively with experimental values.[211] Ion cyclotron resonance spectroscopy on the methylphosphines Me$_n$PH$_{3-n}$, where $n = 1$—3, has been investigated, to define their gas-phase ion chemistry.[212] The proton affinities obtained range from 187.9 kcal mol$^{-1}$ for PH$_3$ to 228.0 kcal mol$^{-1}$ for Me$_3$P, and these values can be fairly closely reproduced by CNDO/2 calculations.[213] The calculated ionization potentials are, however, considerably higher than the experimental values.

Raman spectra of solutions containing MePH$_3^+$ and MePD$_3^+$ have been used to assign the vibrational normal modes and to calculate a force field.[214] Detailed assignments have been given for the vibrational spectra of EtPH$_2$ and EtPD$_2$ in the gaseous, liquid, and solid states.[215]

Optimum geometries have been calculated using the CNDO/2 method for Bu$_2^i$PF and Bu$_2^i$PH, with, in each case, the C—P—C angle being 114°.[216] Bands in the He (I) p.e. spectra for (CF$_3$)$_{3-n}$PH$_n$ and (CF$_3$)$_{3-n}$PCl$_n$, where $n = 0$—2, can be assigned to localized M.O.'s,[217] and the ionization potentials for the phosphorus lone-pair are in the order: (CF$_3$)$_3$P > (CF$_3$)$_2$PX > CF$_3$PX$_2$ > PX$_3$ for X = H or Cl, in agreement with general electronegativity considerations.

The centrosymmetric (CuX)$_4$ core with bidentate diphosphine ligands is present in the X-ray structure of the complex (CuX)$_2$,(Ph$_2$P)$_2$CH$_2$, where X = Br or I.[218] Diarylphosphinous acids are known to add to the CN group of nitriles activated by electronegative substituents, and the kinetics of such reactions between Ph$_2$P(O)H and, for example, p-nitrobenzonitrile and succinonitrile to give Ph$_2$P(O)C(NH)R have been investigated.[219]

New secondary-primary diphosphines can be prepared by reduction of the products from the base-catalysed addition of a phosphine to a vinylphosphorus compound, as shown in equation (32).[220] Methods for the preparation of

$$RPH_2 + CH_2=CHP(O)(OCHMe_2)_2 \longrightarrow RP(H)CH_2CH_2P(O)(OCHMe_2)_2$$

R = Ph, n-hexyl, or neopentyl $\quad\quad\quad\downarrow$ LiAlH$_4$ $\quad\quad\quad$ (32)

$$RPHCH_2CH_2PH_2$$

RP(CH$_2$CH$_2$PH$_2$)$_2$, R$^1$PHCH$_2$CH$_2$PR$_2^2$, and R$_2$PCH$_2$CH$_2$PH$_2$ have also been discussed. Addition to vinylphosphorus compounds is also the basis of a new method for preparing poly(tertiary phosphines).[221] Here the phosphine R$_2$PH reacts with CH$_2$=CHP(S)Me$_2$, giving R$_2$PCH$_2$CH$_2$P(S)Me$_2$, which can be desulphurized by lithium tetrahydroaluminate, providing a ready route to compounds

---

[211] R. K. Safiullin, R. M. Aminova, and Yu. Yu. Samitov, *J. Struct. Chem.*, 1974, **15**, 798.
[212] R. H. Staley and J. L. Beauchamp, *J. Amer. Chem. Soc.*, 1974, **96**, 6252.
[213] C. Leibovici, M. Graffeuil, and J.-F. Labarre, *J. Chim. phys.*, 1975, **72**, 272.
[214] E. Mayer, *J. Mol. Structure*, 1975, **26**, 347.
[215] J. R. Durig and A. W. Cox jun., *J. Chem. Phys.*, 1975, **63**, 2303.
[216] M. Corosine and F. Crasnier, *J. Mol. Structure*, 1975, **27**, 105.
[217] A. H. Cowley, M. J. S. Dewar, and D. W. Goodman, *J. Amer. Chem. Soc.*, 1975, **97**, 3653.
[218] A. Camus, G. Nardin, and L. Randaccio, *Inorg. Chim. Acta*, 1975, **12**, 23.
[219] A. N. Pudovik, T. M. Sudakova, and G. I. Evstaf'ev, *J. Gen. Chem. (U.S.S.R.)*, 1974, **44**, 2371.
[220] R. B. King and J. C. Cloyd jun., *J. Amer. Chem. Soc.*, 1975, **97**, 46.
[221] R. B. King and J. C. Cloyd jun., *J. Amer. Chem. Soc.*, 1975, **97**, 53.

such as $Ph_2PCH_2CH_2PMe_2$ and $PhP(CH_2CH_2PMe_2)_2$. The method can be extended to give penta- and hexa- tertiary phosphines, e.g. $(Me_2PCH_2CH_2)_2PCH_2$-$CH_2PPhCH_2CH_2PMe_2$ and $(Me_2PCH_2CH_2)_2PCH_2CH_2P(CH_2CH_2PMe_2)_2$. Alternative applications of this basic approach, equations (33) and (34), lead to multidentate phosphine ligands containing sulphur or nitrogen atoms.[222] Complexes of the newly synthesized sexidentate ligand (19) have been prepared and characterized.[223]

$$-SH + CH_2=CH-P\big\langle \longrightarrow -SCH_2CH_2P\big\langle \quad (33)$$

$$\big\rangle P-H + CH_2=CHCONH_2 \longrightarrow \big\rangle PCH_2CH_2CONH_2 \quad (34)$$

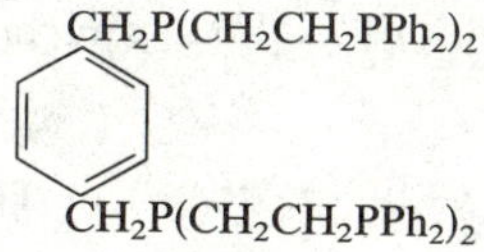

$$\underset{(19)}{\underset{CH_2P(CH_2CH_2PPh_2)_2}{CH_2P(CH_2CH_2PPh_2)_2}}$$

Appel and his co-workers have extended their earlier work on the reaction of $Ph_3P$ and carbon tetrachloride, using such nucleophiles as phenols and thiophenols,[224] phosphoric acid esters,[225,226] phosphinic acids,[225,226] phosphinic amides,[227] substituted sulphur(VI) amides,[228] and ketoximes.[229] The reactions with phenols and thiophenols[224] lead to phosphonium salts, i.e. $[Ph_3PYAr]Cl$ (Y = O or S), while the phosphoric and phosphinic acid reactions[225] are represented by equations (35) and (36). When these reactions are carried out in the presence of triethylamine, the products are (in general) anhydrides, but if a secondary amine is used, intermolecular dehydration, as shown in equations (37) and (38), occurs, and products containing P—N bonds are formed.[226] Similar reactions occur with alcohols [equation (39)], leading to direct ester formation. Phosphinic amides in the presence of triethylamine react to give the corresponding triphenylphosphoranylidene amides $Ph_3P=N(O)PR_2$,[227] but experiments with $(EtO)_2P(O)NH_2$ and $Ph_2P(O)NH_2$ follow an alternative route, leading to cyclophosphazenes, as shown in equation (40). Sulphur amides, i.e. $H_2NSO_2R$ (R = Me, NEt$_2$, NPr$^n_2$, NBu$^n_2$), on the other hand, reacted in the presence of triethylamine in the more usual manner, giving the corresponding phosphoranylidenes $Ph_3P=NSO_2R$.[228] Reactions with ketoximes $R^1R^2C=NOH$ lead to imidoyl chlorides. $R^1CCl=NR^2$, via a Beckman rearrangement,[229] and peptides can be synthesized in reactions with N-protected amino-acids.[230] Attempts to define the

---

[222] D. L. DuBois, W. H. Myers, and D. W. Meek, *J.C.S. Dalton*, 1975, 1011.
[223] M. M. Taqui Khan and A. E. Martell, *Inorg. Chem.*, 1974, **14**, 676.
[224] R. Appel, K. Warning, and K.-D. Ziehn, *Annalen*, 1975, 406.
[225] R. Appel and H. Einig, *Z. anorg. Chem.*, 1975, **414**, 236.
[226] R. Appel and H. Einig, *Z. anorg. Chem.*, 1975, **414**, 241.
[227] R. Appel and H. Einig, *Chem. Ber.*, 1975, **108**, 914.
[228] R. Appel and H. Einig, *Z. Naturforsch.*, 1975, **30b**, 134.
[229] R. Appel and K. Warning, *Chem. Ber.*, 1975, **108**, 1437.
[230] R. Appel, G. Bäumer, and W. Strüver, *Chem. Ber.*, 1975, **108**, 2680.

reactions occurring in these three-component $Ph_3P-CCl_4$-nucleophile systems have involved a detailed examination by quantitative g.l.c.[231]

$$2Ph_3P + 2CCl_4 + ROP(O)(OH)_2 \rightarrow ROP(O)Cl_2 + Ph_3PO + 2CHCl_3 \quad (35)$$
$$R = Ph \text{ or } Pr^i$$

$$Ph_3P + CCl_4 + R_2P(O)OH \rightarrow R_2P(O)Cl + Ph_3PO + CHCl_3 \quad (36)$$
$$R = Me, Ph, \text{ or } OEt$$

$$R^1_2P(O)OH + 2HNR^2R^3 + Ph_3P + CCl_4 \rightarrow$$
$$R^1_2P(O)NR^2R^3 + Ph_3PO + CHCl_3 + R^2R^3NH_2Cl \quad (37)$$

$$R^1OP(O)(OH)_2 + 2Ph_3P + 2CCl_4 + 4HNR_2^2 \rightarrow$$
$$R^1OP(NR_2^2)_2 + 2Ph_3PO + 2CHCl_3 + R_2^2NH_2Cl \quad (38)$$

$$R^1_2P(O)OH + Ph_3P + CCl_4 + R^2OH \xrightarrow{Et_3N} R^1_2P(O)OR^2 + Ph_3PO + CHCl_3 \quad (39)$$

$$R_2P(O)NH_2 + Ph_3P + CCl_4 \rightarrow (R_2PN)_3 + Ph_3PO + CHCl_3 + HCl \quad (40)$$

$$R_2PPR_2 + CCl_4 \rightarrow R_2PCl + R_2PCCl_3 \quad (41)$$
$$R = Ph, \text{ cyclohexyl, } Pr^n, \text{ or } Bu^n$$

These reactions have been extended to diphosphanes, where a preliminary study shows cleavage of the P—P bond in carbon tetrachloride according to equation (41).[232] A similar reaction has been observed with the diphosphane monoxide $Ph_2PP(O)Ph_2$, giving $Ph_2PCl$, $Ph_2PCCl_3$, $Ph_2P(O)Cl$, and $Ph_2P(O)CCl_3$ as products, but with $Me_4P_2$ and $Et_4P_2$ the products were compounds (20) and (21), respectively. When primary or secondary amines were added to the system,

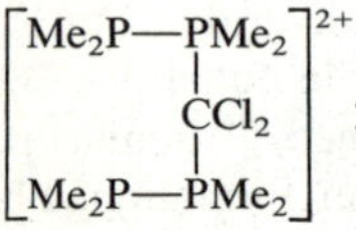

$$\begin{bmatrix} Me_2P-PMe_2 \\ | \\ CCl_2 \\ | \\ Me_2P-PMe_2 \end{bmatrix}^{2+} 2Cl^- \qquad \begin{bmatrix} Et_2P-PEt_2 \\ | \\ CCl_3 \end{bmatrix}^+ Cl^-$$

(20) 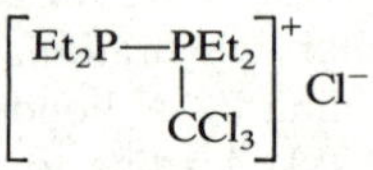                (21)

phosphonium salts with two P—N bonds, e.g. $[R^1_2P(NR^2R^3)_2]^+Cl^-$, were formed,[233] but when 1,2-diphenyl-1,2-diphosphacyclohexane (22) and aniline reacted the product was an unusual seven-membered bis-phosphonium salt (23). Broadly similar results were obtained with cyclophosphanes, where, for example, all the P—P bonds in $Ph_4P_4$ were broken by $CCl_4$, giving $PhPCl(CCl_3)$,[234] but with cyclohexyl-cyclotetraphosphane the substituted diphosphane (24) was obtained. With reactions carried out in the presence of an amine, the products were tris-(amino)phosphonium salts.

*Phosphorus*(v) *Compounds.* The $^1H$ n.m.r. spectrum of the five-co-ordinate compound $Me_4POMe$, which has an axial methyl group, is very complex, as a result of

---

[231] R. Appel and K. Warning, *Chem. Ber.*, 1975, **108**, 606.
[232] R. Appel and R. Milker, *Chem. Ber.*, 1975, **108**, 1783.
[233] R. Appel and R. Milker, *Chem. Ber.*, 1975, **108**, 2349.
[234] R. Appel and R. Milker, *Z. anorg. Chem.*, 1975, **417**, 161.

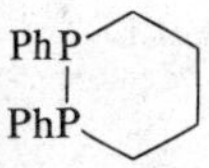

facile proton exchange with methanol and ylidic species such as $Me_3P=CH_2$.[235] At low temperatures, however, the proton-decoupled $^{13}C$ spectra show the expected three methyl environments. The preparation and characterization of the phosphonium perchlorates $Ph_3RP\ ClO_4$, where R = Me, Et, Pr, or $NH_2$, has been announced,[236] and although they are more stable than $PH_4ClO_4$, explosive decomposition is observed in thermal experiments involving more than a few milligrams of pure material. A reaction between phenyl isocyanate and tetrakis-(hydroxymethyl)phosphonium chloride in dry pyridine leads to the carbamic ester $[PhNHCl(O)OCH_2]_4PCl$,[237] and analogous products are obtained from $(HOCH_2)_3P$ and its oxy-derivative.

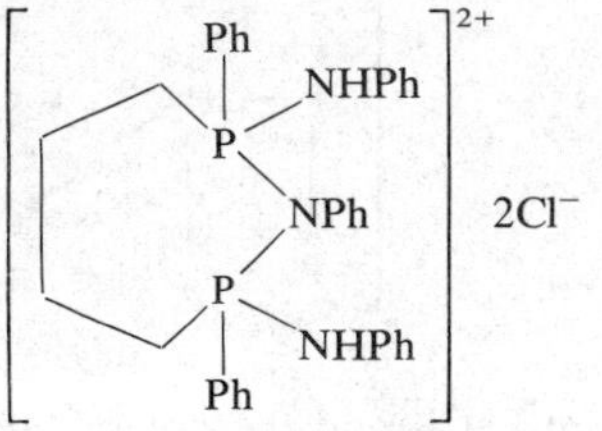

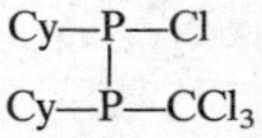

Three new tetra-alkylphosphoranes $Me(CF_3)_3PX$ (X = F, Cl, or OMe) have been synthesized, although previous reports have tended to imply that these compounds are unstable with respect to the phosphonium salt structure.[238] (see also ref. 235). N.m.r. data confirm trigonal-bipyramidal structures and can be interpreted on the basis of a halogen and a $CF_3$ group (or two $CF_3$ groups for X = OMe) in axial positions. Difficulties have also been encountered in isolating penta-alkylphosphoranes owing to their instability with respect to the ylide. Two stable compounds, $(CF_3)_3PMe_2$ and $(CF_3)_2PMe_3$, which have now been prepared by methylation of $(CF_3)_3PCl_2$ or $(CF_3)_2PCl_3$ give n.m.r. spectra at normal temperatures that are interpretable in terms of static trigona-bipyramidal structures.[239] Their stability is attributed to the presence of the highly electron-withdrawing $CF_3$ groups.

$^{31}P$ N.m.r. parameters have been tabulated for compounds in the series $Ph_nR_{3-n}PX$ (n = 0—2; X = O, S, or Se; R = 4-F-, 3-F-, 4-Cl-, or 3-Cl-$C_6H_4$); the chemical shifts appear to give useful information on the $\sigma$- and $\pi$-electron distributions about phosphorus.[240]

A full structure has been reported for the centrosymmetric hydrated 1,6-diphosphacyclodecane derivative (25),[241] in which the ten-membered ring has the normal cyclodecane conformation. The phenyl groups are in *trans*-pseudo-axial positions and the structure is completed by hydrogen bonds between the hydrate water and phosphoryl groups from two molecules.

---

[235] H. Schmidbaur, W. Buchner, and F. H. Köhler, *J. Amer. Chem. Soc.*, 1974, **96**, 6208.
[236] S. R. Jain and P. R. Nambiar, *Indian J. Chem.*, 1974, **12**, 1087.
[237] R. A. Askarova, M. I. Bakhitov, and E. V. Kuznetsov, *J. Gen. Chem. (U.S.S.R.)*, 1974, **44**, 1413.
[238] K. I. The and R. G. Cavell, *J.C.S. Chem. Comm.*, 1975, 279.
[239] K. I. The and R. G. Cavell, *J.C.S. Chem. Comm.*, 1975, 716.
[240] R. F. de Ketelaere and G. P. van der Kelen, *J. Mol. Structure*, 1975, **27**, 363.
[241] M. Dräger, *Chem. Ber.*, 1974, **107**, 3246.

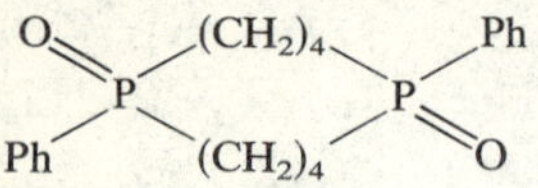

(25) — structure: O=P(Ph)(CH₂)₄–P(Ph)(CH₂)₄=O shown as Ph₂P(=O)(CH₂)₄P(=O)Ph variant

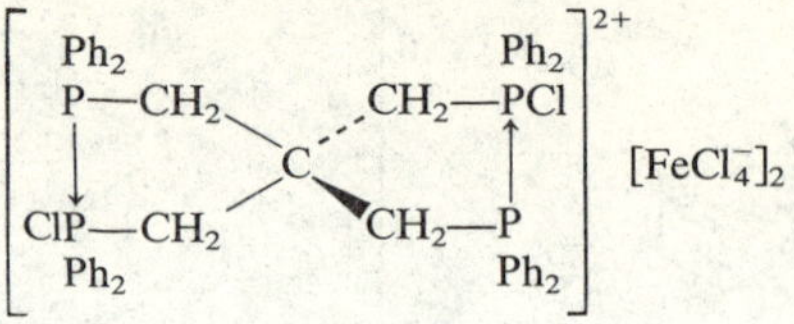

(26) — spirocyclic tetraphosphonium dication [C(CH₂PPh₂Cl)₄]⁴⁺ [FeCl₄⁻]₂

Compound formation between FeCl$_3$ and the polyfunctional donor tetrakis(diphenyldichlorophosphoranomethyl)methane, C(CH$_2$PPh$_2$Cl$_2$)$_4$, gives the tetraphosphonium salt [C(CH$_2$PPh$_2$Cl)$_4$]$^{4+}$ [FeCl$_4^-$]$_4$ at a 4:1 mole ratio, but the spirocyclic compound (26) when the ratio is 2:1.[242a] A similar tetraphosphonium salt results with SbCl$_5$ as the acceptor, but the product with SnCl$_4$ at a 2:1 ratio is surprisingly non-ionic, and it may have a polymeric structure. Similar experiments with the oxy-derivative tetrakis(diphenylphosphorylmethyl)methane, C[CH$_2$P(O)Ph$_2$]$_4$, lead to products with the same stoicheiometries, but the proposed structures are different. For example, the 2:1 and 4:1 complexes with FeCl$_3$ are formulated as (27) and (28), respectively.[242b]

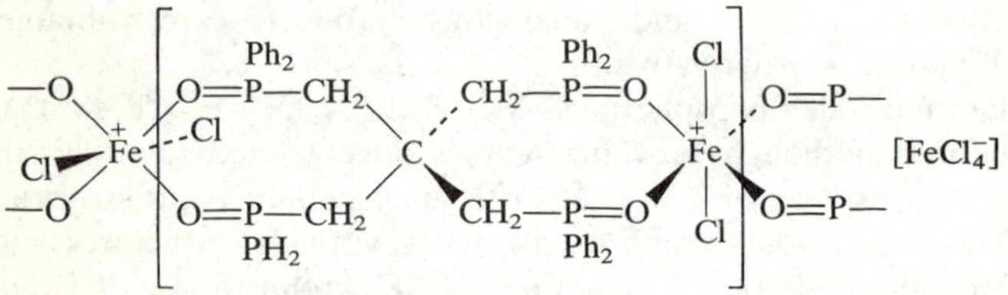

(27)

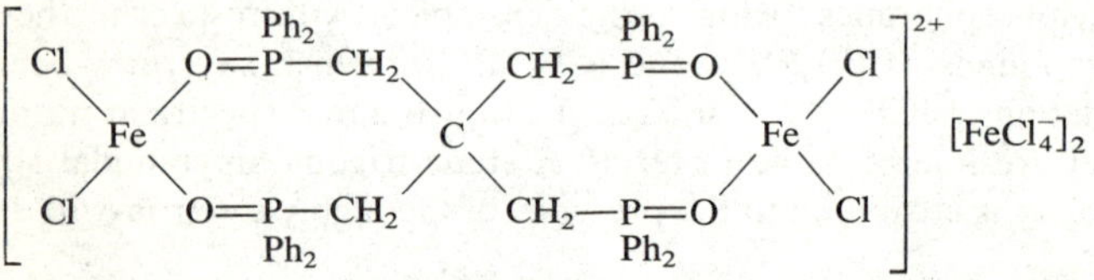

(28)

Factors influencing the stereochemistry of phosphoranyl radicals have been investigated, pointing to trigonal-bipyramidal structures when electron-withdrawing substituents are present, but tetrahedral structures when the substituents are capable of inductive or conjugative stabilization of the charge on phosphorus.[243]

**Bonds to Silicon or Germanium.**—Two new, volatile silicon phosphides, SiF$_3$PH$_2$ and SiF$_2$HPH$_2$, are formed when SiF$_2$ and phosphine are co-condensed, but the products could not be separated.[244] Vibrational spectroscopic data for F$_3$SiPH$_2$,[245] F$_3$SiPD$_2$,[245] (CF$_3$)$_2$PSiMe$_3$,[246] (CF$_3$)$_2$PGeMe$_3$,[246] and

---

[242] J. Ellermann and M. Thierling, *Z. anorg. Chem.*, 1975, **411**, 15, 28.
[243] A. G. Davies, M. J. Parrott, and B. P. Roberts, *J.C.S. Chem. Comm.*, 1974, 973; J. M. F. van Dijk, J. F. M. Pennings, and H. M. Buck, *J. Amer. Chem. Soc.*, 1975, **97**, 4836.
[244] G. R. Langford, D. C. Moody, and J. D. Odom, *Inorg. Chem.*, 1975, **14**, 134.
[245] R. Demuth, *Spectrochim. Acta*, 1975, **31A**, 233.
[246] H. Bürger, J. Cichon, R. Demuth, J. Grobe, and F. Höfler, *Spectrochim. Acta*, 1974, **30A**, 1977.

the arsenic analogues[246] have been recorded and analysed. A detailed investigation of the effect of substituents on the n.m.r. parameters for silylphosphines in the series $Me_nH_{3-n}SiPH_2$, $Me_nH_{3-n}SiPHMe$, $Me_nH_{3-n}SiPMe_2$, $(Me_nH_{3-n}Si)_2PH$, $(Me_nH_{3-n}Si)_2PMe$, and $(Me_nH_{3-n}Si)_3P$ has been carried out, and among the conclusions it was shown possible to treat $\delta_P$ and $J(P—Si)$ in terms of a sum of contributions from individual bond types.[247]

Ready cleavage of the P—Si bond makes compounds such as $(CF_3)_2PSiR_3$ useful reagents for the preparation of new compounds containing the $(CF_3)_2P$ group, and thus provides a direct and efficient route to silylphosphines with R = H or Me.[248] In a similar way $Bu_2^tPSiMe_3$ reacts readily with both $SiCl_4$ and $GeCl_4$ to produce $Bu_2^tPSiCl_3$ or $Bu_2^tPGeCl_3$, but an oxidative cleavage occurs with $SnCl_4$, giving $Bu_2^tCl$, $SnCl_2$, and $Me_3SiCl$.[249] The di-t-butyl compound has also been exploited in reactions with methylchlorogermanes, giving products in the series $Me_nGeCl_{3-n}PBu_2^t$.[250]

1:1 Addition compounds form between aluminium chloride and the silylphosphines $Me_3SiPEt_2$ or $Me_nH_{3-n}SiPMe_2$,[251] which on heating lose methylsilyl chloride, giving trimeric aluminium–phosphorus heterocycles such as (29). The thermolysis product from the adduct $Me_3SiPMe_2,H_2AlCl$ is similar to (29), but the

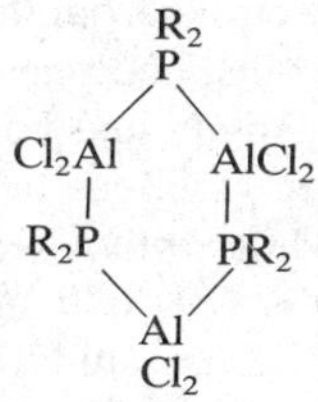

$$
\begin{array}{c}
R_2 \\
P \\
Cl_2Al \quad AlCl_2 \\
| \quad\quad | \\
R_2P \quad\quad PR_2 \\
Al \\
Cl_2
\end{array}
$$

(29)

aluminium atoms now carry a hydrogen and a chlorine atom, while when $HAlMe_2$ is the acceptor a tetrameric thermolysis product, $(Me_2AlPMe_2)_4$, is obtained.

Silylphosphines can add across the C=N bond in aldimines as shown in equation (42) to produce compounds containing an Si—N—C—P atom sequence; further reactions with di-imines and N-acylimines have also been reported.[252]

$$R^1Me_2SiPEt_2 + R^2CH{=}NR^3 \longrightarrow R^1Me_2SiNR^3CR^2HPEt_2 \qquad (42)$$

$R^1$ = Me or H

The reaction between white phosphorus, $Me_3SiCl$, and a mixture of sodium and potassium in either mono- or di-glyme as solvent provides a simple method for preparing the tertiary phosphine $(Me_3Si)_3P$.[253] Scrambling reactions of the latter with $Me_3GeCl$ and $Me_3SnCl$ lead to all ten possible products, which can be identified by $^{31}P$ n.m.r. spectroscopy.[254]

[247] G. Fritz and H. Schäfer, Z. anorg. Chem., 1974, **409**, 137.
[248] L. Maya and A. B. Burg, Inorg. Chem., 1975, **14**, 698.
[249] W.-W. du Mont and H. Schumann, Angew. Chem. Internat. Edn., 1975, **14**, 368.
[250] H. Schumann and W.-W. du Mont, Chem. Ber., 1975, **108**, 2261.
[251] G. Fritz and R. Emül, Z. anorg. Chem., 1975, **416**, 19.
[252] C. Couret, F. Couret, J. Satgé, and J. Escudié, Helv. Chim. Acta, 1975, **58**, 1316.
[253] G. Becker and W. Höderich, Chem. Ber., 1975, **108**, 2478.
[254] H. Schumann, H.-J. Kroth, and L. Rösch, Z. Naturforsch., 1974, **29b**, 608.

PH$_2$ groups and H atoms on germanium are rapidly redistributed, according to a $^1$H n.m.r. study of the compounds MeGe(PH$_2$)$_2$H and Me$_2$Ge(PH$_2$)H,[255] and evidence has been presented for the formation of the new phosphino-germanes GeH$_2$(PH$_2$)$_2$ and GeH(PH$_2$)$_3$ from the thermal redistribution of GeH$_3$PH$_2$. A nitrene insertion reaction is observed when methyl or phenyl azides react with germyl-phosphines or germaphospholans [equation (43)].[256] The initial products

$$\text{Me}_3\text{GePR}^1\text{Me} + \text{R}^2\text{N}_3 \xrightarrow{-\text{N}_2} \text{Me}_3\text{GeNR}^2\text{PR}^1\text{Me} \longrightarrow \text{Me}_3\text{GePMe(NR}^2)\text{R}^1 \quad (43)$$
$$\qquad\qquad\qquad\qquad\qquad\qquad (30) \qquad\qquad\qquad (31)$$

are N-germylated amino-phosphines (30), which isomerize on heating to yield germylated phosphinimines (31); if a second mole of azide is present, further reaction, to give an amino-phosphinimine, e.g. Me$_3$GeNR$^2$PMe(NR$^2$)R$^1$, can occur.

**Bonds to Fluorine.**—*Phosphorus*(III) *Compounds.* Calculations using an *ab initio* pseudopotential SCF method have been carried out for the four phosphorus(III) halides,[257] and it has been possible to derive a basicity scale for a large number of phosphines from CNDO/2 MO calculations.[258] These data are in reasonable agreement with experimental He (I) p.e.s. determinations which are now available[259] for the series R$_2$NPF$_2$, R$_2$NPCl$_2$, and R$_n$PX$_{3-n}$ (X = F, Cl, or H). Variations in the binding energy of the 2p electrons from the X-ray p.e.s. of, *inter alia*, phosphorus(III) halides, phosphonium salts, and phosphorus oxyacids can be discussed in terms of substituent electronegativity and bond ionic character.[260]

High-temperature, high-pressure reactions between PF$_3$ and CO$_2$ or CS$_2$ give, respectively, POF$_3$ and PSF$_3$, while with COS the product is exclusively the thio-compound at lower pressures (*ca.* 680 atm, 300 °C), but POF$_3$ becomes important at 3000 atm.[261] A $^{19}$F n.m.r. study of the PF$_3$–Hg$_2$(AsF$_6$)$_2$ system in liquid sulphur dioxide solution provides new evidence for the species (Hg$_2$PF$_3$)$^{2+}$ and [Hg$_2$(PF$_3$)$_2$]$^{2+}$, with solution Raman data also supporting the former.[262]

A recent investigation into the reaction between allyldifluorophosphine and hydrogen bromide has shown no evidence for the expected radical addition across the double bond, but the two products shown in equation (44) are obtained.[263]

$$\text{CH}_2\text{=CHCH}_2\text{PF}_2 + \text{HBr} \longrightarrow \text{CH}_2\text{=CHCH}_2\text{PBr}_2 + \text{CH}_2\text{=CHCH}_2\text{PHF}_3 \quad (44)$$

The products are consistent with addition of HBr to the difluoride, followed by disproportionation through successive loss of HF and addition of HBr. CNDO/2 calculations on MeOPF$_2$ suggest the existence of two *gauche* equivalent preferred conformations,[264] in disagreement with previous interpretations of microwave spectra. The parameters listed in Table 2 have been determined from electron-diffraction data on F$_2$POPF$_2$,[265] which also suggests the presence of a number of,

[255] A. R. Dahl, C. A. Heil, and A. D. Norman, *Inorg. Chem.*, 1975, **14**, 1095.
[256] J. Escudié, C. Couret, and J. Satgé, *Compt. rend.*, 1975, **280**, C, 783.
[257] C. S. Ewig, P. Coffey, and J. R. van Wazer, *Inorg. Chem.*, 1975, **14**, 1848.
[258] M. Graffeuil, J.-F. Labarre, M. F. Lappert, C. Leibovici, and O. Stelzer, *J. Chim. phys.*, 1975, **72**, 799.
[259] M. F. Lappert, J. B. Pedley, B. T. Wilkins, O. Stelzer, and E. Unger, *J.C.S. Dalton*, 1975, 1207.
[260] E. Fluck and D. Weber, *Z. Naturforsch.*, 1974, **29b**, 603.
[261] A. P. Hagen and B. W. Callaway, *Inorg. Chem.*, 1975, **14**, 1622.
[262] P. A. W. Dean and D. G. Ibbott, *J. Inorg. Nuclear Chem.*, 1974, **11**, 119.
[263] E. R. Falardean, K. W. Morse, and J. G. Morse, *Inorg. Chem.*, 1975, **14**, 1239.
[264] G. Robinet, J.-F. Labarre, and C. Leibovici, *Chem. Phys. Letters*, 1974, **29**, 449.
[265] H. Y. Yow, R. W. Rudolph, and L. S. Bartell, *J. Mol. Structure*, 1975, **28**, 205.

**Table 2** *Parameters for* $F_2POPF_2$
$r(P—O) = 1.631 \pm 0.010$ Å    $\angle O—P—F = 97.6 \pm 1.2°$
$r(P—F) = 1.568 \pm 0.004$ Å    $\angle F—P—F = 99.2 \pm 2.4°$
$\angle P—O—P = 135.2 \pm 1.8°$

probably staggered, conformers. The large P—O—P angle and short P—O bond length both point to the presence of $(p-d)\pi$-bonding.

E.s.r. data are available for the radicals $PH_nF_{4-n}$ ($n = 0$—4),[266,267] $PF_4$,[268,269] $ROPF_3$,[268,270] $PH_2(OH)_2$, and $Me_2P(OH)_2$,[269] while $PR_2$ radicals are considered to be one of the major products in the γ-irradiation of phosphorus(III) compounds.[271] Further data on phosphoranyl radicals are included in ref. 243.

*Phosphorus*(v) *Compounds.* According to CNDO/2 calculations, the only way in which the axial and equatorial fluorines in $PF_5$ can interchange is by the Berry pseudorotation mechanism,[272] and an estimate of the barrier to such an interchange, (1371 cm$^{-1}$ or 3.92 kcal mol$^{-1}$) has been obtained from a detailed Raman study of the $\nu_7$, equatorial bend, of $PF_5$.[273] A new normal-co-ordinate analysis for the $PF_5$ molecule has been published.[274]

$PF_5$ Adducts with, for example, trimethylamine and 3-methylpyridine give $^{19}F$ n.m.r. spectra which cannot be interpreted on a first-order basis, but assignment as an $AB_4X$ system has been confirmed by comparison of the experimental and computer-simulated spectra.[275]

An easy method by which $PF_4Cl$ can be prepared in *ca.* 50% yield is by exchange in the gas phase between $PF_5$ and $BCl_3$,[276] but if the total pressure exceeds 250 Torr the dichloride is also produced. Exchange also occurs between mixtures of phosphorus pentafluoride and the pentachloride in acetonitrile solution, and the $^{19}F$ n.m.r. spectra obtained contain features ascribed to octahedral anions such as $PF_5Cl^-$, *cis*-$PF_4Cl_2^-$, *cis*-$PF_3Cl_3^-$, *cis*-$PF_2Cl_4^-$, and $PFCl_5^-$.[277] The reaction between hydrogen fluoride and organodichlorophosphines in halogenomethanes such as $CFCl_3$ leads to halogen exchange and HF addition, to produce trifluorophosphoranes $RPF_3H$.[278]

Full details are now available on the intramolecular donor–acceptor interaction which occurs in fluorophosphoranes substituted with a 2-methylquinoline group.[279] In the tetrafluoride (32; R = F) the phosphorus is octahedrally co-ordinated, with a P—N bond length of 1.91 Å,[279] while the product from phenyltetrafluorophosphorane (32; R = Ph) contains a plane of symmetry through the

---

[266] A. J. Colussi, J. R. Morton, and K. F. Preston, *J. Phys. Chem.*, 1975, **79**, 1855.
[267] K. Sogabe, A. Hasegewa, T. Komatsu, and M. Miura, *Chem. Letters*, 1975, 663.
[268] A. J. Colussi, J. R. Morton, and K. F. Preston, *J. Phys. Chem.*, 1975, **79**, 651.
[269] Y. I. Gorlov and V. V. Penkovsky, *Chem. Phys. Letters*, 1975, **35**, 25.
[270] I. H. Elson, M. J. Parrott, and B. P. Roberts, *J.C.S. Chem. Comm.*, 1975, 586.
[271] B. W. Fullam, S. P. Mishra, and M. C. R. Symons, *J.C.S. Dalton*, 1974, 2145.
[272] P. Russegger and J. Brickmann, *Chem. Phys. Letters*, 1975, **30**, 276.
[273] L. S. Bernstein, J. J. Kim, K. S. Pitzer, S. Abramowitz, and I. W. Levin, *J. Chem. Phys.*, 1975, **62**, 3671.
[274] T. R. Ananthakrishnan and G. Aruldhas, *J. Mol. Structure*, 1975, **26**, 1.
[275] K.-P. John and R. Schmutzler, *Z. Naturforsch.*, 1974, **29b**, 730.
[276] R. H. Neilson and A. H. Cowley, *Inorg. Chem.*, 1975, **14**, 2019.
[277] Yu. A. Buslyaev, E. G. Il'in, and M. N. Shcherbakova, *Doklady Chem.*, 1974, **217**, 487.
[278] R. Appel and A. Gilak, *Chem. Ber.*, 1975, **108**, 2693.
[279] K.-P. John, R. Schmutzler, and W. S. Sheldrick, *J.C.S. Dalton*, 1974, 1841.

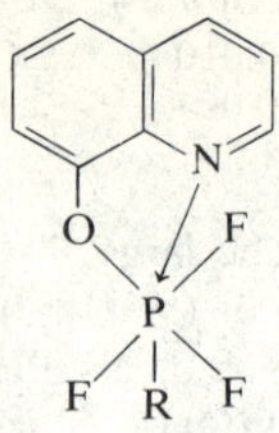

(32)

quinoline group.[280] The phenyl group which is *trans* to the nitrogen of the quinoline is perpendicular to the plane, and here the P—N bond length is 1.980 Å.

Phosphorus pentafluoride will break the Si—C bond in 1,1,3,3-tetramethyl-1,3-disilacyclobutane to give methylenebis(tetrafluorophosphorane) $F_4PCH_2PF_4$, which on hydrolysis gives first the bis(phosphonic difluoride) $CH_2(POF_2)_2$ and finally the bis(phosphonic acid) $CH_2[PO(OH)_2]_2$.[281] New polymethylene-bridged bis(difluorophosphoranes) such as $R_2F_2P(CH_2)_nPF_2R_2$ (R = Me or Ph; $n$ = 1—3) result from hydrogen fluoride reactions with the corresponding silylated iminophosphoranes $R_2P(NSiMe_3)(CH_2)_nP(NSiMe_3)R_2$.[282]

Fluorophosphoranes containing a perfluorinated pinacolyl ring, such as (33), are

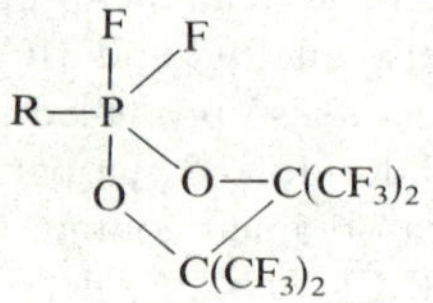

(33)

possible models for the rapid intramolecular exchange in biological systems, and result when a phosphorus(III) fluoride $RPF_2$ (R = Me, Bu, Ph, $Et_2N$, *etc.*) and hexafluoroacetone interact.[283] Conflicts in the simple interpretation of the $^{19}F$ n.m.r. spectrum of $MeSPF_4$, which shows three distinct fluorine environments at −90 °C, have now been resolved by a detailed computer-simulation study,[284] and an ionic formulation, *i.e.* $[Me_2PF_2]^+MF_6^-$, has been assigned to the 1:1 complexes between $Me_2PF_3$ and $PF_5$ or $AsF_5$ on the basis of n.m.r. measurements in acetonitrile solution.[285]

Monoalkoxy-phenyl- and -methyl-trifluorophosphoranes with fluxional trigonal-bipyramidal structures have been prepared from either $PhPF_4$ or $MePF_4$ and an alkoxysilyl ether, and $\Delta G^\ddagger$ values for exchange of the axial and equatorial fluorines have been evaluated.[286] The $^{19}F$ n.m.r. signal of the axial fluorine atoms has been found to be a sensitive probe for investigation of stereochemistry and

---

[280] K.-P. John, R. Schmutzler, and W. S. Sheldrick, *J.C.S. Dalton*, 1974, 2466.
[281] W. Althoff, M. Fild, H. Koop, and R. Schmutzler, *J.C.S. Chem. Comm.*, 1975, 468.
[282] R. Appel and I. Ruppert, *Chem. Ber.*, 1975, **108**, 919.
[283] J. A. Gibson, G.-V. Röschenthaler, and R. Schmutzler, *J.C.S. Dalton*, 1975, 918.
[284] R. B. Johannesen, S. C. Peake, and R. Schmutzler, *Z. Naturforsch.*, 1974, **29b**, 699.
[285] M. Brownstein and R. Schmutzler, *J.C.S. Chem. Comm.*, 1975, 278.
[286] J. G. Riess and D. U. Robert, *Bull. Soc. chim. France*, 1975, 425.

# Elements of Group V

isomerism in asymmetric alkoxyfluorophosphoranes,[287] and its use in the cases of the compounds $R^1PF_3OR^2$, $PhEtPF_2(OCHMeCN)$, and $PhPF_n(OCHMeCN)_{4-n}$ has been described.

Details of the preparation of the monomeric hexafluoroisopropylideneiminophosphoranes $RPF_3[N=C(CF_3)_2]$ (R = Me or Et) from the alkyltetrafluorophosphorane and $(CF_3)_2C=NLi$ have now been reported.[288] A second product, the phosphinimine $RF_2P=NC(CF_3)_2N=C(CF_3)_2$, also results, but with phenyltetrafluorophosphorane only the latter product is obtained. The structure of the trigonal-bipyramidal $(C_6F_5)_3PF_2$ shows a propeller arrangement of the $C_6F_5$ groups in the equatorial plane (mean P—C 1.82 Å), while the axial P—F distances are 1.64 Å.[289]

N-Difluorophosphoryl derivatives of heterocyclic amines such as piperidine and imidazole can be obtained by a number of routes involving $POF_2Cl$ or $POF_3$ and the free amine or its trimethylsilyl derivative; in some cases better yields result from $P_2O_3F_4$ and the silylamine.[290]

**Bonds to Chlorine.**—The reaction between $PCl_3$ and an excess of methanol in the cold has been reinvestigated, to show that more than two moles of HCl are liberated; dimethyl ether can also be detected as a product, but there is no evidence for the phosphonate $MePO(OMe)_2$.[291] A mechanism based on the dealkylation of the assumed intermediate $P(OMe)_3$ has been postulated to account for these observations. Eutectics only are found in the $PCl_3$-$TeCl_4$ and $AsCl_3$-$TeCl_4$ systems, but with $POCl_3$ an incongruently melting 1:1 adduct is formed.[292] Modification of silica surfaces to incorporate $Cr^{III}$-$P^V$-O layers can be achieved by successive treatment with chromyl chloride and phosphorus trichloride.[293]

As part of a general study of reactions of $MePCl_2$ or $MeP(OMe)_2$ to give oligomeric condensation products, exchange in the $MePCl_2$-$Me_nM(OMe)_{4-n}$ systems (M = Si, n = 2 or 3: M = Ge, n = 2) has been investigated.[294] Chloride-methoxide exchange is rapid at room temperature, but this is followed by condensation of the mixed species, as shown in equation (45). In the 'reverse'

$$MeP(OMe)Cl + MeP(OMe)_2 \rightarrow MeP(OMe)P(O)(OMe)Me + MeCl \quad (45)$$

system, i.e. $MeP(OMe)_2$ and $Me_nSiCl_{4-n}$, there is no rapid exchange, and the major observed reaction is the slow condensation of the phosphorus units.

The preferred conformation for both ethyl- and isopropyl-dichlorophosphines is that in which a methyl group and the lone pair are *trans* to each other, according to low-temperature n.m.r. data,[295] and there is evidence for the formation of

---

[287] D. U. Robert, D. J. Costa, and J. G. Riess, *J.C.S. Chem. Comm.*, 1975, 29.
[288] J. A. Gibson and R. Schmutzler, *Z. anorg. Chem.*, 1975, **416**, 222.
[289] W. S. Sheldrick, *Acta Cryst.*, 1975, **B31**, 1776.
[290] E. Fluck and E. Beuerle, *Z. anorg. Chem.*, 1974, **411**, 125.
[291] M. Demarcq and J. Slezionia, *Bull. Soc. chim. France*, 1975, 1605.
[292] A. V. Konov and V. V. Safonov, *Russ. J. Inorg. Chem.*, 1974, **19**, 619.
[293] A. N. Volkova, A. A. Malygin, S. I. Kol'tsov, and V. B. Aleskovskii, *J. Gen. Chem. (U.S.S.R.)*, 1975, **45**, 1.
[294] K. M. Abraham and J. R. Van Wazer, *J. Inorg. Nuclear Chem.*, 1975, **37**, 541.
[295] J. P. Dutasta and J. B. Robert, *J.C.S. Chem. Comm.*, 1975, 747.

non-centrosymmetric dimers in solid dimethylchlorophosphine from detailed vibrational spectroscopy.[296]

Phase studies point to the formation of high-melting 1:1 adducts in the $PCl_5$–$AlCl_3$ and $PCl_5$–$FeCl_3$ systems,[297] while Raman data for the $PCl_5$,$UCl_5$ adduct have been interpreted in terms of the ionic structure $PCl_4^+ UCl_6^-$.[298] Further applications of the phosphorylating action of phosphorus pentachloride have been reported, with substituted vinylphosphonic dichlorides $Cl_2P(O)CR^1=CCl(OR^2)$ and carboxylic acid chlorides $Cl_2P(O)CRClC(O)Cl$ being produced from the esters $R^1CH_2C(O)OR^2$.[299] Similar reactions with tertiary alcohols give an intermediate complex, which with sulphur dioxide leads to 2-chloroalkyl- and 1-alkenyl-phosphonic dichlorides.[300]

The $POCl_3$–$Cl_2$ system shows a peritectic at $-57\,°C$, corresponding to the composition $2POCl_3,Cl_2$,[301] but vapour pressure–composition studies on the $POCl_3$–$WCl_6$ and $POCl_3$–$WOCl_4$ systems indicate no interaction.[302] A yellow amorphous compound with the stoicheiometry $TiP_2O_4Cl_6$ can be obtained from either the reaction of $POCl_3$ with $TiCl_2(OCOR)_2$ or the thermal decomposition of the complex $[TiCl_3(PO_2Cl_2)POCl_3]_2$; from i.r. data it is considered to have a constitution based on $TiOCl(PO_2Cl_2),POCl_3$ units.[303]

Substitution of the chlorine atoms in $POCl_3$ by biphenyl-2-ol and $o$-cyclohexylphenol[304] and in chloromethylphosphonic dichloride, $ClCH_2POCl_2$, by primary aliphatic amines[305] has been investigated. Methylphosphonic dichloride can now be obtained in good yield by a one-step process, i.e. slow addition of $MeP(O)(OMe)_2$ to an excess of refluxing thionyl chloride,[306] and, while the technique can be extended to the ethyl derivative, pure $BuP(O)Cl_2$ could not be prepared.

Vibrational data for $P_2O_3Cl_4$ can be interpreted in favour of $C_s$ symmetry,[307] while there is crystallographic $C_2$ symmetry for the methylene-bridged analogue $(O)PCL_2CH_2P(O)Cl_2$.[308] Molecules of the latter are linked in infinite chains by two C—H$\cdots$O(P) hydrogen bonds (3.23 Å). The P—C—P angle, which is opened to 116.4°, and the shortened P—C bonds (1.804 Å) correlated with the acidity of the methylene protons and their participation in hydrogen bonding.

Two isomeric 1-chlorophospholen 1-oxides (34) and (35), which are separable by fractional distillation, result from the reaction in equation (46).[309] The pyrocatechol chlorophosphate (36) is seen, from electron-diffraction data, to contain a

---

[296] J. R. Durig and J. E. Saunders, J. Mol. Structure, 1975, **27**, 403.
[297] M. K. Chikanova and E. S. Vorontsov, J. Gen. Chem. (U.S.S.R.), 1974, **44**, 2077.
[298] J. Shamir and A. Silberstein, J. Raman Spectroscopy, 1974, **2**, 623.
[299] V. V. Moskva, V. M. Ismailov, S. A. Novruzov, A. I. Razumov, T. V. Zykova, Sh. T. Akhmedov, and R. A. Salakhutdinov, J. Gen. Chem. (U.S.S.R.), 1974, 2574.
[300] V. V. Moskva, L. A. Bashirova, A. I. Razumov, T. V. Zykova, and R. A. Salakhutdinov, J. Gen. Chem. (U.S.S.R.), 1974, **44**, 2578.
[301] M. Michaud and J.-M. Petit, Compt. rend., 1975, **280**, C, 461.
[302] V. I. Trusov and A. V. Suvorov, Russ. J. Inorg. Chem., 1974, **19**, 297.
[303] B. Viard, J. Amaudrut, and C. Devin, Bull. Soc. chim. France, 1975, 1940.
[304] R. J. W. Cremlyn and N. Kishore, Austral. J. Chem., 1975, **28**, 669.
[305] K. Issleib and R. Handke, Z. anorg. Chem., 1975, **413**, 109.
[306] K. Moedritzer and R. E. Miller, Synth. React. Inorg. Metal-Org. Chem., 1974, **4**, 417.
[307] E. Payen and M. Migeon, Compt. rend., 1974, **279**, C, 687.
[308] W. S. Sheldrick, J.C.S. Dalton, 1975, 943.
[309] K. Moedritzer, Synth. React. Inorg. Metal-Org. Chem., 1975, **5**, 45.

$2PCl_3 + P(OCH_2CH_2Cl)_3 + 3C_4H_6 \longrightarrow$ (34) + (35) + $3ClCH_2CH_2Cl$ (46)

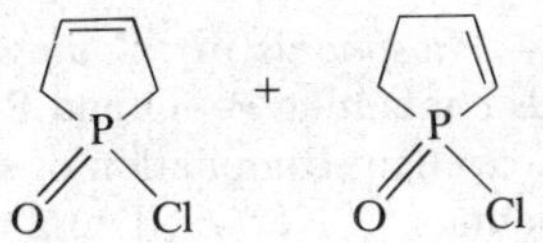

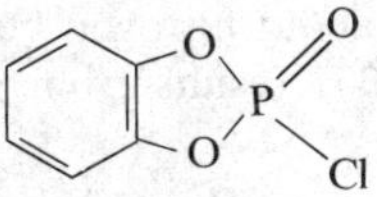

(36)

planar heteroatom ring, with C—O, P=O, P=O distances of, respectively, 1.42, 1.49, and 1.62 Å.[310] A range of radicals of the form $[ROP(O)Cl_2]^-$, generated by γ-irradiation of $ROP(O)Cl_2$, have been investigated by e.s.r. to show that ca. 50% of the spin density is associated with the chlorine atoms.[311]

**Bonds to Bromine or Iodine.**—A reinvestigation of the reaction between hept-1-ene and phosphorus tribromide in the presence of oxygen at 70 °C has shown the formation of 2-bromoheptyl phosphonous dibromide,[312] and not a phosphoryl compound, as previously suggested. Work also continues on reactions of olefins with phosphonium complexes such as $EtPBr_3^+ AlBr_4^-$.[313] With ethylene, there is an exothermic addition, giving an intermediate which with water or ethanol is converted into, respectively, the acid $Et(BrC_2H_4)P(O)OH$ or its ethyl ester. Recent work has led to a room-temperature reaction between the magnesium bromide etherate and the corresponding chloride for the preparation of the phosphorus(III) bromides $(RO)_{3-n}PBr_n$ or $(Et_2N)_{3-n}PBr_n$ which had previously been difficult to obtain.[314]

The following heats of formation at 298.2 K have been obtained from solution calorimetry:[315] $PBr_5(c)$, −60.5(6); $PCl_4BCl_4(c)$, −223.9(6); and $PBr_4BBr_4(c)$, −135.5(7) kcal mol$^{-1}$. A $^{81}Br$ n.q.r. investigation on solid phosphoryl bromide gives little evidence for π-bonding in the P–Br bonds, and points to the solid being orthorhombic, of space group *Pnma*.[316]

More information is now available on the preparation of the substituted phosphorus(III) iodides; for example, good yields of the phosphoramidous di-iodides $R^1R^2NPI_2$ result from substitution of the chlorine in the corresponding dichlorides with sodium iodide in benzene.[317] The products are generally dark red liquids which react readily with both oxygen and water. The recently prepared aryloxy-derivatives $ArOPI_2$ are apparently even less stable and decompose above ca. −50 °C.[318] At 0 °C the disproportionation gives initially (ArO)PI, $PI_3$, and $(ArO)_3P$, but the final products are more complex.

[310] V. A. Naumov and S. A. Shaidulin, *J. Struct. Chem.*, 1974, **15**, 115.
[311] D. J. Nelson and M. C. R. Symons, *J.C.S. Dalton*, 1975, 1164.
[312] S. V. Fridland and M. V. Pobedimskaya, *J. Gen. Chem. (U.S.S.R.)*, 1975, **45**, 227.
[313] R. I. Pyrkin, M. M. Gilyazov, and Ya. A. Levin, *J. Gen. Chem. (U.S.S.R.)*, 1975, **45**, 750.
[314] Z. S. Novikova, M. M. Kabachnik, A. P. Prishchenko, and I. F. Lutsenko, *J. Gen. Chem. (U.S.S.R.)*, 1975, **44**, 1825.
[315] A. Finch, P. J. Gardner, P. N. Gates, A. Hameed, C. P. McDermott, K. K. SenGupta, and M. Stephens, *J.C.S. Dalton*, 1975, 967.
[316] T. Okuda, K. Hosokawa, K. Yamada, Y. Furukawa, and H. Negita, *Inorg. Chem.*, 1975, **14**, 1207.
[317] Zh. K. Gorbatenko, N. G. Feshchenko, and T. V. Kovaleva, *J. Gen. Chem. (U.S.S.R.)*, 1974, **44**, 2311.
[318] N. G. Feshchenko and V. G. Kostina, *J. Gen. Chem. (U.S.S.R.)*, 1975, **45**, 269.

**Bonds to Nitrogen.**—*Phosphorus*(III) *Compounds.* Electron-diffraction data for, *inter alia*, compounds containing P—O and P—N bonds have been discussed and tabulated, giving an excellent compilation of stereochemical data for these important classes of compounds.[319] New CNDO/2 calculations for $PH_2NH_2$ give different results from a previous *ab initio* study with a *gauche*-conformation, with markedly pyramidal geometry about nitrogen being that preferred.[320]

A new azadiphosphilidine (37) results from a gas-phase cyclization reaction

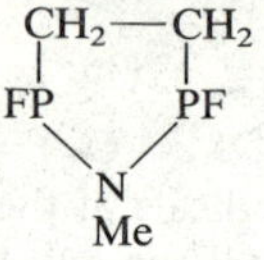

```
     CH₂——CH₂
      |      |
     FP     PF
       \   /
        N
        |
        Me
       (37)
```

between methylamine and 1,2-bis(difluorophosphino)ethane;[321] the corresponding reaction with ammonia gives an analogous product, but in poorer yield and of lower stability, while dimethylamine subsititues one fluorine to give the potentially useful bidentate ligand $Me_2NPFCH_2CH_2PF_2$. Interesting mixed phosphorus-(III)-phosphorus(V) azenes, *e.g.* $RF_2P=NPF_2$, result when the Si—N bond in bis(trimethylsilyl)aminodifluorophosphine is cleaved with either $PF_5$ or $PhPF_4$; reactions with metal carbonyl compounds have also been described.[322]

Cyclic oxa-azaphosphorines can be phosphorylated in the 3-position to give products such as (38) when treated with phosphorus(III) amides of the general type $R^1P(NR_2^2)_2$.[323] Electron-diffraction results for an analogous compound containing a five-membered ring (39) point to an 'envelope' type of ring conformation, with the chlorine atom occupying an axial position.[324]

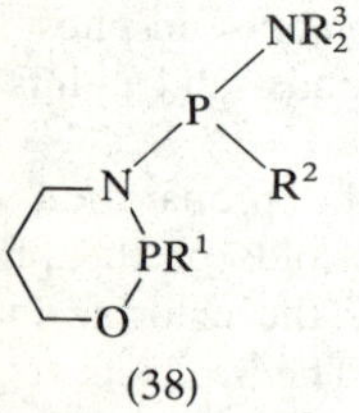

(38) 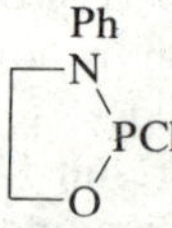   (39)

Although secondary phosphanes such as RP(H)X, where R may be $NHR^1$, $NR_2^1$, $OR^1$, or Cl, are not known, they can be stabilized in the form of metal complexes such as $C_5H_5(CO)_2Mn[PhP(H)NHC_6H_{11}]$.[325] Transamination reactions between $(CF_3)_2PNMe_2$ or the arsenic analogue and secondary amines have been exploited to give a number of new $(CF_3)_2ENR_2$ species.[326]

---

[319] L. Vilkov and L. S. Khaikin, *Topics Current Chem.*, 1975, **53**, 25.
[320] M. Barthelat, R. Mathis, J.-F. Labarre, and F. Mathis, *Compt. rend.*, 1975, **280**, C, 645.
[321] E. R. Falardeau, K. W. Morse, and J. G. Morse, *Inorg. Chem.*, 1975, **14**, 132.
[322] G.-V. Röschenthaler and R. Schmutzler, *Z. anorg. Chem.*, 1975, **416**, 289.
[323] M. A. Pudovik, S. A. Terent'eva, and A. N. Pudovik, *J. Gen. Chem. (U.S.S.R.)*, 1975, **45**, 513.
[324] V. A. Naumov, V. M. Bezzubov, and M. A. Pudovik, *J. Struct. Chem. (U.S.S.R.)*, 1975, **16**, 1.
[325] G. Huttner and H.-D. Müller, *Angew. Chem. Internat. Edn.*, 1975, **14**, 571.
[326] O. Adler and F. Kober, *J. Fluorine Chem.*, 1975, **5**, 231.

Restricted rotation about the P—N bond in $(Me_3C)_2NPX_2$ (X = F, Cl, or Br)[327], and in the silylamino-derivatives $(Me_3Si)_2NP(CF_3)_2$ and $(Me_3Si)Bu^tNP(CF_3)_2$,[328] has been investigated by n.m.r. methods. For the latter compound at room temperature, the data indicate the presence of two rotamers (40) and (41), which,

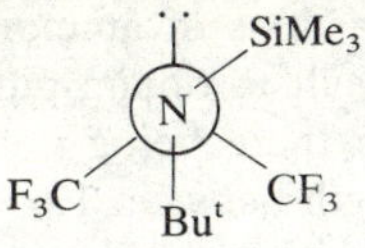

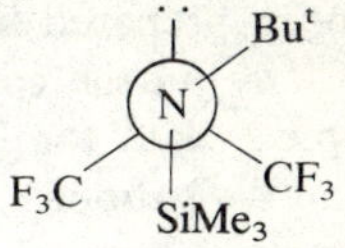

unusually, undergo slow interconversion on the n.m.r. timescale. Coalescence of the two t-butyl signals did not occur unitl 110 °C, giving a value of 20.8 kcal mol$^{-1}$ for $\Delta G_{NP}$.‡

The unusual two- and three-fold co-ordination for, respectively, phosphorus-(III) and -(V) has been achieved previously in trimethylsilyl compounds, but recent work shows that t-butyl groups are also efficient.[329] The reaction in equation (47) gives a 30% yield of the product (42), which with sulphur can be converted into

$$PBr_3 + 2LiNBu^t(SiMe_3) \longrightarrow \quad (42) \quad (47)$$

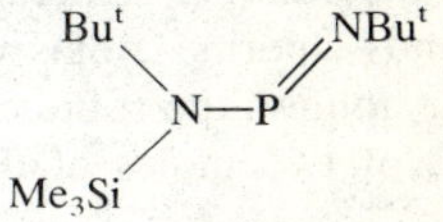

the novel, monomeric, metathiophosphoric acid derivative (43). A reaction of the better known trimethylsilyl-stabilized phosphorus(III) compound (44) with boron halides leads to a new type of heterocycle (45).[330]

A synthesis of a number of new diethoxyphosphino-hydrazines, e.g. $Me_2NNHP(OEt)_2$, $Me_2NN[P(OEt)_2]_2$, and $(EtO)_2PNHNMeP(OEt)_2$, has been reported,[331] while transamination-type reactions between $R^1P(NEt_2)_2$ and substituted amino-hydrazines $H_2NNR^2CH_2CH_2NHR^2$ give derivatives of 1,2,4-triaza-3-phosphacyclohexane (46).[332]

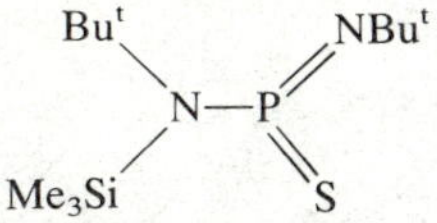

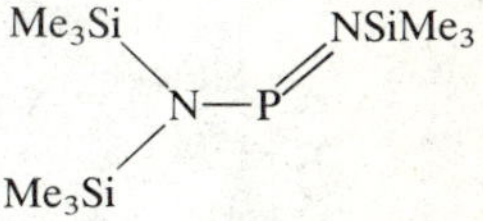

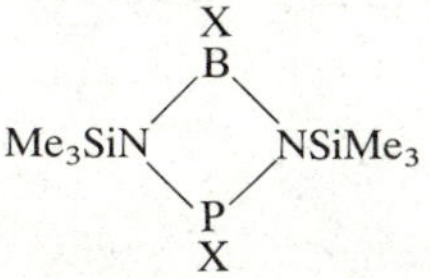

[327] O. J. Scherer and N. Kuhn, Chem. Ber., 1975, **108**, 2478.
[328] R. H. Neilson, R. C. Lee, and A. H. Cowley, J. Amer. Chem. Soc., 1975, **97**, 5302.
[329] O. J. Scherer and N. Kuhn, Angew. Chem. Internat. Edn., 1974, **13**, 811.
[330] E. Niecke and W. Bitter, Angew. Chem. Internat. Edn., 1975, **14**, 56.
[331] D. W. McKennon, G. E. Graves, and L. W. Houk, Synth. React. Inorg. Metal-Org. Chem., 1975, **5**, 223.
[332] G. S. Gol'din, S. G. Fedorov, G. S. Nikitina, and N. A. Smirnova, J. Gen. Chem. (U.S.S.R.), 1974, **44**, 2623.

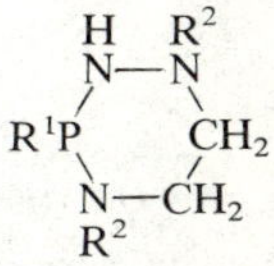

Full n.m.r. data have been obtained for the aminophosphines Pr$^i$NHPR$_2$ (R = Ph or C$_6$F$_5$) and the amino-phosphonium salts derived therefrom;[333] a feature of these spectra is the solvent dependence of the amino-proton resonances. The new bis- and tris-(diflurorophosphino)amines (F$_2$P)$_2$NH and (F$_2$P)$_3$N have recently been prepared from gas-phase reactions of ammonia, trimethylamine, and PF$_2$Cl,[334] but the compounds are difficult to obtain pure.

Crystal and p.e.s. data for P$_4$(NMe)$_6$ point to there being a small degree of $\pi$-bonding in the P—N bonds, which vary in length between 1.65 and 1.74 Å.[335] The P—N—P angles average 124.3°, with the sum of the angles at nitrogen being 356°; the N—P—N angles are 101.7°.

*Phosphorus*(v) *Compounds.* Relationships between the structure of compounds F$_3$P(NH$_2$)$_2$, F$_3$PNPF$_2$, and (CF$_3$)$_2$PNH$_2$ and the magnitude and sign of the J(P–$^{15}$N) coupling constant have been investigated.[336] Dimethylamine reacts with the chlorophosphoranes (CF$_3$)$_3$PCl$_2$ and (CF$_3$)$_2$PCl$_3$ to give the new amine derivatives (CF$_3$)$_3$P(NMe$_2$)$_2$, (CF$_3$)$_3$PClNMe$_2$, and (CF$_3$)$_2$PCl$_2$NMe$_2$.[337] Variable-temperature n.m.r. data indicate the presence of two axial and one equatorial CF$_3$ group in a trigonal-bipyramidal structure for the first compound, while in the other two compounds chlorine atoms occupy axial positions. Vacuum pyrolysis of (CF$_3$)$_3$P(NMe$_2$)$_2$ follows the route observed previously for the difluoride, and on successive loss of two moles of difluorocarbene gives (CF$_3$)$_2$PF(NMe$_2$)$_2$ and CF$_3$PF$_2$(NMe$_2$)$_2$.[338]

The three-fold co-ordination of phosphorus(v) in compound (47) has been confirmed by a full X-ray structure;[339] important parameters are included with the formula, but noteworthy is the shortness of the P—N(imine) bond, in agreement with the presence of (p–p)$\pi$ interaction. Compound (47) is known to be the oxidation product of the phosphorus(III) compound (44) with trimethylsilylazide, but when diazomethane is used the ylide analogue (Me$_3$Si)$_2$-NP(=NSiMe$_3$)(=CH$_2$) cannot be isolated, and the product is the phosphiran (48) formed by a 1,2-cycloaddition.[340] A similar process occurs between diazomethane and (47), when the novel azaphospiridine (49) results.

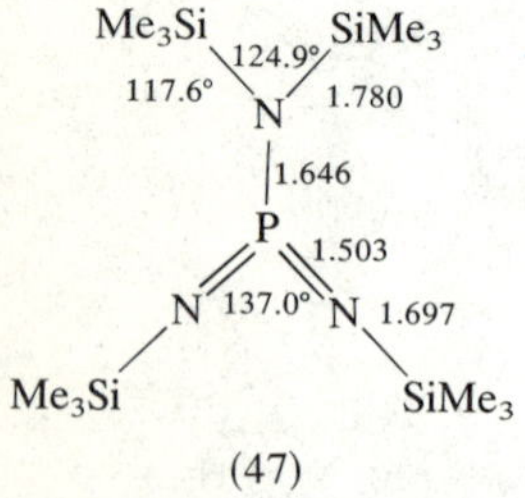

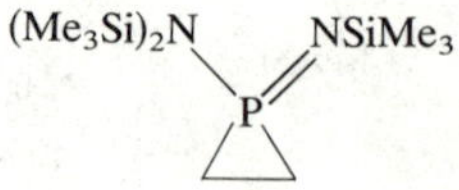

[333] A. H. Cowley, M. Cushner, M. Fild, and J. A. Gibson, *Inorg. Chem.*, 1975, **14**, 1851.
[334] D. E. J. Arnold and D. W. H. Rankin, *J.C.S. Dalton*, 1975, 889.
[335] F. A. Cotton, J. M. Troup, F. Casabianca, and J. G. Riess, *Inorg. Chim. Acta*, 1974, **11**, L33.
[336] J. R. Schweiger, A. H. Cowley, E. A. Cohen, P. A. Kroon, and S. L. Manatt, *J. Amer. Chem. Soc.*, 1974, **96**, 7122.
[337] D. D. Poulin and R. G. Cavell, *Inorg. Chem.*, 1974, **13**, 2324.
[338] D. D. Poulin and R. G. Cavell, *Inorg. Chem.*, 1974, **13**, 3012.
[339] S. Pohl, E. Niecke, and B. Krebs, *Angew. Chem. Internat. Edn.*, 1975, **14**, 261.
[340] E. Niecke and W. Flick, *Angew. Chem. Internat. Edn.*, 1975, **14**, 363.

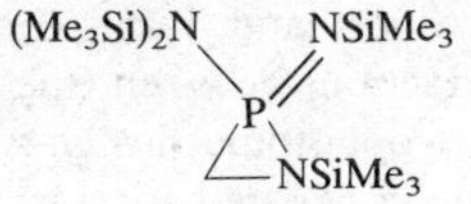

Elements of Group V

(Me$_3$Si)$_2$N\\
\\P—NSiMe$_3$\\
/\\
NSiMe$_3$

(49)

R$_3$P=N—S
N=S
N—S

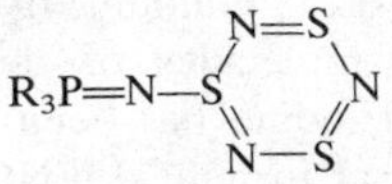

(50)

A series of new phosphazo-phosphonium salts $[(R_3P=N)_n PF_{4-n}]^+ PF_6^-$, where R = Me or Ph and $n$ = 1—3, has been characterized as the product from reactions between the PF$_5$–ether adduct and sililyiminophosphoranes R$_3$P=NSiMe$_3$,[341] while the latter compounds, with S$_4$N$_4$, yield phosphazo-cyclotrithiazenes (50) by elimination of sulphur and (Me$_3$Si)$_2$N$_2$S.[342]

The expected angular structure of the bis(triphenylphosphine)iminium ion $[(Ph_3P)_2N]^+$ has been confirmed in several structures, but in the $[V(CO)_6]^-$ salt the cation lies on an $S_6$ axis and the P—N—P bond is required to be linear.[343] This observation raises problems about bonding, but if the nitrogen is $sp$ hybridized two $(d-p-d)\pi$-bonds could be set up; the shortness of the P—N bonds (1.539 Å) provides some support for this approach.

Trichloromethyl phosphorimidic trichloride, Cl$_3$CN=PCl$_3$, is conveniently prepared from potassium thiocyanate and phosphorus pentachloride in chlorobenzene solution,[344] and new derivatives of the P-trichloromethyl compound (Cl$_3$C)$_2$ClP=NH, in which the hydrogen is replaced by SiCl$_3$, SiMe$_2$Cl, GeBr$_3$, PhCO, etc., have been characterized.[345] Thermal cleavage of (Cl$_3$C)$_2$ClP=NCO-COCl leads to the phosphorane (Cl$_3$C)$_2$PCl$_2$(NCO).[346] One chlorine atom in each trichlorophosphinimine group in SO$_2$(N=PCl$_3$)$_2$ is replaced when two moles of an alcohol are added,[347] and on heating these products the alkyl groups migrate to the nitrogen atoms. Treatment of these latter compounds with (Me$_3$Si)$_2$NR$^1$ then gives the cyclic products (51). The substituted ethane Ph$_3$P=NC$_2$H$_4$N=PCl$_3$

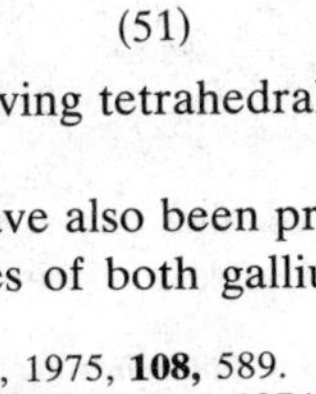

(51)

behaves as a bidentate ligand, giving tetrahedral complexes with Co, Ni, Hg, and Cd halides.[348]

Transition-metal complexes have also been prepared in methanol solution from PO(NH$_2$)$_3$,[349] and 1:1 complexes of both gallium trichloride and bromide form

[341] R. Appel and I. Ruppert, *Chem. Ber.*, 1975, **108**, 589.
[342] I. Ruppert, V. Bastian, and R. Appel, *Chem. Ber.*, 1974, **107**, 3426.
[343] R. D. Wilson and R. Bau, *J. Amer. Chem. Soc.*, 1974, **96**, 7601.
[344] V. Ya. Senenii, A. P. Boiko, G. F. Solodushchenko, N. A. Kirsanova, and V. P. Kukhar', *J. Gen. Chem. (U.S.S.R.)*, 1974, **44**, 1229.
[345] E. S. Kozlov, S. N. Gaidamaka, and L. I. Bobkova, *J. Gen. Chem. (U.S.S.R.)*, 1974, **44**, 1034.
[346] E. S. Kozlov, S. N. Gaidamaka, and L. I. Samarai, *J. Gen. Chem. (U.S.S.R.)*, 1975, **45**, 458.
[347] G. Schönig, U. Klingebiel, and O. Glemser, *Chem. Ber.*, 1974, **107**, 3756.
[348] R. Appel and P. Volz, *Z. anorg. Chem.*, 1975, **413**, 45.
[349] N. Kumagi and H. Mase, *Nippon Kagaku Kaishi*, 1975, 814.

with $PO(NMe_2)_3$.[350] Gallium tri-iodide, on the other hand, forms complexes with one and two molecules of the ligand. Interaction between the hexamethyl derivative and iodine has been examined spectrophotometrically.[351] Complexes with the general formula $Ln(NCS)_3,4L$ have been isolated for $Ln = La$, Ce, Pr, Nd, Sm, Eu, or Gd with $NN$-dimethyldiphenylphosphinamide, $Ph_2P(O)NMe_2$, as the ligand.[352]

The action of hydrogen chloride on $PO(NH_2)_3$ suspended in ether has been re-examined[353] to show the formation of a highly cross-linked product, $(PONH)_n$, by a mechanism involving the formation of either $PO(=NH)(NH_2)$ or $PO(=NH)Cl$ as intermediate. The mixed amino-chloride $MeNHP(O)Cl_2$, according to vibrational spectroscopic and cryoscopic results, exists in the liquid state as a hydrogen-bonded dimer, and, like the parent trichloride, forms adducts with Lewis acids such as $SbCl_5$ and $SnCl_4$.[354] Vibrational spectra have been recorded for the compounds in the series $XP(NMe_2)_nMe_{3-n}$, where $X = O$ or $S$ and $n = 0$—3, and force-constants have been calculated using a simple valence force field.[355] After long periods of heating at high temperatures, mixtures of phenols and $PO(NMe_2)_3$ give the phosphoramidic esters $PO(OAr)_n(NMe_2)_{3-n}$, where $n$ is usually 1 or 2; these compounds will then react with hydrogen chloride to give $POCl(NMe_2)(OAr)$.[356]

The previously unknown bis(difluorophosphoryl)amine $HN(POF_2)_2$ has been synthesized by a reaction between phosphoryl fluoride and the lithium salt of hexamethyldisilazane.[357] Detailed n.m.r. analyses have been reported for compounds containing the structural units $R_2^1P(O)NR^2P(O)R_2^1$ and $R_2^1P(S)NR^2P(S)R_2^1$, where $R^1 = Cl$ or $NMe_2$ and $R^2 = Me$ or Ph, allowing the evaluation of accurate chemical shifts and coupling constants.[358] Preparative methods have been given for the di-thio-compounds mentioned above with, as expected, the non-geminally disubstituted compound $(Me_2N)ClP(S)NRP(S)Cl(NMe_2)$ being produced as a mixture of diastereoisomers.[359] The reaction of the tetrachloride with 6 moles of dimethylamine, however, leads to the cyclic compound (52) rather than a trisubstituted product.

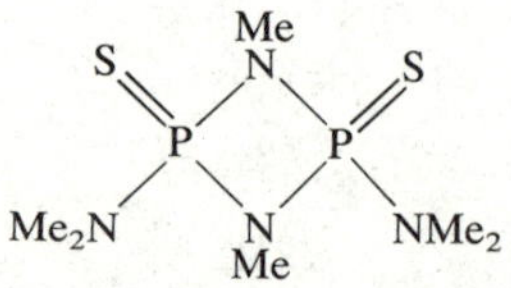

(52)

The course of dimethanolysis of the oxy-sulphur compound $Cl_2P(O)NMeP(S)Cl_2$ is solvent-dependent, with attack occurring initially at the $Cl_2PS$ centre in a

---

[350] J.-C. Couturier and J. Angenault, *Rev. Chim. minérale*, 1974, **11**, 643.
[351] P. Bruno, M. Caselli, and M. DellaMonica, *Inorg. Chim. Acta*, 1974, **10**, 121.
[352] G. Vicentini, L. B. Zinner, and L. Rothschild, *Inorg. Chim. Acta*, 1974, **9**, 213.
[353] L. Riesel and R. Somieski, *Z. anorg. Chem.*, 1975, **415**, 1.
[354] A.-F. Shihada, *Z. anorg. Chem.*, 1975, **411**, 135.
[355] R. Pantzer, W. D. Burkhardt, E. Walter, and J. Goubeau, *Z. anorg. Chem.*, 1975, **416**, 297.
[356] J. Perregaard, E. B. Pedersen, and S.-O. Lawesson, *Rec. Trav. chim.*, 1974, **93**, 252.
[357] E. Fluck and E. Beuerle, *Z. anorg. Chem.*, 1975, **412**, 65.
[358] G. Hägele, R. K. Harris, M. I. M. Wazeer, and R. Keat, *J.C.S. Dalton*, 1974, 1985.
[359] G. Bulloch, R. Keat, and N. H. Tennent, *J.C.S. Dalton*, 1974, 2329.

non-donor solvent but at the alternative $Cl_2PO$ group in ether solution. The linear phosphoryl chlorophosphazenes $Cl_2P(O)[NPCl_2]_nCl$, where $n = 1—3$, react with hexamethyldisilazane[360] and formic acid[361] to give, respectively, new silylamides $Cl_2P(O)[NPCl_2]_n NHSiMe_3$ and $\mu$-imidophosphoric acid chlorides $Cl_2P(O)[NHP(O)Cl]_n Cl$.

New information on the three-membered phosphorus(v) heterocycles which are often postulated as reactive intermediates comes from a study of the reactions of phosphonic diamides $R^1P(O)(NHR^2)_2$.[362] On treatment with t-butyl hypochlorite these are converted into the mono-$N$-chloro-amides, which with base lose hydrogen chloride, forming the diazaphosphoridine oxide (53). The structure is supported by spectroscopic data and by the quantitative formation of the hydrazide (54) on reaction with methanol. New phosphorus(v)-nitrogen-phosphorus(III) systems (55) have been synthesized by transamination reactions between

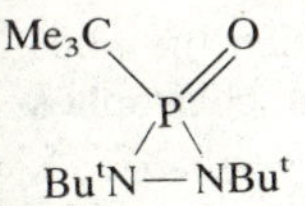

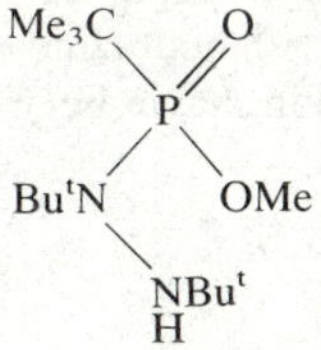

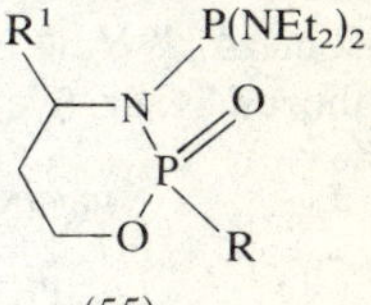

$P(NEt_2)_3$ and 2-substituted-1,3,2-oxazaphosphorine 2-oxides or 2-sulphides.[363] (See also ref. 323).

Preliminary X-ray data on the carcinostat isophosphamide (56; $R = ClCH_2CH_2$) show that the conformation about phosphorus is the same as in its better known isomer phosphamide (57).[364] Important new mechanistic information has been obtained from kinetic measurements on the alkaline hydrolysis of cyclic phosphoramidates (58)[365] and phosphonamidates (59),[366] in the latter the data are

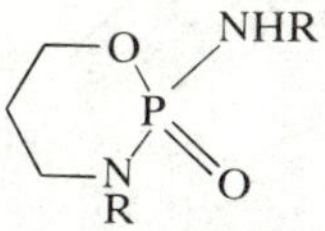

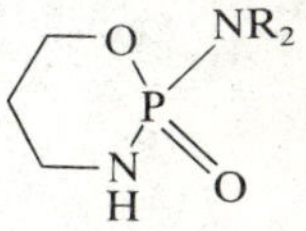

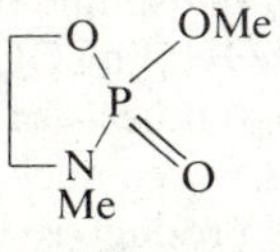

---

[360] L. Riesel and R. Somieski, *Z. anorg. Chem.*, 1975, **411**, 148.
[361] L. Riesel and R. Somieski, *Z. anorg. Chem.*, 1975, **412**, 246.
[362] H. Quast, M. Heuschmann, and M. O. Abdel-Rahman, *Angew. Chem. Internat. Edn.*, 1975, **14**, 486.
[363] M. A. Pudovik, M. D. Medvedev, and A. N. Pudovik, *J. Gen. Chem. (U.S.S.R.)*, 1975, **45**, 683.
[364] H. A. Brassfield, R. A. Jacobson, and J. G. Verkade, *J. Amer. Chem. Soc.*, 1975, **97**, 4144.
[365] C. Brown, J. A. Boudreau, B. Hewitson, and R. F. Hudson, *J.C.S. Chem. Comm.*, 1975, 504.
[366] J. A. Boudreau, C. Brown, and R. F. Hudson, *J.C.S. Chem. Comm.*, 1975, 679.

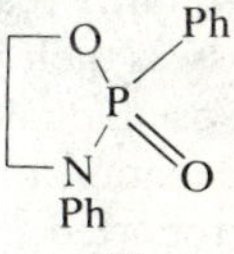

interpreted in terms of square-pyramidal rather than the more usual trigonal-bipyramidal intermediates or transition states.

A tetra-aniline derivative of diphosphoric acid results when ethyl chloride is lost between molecules of $(PhNH)_2POCl$ and $(PhNH)_2P(O)OEt$ in boiling xylene.[367] Further aspects of the aminophosphonate–phosphoramidate rearrangement have been investigated in reactions between diethylhydrogen phosphite and ethyl benzimidate, $EtOC(=NH)Ph$.[368] The initial reaction product is the bis-phosphonate (60), but in the latter stages of the reaction rearrangement gives the phosphoramidate (61). An isomerization reaction also occurs when the phosphoranylidene compound (62), initially obtained from dialkyltrimethylsilylphosphites and methyl diazoacetate, is converted into the phosphinyl derivative (63).[369]

An interesting difference in ring conformation, twist and chair respectively, is displayed by the cis- and trans-isomers of the dithiodihydrazidodimetaphosphoric esters (64), although bond-length and angle data are closely similar for the two isomers.[370] New cyclic phosphorus(v)–hydrazine species (65) in which the phosphorus is in five-fold co-ordination have been prepared from a chlorophos-

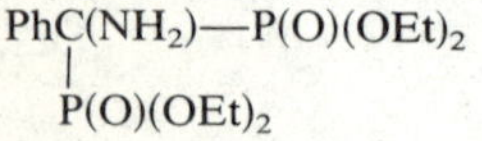

$PhC(NH_2)—P(O)(OEt)_2$
|
$P(O)(OEt)_2$

(60)

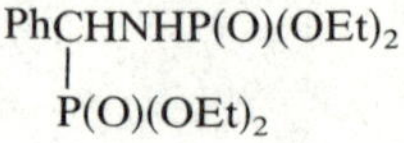

$PhCHNHP(O)(OEt)_2$
|
$P(O)(OEt)_2$

(61)

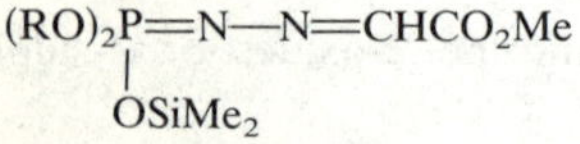

$(RO)_2P=N—N=CHCO_2Me$
|
$OSiMe_2$

(62)

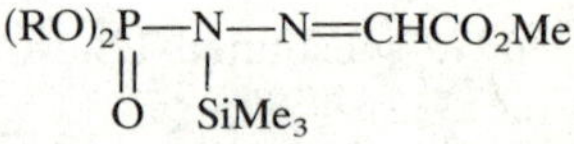

$(RO)_2P—N—N=CHCO_2Me$
‖ |
O SiMe$_3$

(63)

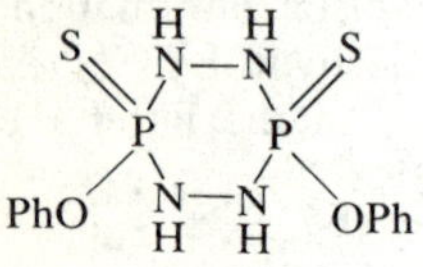

(64)

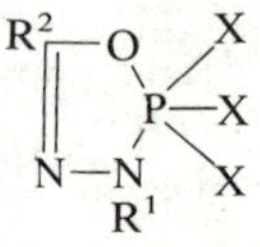

(65) X = Ph or Cl

phorane such as $Ph_3PCl_2$ or $PCl_5$ itself and N-acyl-N'-phenylhydrazine.[371] The compounds decompose fairly readily with loss of the phosphine oxide $X_3PO$, but it has been possible to assess their reactivity toward PhOH, PhCOCl, $PCl_5$, $Et_3N$, etc.

In the phosphorus–pseudohalogen field, the first synthesis of a phosphorylisocyanide (66) by deselenation of the corresponding isoselenocyanate by triethylphosphine has been reported;[372] as expected, the product is very reactive and

---

[367] S. Wagner and H. Scheler, Z. Chem., 1974, **14**, 478.
[368] A. N. Pudovik, M. G. Zimin, I. V. Konovalova, V. M. Pozhidaev, and L. I. Vinogradov, J. Gen. Chem. (U.S.S.R.), 1975, **45**, 26.
[369] A. N. Pudovik and R. D. Gareev, J. Gen. Chem. (U.S.S.R.), 1975, **45**, 220, 221.
[370] U. Engelhardt and H. Hartl, Angew. Chem. Internat. Edn., 1975, **14**, 554; Acta Cryst., 1975, **B31**, 2098.
[371] A. Schmidpeter and J. Luber, Chem. Ber., 1975, **108**, 820.
[372] W. J. Stec, A. Konopka, and B. Uznański, J.C.S. Chem. Comm., 1974, 923.

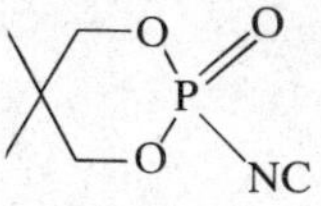

(66)

readily isomerizes to the normal cyanide. Attempts to carry out the same reaction with the analogous phosphorus(III) isoselenocyanatidite using Ph₂POMe to abstract selenium were not successful, and only the normal cyanide was isolated.[373] Mixed chlorideisocyanates were the products when an alkoxy-chloride ROPCl₂ reacted with one mole of sodium cyanate, and although the compounds were unstable they reacted with, respectively, NO₂ and PSCl₃ to form the phosphorus(V) species ROP(O)Cl(NCO) and ROP(S)Cl(NCO), and with antimony trifluoride to give the analogous fluoride.[374]

The azides $P(N_3)_3$, $PO(N_3)_3$, and $P(N_3)_5$ have recently been prepared and characterized in a courageous series of experiments by Buder and Schmidt[375] from the corresponding chloride and sodium azide. The penta-azide must be prepared at $-20\,°C$ to avoid decomposition to the triazide and nitrogen. All are colourless liquids, which, not surprisingly, sometimes explode on warming to room temperature, with $^{31}P$ n.m.r. and vibrational spectra being interpreted to show $C_{3v}$ symmetry for the two triazides and $D_{3h}$ symmetry for the penta-azide. An unstable intermediate azido-species that has been suggested to be $(Me_2N)_3P=NNNR$, where R = Me or Ph, can be isolated from the reaction of tris(dimethylamino)phosphine and either methyl or phenyl azide,[376] while the final products of these and similar reactions with phosphorus(III) compounds are phosphorimidates, according to equations (48) and (49).[377] Low-temperature $^1H$ n.m.r.

$$R^1P(OR^2)_2 + R^3N_3 \longrightarrow R^1P(NR^3)(OR^2)_2 + N_2 \qquad (48)$$

$$R^1P(NR^2_2) + R^3N_3 \longrightarrow R^1P(NR^3)(NR^2_2)_2 + N_2 \qquad (49)$$

spectra point to a barrier of $<8$ kcal mol$^{-1}$ for rotation about the P=N bond in these species and, on heating, rearrangements such as $(MeO)_3P=NMe$ to $(MeO)_2P(O)NMe_2$ sometimes occur.

New non-planar cyclic phosphaborazines (67; X = O or a lone pair of electrons) are the products when lithiated triazaboradecalin reacts with either ClR¹P(O)-NR²P(O)R¹Cl or ClR¹PNR²PR¹Cl,[378] and a P—N—C heterocycle (68) is produced when ammonium chloride and the substituted ethyl phosphorimidic trichloride $CF_3CCl_2N=PCl_3$ interact.[379] The two chlorine atoms can be replaced by aniline residues, and the PCl₂ group is converted into P(O)Cl by carboxylic acids.

Full X-ray structures for two phosphorus–nitrogen-containing heterocycles (69)[380] and (70)[381] have been published, the P—N bond length in the former

[373] W. J. Stec, T. Sudol, and B. Uznański, J.C.S. Chem. Comm., 1975, 467.
[374] V. A. Shokol and L. I. Molyavko, J. Gen. Chem. (U.S.S.R.), 1974, 44, 2615.
[375] W. Buder and A. Schmidt, Z. anorg. Chem., 1975, 415, 263.
[376] H. Goldwhite, P. Gysegem, S. Schow, and C. Swyke, J.C.S. Dalton, 1975, 16.
[377] H. Goldwhite, P. Gysegem, S. Schow, and C. Swyke, J.C.S. Dalton, 1975, 12.
[378] H. Nöth and W. Tinhof, Chem. Ber., 1974, 107, 3806.
[379] V. P. Kukhar' and T. N. Kasheva, J. Gen. Chem. (U.S.S.R.), 1974, 44, 2063.
[380] T. S. Cameron, C. K. Prout, and K. D. Howlett, Acta Cryst., 1975, B31, 2331.
[381] P. A. Tucker and J. C. Van der Grampel, Acta Cryst., 1974, B30, 2795.

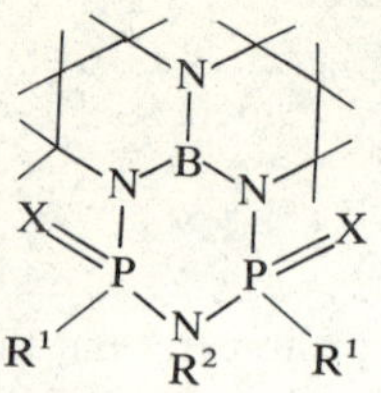

(67) 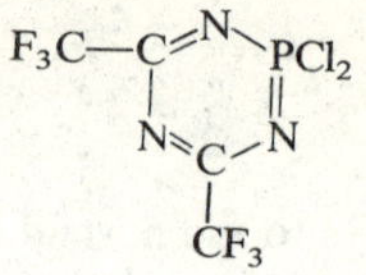    (68)

(1.635 Å) indicating a small degree of $\pi$-bonding. The P—N distance in (70) is much shorter (1.596 Å) and the mean S—N distance is 1.568 Å in the S—N—S segment, but 1.527 Å in the bonds adjacent to phosphorus. The ring conformation, a twisted chair with the oxygen atoms *cis* with respect to the ring, is similar to that in the chlorine analogue.

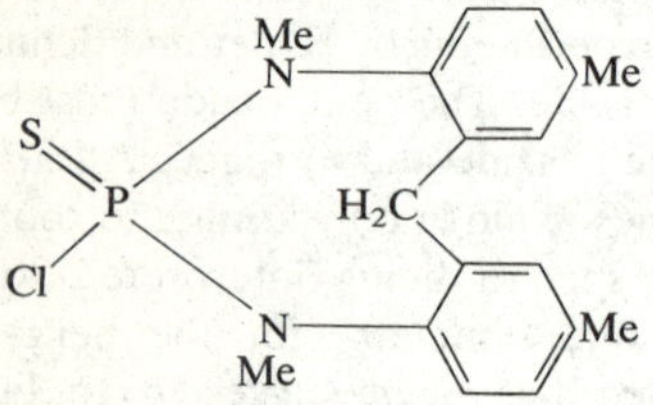

(69) 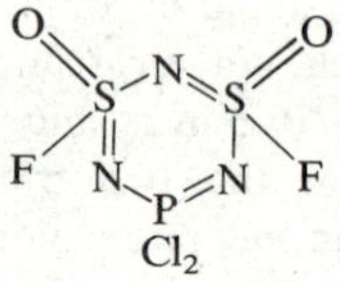    (70)

*Compounds containing* $P_nN_n$ *Rings.* Full structural data are available for five molecules containing $P_2N_2$ ring systems. The nitrogen atoms in the phosphorus-(III) compound (71) lie on a mirror plane, and the molecule has almost exact $C_s$ symmetry; the chlorine atoms are thus mutually *cis*.[382] In the two four-co-ordinate compounds [72(a) X = O, $R^1$ = Cl, $R^2$ = $Bu^t$;[383] (b) X = S, $R^1$ = Ph, $R^2$ = Me[384]] the substituents at phosphorus are in *trans* positions and the ring is planar, with mean P—N bond lengths of 1.661(5) Å for (72a) and 1.688 Å for (72b). The diazadiphosphetidines (73a)[385] and (73b)[386] are also centrosymmetric, with trigonal-bipyramidal geometry at phosphorus, the main features for the former being a long P—C (equatorial) bond [1.888(5) Å] and a N—P—F (equatorial) angle of 134.3(2)°.

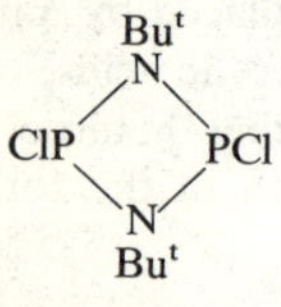

(71) 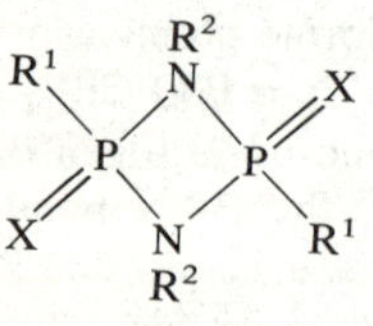    (72) 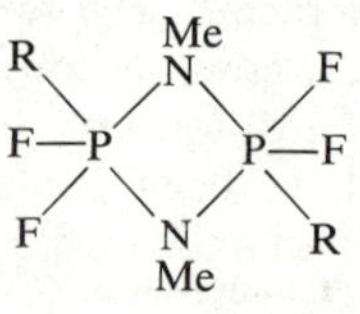    (73)a; R = $CCl_3$; b; R = $C_6F_5$

[382] K. W. Muir, *J.C.S. Dalton*, 1975, 259.
[383] L. Manojlović-Muir and K. W. Muir, *J.C.S. Dalton*, 1974, 2395.
[384] T. S. Cameron, C. K. Prout, and K. D. Howlett, *Acta Cryst.*, 1975, **B31**, 2333.
[385] W. S. Sheldrick and M. J. C. Hewson, *Acta Cryst.*, 1975, **B31**, 1209.
[386] M. Fild, W. S. Sheldrick, and T. Stankiewicz, *Z. anorg. Chem.*, 1975, **415**, 43.

*Elements of Group V* 305

In the preparative area, new cyclo-diphosphazanes (74), where X = O, S, or lone pair and R = Me, Et, or Bu$^t$, result when three moles of t-butylamine react with bis(dichloro-phosphino, -phosphinoyl, or -phosphinothioyl)amine,[387] but attempts to prepare the unknown analogues (ClPNR)$_2$ (R = Me or Et) by a similar route did not give pure products. Continuation of synthetic work by Schmutzler and his co-workers has led to new series of diazadiphosphetidines such as (R$_{3-n}$F$_n$PNMe)$_2$, where R = Me or MeO,[388] and (RF$_2$PNMe)$_2$, where R = ClCH$_2$, Cl$_2$CH, or Cl$_3$C.[389] Interest in these compounds is associated with their complex n.m.r. behaviour, which can often be analysed completely. A zwitterionic structure (75), rather than one involving five-co-ordination at phosphorus, has been

```
      R                              F   Me      Me
   X   N    Cl                    F   N       N—CH₂
    \\P—P//                        \\P—P⁺/     |
   Cl  N   X                     F / \\N/  \\N—CH₂
      Bu^t                          F   Me      Me
      (74)                              (75)
```

proposed for the product obtained from a reaction between (F$_3$PNMe)$_2$ and the di-lithium derivative LiMeNCH$_2$CH$_2$NMeLi on the basis of its n.m.r. and mass spectrometric behaviour.[390]

A new attempt at assigning the vibrational bands in the triphosphazene P$_3$N$_3$Cl$_6$ uses force-constant calculations as the basis and gives improvements, particularly in assignments of the $e^1$ modes.[391] Evidence in favour of the Dewar 'island' model of $\pi$-bonding in phosphazenes has been adduced from magneto-optic measurements on the fluoro-, chloro-, dimethylamino-, and phenoxyderivatives (PNX$_2$)$_{3-5}$.[392] The oligomers (PNCl$_2$)$_{3-7}$ can be separated by t.l.c. on silica gel, and although the $R_f$ values are a linear function of $n$ for $n = 4$—7, that for the trimer is very low.[393]

The reactive intermediates P$_n$N$_n$F$_{2n+1}$N(SnMe$_3$)$_2$, where $n = 3$—5, which result when fully fluorinated cyclophosphazenes react with tris(trimethylstann)amine have been used in reactions with S$_3$N$_3$Cl$_3$,[394] S$_3$N$_2$Cl$_2$,[394] and ClMe$_2$SiN=S=NSiMe$_2$-Cl,[395] giving products such as (76) and (77). An X-ray structure for the mono-

```
                                          Me₂
                                          Si—N
                                         /    \\
  P₄N₄F₇N=S—S            P₃N₃F₅N              S
         |    |                          \\    /
         N    N                          Si—N
          \\S/                            Me₂
          (76)                            (77)
```

[387] G. Bulloch and R. Keat, *J.C.S. Dalton*, 1974, 2010.
[388] R. K. Harris, M. I. M. Wazeer, O. Schlak, and R. Schmutzler, *J.C.S. Dalton*, 1974, 1912.
[389] R. K. Harris, M. Lewellyn, M. I. M. Wazeer, J. P. Woplin, R. E. Dunmur, M. J. C. Hewson, and R. Schmutzler, *J.C.S. Dalton*, 1975, 61.
[390] O. Schlak, R. Schmutzler, H.-M. Schiebel, M. I. M. Wazeer, and R. K. Harris, *J.C.S. Dalton*, 1974, 2153.
[391] R. E. Christopher and P. Gans, *J.C.S. Dalton*, 1975, 153.
[392] J.-P. Faucher, O. Glemser, J.-F. Labarre, and R. A. Shaw, *Compt. rend.*, 1974, **279**, C, 441.
[393] V. Novobilský, V. Kolský, and W. Waněk, *Z. anorg. Chem.*, 1975, **416**, 187.
[394] H. W. Roesky and E. Janssen, *Chem. Ber.*, 1975, **108**, 2531.
[395] H. W. Roesky and B. Kuhtz, *Chem. Ber.*, 1975, **108**, 2536.

amide $P_3N_3F_5NH_2$ shows that the molecules are joined in layers through intermolecular hydrogen bonds to a ring nitrogen atom.[396] The six-membered system is slightly puckered, and the P—N distances to the $PFNH_2$ group (1.58 Å) are significantly longer than the others (1.56 Å).

$^{19}F$ N.m.r. spectra of a number of fluorinated cyclotriphosphazenes, including cis- and trans- (non-geminal) $P_3N_3F_3X_3$, $P_3N_3F_2X_4$ (X = Cl or Br), and the geminal and non-geminal forms of $P_3N_3F_2Cl_2(NMe_2)_2$ have been completely analysed, using an iterative technique.[397] Relationships between structure and the various parameters derived have been considered in general, but particularly interesting is the magnitude of $^4J(FF)$. This has a value of ca. 12 Hz for trans fluorines, but ca. −1 Hz where the alternative cis arrangement is present. Slightly distorted ring conformations, probably associated with crystal-packing effects, have been reported for the $P_3N_3$ systems in cis-$P_3N_3F_3X_3$, where X = Cl or Br.[398] In agreement with the reduced electronegativity, the ring bond-lengths increase from 1.567 Å in the trichloride to 1.580 Å in the bromide. Completing the series of isomeric $P_3N_3Cl_3(NMe_2)_3$ structures, data have now been given for the trans non-geminal isomer;[399] five of the ring atoms are coplanar, but the phosphorus atom carrying the chlorine trans to the other two is 0.15 Å from that plane. According to CNDO/2 calculations, of the three $P_3N_3Cl_3(NMe_2)_3$ isomers, that with the geminal structure appears to have a higher total energy than the cis and trans non-geminal forms.[400]

Partial substitution of $P_3N_3Cl_6$ can be achieved with the sodium salt of OO-diethyl N-methyl aminophosphate and the products $P_3N_3Cl_{6-n}[NMeP(O)(OEt)_2]_n$ (n = 1—3) isolated.[401] Further reactions of these mono- and di-substituted compounds with ammonia at room temperature lead to geminal diamides in both cases, while complete chlorine replacement occurs when liquid ammonia is used.[402] A new series of $^{35}Cl$ n.q.r. spectra of triphosphazenes containing dimethylamine or piperidine groups in addition to chlorine point to the value of this technique in distinguishing positional isomers, but in general the approach does not allow unambiguous assignment of structures to geometrical isomers.[403] Dimethylamino-groups in $P_3N_3(NMe_2)_6$ can be removed in a stepwise manner by treatment with anhydrous hydrogen chloride or bromide.[404] Di- and tri-chlorides can be isolated, but extensive decomposition occurs with hydrogen bromide; with hydrogen iodide the two products obtained were the adducts $P_3N_3(NMe_2)_6,HI$ and $P_3N_3(NMe_2)_6,HI_3$.

Trimeric and tetrameric methyl phosphazenes can be prepared from dimethyltrichlorophosphorane and ammonium chloride while the corresponding pentamer is obtained by treating the fluoride $P_5N_5F_{10}$ with methyl-lithium.[405] The

[396] S. Pohl and B. Krebs, Chem. Ber., 1975, **108**, 2934.
[397] P. Clare, D. B. Sowerby, R. K. Harris, and M. I. M. Wazeer, J.C.S. Dalton, 1975, 625.
[398] P. Clare, T. J. King, and D. B. Sowerby, J.C.S. Dalton, 1974, 2071.
[399] F. R. Ahmed and E. J. Gabe, Acta Cryst., 1975, **B31**, 1028.
[400] J.-P. Faucher, J.-F. Labarre, and R. A. Shaw, J. Mol. Struct., 1975, **25**, 109.
[401] P. Gehlert, H. Schadow, H. Scheler, and B. Thomas, Z. anorg. Chem., 1975, **415**, 51.
[402] P. Gehlert, H. Schadow, H. Scheler, and B. Thomas, Z. anorg. Chem., 1975, **416**, 169.
[403] W. H. Dalgleish, R. Keat, A. L. Porte, D. A. Tong, M. Ul-Hasan, and R. A. Shaw, J.C.S. Dalton, 1975, 309.
[404] S. N. Nabi, R. A. Shaw, and C. Stratton, J.C.S. Dalton 1975, 588.
[405] H. T. Searle, J. Dyson, T. N. Ranganathan, and N. L. Paddock, J.C.S. Dalton, 1975, 203.

compounds are basic and can be quaternized with methyl or ethyl iodide and, in addition, a range of simple salts and complex compounds [such as $P_3N_3Me_6,HCl$; $P_4N_4Me_8,2L$ (L = $HClO_4$ or $HgCl_2$); $P_4N_4Me_8,4L$ (L = $AgNO_3$ or $H_2O$); and $P_5N_5Me_{10},H_2CuCl_4,H_2O$] can be readily prepared. A dipositive cation is considered to be present in the complex $P_4N_4Me_9HZn(NCS)_4$. The quaternized trimeric compound (78) mentioned above undergoes an interesting deprotonation reaction at a *P*-methyl group with the strong base sodium bis(trimethylsilyl)amide, leading to the diazaphosphorine (79), which can be identified through its hydrolysis product (80).[406] Similar reactions are possible with the corresponding tetramer, leading to isolation of (81) and the tetrameric analogue of (80). The octamethyl

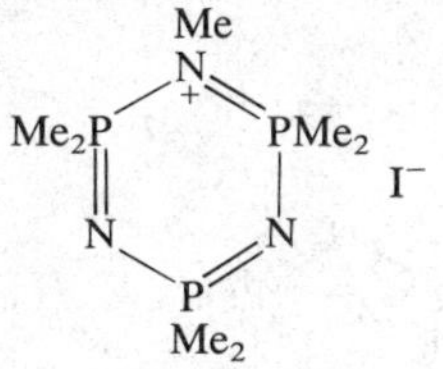

(78) 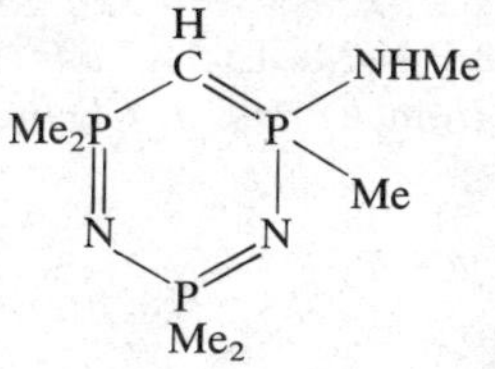        (79) 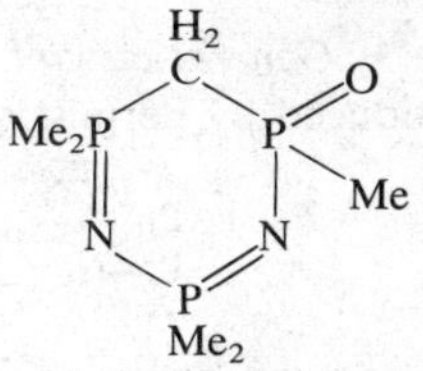        (80)

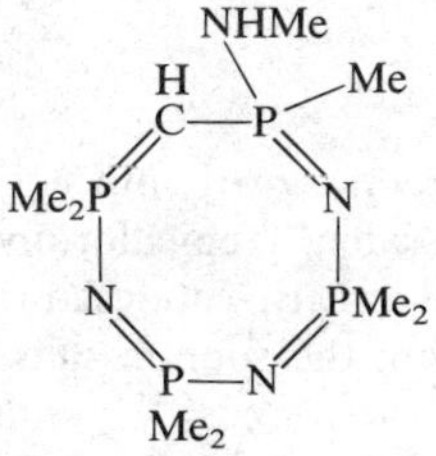

(81)

derivative itself can be deprotonated in ether solution by the action of methyl-lithium, and the isolation of mixed methyl ethyl phosphazenes $P_4N_4Me_{8-n}Et_n$ with $n \leq 4$ after treatment with methyl iodide points to the initial formation of the carbanions $P_4N_4Me_{8-n}(CH_2^-)_n$.[407] One compound isolated during this work is $P_4N_4Me_6Et_2,2HCl$, which from *X*-ray data contains the $P_4N_4Me_6Et_2H_2^{2+}$ cation.[408] The ethyl groups on distant phosphorus atoms are in a *trans* orientation and the $P_4N_4$ ring has a chair conformation (see Figure 1).

A new method for the formation of the $P_3N_3$ ring system uses bis(aminodiphenylphosphorus) nitride chloride [$(H_2NPPh_2)_2N$]Cl and either phosphorus trichloride or $MePCl_2$.[409] With the former, ring closure takes place *via* condensation and a redox disproportionation to give the spirocyclic compound (82), but a monocyclic compound (83) is the product with $MePCl_2$.

Insertion of ethylene oxide into the P—Cl bonds of $P_3N_3Cl_6$ or $P_4N_4Cl_8$ to give

---

[406] H. P. Calhoun, R. T. Oakley, and N. L. Paddock, *J.C.S. Chem. Comm.*, 1975, 454.
[407] H. P. Calhoun, R. H. Lindstrom, R. T. Oakley, N. L. Paddock, and S. M. Todd, *J.C.S. Chem. Comm.*, 1975, 343.
[408] H. P. Calhoun, R. T. Oakley, N. L. Paddock, and J. Trotter, *Canad. J. Chem.*, 1975, **53**, 2413.
[409] A. Schmidpeter and H. Eiletz, *Chem. Ber.*, 1975, **108**, 1454.

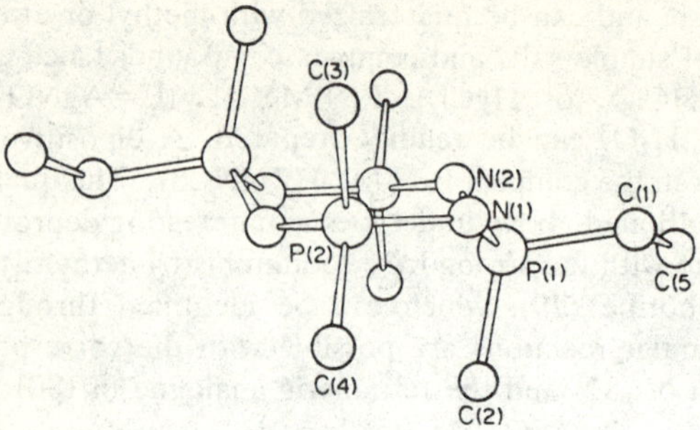

**Figure 1** *General view of the* $P_4N_4Me_6Et_2H_2^{2+}$ *ion* (Reproduced by permission from *Canad. J. Chem.*, 1975, **53**, 2413)

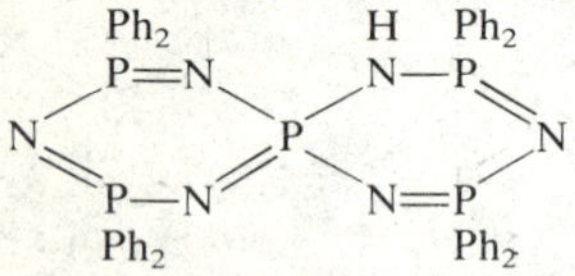

(82) 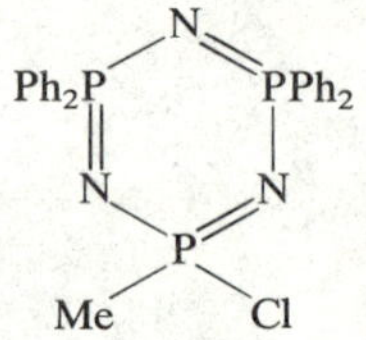     (83)

chloroalkoxy-phosphazenes occurs slowly, but is greatly catalysed by tetra-alkylammonium halides or lithium salts;[410] epichlorohydrin is much less reactive, but up to 70% reaction is achieved in LiBr-catalysed reactions. Complete substitution of $P_3N_3Cl_6$ can be achieved with the sodium salts of *para*-substituted phenols[411] and sodium bis(hydroxymethylbenzenes),[412] and a number of alkoxy-penta-(aryloxy)triphosphazenes have been characterized from reactions of $P_3N_3$-$(OC_6H_4X)_5Cl$.[413] Substitution in the trimeric system by MeSNa has been re-examined to show that the degree of substitution depends mainly on the solvent and reaction temperature.[414]

Pairs of isomeric disubstitution products have been separated from reactions between $P_4N_4Cl_8$ and four moles of ethylamine, t-butylamine, or *N*-methylaniline;[415] with t-butylamine the products are probably the geminal and one of the non-geminal isomers, but probably two of the non-geminal isomers are produced in the other cases.

Structural studies on $Ph_8P_4N_4$ and the corresponding arsenic compound reveal $S_4$ molecular symmetry in each case, with the ring conformation being intermediate between the idealized 'boat' and 'saddle' forms (see Figure 2).[416] In each case the ring bond distances are equal, 1.590 Å for the phosphorus and 1.73 Å for

[410] D. F. Lawson, *J. Org. Chem.*, 1974, **39**, 3357.
[411] R. L. Dieck and M. A. Selvoski, *Inorg. Nuclear Chem. Letters*, 1975, **11**, 313.
[412] M. Kajiwara and H. Saito, *J. Inorg. Nuclear Chem.*, 1975, **37**, 29.
[413] A. A. Volodin, S. N. Zelenetskii, V. V. Kireev, and V. V. Korshak, *J. Gen. Chem.* (*U.S.S.R.*), 1975, **45**, 32.
[414] B. Thomas, H. Schadow, and H. Scheler, *Z. Chem.*, 1975, **15**, 26.
[415] R. Keat, S. S. Krishnamurthy, A. C. Sau, R. A. Shaw, M. N. S. Rao, A. R. V. Murthy, and M. Woods, *Z. Naturforsch.*, 1974, **29b**, 701.
[416] M. J. Begley, D. B. Sowerby, and R. J. Tillott, *J.C.S. Dalton*, 1974, 2527.

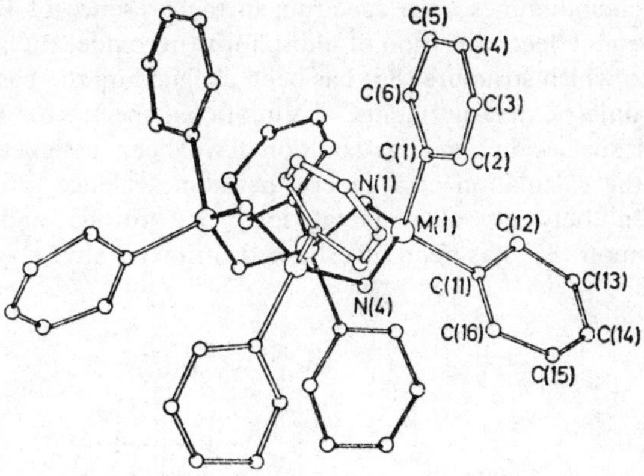

# Elements of Group V

**Figure 2** Ring conformation in $P_4N_4Ph_8$ and $As_4N_4Ph_8$
(Reproduced by permission from J.C.S. Dalton, 1975, 2527)

the arsenic compound, and the angles at nitrogen are small. This is as low as 121.4° for the cyclo-arsazene, probably indicating only a small degree of $4d$-orbital participation. A $D_{2d}$ 'saddle' ring conformation is found in the geminally substituted tetraphenyl-tetrachloro-tetraphosphazene (84), which has overall

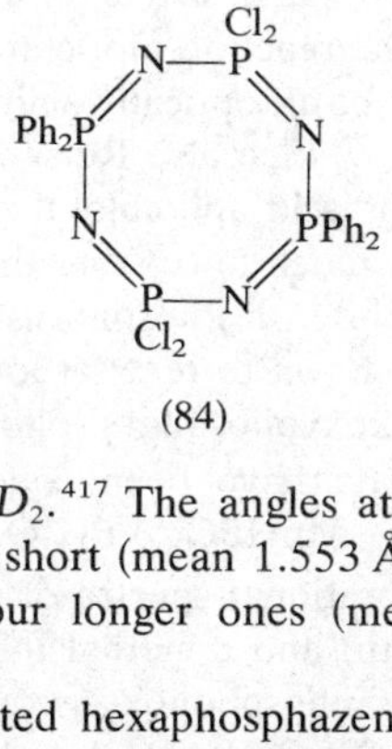

(84)

molecular symmetry close to $D_2$.[417] The angles at nitrogen vary between 128.8 and 135.4° and there are four short (mean 1.553 Å) P—N bonds associated with the $NPCl_2N$ segments and four longer ones (mean 1.591 Å) in the $NPPh_2N$ segments.

The dimethylamine-substituted hexaphosphazene $P_6N_6(NMe_2)_{12}$ more readily behaves as a ligand with transition metals than the corresponding lower oligomers, probably because the twelve-membered system can act as a macrocyclic ligand, accommodating the cations inside the ring.[418] Its behaviour towards a number of metal chlorides and nitrates, forming respectively $2MCl_2,L$ and $M(NO_3)_2,L$, has been defined. The complexes contain the cations $[LMCl]^+$ or $LM(NO_3)]^+$, in which the metal is in trigonal-bipyramidal co-ordination to four nitrogen atoms and the chloride or nitrate group.

**Bonds to Oxygen.**—*Lower Oxidation States.* The origin of the cool green phosphorus flame is an excited-state dimer of phosphorus monoxide, and marked

[17] G. J. Bullen and P. E. Dann, *Acta Cryst.*, 1974, **B30**, 2861.
[18] H. P. Calhoun, N. L. Paddock, and J. N. Wingfield, *Canad. J. Chem.*, 1975, **53**, 1765.

changes in the chemiluminescence spectrum in the presence of $D_2O$ have been observed.[419] Thermal decomposition of phosphorus(III) oxide, $P_4O_6$, leads to good yields of $P_4O_7$, to which structure (85) has been assigned on the basis of $^{31}P$ n.m.r. and mass spectrometric measurements.[420] Vibrational spectra for $H_2PO_2^-$ and the two deuteriated species in aqueous solution have been assigned,[421] and these results used in the calculation of a general harmonic valence force field.[422] The weak interaction between alkali-metal ions or protons and the triphosphate(IV,III,IV) anion (86) has been investigated titrimetrically.[423] γ-Irradiation of

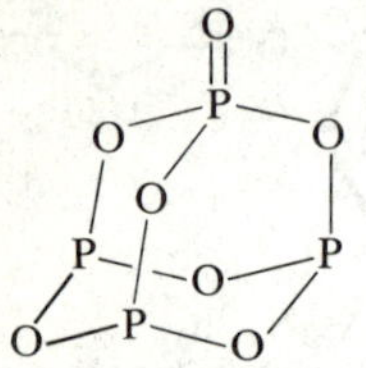

(85) 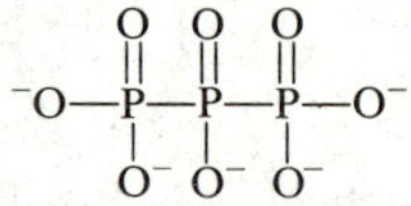 (86)

phosphites such as $P(OMe)_3$ is considered to lead to a radical anion by electron addition followed by protonation at phosphorus;[424] the subsequent loss of an anion or a neutral group can then lead to phosphinyl or phosphoranyl species.

Optically active phosphorus(III) acid esters $PhR^1P^*(OR^2)$ have recently been prepared for the first time by the reaction of a chiral phosphorus chloride and an alcohol or thiol in the presence of an optically active amine, in this case (−)-NN-dimethyl-(1-phenylethyl)amine.[425] (See also Refs. 449 and 450). A kinetic investigation shows that phosphorous acid and chloral react to give 2,2,2-trichloro-1-hydroxyethyl phosphonate at a much lower rate than the corresponding reaction with dimethyl hydrogen phosphite.[426] Phosphorus trichloride and dialkyl phosphites $(RO)_2P(O)H$ have been shown to react at low temperature in the presence of pyridine, producing tris(dialkoxyphosphoryl)phosphines $[(RO)_2P(O)]_3P$.[427]

Thermal analysis and i.r. data have been collected for a number of manganese phosphites, including $MnHPO_3,3.6H_2O$, $Mn_3H_6P_4O_{12},1.5H_2O$, and $MnH_4P_2O_6,H_2O$,[428] and the vibrational spectra of the polymeric hypophosphites $Me_2MO_2PH_2$ (M = Al, Ga, or In) and dimethyl thiophosphinates $Me_2MOSPMe_2$ have been assigned.[429] New phosphonoxy-germoxane compounds such as $(Me_3GeO)_2P(O)H$ and $[Me_2GeOP(O)H—O]_2$ can be isolated from the oxidation

---

[419] R. J. van Zee and A. U. Khan, J. Amer. Chem. Soc., 1974, **96**, 6805; Chem. Phys. Letters, 1975, **36**, 123.
[420] M. L. Walker and J. L. Mills, Synth. React. Inorg. Metal-org. Chem., 1975, **5**, 29.
[421] M. Abenoza and V. Tabacik, J. Mol. Structure, 1975, **26**, 95.
[422] V. Tabacik and M. Abenoza, J. Mol. Structure, 1975, **27**, 369.
[423] R. Šmied, J. Inorg. Nuclear Chem., 1975, **37**, 318.
[424] B. W. Fullam and M. C. R. Symons, J.C.S. Dalton, 1975, 861.
[425] M. Mikolajczyk, J. Drabowicz, J. Omelanczuk, and E. Fluck, J.C.S. Chem. Comm., 1975, 382.
[426] N. A. Vorontsova, V. V. Voronkova, O. N. Vlasov, and N. N. Mel'nikov, J. Gen. Chem. (U.S.S.R.), 1974, **44**, 2598.
[427] D. Weber and E. Fluck, Z. Naturforsch., 1975, **30b**, 60.
[428] M. Ebert and J. Eysseltová, Monatsh., 1974, **105**, 1030.
[429] B. Schaible, K. Roessel, J. Weidlein, and H. D. Hausen, Z. anorg. Chem., 1974, **409**, 176.

# Elements of Group V

products of, respectively, $Me_3GePH_2$ and $Me_2Ge(PH_2)_2$.[430] New reactions of tris-(trimethylsilyl) phosphite $P(OSiMe_3)_3$ have been reported.[431,432] With alkyl bromides, the products are alkyl phosphonates $RP(O)(OSiMe_3)_2$, but when phenacyl bromide is used, the enol phosphonate (87) is initially formed, and this can be readily hydrolysed to the corresponding acid.[431] Silylated ene-diols (88) and quinol-phosphates (89) are the products from α-diketones and p-quinones, respectively.[432]

(87)    (88)    (89)

*Phosphorus*(V) *Compounds.* X-Ray p.e. spectra for many phosphoryl and thiophosphoryl compounds have been analysed to give a series of substituent constants which accurately reproduce the experimental data.[433] There is now strong evidence for a monomeric metaphosphate, isoelectronic with sulphur trioxide, from pyrolysis experiments on methyl but-2-enyl phostonate (90).[434] The

(90)

$MeOPO_2$ species generated in the gas phase can be trapped by N-methylaniline as the N-methyl N-phenyl phosphoramidate, while in the absence of such a reagent the major non-volatile product is methyl polyphosphate.

There is continued general interest in cage structures such as that in $P_4O_{10}$. Previous work had shown the formation of 1,5-μ-oxo-tetrametaphosphate (91) from tetrametaphosphate, but further treatment with dicyclohexyl carbodi-imide leads to the full $P_4O_{10}$ structure, solvated at one apex with a carbodi-imide molecule (92).[435] Similar carbodi-imide-promoted condensations using methylenediphosphonic acid, $(HO)_2OPCH_2PO(OH)_2$, have been followed by $^{31}P$ n.m.r. spectroscopy, pointing to the intermediate formation of the acid (93), the final product being the full anhydride (94).[436]

X-Ray structures have been completed for the caged carbophosphorane (95)[437] and the cyclohexylammonium salt of the di-β-naphthyl ester of diphosphoric acid.[438] The latter is a model compound for conformational studies on physiologically active diphosphates where hydrogen-bonding is probably important. The

---

[30] A. R. Dahl and A. D. Norman, *Inorg. Chem.*, 1975, **14**, 1093.
[31] T. Hata, M. Sekine, and N. Kagawa, *Chem. Letters*, 1975, 635.
[32] T. Hata, M. Sekine, and N. Ishikawa, *Chem. Letters*, 1975, 645.
[33] E. Fluck and D. Weber, *Z. anorg. Chem.*, 1975, **412**, 47.
[34] C. H. Clapp and F. H. Westheimer, *J. Amer. Chem. Soc.*, 1974, **96**, 6710.
[35] T. Glonek, T. C. Myers, and J. R. Van Wazer, *J. Amer. Chem. Soc.*, 1975, **97**, 206.
[36] T. Glonek, J. R. Van Wazer, and T. C. Myers, *Inorg. Chem.*, 1975, **14**, 1597.
[37] H. L. Carrell, H. M. Berman, J. S. Ricci, jun., W. C. Hamilton, F. Ramirez, J. F. Marecek, L. Kramer, and I. Ugi, *J. Amer. Chem. Soc.*, 1975, **97**, 38.
[38] M. K. Wood, M. Sax, and J. Pletcher, *Acta Cryst.*, 1975, **B31**, 76.

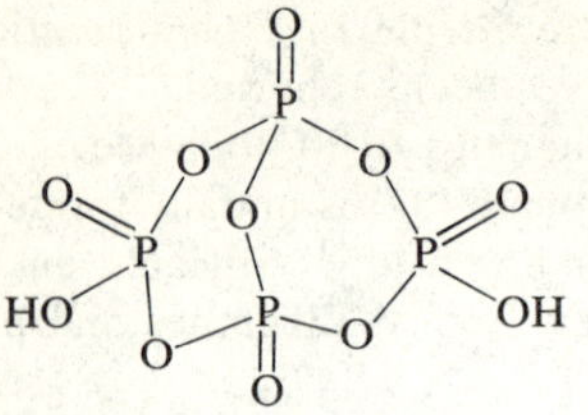

(91) 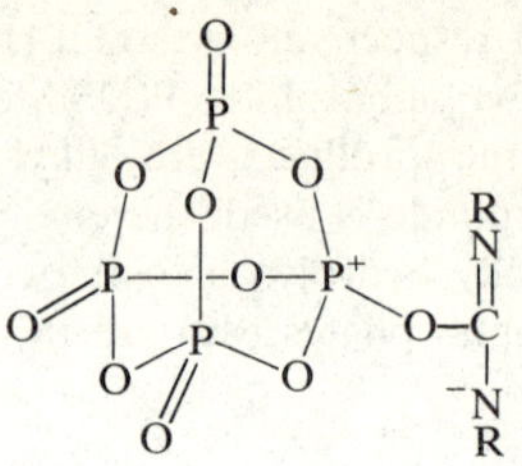 (92) 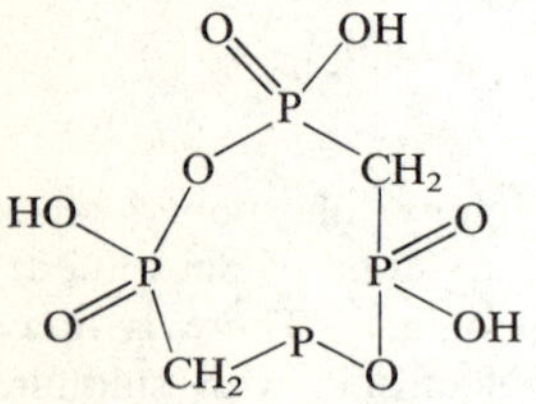 (93) 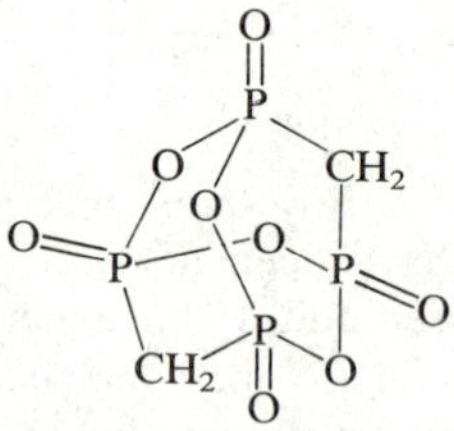 (94) 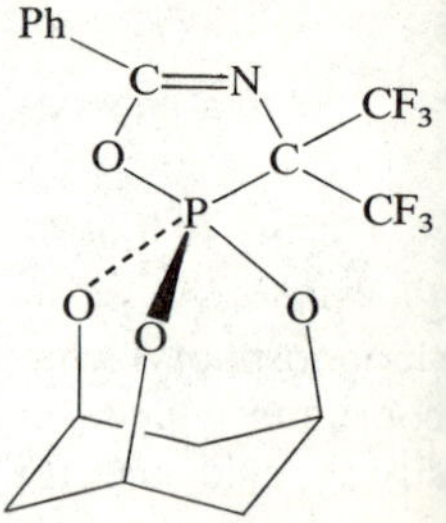 (95)

β-form of aminomethylphosphonic acid, $H_3N^+CH_2PO_3H^-$, contains one long (1.51 Å) and two short (1.49 Å) phosphorus–oxygen bonds, and both N—H···O and O—H···O hydrogen bonds are present.[439] A valence force field has been calculated for this acid, using vibrational data on six isotopically labelled derivatives,[440] and a detailed vibrational analysis is available for $Cl_2HCPO_3H_2$.[441]

Octahedrally co-ordinated solvates of $Al^{3+}$ have been detected by n.m.r. measurements in the presence of trialkyl phosphates, phosphonates and dialkyl hydrogen phosphites,[442] and a number of complexes with first-row transition elements, such as $M[O_2P(OEt)_2]_n$ ($n = 3$ for $V^{III}$ or $Cr^{III}$ and $n = 2$ for $Mn^{II}$ or $Cu^{II}$), can be obtained from the anhydrous chlorides and triethyl phosphate.[443] At temperatures between 180 and 270 °C metal halides react with triethyl or tri-n-butyl thiophosphates to give complexes containing alkyl pyrothiophosphate groups, and although the structures are not known with certainty, spectroscopic measurements indicate the formation of P—S—P rather than P—O—P links in the ligand.[444] New tantalum compounds containing diphenylphosphinate groups in a variety of co-ordination modes have been synthesized.[445]

Different routes leading to the formation of silyl esters of phosphinic acids $R_2P(O)OSiMe_3$ have been investigated and the products characterized, and novel bis(phosphinic acid esters) can be prepared as shown in equation (50).[446] The

$$2Me_2P(X)ONH_4 + Me_2SiCl_2 \rightarrow 2NH_4Cl + Me_2P(X)OSiMe_2OP(X)Me_2 \quad (50)$$

[439] M. Darriet, J. Darriet, A. Cassaigne, and E. Neuzil, *Acta Cryst.*, 1975, **B31**, 469.
[440] C. Garrigou-Lagrange and C. Destrade, *Compt. rend.*, 1975, **280**, C, 969.
[441] B. J. van der Veken and M. A. Herman, *J. Mol. Structure*, 1975, **28**, 371.
[442] J.-J. Delpuech, M. R. Khaddar, A. A. Peguy, and P. P. Rubini, *J. Amer. Chem. Soc.*, 1975, **97**, 3373.
[443] R. C. Paul, V. P. Kapila, R. S. Battu, and S. K. Sharma, *Indian J. Chem.*, 1974, **12**, 827.
[444] C. M. Mikulski, L. L. Pytlewski, and N. M. Karayannis, *Inorg. Chem.*, 1975, **14**, 1559.
[445] H. D. Gillman, *J. Inorg. Nuclear Chem.*, 1975, **37**, 1909.
[446] W. Kuchen and H. Steinberger, *Z. anorg. Chem.*, 1975, **411**, 266.

*Elements of Group V* 313

temperature variation of the dipole moments of the substituted phosphates $(Me_3CO)_3PO$ and $(R_3SiO)_3PO$ (R = Me, Et, or Ph) has been discussed in terms of the changes in proportions of four possible isomeric forms which arise from rotation about the POC(Si) bonds.[447a] All these substituted phosphates form 2:1 complexes with tin(IV) chloride, but no solids were isolated with tellurium tetrachloride.[447b] The synthesis of a new type of silyl ester of a cyclic acyl phosphate (96) has been announced.[448]

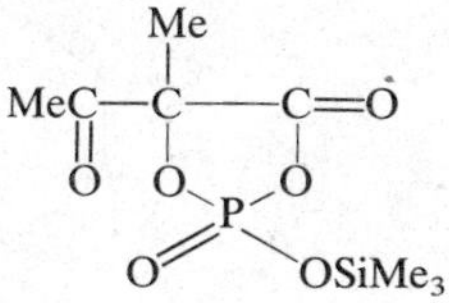

(96)

Stereospecific syntheses have been reported for ethyl isopropyl methyl phosphate, O-ethyl O,S-dimethyl phosphorothioate, and ethyl methyl methylphosphonate,[449a] and optically active alkyl S-methyl methyl phosphonothioates and dialkyl S-methyl methyl phosphorothioates have been prepared from ephedrine–RPSCl$_2$ starting materials.[449b] Finally, optically active phosphines can be obtained when a racemic phosphine oxide such as MePr$^n$PhPO or (97) is reduced with a chiral, non-racemic aluminium hydride.[450]

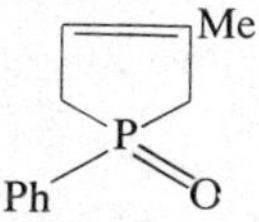

(97) 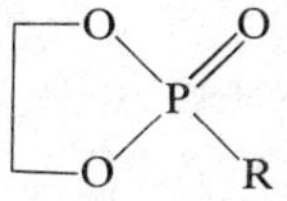        (98) 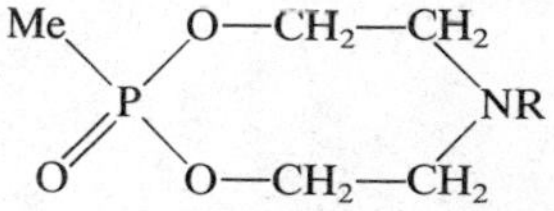        (99)

Hydrogen-bond formation with phenol has been used in an i.r. study of ν-(P=O) in the dioxaphospholans (98) to show that basicity depends on the nature of R and increases in the order OMe < Ph < NMe$_2$.[451] The eight-membered 1,3,6,2-dioxazaphosphocines (99), which are currently of interest for comparison with the better known six-membered analogues, can be obtained from methylphosphonic dichloride and the appropriate 2,2'-iminobis(ethanol).[452]

Two new types of six-co-ordinate phosphorus compound whose exact stereochemistry is as yet unknown are represented by the zwitterion (100), obtained from a spiropentaoxyphosphorane and a substituted pyridine, and (101), with an ion-pair formulation.[453] A further novel feature of the latter is the presence of two bidentate and two unidentate oxygen ligands attached to phosphorus.

[447] (a) Yu. V. Kolodyazhnyi, V. G. Tkalenko, A. P. Sadimenko, N. A. Kardanov, and O. A. Osipov, *J. Gen. Chem. (U.S.S.R.)*, 1975, **45**, 738; (b) *ibid.*, p. 743.
[448] I. Ugi, F. Ramirez, E. V. Hinrichs, P. Lemmen, and J. Firl, *J.C.S. Chem. Comm.*, 1974, 979.
[449] (a) C. R. Hall, T. D. Inch, G. J. Lewis, and R. A. Chittenden, *J.C.S. Chem. Comm.*, 1975, 720; (b) D. B. Cooper, C. R. Hall, and T. D. Inch, *ibid.*, p. 721.
[450] E. Cernia, G. M. Giongo, F. Marcati, W. Marconi, and N. Palladino, *Inorg. Chim. Acta*, 1974, **11**, 195.
[451] M. Revel, R. Pujol, J. Navech, and F. Mathis, *Compt. rend.*, **280**, C, 1297.
[452] N. N. Godovikov, L. A. Vikhreva, and M. I. Kabachnik, *J. Gen. Chem. (U.S.S.R.)*, 1975, **45**, 717.
[453] F. Ramirez, V. A. V. Prasad, and J. F. Marecek, *J. Amer. Chem. Soc.*, 1974, **96**, 7269.

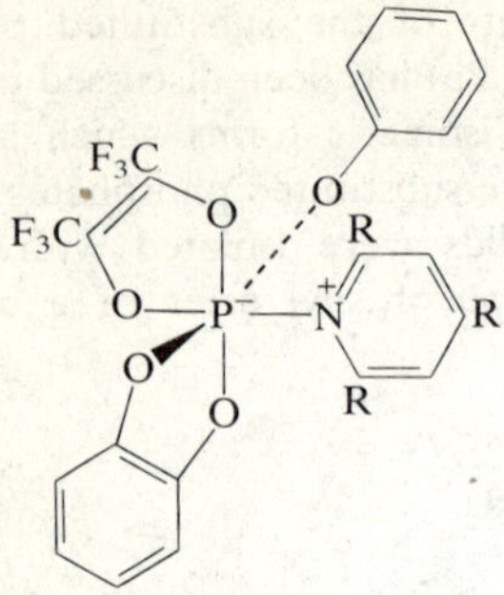

(100)

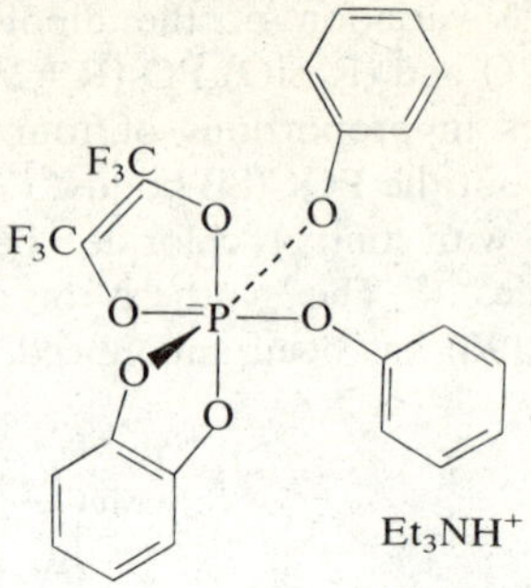

(101) Et$_3$NH$^+$

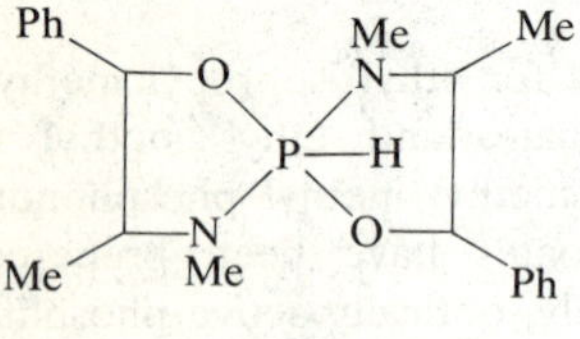

(102)

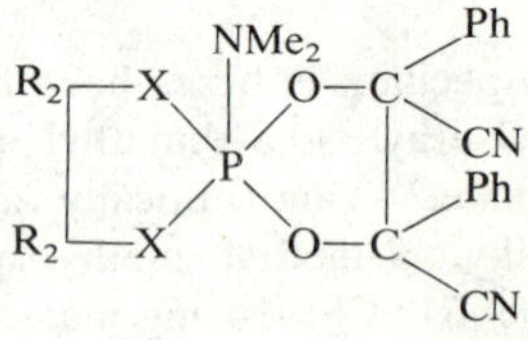

(103) X = O or NMe
R = H or Me

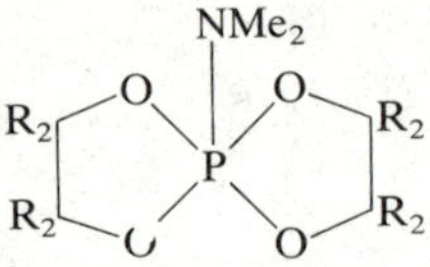

(104)

A number of new five-co-ordinate spiro-phosphoranes such as (102), obtained from P(NMe$_2$)$_3$ and ephedrine,[454] and (103)[455] have also been synthesized. In compound (104), replacement of the dimethylamino-group with both mono- and di-hydroxy species has been studied,[456] and i.r. investigations on a number of similar compounds, but containing P—H bonds point to the establishment of the equilibrium in equation (51) between the phosphorus(v) spirocyclic form and an

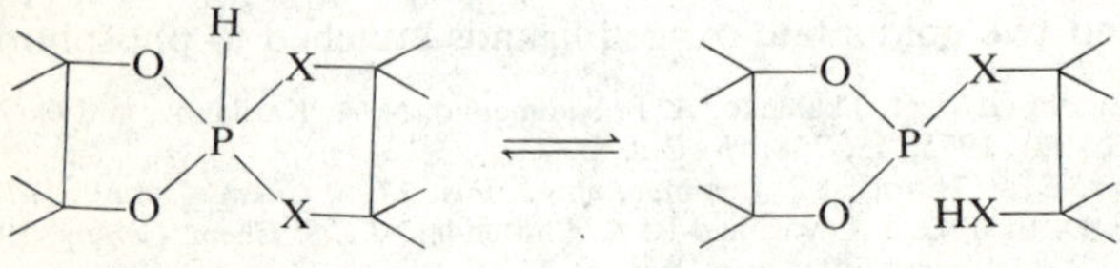

(51)

X = O or NH

[454] M. G. Newton, J. E. Collier, and R. Wolf, *J. Amer. Chem. Soc.*, 1974, **96**, 6888.
[455] M. Willson, R. Burgada, and F. Mathis, *Compt. rend.*, 1975, **280**, C, 225.
[456] D. Bernard and R. Burgada, *Compt. rend.*, 1974, **279**, C, 883.

open phosphorus(III) form.[457,458] Compounds in which a five-co-ordinate phosphorus atom is a bridgehead atom in a bicyclo[3,3,0]octane skeleton, e.g. compounds (105) and (106), are readily obtainable from $R_2PCl$ and $\omega$-hydroxy-

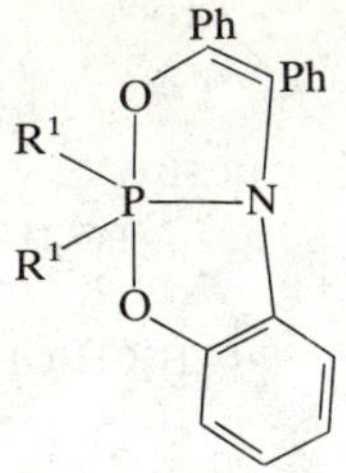

(105) 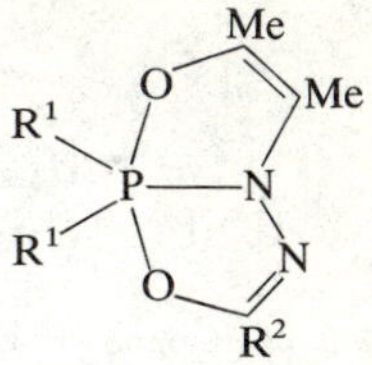          (106)

compounds containing a seven-membered $\alpha,\gamma$-doubly unsaturated chain.[459] The structure of (105; $R^1$ = Ph) contains a trigonal-bipyramidal phosphorus atom, with the oxygen atoms occupying apical positions as shown. The OPO angle is reduced to 171.6° by the five-membered systems, and the two phenyl groups are not equivalent; the structure here is compared with those of similar compounds in which phosphorus is simultaneously part of four- and five- or five- and six-membered rings.[460]

Distorted octahedral structures have been proposed for the 1:1 neutral complexes formed between bivalent transition-metal ions and hydroxymethylphosphonic acid, $HOCH_2PO_3H_2$.[461] A series of new dialkylaminomethylenediphosphonic acids $(HO)_2P(O)CH(NR^1R^2)_2P(O)(OH)_2$ has been prepared from phosphorus trihalides and NN-dialkylformamides.[462] Acid dissociation constants and the complex behaviour toward $Cu^{2+}$ for the phosphinyl-propionic acid $Ph(HO)P(O)CMe(OH)(CO_2H)$ have been evaluated.[463]

New modifications to the edta structure have been reported, in the hope that ligands such as (107)[464] and (108)[465] will have more selective sequestering properties. The work in each case reports the six acid dissociation constants and ligand behaviour with a number of bi- and ter-valent metal ions. A 1:1 complex $Tl(H_5L)$ is formed between $Tl^{III}$ and the $NNN'N'$-tetramethylphosphonic acid analogue of edta $(H_8L)$.[466] Broadly similar investigations have been carried out with nitrilotrismethylenetriphosphonic acid, $N[CH_2P(O)(OH)_2]_3 \equiv (H_6L)$, to show the formation of mono-, di-, and tri-protonated complexes in addition to the normal species $M^{II}L^{4-}$.[467] Compounds with the stoicheiometry $M_3L,3H_2O$ have been isolated, and at higher concentrations polynuclear species such as

[457] R. Mathis, M. Barthelat, Y. Charbonnel, and J. Barrans, *Compt. rend.*, 1975, **280**, C, 809.
[458] A. Munoz, G. Gence, M. Koenig, and R. Wolf, *Compt. rend.*, 1975, **280**, C, 395, 485.
[459] A. Schmidpeter and J. H. Weinmaier, *Angew. Chem. Internat. Edn.*, 1975, **14**, 489.
[460] W. S. Sheldrick, A. Schmidpeter, and J. H. Weinmaier, *Angew. Chem. Internat. Edn.*, 1975, **14**, 490.
[461] G. Brun and G. Jourdan, *Rev. Chim. minérale*, 1975, **12**, 139.
[462] M. Fukuda, Y. Okamoto, and H. Sakurai, *Bull. Chem. Soc. Japan*, 1975, **48**, 1030.
[463] A. Kh. Miftakhova, G. V. Romanov, I. V. Konovalova, V. F. Toropova, and A. N. Pudovik, *J. Gen. Chem. (U.S.S.R.)*, 1974, **44**, 710.
[464] D. T. MacMillan, I. Murase, and A. E. Martell, *Inorg. Chem.*, 1975, **14**, 468.
[465] A. Yu. Kireeva, M. V. Rudomino, and N. M. Dyatlova, *J. Gen. Chem. (U.S.S.R.)*, 1974, **44**, 2594.
[466] V. I. Kornev, N. I. Pechurova, and L. I. Martynenko, *Russ. J. Inorg. Chem.*, 1974, **19**, 146.
[467] L. V. Nikitina, A. I. Grigor'ev, and N. M. Dyatlova, *J. Gen. Chem. (U.S.S.R.)*, 1974, **44**, 1568, 1641.

(107)

(108)

$Ca_5(HL)_2,5H_2O$ can be isolated.[468] I.r. data on these compounds have been interpreted as showing metal–nitrogen interaction in all cases.[469]

*Monophosphates.* X-Ray structural studies on monophosphates and more complex phosphates not mentioned specifically in the text are summarized in Table 3;[470–501] phase-diagram investigations are similarly summarized later (p. 322). The

[468] L. V. Nikitina, N. F. Shugal, A. I. Grigor'ev, and N. M. Dyatlova, *J. Gen. Chem. (U.S.S.R.)*, 1975, **45**, 546.
[469] A. I. Grigor'ev, L. V. Nikitina, and N. M. Dyatlova, *Russ. J. Inorg. Chem.*, 1974, **19**, 1079.
[470] G. Nitsch and H. Schäfer, *Z. anorg. Chem.*, 1975, **417**, 11.
[471] E. Banks, R. Chianelli, and R. Korenstein, *Inorg. Chem.*, 1975, **14**, 1634.
[472] L. W. Schroeder, E. Prince, and B. Dickens, *Acta Cryst.*, 1975, **B31**, 9.
[473] R. Salmon, C. Parent, A. Berrada, R. Brochu, A. Daoudi, M. Vlasse, and G. Le Flem. *Compt. rend.*, 1975, **208**, C, 805.
[474] W. Gebert and E. Tillmanns, *Acta Cryst.*, 1975, **B31**, 1768.
[475] C. Calvo and R. Faggiani, *Canad. J. Chem.*, 1975, **53**, 1849.
[476] H. N. Ng and C. Calvo, *Canad. J. Chem.*, 1975, **53**, 2064.
[477] E. Kostiner and J. R. Rea, *Inorg. Chem.*, 1974, **13**, 2876.
[478] C. Calvo and R. Faggiani, *Canad. J. Chem.*, 1975, **53**, 1516.
[479] I. Tordjman, A. Durif, M. T. Averbuch-Pouchot, and J.-C. Guitel, *Acta Cryst.*, 1975, **B31**, 1143.
[480] M. T. Averbuch-Pouchot, *J. Appl. Cryst.*, 1974, **7**, 511.
[481] A. Whitaker, *Acta Cryst.*, 1975, **B31**, 2026.
[482] J. R. Rea and E. Kostiner, *Acta Cryst.*, 1974, **B30**, 2901.
[483] K. Aurivillius and B. A. Nilsson, *Z. Krist.*, 1975, **141**, 1.
[484] J. Durand, W. Granier, L. Cot, and J. L. Galigné, *Acta Cryst.*, 1975, **B31**, 1533.
[485] W. Granier, J. Durand, and L. Cot, *Rev. Chim. minérale*, 1975, **12**, 147.
[486] N. S. Mandel, *Acta Cryst.*, 1975, **B31**, 1730.
[487] P.-S. Yuen and R. L. Collin, *Acta Cryst.*, 1974, **B30**, 2513.
[488] C.-H. Huang, O. Knop, D. A. Othen, F. W. D. Woodhams, and R. A. Howie, *Canad. J. Chem.*, 1975, **53**, 79.
[489] I. Tordjman, A. Durif, and C. Cavero-Ghersi, *Acta Cryst.*, 1974, **B30**, 2701.
[490] M. Bagieu-Beucher, A. Durif, and J. C. Guitel, *Acta Cryst.*, 1975, **B31**, 2264.
[491] R. Masse, A. Durif, and J.-C. Guitel, *Z. Krist.*, 1975, **141**, 113.
[492] R. Zilber, I. Tordjman, A. Durif, and J.-C. Guitel, *Z. Krist.*, 1974, **140**, 350.
[493] A. G. Nord and K. B. Lindberg, *Acta Chem. Scand. (A)*, 1975, **29**, 1.
[494] M. C. Cavero-Ghersi and A. Durif, *J. Appl. Cryst.*, 1975, **8**, 562.
[495] M.-T. Averbuch-Pouchot and A. Durif, *J. Appl. Cryst.*, 1975, **8**, 564.
[496] Z. Ružic-Toroš, B. Kojić-Prodić, R. Liminga, and S. Popović, *Inorg. Chim. Acta*, 1974, **8**, 273.
[497] M. Laügt and J.-C. Guitel, *Acta Cryst.*, 1975, **B31**, 1148.
[498] M.-T. Averbuch-Pouchot, *J. Appl. Cryst.*, 1975, **8**, 389.
[499] K. Palkina and K.-H. Jost, *Acta Cryst.*, 1975, **B31**, 2281.
[500] K. Palkina and K.-H. Jost, *Acta Cryst.*, 1975, **B31**, 2285.
[501] I. Tordjman, M. Bagieu-Beucher, and R. Zilber, *Z. Krist.*, 1974, **140**, 145.

## Table 3 X-Ray diffraction results

| Compound | Comments | References |
|---|---|---|
| $MBePO_4$ | M = Na, K, or Rb; preparation in molten alkali-metal sulphate medium. The K and Rb salts are isotypic with the $NH_4LiSO_4$ structure but the Na compound is identical with beryllonite. | 470 |
| $MgMPO_4,6H_2O$ | M = K, Rb, Cs, Tl, or $NH_4$; unit-cell dimensions, struvite structure. | 471 |
| $Ca(H_2PO_4)_2,H_2O$ | Neutron-diffraction study; two crystallographically independent $H_2PO_4^-$ ions bonded to each other and to the water molecule. | 472 |
| $Na_3Ln(PO_4)_2$ | Ln = Y, La, Pr, Nd, or Sm—Er, ordered $\beta$-$K_2SO_4$ structure. | 473 |
| $Zr_2O(PO_4)_3$ | Slightly distorted pentagonal-bipyramidal $ZrO_7$ and $PO_4$ tetrahedra share edges to give infinite $(ZrO_3PO_4)^{5-}$ chains. | 474 |
| $BaNaPO_4$ | Glaserite form, phosphate group has $3m$ symmetry with mean P—O distance 1.538 Å; the fourth distance is 1.520 Å. | 475 |
| $FePO_4$ | Isotypic with $AlPO_4$ (berlinite) and related to $\alpha$-quartz structure, $PO_4$ group almost regular. Tetrahedral co-ordination for $Fe^{3+}$. | 476 |
| $Fe_3(PO_4)_2$ | Isotypic with graftonite; contains two almost regular isolated $PO_4$ groups. Two iron atoms have irregular five-co-ordination to oxygen; the third is in highly distorted six-co-ordination. | 477 |
| $Ni_3(PO_4)_2$ | Large distortions in $PO_4$ group, with distances ranging between 1.521 and 1.595 Å; Ni atoms in two types of distorted octahedral sites. | 478 |
| $Zn_2KH(PO_4)_2,\tfrac{5}{2}H_2O$ | Previously reported as a dihydrate. | 479 |
| $Zn(H_2PO_4)_2,2H_2O$ | Powder data. | 480 |
| $ZnH_2P_2O_7$ | Powder data. | 480 |
| $Zn_3(PO_4)_2,4H_2O$ | Hopeite; consists of $ZnO_2(H_2O)_4$ octahedra and $ZnO_4$ and $PO_4$ tetrahedra sharing corners and edges; P—O distances 1.537 Å. | 481 |
| $Cd_2(PO_4)F$ | Isostructural with $Mn_2(PO_4)F$; two types of Cd atoms co-ordinated by four oxygens and two fluorines which are in *cis* positions. | 482 |
| $Hg_3(PO_4)_2$ | Hg in almost linear co-ordination by two oxygens from $PO_4$ tetrahedra. | 483 |
| $NaK_3(PO_3F)_2$ | P—O 1.495(6); P—F 1.630(10) Å. ∠OPO 114.8(7)°; ∠OPF 103.4(3)°. | 484 |
| $M_2PO_3F$ | M = Na, K, Rb, Cs, or $NH_4$; unit-cell dimensions. | 485 |
| $LiMPO_3F$ | M = Na, K, Rb, Cs, or $NH_4$; unit-cell dimensions. | 485 |
| $NaM_3(PO_3F)_2$ | M = K or Rb; unit-cell dimensions. | 485 |
| $Ca_2P_2O_7,2H_2O$ | Six P—O(terminal) distances, with mean value of 1.517 Å; two P—O(bridge) distance are equal (1.623 Å) ∠$O_t$—P—$O_t$ 112°; ∠$O_b$—P—$O_t$ 107°; ∠P—O—P 123.1°. Conformation adopted by $P_2O_7^{4-}$ group related to details of cation environment. | 486 |
| $Na_6Cu(P_2O_7)_2,16H_2O$ | P—O(bridge) 1.620 Å; P—O(terminal) 1.515 Å; ∠POP 129.7°. | 487 |
| $M^{IV}P_2O_7$ | M = Ti, Zr, Hf, Sn, or Pb; powder data. I.r. and Raman spectra also reported. | 488 |
| $Na_2KP_3O_9$ | Single-crystal data. | 489 |
| $Ag_3P_3O_9,H_2O$ | $P_3O_9^{3-}$ anion in chair conformation. | 490 |

**Table 3** (*Continued*)

| Compound | Comments | References |
|---|---|---|
| Ca(NH$_4$)P$_3$O$_9$ | Single-crystal data. | 491 |
| SrNa(P$_3$O$_9$),3H$_2$O | P—O(bridge) 1.60 Å; P—O(terminal) 1.48 Å. | 492 |
| Mg$_2$P$_4$O$_{12}$ | $\bar{1}$ symmetry for P$_4$O$_{12}^{4-}$ ion. P—O(bridge) 1.590—1.599 Å; P—O(terminal) 1.461—1.510 Å; two different ∠POP, of 134.2° and 138.8°. | 493 |
| M$^{II}$M$^{I}$P$_4$O$_{12}$ | M$^{II}$ = Ca or Pb; M$^{I}$ = K, Rb, NH$_4$, or Tl. Preliminary X-ray data, compounds isotypic with SrK$_2$P$_4$O$_{12}$. | 494 |
| Zn$_5$(P$_3$O$_{10}$),17H$_2$O | Powder data. | 495 |
| KThP$_3$O$_{10}$ | Th atoms in 8-fold co-ordination to oxygen atoms from three different triphosphate chains; P$_3$O$_{10}^{5-}$ ion distorted from that in Na$_5$P$_3$O$_{10}$; P—O(bridge) distances are 1.65, 1.55, 1.58, and 1.57 Å; ∠POP 120° and 146°. | 496 |
| Ba$_2$Cu(PO$_3$)$_6$ | Polyphosphate, unit cell contains two (PO$_3$)$_\infty$ chains with period of 12PO$_4$ tetrahedra in *b* direction. | 497 |
| BaM$^{II}$(PO$_3$)$_4$ | Powder data for polyphosphates, with M$^{II}$ = Mn, Cd, Ca, or Hg. | 498 |
| Bi(PO$_3$)$_3$ | Polyphosphate, spiral of PO$_4$ tetrahedra with 6 units per turn; Bi atoms irregularly co-ordinated by 7 oxygen atoms. | 499 |
| BiH(PO$_3$)$_4$ | Polyphosphate with infinite chains of PO$_4$ tetrahedra; the chains show one short bridging P—O bond (1.553 Å) and one long, terminal P—O bond (1.542 Å), suggesting the latter is a P—OH group, Bi in seven-fold co-ordination to oxygen. | 500 |
| CaP$_4$O$_{11}$ | New ultraphosphate structure containing two infinite chains of PO$_4$ tetrahedra with a period of six PO$_4$ units linked by a further P atom. Full structural data reported. | 501 |

X-ray p.e. spectrum of PO$_4^{3-}$ has been assigned as part of a study including the SiO$_4^{4-}$, SO$_4^{2-}$, and ClO$_4^-$ ions, giving data pointing to an increased oxygen 2*s* contribution to bonding with increasing atomic number of the central atom.[502] A potentiometric study of the neutralization of phosphoric acid with KOH in aqueous 3M KCl gives evidence for the formation of a number of dimeric species such as (H$_3$PO$_4$)$_2$, H$_5$P$_2$O$_8^-$, and H$_4$P$_2$O$_8^{2-}$.[503,504]

Controversy associated with the high-energy diphosphate bond has led to a series of *ab initio* M.O. calculations on H$_3$PO$_4$, H$_4$P$_2$O$_7$, H$_2$P$_2$O$_7^{2-}$, the low-energy molecule MeOPO$_3$H$_2$, and acetic anhydride.[505] In agreement with earlier data, the results show that the high-energy nature of the P—O—P linkage is not entirely due to intramolecular electronic effects, and that solvation effects make important contributions.

An unusual tetrameric complex [Al(PO$_4$)(HCl)(EtOH)$_4$]$_4$ obtained from a reaction between anhydrous aluminium chloride and phosphoric acid in ethanol solution has the structure shown in Figure 3.[506] The co-ordination sphere of

---

[502] R. Prins, *J. Chem. Phys.*, 1974, **61**, 2580.
[503] G. Ferroni, J. Galea, G. Antonetti, and R. Romanetti, *Bull. Soc. chim. France*, 1974, 2695.
[504] G. Ferroni, *Bull. Soc. chim. France*, 1974, 2698.
[505] D. M. Hayes, G. L. Kenyon, and P. P. Kollman, *J. Amer. Chem. Soc.*, 1975, **97**, 4762.
[506] J. E. Cassidy, J. A. J. Jarvis, and R. N. Rothon, *J.C.S. Dalton*, 1975, 1497.

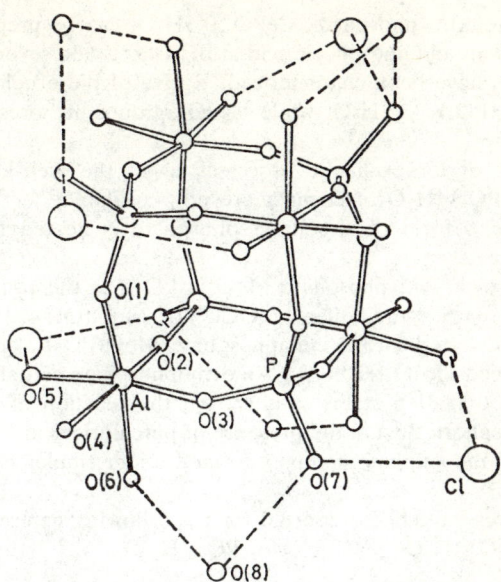

**Figure 3** *Structure of one tetrameric unit* [Al(PO$_4$)(HCl)(EtOH)$_4$]$_4$
(Reproduced by permission from *J.C.S. Dalton*, 1975, 1497)

aluminium contains three oxygen atoms from phosphate groups (Al—O 1.82 Å) and three oxygens from ethanol molecules (Al—O 1.97 Å); the phosphorus atoms are tetrahedrally co-ordinated (P—O 1.50 Å). Four further ethanol molecules are attached *via* hydrogen bonds to the cage, and the fourth oxygen of each phosphate group is hydrogen-bonded to a chlorine atom. An important property of this compound is its ready conversion into glassy films of aluminium phosphate. Products isolated in reactions between aluminium nitrate and ammonium phosphates in aqueous solution depend on the pH and phosphate concentration;[507] two crystalline complexes, (NH$_4$)$_3$H$_6$Al$_4$(PO$_4$)$_7$,16H$_2$O and (NH$_4$)$_2$H$_4$Al$_3$-(PO$_4$)$_5$,12H$_2$O, and a number of amorphous species have been identified. Similar reactions with trisodium and tripotassium phosphates have been carried out.[508]

Lead acetate and nitrate reactions with alkali-metal dihydrogen phosphates have been investigated, showing the formation of Pb$_2$NO$_3$(PO$_4$) from lead nitrate,[509] and recent experiments indicate the precipitation of a similar mixed chloride phosphate Pb$_2$Cl(PO$_4$) from solutions of lead(II) chloride and alkali-metal dihydrogen phosphates.[510] Double decomposition reactions between Pb$_3$(PO$_4$)$_2$ and M$^{II}$TiO$_3$ (M = Ca, Sr, Ba, or Zn) have been investigated by thermogravimetric methods.[511]

---

[507] A. M. Golub and I. I. Boldog, *Russ. J. Inorg. Chem.*, 1974, **19**, 499.
[508] A. M. Golub and I. I. Boldog, *Russ. J. Inorg. Chem.*, 1974, **19**, 955.
[509] E. A. Gyunner, I. S. Vel'mozhnyi, and L. M. Mel'nichenko, *Russ. J. Inorg. Chem.*, 1974, **19**, 326.
[510] I. S. Vel'mozhnyi, E. A. Gyunner, and L. M. Mel'nichenko, *Russ. J. Inorg. Chem.*, 1974, **19**, 821.
[511] I. N. Belyaev, I. I. Belyaeva, and L. N. Aver'yanova, *Russ. J. Inorg. Chem.*, 1974, **19**, 123.

Hexagonal cerium(III) phosphate, $CePO_4,0.5H_2O$, can be prepared from solutions of a cerium salt and phosphoric acid at pH 1–3,[512] and seven new cerium(IV) phosphate species have been characterized.[513] Five of these belong to the series $Ce(OH)_x(PO_4)_x(HPO_4)_{2-2x},yH_2O$, while $H_2PO_4^-$ groups are present in the other two compounds.

Further details of ion-exchange processes using the well-known zirconium phosphate $[Zr(HPO_4)_2,H_2O]$ exchanger are now available,[514,515] and the phases formed when the $\alpha$-form of zirconium phosphate is heated have been identified.[516]

A yellowish brown acid phosphate $MoO(H_2PO_4)_3$ is the product when 40% phosphoric acid reacts with either $MoOCl_5^{2-}$ or $MoO(OH)_3$.[517] On heating to 250 °C, this is converted into a metaphosphate $MoO(PO_3)_3$, and on hydrolysis into the oxo-bridged $Mo_2O_3(H_2PO_4)_4$. An orthophosphate containing platinum in both +2 and +4 oxidation states results from the reaction of $K_2Pt(NO_2)_4$ and $K_2PtCl_6$ with phosphoric acid in the presence of perchloric acid,[518] and a structure similar to that of the sulphato-complex formed under similar circumstances has been suggested.

X-Ray structures have been reported for the following heteropolyacid derivatives: $Na_6Mo_{18}P_2O_{63}(H_2O)_{24}$,[519a] $H_3Mo_{12}PO_{40}(H_2O)_{29-31}$,[519b] and $\alpha$-$Ba_3W_{18}P_2O_{62}(H_2O)_{29}$.[520]

*Apatites.* Solubility data for calcium hydroxyapatite, $Ca_5(OH)(PO_4)_3$, in water containing an indifferent salt have been collected,[521] and the coprecipitation of strontium with the hydroxyapatite has been investigated.[522] Fluoroapatite formation from brushite, $CaHPO_4,2H_2O$, and monetite, $CaHPO_4$, in solutions containing NaF and NaOH is inhibited at higher pH;[523] some hydroxyapatite is formed, and fluoride incorporation into this occurs at pH below 9.5. Decomposition of carbonated apatites at 1000 °C gives hydroxyapatite, sodium oxide, and lime,[524] and the B-type carbonate apatites $Ca_{10-x}Na_{2x/3}(PO_4)_{6-x}(CO_3)_x(H_2O)_yZ_{2-(x/3)}$ (Z = OH or F, $0 \le x \le 3$, $0 \le y \le x$) are isomorphous with the stoicheiometric hydroxy- or fluoro-apatites.[525] High-temperature conditions under which oxide ions in oxyapatite $Ca_{10}O(PO_4)_6$ are substituted by sulphide ions (from sulphur vapour or calcium sulphide) have been defined.[526]

Hydrothermally grown manganese chloroapatite, $Mn_5(PO_4)_3Cl_{0.9}(OH)_{0.1}$, has been shown to be isostructural with calcium fluoroapatite but the phosphate

---

[512] K. Hukuo and Y. Hikichi, *Nippon Kagaku Kaishi*, 1975, 622.
[513] R. G. Herman and A. Clearfield, *J. Inorg. Nuclear Chem.*, 1975, **37**, 1697.
[514] Y. Hasegawa and I. Tomita, *Bull. Chem. Soc. Japan*, 1974, **47**, 2389.
[515] S. Ahrland, N.-O. Björk, R. Blessing, and R. Herman, *J. Inorg. Nuclear Chem.*, 1974, **36**, 2377.
[516] A. Clearfield and S. P. Pack, *J. Inorg. Chem.*, 1975, **37**, 1283.
[517] H. K. Saha, S. S. Mandal, and T. R. Chaudhuri, *J. Inorg. Nuclear Chem.*, 1975, **37**, 840.
[518] S. I. Ginsburg, M. I. Yuz'ko, and Yu. Ya. Kharitonov, *Russ. J. Inorg. Chem.*, 1974, **19**, 251.
[519] (a) R. Strandberg, *Acta Chem. Scand.* (A), 1975, **29**, 350; (b) *ibid.*, p. 359.
[520] K. Y. Matsumoto and Y. Sasaki, *J.C.S. Chem. Comm.*, 1975, 691.
[521] H. E. L. Madsen, *Acta Chem. Scand.* (A), 1975, **29**, 745.
[522] O. Fujino, *Bull. Chem. Soc. Japan*, 1975, **48**, 1455.
[523] E. J. Duff, *J. Inorg. Nuclear Chem.*, 1975, **37**, 503.
[524] C. Vignoles, J.-C. Trombe, G. Bonel, and G. Montel, *Compt. rend.*, 1975, **280**, C, 275.
[525] C. Vignoles, G. Bonel, and G. Montel, *Compt. rend.*, 1975, **280**, C, 361.
[526] J.-C. Trombe and G. Montel, *Compt. rend.*, 1975, **280**, C, 567.

groups are more distorted than previously found in these structures.[527] The structure of $Sr_{9.402}Na_{0.209}(PO_4)_6B_{0.996}O_2$, prepared by heating a mixture of $2(NH_4)_2HPO_4, 3SrCO_3$, and $6Na_2B_4O_7, 10H_2O$ to 1450 °C, is apatitic but contains the previous unobserved linear O—B—O grouping.[528]

*Diphosphates.* Changes in the configurational entropy for the anhydrous sodium diphosphates have been calculated from observed crystal structures,[529] and factors such as water content of the atmosphere influencing the formation of $\alpha$- or $\beta$-$Na_2H_2P_2O_7$ from $NaH_2PO_4$ by thermolysis have been elucidated.[530] The interaction between $Mg^{2+}$ and diphosphoric acid in water and $D_2O$ has been examined kinetically by the temperature-jump method.[531]

Thermal decomposition of $CaH_2P_2O_7$ at 400—450 °C gives the metaphosphate $\beta$-$Ca(PO_3)_2$, but material that has been heat-treated at 550 °C shows evidence for phosphates of lower molecular weight also.[532] Several products result when indium trichloride and $Cs_4P_2O_7$ react in water at 25 °C, e.g. $In_4(P_2O_7)_3, 14H_2O$, $CsInP_2O_7, 3H_2O$, $Cs_2In(OH)P_2O_7, 3H_2O$, and basic diphosphates of variable composition.[533] A rubidium compound $RbInP_2O_7, 4H_2O$ has also been obtained, which, together with the caesium compounds above, has been investigated by thermogravimetric and X-ray methods. A manganese salt, $Rb_2MnP_2O_7, 3H_2O$, has also been prepared and characterized.[534]

*Meta- and Poly-phosphates.* Radical formation on $\gamma$-irradiation of $K_3P_3O_9$ has been investigated,[535] as has the breakdown of the tri- and tetra-metaphosphate structures on heating to 450 °C with disodium cyanamide, $Na_2NCN$.[536] Two of the products are $Na_3PO_3N_2C$ and $NaPO_2(N_2C)_2$, which result from the initial cleavage of the ring by the $N_2C^{2-}$ ion. When barium tetrametaphosphate and barium difluoride in 1:1 ratio are heated to between 400 and 1200 °C the identified product is $BaPO_3F$.[537] The ammonium hexa- and octa-metaphosphates $(NH_4)_2P_6O_{18}, H_2O$ and $(NH_4)_8P_8O_{24}, 3H_2O$, respectively, have been obtained from the corresponding sodium salts and characterized.[538]

Dehydration of $Na_5P_3O_{10}, 6H_2O$ in an atmosphere of water has been studied,[539] and the conditions under which phosphoric acid and either $Al_2O_3$ or $Al(OH)_3$ react to give the I and II forms of $AlH_2P_3O_{10}$ and the A and B types of $Al_4(P_4O_{12})_3$ have been established.[540] The complex ions $TlP_3O_{10}^{4-}$ and

[527] G. Engel, J. Pretzsch, V. Gramlich, and W. H. Baur, *Acta Cryst.*, 1975, **B31**, 1854.
[528] C. Calvo, R. Faggiani, and N. Krishnamachari, *Acta Cryst.*, 1975, **B31**, 188.
[529] K. Y. Leung, *Canad. J. Chem.*, 1975, **53**, 1739.
[530] A. de Sallier-Dupin and P. Dugleux, *Bull. Soc. chim. France*, 1975, 1105.
[531] H. Silber and P. Wehner, *J. Inorg. Nuclear Chem.*, 1975, **37**, 1025.
[532] M. N. Nabiev, M. Adylova, M. T. Saibova, I. Nishanov, and O. Momot, *Russ. J. Inorg. Chem.*, 1974, **19**, 15.
[533] E. N. Deichman, I. V. Tananeva, Zh. A. Ezhova, *Russ. J. Inorg. Chem.*, 1974, **19**, 139.
[534] M. V. Goloshchapov and B. V. Martynenko, *Russ. J. Inorg. Chem.*, 1974, **19**, 611.
[535] Y. Kobayashi, N. Matsuura, and A. Indelli, *Chem. Letters*, 1975, 1.
[536] H. Köhler, U. Lange, and R. Uebel, *Z. anorg. Chem.*, 1975, **413**, 119.
[537] G. P. Nikolina, M. L. Petrovskaya, D. M. Yudin, A. V. Shendrik, and B. G. Fedorushkov, *Russ. J. Inorg. Chem.*, 1974, **19**, 332.
[538] S. I. Vol'fkovich, L. V. Kubasova, M. L. Koz'mina, and T. L. Roslaya, *Doklady Chem.*, 1975, **220**, 116.
[539] L. A. Lesnikovich, M. M. Pavlyuchenko, and E. A. Prodan, *Russ. J. Inorg. Chem.*, 1974, **19**, 208.
[540] M. Tsuhako, K. Hasegawa, T. Matsuo, I. Motooka, and M. Kobayashi, *Bull. Chem. Soc. Japan*, 1975, **47**, 1830.

$[Tl(P_3O_{10})_2]^{9-}$ have been identified as important species in solubility studies on $TlIO_3$ and $Tl_2CrO_4$ in sodium triphosphate solutions.[541]

High-temperature calorimetry has led to the heat of formation data given in Table 4 on the interconversion of sodium trimetaphosphate, Kurrol's salt, and

**Table 4** Heats of formation of various salts

| Process of formation | $\Delta H°/kJ$ (mole $NaPO_3)^{-1}$ |
|---|---|
| $\frac{1}{3}Na_3P_3O_9 \rightarrow$ Maddrell's salt | $-21.5 \pm 1.9$ |
| $\frac{1}{3}Na_3P_3O_9 \rightarrow$ Kurrol's salt | $-12.4 \pm 1.5$ |
| Kurrol's salt $\rightarrow$ Maddrell's salt | $-9.03 \pm 1.5$ |

Maddrell's salt,[542] which are interpreted as showing that kinetic effects are responsible for difficulties in converting the cyclic trimer into the linear polymeric forms. The potassium polyphosphate samples (Kurrol's salt) obtained from the fusion of potassium sulphate and phosphoric acid have chains terminated by $OSO_3^-$ groups, according to i.r. spectra.[543] Kinetic studies show a first-order dependence of polyphosphate concentration in the hydrolysis of Graham's salt to give trimetaphosphate and short-chain species.[544] The vaporization of sodium metaphosphate from a Knudsen cell leads to predominantly monomeric, with some dimeric, species, according to mass-spectrometric data.[545]

*Phase Studies.* The following systems have been investigated, formulae in parentheses being identified phases. $Tl_3PO_4$–$TlH_2PO_4$–$H_2O$ ($Tl_2HPO_4$),[546] $TlH_2PO_4$–$H_3PO_4$–$H_2O$ [$TlH_5PO_4)_2$],[547] $NaH_2PO_4$–$H_2O$–$Me_2CO$,[548] $Na_2HPO_4$–$H_2O$–$Me_2CO$,[548] $(NH_4)_4P_2O_7$–$Cd_2P_2O_7$–$H_2O$ [$(NH_4)_2CdP_2O_7,3H_2O$ and $(NH_4)_2CdP_2O_7,H_2O$],[549] $NaPO_3$–$RbPO_3$ ($Na_2RbP_3O_9$),[550] $(NH_4)_6P_4O_{13}$–$InCl_3$–$H_2O$ [$(NH_4)_3In_3(P_4O_{13})_2,15H_2O$ and $(NH_4)_3InP_4O_{13},3H_2O$],[551] $Al(PO_3)_3$–$Ba(PO_3)_2$.[552]

**Bonds to Sulphur or Selenium.**—Fluorination of $P_4S_3$ or $P_4S_{10}$ with nitrogen trifluoride depends primarily on the temperature and flow rate of $NF_3$, with $PSF_3$ and $PF_3$ being the products at temperatures around 350 °C.[553] At *ca.* 200 °C, however, the fluorophosphazenes $(PNF_2)_{3-9}$ result, in good yield, while substitution of the phosphorus sulphide with calcium phosphide leads to the ionic species $F_3P(NPF_2)_xNPF_3^+PF_6^-$ in addition to $PF_5$ and the phosphazenes. A new mixed arsenic–phosphorus cage sulphide $P_2As_2S_3$ (109) has been identified as a product

---

[541] K. G. Burtseva and S. L. Saval'skii, *Russ. J. Inorg. Chem.*, 1974, **19**, 841.
[542] J. H. E. Jeffes and A. E. M. Warner, *J.C.S. Faraday I*, 1975, **71**, 670.
[543] B. A. Beremzhanov, Yu. A. Pokrovskaya, D. Z. Serazetdinov, and E. V. Poletaev, *Russ. J. Inorg. Chem.*, 1974, **19**, 397.
[544] M. Watanabe, S. Sato, and H. Saito, *Bull. Chem. Soc. Japan*, 1975, **48**, 896.
[545] A. V. Steblevskii, A. S. Alikhanyan, I. D. Sokolova, and V. I. Gorgoraki, *Russ. J. Inorg. Chem.*, 1974, **19**, 789.
[546] Y. Oddon, L. Porte, G. Coffy, and A. Tranquard, *Bull. Soc. chim. France*, 1975, 1484.
[547] G. Coffy, Y. Oddon, M.-J. Boinon, and A. Tranquard, *Compt. rend.*, 1975, **280**, C, 1301.
[548] G. Ferroni, J. Galea, and G. Antonetti, *Bull. Soc. chim. France*, 1974, 2731.
[549] G. A. Selivanova and N. T. Kudryavtsev, *Russ. J. Inorg. Chem.*, 1974, **19**, 1067.
[550] C. Cavero-Ghersi and A. Durif, *Compt. rend.*, 1975, **280**, C, 579.
[551] G. V. Rodicheva, E. N. Deichman, I. V. Tananaev, and Zh. K. Shaidarbekova, *Russ. J. Inorg. Chem.*, 1974, **19**, 814.
[552] M. I. Kuz'menkov, S. V. Plyshevskii, and V. V. Pechkovskii, *Russ. J. Inorg. Chem.*, 1974, **19**, 881.
[553] A. Tasaka and O. Glemser, *Z. anorg. Chem.*, 1974, **409**, 163.

from a sealed-tube reorganization reaction between $P_4S_3$ and $As_2S_3$ at 500 °C.[554] A non-polymeric structure is expected on the basis of its solubility, and vibrational data point to the presence of both As—As and P—As bonds. No mixed P–S–Se compounds were identified in a number of systems investigated, including $P_4Se_3$–$P_4S_3$, $P_4Se_4$–$P_4S_7$, $P_4Se_{10}$–$P_4S_{10}$, and $P_4S_{10}$–Se.[555]

A value of 0.91 D has been calculated for the free electron-pair moment in $P^{III}$ compounds, using dipole-moment data for $P_4S_3$ and the phosphabicyclo-octane derivative (110).[556] Calculations for the compounds in the series $PH_nMe_{3-n}$ with $n = 0$—3 point to variation in the free electron-pair moment as being the most important factor in determining the observed dipole moments; extension to a variety of other species allows the calculation of new values for the moments of typical bonds to phosphorus.

In addition to replacement of the oxygen atom in phosphorodiamidous acids $(R_2N)_2P(O)H$ by sulphur, the exothermic reaction with $P_2S_5$ also leads to insertion of a sulphur atom into the P—H bond, to give derivatives of the dithio-acid $(R_2N)_2P(S)SH$.[557] Five-membered-ring systems (111) rather than the previously

(109)

(110)

(111) R = Ph or p-FC$_6$H$_4$

considered six-membered P—S systems are present in $Ph_3P_3S_3$[558] and (p-FC$_6$H$_4$)$_3$P$_3$S$_3$.[559] The former compound can be obtained by two new routes, i.e. the reaction of $S_3Cl_2$ with the dipotassium salt of triphenyltriphosphine or that between phenylphosphine and sulphur dichloride;[558] the p-fluoro-derivative has advantages over the phenyl compounds as it is more soluble, and from $^{19}F$ and $^{31}P$ n.m.r. measurements it has been shown that there are three different fluorine and phosphorus environments.

The tautomeric equilibrium between secondary phosphane sulphides $HR_2P=S$ and thiophosphinic acids $R_2PSH$ lies completely on the side of the former at room temperature, and P—H bond cleavage does not occur until over 180 °C.[560] The proton-transfer reaction can, however, be brought about much more readily by co-ordination to transition metals, which initially takes place via the sulphur atom. On raising the temperature, quantitative conversion into the phosphorus-bonded form occurs.

---

[554] A.-M. Leiva, E. Fluck, H. Müller, and G. Wallenwein, Z. anorg. Chem., 1974, **409**, 215.
[555] Y. Monteil and H. Vincent, Bull. Soc. chim. France, 1975, 1025.
[556] E. I. Matrosev, G. M. Petov, and M. I. Kabachnik, J. Struct. Chem., 1974, **15**, 229, 234.
[557] E. E. Nifant'ev, I. V. Shilov, V. S. Blagoveshchenskii, and I. V. Komlev, J. Gen. Chem. (U.S.S.R.), 1975, **45**, 282.
[558] M. Baudler, D. Koch, Th. Vakratsas, E. Tolls, and K. Kipker, Z. anorg. Chem., 1975, **412**, 239.
[559] M. R. Le Geyt and N. L. Paddock, J.C.S. Chem. Comm., 1975, 20.
[560] E. Lindner and H. Dreher, Angew. Chem. Internat. Edn., 1975, **14**, 416.

A unique reaction in which a sulphur atom is transferred between two phosphorus atoms in the same molecule has been observed when $Ph_2P(S)CH_2PMe_2$ is heated.[561] Transfer occurs to the more basic phosphorus atom, giving $Ph_2PCH_2$-$P(S)Me_2$, and the observation of both $Ph_2P(S)CH_2P(S)Me_2$ and $Ph_2PCH_2PMe_2$ as intermediates implies a complex mechanism. Two new isomeric phosphorus compounds containing the element in both the +3 and +5 oxidation states that have recently been prepared are $F_2P(S)SP(CF_3)_2$ and $F_2PSP(S)(CF_3)_2$, which on mixing equilibrate to give the symmetrical compounds $F_2P(S)SPF_2$ and $(CF_3)_2P(S)SP(CF_3)_2$;[562] details of temperature-dependent n.m.r. spectra and a number of chemical reactions have been discussed. Good yields of the $N$-silyl derivatives $R_2P(S)NHSiMe_3$ have been reported from the reaction of hexamethyldisilazane and dialkylthiophosphoryl halides $R_2P(S)X$.[563] Both mono- and di-esters, e.g. $CCl_3P(S)Cl(OR)$ and $CCl_3P(S)(OR)_2$, can be obtained from (trichloromethyl)phosphonothioic dichlorides.[564] The thiophosphoramidate ion $(EtO)_2P(S)$-$NR^-$, which can be prepared from butyl-lithium and $N$-alkylamidothioates, is ambidentate, reacting at both the nitrogen and sulphur atoms.[565]

Ionization energies have been determined by p.e.s. for compounds in the two series $(Me_2N)_{3-n}Cl_nPS$ and $(EtO)_{3-n}Cl_nPS$ ($n = 0$—3), and the first ionization potential has been shown to correlate linearly with $\Delta G°$ values for complex formation with iodine in carbon tetrachloride.[566] Re-examination of the reaction between tetramethyldiphosphine disulphide and hydrated copper(II) chloride in ethanol shows the formation of a dinuclear copper(I) complex $[(Me_4P_2S_2)CuCl]_2$ as the major product, while its precursor, $(Me_4P_2S_2)CuCl_2$, is obtained in minor amounts.[567] X-Ray structure determinations have been carried out on both compounds.

$O$-Alkyl phosphenodithioates, probably as cyclic trimers $(PS)_3(OR)_3S_3$, are isolable as dark green liquids when anhydrous hydrogen chloride is passed through a suspension of the trithioates $RPO(S)(SK)_2$ in chloroform.[568] The compounds react with hydroxy-derivatives to give the known dithioates $R^1O(R^2O)P(S)SH$.

An analysis of the $^1H$ n.m.r. spectrum of 2-phenyl-1,3,2-dithiaphosphorinan (112; $R^1 = Ph$, $R^2 = H$) has been reported,[569] and the phosphorinan (112; $R^1 = H$, $R^2 = Me$) has been synthesized by reduction of the corresponding chloride.[570] The new cyclic dithioates (113; R = H or Me) can be prepared by the reaction in equation (52).[571]

---

[561] S. O. Grim and J. D. Mitchell, *J.C.S. Chem. Comm.*, 1975, 634.
[562] L. F. Doty and R. G. Cavell, *Inorg. Chem.*, 1974, **13**, 2722.
[563] H. Steinberger and W. Kuchen, *Z. Naturforsch.*, 1974, **29b**, 611.
[564] O. N. Grishina and L. M. Kosova, *J. Gen. Chem. (U.S.S.R.)*, 1975, **45**, 276.
[565] M. Dreux and P. Savignac, *Compt. rend.*, 1975, **280**, C, 297.
[566] F. I. Vilesov, S. N. Lopatin, V. I. Vovna, R. Paetzold, and K. Niendorf, *Z. phys. Chem. (Leipzig)*, 1974, **255**, 661.
[567] F. A. Cotton, B. A. Frenz, D. L. Hunter, and Z. C. Mester, *Inorg. Chim. Acta*, 1974, **11**, 111, 119.
[568] N. A. Kolesnikova, *J. Gen. Chem. (U.S.S.R.)*, 1974, **44**, 2628.
[569] K. Bergesen, *Acta Chem. Scand. (A)*, 1975, **29**, 567.
[570] E. E. Nifant'ev, A. A. Borisenko, A. I. Zavalishina, and S. F. Sorokina, *Doklady Chem.*, 1974, **219**, 839.
[571] D. A. Predvoditelev, D. N. Afanas'eva, and E. E. Nifant'ev, *J. Gen. Chem. (U.S.S.R.)*, 1974, **44**, 2586.

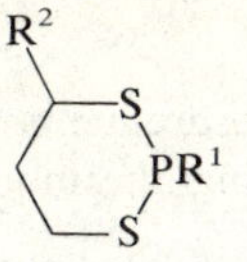

# Elements of Group V

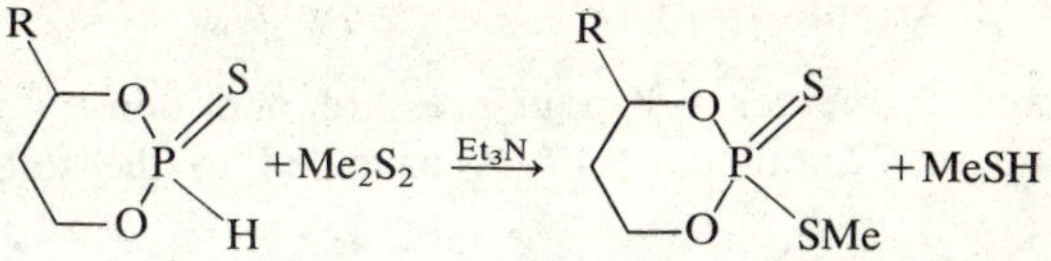

(112) $R^2\!\!-\!\!S\!-\!PR^1\!-\!S$

(113) $R\!-\!O\!-\!P(\!=\!S)(OH)$ + Me$_2$S$_2$ $\xrightarrow{\text{Et}_3\text{N}}$ $R\!-\!O\!-\!P(\!=\!S)(OSMe)$ + MeSH (52)

An alternative preparative route to the unstable phosphorus thiocyanates uses Me$_3$SiCN in reaction with sulphenyl chlorides $R^1R^2P(O)SCl$, where $R^1 = R^2 =$ Me$_3$CCH$_2$O or Me$_2$CHO, and $R^1 = $Me$_3$C, $R^2 = $Ph.[572] Because of steric effects, these products are more stable than those previously isolated.

The structure of BiPS$_4$ consists of a network of sulphur tetrahedra, every second one of which is occupied by a phosphorus atom;[573] while in PdPS the non-metal atoms can be considered as forming (SPPS)$^{4-}$ ions.[574]

The selenide P$_4$Se$_4$ has been described for the first time and was obtained from a reaction between the elements or by heating a mixture of P$_4$Se$_3$ and selenium.[575] The structure (114) has been suggested on the basis of X-ray powder data and

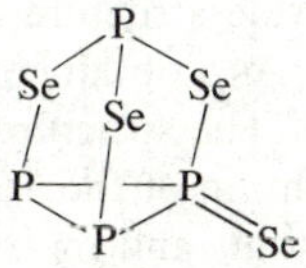

(114)

spectroscopic measurements. Chloro-phosphines (or -arsines) readily cleave the Se—Si bond in methylseleno(trimethyl)silane, giving a general, high-yield method for preparing compounds such as $R_{3-n}P(SeMe)_n$ ($n = 1—3$) for R = Ph, and $n = 1$ for R = OEt.[576] The simple reaction of a phosphorus(III) compound with elemental selenium to give a P=Se link has been exploited in the formation of a number of new $OO$-dialkyl phosphorochloridoselenoates (RO)$_2$P(Se)Cl.[577] Potassium $OO$-diaryl phosphorodiselenoates (ArO)$_2$P(Se)SeK undergo alcoholysis on dissolution in ethyl or propyl alcohols in the presence of traces of hydrogen chloride to give phenols and the corresponding dialkyl compounds.[578]

A chemical-transport method, using chlorine gas, has been used to prepare the iron(II) triselenide FePSe$_3$, which has a structure related to that of CdI$_2$.[579] The octahedral sites in the hexagonal close-packed array of selenium atoms are occupied by Fe and P$_2$ units in the ratio 2:1.

---

[572] A. Łopusiński, J. Michalski, and W. J. Stec, *Angew. Chem. Internat. Edn.*, 1975, **14**, 108.
[573] H. Zimmermann, C. D. Carpenter, and R. Nitsche, *Acta Cryst.*, 1975, **B31**, 2003.
[574] W. Jeitschko, *Acta Cryst.*, 1974, **B30**, 2565.
[575] Y. Monteil and J. E. Guerchais, *Z. anorg. Chem.*, 1975, **416**, 181.
[576] J. W. Anderson, J. E. Drake, R. T. Hemmings, and D. L. Nelson, *Inorg. Nuclear Chem. Letters*, 1975, **11**, 233.
[577] I. A. Nuretdinov, L. K. Nikonorova, N. P. Grechkin, and R. G. Gainullina, *J. Gen. Chem. (U.S.S.R.)*, 1975, **45**, 526.
[578] N. I. Zemlyanskii, L. M. Dzikovskaya, and A. P. Vas'kov, *J. Gen. Chem. (U.S.S.R.)*, 1975, **45**, 463.
[579] B. Taylor, J. Steger, A. Wold, and E. Kostiner, *Inorg. Chem.*, 1974, **13**, 2719.

### 3 Arsenic

**Arsenic and Arsenides.**—Vapour-pressure and density measurements on the element up to 77 atm and 1400 K have led to the following thermodynamic data:[580]

$$\Delta H_f^\circ(As_4, g, 298.15 \text{ K}) = 37.34 \pm 0.2 \text{ kcal mol}^{-1}$$
$$\Delta H_f^\circ(As_2, g, 298.15 \text{ K}) = 45.54 \pm 0.5 \text{ kcal mol}^{-1}$$

The mixed phosphine-arsine $(CF_3)_2PAsH_2$ is the product when trimethyliodosilane is displaced from a mixture of $(CF_3)_2PI$ and $Me_3SiAsH_2$.[581] Although the compound decomposes to $(CF_3)_2PH$ and a polymeric arsenic hydride on standing, it has been characterized and fully investigated. The diarsine analogue $(CF_3)_2AsAsH_2$ has similarly been prepared but cannot be isolated because of lower stability.

The compound $Sn_4As_3$ has been identified in the SnAs–Sn system.[582] The high-temperature modification of $V_4As_3$, according to powder diffraction data, is isotypic with $Cr_4As_3$,[583] while the americium binary compounds AmAs, AmSb, and AmBi have the sodium chloride structure.[584]

The ternary phase $Ca_{10+x}Si_{12-2x}As_{16}$ obtained from the elements at 1000 °C has values of $x$ between zero and 4.[585] The structure contains arsenic atoms in slightly distorted cubic close packing, with ten of the 16 octahedral sites in the molecular unit being occupied by calcium atoms and the remainder by either calcium or $Si_2$ units.[586] A phosphorus analogue can be prepared which is isotypic. Direct synthesis from the elements leads to the non-stoicheiometric arsenide $Ni_xMo_2As_3$, where $0.7 \leq x \leq 1$, which has a structure based on that of NiAs.[587] In the $Cd_4As_2I_3$ structure the cadmium atoms form a pseudo-cubic face-centred structure, with the octahedral sites occupied by iodine atoms or $As_2$ groups (As—As 2.40 Å).[588]

**Bonds to Carbon.**—Trimethylarsine forms stable 1:1 addition compounds with boron tribromide and tri-iodide, but the corresponding trimethylantimony derivatives decompose slowly at room temperature to give $Me_3SbX_2$ (X = Br or I).[589] The crystalline 1:2 adduct of dimethylchloroarsine and thiourea from single-crystal data should be formulated as the ionic compound $[Me_2AsSC(NH_2)_2]^+$-$Cl^-,SC(NH_2)_2$;[590] the As—S distance is 2.320(2) Å.

A new chloro-arsine, $(1\text{-naphthyl})_2AsCl$, can be prepared by the reaction of

---

[580] H. Rau, *J. Chem. Thermodynamics*, 1975, **7**, 27.
[581] A. Breckner and J. Grobe, *Z. anorg. Chem.*, 1975, **414**, 269.
[582] T. Z. Vdovina and Z. S. Medvedeva, *Russ. J. Inorg. Chem.*, 1974, **19**, 1234.
[583] R. Berger, *Acta Chem. Scand.* (A), 1974, **28**, 771.
[584] J. W. Roddy, *J. Inorg. Nuclear Chem.*, 1974, **36**, 2531.
[585] M. Hamon, J. Guyader, and J. Lang, *Rev. Chim. minérale*, 1975, **12**, 1.
[586] M. Hamon, J. Guyader, P. L'Haridon, and Y. Laurent, *Acta Cryst.*, 1975, **B31**, 445.
[587] A. Guérin, M. Potel, and M. Sergent, *Compt. rend.*, 1974, **279**, C, 517; *Rev. Chim. minérale*, 1975, **12**, 335.
[588] J. Gallay, G. Allais, and A. Deschanvres, *Acta Cryst.*, 1975, **B31**, 2274.
[589] M. L. Denniston and D. R. Martin, *J. Inorg. Nuclear Chem.*, 1974, **36**, 2175.
[590] P. H. Javora, R. A. Zingaro, and E. A. Myers, *Cryst. Struct. Comm.*, 1975, **4**, 67.

As$_2$O$_3$ with the Grignard reagent from 1-naphthyl bromide followed by decomposition with hydrochloric acid.[591] The resolution into optically active isomers of the arsenic(III) compounds (115) as hydrogen tartrates and the arsonium salts (116) has been achieved.[592]

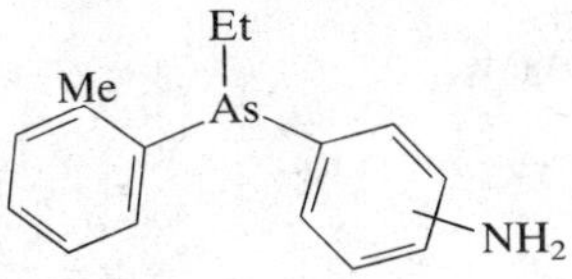

(115)

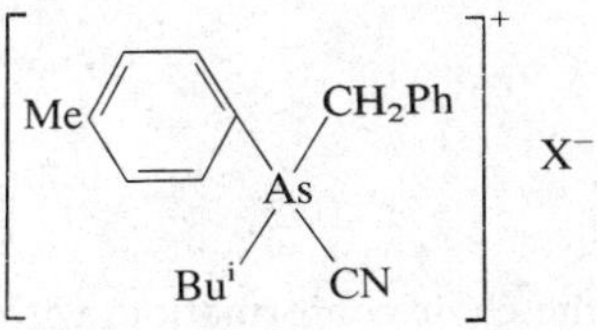

(116) X = Br or ClO$_4$

The barrier to pyramidal inversion at arsenic has been calculated for a variety of compounds, and where experimental data are available there is satisfactory agreement.[593] The low barrier to inversion in the 6π-electron arsoles (117) is noteworthy, pointing to extensive delocalization in the planar transition state and lending support to the contention that cyclic delocalization also occurs in the pyramidal ground state. The six-membered arsabenzene molecule (118; R = H) is

(117) 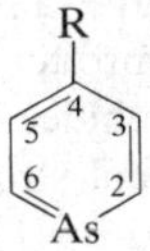    (118)

strictly planar, from recent electron-diffraction measurements, and has the molecular parameters shown below:[594]

| | | | |
|---|---|---|---|
| $r$(C—As) | 1.850(3) Å | ∠CAsC | 97.3(1.7)° |
| $r$[C(2)—C(3)] | 1.39(3) Å | ∠AsCC | 125.1(2.8)° |
| $r$[C(3)—C(4)] | 1.40(3) Å | ∠C(2)C(3)C(4) | 124.9(2.9)° |

Substituted arsabenzene molecules with alkoxy- or aldehyde groups in the 4-position (118; R = OMe, OEt, or CHO) have been prepared for the first time.[595]

Relatively little is known about the arsenic analogues of nitrenes, but a complex of the four-electron ligand PhAs has been prepared via demetallation of (CO)$_5$CrAsPhLi$_2$.[596] The product, PhAs[Cr(CO)$_5$]$_2$, from an X-ray determination, contains arsenic in trigonal planar co-ordination, and the short (2.38 Å) Cr—As bond distance points to π-bonding, probably of the (d–p) type. A new cyclic

---

[591] R. D. Gigauri, M. A. Indzhiya, B. D. Chernokal'skii, M. M. Ugulava, and Ts. A. Sidamashvili, *J. Gen. Chem. (U.S.S.R.)*, 1974, **44**, 1511.
[592] F. D. Yambushev, Yu. F. Gatilov, N. Kh. Tenisheva, and V. I. Savin, *J. Gen. Chem. (U.S.S.R.)*, 1974, **44**, 2458; L. B. Ionov, Yu. F. Gatilov, I. P. Mukanov, and L. G. Kokorina, *ibid.*, p. 2461.
[593] J. D. Andose, A. Rauk, and K. Mislow, *J. Amer. Chem. Soc.*, 1974, **96**, 6904.
[594] T. C. Wong, A. J. Ashe'tert., and L. S. Bartell, *J. Mol. Structure*, 1975, **25**, 65.
[595] G. Märkl and F. Kneidl, *Angew. Chem. Internat. Edn.*, 1974, **13**, 667, 668.
[596] G. Huttner and H.-G. Schmid, *Angew. Chem. Internat. Edn.*, 1975, **14**, 433; G. Huttner, J. von Seyerl, M. Marsili, and H.-G. Schmid, *ibid.*, p. 434.

system (119) is generated in the reaction of cacodylic acid and $Cr(CO)_6$;[597] the

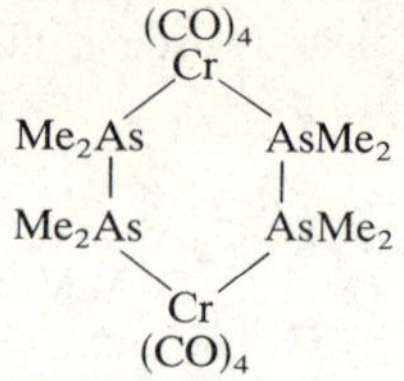

$$\begin{array}{c} (CO)_4 \\ Cr \\ Me_2As \diagup \quad \diagdown AsMe_2 \\ | \qquad\qquad | \\ Me_2As \diagdown \quad \diagup AsMe_2 \\ Cr \\ (CO)_4 \\ (119) \end{array}$$

ring is in the chair conformation, with Cr—As and As—As distances of 2.480(1) and 2.442(1) Å, respectively. Stabilization of a nine-membered cyclo-arsine $(AsMe)_9$ in the compound $Cr_2(CO)_6(AsMe)_9$ has recently been reported, while an open-chain $(AsPr^n)_8$ species is present in the molybdenum compound $Mo_2(CO)_6(AsPr^n)_8$.[598]

Hydrolysis of the cyano-arsine PhRAsCN (R is an alkyl group) leads to the formic acid derivative $PhRAsCO_2H$,[599] while esters of this acid can be prepared by treating the Grignard reagent $PhR^1AsMgX$ with $ClCO_2R^2$.[600] The (ethylphenylarsino)formic acid $EtPhAsCO_2H$ can be resolved by fractional crystallization of the quinine salts, giving pure samples of one diastereoisomer, but in moist air the racemic oxide is rapidly produced.[601] The enantiomer can, however, be converted into the analogous sulphide, which is of high optical stability.

A non-ionic, trigonal-bipyramidal structure with fluorine atoms in apical positions [1.834(7) Å] has been found for $Ph_3AsF_2$;[602] idealized $D_3$ symmetry is not achieved, owing to unequal rotation of the phenyl rings about the As—C bond.

The arsenic ylide $Ph_3As=CH_2$ has now been isolated as an unstable solid from a reaction between $Ph_3AsMeBr$ and sodium amide in THF,[603] and the analogous methyl ylide $Me_3As=CH_2$ is the immediate precursor of an unusual five-co-ordinate tetra-alkylarsonium compound.[604] Methanol adds to the ylide to give $Me_4AsOMe$, which from low-temperature n.m.r. spectra has the structure (120).

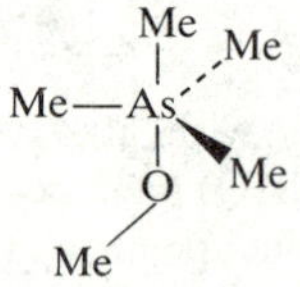

$$\begin{array}{c} Me \\ | \quad Me \\ Me—As \\ | \quad \diagdown Me \\ O \\ | \\ Me \\ (120) \end{array}$$

**Bonds to Halogens.**—From new data on the heat of hydrolysis, the standard heat of formation of $AsF_3$ has been revised to $-205.1$ kcal mol$^{-1}$ (858.14 kJ mol$^{-1}$), giving

---

[597] F. A. Cotton and T. R. Webb, *Inorg. Chim. Acta*, 1974, **10**, 127.
[598] P. S. Elmes, B. M. Gatehouse, D. J. Lloyd, and B. O. West, *J.C.S. Chem. Comm.*, 1974, 953.
[599] L. B. Ionov, Yu. F. Gatilov, I. P. Mukanov, and L. G. Kokorina, *J. Gen. Chem. (U.S.S.R.)*, 1974, **44**, 1704.
[600] Yu. F. Gatilov, L. B. Ionov, L. G. Kokorina, and I. P. Mukanov, *J. Gen. Chem. (U.S.S.R.)*, 1974, **44**, 1727.
[601] Yu. F. Gatilov, I. P. Mukanov, and L. G. Kokorina, *J. Gen. Chem. (U.S.S.R.)*, 1974, **44**, 1841.
[602] A. Augustine, G. Ferguson, and F. C. Marsh, *Canad. J. Chem.*, 1975, **53**, 1647.
[603] Y. Yamamoto and H. Schmidbaur, *J.C.S. Chem. Comm.*, 1975, 668.
[604] H. Schmidbaur and W. Richter, *Angew. Chem. Internat. Edn.*, 1975, **14**, 183.

a value of 108.5 kcal mol$^{-1}$ as the As—F bond energy.[605] The AsF$_4$ radical may be generated by γ-irradiation of AsF$_3$ in sulphur hexafluoride, according to e.s.r. data,[606] and M.O. calculations on AsF$_3^-$ and AsCl$_3^-$ species by an iterative extended Hückel procedure give reasonable results on the shapes and unpaired-spin densities.[607]

I.r. data for the pyridine, bipyridyl, and terpyridyl complexes of arsenic, antimony, and bismuth(III) halides have been interpreted in terms of halogen bridging in the bipyridyl species, but the terpyridyl compounds are formulated as [SbX$_2$(terpy)]$_2^+$ SbX$_5^{2-}$, [AsCl$_2$(terpy)]$^+$ AsCl$_4^-$, and [AsBr$_2$(terpy)$^+$]Br$^-$.[608] The bismuth adduct BiCl$_3$,terpy, on the other hand, is covalent.

Arsenic pentafluoride is readily intercalated into graphite at room temperature to give a blue-black product with the stoicheiometry C$_{10}$AsF$_5$.[609] On heating, weight loss begins at ca. 60 °C, and decomposition occurs in several stages, giving species with the composition C$_{15}$AsF$_5$ and C$_{30}$AsF$_5$. Arsenic pentafluoride in liquid sulphur dioxide solution vigorously oxidizes a number of transition metals, giving, for example, Mn(AsF$_6$)$_2$ and the compound MF(AsF$_6$) for M = Fe or Ni.[610] With cobalt, the product is a mixture of CoF$_2$ and CoF(AsF$_6$), and a similar mixture is obtained when antimony pentafluoride reacts with cobalt. Manganese, iron, and nickel, on the other hand, give the normal bivalent hexafluoro-antimonates.

Vibrational data for AsF$_5$,MeCN and SbCl$_5$,MeCN have been analysed, yielding stretching force constants of 1.9 and 1.1 mdyn Å$^{-1}$ for the As—N and Sb—N bonds, respectively;[611] polymorphism in the KAsF$_6$ system has been investigated by vibrational spectroscopy.[612]

**Bonds to Nitrogen.**—Arsenic trifluoride reacts with lithium-substituted silylamines, giving amino-derivatives AsF$_n$[N(R)SiMe$_3$]$_{3-n}$ (R = SiMe$_3$ or Bu$^t$; $n$ = 1 or 2),[613] and on heating to 150 °C the monofluoride AsF[N(SiMe$_3$)$_2$]$_2$ loses fluorotrimethylsilane, forming the arsenic heterocycle (121). Secondary amines readily replace the chlorine atoms in methanebis(methylchloroarsine) CH$_2$(AsMeCl)$_2$, giving the expected products, which on hydrolysis yield the eight-membered heterocycle (122).[614] The dimethylamino-groups in As(NMe$_2$)$_3$ can be displaced by oxime groups in a step-wise manner to give the products (Me$_2$N)$_{3-n}$-As (ONCR$^1$R$^2$)$_n$ ($n$ = 1—3),[615] and aminolysis of the dialkyl arsonates R$^1$As(O)-(OR$^2$)$_2$ leads to two series of arsonamidic esters, i.e. R$^1$As(O)(OR$^2$)(NR$_2^3$) and R$^1$As(O)(NR$_2^2$).[616]

---

[605] A. A. Woolf, J. Fluorine Chem., 1975, **5**, 172.
[606] A. J. Colussi, J. R. Morton, and K. F. Preston, Chem. Phys. Letters, 1975, **30**, 317.
[607] P. K. Mehrotra, J. Chandrasekhar, S. Subramanian, and P. T. Manoharan, Chem. Phys. Letters, 1974, **28**, 402.
[608] A. M. Brodie and C. J. Wilkins, Inorg. Chim. Acta, 1974, **8**, 13.
[609] L. Chun-Hsu, H. Selig, M. Rabinovitz, I. Agranat, and S. Sarig, Inorg. Nuclear Chem. Letters, 1975, **11**, 601.
[610] P. A. W. Dean, J. Fluorine Chem., 1975, **5**, 499.
[611] D. M. Byler and D. F. Shriver, Inorg. Chem., 1974, **13**, 2697.
[612] A. M. Heyns and C. W. F. T. Pistorius, Spectrochim. Acta, 1975, **31A**, 1293.
[613] E. Niecke and W. Bitter, Synth. React. Inorg. Metal-org. Chem., 1975, **5**, 231.
[614] F. Kober, Z. anorg. Chem., 1975, **412**, 202.
[615] J. Kaufmann and F. Kober, Z. anorg. Chem., 1975, **416**, 152.
[616] V. S. Gamayurova, Z. G. Daineko, B. D. Cherokal'skii, R. R. Shagidullin, I. A. Lamanova, and L. V. Avvakumova, J. Gen. Chem. (U.S.S.R.), 1974, **44**, 1506.

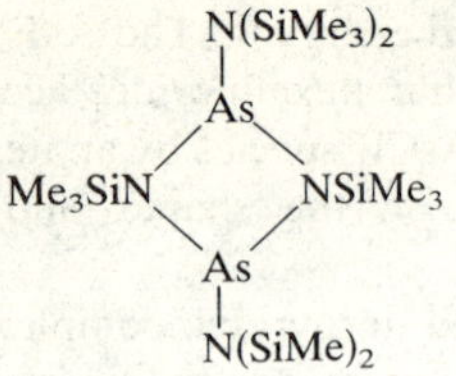

(121) 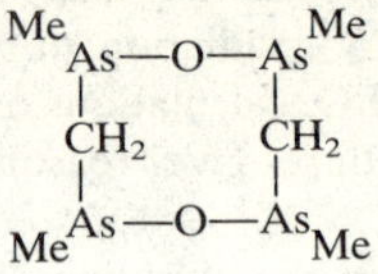    (122)

When pentaphenylarsenic reacts with potassium amide in ammonia, the product depends markedly on the temperature with, for example, the products being K[Ph$_3$AsN],NH$_3$, K$_2$[PhAs(NH)$_3$], and K$_3$[HNAs(NH)$_3$] at respectively 0, 60, and 120 °C.[617] Similar reactions have been described using iminotribenzylarsorane, when products such as K[(PhCH$_2$)$_3$AsN], K$_2$[PhCH$_2$As(NH)$_3$], and K$_3$-[HNAs(NH)$_3$] are obtained at −70, 0, and 80 °C.

**Bonds to Oxygen.**—The monoclinic form of arsenic(III) oxide, claudetite II, has been examined by X-ray methods, showing two crystallographically different arsenic atoms in trigonal-pyramidal co-ordination to oxygen.[618] Silver trifluoroacetate reacts with arsenic trichloride in dichloromethane to give As(O$_2$CCF$_3$)$_3$ as a low-melting solid,[619] and the formation of addition compounds with bipyridyl and boron tribromide has also been described. $^1$H N.m.r. data for a series of 1,3,2-dioxarsolans (123)[620] and the six-membered cyclic arsenites (124)[621] have been analysed in detail.

A new crystal-structure determination for the oxychloride (125) has not confirmed the previously reported large value for the As—O—As angle, but gives a

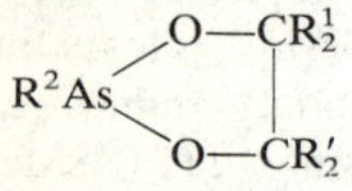

(123) R$^1$ = H or Me
R$^2$ = Cl or Ph

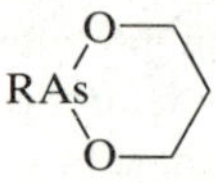

(124) R = Cl, Br, OPh, or OMe

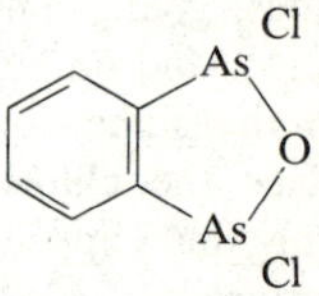

(125)

value close to 122° for both the chloride and the isomorphous bromide.[622] The angle is in fact smaller than any previously reported, and the As—O distance (1.787 Å) is correspondingly longer. Rearrangement of arsine oxides in a direction contrary to that of the Arbuzov reaction occurs with ethyldiphenylarsine oxide in the presence of alkyl halides,[623] the major product being the arsinous ester Ph$_2$AsOR, where the R substituent is derived from the added alkyl halide.

New heteropolyanions containing dialkyl or diaryl arsinate groups, e.g. [R$_2$As-Mo$_4$O$_{14}$(OH)]$^{2-}$, have been prepared, and the structure of that with R = Me has

[617] B. W. Ross and W. B. Marzi, *Chem. Ber.*, 1975, **108**, 1518.
[618] F. Pertlik, *Monatsh.*, 1975, **106**, 755.
[619] C. D. Garner and B. Hughes, *Inorg. Chem.*, 1975, **14**, 1722.
[620] Yu. Yu. Samitov, N. K. Tazeeva, and N. A. Chadaeva, *J. Struct. Chem.*, 1975, **16**, 29.
[621] D. W. Aksnes, *Acta Chem. Scand.* (A), 1974, **28**, 1175.
[622] J. C. Dewan, K. Henrick, A. H. White, and S. B. Wild, *Austral. J. Chem.*, 1975, **28**, 15.
[623] Yu. F. Gatilov, B. E. Abalonin, and Z. M. Izmailova, *J. Gen. Chem.* (*U.S.S.R.*), 1975, **45**, 42.

been determined.[624] Conditions for obtaining tungstoarsen(III)ates with As:W ratios of 2:21, 2:20, 4:40, and 1:18 have been defined.[625]

A number of hydroxyarsonium salts $R^1R^2_2$AsOHX have been prepared from the corresponding oxide and a hydrohalic acid; the thermal decomposition of such species leads to a mixture of products resulting from loss of both an alcohol and an alkyl halide.[626] Normal-co-ordinate analyses have been performed for the molecules $Me_3AsO$ and $(CD_3)_3AsO$,[627] and the donor ability of arsine oxide to phenol has been assessed by measuring the shift in $v(O—H)$ for the latter.[628] The correct formulation of the product obtained by treatment of the hydrogen-bonded species $Ph_3AsOHCl$ with iodine monochloride in acetonitrile has been determined as (126) from an X-ray investigation.[629] The hydrogen atoms were located,

$$[Ph_3As—O—H \cdots Cl \cdots H—O—AsPh_3]^+ \; ICl_2^-$$
(126)

leading to values of 0.89 Å and 2.13 Å for the O—H and H $\cdots$ Cl distances, respectively.

The electronic structures of $AsO_4^{3-}$, $AsO_6^{7-}$, and the $Se^{VI}$ and $Br^{VII}$ analogues have been investigated by the neglect of differential overlap M.O. method.[630] New phases have been characterized in the following systems: $Na_2O–As_2O_5–H_2O$,[631] $K_3AsO_4–M_3(AsO_4)_2–H_2O$ (M = Sr or Ba),[632] and $Sc(NO_3)_3–H_3AsO_4–H_2O$.[633] Amorphous indium arsenate $InAsO_4,H_2O$ is precipitated from acid media, but crystallization as the dihydrate occurs on warming.[634] Thermogravimetric results have been reported for the newly synthesized tin arsenates $H_2[Sn(AsO_4)_2],H_2O$, $H_2[Sn(AsO_4)_2],0.5H_2O$, $Sn(OH)AsO_4,xH_2O$, and $H[Sn(OH)ClAsO_4],2H_2O$;[635] the monohydrate mentioned first is isotypic with the zirconium analogue and is a cation-exchanger, while the hydroxyarsenate is an isotype of $Ge(OH)PO_4$.[636] Heating of $Zr(HAsO_4)_2,H_2O$ leads to two forms of the anhydrous salt, two of the diarsenates $ZrAs_2O_7$, with $Zr_3(AsO_4)_4$ being the final product at 900 °C.[637] Tantalum arsenate is a new inorganic ion-exchanger with analytical importance in a number of separations of metal ions.[638] The double arsenates $Na_2Th(AsO_4)_2$ and $Na_2Np(AsO_4)_2$ have been characterized.[639]

[624] K. M. Barkigia, L. M. Rajković, M. T. Pope, and C. O. Quicksall, J. Amer. Chem. Soc., 1975, **97**, 4146.
[625] J. Martin-Frère and G. Hervé, Compt. rend., 1974, **279**, C, 895.
[626] B. E. Abalonin, Yu. F. Gatilov, and Z. M. Izmailova, J. Gen. Chem. (U.S.S.R.), 1974, **44**, 1513.
[627] F. Watari, Spectrochim. Acta, 1975, **31A**, 1143.
[628] B. N. Laskorin, V. V. Yakshin, L. A. Fedorova, and N. A. Lyubosvetova, Doklady Chem., 1975, **219**, 853.
[629] F. C. March and G. Ferguson, J.C.S. Dalton, 1975, 1381.
[630] B. F. Shchegolev and M. E. Dyatkina, J. Struct. Chem., 1974, **15**, 304.
[631] T. Jouini and H. Guérin, Bull. Soc. chim. France, 1975, 973.
[632] N. Ariguib-Kbir and R. Stahl-Brasse, Bull. Soc. chim. France, 1974, 2343.
[633] N. A. Chernova, L. N. Komissarova, G. Ya. Pushkina, and N. P. Khrameeva, Russ. J. Inorg. Chem., 1974, **19**, 812.
[634] E. N. Deichman, Zh. A. Ezhova, I. V. Tananaev, and Yu. Ya. Kharitonov, Russ. J. Inorg. Chem., 1974, **19**, 19.
[635] A. Démaret, F. d'Yvoire, and H. Guérin, Bull. Soc. chim. France, 1974, 2334.
[636] F. d'Yvoire and A. Démaret, Bull. Soc. chim. France, 1974, 2340.
[637] N. G. Chernorukov, I. A. Korshunov, M. I. Zhuk, and N. P. Shuklina, Russ. J. Inorg. Chem., 1974, **19**, 1139.
[638] J. P. Rawat and S. Q. Mujtaba, Canad. J. Chem., 1975, **53**, 2586.
[639] W. Freundlich, A. Erb, and M. Pagès, Rev. Chim. minérale, 1974, **11**, 598.

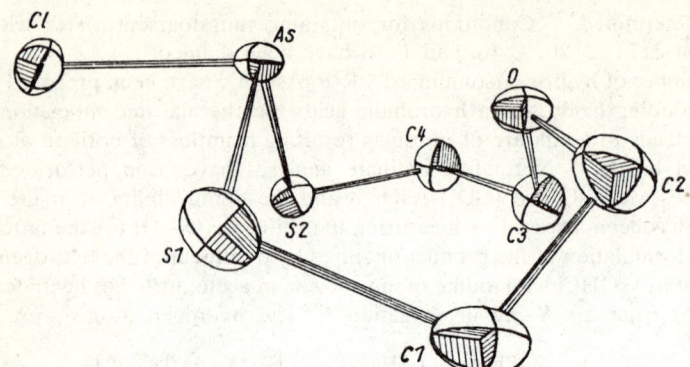

**Figure 4** *Structure of 5-chloro-1-oxa-4,6-dithia-5-arsaoctane*
(Reproduced by permission from Z. anorg. Chem., 1975, **411**, 79)

**Bonds to Sulphur or Selenium.**—Pseudo-trigonal-bipyramidal co-ordination about arsenic is found in 5-chloro-1-oxa-4,6-dithia-5-arsaoctane (see Figure 4), with the eight-membered ring stabilized in a boat–chair conformation by 1,5 As···O transannular interaction.[640] The reactions of alkyl halides and thiobis(diorganoarsines) have now been described as an extension of earlier work on the oxy-derivatives;[641] ethyl bromide and $(Ph_2As)_2S$ give $Ph_2AsBr$ in good yield at temperatures above 200 °C but with alkyl-arsines the reaction is more complex, involving, *inter alia*, the formation of arsonium salts.

Bromination of arsenic, antimony, and bismuth tris(dialkyldithiocarbamates) leads to displacement of one ligand and the formation of a series of bromobis-(dialkyldithiocarbamates).[642] The products are monomers for arsenic and antimony but there is a tendency to dimerization with bismuth. Similar monohalogeno-derivatives have been identified in reactions of chlorine, bromine, or iodine with the newly synthesized complexes of piperidyldiselenocarbamate.[643] Tertiary arsine sulphides $R_3AsS$ (R = Ph or cyclohexyl) react with dimethyl sulphate to give the arsonium compounds $[R_3AsSMe]MeSO_4$,[644] while similar alkylations can be achieved with the cyclohexyl compound using $Et_3O^+ BF_4^-$ or MeI.

Four independent $AsS_4$ tetrahedra are present in the structure of $K_3AsS_4$ recently determined, mean values for As—S, S—S, and ∠SAsS being 2.165 Å, 3.528 Å, and 109.3°.[645]

A good quality $As_2Se_3$ glass with a boiling point of 846 °C can be prepared from the elements in a radio-frequency furnace under pressure,[646] and two new arsenic–selenium compounds (127) and (128) result when selenium reacts with,

---

[640] M. Dräger, Z. anorg. Chem., 1975, **411**, 79.
[641] B. E. Abalonin, Yu. F. Gatilov, and G. I. Vasilenko, J. Gen. Chem. (U.S.S.R.), 1975, **45**, 763.
[642] G. E. Manoussakis, C. A. Tsipis, and C. C. Hadjikostas, Canad. J. Chem., 1975, **53**, 1530.
[643] G. E. Manoussakis, C. A. Tsipis, and A. G. Christophides, Z. anorg. Chem., 1975, **417**, 235.
[644] B. D. Chernokal'skii, I. B. Levenshtein, R. R. Shagidullin, S. V. Izosimova, and I. A. Lamanova, J. Gen. Chem. (U.S.S.R.), 1974, **44**, 1838.
[645] M. Palazzi, S. Jaulmes, and P. Laruelle, Acta Cryst., 1974, **B30**, 2378.
[646] E. H. Baker, J.C.S. Dalton, 1975, 1589.

# Elements of Group V

(127) MeAs, AsPh with Se—Se bridge and Se

(128) MeAs, AsMe with Se, Se, As-Me

respectively, (PhAs)$_6$ and (MeAs)$_5$.[647] The As–Te diagram has been determined.[648]

## 4 Antimony

**Antimonides.**—The crystal structure of Ca$_2$Sb, which results when Ca$_2$Sb$_{1-x}$I$_x$ is heated, consists of cubes of antimony atoms face-centred by calcium atoms,[649] and the compounds Ba$_2$Sb, Ba$_5$Sb, Sr$_2$Bi, Sr$_5$Bi$_3$, and Ba$_5$Bi$_3$ have been synthesized and investigated structurally.[650] $^{121}$Sb Mössbauer spectra have been reported for NbSb$_2$, TaSb$_2$,[651] Mo$_3$Sb$_7$, and Nb$_3$Sb$_2$Te$_5$.[652] Re-examination of the Li–Au–Sb system has led to identification of a cubic phase LiAuSb,[653] and the phases Sb$_{0.07}$WO$_3$ and Sb$_{0.15}$WO$_3$ have been identified in the bronzes obtained when antimony and WO$_3$ react at 900 °C.[654]

**Bonds to Carbon.**—The ligand behaviour of Ph$_2$SbCH$_2$CH$_2$SbPh$_2$ toward carbonyls[655] and Pd$^{II}$, Pt$^{II}$, and Rh$^{III}$ [656] has been investigated. A crystal-structure determination reveals a dimeric structure with a double chlorine bridge (129) for anhydrous diphenylantimony trichloride in the solid state,[657] and compounds formulated as the tetramethylammonium or tetraphenylarsonium salts of the anions [Ph$_2$SbCl$_3$X]$^-$ (X = Cl, Br, N$_3$ or NCS) have been prepared.[658] The predicted sign reveral of the quadrupole coupling constant has been observed in the $^{121}$Sb Mössbauer spectra of Ph$_2$SbCl$_2$(acac), with a *trans* octahedral structure, and Ph$_2$SbCl$_2$(oxine), with the *cis* configuration.[659]

I.r. and Raman data are now available for (C$_6$F$_5$)$_3$Sb and the dichloride and bromide,[660] and structural studies have been reported for three $\mu$-oxo-compounds (Me$_3$SbX)$_2$O (X = Cl, ClO$_4$, or N$_3$).[661] All three species have no symmetry and are extensively disordered, but the antimony atoms have distorted trigonal-bipyramidal co-ordination. In the chloride and perchlorate, however, these groups

---

[647] D. Herrmann, *Z. anorg. Chem.*, 1975, **416**, 50.
[648] R. Blachnik, A. Jäger, and G. Euninga, *Z. Naturforsch.*, 1975, **30b**, 191.
[649] C. Hamon, R. Marchand, P. L'Haridon, and Y. Laurent, *Acta Cryst.*, 1975, **B31**, 427.
[650] B. Eisenmann and K. Deller, *Z. Naturforsch.*, 1975, **30b**, 66.
[651] L. Brattås, J. D. Donaldson, A. Kjekshus, D. G. Nicholson, J. T. Southern, *Acta Chem. Scand. (A)*, 1975, **29**, 217.
[652] J. D. Donaldson, A. Kjekshus, D. G. Nicholson, and J. T. Southern, *Acta Chem. Scand. (A)*, 1974, **28**, 866.
[653] H.-U. Schuster and W. Dietsch, *Z. Naturforsch.*, 1975, **30b**, 133.
[654] M. Parmentier, A. Courtois, and C. Gleitzer, *Compt. rend.*, 1974, **279**, C, 899.
[655] T. W. Beall and L. W. Houk, *Inorg. Chem.*, 1974, **13**, 2549.
[656] W. Levason and C. A. McAuliffe, *Inorg. Chem.*, 1974, **13**, 2765.
[657] J. Bordner, G. O. Doak, and J. R. Peters, jun., *J. Amer. Chem. Soc.*, 1974, **96**, 6763.
[658] N. Bertazzi, L. Pellerito, and G. C. Stocco, *Inorg. Nuclear Chem. Letters*, 1974, **10**, 855.
[659] J. N. R. Ruddick and J. R. Sams, *Inorg. Nuclear Chem. Letters*, 1975, **11**, 229.
[660] B. A. Nevett and A. Perry, *Spectrochim. Acta*, 1975, **31A**, 101.
[661] G. Ferguson, F. C. March, and D. R. Ridley, *Acta Cryst.*, 1975, **B31**, 1260.

are attached by long bonds (2.71 and 2.60 Å, respectively). A non-ionic trigonal-bipyramidal structure is also found for the di-isocyanate $Ph_3Sb(NCO)_2$, which has the nitrogen atoms in the apical positions.[662]

Organometallic halides such as $Ph_4SbCl$, $Me_4SbCl$, and $Ph_3TeCl$ react with the silver salts of acetic and benzoic nitrosolates to give products such as (130), for which a full structure is available.[663]

(129)

(130)

**Bonds to Halogens.**—*Antimony*(III) *Compounds.* The formation of $SbF_3$ and $Sb_{11}F_{43}$ in the incomplete fluorination of antimony in a flow system has been confirmed,[664] and Raman spectra of the latter show peaks associated with $SbF_2^+$, $Sb_2F_5^+$ units and the long Sb $\cdots$ F bridges in the $Sb_3F_{13}^{5+}$ cation. Discrete $SbF_5^{2-}$ anions are present in $Na_2SbF_5$, according to X-ray data;[665] the ions have a distorted square-pyramidal structure, showing stereochemical activity of the lone electron pair, but the structure is not isotypic with that of the ammonium salt. The structures of three new heptafluoroantimonates $MSb_2F_7$ (M = Rb, $NH_4$, or Tl) have been discussed in relation to those of the potassium and caesium salts,[666] and characterization of a number of $MSb_3F_{10}$ species has been reported.[667] The structure of the sodium salt has been determined, showing the presence of $[(SbF_3)_3F]^-$ units which are linked together through three equatorial fluorine bridges to form infinite $(Sb_3F_{10})_n^{n-}$ layers (see Figure 5).[668] The bonding of a fluoride ion to three $SbF_3$ groups presents interesting possibilities of interaction between a filled fluorine $p_z$-orbital and 5d-orbitals on antimony.

Photoelectron spectra for the antimony trihalides[669] and $^{121}$Sb, $^{35}$Cl, and $^{79}$Br n.q.r. spectra[670] for a number of adducts of the trichloride and bromide have been reported. The complex $SbCl_3,GaCl_3$ has been isolated, whereas a simple eutectic only is observed in the $SbCl_3$–$AlCl_3$ system.[671] Antimony and bismuth halide complexes with substituted 1,2-dithiol-3-thiones can be obtained,[672] and the

---

[662] G. Ferguson, R. G. Goel, and D. R. Ridley, *J.C.S. Dalton*, 1975, 1288.
[663] J. Kopf, G. Vetter, and G. Klar, *Z. anorg. Chem.*, 1974, **409**, 285.
[664] A. J. Hewitt, J. H. Holloway, and B. Frlec, *J. Fluorine Chem.*, 1975, **5**, 169.
[665] R. Fourcade, G. Mascherpa, E. Philippot, and M. Maurin, *Rev. Chim. minérale*, 1974, **11**, 481.
[666] N. Habibi, B. Ducourant, R. Fourcade, and G. Mascherpa, *Bull. Soc. chim. France*, 1974, 2320.
[667] B. Ducourant, B. Bonnet, R. Fourcade, and G. Mascherpa, *Bull. Soc. chim. France*, 1975, 1471.
[668] R. Fourcade, G. Mascherpa, and E. Philippot, *Acta Cryst.*, 1975, **B31**, 2322.
[669] D. G. Nicholson and P. Rademacher, *Acta Chem. Scand.* (A), 1974, **28**, 1136.
[670] A. D. Gordeev and I. A. Kyuntsel', *J. Struct. Chem.*, 1974, **15**, 827; T. Okuda, K. Yamada, H. Ishihara, and H. Negita, *Chem. Letters*, 1975, 785.
[671] L. A. Nisel'son, Z. N. Orshanskaya, and K. V. Tret'yakova, *Russ. J. Inorg. Chem.*, 1974, **19**, 580.
[672] F. Petillon and J. E. Guerchais, *J. Inorg. Nuclear Chem.*, 1975, **37**, 1863.

# Elements of Group V

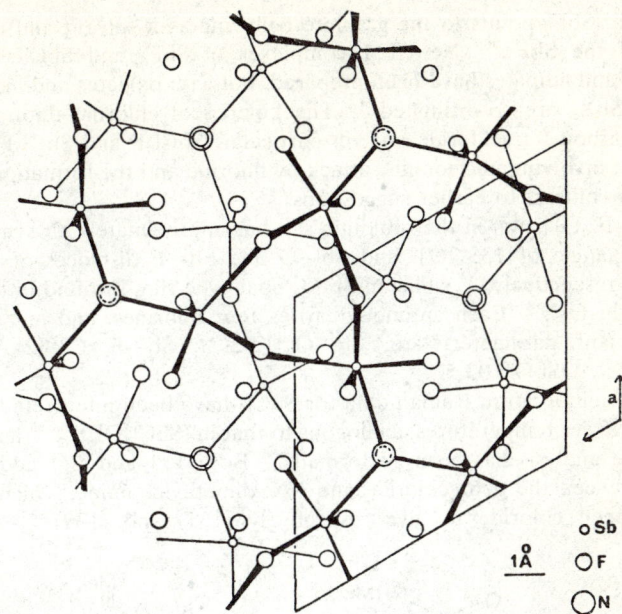

**Figure 5** *Projection of the structure of* Na[(SbF$_3$)$_3$F] *down the z-axis*
(Reproduced by permission from *Acta Cryst.*, 1975, **B31**, 2322)

structure of the newly prepared 1:1 compound of SbCl$_3$ with 1,4-dithian shows co-ordination of antimony by three chlorines and two sulphur atoms.[673]

U.v. and Raman measurements on solutions of Sb$^{III}$ in hydrochloric acid confirm the presence of mixtures of SbCl$_4^-$ and SbCl$_6^{3-}$ ions while the equilibria also involve SbCl$_3$(HSO$_4$)$^-$ when sulphuric acid is added.[674] Complex antimony(III)-bromides R$_x$Sb$_y$Br$_z$, where R is an aromatic amine, have been formed with $x:y:z$ ratios of 1:1:4, 2:1:5, and 3:2:9,[675] the products being intensely coloured when halogeno-pyridines are used.

Standard enthalpies and free energies of formation for Cs$_2$SbBr$_6$ and Cs$_2$SbBr$_5$ are, respectively, −268.3, −253.3 and −264.5, −251.4 kcal mol$^{-1}$.[676]

*Antimony(v) Compounds.* The standard heat of formation of SbF$_5$ from fluorine combustion experiments is −319.5 kcal mol$^{-1}$,[677] and the intercalation of this compound into graphite has been discussed in two reports. The first[678] considers the formation of C$_{6.5}$SbF$_5$ as the first product while the second[679] reports the existence of C$_n$SbF$_5$, where $n = 4$, 6, or 12. A wide-line n.m.r. investigation of

[673] G. Kiel and R. Engler, *Chem. Ber.*, 1974, **107**, 3444.
[674] J. Milne, *Canad. J. Chem.*, 1975, **53**, 888.
[675] J. M. Stewart, K. L. McLaughlin, J. J. Rossiter, J. R. Hurst, R. G. Haas, V. J. Rose, B. E. Ciric, J. A. Murphy, and S. L. Lawton, *Inorg. Chem.*, 1974, **13**, 2767.
[676] S. H. Lee, R. M. Murphy, and C. A. Wulff, *J. Chem. Thermodynamics*, 1975, **7**, 33.
[677] J. Bousquet, J. Carré, M. Kollmannsberger, and P. Barbieri, *J. Chim. phys.*, 1975, **72**, 280.
[678] J. Melin and A. Hérold, *Compt. rend.*, 1975, **280**, C, 641.
[679] A. A. Opalovskii, A. S. Nazarov, and A. A. Uminskii, *Russ. J. Inorg. Chem.*, 1974, **19**, 827.

intercalated $SbF_5$ points to the graphite behaving as a solvent and reducing the activity of the $SbF_5$.[680] New 1:1 complexes of $SbF_5$ and $SbCl_5$ with dialkyl succinates and adipates have been prepared, but with oxalates and malonates two moles of $SbX_5$ are co-ordinated.[681] The kinetics of chlorine–fluorine exchange between carbon tetrachloride and, in particular, $SbCl_4F$ and $SbCl_2F_3$ point to a mechanism involving nucleophilic attack by fluorine and the formation of fluorine bridges, in contrast to earlier suggestions.[682]

The $(Sb_2F_{10}O)^{2-}$ ion in the rubidium salt has approximately $C_s$ symmetry, with an SbOSb angle of 133.2(7)° and Sb—O and Sb—F distances of 1.91(2) and 1.88(2) Å, respectively,[683] while an $Sb_3O_3$ boat-type ring is found in the structure of $Cs_3(Sb_3F_{12}O_3)$.[684] Each antimony carries four fluorines, and mean values for the important parameters are: Sb—O 1.92(2); Sb—F 1.88(2) Å; ∠SbOSb 133.5(1.0)°; ∠OSbO 103.5°.

Variable-temperature Raman data for $SbCl_5$ have been interpreted in terms of a dimer at lower temperatures, analogous to that in $(SbCl_4OEt)_2$,[685] and a normal-co-ordinate analysis has been performed.[686] Both 1:1 and 2:1 adducts can be formed between the pentachloride and $NN'$-dimethyloxamide, which on heating lose hydrogen chloride to give respectively (131) and (132).[687] A series of

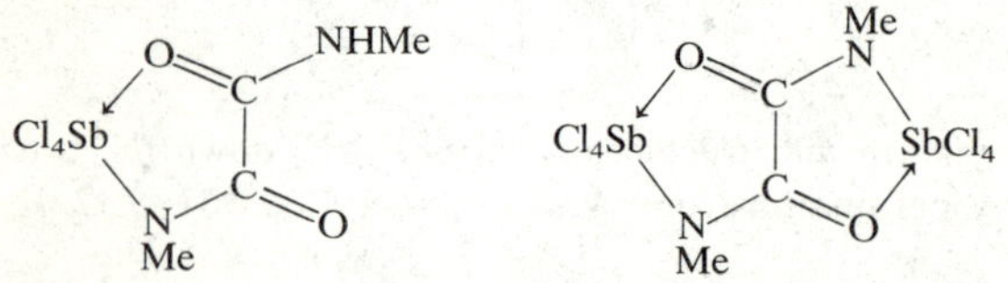

(131)             (132)

complex hexachloroantimonates containing metal cations solvated with a nitro-compound (L) such as nitromethane or nitrobenzene can be prepared with stoicheiometries such as $M^IL(SbCl_6)$, $M^IL_2(SbCl_6)$, $M^{II}L_n(SbCl_6)_2$ ($n = 4$ or 6), and $M^{III}L_n(SbCl_6)_3$ ($n = 3$ or 6).[688] Ionization of $SbCl_5$ in the presence of either tetramethyleneurea or DMSO (L) in 1,2-dichloroethane solution has been confirmed by the observation of a $^{121}Sb$ n.m.r. signal from the $SbCl_6^-$ ion and $^1H$ n.m.r. data on the $L_2SbCl_4^+$ ion.[689] The n.m.r. activity of $^{121}Sb$ has been generally surveyed, and data have been collected for a number of different compound types, including $MSbCl_6$, $M_2SbBr_9$, $MSb(OH)_6$, $M_2H_2Sb_2O_7$, and $MSbS_4$, where M is a univalent cation.[690]

---

[680] L. B. Ebert, R. A. Huggins, and J. I. Brauman, *J.C.S. Chem. Comm.*, 1974, 924.
[681] K. C. Malhotra and S. M. Sehgal, *Indian J. Chem.*, 1974, **12**, 982.
[682] L. Kolditz and S. Schultz, *J. Fluorine Chem.*, 1975, **5**, 141.
[683] W. Haase, *Acta Cryst.*, 1974, **B30**, 2508.
[684] W. Haase, *Acta Cryst.*, 1974, **B30**, 2465.
[685] R. Heimburger and M. J. F. Leroy, *Spectrochim. Acta*, 1975, **31A**, 653.
[686] W. Brockner, S. J. Cyvin, and H. Hovdan, *Inorg. Nuclear Chem. Letters*, 1975, **11**, 171.
[687] R.-A. Laber and A. Schmidt, *Z. anorg. Chem.*, 1975, **416**, 32.
[688] C. Dragulescu, E. Petrovici, and I. Lupu, *Monatsh.*, 1974, **105**, 1170, 1176, 1184.
[689] P. Stilbs and G. Olofsson, *Acta Chem. Scand. (A)*, 1974, **28**, 647.
[690] R. G. Kidd and R. W. Matthews, *J. Inorg. Nuclear Chem.*, 1975, **37**, 661.

**Bonds to Oxygen.**—*Antimony*(III) *Compounds.* The structure of cubic $Sb_2O_3$ has been refined and the $Sb_4O_6$ structure confirmed;[691] each antimony is co-ordinated to three oxygens [Sb—O 1.977(1) Å] with the lone pair completing a tetrahedral arrangement. Structures have also been reported for $Sb_2O_3,2SO_3$, which contains two $SO_4$ tetrahedra and two $SbO_3$ pyramids sharing edges to give $(Sb_2O)(SO_4)_2$ as shown in (133),[692] and the compound previously described as $4Sb_2O_3,Cl_2O_7,2H_2O$, but now known to be $Sb_4O_5(OH)ClO_4,\frac{1}{2}H_2O$.[693] This species is one of those produced in the hydrolysis of $Sb^{III}$ in perchloric acid solutions.[694] Reactions between potassium, rubidium, or caesium superoxide and $Sb_2O_3$ lead to the antimonates $MSbO_3$ and $M_3SbO_4$,[695] while a tungstate $Sb_2WO_6$ is produced on heating mixtures of the oxide and $WO_3$ to 750 °C.[696] Structures have been proposed for the alkali-metal and ammonium antimony fluorosulphates, $M_2Sb(SO_4)F_3$, $M_6Sb_4(SO_4)_3F_{12}$, and $M_2Sb_2(SO_4)F_6$, on the basis of spectroscopic measurements.[697]

A number of new 1,3,2-dioxastibinans (134), where R = H or Me and X = Cl or

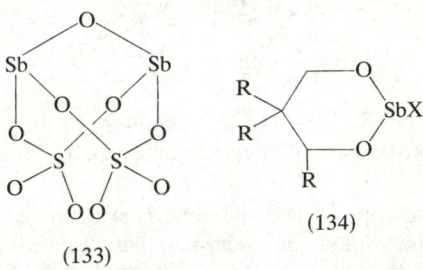

(133)   (134)

OMe, have been synthesized.[698] The structure of tartar emetic, $K_2Sb_2(d\text{-}C_4H_2O_6)_6,3H_2O$, is illustrated in Figure 6, showing co-ordination of each antimony to a carboxyl oxygen and an α-hydroxyl atom from two tartrate groups.[699]

The α-form of $Sb_5O_7I$ consists of pseudo-hexagonal sheets with the composition $Sb_2[Sb_3O_7]_\infty^+$ connected by iodide ions,[700] while both $Sb^{III}$ and $Sb^V$ in respectively four- and six-fold co-ordination are present in α-$Sb_2O_4$ (cervantite).[701] Three new mixed oxidation state species have recently been prepared, *i.e.* $Ca_6Sb^{III}Sb^VO_{10}$, $Ca_4Sb^{III}Sb_7^VO_{23}$ and $Ca_{10}Sb^{III}Sb_9^VO_{34}$.[702]

*Antimony*(V) *Compounds.* The antimony(V) alcoholates $NaSb(OEt)_6$, $[Sb(OEt)_5]_2$, $(SbCl_4OEt)_2$, etc., reported last year, have been characterized further by $^1H$ n.m.r.

---

[691] C. Svensson, *Acta Cryst.*, 1975, **B31**, 2016.
[692] R. Mercier, J. Douglade, and F. Theobald, *Acta Cryst.*, 1975, **B31**, 2081.
[693] J.-O. Bovin, *Acta Chem. Scand.* (A), 1974, **28**, 723.
[694] S. Ahrland and J.-O. Bovin, *Acta Chem. Scand.* (A), 1974, **28**, 1089.
[695] G. Duquenoy, *Rev. Chim. minérale*, 1974, **11**, 474.
[696] M. Parmentier, A. Courtois, and C. Gleitzer, *Compt. rend.* 1975, **280**, C, 985.
[697] R. L. Davidovich, V. I. Sergienko, L. A. Zemnukhova, Yu. Ya. Kharitonov, and V. I. Kostin, *Russ. J. Inorg. Chem.*, 1974, **19**, 698.
[698] B. A. Arbuzov, Yu. Yu. Samitov, and Yu. M. Mareev, *Doklady Chem.*, 1974, **219**, 821.
[699] M. E. Gress and R. A. Jacobson, *Inorg. Chim. Acta*, 1974, **8**, 209.
[700] V. Krämer, *Acta Cryst.*, 1975, **B31**, 234.
[701] P. S. Gopalakrishnan and H. Manohar, *Cryst. Struct. Comm.*, 1975, **4**, 203.
[702] N. Zenaidi, R. Renaud, and F.-A. Josien, *Compt. rend.*, 1975, **280**, C, 1029.

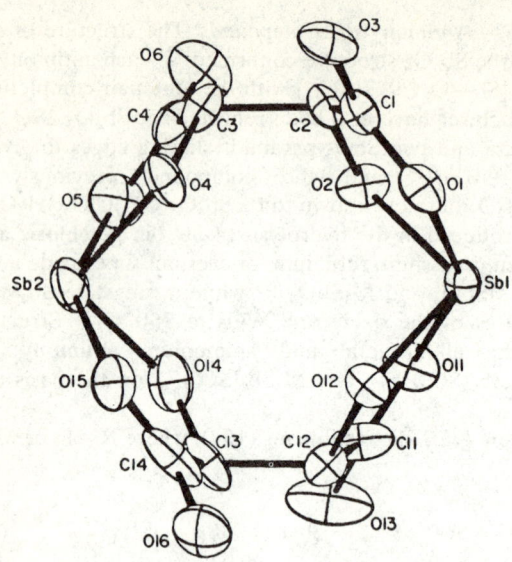

**Figure 6** *Structure of the* $Sb_2(d\text{-}C_4H_2O_6)_2^{2-}$ *anion in tartar emetic*
(Reproduced by permission from *Inorg. Chim. Acta*, 1974, **8**, 209)

and vibrational spectroscopy,[703] and the dimeric structure of the mono-alkoxides has been shown to be broken on treatment with Lewis bases to give six-co-ordinate 1:1 adducts, probably with a *cis* configuration.[704,705] Vibrational data for bis[tetrachloroantimony(v)] oxalate have been discussed in terms of structure (135), which contains five-membered ring systems, rather than the alternative

$$Cl_4Sb\begin{matrix}O\\ \\O\end{matrix}\begin{matrix}C\\ \\C\end{matrix}\begin{matrix}O\\ \\O\end{matrix}SbCl_4$$

(135)

with four-membered rings,[706] and the oxalate group also behaves as a bidentate ligand in $Sb(C_2O_4)OH$, obtained by treating $Sb_2O_3$ with aqueous oxalic acid.[707] Further examples of monocarboxylates $SbCl_4O_2CR$ have been prepared using sodium salts of propionic, isobutyric, and benzoic acids, and monomeric structures with bidentate carboxyl-groups have been postulated.[708]

Mössbauer data have been reported for $Me_3Sb(O_2CR)_2$[709] and for a series of

---

[703] R.-A. Laber and A. Schmidt, *Z. anorg. Chem.*, 1974, **409**, 129.
[704] R.-A. Laber and A. Schmidt, *Z. anorg. Chem.*, 1975, **416**, 41.
[705] R. C. Paul, H. Madan, and S. L. Chadha, *J. Inorg. Nuclear Chem.*, 1975, **37**, 447.
[706] R.-A. Laber and A. Schmidt, *Chem. Ber.*, 1975, **108**, 1125.
[707] S. Ambe, *J. Inorg. Nuclear Chem.*, 1975, **37**, 2023.
[708] R.-A. Laber and A. Schmidt, *Z. anorg. Chem.*, 1975, **414**, 261.
[709] R. G. Goel, J. N. R. Ruddick, and J. R. Sams, *J.C.S. Dalton*, 1975, 67.

# Elements of Group V

metal antimonates containing octahedral $SbO_6$ units.[710] An interesting feature in the structure of $Tl_5SbO_5$ is the presence of isolated $Sb_2O_{10}$ units, formed by edge sharing of two $SbO_6$ octahedra, observed here for the first time.[711] The formation of $Cd_2Sb_2O_7$ has been confirmed, but under somewhat different conditions it is possible to obtain a metastable, non-stoicheiometric pyrochlore $Cd_{1.90}Sb_2O_{6.90}$.[712] A number of mixed species, e.g. $Cd_{2-x}Bi_x(Sb_{2-x}Ti_x)O_7$,[712] $Cd_2(Sb_{2-x}M_x)O_7$ (M = Nb or Ta),[713] and $Cd_{2-x}Bi_x(Sb_{2-x}Sn_x)O_7$,[713] have also been synthesized, along with the tungsten species $M^1M^2WO_6$ ($M^1$ = Li or Na; $M^2$ = Sb, Nb, or Ta).[714] X-Ray and thermal decomposition data have been reported for zirconium antimonate.[715]

**Bonds to Sulphur or Selenium.**—A polymeric anion (136) formed from edge-sharing pseudo-trigonal-bipyramidal units is present in the red thioantimonite

(136)

KSbS$_2$ obtained under mild hydrothermal conditions from potassium hydrosulphide and $Sb_2S_3$,[716] while an unusual structure containing $Sb_2O_3$ tubes of condensed $SbO_3$ pyramidal units is found in the compound $K_3SbS_3,3Sb_2O_3$.[717] Positions between the tubes are occupied by $SbS_3^{3-}$ pyramids. Powder X-ray data are available for $NaSbS_2$ and $NaSbSe_2$.[718]

Members of the $Pb_{3+2n}Sb_8S_{15+2n}$ family of phases from the $PbS-Sb_2S_3$ system have been investigated by X-ray methods. For $n = 0$, the compound contains two kinds of interlocked complexes, i.e. $Pb_2Sb_4S_6$ and $PbSb_4S_9$,[719] while for the second member the structure contains four independent antimony atoms, three of which have square-pyramidal co-ordination, but the fourth is trigonal-pyramidal.[720] Prediction of the structures of higher members is considered possible.[721] Structures have also been reported for zinckenite $Pb_6Sb_{14}S_{27}$[722] and samsonite $Ag_4MnSb_2S_6$.[723]

$^{121}$Sb Mössbauer data for the compounds SbSBr, SbSeBr, the corresponding iodides, and SbTeI show the presence of one antimony site only, but indicate that

---

[710] J. B. Wooten, G. G. Long, and L. H. Bowen, *J. Inorg. Nuclear Chem.*, 1974, **36**, 2177.
[711] M. Bouchama and M. Tournoux, *Rev. Chim. minérale*, 1975, **12**, 93.
[712] G. Desgardin, G. Jeanne, and B. Raveau, *Compt. rend.*, 1974, **279**, C, 565.
[713] C. Hénault, G. Desgardin, and B. Raveau, *Rev. Chim. minérale*, 1975, **12**, 247.
[714] C. Michel, D. Groult, and B. Raveau, *J. Inorg. Nuclear Chem.*, 1975, **37**, 247; C. Michel, D. Groult, A. Deschanvres, and B. Raveau, *ibid.*, p. 251.
[715] A. N. Lapitskii and V. F. Tikavyi, *Russ. J. Inorg. Chem.*, 1974, **19**, 818.
[716] H. A. Graf and H. Schäfer, *Z. anorg. Chem.*, 1975, **414**, 211.
[717] H. A. Graf and H. Schäfer, *Z. anorg. Chem.*, 1975, **414**, 220.
[718] V. G. Kuznetsov, A. S. Kanishcheva, and A. V. Salov, *Russ. J. Inorg. Chem.*, 1974, **19**, 696.
[719] E. W. Nuffield, *Acta Cryst.*, 1975, **B31**, 151.
[720] S.-A. Cho and B. J. Wuensch, *Z. Krist.*, 1974, **139**, 351.
[721] J. J. Kohatsu and B. J. Wuensch, *Acta Cryst.*, 1974, **B30**, 2935.
[722] J. C. Portheine and W. Nowacki, *Z. Krist.*, 1975, **141**, 79.
[723] A. Edenharter and W. Nowacki, *Z. Krist.*, 1974, **140**, 87.

the chalcogen atoms have a greater effect on the antimony valence electrons than have the halogens.[724]

The compounds $Sn_2Sb_6Se_{11}$ and $Sn_2Sb_2Se_5$ have been identified in the $Sb_2Se_3$–SnSe system,[725] and the $NaSbSe_2$–$Sb_2Se_3$ system has been investigated.[726]

## 5 Bismuth

The formation of a co-ordinate bond between bismuth and a transition metal in a positive oxidation state has been observed for the first time in a square-pyramidal nickel complex of tris(o-dimethylarsinophenyl)bismuthine (137).[727] Both bismuth atoms in μ-oxo-bis(perchloratotriphenylbismuth) (138) are trigonal-bipyramidally

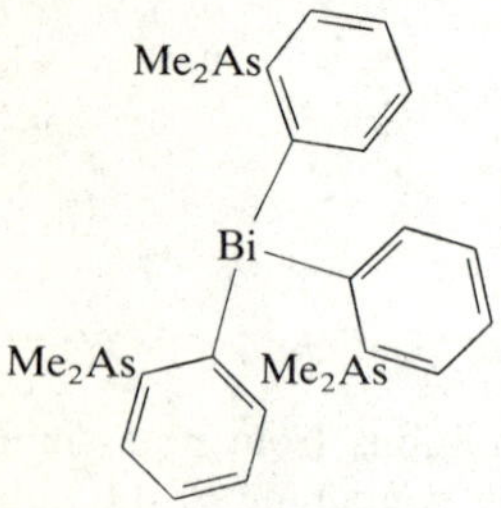

(137) 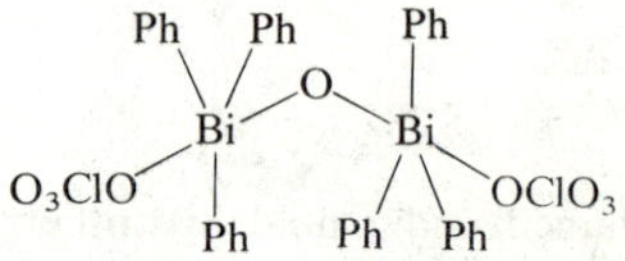   (138)

co-ordinated, with equatorial phenyl groups and weakly co-ordinated perchlorate groups [Bi—O 2.65(1) Å] in apical positions.[728] The Bi—O bridge bond is 2.07(1) Å and the BiOBi angle 142.4(7)°.

An improvement in the synthesis of the chloride $Bi_{12}Cl_{14}$, based on crystallization from a reduced $KCl$–$BiCl_3$ melt, has been reported, together with new X-ray diffraction data.[729] Bismuth(III) chloride forms a 1:1 adduct with $SnCl_2$,[730] and a structure determination for the thiourea adduct $3BiCl_3,7SC(NH_2)_2$ shows the presence of the dinuclear cation $[Bi_2Cl_4(tu)_6]^{2+}$ and the anion $[BiCl_5(tu)]^{2-}$; all bismuth atoms are in octahedral co-ordination.[731] Stability constants for all possible bismuth(III) anions $BiCl_n(NO_3)_m^{3-n-m}$, where $n + m = 2$—4, have been estimated by Sillèn's method, and although species containing five groups occur at high ionic strengths, no mixed hexa-co-ordinated species were identified.[732]

Bismuth(III) chloride reacts with $LiNR_2$ for R = Me, Et, or Pr to give the

---

[724] L. Brattås, A. Kjekshus, D. G. Nicholson, and J. T. Southern, Acta Chem. Scand. (A), 1975, 29, 220.
[725] G. G. Gospodinov, I. N. Odin, and A. V. Novoselova, Russ. J. Inorg. Chem., 1974, 19, 895.
[726] V. A. Bazakutsa, N. I. Gnidash, A. B. Lazarev, E. I. Rogacheva, A. V. Salov, S. I. Berul', and M. B. Vasil'eva, Russ. J. Inorg. Chem., 1974, 19, 265.
[727] W. Levason, C. A. McAuliffe, and S. G. Murray, J.C.S. Chem. Comm., 1975, 164.
[728] F. C. March and G. Ferguson, J.C.S. Dalton, 1975, 1291.
[729] R. M. Friedman and J. D. Corbett, Inorg. Chim. Acta, 1973, 7, 525.
[730] N. V. Karpenko, Russ. J. Inorg. Chem., 1974, 19, 876.
[731] L. P. Battaglia, A. B. Corradi, G. Pellizzi, and M. E. V. Tani, Cryst. Struct. Comm., 1975, 4, 399.
[732] V. A. Fedorov, T. N. Kalosh, and L. I. Shmyd'ko, Russ. J. Inorg. Chem., 1974, 19, 991.

# Elements of Group V

tris(dialkylamino)bismuthines, which are stable at low temperature and have properties similar to those of the antimony and arsenic analogues.[733]

The compound $Bi_6O_7FCl_3$, which contains infinite layers based on $(Bi_6O_7F^{3+})_n$ ions and columns of chloride ions, has been identified in the $BiOCl$–$BiOF$–$Bi_2O_3$ system,[734] and the $Bi_2O_3$–$CaO$ system shows the presence of $Bi_{14}Ca_5O_{26}$, $Bi_2CaO_4$, $Bi_{10}Ca_7O_{22}$, and $Bi_6Ca_7O_{16}$.[735]

Methods for the preparation of the selenate $Bi_2(SeO_4)_3$ have been investigated and the compound has been characterized.[736] Powder X-ray data have been collected for the pyrochlore $Bi_2Ru_2O_7$,[737] $Bi_2UO_6$,[738] and $Bi_{19}S_{27}Cl_3$.[739]

The compounds $TlBiS_2$ and $Tl_4Bi_2S_5$ have been identified in the $Tl_2S$–$Bi_2S_3$ system[740] and three new non-stoicheiometric phases in the $CdS$–$Bi_2S_3$ system.[741] Chemical transport methods have been used to synthesize a number of mixed sulphide species, e.g. $(Pb_{0.12}Bi_{0.88})Bi_2Cu_{3.12}S_5I_2$,[742] $Cu_4Bi_5S_{10}$,[743] and $Bi_2Cu_3S_4Br$,[744] which have been examined by X-ray methods. The structure of cosalite, $Pb_2Bi_2S_5$, has also been determined.[745]

---

[733] F. Ando, T. Hayashi, K. Ohashi, and J. Koketsu, *J. Inorg. Nuclear Chem.*, 1975, **37**, 2011.
[734] F. Hopfgarten, *Acta Cryst.*, 1975, **B31**, 1087.
[735] P. Conflant, J.-C. Boivin, and G. Tridot, *Compt. rend.*, 1974, **279**, C, 457.
[736] Z. Mladenova-Bontschewa and G. Georgiew, *Monatsh*, 1975, **106**, 283.
[737] F. Abraham, G. Nowogrocki, and D. Thomas, *Compt. rend.*, 1975, **280**, C, 279.
[738] A. S. Koster, J. P. P. Renaud, and G. D. Rieck, *Acta Cryst.*, 1975, **B31**, 127.
[739] V. Krämer, *Z. Naturforsch.*, 1974, **29b**, 688.
[740] M. Julien-Pouzol and M. Guittard, *Bull. Soc. chim. France*, 1975, 1037.
[741] C. Bronty, P. Spinat, and P. Herpin, *Rev. Chim. minérale*, 1975, **12**, 60.
[742] M. Ohmasa and K. Mariolacos, *Acta Cryst.*, 1974, **B30**, 2640.
[743] K. Mariolacos, V. Kupčík, M. Ohmasa, and G. Miehe, *Acta Cryst.*, 1975, **B31**, 703.
[744] K. Mariolacos and V. Kupčík, *Acta Cryst.*, 1975, **B31**, 1762.
[745] T. Srikrishnan and W. Nowacki, *Z. Krist.*, 1974, **140**, 114.

# 6
# Elements of Group VI

BY M. G. BARKER

Several reviews of the chemistry of the Group VI elements have been published in the past year. The M.T.P. second series includes articles on inorganic sulphur chemistry,[1] cyclic, sulphur–nitrogen compounds,[2] inorganic selenium chemistry,[3] polyatomic cations of S, Se, and Te,[4] and the inorganic chemistry of tellurium.[5] The Group VI elements also feature in two general reviews on The Typical Elements[6] and Non-Metallic Elements.[7] The chemistry of oxygen[8] and the remaining elements of the Group[9] also appear in a series published by Pergamon Press. The reactions of inorganic compounds and complexes involving oxygen and hydrogen peroxide are covered in an article[10] which places emphasis on mechanisms and kinetics. A review[11] with over 400 references deals with ozone in inorganic chemistry, and reactions involving water are covered in a Japanese paper.[12] The stereochemistry of compounds containing S—N or S—O bonds forms part of a larger review,[13] and literature up to the middle of 1974 on sulphur atoms as ligands in metal complexes has been surveyed.[14] Finally, the electrochemistry of Se and Te is dealt with in detail in a review[15] of 178 papers.

## 1 Oxygen

**The Element.**—The production of singlet oxygen by three different methods has been announced. The first article describes a relatively cheap method of producing excited oxygen by subjecting ordinary oxygen, under low pressure, to a

---

[1] B. Meyer and M. Schmidt, in 'M.T.P. International Reviews of Science, Inorganic Chemistry, Series. Two', Butterworths, London, 1975, Vol. 3, p. 1.
[2] A. J. Banister in ref. 1, p. 41.
[3] K. Dostal in ref. 1, p. 85.
[4] R. J. Gillespie and J. Passmore in ref. 1, p. 121.
[5] A. Engelbrecht and F. Sladky, in ref. 1, p. 137.
[6] D. Millington, J. M. Winfield, and M. G. H. Wallbridge, *Ann. Reports* (A), 1974, **70**, 279.
[7] J. Burgess, in 'Inorganic Reaction Mechanisms', ed. J. Burgess (Specialist Periodical Reports), The Chemical Society, London, 1974, Vol. 3, p. 117.
[8] E. A. V. Ebsworth, J. A. Connor, and J. J. Turner, 'The Chemistry of Oxygen', Pergamon Press, New York, 1975.
[9] M. Schmidt, W. Siebert, and K. W. Bagnall, 'The Chemistry of Sulphur, Selenium, Tellurium and Polonium', Pergamon Press, New York, 1975.
[10] A. McAuley in ref. 7, p. 98.
[11] J. O. Edwards, *Ozone Chem. Technol.*, 1975, 185.
[12] Y. Otsuji, *Kagadu (Kyoto), Zokan*, 1974, **63**, 75.
[13] L. V. Vilkov and L. S. Khaikin, *Topics Current Chem.*, 1975, **53**, 25.
[14] H. Vahrenkamp, *Angew. Chem. Internat. Edn.*, 1975, **14**, 322.
[15] A. I. Alekperov, *Russ. Chem. Rev.*, 1974, **43**, 235.

high-frequency discharge.[16] The singlet oxygen produced by the apparatus was shown to be suitable for the chemical oxidation of substrates. The aqueous decomposition of potassium perchromate has been shown[17] to generate singlet oxygen, with an upper limit of the yield at 6%. Thermal decomposition was, however, shown to give $K_2CrO_4$, $KO_2$, and $O_2$, with no production of singlet oxygen. Phthaloyl peroxide (PPO) has been shown[18] to be an efficient chemical generator of singlet oxygen since the reaction of PPO with 1,3-diphenylisobenzofuran affords the photo-oxidation product o-dibenzoylbenzene in 60% yield. The temperature dependence of the absolute rate constant for the reaction of oxygen ($^3P$) atoms with a series of aromatic hydrocarbons has been measured[19] over the temperature range 299—392 K.

A wide variety of reactions of oxygen involved in biological systems has been discussed[20] and an attempt made to group the reactions into a set of rather simple reaction patterns, in the hope that a coherent understanding of the reaction mechanisms of oxygen could be gained. A polarographic investigation of dissolved oxygen in non-aqueous solvents has shown[21] that in an aprotic solvent, the dissolved oxygen is reduced to superoxide, and then to peroxide, with alkylammonium salt as the supporting electrolyte. With alkali metal as the supporting electrolyte, it is reduced further to form a third wave. The photoion and photoelectron spectroscopic kinetic energy distributions of oxygen have been measured [22] at a number of incident monochromatic wavelengths down to 304 Å. Molecular oxygen in an argon matrix has been bombarded[23] with 2.0 keV protons and deuterons during condensation at 15 K. Bands due to $O_3$ and $HO_2$ were observed, but of most interest was the band of isolated $O_3^-$ at 804.3 cm$^{-1}$. Studies of the reactions of molecular oxygen with several alkaline-earth metal atoms in Ar and $N_2$ matrices have enabled a mechanism for the formation of the metal oxides to be proposed.[24] A similar study[25] using ozone has shown it to be an excellent precursor for oxygen-abstraction reactions in nitrogen matrices.

*Ozone.* The rate of formation of $O_3$ on a Pt anode has been determined[26] in concentrated solutions of pure $H_2SO_4$ and in mixtures of $H_2SO_4$ and $(NH_4)_2SO_4$. The dependence of the yield of ozone on the current density was evaluated and an expression given. The decomposition of $K_2S_2O_8$ by acids yields ozone. Isotopic studies and chemical analyses indicate[27] that the ozone is formed from the reaction of $S_2O_8^{2-}$ with $H_2O_2$, which is formed from the interaction of $H_2S_2O_8$ with $H_2O$. It was shown that one oxygen atom in the ozone comes from $S_2O_8^{2-}$, and two atoms from the $H_2O_2$.

---

[16] R. Chapelon and B. Pouyet, *Bull. Soc. chim France*, 1974, 2367.
[17] J. W. Peters, P. J. Bekowies, A. M. Winer, and J. N. Pitts, *J. Amer. Chem. Soc.*, 1975, **97**, 3299.
[18] K.-D. Gundermann and M. Steinfatt, *Angew. Chem. Internat. Ed.*, 1975, **14**, 560.
[19] R. Atkinson and J. N. Pitts, *J. Phys. Chem.*, 1975, **79**, 295.
[20] E. I. Ochiai, *J. Inorg. Nuclear Chem.*, 1975, **37**, 1503.
[21] T. Fujinaga and S. Sakura, *Bull. Chem. Soc. Japan*, 1974, **47**, 2781.
[22] J. L. Gardner and J. A. R. Samson, *J. Chem. Phys.*, 1975, **62**, 4460.
[23] L. Andrews, B. S. Ault, J. M. Grzybowski, and R. O. Allen, *J. Chem. Phys.*, 1975, **62**, 2461.
[24] B. S. Ault and L. Andrews, *J. Chem. Phys.*, 1975, **62**, 2312.
[25] B. S. Ault and L. Andrews, *J. Chem. Phys.*, 1975, **62**, 2320.
[26] J. Balej and M. Thumova, *Coll. Czech. Chem. Comm.*, 1974, **39**, 3409.
[27] A. Blums and J. Sauka, *Tezisy Dokl. Vses. Soveshch. Khim. Neorg. Perekisnykh Soedin.*, 1973, 111.

A theoretical investigation[28] has shown the cyclic conformer of $O_3$ to be 16 kcal mol$^{-1}$ above the preferred open-chain form. The possible experimental significance of such a relatively stable ring conformer was assessed. Further studies on the ring state have been reported[29] in a study of the ground and excited states of ozone. Potassium ozonide dissolved in liquid $NH_3$ absorbs fairly strongly in the u.v. spectrum from 300 nm, in addition to the strong band in the violet–blue region. Previous conflicting results have been shown[30] to be due to interference from other species, such as $O_3, O_3^-$, and $HO_2$.

The reaction of ozone with $H_2S$ has been studied[31] from 25 to 70 °C over a range of pressures. Previous work on this reaction employed ozonized $O_2$, and the stoicheiometry of the reaction was reported as:

$$O_3 + H_2S = SO_2 + H_2O \tag{1}$$

The present investigation found that oxygen was the most important product of the reaction, and that the ($O_2$ formed)/($O_3$ used) ratio approaches 1.5. The $H_2O/SO_2$ ratio also varied considerably from unity. Two alternative mechanisms were proposed both of which account for the observed product ratios and the observed rate law. Both mechanisms were thought to include an initial step which took place *via* a five-membered ozonide ring (1). Further reactions of ozone which have been studied are those with $C_2Cl_4$,[32] $NO_2$ and $SO_2$,[33] $CH_3$, $CH_3O$, and $CH_3O_2$,[34] and at low pressures with $Fe(CO)_5$ and $Ni(CO)_4$.[35]

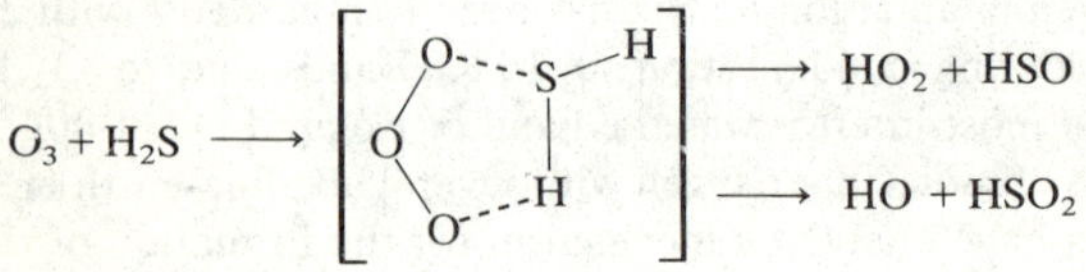

$$O_3 + H_2S \longrightarrow \begin{bmatrix} \text{five-membered ring} \end{bmatrix} \begin{array}{l} \longrightarrow HO_2 + HSO \\ \longrightarrow HO + HSO_2 \end{array}$$

(1)

**Oxides, Peroxides, and Superoxides.**—The electron affinity of $O^{2-}$ has been computed[36] from interaction-potential functions applied to the oxides MO (M = Be, Mg, Ca, Sr, or Ba). The value (728.7 kJ mol$^{-1}$) obtained agrees well with those derived from other methods.

The Raman spectra of peroxides of alkali, alkaline earth, and some transition metals have been measured.[37] The O—O stretching vibration produces an intense band in the expected frequency region of 750—850 cm$^{-1}$. The existence of $BeO_2$ was inferred from the observation of a weak band at 815 cm$^{-1}$, and frequency shifts in the hydrates of alkali-metal and alkaline-earth peroxides were thought to indicate differences of water association.

[28] S. Shih, R. J. Buenker, and S. D. Peyerimhoff, *Chem. Phys. Letters*, 1974, **28**, 463.
[29] P. J. Hay, T. H. Dunning, and W. A. Goodard, *J. Chem. Phys.*, 1975, **62**, 3912.
[30] P. A. Giguere and K. Herman, *Canad. J. Chem.*, 1974, **52**, 3941.
[31] S. Glavas and S. Toby, *J. Chem. Phys.*, 1975, **79**, 779.
[32] E. Mathias, E. Sanhueza, I. C. Hisatune, and J. Heicklen, *Canad. J. Chem.*, 1974, **52**, 3852.
[33] D. D. Davis, J. Prusazcyk, M. Dwyer, and P. Kim, *J. Phys. Chem.*, 1974, **78**, 1775.
[34] R. Simonaitis and J. Heicklen, *J. Phys. Chem.*, 1975, **79**, 298.
[35] W. Groth, U. Schurath, and K. Weber, *Z. phys. Chem. (Frankfurt)*, 1974, **93**, 159.
[36] K. P. Thakur, *J. Inorg. Nuclear Chem.*, 1974, **36**, 2171.
[37] H. H. Eysel and S. Thym, *Z. anorg. Chem.*, 1975, **411**, 97.

A series of papers on the behaviour of alkali-metal peroxides has been published; topics covered include their decomposition in a lithium perchlorate melt,[38] their reactions with fused alkali-metal nitrates,[39] and the thermal decomposition of several metal peroxides.[40] Also included are the reactions of alkali-metal hydroxides with $H_2O_2$ and $H_2O$,[41] and the reaction of $H_2O_2$ with some Group I metal fluorides.[42] Several papers have also dealt with the decomposition of hydrogen peroxide, either spontaneously[43] or on catalysts such as NiO–ZnO,[44,45] NiO–CuO,[45] $V^{5+}$ with $H_2SO_5$ and $H_2SO_4$ also in the system,[46] and lead dioxide in acid and alkaline solutions.[47] The photo-oxidation of sulphur to $H_2SO_4$ by $H_2O_2$ has been shown[48] to be inhibited by iodine, which seems to act as a trap for $H_2O_2$ molecules as well as H and OH radicals. The photoelectron spectra of $H_2O_2$ and $H_2S_2$ have been measured[49] and *ab initio* calculations performed.

Solutions of dicyclohexyl-18 crown 6 in DMSO have been used[50] to prepare pale yellow 0.15 mol l$^{-1}$ solutions of $KO_2$ which contain the $O_2^-$ anion in approximately the same concentration. The specificity of the superoxide dismutase enzyme was used to show that the solution did indeed contain the superoxide ion in solution. The redox potentials of the superoxide and hydroperoxy free radicals, $O_2^-$ and ·$HO_2$, have been measured[51] by the fast reaction technique of pulse radiolysis and kinetic absorption spectrophotometry. A d.t.a. study[52] has shown that, during the heating of $LiClO_4$–$KO_2$ mixtures containing <30% $KO_2$, a eutectic melt occurred at 100—250 °C with the loss of superoxide oxygen. At 250—300 °C the mixture melted with loss of peroxide oxygen; and at 360—500 °C the perchlorate decomposed with the loss of all the perchlorate oxygen.

The kinetics of oxygen evolution, on warming the trapped products (at −196 °C) from water or $H_2O_2$ vapour dissociated in a glow discharge, have been studied[53] by the manometric method. Under closely controlled conditions it was possible to distinguish clearly the decomposition of the two intermediates $H_2O_3$ and $H_2O_4$. The latter begins to decompose measurably at about −115 °C; the trioxide decomposes readily between −50 and −35 °C. Less water, and more of the higher oxides, were obtained from dissociated $H_2O_2$ than from water vapour.

---

[38] D. G. Lemesheva and V. Ya. Rosolovskii, *Tezisy Dokl. Vses. Soveshch. Khim. Neorg. Perekisnykh Soedin.*, 1973, 88.
[39] A. Salta, V. Bruners, and Dz. Peica, *Tezisy Dokl. Vses. Soveshch. Khim. Neorg. Perekisnykh Soedin.*, 1973, 89.
[40] V. Z. Kuprii and V. A. Lunenok-Burmakina, *Tezisy Dokl. Vses. Soveshch. Khim. Neorg. Perekisnykh Soedin.*, 1973, 90.
[41] T. A. Dobrynina, *Tezisy Dokl. Vses. Soveshch. Khim. Neorg. Perekisnykh Soedin.*, 1973, 10.
[42] T. A. Dobrynina, A. Bekmuratov, and N. A. Akhapina, *Tezisy Dokl. Vses. Soveshch. Khim. Neorg. Perekisnykh Soedin.*, 1973, 39.
[43] I. A. Kazarnovskii, *Doklady Chem.*, 1975, **221**, 200.
[44] V. Mucka and J. Cabicar, *Coll. Czech. Chem. Comm.*, 1975, **40**, 947.
[45] V. Mucka, J. Cabicar, and A. Motl, *Coll. Czech. Chem. Comm.*, 1975, **40**, 340.
[46] H. Guiraud, *Bull. Soc. chim. France*, 1975, 521.
[47] A. A. Konoplina, M. M. Andrusev, and L. A. Nikolaev, *Russ. J. Inorg. Chem.*, 1975, **49**, 203.
[48] I. N. Barshchevskii, *Russ. J. Phys. Chem.*, 1975, **49**, 422.
[49] D. W. Davies, *Chem. Phys. Letters*, 1974, **28**, 520.
[50] J. S. Valentine and A. B. Curtis, *J. Amer. Chem. Soc.*, 1975, **97**, 224.
[51] P. S. Rao and E. Hayon, *J. Phys. Chem.*, 1975, **79**, 397.
[52] V. Bruners, A. Salta, and I. I. Vol'nov, *Tezisy Dokl. Vses. Soveshch. Khim. Neorg. Perekisnykh Soedin.*, 1973, 87.
[53] J. L. Arnau and P. Giguere, *Canad. J. Chem.*, 1975, **53**, 2490.

**Oxygen Halides.**—A new method for the preparation of $O_2F_2$ has been announced. Normally the compound is prepared by the electric discharge of a mixture of oxygen and fluorine at −183 °C and 15—20 mmHg; in the new method[54] the same mixture, in liquid form, is subjected to near-u.v. light at −196 °C. Under these conditions oxygen, as well as oxygen difluoride, reacts with fluorine, and dioxygen difluoride is formed. The reaction is quantitative and it can also be used as an efficient and simple method for the removal of oxygen and/or oxygen difluoride from impure elemental fluorine.

An alternative approach to the VSEPR method for the prediction of bond angles in $OF_2$ and $O_2F_2$ has been put forward.[55] Bond angles predicted by this method, which is based on geometrical rather than electronic factors, are in very reasonable agreement with experimental values (Table 1).

**Table 1** *Observed and calculated bond angles in some halides and hydrides of oxygen*

| Compound | Observed angle | Calculated angle |
|---|---|---|
| $OF_2$ | 101.3° | 100.5° |
| $OCl_2$ | 110.9° | 115.8° |
| $O_2F_2$ | 109.5° | 108.1° |
| $O_2H_2$ | 96.9° | 98.2° |

The photolysis of oxygen difluoride in the presence of carbonyl sulphide and carbonyl sulphide–oxygen mixtures has been studied.[56] No change in the total pressure occurs during the course of the reaction, the reaction products being $F_2SO$ and CO. Sulphur dioxide was not detected in reactions carried out with oxygen-free $OF_2$. If irradiation was continued after total conversion of SCO, the further products $F_4SO$, $F_2SO_2$, $F_2CO$, $CF_3OF$, and $(CF_3)_2O_2$ were formed. Because of the discrepancies in the e.s.r. spectra of $O_2^+$, and for the purpose of understanding the behaviour of $O_2^+$ in a crystalline environment, the e.s.r. spectra of this ion in the presence of several counterions have been investigated.[57] An improved synthesis of $O_2^+ BF_4^-$ was described involving low-temperature u.v. photolysis of a $BF_3$–$F_2$–$O_2$ mixture.

## 2 Sulphur

**The Element.**—CNDO/2 MO calculations[58] have predicted that the boat form of cyclohexasulphur exists, and that its potential energy is *ca.* 4 kcal mol$^{-1}$ less than that of the chair form which is found in the rhombohedral crystal. The interconversion has a barrier of 23 kcal mol$^{-1}$ and is not forbidden by symmetry rules. A Mulliken population analysis gave a picture of the bonding in cyclohexasulphur which was consistent with thermodynamic data. Calculations have also been

---
[54] A. Smalc, K. Lutar, and J. Slivnik, *J. Fluorine Chem.*, 1975, **6**, 287.
[55] C. Glidewell, *Inorg. Nuclear Chem. Letters*, 1975, **11**, 353.
[56] D. Soria, O. Salinovich, E. Ramondelli de Staricco, and E. H. Staricco, *J. Fluorine Chem.*, 1974, **4**, 437.
[57] I. B. Goldberg, K. O. Christe, and R. D. Wilson, *Inorg. Chem.*, 1975, **14**, 152.
[58] Z. S. Herman and K. Weiss, *Inorg. Chem.*, 1975, **14**, 1592.

performed[59] on $S_8$; *ab initio* methods were used, but without the inclusion of *d*-orbitals.

The vacuum-u.v. photoelectron spectrum of the $S_2$ ($^3\Sigma_g$) molecule has been investigated.[60] Eleven ionic states of the disulphur molecule were identified and good correlation was found between the position of these states and those observed previously for $O_2^+$ and $SO^+$. Calculated bond lengths and measured vibrational level separations were used, along with the known ionization potentials of atomic sulphur, to construct a potential-energy diagram for $S_2^+$. The u.v. and X-ray photoelectron spectra of the $S_8$ molecule have been obtained[61] and a calculation of the electronic structure has been carried out which enabled a tentative assignment of the valence-region spectrum to be proposed. The pure rotational Raman spectrum of $S_2$ has been measured and rotational constants have been derived.[62] Vibrational assignments and normal-co-ordinate treatments have been made for $S_8O$,[63] and $S_{12}$,[64] using a modified Urey–Bradley force field with 12 force-constants. The valence force-constants obtained have been compared[65] with those of $S_6$, $S_7$, and $S_8$, and the dependence of some normal vibrations on ring size has been discussed. Non-linear relationships were established between S—S stretching frequencies and force-constants, respectively, and S—S bond distances. The Raman intensity of S—S stretching vibrations was found to decrease with increasing bond order. The spectral properties of $S_6$, $S_8$, and $S_{12}$ have also been studied[66] from a different aspect, namely the so-called atomic force-constants. For sulphur in the above configurations the atomic force-constants were found to be $475 \pm 2$ nm$^{-1}$, and to be transferable between the three compounds.

Curves describing the intensity of the X-rays scattered by liquid sulphur at 100—250 °C have been obtained.[67] A difference was observed between the diffraction patterns for three temperature ranges (100—130, 170—200, and 250 °C). A comparison of the structure factors for $S_8$ rings and chains with those for sulphur at 100 °C showed that at low temperatures the structural units in the melt are $S_8$ rings in a largely coplanar arrangement. From radial distribution curves for atoms in liquid sulphur it was suggested that at 170—200 °C the rings are vigorously decomposed and at 250 °C the structure of the melt consists of macromolecules.

Electrochemical measurements[68] have shown the limiting stage in the reaction of sulphur with liquid ammonia at −40 °C to be the physical dissolution of the element. The number of moles of solid S dissolving equalled the number of $NH_4^+$ SCN$^-$ ion pairs formed in the presence of KSCN. The principal reaction taking place was:

$$3S + 2NH_3 = N_2 + 3H_2S \tag{2}$$

[59] G. L. Carlson and L. G. Pederson, *J. Chem. Phys.*, 1975, **62**, 4567.
[60] J. M. Dyke, L. Golob, N. Jonathan, and A. Morris, *J.C.S. Faraday II*, 1975, **71**, 1026.
[61] N. V. Richardson and P. Weinberger, *J. Electron Spectroscopy*, 1975, **6**, 109.
[62] P. A. Freedman, W. J. Jones, and A. Rogstad, *J.C.S. Faraday II*, **71**, 286.
[63] R. Steudel and D. F. Eggers, *Spectrochim. Acta*, 1975, **31A**, 871.
[64] R. Steudel and D. F. Eggers, *Spectrochim. Acta*, 1975, **31A**, 879.
[65] R. Steudel, *Spectrochim. Acta*, 1975, **31A**, 1065.
[66] W. T. King, *Spectrochim. Acta*, 1975, **31A**, 1421.
[67] Yu. G. Poltavtsev and Yu. V. Titenko, *Russ. J. Inorg. Chem.*, 1975, **49**, 178.
[68] R. Guiraud, M. Aubry, and B. Gilot, *Bull. Soc. chim. France*, 1975, 490.

Some preliminary results[69] on the electrochemical behaviour of elemental sulphur (and selenium) in $AlCl_3$–NaCl melts have been reported. The results for both elements were difficult to interpret, due in part to the low solubility of both elements in the melt and the modification of the electrode surfaces by the products of electrode reactions. In basic melts it was suggested that an $S_n^{2+}$ cation is formed, in agreement with spectral studies. In acidic melts the results were more complex, and no exact conclusions could be drawn as to the nature of the species. A spectrophotometric study of the reactions of sulphur, selenium, and tellurium with aqueous solutions of NaOH at 150 °C (S, Se) and 300 °C (Te) has been carried out.[70] From the experimental data the formation of the chain-like structure of polychalcogenide compounds was presumed. The spectra confirmed previous results concerning the mechanism of interaction of elemental chalcogens with aqueous NaOH solutions.

The photo-oxidation of sulphur to $H_2SO_4$ by hydrogen peroxide takes place in aqueous tetrachloromethane and is intensified by methanol.[71] The reaction of potassium tetracyanonickelate(II) with molten sulphur, and with potassium tetrasulphide in molten sulphur, has been studied.[72] The physical adsorption of $SO_2$, $SF_6$, $CO_2$, $N_2$, Ar, and neopentane on rhombic sulphur has been reported.[73] The chemical forms of trace quantities of sulphur in arsenic trichloride have been identified[74] by radiochemical methods as elemental sulphur and $SO_2Cl_2$. The reactions of sulphur vapour with lanthanide oxides (at 1050—1120 °C), oxycarbonates, or oxalates (at 700 °C) has been shown[75] to give the oxysulphide. The reactions are thought to proceed *via* the lanthanyl ion $[LnO]^+$:

$$4(LnO)2O + 3S_2(\text{vapour}) = 4(LnO)_2S + 2SO_2 \quad (3)$$

This method allows the preparation of the oxysulphides in a non-reducing atmosphere. The sulphides $Ln_2S_3$ are obtained above 1200 °C.

The preparation of $S_8O$ by the oxidation of $S_8$ has been announced.[76]

$$S_8 + CF_3CO_3H = S_8O + CF_3SO_2H \quad (4)$$

The reaction involves the use of perfluoroacetic acid as the oxidizing agent and gives $S_8O$ in a 45% yield.

**Sulphur–Halogen Compounds.**—The available data for the physicochemical and, primarily, chemical properties of $SF_6$ have been collated in a review.[77] The variation in the melting point of $SF_6$ with pressure has been calculated[78] by means of the Clausius–Clapeyron equation as 0.08—0.1 K atm$^{-1}$, which has been confirmed by experiment. The reactions of $SF_6$ with various container materials have been studied.[79] From analytical results and i.r. absorption frequencies, the

---

[69] R. Marassi, G. Mamantov, and J. Q. Chambers, *Inorg. Nuclear Chem. Letters*, 1975, **11**, 245.
[70] M. Z. Ugorets. K. T. Rustembekov, and Y. A. Kushnikov, *Tr. Khim.-Metall. Inst., Akad. Nauk Kazakh. S.S.R.*, 1974, **24**, 48.
[71] I. N. Barshchevskii, *Russ. J. Phys. Chem.*, 1975, **49**, 424.
[72] S. B. Cole and J. Kleinberg, *Inorg. Chim. Acta*, 1975, **14**, 111.
[73] F. Sollberger and F. Stoeckli, *Helv. Chim. Acta*, 1974, **57**, 2327.
[74] A. A. Efremov, V. A. Fedorov, E. A. Efremov, D. M. Mgaloblishvili, and S. A. Fedotikov, *Izvest. Akad. Nauk S.S.S.R., Neorg. Materials*, 1975, **11**, 400.
[75] R. Heindl and J. Loriers, *Bull Soc. chim. France*, 1974, 377.
[76] R. Steudel and J. Latte, *Angew. Chem. Internat. Edn.*, 1974, **13**, 603.
[77] A. A. Opalovskii and E. U. Lobkov, *Russ. Chem. Rev.*, 1975, **44**, 97.
[78] A. I. Semenova and D. S. Tsiklis, *Russ. J. Phys. Chem.*, 1975, **49**, 778.
[79] D. K. Padma and A. R. Vasudeva Murthy, *J. Fluorine Chem.*, 1975, **5**, 181.

reaction between quartz and sulphur hexafluoride can be expressed in terms of the equation:

$$2SF_6 + SiO_2 = SiF_4 + SOF_2 + OF_2 \quad (5)$$

A complete quantitative estimate of the $SiF_4$ and $OF_2$ content could not be obtained because of secondary reactions. In stainless steel and copper reaction tubes, $SF_6$ decomposed to $SF_4$ and the metal fluoride. Thermodynamic calculations have shown that at temperatures up to 1500 °C $SF_6$ dissociates principally to $SF_4$; these results therefore lend further support to the calculations. Metal fluorides have also been observed[80] as the products of the exothermic reactions between $SF_6$ and some oxides of Group II and Group III elements at 500—800 °C.

Several new derivatives of sulphur hexafluoride have been prepared. $NSF_3$ reacts[81] with $OSF_4$ to give pentafluorosulphur oxide difluoride imide (2), which hydrolyses with alkali to the ion $[F_5S\text{—}NSO_2F]^-$.

$$NSF_3 + OSF_4 \xrightarrow[HF]{85\,°C} F_5S\text{—}NS(O)F_2 \quad (6)$$
$$(2)$$

$$F_5S\text{—}NS(O)F_2 + H_2O \xrightarrow{Ph_4AsCl\text{—}OH^-} [F_5S\text{—}NSO_2F]^- [AsPh_4]^+ + HF \quad (7)$$

From $NSF_2\text{—}NMe_2$ and $SF_4$ or $OSF_4$, only $F_5S\text{—}NMe_2$ is formed, together with the decomposition products NSF and $(NSOF)_3$, respectively. New amino-derivatives of $SF_6$ of the type (3) and (4) have been prepared[82] by Si—N bond

$$SF_5\text{—}N{=}SF_2 + (Me)_3Si\text{—}NR_2 \longrightarrow SF_5\text{—}N{=}S\begin{array}{c}F\\\diagdown\\NR_2\end{array} + Me_3SiF \quad (8)$$

R = Me or Et.  (3)

$$SF_5\text{—}N{=}SF_2 + 2Me_3Si\text{—}NR_2 \longrightarrow SF_5\text{—}N{=}S\begin{array}{c}NR_2\\\diagdown\\NR_2\end{array} + 2Me_3SiF \quad (9)$$

R = Me or Et.  (4)

cleavage reactions of silylamines with pentafluorosulphur-sulphur difluoride imide. The reaction of the latter with $PCl_5$ yields $SF_5\text{—}N{=}SCl_2$. The mass, i.r., and n.m.r. spectra of the new compounds were reported and compared with those of other derivatives.

An extensive range of reactions of $SF_5Cl$ and $SF_5Br$ with a variety of methyl-, halogeno-, and di-silanes showed[83] only that $SF_5Cl$ and $SF_5Br$ functioned mainly as fluorinating agents. However, the addition of $SF_5Cl$ and $SF_5Br$ to the olefinic double bonds of several vinylsilanes gave the following reactions:

$$SF_5Cl + Me_3SiCH{=}CH_2 \xrightarrow{100\,°C} Me_3SiC_2H_3ClSF_5 \quad (10)$$
$$SF_5Br + Me_3SiCH{=}CH_2 \xrightarrow{25\,°C} Me_3SiC_2H_3BrSF_5 \quad (11)$$
$$SF_5Br + Cl_3SiCH{=}CH_2 \xrightarrow{0\,°C} Cl_3SiC_2H_3BrSF_5 \quad (12)$$

[80] A. A. Opalovskii, E. U. Lobkov, V. N. Lyubimov, Yu. V. Zakhar'ev, V. N. Grankin, and V. N. Mit'kin, *Zhur. neorg. Khim.*, 1975, **20**, 1179.
[81] R. Höfer and O. Glemser, *Z. Naturforsch.*, 1975, **30b**, 458.
[82] R. Höfer and O. Glemser, *Z. anorg. Chem.*, 1975, **416**, 263.
[83] A. D. Berry and W. B. Fox, *J. Fluorine Chem.*, 1975, **6**, 175.

These products were thought to be the first reported examples of molecules incorporating both pentafluorosulphur- and silyl-groupings.

Neutralization processes of ions in the radiolysis of ethane or ethylene with $SF_6$ have been shown[84] to lead to the formation of the $SF_5$ radical. Compounds of the type $RSF_5$ are formed as a result of the recombination reactions with hydrocarbon radicals. A kinetic e.s.r. study[85] of the self-reaction of $SF_5$, and a spectroscopic and kinetic e.s.r. study of its reaction with 1,1-di-t-butylethylene, have been reported. The radical undergoes self-reaction by a second-order process and adds to 1,1-di-t-butylethylene to give $Bu_2^tCCH_2SF_5$, which decomposes by a first-order process.

On cooling sulphur-fluorine compounds with three- or four-co-ordinate sulphur atoms, changes have been observed[86] in the chemical shifts of fluorine atoms and in spin coupling constants, which suggests that fluorine bridges are formed. In solutions of $SF_4$, the presence of associated species is indicated by a decrease in the distance between the two triplets. In the absence of HF, the fluorine exchange in $SF_4$ proceeds, in the liquid state, in solution, and in the gaseous state, at almost the same rate.[87] Evidently, in these cases, the reaction is intramolecular. The reaction of $SF_4$ with pyridine has been shown[88] to give the adducts $SF_4, xpy$, where $x = 1$, 2, 4, and 8. The adducts are stable only below $-30\,°C$, and above this temperature they undergo partial decomposition to $SF_4$, a brown viscous liquid in the case of $SF_4$,py and white solids in the case of the other three adducts. The brown liquid corresponds to the molecular formula $SF_2py$, its i.r. spectrum indicating co-ordination through the nitrogen of the amine and S—F bond frequency of ca. $840\,cm^{-1}$.

The reaction of the recently reported compound $[SF_2NCO]^+[AsF_6]^-$ with NOCl takes place[89] through nucleophilic attack of $Cl^-$ at the C atom of the isocyanate group, and results in the formation of chloroformyl sulphur difluoride imide (5).

$$[SF_2NCO]^+[AsF_6]^- + NOCl = ClC(O)NSF_2 + [NO]^+[AsF_6]^- \quad (13)$$
$$(5)$$

The compound is unstable, and isomerizes rapidly, even in the gaseous state at low pressure, to give the chlorofluoro-imide (6). It is not therefore possible to isolate (5) in the pure state. The chlorofluoro-imide is a colourless liquid, stable

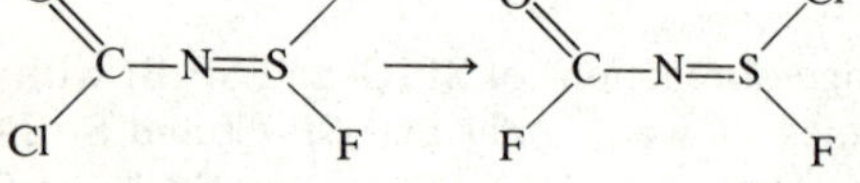

only at low temperatures; at room temperature it rapidly dismutates:

$$2\,COFNSClF \rightarrow COFNSCl_2 + COFNSF_2 \quad (14)$$

---

[84] J. Gawlowski and J. A. Herman, *Canad. J. Chem.*, 1974, **52**, 3631.
[85] J. C. Tait and J. A. Howard, *Canad. J. Chem.*, 1975, **53**, 2361.
[86] W. Gombler and F. Seel, *J. Fluorine Chem.*, 1974, **4**, 333.
[87] F. Seel and W. Gombler, *J. Fluorine Chem.*, 1974, **4**, 327.
[88] D. K. Padma, *J. Fluorine Chem.*, 1974, **4**, 441.
[89] R. Mews, *J. Fluorine Chem.*, 1974, **4**, 445.

# Elements of Group VI

The reaction of $[OSF_2NCO]^+[AsF_6]^-$ with NOCl gives (7), which shows no tendency to isomerize, probably because of the presence of the more stable $S^{VI}$—F bond.

$$[OSF_2NCO]^+[AsF_6]^- + NOCl = ClC(O)NSOF_2 + [NO]^+[AsF_6]^- \quad (15)$$
$$(7)$$

Aryl oxysulphur(IV) fluorides may be prepared from the interaction[90] between aryl silyl ethers and $SF_4$ and its derivatives. The compounds prepared, $ArOSF_2CF_3$, $ArOSF_2CF_3$, $(ArO)_2SF_2$, $(ArO)_3SF$, and $ArOSF_2C_3F_7$, have been examined spectroscopically, and in all cases the structures appear to be based on a trigonal-bipyramidal arrangement about the sulphur atom, with the lone pair equatorial and the fluorine atoms apical. A study[91] of the $^{19}F$ n.m.r. spectrum of $CF_3SF_2CF_2CF_3$ has shown that the geminal fluorine atoms of the methylene group as well as the geminal fluorine atoms bonded to the suphur are magnetically non-equivalent. Thus this molecule may be described in terms of an $AA'XX'Y_3Z_3$ system.

The vapour-state Raman spectra and thermodynamic properties of sulphur dichloride and disulphur dichloride have been determined[92] over the temperature range 300—700 K.

The nature of mixed cations of sulphur in $H_2S_2O_7$ has been studied.[93] Elemental sulphur, SCl, and $SCl_2$ were added to disulphuric acid, sometimes in the presence of halogens. Cryoscopic and conductance measurements enabled equations to be proposed to account for all the observed reactions. The blue coloration due to the $S_8^{2+}$ ion was observed when SCl and $SCl_2$ were dissolved in $H_2S_2O_7$:

$$16SCl_2 + 30H_2S_2O_7 \rightarrow S_8^{2+} + 8SCl_3^+ + 10HS_3O_{10}^- + 8HSO_3Cl + SO_2 + 21H_2SO_4 \quad (16)$$

In the presence of halogens, $SCl_2$ forms $SX_3^+$, whilst SCl forms $S_2X_3^+$, which disproportionates to $SX_3^+$ and S. The $SX_3^+$ ions are also formed when elemental sulphur is dissolved in $H_2S_2O_7$ containing Cl or Br, or when halogens are added to the colourless solution of $S_4^{2+}$ ions in disulphuric acid. The vibrational spectra of MeSCl have been investigated,[94] and all fundamentals identified, as well as several combination bands.

The use of $NSF_3$ as a ligand in transition-metal complexes has been demonstrated by the preparation[95] of $[Re(CO)_5(NSF_3)][AsF_6]$. The cation has $C_{3v}$ symmetry and the $NSF_3$ ligand is bonded to rhenium through nitrogen. Imidosulphenylhalides $(CF_3)_2C=N-SX$ (X = Cl or F) have been prepared[96] from $(CF_3)_2CN_2$ and $(NSCl)_3$ and $Hg(NSF_2)_2$, respectively:

$$(NSCl)_3 + 3(CF_3)_2CN_2 \rightarrow 3(CF_3)_2C=N-SCl + 3N_2 \quad (17)$$

---

[90] J. I. Darragh, S. F. Hossain, and D. W. A. Sharp, *J.C.S. Dalton*, 1975, 218.
[91] D. T. Sauer, S.-L. Yu, W. N. Shepard, J. A. Magnuson, R. A. Porter, and J. M. Shreeve, *Inorg. Chem.*, 1957, **14**, 1228.
[92] S. G. Frankiss and D. J. Harrison, *Spectrochim. Acta*, 1975, **31A**, 161.
[93] A. Bali and K. C. Malhotra, *Austral. J. Chem.*, 1975, **28**, 983.
[94] F. Winther, A. Guarnieri, and O. Faurskov Nielsen, *Spectrochim. Acta*, 1975, **31A**, 689.
[95] R. Mews and O. Glemser, *Angew. Chem. Internat. Edn.*, 1975, **14**, 186.
[96] J. Varwig, H. Steinbeisser, R. Mews, and O. Glemser, *Z. Naturforsch.*, 1974, **29b**, 813.

Structural and dynamic parameters for trifluoromethyliminosulphur difluoride [$F_3CNSF_2$], in the gas phase, have been determined[97] from molecular-electron scattering functions. The parameters are C—F = 1.332, [S=N] = 1.447, N—C = 1.469, S—F = 1.583 Å, ∠CNS = 130.4°, ∠NSF = 112.6°, ∠FSF = 81.1°, and ∠NCF = 110.3°. A wide range of models were tested, covering many positional isomers about the S=N and N—C bonds. The bisector of the $SF_2$ angle is essentially *cis* to the N—C bond; the thermal average position of the $CF_3$ group is approximately *gauche* with respect to the N—S bond (Figure 1).

By their reactions with $NaN(SiMe_3)_2$, the chlorides $C_6F_5SCl$, $C_6F_5S(O)Cl$, and $C_6F_5SO_2Cl$ have been converted[98] into $C_6F_5SN(SiMe_3)_2$, $C_6F_5S(O)N(SiMe_3)_2$, and $Na^+[C_6F_5SO_2]^-$, respectively. In this instance the reactivity pattern of the pentafluorophenyl compounds does not differ from that of similar phenyl derivatives. Fluoro(trifluoromethyl)disulphane results from the reaction of gaseous chloro(trifluoromethyl)disulphane with active KF. The new compound was

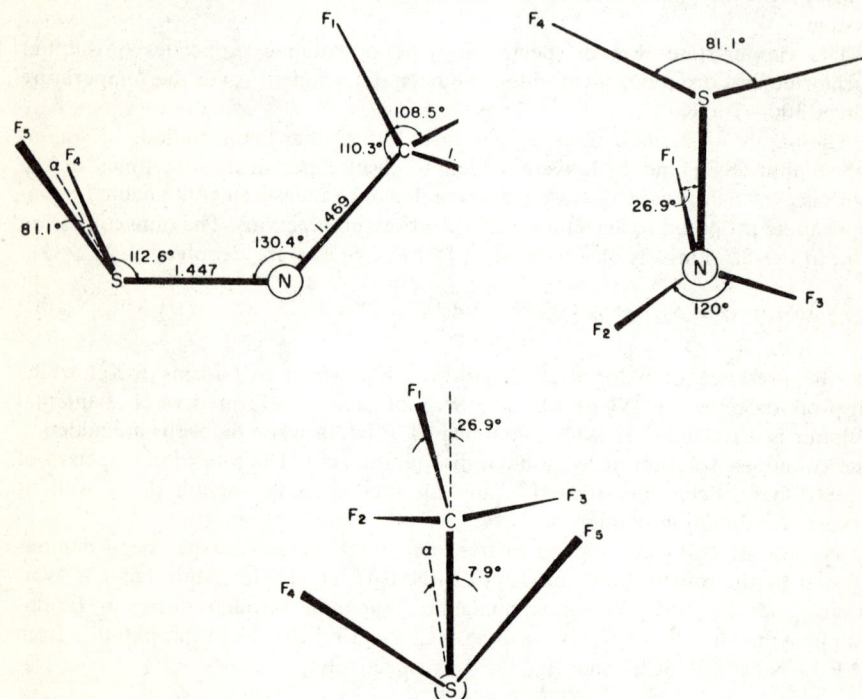

**Figure 1** *Projected structures for* $F_3CNSF_2$: (a) *side view*; (b) *projected view along the CN bond*; (c) *projected view along the SN bond*
(Reproduced by permission from *Inorg. Chem.*, 1975, **14**, 1859)

[97] R. R. Karl and S. H. Bauer, *Inorg. Chem.*, 1975, **14**, 1859.
[98] D. Rinne and A. Blaschette, *Z. Naturforsch.*, 1975, **30b**, 323.

characterized[99] by n.m.r., i.r., and mass spectra and by its volatility; a new synthesis for $CF_3SSCl$ was also reported.

**Sulphur–Oxygen–Halogen Compounds.**—Ebulliometry of sulphuryl chloride–thionyl chloride mixtures has been carried out at atmospheric pressure.[100] The system can be considered as being ideal. The weak complex species $SO_2X^-$, $SOCl_2X^-$, and $SO_2Cl_2X^-$ (X = Cl, Br, or I) have been studied[101] in acetonitrile, DMSO, and water (for $SO_2X^-$ only). It appears that solvation of the halide ions by DMSO is important in deciding the stabilities of the different species. The reactions of $H_2S_2$, $H_2S_5$, and crude sulphane with thionyl chloride at low temperatures have been reported.[102] Of particular interest is the formation of $S_8O$ from $SOCl_2$ and $H_2S_x$ at $-40\,°C$ in $CS_2$–$Me_2O$. The solubility, density, stability, mass spectrum, and some simple reactions of $S_8O$ were also described. Experiments on the oxidative addition of thionyl halides to transition-metal complexes have shown[103] that the complexes $[(Ph_3P)_4Pd]$ and $[(Ph_3P)_2Ni(NO)Cl]$ react with $SOX_2$ to give $trans$-$[(Ph_3P)_2PdX_2]$ (X = Cl or Br) and $[(Ph_3PX)(NiX_3)]$, respectively. In contrast, $SOCl_2$ is added oxidatively by $[(Ph_3P)_2Ir(CO)Cl]$ to give the complex $[(Ph_3P)_2Ir(CO)(Cl_2)SOCl]$.

Three different methods for the preparation of peroxydisulphuryl difluoride have been described. In the first preparation, $S_2O_6F_2$ was prepared[104] by mixing $HSO_3F$ and $O_2^+[AsF_6]^-$ at low temperatures. The latter compound was prepared photochemically from $AsF_5$, $O_2$, and $F_2$. The second preparation involves the quantitative reaction of $CrF_5$ with $SO_3$ according to the equation:

$$CrF_5 + 5SO_3 \rightarrow S_2O_6F_2 + Cr(SO_3F)_3 \qquad (18)$$

Pure $S_2O_6F_2$ may be removed from the reaction mixture at $-22\,°C$, and the other reaction product, $Cr(SO_3F)_3$, which had not been isolated previously, was also studied in some detail.[105] The third paper[106] describes the preparation of peroxydisulphuryl difluoride, fluorine fluorosulphate, and peroxysulphuryl difluoride by the photochemical reaction of fluorine and $SO_3$:

$$2SO_3 + F_2 \rightarrow S_2O_6F_2 \qquad (19)$$

fluorine and $S_2O_6F_2$:

$$S_2O_6F_2 + F_2 \rightarrow 2SO_3F_2 \qquad (20)$$

and $OF_2$ and $SO_3$, respectively:

$$SO_3 + OF_2 \rightarrow FSO_4F \qquad (21)$$

$FSO_4F$ is a strong oxidizing agent and caused *explosions* when being warmed rapidly from $-196$ to $-20\,°C$, and when it reacted with organic matter.

---

[99] W. Gombler, *Z. anorg. Chem.*, 1975, **416**, 235.
[100] M. Aubry and F. Guiraud, *Bull. Soc. chim. France*, 1974, 1857.
[101] S. B. Salama and S. Wasif, *J.C.S. Dalton*, 1975, 151.
[102] R. Steudel and M. Rebsch, *Z. anorg. Chem.*, 1975, **413**, 252.
[103] G. Schmid and G. Ritter, *Z. anorg. Chem.*, 1975, **415**, 97.
[104] A. Smalc, *Vestn. Slov. Kem. Drus.*, 1974, **21**, 5.
[105] S. D. Brown and G. L. Gard, *Inorg. Nuclear Chem. Letters*, 1975, **11**, 19.
[106] M. Gambaruto, J. E. Sicre, and H. J. Schumacher, *J. Fluorine Chem.*, 1975, **5**, 175.

Specific conductance, i.r., and $^1$H n.m.r. spectra of $HN(SO_2Cl)_2$ have been reported.[107] It behaves as a 1:1 electrolyte in $PhNO_2$ and forms a 1:1 adduct with pyridine; i.r. spectra of the latter indicate the presence of $pyH^+$ and $[N(SO_2Cl)_2]^-$ ions. Further data was obtained to support the autoionization:

$$2HN(SO_2Cl)_2 \rightleftharpoons [H_2N(SO_2Cl)_2]^+ + [N(SO_2Cl)_2]^- \quad (22)$$

$HNSOF_2$ has been shown[108] to react with metal carbonates to form the salts $M^+NSOF_2^-$, where M = Li, Cs, or Ag. The reaction of $Ag^+[NS(O)F_2]^-$ with the phosphoryl halides $OPF_2Cl$, $OPFCl_2$, and $OPCl_3$ has been shown[109] to give a series of phosphorylsulphuroxide fluoride imides, whilst reaction with $SPCl_3$ yields the thiophosphoryl halides $SPCl_{3-n}(NSOF_2)_n$. Decomposition of $SP(NSOF_2)_3$ takes place according to the equation:

$$SP(NSOF_2)_3 \rightarrow SPF(NSOF_2)_2 + 1/3(NSOF)_3 \quad (23)$$

The reactions of $FSO_2NCO$ with some metal oxides proceed[110] as shown in the following equations:

$$HgO + FSO_2NCO \rightarrow HgNSO_2F, MeCN + CO_2 \quad (24)$$

$$HgO + 2FSO_2NCO \rightarrow Hg[N(SO_2F)C(O)NSO_2F], MeCN + CO_2 \quad (25)$$

$$Ag_2O + 2FSO_2NCO \rightarrow Ag_2[N(SO_2F)C(O)NSO_2F], xMeCN + CO_2 \quad (26)$$

It is thought that unsolvated $HgNSO_2F$ may exist as a chain polymer, with a backbone (8) formed of linear N—Hg—N bonds.

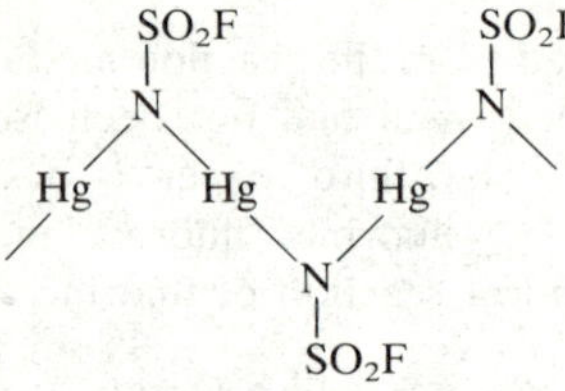

(8)

A series of imidosulphuryl difluorides has been prepared[111] from the reaction of $Hg(NSOF_2)_2$ and the corresponding halide:

$$ZCl_n + \frac{n}{2}Hg(NSOF_2)_2 \longrightarrow Z(NSOF_2)_n + \frac{n}{2}HgCl_2 \quad (27)$$

**Sulphur–Nitrogen Compounds.**—*Linear Compounds.* The crystal structures of the polymeric metal $(SN)_x$, polythiazyl, prepared by the solid-state, room-temperature polymerization of $S_2N_2$ over 3 days, followed by heating at 75 °C for 2 h, and of $S_2N_2$ have been determined.[112] Polythiazyl consists of an almost planar chain of

[107] R. C. Paul, P. Kapoor, R. Kapoor, and R. D. Verma, *Indian J. Chem.*, 1975, **13**, 619.
[108] M. Feser, R. Höfer, and O. Glemser, *Z. Naturforsch.*, 1974, **29b**, 716.
[109] M. Feser, R. Höfer, and O. Glemser, *Z. Naturforsch.*, 1975, **30b**, 327.
[110] R. E. Noftle and J. Crews, *Inorg. Chem.*, 1974, **13**, 3031.
[111] C. Jäckh, A. Roland, and W. Sundermeyer, *Chem. Ber.*, 1975, **108**, 2580.
[112] A. G. MacDiarmid, C. M. Mikulski, P. J. Russo, M. S. Saran, A. F. Garito, and A. J. Heeger, *J.C.S. Chem. Comm.*, 1975, 476.

alternating sulphur and nitrogen atoms, with an electrical conductivity along the chain comparable to that of Hg. The refined structure ($R = 0.11$) gave the following parameters for the $(SN)_x$ chain (9). S(a)—N(a) = 1.593, S(a)—N(b) = 1.628, S(a)—S(b) = 2.789, N(a)—N(b) = 2.579, and S(a)—N(c) = 2.864 Å: bond

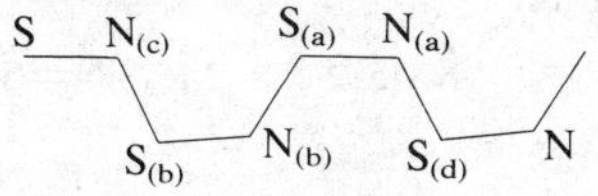

$$\begin{array}{c} S \quad N_{(c)} \quad S_{(a)} \quad N_{(a)} \\ S_{(b)} \quad N_{(b)} \quad S_{(d)} \quad N \end{array}$$

(9)

angles SNS = 119.9 and NSN = 106.2°. These $X$-ray diffraction results are markedly different from those obtained in an earlier electron-diffraction study. The molecule $S_2N_2$ is square planar, with essentially equal S—N bond lengths (1.651 and 1.657 Å), and with SNS and NSN bond angles of 90.42 and 89.58°, respectively.

The i.r. spectrum and force fields of matrix-isolated cis-thionylimide, HNSO, have been studied.[113] Photolysis of cis-HNSO leads to the formation of four products; these have been identified,[114] and the following reaction scheme was proposed for the photolytically induced process: cis-HNSO → cis-HOSN → cis-HSNO + trans-HSNO → SNO. It has also been shown[115] that trans-HNSO may be obtained by the low-energy (>3000 Å) photolysis of cis-HNSO in an argon matrix; i.r. spectra of the two isomers were compared and a further photolysis product was thought to be the radical NSO.

The unstable compound silylsulphinylamine, $SiH_3NSO$, has been prepared[116] by the reaction of $Me_3GeNSO$ or $Bu_3SnNSO$ with $SiH_3Br$, or of $(H_3Si)_3N$ with $SOCl_2$. Electron-diffraction studies show that the skeleton is bent at nitrogen and at sulphur, has an unusually long Si—N bond, and is non-planar.

The compound $S(=NSiMe_3)_3$ has been shown[117] to react with trifluoroacetic anhydride to form (10) and (11), which are nitrogen analogues of sulphur trioxide.

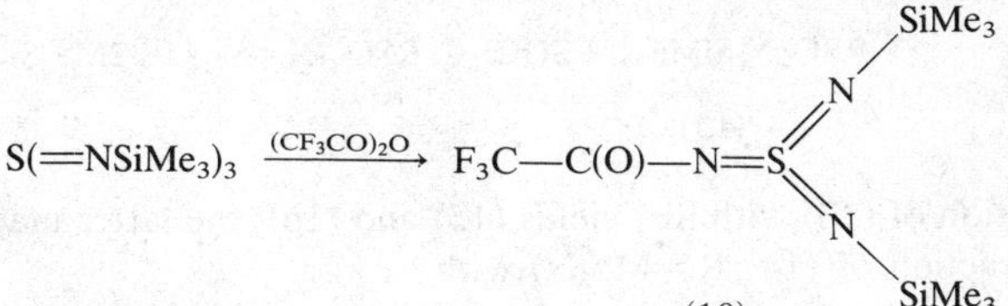

$$S(=NSiMe_3)_3 \xrightarrow{(CF_3CO)_2O} F_3C-C(O)-N=S\begin{array}{c} N-SiMe_3 \\ N-SiMe_3 \end{array}$$ (28)

(10)

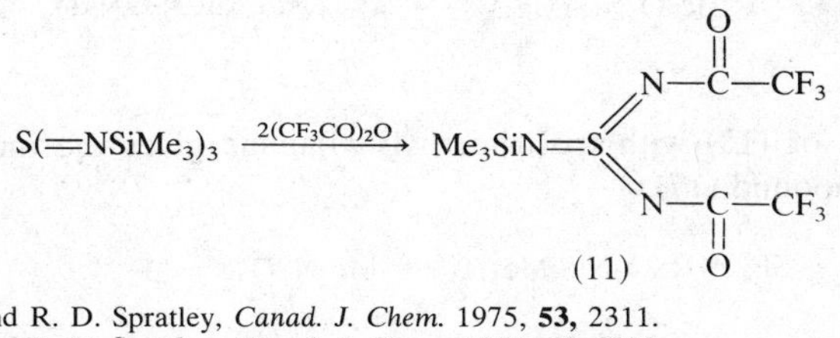

$$S(=NSiMe_3)_3 \xrightarrow{2(CF_3CO)_2O} Me_3SiN=S\begin{array}{c} N-C(O)-CF_3 \\ N-C(O)-CF_3 \end{array}$$ (29)

(11)

[113] P. O. Tchir and R. D. Spratley, *Canad. J. Chem.* 1975, **53**, 2311.
[114] P. O. Tchir and R. D. Spratley, *Canad. J. Chem.*, 1975, **53**, 2318.
[115] P. O. Tchir and R. D. Spratley, *Canad. J. Chem.*, 1975, **53**, 2331.
[116] S. Cradock, E. A. V. Ebsworth, G. D. Meikle, and D. W. H. Rankin, *J.C.S. Chem. Comm.*, 1975, 805.
[117] R. Höfer and O. Glemser, *Z. Naturforsch.*, 1975, **30b**, 460.

The nitrogen analogue of sulphuric acid (12) may be prepared by the reaction of $F_2S(=NSiMe_3)_2$ and $(Me_3Sn)_2NMe$.

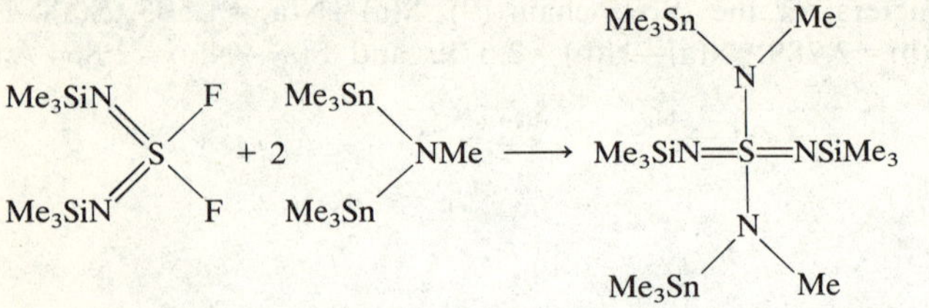

$$Me_3SiN{=}\underset{Me_3SiN}{\overset{F}{\underset{F}{S}}} + 2\ \underset{Me_3Sn}{\overset{Me_3Sn}{N}}{-}NMe \longrightarrow Me_3SiN{=}\underset{\underset{Me_3Sn}{N}{\diagdown}Me}{\overset{\overset{Me_3Sn}{N}{\diagup}Me}{S}}{=}NSiMe_3 \quad (30)$$

(12)

A large number of $N$-phoshoranylidene derivatives have been prepared[118] by the general reaction:

$$Ph_3P + CCl_4 + H_2NSO_2R + Et_3N \rightarrow Ph_3P{=}N{-}SO_2R + HCCl_3 + Et_3NH^+Cl^- \quad (31)$$

$(R = Me, NEt_2, NPr_2, NBu^n_2, OMe, OEt, or F)$

The preparation of $N$-(triphenylphosphoranylidene) sulphamoyl azide, $Ph_3P:NSO_2N_3$, has been described,[119] being by reduction of the hydrazide with nitrous acid:

$$Ph_3P:NSO_2NHNH_2 + NaNO_2 + HCl \rightarrow Ph_3P:NSO_2N_3 + NaCl + 2H_2O \quad (32)$$

or by the reaction of sodium azide with the chloride in methyl ethyl ketone:

$$Ph_3P:NSO_2Cl + NaN_3 \rightarrow Ph_3P:NSO_2N_3 + NaCl \quad (33)$$

Reactions of the azide with phosphites and thiophosphites were investigated. The methods of preparation of silylated amidosulphuric acid derivatives have been described;[120] some of the silylated compounds obtained (13) were treated with thionyl chloride to produce the $N$-sulphinylamides (14).

$$RSO_2N[SiMe_3]_2 + SOCl_2 \rightarrow RSO_2{-}N{=}S{=}O + 2Me_3SiCl \quad (34)$$
$\phantom{RSO_2N[SiMe_3]_2 + SOCl_2 \rightarrow\ }$ (13) $\phantom{+ SOCl_2 \rightarrow\ \ }$ (14)

The reaction of (13) with $SF_4$ yields (15) and (16); the latter may also be obtained by the reaction of (14; $R = Me_2N$) with $SF_4$.

$$SF_4 + 2RSO_2N(SiMe_3)_2 \longrightarrow Me_2NSO_2N{=}S{=}NSO_2$$
$$(15)$$

The reaction of (15) with tris(trimethylsilyl)amine gives the unsymmetrically substituted compound (17).

$$SF_4 + RSO_2N(SiMe_3)_2 \longrightarrow Me_2NSO_2N{=}SF_2$$
$$(16)$$

---

[118] R. Appel and H. Einig, *Z. Naturforsch.*, 1975, **30b**, 134.
[119] D. E. Arrington, *J.C.S. Chem. Comm.*, 1975, 1221.
[120] R. Appel and M. Montenarh, *Chem. Ber.*, 1975, **108**, 2340.

$$Me_2NSO_2N{=}S{=}NSO_2 + N(SiMe_3)_3 \longrightarrow Me_2NSO_2N{=}S{=}NSiMe_3$$
$$(15) \hspace{4cm} (17)$$

Reactions of $R_2NSCl$ with $Hg(SCF_3)_2$ have been shown[121] to result in $R_2NSSCF_3$ (R = Me or Et), which can be cleaved with HX to give $CF_3SSX$ (X = Cl or Br), $CF_3SSCl$ reacts with $EtOC(S)SK$ or $AgOCN$ to yield $EtOC(S)SSSCF_3$ or $CF_3SSNCO$, respectively. The latter hydrolyses to $CF_3SSNHC(O)NHSSCF_3$ and $CO_2$. N-Triorganostannyl sulphinamides of the type (18) have been prepared[122]

$$R^1S{-}NR^2H + Me_2NSnR_3^3 \longrightarrow R^1S{-}N\begin{matrix}R^2\\ \\SnR_3^3\end{matrix} \quad (35)$$
$$\|\hspace{3.5cm}\|$$
$$O \hspace{3.5cm} O$$
$$(18)$$

$$R^1 = p\text{-}MeC_6H_4 \text{ or Me}; \quad R^2 = Me, Pr^i, \text{ or } C_6H_4$$
$$R^3 = Me, Bu_i, \text{ or Ph}$$

by a transamination reaction from $R_3^3SnNMe_2$ and $R^1S(O){-}N(R^2)H$. Thiourea has been shown[123] to react rapidly with nitrous acid in aqueous solution to form an equilibrium concentration of a coloured S-nitroso-compound which is thought to be $(NH_2)_2C{=}S{-}N{=}O$. The kinetics of the much slower reaction to form nitrogen and thiocyanic acid have also been investigated; the mechanism proposed is a rate-determining S-to-N migration of the nitroso-group in the conjugate base of the S-nitroso-compound. Chlorosulphonyl isocyanate has been shown to react[124] with NN-dialkyl- or NN-diaryl-sulphamides to give the previoulsy unknown N-chlorosulphonyl-N'-(dialkyl- and -diaryl-sulphamoyl)ureas (19), which may be hydrolysed to the corresponding N-(dialkyl- and -diaryl-sulphamonyl)ureas.

$$R_2NSO_2{-}NH_2 + ClSO_2{-}NCO \longrightarrow R_2NSO_2{-}NH{-}CO{-}NH{-}SO_2Cl \quad (36)$$
$$(19)$$

$$R = Me, Et, Ph, {-}[CH_2]_5{-}, \text{ or } {-}(CH_2)_2 O(CH_2)_2{-}$$

A crystal-structure determination[125] on the compound $[(Me_2S)_2N]^+Br^-,H_2O$ has shown that the cation possesses a bent S—N—S unit (Figure 2), and approximately tetrahedral angles at sulphur, with average bond lengths and angles: S—N 1.64, S—C 1.80 Å, $\angle$SNS 110.8, $\angle$NSC 102.9°, and $\angle$CSC 99.1°.

Compounds of the type $Ph_3PNSO_2OR$ and $Ph_3PNSO_2NHR$ (R = Me, Et, $Pr^n$, $Bu^n$, or o-, m-, or $p$-$BrC_6H_4$) have been prepared[126] in good yield by the reaction of (triphenylphosphoranylidene)sulphamoyl chloride, $Ph_3PNSO_2Cl$, with alcohols in pyridine or with amines in chloroform; derivatives of secondary amines may also be prepared by the latter method. Heating the methyl ester in pyridine produces

[121] F. Bur-Bur, A. Haas, and W. Klug, Chem. Ber., 1975, **108**, 1365.
[122] E. Wenschuh, W. D. Riedmann, L. Korecz, and K. Burger, Z. anorg. Chem., 1975, **413**, 143.
[123] K. A.-Mallah, P. Collings, and G. Stedman, J.C.S. Dalton, 1974, 2469.
[124] R. Appel and M. Montenarh, Chem. Ber., 1975, **108**, 618.
[125] A. M. Griffin and G. M. Sheldrick, Acta Cryst., 1975, **B31**, 893.
[126] D. E. Arrington, Inorg. Chem., 1975, **14**, 1236.

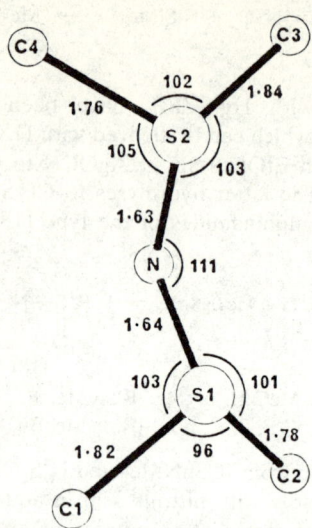

**Figure 2** *The* [Me$_2$SNSMe$_2$] *cation, showing bond lengths and angles* (Reproduced by permission from *Acta Cryst.*, 1975, **B31**, 893)

a compound containing the Ph$_3$PNSO$_2$O$^-$ ion, which gives the free acid on acidification.

*Cyclic Compounds.* The X-ray photoelectron spectra of cyclo-octasulphur and all the known cyclic sulphur imides S$_{8-x}$(NH)$_x$ have been examined[127] with the object of determining, from chemical shifts, the charges on the sulphur and nitrogen atoms. Small positive charges were found on the sulphur atoms adjacent to nitrogen, the other sulphur atoms being practically neutral. Core electron binding energies (/eV) were as follows: N$_{1s}$ = 399.1, S—S—S = 163.9, S—S—N = 164.6, and N—S—N = 164.8.

The reaction of azide ion with cyclo-octasulphur in tris(dimethylamino)-phosphine oxide produces[128] the [S$_4$N]$^-$ ion, whereas Li$_3$N gives predominantly the S$_3^-$ radical ion. The azide reaction was found to be a convenient and direct synthesis of S$_7$NH. Equilibria of the type:

$$[S_7NH]^- \rightleftharpoons [S_4N]^- + 3/8 S_8 \rightleftharpoons \tfrac{1}{2}[S_6N_2]^{2-} + \tfrac{1}{2}S_8 \qquad (37)$$

were invoked to explain the formation and relatively low yields of S$_6$(NH)$_2$ isomers in these syntheses.

The first oxide of a five-membered sulphur–nitrogen ring, N$_2$S$_3$O, has been synthesized[129] by the reaction of a suspension of (20) in CH$_2$Cl$_2$ with an excess of SOF$_2$ at room temperature. N$_2$S$_3$O (21) was isolated as a red liquid which could be distilled under vacuum without decomposition (b.p. 50 °C at 0.01 Torr), and

---

[127] A. Barrie, H. Garcia-Fernandez, H. G. Heal, and R. J. Ramsey, *J. Inorg. Nuclear Chem.*, 1975, **37**, 311.
[128] J. Bojes and T. Chivers, *J.C.S. Dalton*, 1975, 1715.
[129] H. W. Roesky and H. Wiezer, *Angew. Chem. Internat. Edn.*, 1975, **14**, 258.

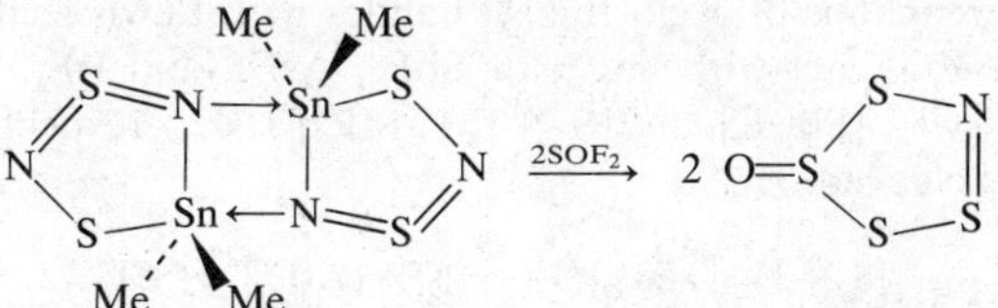

*Elements of Group VI* 359

$$(20) \xrightarrow{2SOF_2} 2 \; (21) \quad (38)$$

which showed no signs of decomposition even after several weeks storage at room temperature.

The new sulphur–nitrogen ring system (22), containing two sulphanuric fluoride and one alkylsulphimide groups, has been prepared[130] by the reation of $Ag^+(NSOF)_2(NSO_2)^-$ with alkyl iodides at room temperature in $CH_2Cl_2$. The structure shown was confirmed against other possible isomers by n.m.r. spectroscopy.

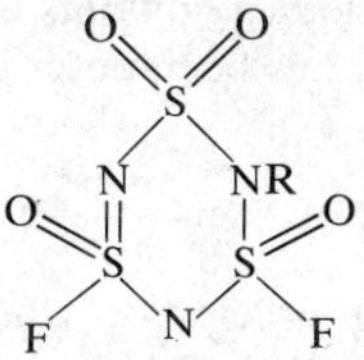

(22)

Sulphanuric fluoride (23) has been shown to react[131] with MeOH and $Me_3N$ to give the $Me_4N^+$ salt of (24). The free acid (24) was obtained by ion exchange and isolated as the monohydrate; the silver salt could be prepared by the reaction of (24) with $Ag_2CO_3$. Thiols react with (23) and $NMe_3$ with reduction to give the $Me_3NH^+$ salt of (25), from which the $Ph_4P^+$ and $Ph_4As^+$ salts were prepared. The $Me_3NH^+$ salt reacts with $PCl_5$ and $PF_5$ to form (26) with X = Cl and F, respectively.

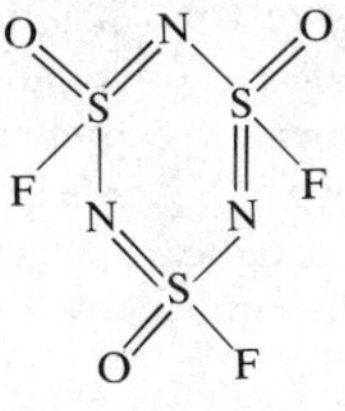

(23) 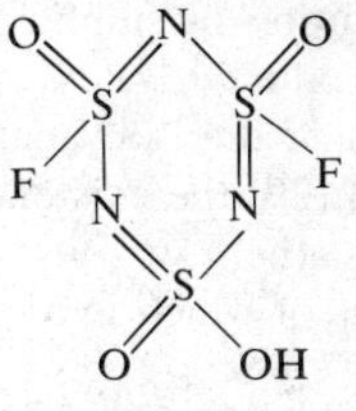  (24)

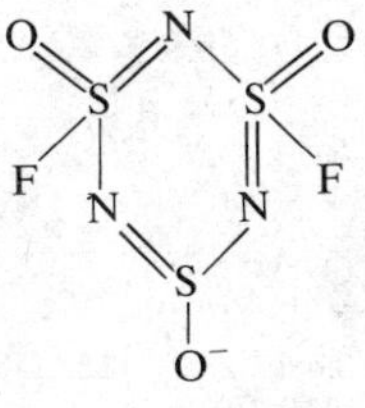

(25) 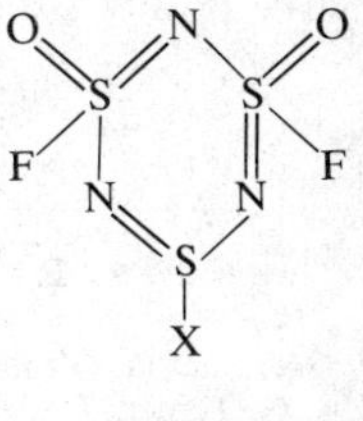  (26)  X = Cl or F

[130] D.-L. Wagner, H. Wagner, and O. Glemser, *Z. Naturforsch.*, 1975, **30b**, 279.
[131] D.-L. Wagner, H. Wagner, and O. Glemser, *Chem. Ber.*, 1975, **108**, 2469.

A study[132] of reactions of cyclo-thiazyl halides with Lewis acids has shown that $(SNCl)_3$ and $(SNF)_3$ react with $SbCl_5$ or $SbF_5$, $AsF_5$, and $BF_3$, to give the ionic compounds $[S_3N_3Cl_2]^+[SbCl_6]^-$ and $[S_3N_3F_2]^+[MF_6]^-$ ($BF_4^-$), respectively. On heating, the $(NS)_3$ ring is cleaved, e.g.

$$[S_3N_3F_2]^+[AsF_6]^- \xrightarrow{85\,°C} [NS]^+[AsF_6]^- + NSF \qquad (39)$$

The reaction of $AsF_5$ in liquid $SO_2$ with the $(NSF)_4$ ring indicates the lower stability of this ring, since the reaction products are a mixture of $[NS]^+[AsF_6]^-$ and $[S_3N_3F_2]^+[AsF_6]^-$.

High-purity solutions of $S_4N_2$ may be prepared[133] by a chromatographic method. The compound was prepared by refluxing $S_4N_4$ and sulphur in PhMe, cooling, removing PhMe with a stream of dry $N_2$, dissolving the $S_4N_2$ in $CCl_4$ and hexanes, filtering, and passing through a column packed with silica gel powder in hexanes. It is claimed that the product contains no $S_8$ nor any other contaminant. A new safe preparation for $S_4N_2$ has been described.[134] The reaction involves the preparation of $Hg(S_7N)_2$ by:

$$S_7NH + Hg(OAc)_2 \xrightarrow[-10\,°C]{MeOH} Hg(S_7N)_2 \qquad (40)$$

The compound is obtained as a precipitate which decomposes over three days under $N_2$ at room temperature to form $S_4N_2$, which was extracted, from the other reaction products [$S_4N_4$ and $(S_7N)_2S_x$], by gel-permeation chromatography.

The energies calculated for $S_4N_4$ by a theoretical study[135] favour a structure with coplanar nitrogen atoms rather than coplanar sulphur atoms. The results show electron delocalization, a bent S—N bond involving pure p-orbitals on the sulphur and nitrogen atoms, and a pure p-bent bond between the sulphur atoms on the same side of the coplanar nitrogen atoms. The solubilities of $S_4N_4$ in some thirteen organic solvents have been measured.[136]

The cyclic addition of sulphonyl isocyanates to $S_4N_4$ has been shown.[137] The monoaddition product (27) is formed without exception and is explained by 1,4-cycloaddition of the isocyanate to $S_4N_4$. During the preparation and thermal decomposition of (27) the existence of the known $S_2N_2^{+\cdot}$ radical species was shown by e.s.r. methods; there was also evidence for a radical species with three active coupling N-atoms of which two are magnetically equivalent.

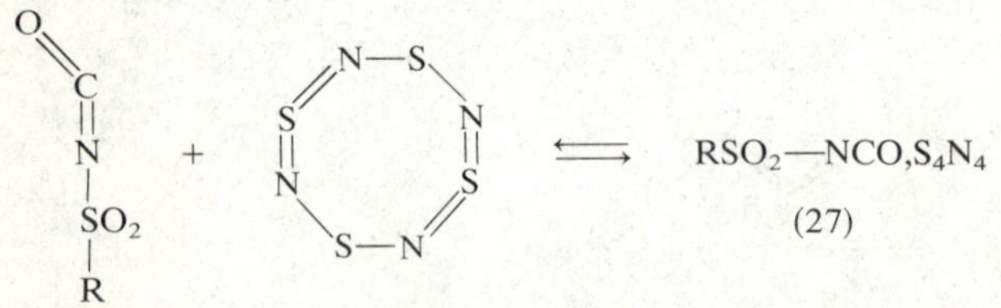

R = F, Cl, or NCO

[132] R. Mews, D.-L. Wagner, and O. Glemser, Z. anorg. Chem., 1975, **412**, 148.
[133] R. R. Adkins and A. G. Turner, J. Chromatog., 1975, **110**, 202.
[134] H. G. Neal and R. J. Ramsey, J. Inorg. Nuclear Chem., 1975, **37**, 286.
[135] M. S. Gopinathan and M. A. Whitehead, Canad. J. Chem., 1975, **53**, 1343.
[136] S. Hamada, Y. Kudo, and M. Kawano, Bull. Chem. Soc. Japan, 1975, **48**, 719.
[137] R. Appel, M. Montenarh, and I. Ruppert, Chem. Ber., 1975, **108**, 582.

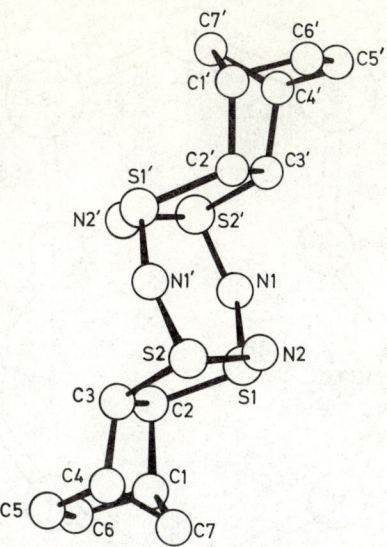

**Figure 3** *One molecule of the adduct* $C_{14}H_{16}N_4S_3$
(Reproduced by permission from *Acta Cryst.*, 1975, **B31**, 895)

The crystal structure of the bis norbornadiene adduct of $S_4N_4$ has been determined.[138] One C=C bond in each diene molecule adds across an S—N—S unit to give a five-membered C—S—N—S—C ring (Figure 3). The adduct retains the eight-membered $(NS)_4$ ring, and possesses a crystallographic centre of symmetry. Some mean bond lengths and angles are: S—N 1.62, S—C 1.84 Å; ∠NSN 118.0; ∠SNS 126.4° (eight-membered ring) and 118.5° (both rings).

The tetrasulphur pentanitride anion has been prepared[139] by a sequence of reactions starting with that of $Me_3CN=SCl_2$ in pentane with $(Me_3Si)_2NLi$ at $-78$ °C to give $Me_3CN=S=NSiMe_3$, which undergoes methanolysis to give $[Me_3CNH_3]^+[S_4N_5]^-$. Of the possible alternatives, structure (28) was preferred since it formally corresponds to an $S_4N_4$ molecule after insertion of an additional N atom into a transannular S—S bond.

(28)

Two compounds with the general formula $S_4N_4O_2$ have been isolated[140] from the reaction between $Me_3SiN=S=NSiMe_3$ and $FSO_2N=S=O$ in $CH_2Cl_2$. An

---

[138] A. M. Griffin and G. M. Sheldrick, *Acta Cryst.*, 1975, **B31**, 895.
[139] O. J. Scherer and G. Wolmershäuser, *Angew. Chem. Internat. Edn.*, 1975, **14**, 485.
[140] H. W. Roesky, W. G. Böwing, I. Rayment, and H. M. M. Shearer, *J.C.S. Chem. Comm.*, 1975, 735.

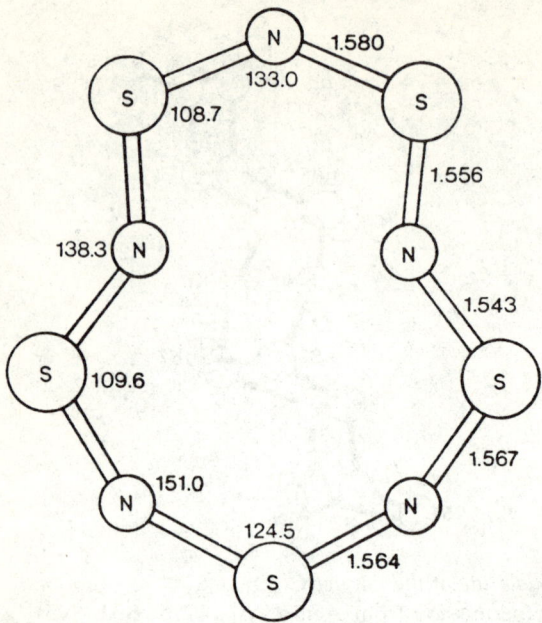

**Figure 4** *The $[S_5N_5]^+$ cation*
(Reproduced from *J.C.S. Chem. Comm.*, 1975, 735)

X-ray crystal structure of the first compound showed it to be ionic, with the formulation $[S_5N_5]^+[S_3N_3O_4]^-$. The shape of the $[S_5N_5]^+$ cation (Figure 4), differs greatly from the heart-shaped ion reported for $[S_5N_5]^+[AlCl_4]^-$. The ring is almost planar, and the bond distances across the ring are the same. The S—N bond lengths are normal for a delocalized S—N system, lying in the range 1.543—1.580 Å. The $[S_3N_3O_4]^-$ anion has not previously been reported; as shown in Figure 5, it contains an $S_3N_3$ ring with bond distances in the di-imide N—S—N

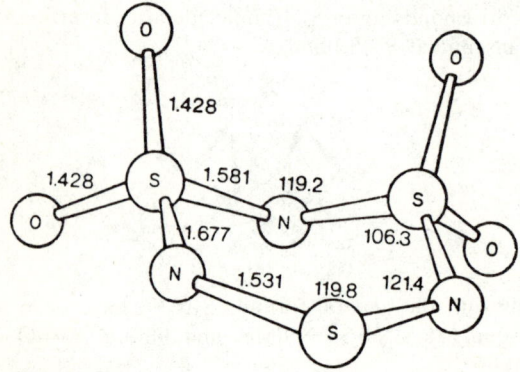

**Figure 5** *The $[S_3N_3O_4]^-$ anion*
(Reproduced from *J.C.S. Chem. Comm.*, 1975, 735)

unit which are very short. This indicates substantial multiple bonding in this part of the ring, whereas the bond joining this unit to the SO$_2$—N—SO$_2$ portion of the ring lie in the range of S—N single bonds.

Thionyl chloride reacts with both S$_7$NH and C$_6$H$_5$HgNS$_7$ to form[141] colourless crystals of (S$_7$N)$_2$SO. The reaction of S$_4$N$_4$ and S$_3$N$_2$Cl$_2$ with Me$_2$Si(NMe$_2$)$_2$ has been show[142] to give Me$_2$Si(NMe$_2$)N:SN–SN(Me)$_2$ and Me$_2$Si(Cl)N:S:N—SNMe$_2$ respectively.

**Other Sulphur-containing Ring Compounds.**—The amines (29) have been shown[143] to react with trithiadiazolium chloride to give the compound (30), which has a five-membered S—N ring as the substituent. In contrast to this, the six-membered-ring compound F$_5$P$_3$N$_3$-N(SnMe$_3$)$_2$ gave, with 2,4,6-trichloro-S-trithiatriazine, the compound (31), for which a horseshoe-like N—S—N—S structure was proposed. The synthesis of the heterocycle (32) has been described.[144] This ring can also act as a substituent to the six-membered F$_5$P$_3$N$_3$. A series of reactions have been described[145] which lead to the formation of (33a), which reacts with (Me$_3$Si)$_2$NMe to form the heterocycle (33b).

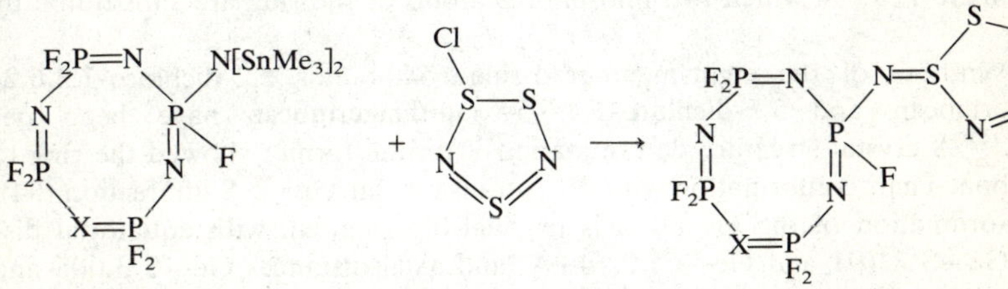

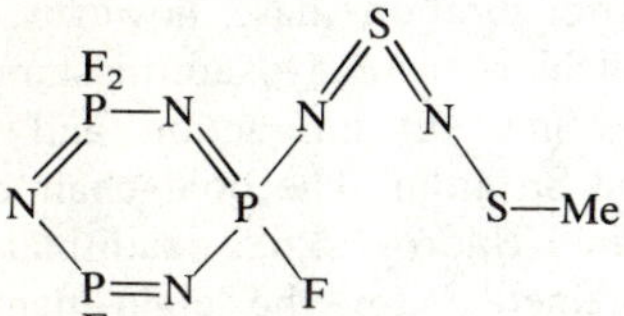

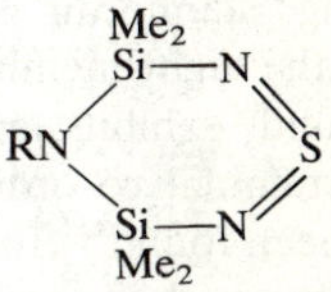

R = Me$_3$Sn, H, or Me$_2$N—SO$_2$

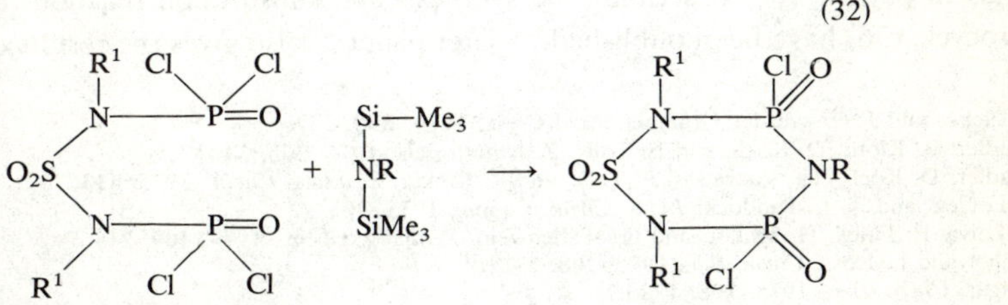

[141] R. Steudel and F. Rose, *Z. Naturforsch.*, 1975, **30b**, 810.
[142] H. W. Roesky, W. Schaper, W. Grosse-Böwing, and M. Dietl, *Z. anorg. Chem.*, 1975, **416**, 306.
[143] H. W. Roesky and E. Janssen, *Chem. Ber.*, 1975, **108**, 2531.
[144] H. W. Roesky and B. Kuhtz, *Chem. Ber.*, 1975, **108**, 2536.
[145] G. Schoning, U. Klingebiel, and O. Glemser, *Chem. Ber.*, 1974, **107**, 3756.

A structure determination[146] of $NPCl_2(NSOF)_2$ at $-160\,°C$ has been carried out in order to give further information on the $p_\pi$–$d_\pi$ bonding and degree of ring planarity of the compound. The six-membered ring has a twisted chair conformation, with oxygen atoms *cis* with respect to the ring plane. The N—S bonds in the S—N—S section of the ring have a mean length of 1.568 Å whilst the mean value of the other N—S bond lengths is 1.527 Å.

A study of the $^{31}P$ n.m.r. spectrum of $\alpha$-$P_4S_3I_2$ has shown[147] the compound to possess the same bicyclic structure in solution as in the solid state.

N.m.r. data have also been used to show that the compound $(C_6H_6PS)_3$ has the heterocyclophosphane structure with a five-membered $P_3S_2$ ring containing isolated sulphur atoms. One of the directly connected phosphorus atoms is linked to another sulphur atom in an *exo*-position, in addition to the phenyl group. The above structure has also been confirmed by a second study.[149]

The reaction between $P_4S_3$ and $As_2S_3$ at 500 °C in a sealed tube has been reported.[150] From the reaction products, the new compound $P_2As_2S_3$ was isolated. According to its properties and vibrational spectra, it has a cage structure, analogous to $P_4S_3$, in which two phosphorus atoms of the ring are substituted by arsenic.

The synthesis of the eight-membered-ring compounds 2,2-dichloro-1,3,6,2-trithiagermocan and 5,5-dichloro-1,4,6,5-oxadithiagermocan have been described.[151] A crystal structure determination[152] on the former showed the ring to have a boat–chair conformation with a 1,5-transannular Ge · · · S interaction (34). The co-ordination of the Ge atom is trigonal-bipyramidal, with equatorial distances Ge—S 2.191 and Ge—Cl 2.149 Å, and axial distances Ge—S 3.005 and Ge—Cl 2.208 Å. An almost identical structure has also been found[153] for the corresponding stannocan. A chair–chair configuration has, however, been found[154] for the eight-membered ring in 5,5-dichloro-1,4,6,5-oxadithiastannocan. The compound exhibits a 1,5-transannular Sn · · · O interaction and has a trigonal-bipyramidal co-ordination around the Sn atom. The boat–chair configuration has been found[155] for the ring system in 5-chloro-1,4,6,5-oxadithiarsocan. Crystal structure data have also been obtained[156] for the eight-membered heterocycle $(Me_2Si)_2S_2N_2$ (35).

A series of papers[157–160] describing the synthesis and substitution reactions of the heterocycle (36) have been published. A later paper[161] also gives the results of

---

[146] P. A. Tucker and J. C. van der Grampel, *Acta Cryst.*, 1974, **B30**, 2795.
[147] M. Baudler, B. Kloth, D. Koch, and E. Tolls, *Z. Naturforsch.*, 1975, **30b**, 340.
[148] M. Baudler, D. Koch, Th. Vakratsas, E. Tolls, and K. Kipker, *Z. anorg. Chem.*, 1975, **413**, 239.
[149] M. R. LeGeyt and N. L. Paddock, *J.C.S. Chem. Comm.*, 1975, 20.
[150] A.-M. Leiva, E. Fluck, H. Muller, and G. Wallenwein, *Z. anorg. Chem.*, 1974, **409**, 215.
[151] M. Dräger and L. Ross, *Chem. Ber.*, 1975, **108**, 1712.
[152] M. Dräger, *Chem. Ber.*, 1975, **108**, 1723.
[153] M. Dräger and R. Engler, *Chem. Ber.*, 1975, **108**, 17.
[154] M. Dräger and R. Engler, *Z. anorg. Chem.*, 1975, **413**, 229.
[155] M. Dräger, *Z. anorg. Chem.*, 1975, **411**, 79.
[156] G. Ertl and J. Weiss, *Z. Naturforsch.*, 1974, **29b**, 803.
[157] W. Schramm, G. Voss, M. Michalik, G. Rembarz, and E. Fischer, *Z. Chem.*, 1975, **15**, 19.
[158] W. Schramm, G. Voss, G. Rembarz, and E. Fischer, *Z. Chem.*, 1974, **14**, 471.
[159] W. Schramm, G. Voss, G. Rembarz, and E. Fischer, *Z. Chem.*, 1975, **15**, 57.
[160] W. Storek, W. Schramm, G. Voss, G. Rembarz, and E. Fischer, *Z. Chem.*, 1975, **15**, 104.
[161] G. Leonhardt. R. Scheibe, W. Schramm, G. Voss, and E. Fischer, *Z. Chem.*, 1975, **15**, 193.

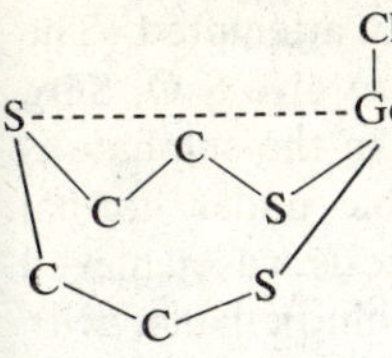

an ESCA study from which the binding energies of the sulphur(IV) and sulphur(II) atoms in the compounds were derived.

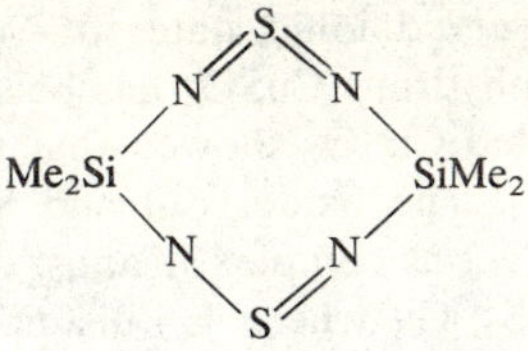

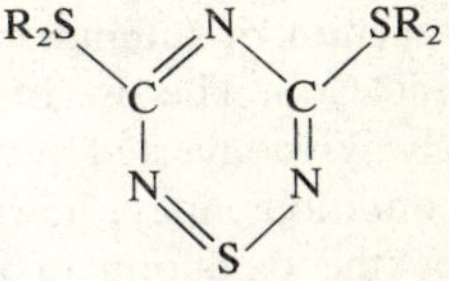

$NN'$-Bis(trimethylsilyl)sulphur(IV) di-imide and bis(trimethylsilylimido)-sulphoxide have been shown[162] to react with chlorosulphonic isocyanate to give the heterocycles (37) and (38), respectively.

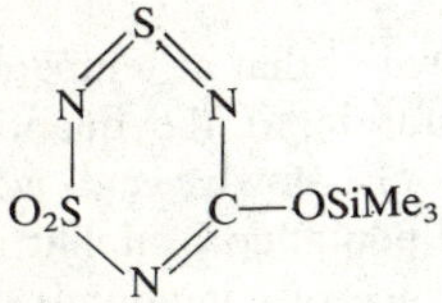

The five-membered heterocyclic ketone (39) has been prepared[163] by the reaction of dimeric 5,5-dimethyl-1,3$\lambda^4$,2,4,5-dithiadiazastannole (40), obtained by the reaction of $S_4N_4$ with tris(trimethylstannyl)amine, with an excess of carbonyl difluoride at room temperature. The cyclic ketone is the first representative of a new class of compounds that have become readily accessible by this method.

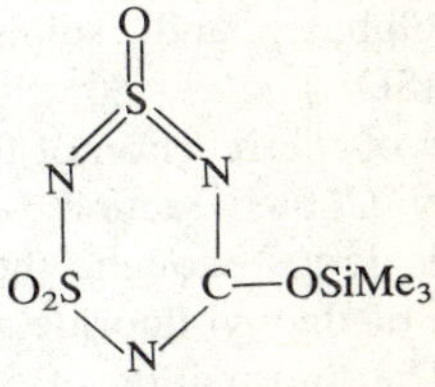

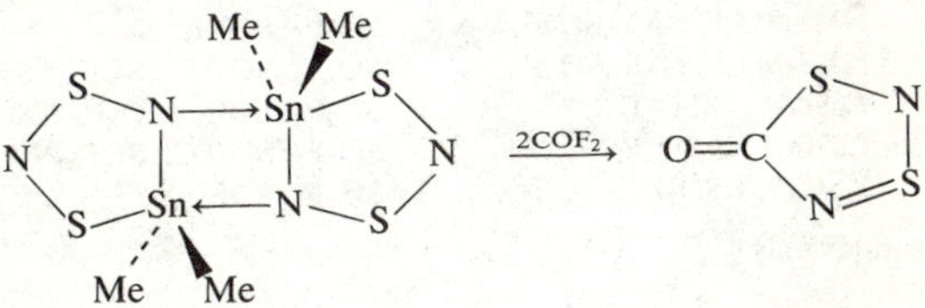

The complexing power of 1,2,3,4-thiatriazole-5-thiolate, $CS_2N_3$, with several metallic cations has been investigated[164] by polarography. The crystal structure of $(BrBS)_3$ has been reinvestigated.[165]

**Sulphur–Oxygen Compounds.**—The i.r. spectra of matrix-isolated $S_2O$ isotopes have been measured.[166] The photoelectron spectra of $S_2O$, $O_3$, and $SO_2$ have been compared, and further experimental details of the spectrum of $S_2O$ presented.[167]

---

[162] R. Appel, H. Uhlenhaut, and M. Montenarh, Z. Naturforsch., 1974, **29b**, 799.
[163] H. W. Roesky and E. Wehner, Angew. Chem. Internat. Edn., 1975, **14**, 498.
[164] E. A. Neves and D. W. Franco, J. Inorg. Nuclear Chem., 1975, **37**, 277.
[165] W. Schwarz, H.-D. Hausen, and H. Hess, Z. Naturforsch., 1974, **29b**, 596.
[166] A. G. Hopkins, F. P. Daly, and C. W. Brown, J. Phys. Chem., 1975, **79**, 1849.
[167] P. Rosmus, P. D. Dacre, B. Solouki, and H. Bock, Theor. Chim. Acta, 1974, **35**, 129.

An assignment of the observed ionic states of $S_2O$ was also attempted. The reaction of sulphur with anhydrous $CuSO_4$ has been shown[168] to give $S_2O$, $SO_2$, and CuS. The use of labelled $CuSO_4$ showed that the sulphur in the sulphate is always converted into $SO_2$. The oxides SO and $S_2O_2$, unstable under normal conditions, may, however, act as complex ligands.[169] The synthesis was achieved by the oxidation of $[IrL_2(S_2)]Cl$, where L = bis(diphenylphosphino)ethane, with sodium periodate. The reaction gave the two compounds cis-$[IrL_2(S_2O_2)]Cl$ and trans-$[IrL_2(SO)_2]Cl$; the structure of the former was established by X-ray diffraction, and that of the latter by i.r. spectra.

The vibrational spectrum of $SO_2$ in various matrices has been measured.[170] Sulphur dioxide has been[171] photochemically oxidized to $SO_3^-$ on the surface of MgO in the presence of water vapour, oxygen, or $N_2O$. The complex that gives rise to the transient yellow colour which is formed on mixing equimolar solutions of thiosulphate and sulphur dioxide has been identified[172] as being $Na_2[S_2O_3(SO)_2]$.

A study of the reaction of fluorine with $SO_2$ has shown[173] that at temperatures up to 500 °C the reaction rapidly gives sulphuryl fluoride. If the quantity of fluorine is large enough, then $SF_6$ is formed, but only slowly, and via the formation of thionyl fluoride, thionyl tetrafluoride, and pentafluorosulphur hypofluorite. The fluorination of $SO_2$ by $XeF_2$, in the presence of compounds of the type MX (M = $NMe_4$, Cs, or K: X = F or Cl), has been studied.[174] A variety of products were observed (Table 2) and reaction mechanisms were proposed in which the $XeF_2$ functions as a weak Lewis acid.

**Table 2** Products of the reactions $MX + XeF_2 + SO_2$

| MX | Solid products | Gaseous products |
| --- | --- | --- |
| $NMe_4F$ | $NMe_4SO_2F$; $NMe_4SO_3F$ | $SO_2F_2$; $SOF_2$;[a] Xe |
| $NMe_4Cl$ | $NMe_4SO_2F$; $NMe_4SO_3F$ | $SO_2F_2$; $SO_2FCl$; $SOF_2$;[a] $S_2O_5F_2$;[a] Xe |
| CsF | $CsSO_3F$; $CsSO_2F$ | $SO_2F_2$; $SOF_2$; $S_2O_5F_2$;[a] Xe |
| CsCl | $CsSO_3F$; $CsSO_2F$ | $SO_2F_2$; $SO_2FCl$; $SOF_2$; $S_2O_5F_2$;[a] Xe |
| KF | $KSO_3F$; $KSO_2F$ | $SO_2F_2$; $SOF_2$; Xe |

[a] Trace quantities only.

The crystal structure of the 1:2 complex formed between $NNN'N'$-tetramethyl-p-phenylenediamine and $SO_2$ has been determined.[175] The structure consists of discrete molecular complexes containing one TMPD molecule and two $SO_2$ molecules attached through two N—S bonds of equal lengths (2.340 and 2.337 Å). The S—O distances are essentially the same as those found in solid and gaseous $SO_2$, but the O—S—O angles are considerably contracted.

Three papers have been published on the $NH_3$–$SO_2$–$H_2O$ system. The enthalpy

---

[168] K. R. Nair and A. R. V. Murthy, J. Indian Inst. Sci., 1974, **56**, 302.
[169] G. Schmid and G. Ritter, Chem. Ber., 1975, **108**, 3008.
[170] D. Maillard, M. Allavena, and J. P. Perchard, Spectrochim. Acta, 1975, **31A**, 1523.
[171] M. J. Lin and J. H. Lunsford, J. Phys. Chem., 1975, **79**, 892.
[172] A. F. Ryabinina and V. A. Oshman, Russ. J. Phys. Chem., 1975, **49**, 154.
[173] A. Vanderchmitt, Report CEA-R-4613, 1974, France.
[174] I. L. Wilson, J. Fluorine Chem., 1975, **5**, 13.
[175] J. D. Childs, D. van der Helm, and S. D. Christian, Inorg. Chem., 1975, **14**, 1386.

and entropy of the reaction have been determined,[176] and the temperature dependence of the equilibrium constant of the formation of $(NH_4)_2SO_3$ has been calculated from available experimental data.[177] The third study[178] investigated the solid-state reaction by warming a low-temperature matrix of $SO_2$ with excess $NH_3$. An adduct $(NH_3)_2,SO_2$ was observed at $-90\,°C$ which decomposed when the temperature rose to $-50\,°C$ to $NH_3,SO_2$. With excess $SO_2$ only the latter appears at $-150\,°C$, and with $H_2O$ in the system it is converted at $-80\,°C$ into ammonium sulphite, which is stable to room temperature.

The equilibrium:
$$2La_2O_3 + 3SO_2 \rightleftharpoons 2La_2O_2SO_4 + \tfrac{1}{2}S_2 \qquad (44)$$

has been studied[179] in the range 700—1200 °C. The enthalpy and entropy of the reaction, and the previously unknown values of $\Delta H°_{298}$ and $\Delta S°_{298}$ for the compound $La_2O_2SO_4$, were calculated. The oxidation of sodium and potassium by sulphur dioxide has been studied[180] in the temperature ranges 110—200 and 25—130 °C, respectively. From the data obtained it could be deduced that the oxidation of Na gave sodium sulphide whilst that of potassium gave potassium dithionate. The reactivity of $SO_2$ twoards commercial limestones has been investigated.[181]

The Raman spectrum of matrix-isolated $SO_3$ has been measured,[182] along with spectra of pure gaseous $SO_3$. The results give an unambiguous confirmation of the assignments made previously by Kaldor *et al.* (A. Kaldor, A. G. Maki, A. J. Dorney, and I. M. Mills, *J. Mol. Spectroscopy*, 1973, **45**, 247).

Under a moderate temperature increase the reaction of $SO_3$ and some chlorides or chlorosulphates has been shown[183] to give condensed polysulphates and $S_2O_5Cl_2$. On increasing the temperature, the degree of condensation of the polysulphates decreases, with disulphate or even sulphate being formed with $SO_2Cl_2$. The liquid–solid equilibria in the system $B_2O_3$–$SO_3$–$H_2O$ have been studied at 25 °C, and several compounds were isolated and characterized by X-ray diffraction.[184]

The structure of $Sb_2O_3,2SO_3$ has been determined.[185] The structure consists of molecular units built of two $SO_4$ tetrahedra and two $SbO_3$ pyramids sharing corners.

*Sulphates.* The $BaSO_4$–$SrSO_4$ system has been studied[186] by the preparation of mixed crystals. Coprecipitation and heat-treatment methods were compared, and it was found that the samples prepared by heat treatment were more homogeneous. I.r. frequency shifts of the sulphate absorption bands were studied as a function of composition, with the $v_1(SO_4^{2-})$ vibration being found to be the least

---

[176] R. Landreth, R. G. de Pena, and J. Heicklen, *J. Phys. Chem.*, 1975, **79**, 1785.
[177] J. Czarnecki, J. Haber, J. Pawlikowska-Czubak, and A. Pomianowski, *Z. anorg. Chem.*, 1974, **410**, 213.
[178] I. C. Hisatsune and J. Heicklen, *Canad. J. Chem.*, 1975, **53**, 2646.
[179] A. A. Grizik, I. G. Abdullina, and N. M. Garifdzhanova, *Russ. J. Inorg. Chem.*, 1974, **19**, 1412.
[180] Ph. Touzain, M.-F. Ayedi, and J. Besson, *Bull. Soc. chim. France*, 1974, 421.
[181] M. Hartman, *Coll. Czech. Chem. Comm.*, 1975, **40**, 1466.
[182] S.-Y. Tana and C. W. Brown, *J. Raman Spectroscopy*, 1975, **3**, 387.
[183] E. Puskaric, R. De Jaeger, and J. Heubel, *Rev. Chim. minérale*, 1975, **12**, 374.
[184] I. Nowdjavan, P. Vitse, and J. Potier, *Compt. rend.*, 1975, **280**, C, 755.
[185] R. Mercier, J. Douglade, and F. Theobald, *Acta Cryst.*, 1975, **B31**, 2081.
[186] I. Krivy, J. Moravec, and O. Vojtech, *Coll. Czech. Chem. Comm.*, 1974, **39**, 3603.

sensitive to lack of homogeneity. A similar study of the $BaSO_4$–$CaSO_4$ system has also been carried out.[187] The vibrational spectra of single crystals of ammonium sulphate[188] and of potassium sulphate[189] have been studied in detail.

Several studies of the physical properties of sulphates have been carried out; these will not be treated in detail but are listed as follows: a determination of the dissociation constants of some univalent sulphate ion-pairs,[190] the electrostriction of ammonium sulphate,[191] dielectric and n.m.r. investigations of phase transitions in lithium ammonium sulphate,[192] the surface structure of barium sulphate crystals in aqueous solution,[193] optical activity and the electro-optical effect in crystals of $Cd_2(NH_4)_2(SO_4)_3$,[194] apparent molal volumes and heat capacities of $Na_2SO_4$, $K_2SO_4$, and $MgSO_4$ in water,[195] and densities, heats of fusion, and refractive indices of double sulphates of univalent metals.[196]

Following the practice of previous years, the studies of sulphate phase diagrams are presented in tabular form (Table 3[197—205]).

**Table 3** Sulphate phase systems that have been studied

| System | Reference |
|---|---|
| $BeSO_4$–$Li_2SO_4$ | 197 |
| Cs, Na‖$Cl^-$, $SO_4^{2-}$, $H_2O$ | 198 |
| K, Rb, Sr‖$SO_4^{2-}$ | 199 |
| Cs, Rb, Sr‖$SO_4^{2-}$ | 200 |
| $MgSO_4$–thiourea–$H_2O$ | 201 |
| $CdSO_4$, $MSO_4$, $H_2O$ | 202 |
| $MnSO_4$–$CaSO_4$–$H_2O$ | 203 |
| Li, Rb,‖$Cl^-$, $SO_4^{2-}$ | 204 |
| $(NH_4)_2SO_4$–$Cs_2SO_4$–$H_2O$ | 205 |

Analysis of the water evolved in the thermal decomposition of calcium sulphate hydrates has shown[206] that below $p_{H_2O} = 2.6 \times 10^3$ N m$^{-2}$, $CaSO_4,\frac{1}{2}H_2O$ has a lower hydration temperature that $CaSO_4,2H_2O$, and two-stage dehydration of the latter

---

[187] J. Moravec, V. Sara, and O. Vojtech, *Coll. Czech. Chem. Comm.*, 1975, **40**, 815.
[188] P. Vankateswarlu, H. D. Bist, and Y. S. Jain, *J. Raman Spectroscopy*, 1975, **3**, 143.
[189] F. Meserole, J. C. Decius, and R. E. Carlson, *Spectrochim. Acta*, 1974, **30A**, 2179.
[190] E. J. Readon, *J. Phys Chem.*, 1975, **79**, 422.
[191] M. P. Zaitseva and G. P. Rozhnova, *Soviet Phys. Cryst.*, 1975, **20**, 105.
[192] V. I. Yuzvak, L. I. Zherebtsova, V. B. Shkuryaeva, and I. P. Aleksandrova, *Soviet Phys. Cryst.*, 1975, **19**, 480.
[193] I. V. Melikhov, B. D. Nebylitsyn, and V. N. Rudin, *Soviet Phys. Cryst.*, 1975, **19**, 521.
[194] N. R. Ivanov and C. Koniak, *Soviet Phys. Cryst.*, 1975, **19**, 755.
[195] G. Perron, J. E. Desnoyers, and F. J. Millero, *Canad. J. Chem.*, 1975, **53**, 1134.
[196] G. A. Bukhalov, E. L. Kozachenko, D. V. Sementsova, and V. V. Keropyan, *Russ. J. Inorg. Chem.*, 1975, **20**, 343.
[197] V. G. Vasilev, V. S. Markov, E. G. Teterin, and O. N. Vtkina, *Russ. J. Inorg. Chem.*, 1974, **19**, 871.
[198] I. F. Poletaev, V. E. Plyushchev, and A. P. Lyudomirskaya, *Russ. J. Inorg. Chem.*, 1975, **19**, 926.
[199] N. V. Maksina, N. N. Polivanova, and N. A. Finkelshtein, *Russ. J. Inorg. Chem.*, 1974, **19**, 1053.
[200] N. V. Maksina, N. N. Polivanova, and N. A. Finkelshtein, *Russ. J. Inorg. Chem.*, 1974, **19**, 1053.
[201] V. G. Skvortsov, *Russ. J. Inorg. Chem.*, 1974, **19**, 1089.
[202] A. S. Karnaukhov, T. P. Fedorenko, M. I. Vaisfeld, M. K. Onishchenko, and V. G. Shevchuk, *Russ. J. Inorg. Chem.*, 1974, **19**, 1086.
[203] B. I. Zhelnin and G. I. Gorshtein, *Russ. J. Inorg. Chem.*, 1975, **20**, 142.
[204] E. A. Akopov and Zh. G. Moiseenko, *Russ. J. Inorg. Chem.*, 1975, **20**, 283.
[205] J. Balej and V. G. Sevcuk, *Coll. Czech. Chem. Comm.*, 1974, **39**, 3423.
[206] T. P. Lees, *Proc. Soc. Analyt. Chem.*, 1974, **11**, 94.

is not possible. $\Delta H$ Values for both hydrates were calculated to be 38.5 kJ mol$^{-1}$ for $\frac{1}{2}$H$_2$O and 343 or 259 kJ mol$^{-1}$ for 2H$_2$O. The thermolysis of manganese hydrogenosulphate has been shown[207] to give sulphate and H$_2$SO$_4$, which is in accordance with thermodynamic calculations. The dehydration processes of ZnSO$_4$,7H$_2$O and NiSO$_4$,6H$_2$O have been shown[208] to follow a very similar sequence of reactions. Thermal decomposition studies[209] on basic cadmium sulphates have shown that the intermediate product of the thermolysis is the dioxy salt CdSO$_4$,2CdO, for which crystal data were given. The compositions of the pure basic sulphates of copper and cobalt, obtained at 20 °C by addition of NaOH to CuSO$_4$ and CoSO$_4$ solutions, have been determined.[210]

Chlorine fluorosulphate has been shown[211] to react with HCl at 20 °C to give Cl$_2$ and pure HOSO$_2$F, and with N$_2$O$_4$ to give NO$_2$OSO$_2$F and ClNO$_2$. Reaction with CO gave ClC(O)OSO$_2$F, which reacted with additional ClOSO$_2$F to give S$_2$O$_5$F$_2$. Lead(IV) fluorosulphate, which is surprisingly stable at room temperature, has been prepared[212] in quantitative yield by the reaction:

$$Pb(CF_3COO)_4 + 4HSO_3F = Pb(SO_3F)_4 + 4CF_3CO_2H \qquad (45)$$

The compound decomposes slowly at 150 °C and rapidly at 210 °C to yield lead(II) fluorosulphate and probably S$_2$O$_5$F$_2$ and O$_2$. Vibrational spectra indicated a polymeric structure with distorted octahedral configuration of fluorosulphate groups about the central lead atom, a structure similar to that of the analogous tin compound.

Three studies on the oxidation of sulphite have been published. A kinetic expression for the oxidation of solid sodium sulphite by oxygen was the result of the first study,[213] whilst the effects of copper ions[214] and ferric ions[215] on the autoxidation of sulphite were the subject of the remaining papers.

*Thiosulphates and Thionates.* A neutron-diffraction study[216] has shown that in barium thiosulphate monohydrate the tetrahedral S$_2$O$_3^{2-}$ anion has the bond lengths S—S = 1.979 and S—O = 1.472–1.483 Å; the Ba—S distances are 3.355 and 3.424 Å. A crystal structure determination of bis(ethylenethiourea)zinc(II) thiosulphate has shown[217] that each zinc atom is tetrahedrally surrounded by three sulphur atoms (two from etu and one from the thiosulphate group, with the mean distance Zn—S = 2.320 Å) and one oxygen atom from the S$_2$O$_3$ group. Crystalline Zn(NH$_3$)$_3$(S$_2$O$_3$),H$_2$O and Cd(NH$_3$)$_3$(S$_2$O$_3$) have been prepared[218] and characterized; spectral data indicate that the S$_2$O$_3^{2-}$ ion is unidentate and is bonded to the metal through sulphur.

[207] G. Palavit and S. Noel, *Bull. Soc. chim. France*, 1975, 1040.
[208] G. Rabbering, J. Wanrooy, and A. Schuijff, *Thermochim. Acta*, 1975, **12**, 57.
[209] L. Walter-Levy, D. Groult, and J. W. Visser, *Bull. Soc. chim. France*, 1974, 383.
[210] J. Glibert, *Bull. Soc. Chim. France*, 1975, 459.
[211] A. V. Fokin, A. D. Nikolaeva, Yu. N. Studnev, A. I. Rapkin, N. A. Proshin, and L. D. Kuznetsov, *Izvest. Akad. Nauk. S.S.S.R., Ser. khim.*, 1975, 1000.
[212] H. A. Carter, C. A. Milne, and F. Aubke, *J. Inorg. Nuclear Chem.*, 1975, **37**, 282.
[213] J. Marecek, J. Bares, and E. Erdos, *Coll. Czech. Chem. Comm.*, 1975, **40**, 827.
[214] J. Veprek-Siska and S. Lunak, *Z. Naturforsch.*, 1974, **29b**, 689.
[215] J. Veprek-Siska, S. Lunak, and A. El-Wakil, *Z. Naturforsch.*, 1974, **29b**, 812.
[216] L. Manojlović-Muir, *Acta Cryst.*, 1975, **B31**, 135.
[217] S. Baggio, R. F. Baggio, and P. K. De Perazzo, *Acta Cryst.*, 1974, **B30**, 2166.
[218] Z. Gabelica, *Ann. Soc. Sci. Bruxelles, Ser.* 1, 1975, **89**, 149.

The stoicheiometry and kinetics of reaction of aqueous ammonia solutions of copper(II) ions with thiosulphate ion in the presence of oxygen have been examined.[219] The amount of oxygen consumed and the relative amounts of the final sulphur products, namely trithionate and sulphate ions, are dependent on the initial $S_2O_3^{2-}$ concentration and pH. The most active species for $S_3O_6^{2-}$ formation is a tetra-amminecopper(II) complex having one axial $S_2O_3^{2-}$ and one axial $O_2$ ligand. A complex having both axial and equatorial $S_2O_3^{2-}$ ligands as well as an axial $O_2$ was suggested as the reactive intermediate for sulphate formation.

A specific rate constant for the rapid complexation of $Ag^+$ ion by thiosulphate ion in aqueous solution has been determined,[220] using the recently developed laser optical–acoustic technique for measurement of sound absorption. In aqueous solution the predominant complexed species is the 1:2 ion $[Ag(S_2O_3)_2]^{3-}$.

The stoicheiometry of the oxidation of thiosulphate by the tetramminegold(III) ion in acid solution is:

$$[Au(NH_3)_4]^{3+} + 4S_2O_3^{2-} + 4H^+ \rightarrow [Au(S_2O_3)_2]^{3-} + S_4O_6^{2-} + 4NH_4^+ \quad (46)$$

Reaction-kinetic and equilibrium studies of the system have been carried out,[221] and a series of consecutive reactions has been proposed. The cyanolysis and photometric determination of thiosulphate, trithionate, and tetrathionate have been investigated[222] under a variety of conditions. The Raman spectra of $K_2S_2O_6$[223] and $NH_3SO_3$[224] have been measured.

*Sulphuric Acid and Related Compounds.* The vibrational spectra of $H_2SO_4$ and $D_2SO_4$ in the crystalline state have been measured[225] at various temperatures. The cation-transference numbers of Li, Na, K, $NH_3$, Ag, and Ba hydrogen sulphates, and of solutions of water and acetic acid, have been measured[226] in 100% $H_2SO_4$ at 25 °C. The numbers obtained were of the order of 0.005, much smaller than was previously believed. The behaviour of hafnium dioxide[227] and copper(II) oxide[228] in sulphuric acid has been studied and various compounds have been isolated.

The structure of $KHSO_4$ contains two $HSO_4^-$ ions in the asymmetric unit,[229] one type being linked into a polymeric chain by hydrogen-bonding along a glide plane, the other forming a dimer across a centre of symmetry. A Raman spectral study has been made[230] of the dissociation of the $HSO_4^-$ ion at 25 °C in $H_2SO_4$, $NH_4SO_4$, and $LiHSO_4$ over a wide concentration range. Evidence for an ion pair

---

[219] J. J. Byerley, S. A. Fouda, and G. L. Rempel, *J.C.S. Dalton*, 1975, 1329.
[220] M. M. Farrow, N. Purdie, and E. M. Eyring, *Inorg. Chem.*, 1975, **14**, 1584.
[221] G. Nord, L. H. Skibsted, and A. S. Halonin, *Acta Chem. Scand.* (A), 1975, **29**, 505.
[222] T. Mizoguchi and T. Okabe, *Bull. Chem. Soc. Japan*, 1975, **48**, 1799.
[223] P. Dawson, M. M. Hargreave, and G. R. Wilkinson, *Spectrochim. Acta*, 1975, **31A**, 1533.
[224] G. Lucazeau, A. Lautie, and A. Novak, *J. Raman Spectroscopy*, 1975, **3**, 161.
[225] A. Goypiron, J. De Villepin, and A. Novak, *Spectrochim. Acta*, 1975, **31A**, 805.
[226] D. P. Sidebottom and M. Spiro, *J. Phys. Chem.*, 1975, **79**, 943.
[227] M. M. Godneva, D. L. Motov, R. F. Okhrimenko, and S. A. Kobycheva, *Russ. J. Inorg. Chem.*, 1974, **19**, 1646.
[228] N. P. Shevelev, I. G. Gorichev, N. G. Klyuchnikov, and R. I. Nazarova, *Russ. J. Inorg. Chem.*, 1974, **19**, 931.
[229] F. A. Cotton, B. A. Frenz, and D. L. Hunter, *Acta Cryst.*, 1975, **B31**, 302.
[230] D. J. Turner, *J.C.S. Faraday I*, 1974, **70**, 1346.

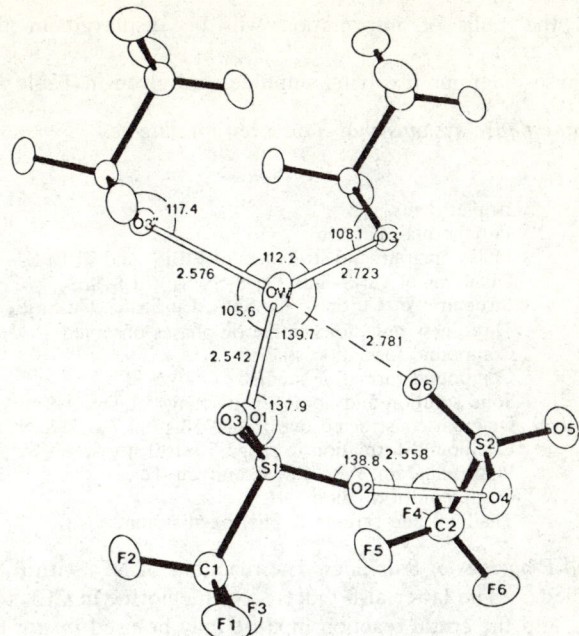

**Figure 6** *The structure of* $2CF_3SO_3H, H_2O$ *at 85 K*
(Reproduced by permission from *Acta Cryst.*, 1975, **B31**, 2208)

$H_3O^+, SO_4^{2-}$ was obtained, and it was suggested that electrochemical and spectrophotometric measurements of the dissociation constant of $HSO_4^-$ should take into account the presence of such ion pairs.

Compounds with the composition $Ti(MeSO_3)_n Cl_{4-n}$ ($n = 1$—4) have been prepared by the action of methanesulphonic acid on titanium(IV) chloride.[231]

The crystal structure of trifluoromethanesulphonic acid hemihydrate has been determined.[232] The structure consists of $H_3O^+$ ions, $CF_3SO_3^-$ ions, and $CF_3SO_3H$ molecules hydrogen-bonded together to form double layers, with the oxonium ion hydrogen-bonded to three different sulphonate ions in an asymmetric pyramidal arrangement (Figure 6). The crystal structure of the dihydrate[233] consists of $H_5O_2^+$ ions and $CF_3SO_3^-$ ions hydrogen-bonded together to form double layers which are held together by van der Waals forces. Two water molecules are held by a short hydrogen bond to form $H_5O_2^+$ ions of the asymmetric type with a *gauche* conformation. The $CF_3SO_3^-$ ion has a staggered conformation with approximate $C_{3v}$ symmetry.

**Sulphides.**—Since many of the sulphides treated in this section will have been dealt with in detail elsewhere in this volume, only features of interest will be

---

[31] R. C. Paul, V. P. Kapila, and S. K. Sharma, *J. Inorg. Nuclear Chem.*, 1974, **36**, 1933.
[32] R. G. Delaplane, J.-O. Lundgren, and I. Olovsson, *Acta Cryst.*, 1975, **B31**, 2208.
[33] R. G. Delaplane, J.-O. Lundgren, and I. Olovsson, *Acta Cryst.*, 1975, **B31**, 2202.

outlined, and the bulk of information will be displayed in the form of Tables.

Published phase diagrams involving sulphides are given in Table 4.[234–246]

**Table 4** *Sulphide phase systems that have been investigated*

| System | Notes | Ref. |
|---|---|---|
| P–S–Se | Isopleth cuts | 234 |
| P–S–Se | Polythermal diagram | 235 |
| $Tl_2S–Bi_2S_3$ | Phase diagram and structures of $TlBiS_2$ and $Tl_4Bi_2S_5$ | 236 |
| Cu–Sn–S | Diagrams of $Cu_2S–SnS$, $Cu_2S–Sn_2S_3$, $Cu_2S–SnS_2$ Structures of $Cu_4SnS_4$, $Cu_2SnS_3$, $Cu_2Sn_4S_9$, $Cu_4Sn_3S_8$ | 237 |
| $Bi_2S_3$–CdS | Three new non-stoicheiometric phases observed | 238 |
| $In_2S_3$–$Tl_2S_3$ | Compound formation studied | 239 |
| $As_2S_3$–$In_2S_3$ | Compound formation studied | 240 |
| Dy–O–S | Solid solution and homogeneity range of $Dy_2O_2S$ | 241 |
| U–S | Diagram constructed over range 53.1–64.7 at % S | 242 |
| $KSbS_2$–$Sb_2S_3$ | Compound formation in range 50–100 mol % $Sb_2S_3$ | 243 |
| Cu–S | Phase diagram. Also Cu–Se and Cu–Te | 244 |
| Sc–S | Compound formation | 245 |
| $Tl_2X$–$L_2X_3$ | Thallium and rare-earth chalcogen systems | 246 |

*Preparation and Properties of Sulphides.* The reactions of $NF_3$ with $P_4S_3$ and $P_4S_{10}$ have been studied.[247] The latter also reacts[248] with chlorine in $CCl_4$ to give $S_2Cl_2$, $SCl_2$, and $PCl_5$, and the crude reaction mixture may be used *in situ* for reactions requiring $S_2Cl_2$. Two new mixed-valence phosphorus compounds $F_2P(S)SP(CF_3)_2$ and $(CF_3)_2P(S)SPF_2$ are formed[249] by the reaction of the acids $F_2PS_2H$ or $(CF_3)_2PS_2H$ with the aminophosphines $(CF_3)_2PN(CH_3)_2$ or $F_2PN(CH_3)_2$.

A study has been made[250] of a number of methods for the preparation of halide sulphides and selenides of W and Mo in several oxidation states. The method of most general application is the reaction of $Sb_2S_3$ or $Sb_2Se_3$ with the appropriate halide. Sulphides and selenides of the type $In_{3-x}Ga_xX_3Z$ (X = S or Se; Z = P or As; x = 1 or 2.) have been prepared[251] and structure types determined by X-ray diffraction. Replacement of the $O^{2-}$ ion by $S^{2-}$ in oxyapatite, $Ca_{10}(PO_4)_6O$, may

[234] Y. Monteil and H. Vincent, *Bull. Soc. chim. France*, 1975, 1025.
[235] Y. Monteil and H. Vincent, *Bull. Soc. chim. France*, 1975, 1029.
[236] M. Julien-Pouzol and M. Guittard, *Bull. Soc. chim. France*, 1975, 1037.
[237] M. Khanafer, J. Rivet, and J. Flahaut, *Bull. Soc. chim. France*, 1974, 2670.
[238] C. Brouty, P. Spinat, and P. Herpin, *Rev. Chim. minérale*, 1975, **12**, 60.
[239] N. M. Kompanichenko, I. S. Chaus, L. E. Demchenko, V. D. Sukhenko, and I. A. Sheka, *Russ. J. Inorg. Chem.*, 1975, **20**, 422.
[240] L. E. Demchenko, I. S. Chaus, N. M. Kompanichenko, and V. D. Sukhenko, *Russ. J. Inorg. Chem.*, 1975, **20**, 274.
[241] A. A. Grizik, A. A. Eliseev, G. P. Borodulenko, and Yu. F. Streltsova, *Russ. J. Inorg. Chem.*, 1975, **20**, 124.
[242] G. V. Ellert, V. G. Sevasyanov, and V. K. Slovyanskikh, *Russ. J. Inorg. Chem.*, 1974, **19**, 1699.
[243] V. I. Bazakutsa, N. I. Gnidash, L. N. Sukhorukova, E. I. Rogacheva, A. V. Salov, S. I. Berul, and V. B. Lazarev, *Russ. J. Inorg. Chem.*, 1974, **19**, 1387.
[244] B. P. Burylev, N. N. Fedorova, and L. Sh. Tsemekhman, *Russ. J. Inorg. Chem.*, 1974, **19**, 1249.
[245] V. Brozek, J. Flahaut, M. Guittard, M. Julien-Pouzol, and M. P. Pardo, *Bull. Soc. chim. France*, 1740.
[246] M. S. Kabre, M. Julien-Pouzol, and Guittard, *Bull. Soc. chim. France*, 1974, 1881.
[247] A. Tasker and O. Glemser, *Z. anorg. Chem.*, 1974, **409**, 163.
[248] Z. Zur and E. Dykman, *Chem. and Ind.*, 1975, 436.
[249] L. F. Doty and R. G. Cavell, *Inorg. Chem.*, 1974, **13**, 2722.
[250] D. Britnell, G. W. A. Fowles, and D. A. Rice, *J.C.S. Dalton*, 1974, 2191.
[251] M. Robbins and V. G. Lambrecht, *Materials Res. Bull.*, 1975, **10**, 331.

be achieved by the reaction of the latter with CaS or by the reaction of $Ca_3(PO_4)_2$ with CaS.[252] The product in both cases contains CaO, but a pure sample may be prepared by the use of sulphur vapour. The dissolution of ethyltin sesquisulphide in aqueous solution, in the presence of sulphide and at pH values between 8 and 11, is due to the formation of two complexes,[253] $[EtSnS_3]^{3-}$ and $[(EtSn)_3(OH)_6(HS)_8]^{5-}$; in the latter complex, the complexed sulphides are oxidized with greater difficulty than the free sulphides.

I.r. spectra of some lanthanide disulphides have been measured[254] and the presence of —S—S groups has been demonstrated. Values for the electron affinities of $S^{2-}$, $Se^{2-}$, and $Te^{2-}$ have been calculated[255] as part of a study of the properties of alkaline-earth sulphides, selenides, and tellurides. Thermodynamic data for the sulphides of Fe, Co, and Ni have been related[256] to the pH of their precipitation. The standard enthalpy of formation for the non-stoicheiometric sulphide $Fe_{1-x}S$ has been deduced.[257] Thermodynamic data for some lead chalcogenides have been determined.[258]

High-temperature investigations by $X$-ray diffraction, d.t.a, and quenching experiments have been performed[259] for $FeS_2$, $FeSe_2$, $FeTe_2$, $CoTe_2$, and $CuSe_2$, and details of the synthesis of $OsTe_2$ are included.

Structural data have been published for a considerable number of sulphides; these have been collected in Table 5.[260—286]

[252] J. C. Trombe and G. Montel, *Compt. rend.*, 1975, **280**, C, 567.
[253] M.-Cl. Langlois and M. Devaud, *Bull. Soc. chim. France*, 1974, 789.
[254] Yu. M. Golovin, K. I. Petrov, E. M. Loginova, A. A. Grizik, and N. M. Ponomarev, *Russ. J. Inorg. Chem.*, 1975, **20**, 155.
[255] K. P. Thakur and J. D. Pandey, *J. Inorg. Nuclear Chem.*, 1975, **37**, 645.
[256] N. Blaton and J. Glibert, *Bull. Soc. chim. France*, 1975, 1527.
[257] G. Bugli, L. Abello, and G. Pannetier, *Bull. Soc. chim. France*, 1975, 2019.
[258] R. Blachnik and R. Igel, *Z. Naturforsch.*, 1974, **29b**, 625.
[259] A. Kjekshus and T. Rakke, *Acta Chem. Scand. (A)*, 1975, **29**, 443.
[260] B. Leclerc and T. S. Kabre, *Acta Cryst.*, 1975, **B31**, 1675.
[261] T. J. Isaacs, *Z. Krist.*, 1975, **141**, 104.
[262] D. Schmitz and W. Bronger, *Z. Naturforsch.*, 1975, **30b**, 491.
[263] K.-J. Range and G. Mahlberg, *Z. Naturforsch.*, 1975, **30b**, 81.
[264] H. A. Graf and H. Schäfer, *Z. anorg. Chem.*, 1975, **414**, 211.
[265] H. A. Graf and H. Schäfer, *Z. anorg. Chem.*, 1975, **414**, 220.
[266] E. W. Nuffield, *Acta Cryst.*, 1975, **B31**, 151.
[267] J. C. Portheine and W. Nowacki, *Z. Krist.*, 1975, **141**, 79.
[268] T. Srikrishnan and W. Nowacki, *Z. Krist.*, 1975, **141**, 174.
[269] A. Edenharter and W. Nowacki, *Z. Krist.*, 1974, **140**, 87.
[270] V. Krämer, *Z. Naturforsch.*, 1974, **29b**, 688.
[271] T. Srikrishnan and W. Nowacki, *Z. Krist.*, 1974, **140**, 114.
[272] Y. Takeuchi and T. Ozawa, *Z. Krist.*, 1975, **141**, 217.
[273] Y. Takeuchi, J. Takagik and T. Yamanaka, *Z. Krist.*, 1974, **140**, 249.
[274] G. Dittmar and H. Schäfer, *Acta Cryst.*, 1975, **B31**, 2060.
[275] J. C. Jumas, E. Philippot, and M. Maurin, *J. Solid State Chem.*, 1975, **14**, 152.
[276] W. Schiwy, Chr. Blutau, D. Gäthje, and B. Krebs, *Z. anorg. Chem.*, 1975, **412**, 1.
[277] S. Furuseth, L. Brattas, and A. Kjekshus, *Acta Chem. Scand. (A)*, 1975, **29**, 623.
[278] J. T. Szymanski, *Z. Krist.*, 1974, **140**, 218.
[279] J. T. Szymanski, *Z. Krist.*, 1974, **140**, 240.
[280] A. Katori, S. Anzai, and T. Yoshida, *J. Inorg. Nuclear Chem.*, 1975, **37**, 324.
[281] C. Riedel and R. Schoellhorn, *Materials Res. Bull.*, 1975, **10**, 629.
[282] I. E. Grey, *Acta Cryst.*, 1975, **B31**, 45.
[283] V. Rajami and C. T. Prewitt, *Canad. Mineralogist*, 1975, **13**, 75.
[284] W. Jeitschko and P. C. Donohue, *Acta Cryst.*, 1975, **B31**, 1890.
[285] G. Collin, C. Dagron, and F. Thevet, *Bull. Soc. chim. France*, 1974, 418.
[286] L. E. Drafall, G. J. McCarthy, C. A. Sipe, and W. B. White, *Proc. Rare Earth Res. Conf.*, 11th, 1974, **2**, 954.

**Table 5** Structural data of binary and ternary sulphides

| Compound | Space group | Structural information | Ref. |
|---|---|---|---|
| $Tl_2S_5$ | $P2_12_12_1$ | $S_5^{2-}$ ion with negative charge on end atoms; pyramidal co-ordination round thallium | 260 |
| $TlInS_2$ | — | Four new phase modifications observed from melts, precession data | 261 |
| $CsGaS_2$ | $C2/c$ | Isotypical with $RbFeS_2$ | 262 |
| $AInS_2$ | — | A = K, Rb, Cs. High-pressure modifications | 263 |
| $KSbS_2$ | $C2/c$ | $SbS_2^-$ chains built up by trigonal bipyramids sharing edges | 264 |
| $K_3SbS_3, 3Sb_2O_3$ | $P6_3$ | $Sb_2O_3$ tubes containing $K^+$, charge neutralized by $SbS_3^{2-}$ pyramids between tubes | 265 |
| $Pb_3Sb_8S_{15}$ | | $Pb_2Sb_4S_6$ and $PbSb_4S_9$ units interlocked | 266 |
| $Pb_6Sb_{14}S_{27}$ | $P6_3$ | $SbS_3$ groups joined in endless chains forming spirals held together by Pb ions | 267 |
| $HgSb_4S_8$ | $A2/a$ | Formula $HgSb_4S_8$ confirmed. $S_2$ group with S—S = 2.06 Å joining two $Sb_2S_4$ double chains | 268 |
| $MnAg_4Sb_2S_6$ | $P2_1/n$ | $MnS_6$ octahedral units, $SbS_3$ pyramidal | 269 |
| $Bi_9S_{27}Cl_3$ | — | Powder data | 270 |
| $Pb_2Bi_2S_5$ | $Pbna$ | Bi—S = 2.57—3.45, Pb—S = 2.72—3.47, Cu—S = 2.21—2.61 Å | 271 |
| $Cu_4Bi_4S_9$ | $Pbnm$ | $S_2$ group with S—S = 2.095 Å | 272 |
| $Pb_{1-x}Bi_{2x/3}S, 2Bi_2S_3$ | | Three phases studied in the range $x = 0$—0.266 | 273 |
| $GeS_2$ | $P2_1/c$ | High-temperature form with $GeS_4$ tetrahedra | 274 |
| $Na_4Sn_3S_8$ | $C2/c$ | $SnS_4$ tetrahedra linked with $SnS_5$ trigonal bipyramids | 275 |
| $K_2SnS_3, 2H_2O$ | $Pnma$ | $SnS_4$ tetrahedra sharing corners to give an endless chain Sn—S = 2.345—2.458 Å | 276 |
| $TiS_3, ZrS_3, ZrSe_3, ZrTe_3, HfS_3, HfSe_3$ | — | Two variants of the $ZrSe_3$ structure type found | 277 |
| $CuFe_2S_3$ | $Pcmn$ | Cu—S = 2.276—2.326, Fe—S = 2.258—2.304 Å | 278 |
| $CuFe_2S_3$ | $F\bar{4}3m$ | High-temperature form. Complete disorder of Cu and Fe between metal sites | 279 |
| $(Cr_{1-x}V_x)_5S_6$ | | Lattice parameters of solid solutions | 280 |
| $Ti_{1.023}S_2$ | $P\bar{3}m1$ | Refinement of structure | 281 |
| $Ba_5Fe_9S_{18}$ | $P4/ncc$ | Chains of edge-shared Fe—S tetrahedra with Ba atoms between chains | 282 |
| $Co_9S_8$ | $Fm3m$ | Co—S = 2.359 Å | 283 |
| $Y_2HfS_5$ | $Pnma$ | $U_3Se_5$ type with Hf substituting for U in an ordered manner, $[HfS_5]^{6-}$ polyanion | 284 |
| $LuSBr$ | $Pmmn$ | Powder data | 285 |
| Rare-earth sulphides | — | Powder data for sulphides, oxysulphides, and polysulphides | 286 |

# Elements of Group VI

Several studies of the chemistry of layered transition-metal disulphides have been published[287-293]. Of most interest is the report[293] of the first superconducting polyelectrolytes, which are the hydrated ternary layered sulphides with the formulae $[(A_x)^{x+}(H_2O)_y TaS_2]^{x-}$. For $A^+$ = alkali metal the transition temperatures range from 2.8 to 5.45 K, with the temperature being dependent on the cation radius.

**Hydrogen Sulphide.**—Ion-molecule reactions in methanol and $H_2S$ have been studied[294] using a photoionization quadrupole mass spectrometer; the proton affinity of $H_2S$ was determined. The sulphur $K_\beta$ emission and $K$ absorption spectra from gaseous $H_2S$ have been measured.[295] The $K_\beta$ spectrum comprises four peaks and the absorption spectrum three peaks that are of decreasing energy with increasing photon energy.

Hydrogen sulphide has been shown to react[296] with bromide ions in a eutectic melt of lithium and potassium bromides according to the equation:

$$H_2S + Br^- = HS^- + HBr \quad (47)$$

The reaction follows closely that observed for chlorine with $H_2S$, the only difference being that the equilibrium constants and reaction rates are lower in the case of $Br^-$. A study of the kinetics of the reaction between $H_2S$ and $Pb^{2+}$ ions in a eutectic melt of LiCl and KCl has allowed[297] the following reaction scheme to be suggested:

$$H_2S + Cl^- \rightarrow HS^- + HCl \quad (48)$$
$$2HS^- + Pb^{2+} \rightarrow PbS + H_2S \quad (49)$$
$$PbS \rightarrow Pb^{2+} + S^{2-} \quad (50)$$
$$S^{2-} + H_2S \rightarrow 2HS^- \quad (51)$$
$$S^{2-} + HCl \rightarrow HS^- + Cl^- \quad (52)$$
$$H_2S + 2Cl^- + Pb^{2+} \rightarrow PbS + 2HCl \quad (53)$$

The oxidation of $H_2S$ (by $O_2$) over microporous carbons has been studied.[298] The predominant reaction:

$$H_2S + \tfrac{1}{2}O_2 \rightarrow \tfrac{1}{2}S_2 + H_2O \quad (54)$$

was first-order in $H_2S$ concentration, and independent of $O_2$ concentration. The adsorption of $H_2S$ on microporous carbons has also been studied.[299] The interactions of oxygen with $H_2S$, of $O_2$ with $H_2$, and of $H_2S$ with $H_2$ have also been

---

[287] R. Schöllhorn and A. Lerf, *J. Less-Common Metals*, 1975, **42**, 89.
[288] G. V. Subba Rao, M. W. Shafer, and J. C. Tsang, *J. Phys. Chem.*, 1975, **79**, 553.
[289] G. V. Subba Rao and M. W. Shafer, *J. Phys. Chem.*, 1975, **79**, 557.
[290] M. Dines and R. Levy, *J. Phys. Chem.*, 1975, **79**, 1979.
[291] R. Schöllhorn and A. Lerf, *Z. Naturforsch.*, 1974, **29b**, 804.
[292] R. R. Chianelli, J. C. Scanlon, M. S. Whittingham, and F. R. Gamble, *Inorg. Chem.*, 1975, **14**, 1691.
[293] R. Schöllhorn, A. Lerf, and F. Sernetz, *Z. Naturforsch.*, 1974, **29b**, 810.
[294] L. Y. Wei and L. I. Bone, *J. Phys. Chem.*, 1974, **78**, 2527.
[295] R. E. LaVilla, *J. Chem. Phys.*, 1975, **62**, 2209.
[296] J. Mala and I. Slama, *Coll. Czech. Chem. Comm.*, 1975, **40**, 263.
[297] J. Mala and I. Slama, *Coll. Czech. Chem. Comm.*, 1975, **40**, 36.
[298] O. C. Carihso and P. L. Walker, *Carbon*, 1975, **13**, 233.
[299] O. C. Carihso and P. L. Walker, *Carbon*, 1975, **13**, 241.

investigated, using a $Ni/Al_2O_3$ catalyst.[300] A thermodynamic study[301] on the nucleation phenomenon when iron reacts with $H_2$–$H_2S$ mixtures has been carried out.

A determination[302] of the structure of hydrazinium(1+) hydrogen sulphide has shown it to contain $N_2H_5^+$ and $HS^-$ ions. The —$NH_3$ part of each $N_2H_5^+$ ion is connected by three weak hydrogen bonds with different $HS^-$ ions. The relationship between dihedral angle and bond length in HSSH has been studied[303] by the CNDO/2 MO method. The inclusion of sulphur $d$-orbitals improved the agreement between calculated and experimental results.

**Polysulphides.**—The trisulphur radical ion $S_3^-$ has been the feature of interest in several papers. The colour of alkali-metal polysulphides (blue) in solvents that are electron-pair donors and of ultramarine have been attributed to the $S_2^-$ ion in the past, but it now seems certain that it is the $S_3^-$ ion that is responsible.[304] The species, which may be readily identified by its spectroscopic properties, is encountered in a variety of situations, and may have a possible role as an intermediate in transformations involving elemental sulphur or sulphide ions. An e.s.r. study[305] of blue solutions of polysulphides in DMF also concludes that the signal with $g = 2.0280$ arises from the $S_3^-$ ion. The Raman spectra of rhombic sulphur dissolved in several primary amines have been stuided in detail.[306] Several corrections to previous assignments (*J. Phys. Chem.*, 1973, **77**, 1859) were made, and it is now thought that the bands observed at 535 cm$^{-1}$ in the Raman spectra and at 585 cm$^{-1}$ in the i.r. are due to the symmetric and antisymmetric stretching vibrations of $S_3^-$, respectively. The bands previously reported at 400, 440, and 510 cm$^{-1}$ are now assigned to the polysulphides $S_4^{2-}$ and $S_8^{n-}$. A linear relationship has been established[307] between S—S bond distances and both S—S force-constants and S—S stretching frequencies from vibrational spectra. Bond distances were derived for $S_2^-$ (2.00 Å), $S_3^-$ (1.95 Å), and $S_4^{2+}$ (2.00 Å).

The preparation of the polysulphides $K_2S_n$ ($n = 3$, 4, 5, or 6) in ethanol and in liquid ammonia have been described,[308] and lattice parameters calculated from powder diffraction data.[309] A crystal-structure determination[310] of the polysulphides $Ba_2S_3$ and $BaS_3$ has shown that $Ba_2S_3$ contains a sulphide ion and a $S_2^{2-}$ ion with a S—S distance of 2.32 Å. In $BaS_3$ the polysulphide anion is $S_3^{2-}$, with S—S = 2.074 Å, and ∠SSS is 114.8°.

**Other Sulphur-containing Compounds.**—The stable complex $[Pt(WS_4)_2]^{2-}$ has been prepared as the tetraphenylphosphonium salt by the reaction[311] of $K_2[PtI_6]$, $(NH_4)_2WS_4$, and $Ph_4PCl$. The $[Pt(WS_4)_2]^{2-}$ ion has $D_{2h}$ symmetry, with a planar

---

[300] A. Rudajevova and V. Pour, *Coll. Czech. Chem. Comm.*, 1975, **40**, 597.
[301] B. Blaise, A. Genty, and J. Bardolle, *Bull. Soc. chim. France*, 1974, 1229.
[302] F. Lazarini and M. Varfan-Jarec, *Acta Cryst.*, 1975, **B31**, 2355.
[303] L. J. Saethre, *Acta Chem. Scand.* (A), 1975, **29**, 558.
[304] T. Chivers, *Nature*, 1974, **252**, 32.
[305] F. Seel and H.-J. Güttler, *Z. Naturforsch.*, 1975, **30b**, 88.
[306] F. P. Daly and C. W. Brown, *J. Phys. Chem.*, 1975, **79**, 350.
[307] R. Steudel, *Z. Naturforsch.*, 1975, **30b**, 281.
[308] J.-M. Letoffe, J.-M. Blanchard, and J. Bousquet, *Bull. Soc. chim. France*, 1975, 485.
[309] J.-M. Letoffe, J.-M. Blanchard, M. Prost, and J. Bousquet, *Bull. Soc. chim. France*, 1975, 148.
[310] S. Yamaoka, J. T. Lemley, J. M. Jenks, and H. Steinfink, *Inorg. Chem.*, 1975, **14**, 129.
[311] A. Müller, M. C. Chakravorti, and H. Dornfeld, *Z. Naturforsch.*, 1975, **30b**, 162.

PtS$_4$ chromophore. The trithiotungstates (Ph$_4$P)$_2$W$_3$S$_9$ and (Ph$_4$As)$_2$W$_3$S$_9$) have been prepared.[312] A structure determination of the former has shown the isolated W$_3$S$_9^{2-}$ ion to have the structure (41), with the terminal tungsten atoms in

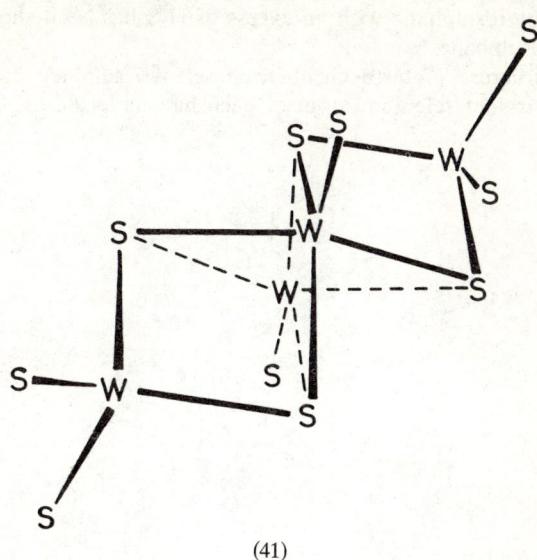

(41)

tetrahedral co-ordination and the central tungsten atom in a strongly distorted square-pyramidal arrangement. A crystal-structure determination[313] on Cs$_8$Sn$_{10}$O$_4$S$_{20}$,13H$_2$O has shown it to contain the new type of polyanion Sn$_{10}$O$_4$S$_{20}^{8-}$. The thio-oxo-anion has idealized $T_d$ symmetry and can be described as an arrangement of ten corner-linked SnS$_4$ tetrahedra with the octahedral sites lying between these tetrahedra occupied by oxygen atoms. The species, which is also stable in solution, it thought to represent the largest thio-anion to be discovered so far.

The salt K$_2$CS$_4$ and its addition compounds K$_2$CS$_4$,$n$L (L = MeOH, $n$ = 1; L = H$_2$O, $n = \frac{1}{2}$) have been prepared and characterized.[314] The nature of the addition compounds as inclusion compounds was discussed in terms of the geometry and crystal arrangement of the CS$_4^{2-}$ ions, which allow the inclusion of MeOH but not of more voluminous molecules. The reaction of sulphur with (Me$_4$N)CS$_3$ in MeOH solution has been shown[315] to give (Me$_4$N)CS$_4$,MeOH, which on heating to 70 °C gives (Me$_4$N)$_2$CS$_4$. The preparation and structure of (Me$_4$N)$_2$C$_2$S$_6$,$\frac{1}{2}$CS$_2$ have been reported.[316] The C$_2$S$_6^{2-}$ anion consists of two planar CS$_3$ groups connected by an S$_2$ bridge. The crystal structures of potassium dithioacetate[317] and potassium dithio-oxalate[318] have been reported.

[312] E. Koniger-Ahlborn and A. Müller, *Angew. Chem. Internat. Edn.*, 1974, **14**, 573.
[313] W. Schiwy and B. Krebs, *Angew. Chem. Internat. Edn.*, 1975, **14**, 436.
[314] M. Abrouk, *Rev. Chim. minérale*, 1974, **11**, 726.
[315] M. Robineau and D. Zins, *Compt. rend.*, 1975, **280**, C, 759.
[316] P. Silber, M. Robineau, D. Zins, and M. C. Brianso-Perucaud, *Compt. rend.*, 1975, **280**, C, 1517.
[317] M. M. Borel and M. Ledersert, *Z. anorg. Chem.*, 1975, **415**, 285.
[318] R. Mattes, W. Meschede, and W. Stork, *Chem. Ber.*, 1975, **108**, 1.

The electronic structures of $SH_2$, $SH_4$, and $SH_6$ have been investigated[319] by *ab initio* theoretical methods. It was concluded that $SH_4$ lies energetically above $SH_2 + H_2$ and that $SH_6$ lies at least 36 kcal mol$^{-1}$ above $SH_2 + 2H_2$. The reaction of trifluoromethylchlorosulphane with an excess of $H_2S$ has been shown[320] to yield trifluoromethyldisulphane.

The crystal structure of bis(*o*-nitrobenzeneselenyl) sulphide has been determined.[321] The bivalent selenium atom of each half molecule (42) occurs as the

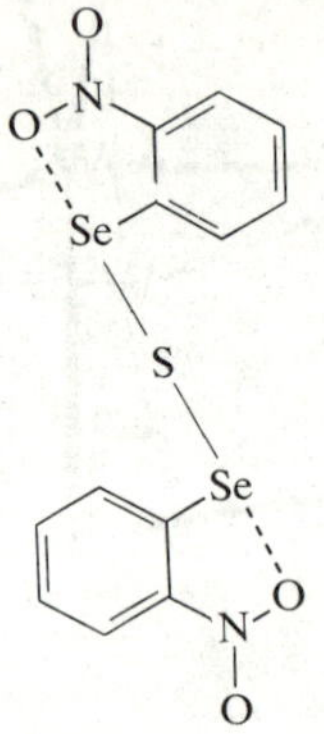

(42)

central atom in a practically planar three-co-ordinated system, with S and O as *trans* ligand atoms and a benzene carbon atom as the third ligand atom. The bond lengths from the Se atom to S, O, and C are 2.202, 2.574, and 1.918 Å, respectively. A loose five-membered ring is thus formed.

## 3 Selenium

**The Element.**—X-Ray diffraction studies[322] have been made on vitreous and liquid selenium at 20, 230, 300, and 430 °C. At 300 °C the dominant structural elements of the melt are $Se_8$ rings. Near the melting point and at 430 °C the melt consists of chains and species of low molecular weight. On the basis of model concepts for the structure of liquid selenium, the temperature variation of its conductivity has been accounted for.

γ-Ray emissions from $^{84}Se$, $^{85}Se$, $^{86}Se$, and $^{87}Se$ have been studied[323] over the energy range 300—7100 keV; the isotopes were separated from a mixture of fission products using high-speed radiochemical methods. The decay and isomerism of $^{70}Se$ have also been measured.[324]

---

[319] G. M. Schwenzer and H. F. Schaefer, *J. Amer. Chem. Soc.*, 1975, **97**, 1393.
[320] W. Gombler and F. Seel., *Z. Naturforsch.*, 1975, **30b**, 169.
[321] R. Eriksen, *Acta Chem. Scand.* (A), 1975, **29**, 517.
[322] Yu. G. Poltavtsev, *Russ. J. Phys. Chem.*, 1975, **49**, 182.
[323] M. H. Hurdus and L. Tomlinson, *J. Inorg. Nuclear Chem.*, 1975, **37**, 1.
[324] J. J. LaBreque, I. L. Preiss, H. Bakhru, and R. I. Morse, *J. Inorg. Nuclear Chem.*, 1975, **37**, 623.

# Elements of Group VI

I.r. and radioactive tracer studies[325] have shown that Se and Sb trace impurities in $AsCl_3$ are in the form of $SeOCl_2$ and $SbCl_3$, respectively. The literature concerning the stereochemical behaviour of $Se^{IV}$ oxides, fluorides, and oxyfluorides has been surveyed.[326] Several other elements, including $Te^{IV}$ were also dealt with.

**Selenium–Halogen Compounds.**—The reaction of selenium (and tellurium) chlorides with sulphur chloride ($S_2Cl_2$) has been studied.[327] $Se_2Cl_2$ reacted with $S_2Cl_2$ in a 1:3 molar ratio to give $SeCl_4$ and sulphur; with increasing temperature the amount of $SeCl_4$ formed decreased. $SeCl_4$ is insoluble in $S_2Cl_2$ at room temperature, and at higher temperatures it sublimes. No reaction was observed between $S_2Cl_2$ and $TeCl_4$. A study[328] of the extraction of $Se^{IV}$ halides in organic solvents has shown that no correlation exists between the extractability of the element and the ionization potentials of the solvents. A donor–acceptor mechanism is not therefore applicable in the case of $Se^{IV}$, although such a mechanism has been shown to hold for the extraction of $Sb^{III}$ in such solvents. The fluorine exchange in selenium tetrafluoride has been shown[329] to be suppressed at $-140\,°C$ in a solution in methyl fluoride. $SeF_4$ is a monomer species under these conditions. The reciprocal displacement system $SeCl_4 + 2Pb \rightleftharpoons 2PbCl_2 + Se$ has been investigated by studying[330] the $PbCl_2$–$SeCl_4$ binary system and the $2Pb$–$SeCl_4$ and $2PbCl_2$–$Se$ diagonal sections. The equilibrium in the system is irreversibly shifted to the side of the pair $PbCl_2 + Se$.

**Selenium–Oxygen Compounds.**—A study of the Raman spectra of molten and gaseous selenium dioxide has shown them to be nearly identical.[331] The number of bands indicated the existence of a temperature-dependent equilibrium of monomeric and dimeric $SeO_2$ in the molten and gaseous states. A chain structure, as in the crystal, could not be found, and an oxygen-bridged planar ring (43) seemed most likely for the structure of the dimer. Broadening of the bands

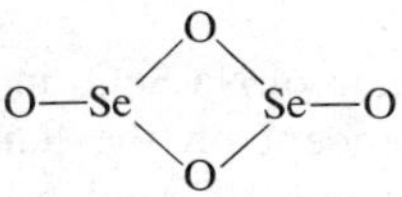

$$O-Se\begin{array}{c}O\\ \\O\end{array}Se-O$$

(43)

and two intense, polarized lines at $\nu = 955$ and $966\,cm^{-1}$ indicated an equilibrium of the *trans*- ($C_{2h}$) with the *cis*-structure ($C_{2v}$).

The reaction of $SeO_2$ with ClF has been shown[332] to give $SeOF_2$ and, with excess ClF, $SeF_4$; the latter is also formed in the reaction of $SeCl_4$ with ClF. The

---

[325] E. A. Efremov, V. A. Fedorov, A. A. Efremov, V. A. Efremov, and S. A. Fedotikov, *Zhur. priklad. Khim.*, 1975, **48**, 1127.
[326] J. Galy, G. Meunier, S. Andersson, and A. Astrom, *J. Solid State Chem.*, 1975, **13**, 142.
[327] N. I. Timoshchenko, N. Fortunatov, and Z. A. Fokina, *Nauchn. Tr. Gos. Nauchno-Issled. Proektn. Inst. Redkomet. Prom.-sti.* 1974, **58**, 119.
[328] S. Alian, W. Sanad, M. Nofal, and H. Khalifan, *J. Inorg. Nuclear Chem.*, 1975, **37**, 297.
[329] K. Seppelt, *Z. anorg. Chem.*, 1975, **416**, 12.
[330] Yu. P. Afinogenov, *Russ. J. Inorg. Chem.*, 1975, **20**, 312.
[331] H. Ziemann and W. Bues, *Z. anorg. Chem.*, 1975, **416**, 348.
[332] C. Lau and J. Passmore, *J. Fluorine Chem.*, 1975, **6**, 77.

solid-state reaction of $SeO_2$ with $VO_2$ at 400 °C gives $VSe_2O_6$, which has been shown[333] to have a structure comprising three independent $[(VO_5)_n]^{6n-}$ strings of octahedra sharing corners. The strings are connected to each other by $[Se_2O]^{6+}$ groups, which have also been observed in the structure[334] of $ZnSe_2O_5$ as prepared by the reaction of ZnO with $SeO_2$.

The uncertainty about the constitution of the vapour of $SeO_3$ may have been settled. It has been claimed[335] that only tetramer molecules are present. A redetermination of the vapour pressure gave a straight line in the Van't Hoff plot, indicating that there is no anomalous behaviour for $SeO_3$. New values for the heat of sublimation and heat of evaporation were found. Freezing points have been determined[336] in the $SeO_3$—$H_2O$ system, and a maximum freezing point of 61.950 °C was found for the pure acid. Water behaves as a strong electrolyte and $SeO_3$ as a weak electrolyte in this system.

The existence of two $SeO_3$–$AsCl_3$ complexes with the molar ratios of 1:1 and 2:1 has been detected[337] conductimetrically in liquid $SO_2$ solution. Both complexes were isolated and a reaction scheme was suggested (Scheme 1). Both complexes have $SeO_3$ as a donor, which is bonded to the arsenic atom by one of its oxygen atoms, and both are thermally unstable, decomposing rapidly at room temperature. The reaction of $SeO_3$ with pyridine in liquid $SO_2$ has been shown[338] to give the compound $(py)Se_2O_6$ in addition to $(py)SeO_3$ (Scheme 2). Partial hydrolysis of $(py)Se_2O_6$ under controlled conditions leads initially to polymerization, but the final product consists of an equimolar mixture of selenic acid and pyridinium hydrogen selenate. $(py)SeO_3$ was shown to exist in two crystal modifications, with a reversible transition; only one form is stable at room temperature. Polymeric $(py)_2Se_2O_5$ may be precipitated during the reaction of $(Se_2O_5)_n$ with pyridine at room temperature.[339] Vibrational spectra indicate that the complex forms a linear polymer. The selenium atom has trigonal-bipyramidal co-ordination, with Se—O—Se bridge bonds and one terminal oxygen atom in the equatorial position.

*Selenates.* The degree of reduction of $Na_2SeO_4$ in the $Na_2SeO_4$–Se–NaOH system has been shown[340] to be dependent on the temperature and alkalinity of the solution. At 280 and 300 °C, and 100 and 50 g l$^{-1}$ NaOH, the selenate was completely reduced. Studies[341,342] of the Raman spectra of $K_2SeO_4$ in the temperature range 87—293 K have shown no significant differences to exist between the paraelectric phase at 293 K and the ferroelectric phase at 87 K; it would seem that only minor lattice distortions are taking place. A study of the Raman spectra

---

[333] G. Meunier, M. Bertaud, and J. Galy, *Acta Cryst.*, 1974, **B30**, 2834.
[334] G. Meunier and M. Bertaud, *Acta Cryst.*, 1974, **B30**, 2840.
[335] R. H. T. Bleijerveld and F. C. Mijlhoff, *Rec. Trav. chim.*, 1975, **94**, 190.
[336] S. Wasif, M. M. Novr, and M. A. Hussein, *J.C.S. Farady I*, 1974, **70**, 929.
[337] J. Touzin, P. Bauer, and M. Jaros, *Coll. Czech. Chem. Comm.*, 1975, **40**, 1322.
[338] J. Touzin and P. Bauer, *Coll. Czech. Chem. Comm.*, 1975, **40**, 1296.
[339] J. Touzin and E. Chrobokova, *Coll. Czech. Chem. Comm.*, 1975, **40**, 1316.
[340] R. Pelyukpashidi, A. V Emelina, and A. A. Kozhakova, *Tr. Khim.-Metall. Inst., Akad. Nauk Kazakh. S.S.R.*, 1974, **24**, 58.
[341] V. Fawcett, R. J. B. Hall, D. A. Long, and V. N. Sankaranarayanan, *J. Raman Spectroscopy*, 1974, **2**, 629.
[342] V. Fawcett, R. J. B. Hall, D. A. Long, and V. N. Sankaranarayanan, *J. Raman Spectroscopy*, 1975, **3**, 229.

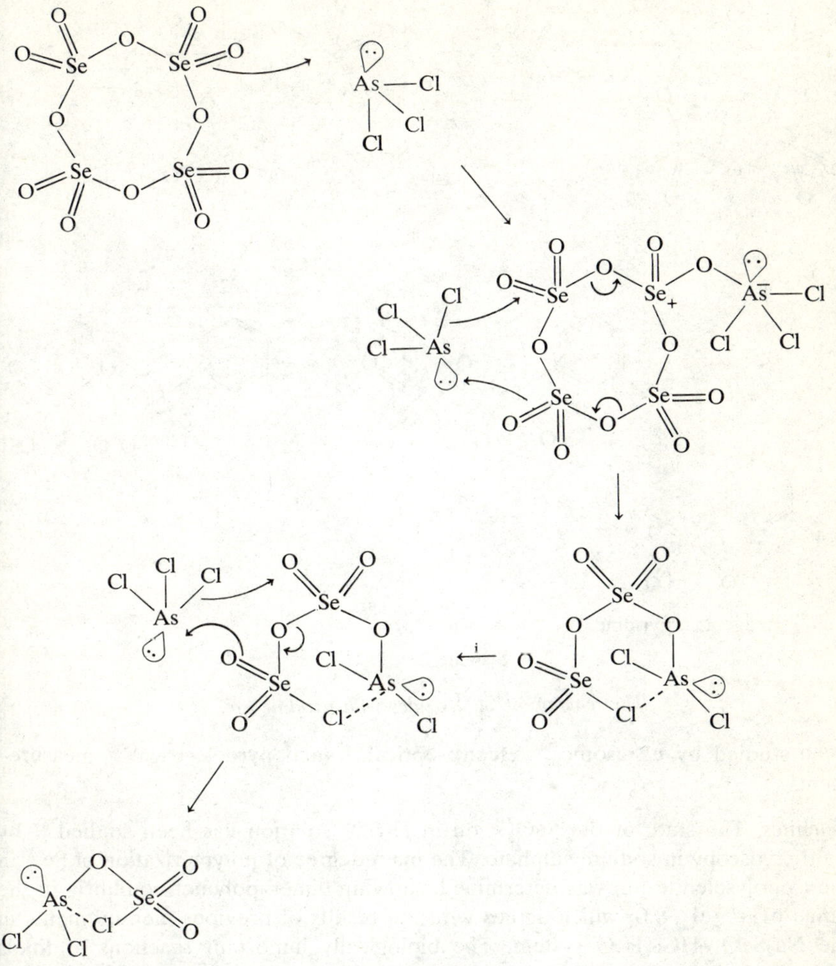

**Scheme 1**

*The reaction of Se$^{VI}$ oxide with AsCl$_3$*

Reagent: i, 2AsCl$_3$

of single crystals of NaNH$_4$SeO$_4$,2H$_2$O in various orientations[343] supports the evidence from n.m.r. studies that the phase transition at 180 K is of a displacive type, with ordering of the NH$_4^+$ ions, but not of the complete crystal, at the transition temperature. This result compares with those from NaNH$_4$SO$_4$,2H$_2$O which suggest[344,345] that a large lattice distortion takes place in this compound at the transition temperature. The phase transition in NaNH$_4$SeO$_4$,2H$_2$O has also

---

[343] V. Fawcett, D. A. Long, and V. N. Sankaranarayanan, *J. Raman Spectroscopy*, 1975, **3**, 177.
[344] V. Fawcett, D. A. Long, and V. N. Sankaranarayanan, *J. Raman Spectroscopy*, 1975, **3**, 197.
[345] V. Fawcett, D. A. Long, and V. N. Sankaranarayanan, *J. Raman Spectroscopy*, 1975, **3**, 217.

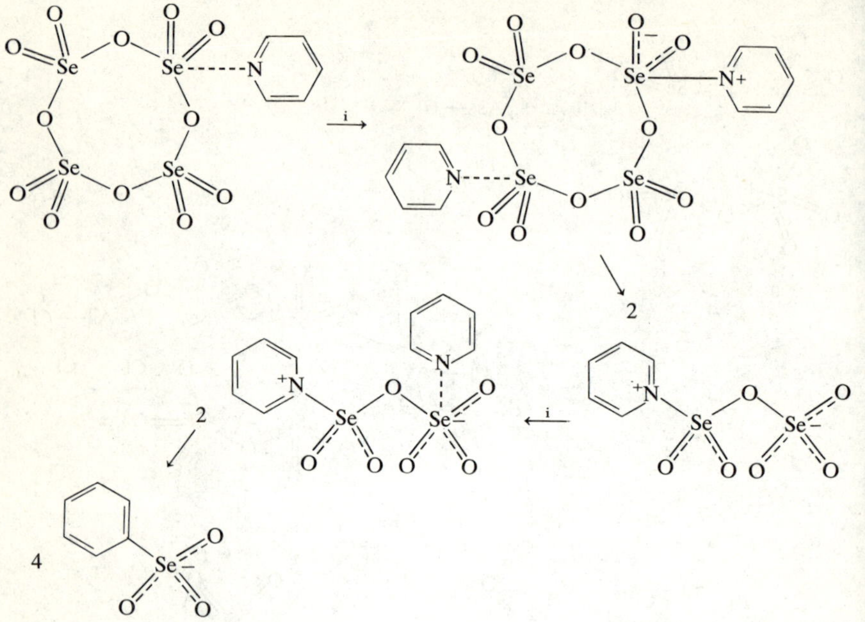

Reagent: i, pyridine

**Scheme 2**

*The reaction of Se$^{VI}$ oxides with pyridine*

been studied by ultrasonic,[346] electro-optical,[347] and pyroelectrical[348] measurements.

*Selenites.* The state of the SeO$_3^{2-}$ ion in H$_2$SO$_4$ solution has been studied[349] by 'salt' cryoscopy in sodium sulphate. The main degree of polymerization of Se$^{IV}$ in the isopolyselenite ion was determined, showing dimer–polymer equilibria in the range pH = 7.0—8.0, which agrees with the results of previous measurements in the Na$_2$SeO$_4$–HCl–H$_2$O system. The biologically important reactions of thiols with Na$_2$SeO$_3$ to give RSSeSR and RSSR have been studied at pH 7.4 in phosphate buffer, using the thiols cysteine, glutathione, $\alpha$-thioglycerol, and $\beta$-mercaptoethanol.[350] At least two intermediates are formed in the reactions, one of which is purely inorganic. The behaviour of crystalline Rb$_2$SeO$_3$ on heating to 1000 °C has been studied.[351] At 180 °C in air a very small amount of the selenite undergoes disproportionation, with the formation of elemental selenium. At about 310 °C stepwise oxidation begins, and leads to the formation of a product

---

[346] A. I. Krupnyi, L. A. Shabanova, and K. S. Aleksandrov, *Soviet Phys. Cryst.*, 1975, **20**, 218.
[347] A. T. Anistratov and S. V. Mel'nikova, *Soviet Phys. Cryst.*, 1975, **19**, 504.
[348] L. I. Zherebtsova and A. V. Shtain, *Soviet Phys. Cryst.*, 1975, **20**, 103.
[349] V. P. Kuzmicheva and E. Sh. Ganelina, *Russ. J. Inorg. Chem.*, 1974, **19**, 1156.
[350] J. Wafflart, Y. Bardoux, and J. Hladik, *Compt. rend.*, 1975, **280**, C, 617.
[351] T. V. Klushina, O. N. Evstafeva, N. M. Selivanova, Yu. M. Khozhainov, and A. Ya. Monosova, *Russ. J. Inorg. Chem.*, 1975, **20**, 160.

# Elements of Group VI

whose composition corresponds to the formula $Rb_2SeO_3,3Rb_2SeO_4$; the product melts at 990 °C without decomposition.

A neutron-diffraction study[352] of $NH_4H_3(SeO_3)_2$ has confirmed the results of an earlier (1972) $X$-ray-diffraction study, including the predicted positions of hydrogen atoms. The $H_2SeO_3$ molecules and the $HSeO_3^-$ ions in the structure are each connected *via* hydrogen bonds to form two different types of zig-zag chains running perpendicular to one another. These are interlinked by a third similar hydrogen bond to form a three-dimensional network. The $NH_4^+$ ions further stabilize the structure by forming four N—H $\cdots$ O bonds to oxygen atoms in different chains. Neutron diffraction has also been used[353] to study the arrangement of the hydrogen isotopes in sodium trihydrodeuterioselenites. The phase transitions in crystals of sodium, rubidium, and potassium trihydroselenites have been studied[354] by $X$-ray diffraction, and currently known experimental data on the anomalies of the physical properties of $RbH_3(SeO_3)_2$ which accompany a phase transition have been summarized.[355] The phase diagram of the $Na(D_{0.13}H_{0.87})_3(SeO_3)_2$ crystal in temperature–radiation dose co-ordinates has been determined.[356]

Spectroscopic and magnetic measurements[357] on the compound $VO(O_2SeOEt)_2$, prepared by the reaction of $VOCl_2$ with $H_2SeO_3$ in EtOH, suggest the monomeric structure (44). Pyrolysis of the compound proceeds by a complex mechanism; at about 200 °C the compound decomposes to $VO_2$ and $SeO_2$, with partial reduction of $Se^{IV}$ to $Se^0$. Raising the temperature leads to oxidation of vanadium and the formation of vanadyl(v) selenite. Complete dehydration of $NiSeO_3,2H_2O$ has been shown[358] to take place in two stages in the temperature range 180—550 °C, and is accompanied by the partial reduction of the selenite. The intermediates were amorphous towards $X$-rays and could not be positively identified.

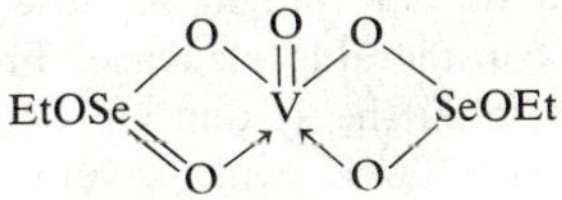

**Selenides.**—The photochemical decomposition of hydrogen selenide has been studied;[359] it was concluded that the only primary step operative in the $\lambda >$ 200.0 nm photolysis is the free-radical mode of decomposition:

$$H_2Se + h\nu \rightarrow H + SeH \qquad (55)$$

---

[352] R. Tellgren and R. Liminga, *Acta Cryst.*, 1974, **B30**, 2497.
[353] Yu. Z. Nozik, *Soviet Phys. Cryst.*, 1975, **20**, 98.
[354] S. Kh. Aknazarov, V. Sh. Shekhtman, and L. A. Shuvalov, *Soviet Phys. Cryst.*, 1975, **19**, 804.
[355] L. A. Shuvalov, N. R. Ivanov, A. M. Shirokov, A. A. Boiko, L. F. Kirpichnikova, N. V. Gordeeva, A. I. Baranov, A. Fouskova, V. S. Ryabkin, V. A. Babayants, and S. Kh. Aknazarov, *Soviet Phys. Cryst.*, 1975, **20**, 206.
[356] E. V. Peshikov and L. A. Shuvalov, *Soviet Phys. Cryst.*, 1975, **19**, 679.
[357] A. A. Kuznetsova, V. V. Kovalev, and V. T. Kalinnikov, *Russ. J. Inorg. Chem.*,
[358] V. N. Makatun, V. V. Pechkovski, R. Ya. Mel'nikova, and M. N. Ryer, *Russ. J. Inorg. Chem.*, 1974, **19**, 1851.
[359] D. C. Dobson, F. C. James, I. Safarik, H. E. Gunning, and O. P. Strausz, *J. Phys. Chem.*, 1975, **18**, 771.

with a quantum efficiency of unity. The intermediacy of H and SeH in the reaction, and the presence of Se and $Se_2$ originating from a secondary reaction, was demonstrated. The SeH radicals decay largely by the disproportionation reaction:

$$2SeH \rightarrow H_2Se + Se(^3P) \tag{56}$$

The complete $X$-ray powder pattern of InSe, not previously published, has been indexed[360] on the basis of a layer structure in the space group $R3m$, with $a_0 = 4.0046$, $c_0 = 25.960$ Å. The effects of various heat treatments on the powder patterns were discussed. A crystal-structure determination[361] has given the same space group, with the lattice constants $a_0 = 4.00$ and $c_0 = 25.32$ Å. The structure can be considered as being formed of double layers of selenium atoms, parallel to the (001) plane, between which occur pairs of indium atoms.

The thermal expansion of GeSe has been studied[362] from 20 to 760 °C by $X$-ray diffraction. The relative changes of the crystallographic axes indicate a rearrangement of the orthorhombic structure towards a NaCl-type structure with increasing temperature; a transition temperature of 651 °C was found. The sublimation kinetics of GeSe single crystals have been measured.[363] Single crystals belonging to a new modification of $GeSe_2$ have been prepared by several methods.[364] This new form appears orthorhombic, with $a = 7.037$, $b = 11.826$, and $c = 16.821$ Å. The space group could not be determined because cutting samples from large as-grown crystals induced considerable cleavage. Free-volume theory has been applied[365] to glasses and melts in the system Ge–Se.

The reaction of $P_2Se_5$ at 700 °C with several metal selenides has been studied,[366] and several new compounds with the compositions $SnPSe_3$, $PbPSe_3$, $BiPSe_4$, and $In_2P_3Se_9$ were isolated. The tetraphosphorus tetraselenide $P_4Se_4$ has been prepared[367] by heating a mixture of $P_4Se_3$ and Se between 250 and 300 °C. It has been suggested that the new compound, which exists in two allotropic forms with a transition temperature of 300 °C, has a structure derived from the tetrahedral structure of white phosphorus.

Arsenic triselenide glass of good quality has been prepared[368] by the reaction of the elements under argon in a high-pressure r.f. furnace. The boiling-point relation at elevated pressures was determined. $^{121}Sb$ Mössbauer spectra of SbSeBr and SbSeI have been obtained[369] at 4.2 K and have been compared to those of similar $Sb^{III}$ chalcogenides.

The decomposition of $CuAlX_2$ (X = X or Se) at high pressures and temperatures has been shown[370] to give $CuX_2$ and the spinel phases $CuAl_5X_8$, with

---

[360] K. C. Nagpal and S. Z. Ali, *Indian J. Chem.*, 1975, **13**, 258.
[361] A. Likforman, D. Carre, J. Etinne, and B. Bachet, *Acta Cryst.*, 1975, **B31**, 1252.
[362] H. Wiedermeier and P. A. Siemers, *Z. anorg. Chem.*, 1975, **411**, 90.
[363] E. A. Irene and H. Wiedemeier, *Z. anorg. Chem.*, 1975, **411**, 182.
[364] J. Burgeat, G. Le Roux, and A. Brenac, *J. Appl. Cryst.*, 1975, **8**, 325.
[365] A. Feltz, *Z. anorg. Chem.*, 1975, **412**, 20.
[366] A. I. Brusiloverts and N. V. Teplyakova, *Russ. J. Inorg. Chem.*, 1974, **19**, 1732.
[367] Y. Monteil and H. Vincent, *Z. anorg. Chem.*, 1975, **416**, 181.
[368] E. H. Baker, *J.C.S. Dalton*, 1975, 1589.
[369] J. D. Donaldson, A. Kjekshus, D. G. Nicholson, and J. T. Southern, *Acta Chem. Scand.* (A), 1975, **29**, 220.
[370] K. J. Range, G. Engert, and M. Zabel, *Z. Naturforsch.*, 1974, **29b**, 807.

tetrahedral co-ordination of the copper atoms. The new compound $CuAl_5Se_8$ crystallizes in the cubic space group $Fd3m$, with $a_0 = 10.53$ Å. Iron phosphorus triselenide, $FePSe_3$, has been prepared[371] by chemical vapour transport in chlorine. The structure of $FePSe_3$ is related to that of $CdI_2$; selenium atoms are in a hexagonal, close-packed array, with iron atoms and phosphorus–phosphorus pairs occupying the trigonally distorted octahedral holes in an ordered arrangement. Magnetic and Mössbauer measurements were also carried out. A phase with the composition $NbSe_3$ has been found[372] in the Nb–Se system, having the monoclinic unit-cell dimensions $a = 10.006$, $b = 3.478$, $c = 15.626$ Å, $\beta = 109.30°$, and with the space group $Pm$.

The system Tl–Ta–Se, at low thallium concentrations, gives rise to a phase with the empirical formula $Tl_xTaSe_2$ in which $x$ has values between 0.33 and 0.5. A structural study[373] shows the compound to consist of $TaSe_2$ layers between which the thallium atoms are arranged in such a way that both metals attain a trigonal-prismatic co-ordination concerning the selenium atoms. The relevant bond lengths are compared with those of the two structural modifications of $TaSe_2$ in Table 6.

**Table 6** *Bond lengths in $Tl_{0.33}TaSe_2$ compared to the two structural modifications of $TaSe_2$*

| Bond type | Bond lengths/Å | | |
|---|---|---|---|
| | $Tl_{0.33}TaSe_2$ | $2s\,TaSe_2$ | $4s\,TaSe_2$ |
| Ta—Se | 2.603 | 2.59 | 2.60 |
| Se—Se within the layer | 3.319 | 3.35 | 3.32 |
| Se—Se between layers | 5.074 | 3.59 | 3.60 |
| Se—Se in the horizontal | 3.473 | 3.43 | 3.45 |
| Tl—Se | 3.234 | — | — |

The crystal structure of $Ni_6Se_5$ quenched from 420 °C has been determined.[374] The selenium atoms form a zig-zag pattern, with the nickel atoms in deformed tetrahedral, octahedral, or pyramidal positions in the selenium lattice. $Ag_2Se$ is formed during heating a suspension of Ag powder and selenium in alkaline solution. The temperature of reaction was found[375] to decrease with increasing alkali concentration because of activation of selenium by hydroxide ion. The ternary selenides $Cs_2Pd_3Se_4$, $Rb_2Pd_3Se_4$, and $K_2Pd_3Se_4$ have been prepared[376] by fusion of alkali-metal carbonates with palladium and selenium at 850 °C. The following phase systems have been studied: Cu–Se,[377] AsS–AsSe.[378]

---

[371] B. Taylor, J. Steger, A. Wold, and E. Kostiner, *Inorg. Chem.*, 1974, **13**, 2719.
[372] A. Meerschaut and J. Rouxel, *J. Less-Common Metals*, 1975, **39**, 197.
[373] D. Muller, F. E. Poltmann, and H. Hahn. *Z. anorg. Chem.*, 1974, **410**, 129.
[374] G. Akesson and E. Roest, *Acta Chem. Scand.* (A), 1975, **29**, 236.
[375] M. Z. Ugorets, A. I. Kostikov, and T. I. Glazkova, *Tr. Khim.-Metall. Inst., Akad. Nauk Kazakh. S.S.R.*, 1974, **24**, 37.
[376] J. Huster and W. Bronger, *Z. Naturforsch.*, 1974, **29b**, 594.
[377] R. M. Murray and R. D. Heyding, *Canad. J. Chem.*, 1975, **53**, 878.
[378] E. G. Zhukov, O. I. Dzhaparidze, and S. A. Dembovskii, *Russ. J. Inorg. Chem.*, 1974, **19**, 933.

$Sn_3S_3 + Sb_2Se_3 - Sn_3Se_3 + Sb_2S_3$,[379] $KSbSe_2 - Sb_2Se_3$,[380] Se–U and Te–U,[381] and Ga–Ge–Se.[382]

**Other Compounds of Selenium.**—The crystal structure of $KSe(SCN)_3, \frac{1}{2}H_2O$ has been determined.[383] The compound was prepared by the reaction:

$$Se + Br_2 + 3SCN^- \rightarrow Se(SCN)_3^- + 2Br^- \qquad (57)$$

In the crystals, the selenium trithiocyanate ions are dimerized (45), with the two selenium atoms and the six sulphur atoms approximately coplanar. The six cyano-groups are located on the same side of the plane and each of the two selenium atoms has an approximately square-planar co-ordination. An almost identical arrangement was found[384] for the crystal structure of $KSe(SeCN)_3, \frac{1}{2}H_2O$ in which eight selenium atoms participate in the dimerized unit (46). The unit can

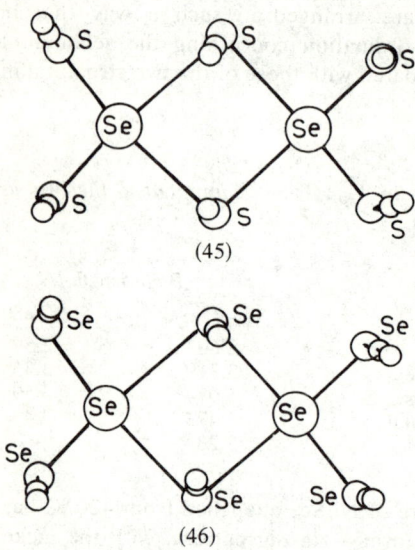

(45)

(46)

be looked upon as built up of two selenium diselenocyanate molecules, $Se(SeCN)_2$, bridged together through the selenium atoms of two selenocyanate ions. The structure of the triselenocyanate ion has also been studied in the structure[385] of $Cs(SeCN)_3$. The sequence of three selenium atoms in the ion (47) is very nearly linear, with ∠SeSeSe = 178.31° and Se—Se bond lengths of 2.650 Å, which is 0.31 Å longer than single covalent selenium–selenium bonds. An almost linear three-selenium system (∠SeSeSe = 173.8°) has also been found[386] in the

---

[379] G. G. Gospodinov, I. N. Odin, and A. V. Novoselova, *Russ. J. Inorg. Chem.*, 1974, **19**, 895.
[380] V. A. Bazakutsa, N. I. Gnidash, L. N. Sukhorukova, M. P. Vasileva, E. I. Rogacheva, S. I. Berul, A. V. Salov, and V. B. Lazarev, *Russ. J. Inorg. Chem.*, 1974, **19**, 1599.
[381] G. V. Ellert, V. G. Sevastyanov, and V. K. Slovyanskikh, *Russ. J. Inorg. Chem.*, 1975, **20**, 120.
[382] Cl. Thiebault, L. Guen, R. Eholie, and J. Flahaut, *Bull. Soc. chim. France*, 1975, 967.
[383] S. Hauge and P. A. Henriksen, *Acta Chem. Scand.* (A), 1975, **29**, 778.
[384] S. Hauge, *Acta Chem. Scand.* (A), 1975, **29**, 771.
[385] S. Hauge, *Acta Chem. Scand.* (A), 1975, **29**, 163.
[386] S. Hauge, D. Opedal, and J. Askog, *Acta Chem. Scand.* (A), 1975, **29**, 225.

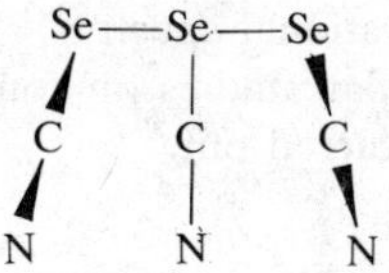

Elements of Group VI                                                          387

```
       Se ── Se ── Se
      ╱       │       ╲
     C        C        C
    ╱         │         ╲
   N          N          N
```
(47)

tris(selenourea) ion, as found in the dibromide and dichloride hydrates. The structure of selenium bis(1-pyrrolidine-carbodiselenoate) has been determined.[387] There are two types of crystallographically independent molecules in the unit cell but both are very similar to each other. The monomeric compounds can be regarded as four-co-ordinated trapezoid planar selenium(II) complexes with two bidentate diselenocarbamate ligands. Average bond lengths from the central selenium atoms to the ligand selenium atoms are 2.445 and 2.868 Å, respectively, for the short and long bonds which are *trans* to each other.

Dimethyltin selenide has been shown[388] to react with diorganyliodoboranes to give $R_2B$—Se—$BR_2$. The compounds with $R = C_6H_{11}$ and $C_6H_5$ are stable, but with $R = CH_3$ and $C_4H_9$ decomposition to polymeric $(RBSe)_n$ was observed. The reaction of 3,4-xylenyl-1,1-di-iodoborane with polymeric iodoboron selenide at 180 °C has been shown[388a] to yield xyleno-1,2,5-selenadiborolen (48).

The compounds (49) and (50) have been prepared[389] by the reaction of the

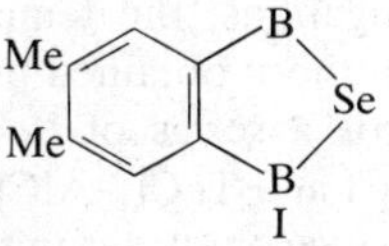

(48) 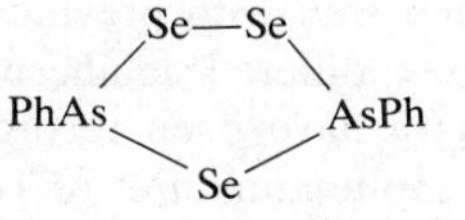   (49) 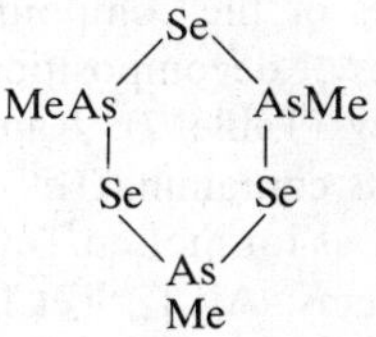   (50)

corresponding organocycloarsine and elemental selenium. The structures were concluded from u.v., i.r., $^1H$ n.m.r., and mass spectrometric measurements. The reactions of esters of selenophosphorus acids with sulphuryl chloride have also been studied.[390]

Potassium monoselenotetrathionate, $K_2SeS_3O_6 \cdot \frac{1}{2}H_2O$, has been prepared[391] by the reaction of the selenotrithionate and selenopentathionate in an acid medium, and a number of oxidation and decomposition reactions have been studied.

$$SeS_2O_6^{2-} + SeS_4O_6^{2-} \rightarrow 2SeS_3O_6^{2-} \qquad (58)$$

---

[387] S. Esperas, S. Husebye, and A. Rolandsen, *Acta Chem. Scand.* (A), 1975, **29**, 608.
[388] F. Riegel and W. Siebert, *Z. Naturforsch.*, 1974, **29b**, 719.
[388a] W. Siebert and B. Asgarouladi, *Z. Naturforsch.*, 1975, **30b**, 647.
[389] D. Herrmann, *Z. anorg. Chem.*, 1975, **416**, 50.
[390] I. A. Nuretdinov, R. Kh. Giniyatullin, and E. V. Bayandina, *Izvest. Akad. Nauk S.S.S.R. Ser. khim.*, 1975, 726.
[391] V. Zelionkaite and V. Janickis, *Russ. J. Inorg. Chem.*, 1974, **19**, 1762.

The preparation[392] and vibrational spectra[393] of several derivatives of $CF_3Se$ have been described, and similar studies on some transition-metal pentafluoro-orthochalogenides have been carred out.[394]

## 4 Tellurium

**The Element.**—An $X$-ray diffraction study[395] of molten tellurium has shown the atomic radial distribution curves to have $r_1 = 2.95$ and $r_2 = 4.16$ Å at 450 °C, and 2.95 and 4.18 Å at 510 °C. The transport of tellurium nuclides from fission products, using an oxygen–nitrogen mixture, in the gas phase, has been investigated.[396]

**Tellurium–Halogen Compounds.**—Enthalpies and entropies of evaporation of $TeCl_4$ and of sublimation of $TeBr_4$ have been measured.[397] Using the above data, the equilibrium constant of the reaction:

$$TeX_4(g) \to TeX_2(g) + X_2(g) \tag{59}$$

was calculated, and enthalpies and entropies of reaction were determined. The solubility of $TeCl_4$ in tin tetrachloride has been measured[398] at various temperatures between 18 and 140 °C; values found varied from 0.30 mole % at the lower temperature to 1.97 mole % at 140 °C. The thermal stabilities of $TeI_4$ and hexaiodotellurites of K,Rb, and Cs have been studied[399] by tensimetric and thermogravimetric analyses. In the temperature ranges studied the vapour pressures of the compounds agreed well with previous investigations; the temperatures of decomposition, however, differed significantly from those obtained previously (Table 7). Raman spectra have been recorded[400] for a series of $KAlCl_4$ melts containing $Te^{IV}$ and for molten mixtures of $TeCl_4$–$KCl$ and $TeCl_4$–$AlCl_3$, as well as for molten $TeCl_4$. Bands were assigned for the following complexes in these systems: $AlCl_4^-$, $TeCl_4$, $TeCl_6^{2-}$, $TeCl_5^-$, and $TeCl_3^+$. The spectra of $TeCl_4$ change little in going from solutions in $KAlCl_4$ to pure molten $TeCl_4$, which indicates that polymeric units such as $Te_4Cl_{16}$, found in solid $TeCl_4$, are not present in these melts. The structure of $TeCl_4$ in the molten state is now thought to be $Cl_3Te$—$Cl$, of $C_{3v}$ symmetry, with one weakly bound and three strongly bound chlorides, in constrast to the $C_{2v}$ molecule observed in gaseous $TeCl_4$. Studies have also been carried out on the eutectic system $BiCl_3$–$TeCl_4$,[401] and interactions in the systems $Te^{4+} \parallel Br,Cl,I$ and $Cs,K,Rb,Te^{4+} \parallel I^-$ have been investigated.[402,403]

---

[392] C. J. Marsden, *J. Fluorine Chem.*, 1975, **5**, 423.
[393] C. J. Marsden, *J. Fluorine Chem.*, 1975, **5**, 401.
[394] K. Seppelt, *Chem. Ber.*, 1975, **108**, 1823.
[395] Yu. G. Poltavsev, *Russ. J. Phys. Chem.*, 1975, **49**, 429.
[396] W. Bogle, K. Bachmann, and V. Matschoss, *J. Inorg. Nuclear Chem.*, 1975, **37**, 1577.
[397] H. Oppermann, G. Stover, and E. Wolf, *Z. anorg. Chem.*, 1974, **410**, 179.
[398] Zh. K. Feskova, V. V. Safonov, and V. I. Ksenzenko, *Russ. J. Inorg. Chem.*, 1974, **19**, 1083.
[399] V. V. Safonov, O. V. Lemenshko, S. M. Chernylch, and B. G. Korshunov, *Russ. J. Inorg. Chem.*, 1974, **19**, 791.
[400] F. W. Poulsen, N. J. Bjerrum, and O. F. Nielsen, *Inorg. Chem.*, 1974, **13**, 2693.
[401] L. A. Niselson, V. P. Borisova, and K. V. Tret'yakova, *Russ, J. Inorg. Chem.*, 1974, **19**, 1544.
[402] V. V. Safonov, N. F. Kuklina, and B. G. Korshunov, *Russ. J. Inorg. Chem.*, 1975, **20**, 309.
[403] V. V. Safonov, N. F. Kuklina, and B. G. Korshunov, *Russ. J. Inorg. Chem.*, 1974, **19**, 1744.

**Table 7** *Thermal decomposition temperatures of some hexaiodotellurites of K, Rb, and Cs, and of Te$^{IV}$ iodide*

| | Temperature/°C | | |
|---|---|---|---|
| | Tensimetric | T.g.a. | |
| Compound | at $p = 760$ mmHg | Start | Maximum |
| TeI$_4$ | 290 | 187 | 267 |
| | 313$^a$ | 118$^b$ | 190$^b$ |
| K$_2$TeI$_6$ | 373 | 149 | 347 |
| | | 130$^{b,c}$ | 235,$^b$ 245$^c$ |
| Rb$_2$TeI$_6$ | 422 | 315 | 440 |
| | | 185$^{b,c}$ | 290$^{b,c}$ |
| Cs$_2$TeI$_6$ | 465 | 355 | 503 |
| | | 225$^{b,c}$ | 335$^{b,c}$ |

$^a$ S. A. Ivashin and E. S. Petrov, *Izvest. sibirsk. Otdel. Akad. Nauk S.S.S.R., Ser. khim. Nauk*, 1971, No. 4, p. 28; $^b$ G. R. Allakhverdol, *Zhur. neorg. Khim.*, 1968, **13**, 3377; $^c$ B. D. Stepin, *Rare Alkali Elements*, 1969, 80.

Infrared and Raman spectra of TeF$_5$OCs and TeF$_5$OAg in MeCN solutions and in the solid state have been reported.[404] The vibrations of the TeF$_5$O$^-$ ion could be assigned on the basis of $C_{4v}$ symmetry, and contact ion-pair formation was observed in concentrated solutions which gave rise to two new bands which were assigned to the Te—O and the axial Te—F stretching modes in the complex. A normal-co-ordinate analysis of the TcF$_5$O$^-$ and SeF$_5$O$^-$ ions was also carried out. The *cis-* influence in substituted TeF$_6$ compounds has been studied by a MO method.[405] The calculations confirm the experimental observations that the fluorine atoms *cis* to a ligand are more labile and hence more susceptible to replacement. Ligands could be placed in order of *cis* influence for the monosubstituted TeF$_5$X. The equilibrium:

$$\text{Te(OH)}_6 + n\text{HF} \rightleftharpoons \text{Te(OH)}_n\text{F}_{6-n} + n\text{H}_2\text{O} \qquad (60)$$

has been studied,[406] mainly by $^{19}$F n.m.r. spectroscopy. In the solvolysis of Te(OH)$_6$ in HF, products up to $n = 3$ have been identified, while in the hydrolysis of TeF$_6$, products with $n = 0$—5 have been observed. Solvolysis of selenic acid in HF was shown to lead to pentafluoro-orthoselenic acid and two other products, one of which is probably HSeO$_3$F. Most of the substituted derivatives Te(OMe)$_n$F$_{6-n}$ have now been observed in a study[407] of the solvolysis of Te(OMe)$_6$ in anhydrous HF. The reaction of tellurium hexafluoride with ethylene glycol and other polyhydric alcohols has been studied.[408] A large number of products were characterized and the reactions were thought to be step-wise, with the relative proportions of products largely determined by the equilibrium shown in Scheme 3.

[404] E. Mayer and F. Sladky, *Inorg. Chem.*, 1975, **14**, 589.
[405] D. R. Armstrong, G. W. Fraser, and G. D. Meikle, *Inorg. Chim. Acta*, 1975, **15**, 39.
[406] U. Elgad and H. Selig, *Inorg. Chem.*, 1975, **14**, 140.
[407] I. Agranat, M. Rabinovitz, and H. Selig, *Inorg. Nuclear Chem. Letters*, 1975, **11**, 185.
[408] G. W. Fraser and G. D. Meikle, *J.C.S. Dalton*, 1975, 1033.

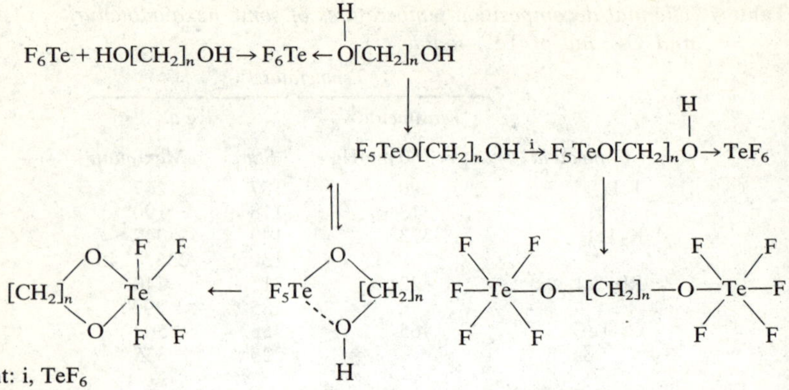

**Scheme 3**
*The reaction of TeF$_6$ with polyhydric alcohols*

Reagent: i, TeF$_6$

The heats of formation of potassium, rubidium, and caesium hexachlorotellurates(IV) have been found[409] from their heats of solution, and those of crystalline TeCl$_4$ and the metal chlorides determined from isothermal calorimetry. Thermal and elastic properties of the alkali-metal hexachlorotellurates have also been investigated.[410] Spectroscopic studies on a series of dialkyltellurium tetraiodides have been carried out.[411]

Isotopically pure samples of Ph$_2$TeCl$_2$, Ph$_2$TeBr$_2$, and (p-MeC$_6$H$_4$)$_2$TeBr$_2$ have been prepared and their spectra examined, so that a definitive assignment of the Te—C stretching vibration could be made.[412] Some $^{125}$Te Mössbauer parameters have been reported[413] for a number of aryltellurium-(II) and -(IV) compounds. The crystal structures of several organotellurium halides have been determined; relevant crystal data[414—418] are given in Table 8.

**Tellurium–Oxygen Compounds.**—Single crystals of Te$_4$O$_9$ have been prepared[419] by hydrothermal synthesis. The crystals have a hexagonal unit cell and exhibit disorder. The structure is built up of covalent (TeO$_3$,3TeO$_2$)$_n$ layers and the disorder arises since these layers can be stacked either normally or inverted. The normally stacked structure revealed the expected Te$^{IV}$ and Te$^{VI}$ co-ordination in the layers; Te$^{VI}$ is octahedrally co-ordinated by oxygen atoms, with Te—O = 1.903 and 1.948 Å, whilst the four strong Te$^{IV}$—O bonds of 1.883, 1.902, 2.020, and 2.144 Å are directed towards four corners of a trigonal bipyramid.

[409] Yu. M. Golutvin, E. G. Maslennikova, A. N. Li, V. V. Safonov, and B. G. Korshunov, *Russ. J. Phys. Chem.*, 1974, **48**, 1604.
[410] N. L. Rapoport, O. V. Ioffe, G. R. Allakhverdov, and B. D. Stepin, *Russ. J. Phys. Chem.*, 1974, **48**, 1165.
[411] K. V. Smith and J. S. Thayer, *Inorg. Chem.*, 1974, **13**, 3021.
[412] N. S. Dance and W.-R. McWhinnie, *J.C.S. Dalton*, 1975, 43.
[413] F. J. Berry, E. H. Kustan, and B. C. Smith, *J.C.S. Dalton*, 1975, 1323.
[414] J. D. McCullough, *Inorg. Chem.*, 1975, **14**, 1142.
[415] K. S. Fredin, K. Maroy, and S. Slogvik, *Acta Chem. Scand.* (A), 1975, **29**, 212.
[416] O. Vikane, *Acta Chem. Scand.* (A), 1975, **29**, 787.
[417] O. Vikane, *Acta Chem. Scand.* (A), 1975, **29**, 738.
[418] O. Vikane, *Acta Chem. Scand.* (A), 1975, **29**, 763.
[419] O. Lindqvist, W. Mark, and J. Moret, *Acta Cryst.*, 1975, **B31**, 1255.

**Table 8** Crystal data for some organo-tellurium halides

| Compound | Co-ordination, and bond distances/Å | Ref. |
|---|---|---|
| $C_{12}H_8TeI_2$ | Te octahedral, with Te—I = 2.944, 2.928, 3.717, and 3.696; Te—C = 2.111 and 2.113 | 414 |
| $Te(trtu)_2Cl_2$ | Te square-planar, with Te—Cl = 2.964 and Te—S = 2.465 | 415 |
| $Te(trtu)_2Br_2$ | Te square-planar, with Te—Br = 2.994 and Te—S = 2.499 | 415 |
| $C_6H_5Te(etu)I$ | Te three-co-ordinate, with Te—I = 3.0033, Te—S = 2.614, and Te—C = 2.124 | 416 |
| $C_6H_5Te(esu)I$ | Te three-co-ordinate, with Te—I = 3.0951, Te—Se = 2.6791, and Te—C = 2.112 | 416 |
| $C_6H_5Te(etu)Br$ (1st type) | Te three-co-ordinate, with Te—Br = 2.8348, Te—S = 2.556, and Te—C = 2.116 | 417 |
| $C_6H_5Te(etu)Br$ (2nd type) | Te four-co-ordinate, with Te—Br = 2.9694 and 3.831, Te—S = 2.5231, and Te—C = 2.123 | 417 |
| $C_6H_5Te(etu)Cl$ | Te three-co-ordinate, with Te—Cl = 2.8486, Te—S = 2.5211, Te—C = 2.120; possible fourth position with Te—Cl = 3.7401 | 418 |
| $C_6H_5Te(esu)Br$ | Te three-co-ordinate, with Te—Br = 3.0537, Te—S = 2.5211, Te—C = 2.120; possible fourth position with Te—Br = 3.8490. | 418 |

Phase analysis of the mixed oxide system $TeO_2$–$MoO_3$ has indicted[420] the existence of a new phase, $\alpha$-$Te_2MoO_7$, that is stable at room temperature. Below 500 °C, mixtures of crystalline products were obtained or complete devitrification could be induced; above 500 °C a tendency to glass formation was observed. Quenching of a melt of $Te_2MoO_7$ yielded a dark yellow glass, $\beta$-$Te_2MoO_7$.

Raman spectra have indicated[421] that aqueous hydrofluoric acid solutions of $TeO_2$ contain only $TeF_5^-$ and $[Te(OH)F_4]^-$ ions; the latter was characterized by Raman and i.r. spectra in solid $KTe(OH)F_4$ and in solution The OH group is in the apical position, according to the spectra, but there was considerable departure from the spectrum expected for a $C_{4v}$ ion. The interactions of the components in the $TeO_2$–$TeCl_4$–$TeBr_4$ and $TeO_2$–$TeCl_4$–$TeI_4$ systems have been investigated.[422]

The crystal structure of $BaTeO_3$ has been studied.[423] In the compound, tellurium forms pyramidal $TeO_3$ groups which are not linked with one another, but are joined to $BaO_9$ polyhedra by common edges; the latter are linked one with another by common edges only. A similar type of co-ordination was observed[424] in $Li_2TeO_3$, in which the $TeO_3$ pyramids are linked to corners of $LiO_4$ groups which are joined to other $LiO_4$ groups by both common corners and edges. The new tellurates $Bi_6Te_2O_{15}$ and $Bi_2TeO_6$ have been prepared[425] by the reversible reaction of oxygen with bismuth tellurites having the composition $Bi_{1-x}Te_xO_{(3+x)/2}$. $Bi_2TeO_6$ was shown to have structural similarities to the compounds $Bi_2MoO_6$ and $Bi_2WO_6$, but the possibility of ferroelectricity, common in this type of compound, was ruled out.

---

[420] J. C. J. Bart, G. Petrini, and N. Giordano, *Z. anorg. Chem.*, 1975, **412**, 258.
[421] J. B. Milne and D. Moffett, *Inorg. Chem.*, 1974, **13**, 2750.
[422] V. V. Safonov, V. S. Nikulenko, and B. G. Korshunov, *Russ. J. Inorg. Chem.*, 1975, **20**, 640.
[423] F. Folger, *Z. anorg. Chem.*, 1975, **411**, 111.
[424] F. Folger, *Z. anorg. Chem.*, 1975, **411**, 103.
[425] B. Frit and M. Jaymes, *Bull. Soc. chim. France*, 1974, 402.

A structural analysis[426] of $(NH_4)_6[TeMo_6O_{24}], Te(OH)_6, 7H_2O$ has shown that the heteropolyanion $[TeMo_6O_{24}]^{6-}$ consists of $TeO_6$ octahedra surrounded by six $MoO_6$ octahedra that are condensed in a flat arrangement. Interactions in the systems $AlCl_3-K_4H_2TeO_6-H_2O$[427] and $MCl_3-H_6TeO_6-KOH-H_2O$ (M = Al, Ga, or In)[428] have been studied.

**Tellurides.**—The crystal structure of TlTe has been redetermined[429] and the tellurium sublattice corrected. Short Te—Te atomic distances were observed in the structure which are comparable with covalent bond distances. The ternary telluride $MnIn_2Te_4$, prepared by direct synthesis from the elements at 800 °C, has been shown[430] to have a structure derived from that of zinc blende. The same type of structure has also been found[431] for the ternary tellurides $MgAl_2Te_4$, $MgGa_2Te_4$, and $MgIn_2Te_4$. The intermediate phases in the $Tl_2X-L_2X_3$ system (X = S, Se, or Te; L = rare-earth element) have been prepared and a structural study has been carried out.[432] Two types of structure were identified, definite rhombohedral compounds of the $\alpha$-$NaFeO_2$ type and face-centred-cubic solid solutions of the $Th_3P_4$ type.

The vapour pressures of tellurium in iron–tellurium alloys have been measured[433] at temperatures between 550 and 900 °C. The activities and partial molar enthalpies of tellurium were calculated and integral free energies of formation obtained. Magnetic measurements have also been made in the same system,[434] for which the phase diagram has been determined.[435] The enthalpies of formation of solid Co–Te alloys have been determined.[436] The following phase diagrams involving tellurium have been studied: As–Te,[437] $As_2Te_3$–PbTe,[438] Cu–Sn–Te,[439] Pb–Sn–Te,[440] and S–Te–U.[441]

**Other Compounds of Tellurium.**—The tris(o-ethylxanthato)tellurate(II) anion, as its tetraethylammonium salt, has been shown[442] by X-ray methods to have the planar pentagonal structure shown in Figure 7. The environment of the tellurium atom is asymmetric, as two of the Te—S bonds (3.053 and 3.058 Å) are longer than the other three (2.645, 2.676, and 2.497 Å). Tetrakis(4-morpholine-carbodithioato)tellurium(IV) has been shown[443] to contain a central tellurium atom bonded to all eight sulphur atoms in a slightly distorted dodecahedral $C_{2d}$ configuration. The Te—S bond lengths were found to vary between 2.672 and

---

[426] H. T. Evans, *Acta Cryst.*, 1974, **B30**, 2095.
[427] P. K. Bol'shakova and A. A. Kudinova, *Russ. J. Inorg. Chem.*, 1975, **20**, 960.
[428] N. K. Bol'shakova and A. A. Kudinova, *Russ. J. Inorg. Chem.*, 1975, **20**, 675.
[429] J. Weis, H. Schafer, B. Eisenmann, and G. Schön, *Z. Naturforsch.*, 1974, **29b**, 585.
[430] K.-J. Range and H.-J. Hubner, *Z. Naturforsch.*, 1975, **30b**, 145.
[431] P. Dotzel, E. Franke, H. Schafer, and G. Schön, *Z. Naturforsch.*, 1975, **30b**, 179.
[432] M. S. Kabre, M. Julien-Pouzol, and M. Guittard, *Bull. Soc. chim. France*, 1974, 1881.
[433] H. Ipser and K. L. Komarek, *Monatsh.*, 1974, **105**, 1344.
[434] K. L. Komarek and P. Terzieff, *Monatsh.*, 1975, **106**, 145.
[435] H. Ipser, K. L. Komarek, and H. Mikler, *Monatsh.*, 1974, **105**, 1322.
[436] K. L. Komarek, P. Matyas, and J. Mikler, *Monatsh.*, 1975, **106**, 73.
[437] R. Blachnik, A. Jäger, and G. Enninga, *Z. Naturforsch.*, 1975, **30b**, 191.
[438] C. Carcaly, J. Rivet, and J. Flahaut, *J. Less-Common Metals*, 1974, **38**, 245.
[439] A. Gaumann, *J. Less-Common Metals*, 1974, **38**, 245.
[440] V. P. Zlomanov, L. Kuan Fu, A. M. Gas'kov, and A. V. Novoselova, *Russ. J. Inorg. Chem.*, 1974, **19**, 1385.
[441] G. V. Ellert and V. K. Slovyanskikh, *Russ. J. Inorg. Chem.*, 1975, **20**, 280.
[442] B. F. Hoskins and C. D. Pannan, *J.C.S. Chem. Comm.*, 1975, 408.
[443] S. Esperas and S. Husebye, *Acta Chem. Scand. (A)*, 1975, **29**, 185.

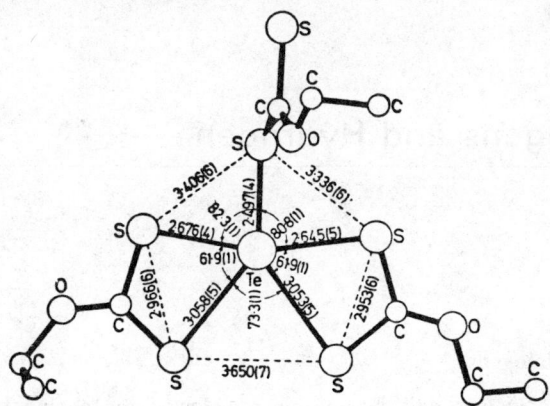

**Figure 7** *The structure of the* [Te(Etxan)$_3$]$^-$ *anion, showing the pentagonal-planar co-ordination about the tellurium atom*
(Reproduced from *J.C.S. Chem. Comm.*, 1975, 408)

2.824 Å. The structures of the phenyldithiocyanatotellurate(II) ion and the phenyldiselenocyanatotellurate(II) ion have been shown[444] to be based on square-planar co-ordination, with one position, *trans* to the phenyl group, vacant. Each tellurium atom is bonded to a phenyl carbon atom and, in a direction perpendicular to the Te—C bond, to two thiocyanate sulphur atoms or to two selenocyanate selenium atoms (51).

(Se)S—Te—S(Se)

(51)

The $^{125}$Te Mössbauer spectra of a number of Te$^{II}$ and Te$^{IV}$ complexes with sulphur-containing ligands have been measured.[445] The Te$^{II}$ compounds exhibited large quadrupole splittings of 12.4—15.2 mm s$^{-1}$, and isomer shifts of +0.33 to +0.91 mm s$^{-1}$, which was thought to indicate the incorporation of some 5s character in the bonding scheme for several of these compounds. The preparation of several complexes of Te$^{II}$ by the reactions of diphenyl ditelluride with halogens in the presence of substituted selenoureas has been described.[446]

## 5 Polonium

The reactions of polonium vapour with both Main-group and Transition-group IV elements have been studied.[447] Polonium vapour did not react with Si, Ta, or Zr carbides, but did react with tin at 370 °C to give a tin polonide, which dissociated at 670 °C to the elements.

[444] S. Hauge and O. Vikane, *Acta Chem. Scand.* (A), 1975, **29**, 755.
[445] R. M. Cheyne and C. H. W. Jones, *Canad. J. Chem.*, 1975, **53**, 1855.
[446] O. Vikane, *Acta Chem. Scand.* (A), 1975, **29**, 152.
[447] A. S. Abakumov and Z. V. Ershova. *Radiokhimiya*, 1974, **16**, 879.

# 7
# The Halogens and Hydrogen

BY M. F. A. DOVE

## 1 Halogens

**The Elements.**—Downs and Adams have reviewed the chemistry of chlorine, bromine, iodine, and astatine.[1] The role of fluorine in our environment and the use of lichens as monitors for air pollution have been discussed by Dobbs.[2] Particulate rather than gaseous fluorides were shown to be the pollutants in the vicinity of one aluminium refining plant.[3] Bronchitis, pneumonia, nervous disorders, and skin diseases were said[4] to occur more frequently in workers at a Russian $AlF_3$ production plant than in a control group.

Improved methods for producing $^{18}F$ have been described. Using an enriched $^6Li_2CO_3$ target, 65—75 mCi of usable carrier-free material can be obtained after a 3 h irradiation at $4.2 \times 10^{13}$ neutrons $cm^{-2}$ $s^{-1}$.[5] Yields of $^{18}F$ as high as 98% were realized by Parker *et al.*, and in addition they reduced the separation time to 7—15 minutes.[6] Bombardment of a target system with $\tau(^3He)$-particles consisting of $^{20}Ne$ and $H_2^{16}O$, in that order, separated by a thin foil has yielded simultaneously anhydrous and aqueous $^{18}F$, in good yields:[7] neon has a maximum cross-section for nuclear reaction at higher energy than has $^{16}O$ and hence the method yields greater amounts of $^{18}F$ than would be obtained from either target separately.

The reactions of $F(^2P)$ atoms have been studied kinetically in a discharge–flow apparatus at 297 K.[8] The major reaction channel for the $F + ICl$ reaction forms $IF + Cl$, although a minor process yields $ClF + I$. Fluorine atoms react with CO and Xe slowly to yield $COF_2$ and $XeF_2$, respectively, but no reaction was detected with either He or Kr. Fluorine atoms abstracted hydrogen from HOF, generating OF radicals. The mechanisms of the reactions of F and OF radicals with NOF have been investigated by Schumacher *et al.*[9] Quantum yields were shown to be

---

[1] A. J. Downs and C. J. Adams, 'The chemistry of Chlorine, Bromine, Iodine and Astatine', Pergamon Press, Elmsford, N.Y., 1975.
[2] C. G. Dobbs, *Fluoride*, 1974, **7**, 123.
[3] T. Okita, K. Kaneda, T. Yanaka, and R. Sugai, *Atmos. Environ.*, 1974, **8**, 927.
[4] Yu. I. Solov'eva, *Gig. Tr. Prof. Zabol.*, 1974, 7.
[5] P. K. H. Chan, G. Firnau, and E. S. Garnett, *Radiochem. Radioanalyt. Letters*, 1974, **19**, 237.
[6] W. C. Parker, C. P. G. DaSilva, and W. H. G. Francis, Report 1973, IEA-302 (*Chem. Abs.*, 1975, **82**, 23234).
[7] T. Nozacki, M. Iwamoto, and T. Ido, *Internat. J. Applied Radiation Isotopes*, 1974, **25**, 393; Jap. P. 75 54796 (*Chem. Abs.*, 1975, **83**, 169980).
[8] E. H. Appelman and M. A. A. Clyne, *J.C.S. Faraday I*, 1975, **71**, 2072.
[9] P. J. Bruna, J. E. Sicre, and H. J. Schumacher, *Anales Soc. Cient. Argentina*, 1972, **194**, 69.

less than 1 for the reactions:

$$F_2 + NOF + h\nu \rightarrow NOF_3; \quad OF_2 + NOF + h\nu \rightarrow NO_2F + F_2$$

Axworthy et al.[10] have reported the results of calculations on the combustion generation of F atoms in reactions of potential use in an HF continuous-wave laser. Various combinations of oxidizers ($F_2$, $N_2F_4$, $NF_3$, $ClF_5$, $ClF_3$) and fuels were considered.

Argon matrix reactions of alkali-metal atoms with $F_2$ have been studied, using laser Raman and i.r. spectroscopy.[11] The F—F stretching motion in the $F_2^-$ ion occurs at ca. 460 cm$^{-1}$, compared with 892 cm$^{-1}$ for the neutral molecule; the latter band is very close to that for the gaseous molecule, and this fact clearly shows that intermolecular fluorine bonding is very feeble. Absolute measurements of the coefficient of shear viscosity of compressed liquid $F_2$ at 90—300 K and for pressures up to 20 MPa and of saturated liquid $F_2$ at 70—144 K have been reported.[12]

The direct fluorination of $C_2B_5H_7$ under carefully controlled conditions has been shown to yield a number of novel, fluorine-containing carbaboranes as well as $CH_2(BF_2)_2$.[13] A new mild fluorinating agent, $(CF_2NCl)_3$, reacts as if it contains positively charged chlorine;[14] it is an involatile, colourless liquid which may be stored in Pyrex vessels.

Equilibrium distributions have been reported for the scrambling of fluorine and chlorine between pairs of the following centres: $Me_2C$, $Me_2Si$, $Me_3Si$, $Me_2Ge$, $Me_3Ge$, MePO, and $Me_2As$.[15] In Group IV there is a strong preference for F to be bound to silicon rather than to comparable carbon or germanium centres. There is a slight preference for $Me_2PO$ rather than for $Me_2Si$, but changes in substituents can reverse this; F prefers $Me_2As$ to $Me_3Ge$. Overall, mixed fluoro-chloro-species are less abundant at equilibrium than is to be expected statistically.

Estimates have been made of the annual emissions of HCl, HF, and $SO_2$ from volcanic eruptions to the tropo- and strato-spheres.[16] The results indicate that man-made chlorofluorocarbons are potentially more important in stratospheric chemistry than halides of volcanic origin. Lovelock[17] has listed the concentration and concentration profiles of halocarbons in the troposphere and the lower stratosphere over the U.K. and the mid-Atlantic: he has also estimated the total quantity of chloro-species transferred to the stratosphere by halocarbon sources. The interest in such materials, and especially in chlorofluorocarbons, e.g. $CFCl_3$ and $CF_2Cl_2$, used as propellants and refrigerants, is that they are photolysed to give chlorine atoms in the stratosphere; the Cl then destroys ozone by reactions (part of the $ClO_x$ cycle) such as (1).

---

[10] A. E. Axworthy, J. Q. Weber, E. C. Curtis, and C. Selph, *West. States Sect., Combust. Inst.*, 1974, WSS/CI-74-13 (*Chem. Abs.* 1975, **82**, 8278).
[11] W. F. Howard and L. Andrews, *Inorg. Chem.*, 1975, **14**, 409.
[12] W. M. Haynes, *Physica*, 1974, **76**, 1.
[13] N. J. Marachin and R. J. Lagow, *Inorg. Chem.*, 1975, **14**, 1855.
[14] R. L. Kirchmeier, G. H. Sprenger, and J. M. Shreeve, *Inorg. Nuclear Chem. Letters*, 1975, **11**, 699.
[15] S. C. Pace and J. G. Riess, *J. Organometallic Chem.*, 1974, **76**, 325.
[16] R. D. Cadle, *J. Geophys. Res.*, 1975, **80**, 1650.
[17] J. E. Lovelock, *Nature*, 1974, **252**, 292.

$$Cl + O_3 \rightarrow ClO + O_2; \quad ClO + O \rightarrow Cl + O_2 \tag{1}$$

It is thought that the halogenomethanes may remain at altitudes of 20—40 km for 40—150 yr and may well reach saturation values of 10—30 times the present levels.[18a,b] According to Wofsy et al.[19] the reduction in the ozone conentration in the stratosphere caused by the chloromethanes could be as much as 3% by 1980 and 16% by the year 2000, if the chloromethane consumption were to grow at the rate of 10% $yr^{-1}$. Even if they were banned in 1990, the residual effect could be significant, and might endure for several hundred years. Other investigators, however, take a less pessimistic view (*e.g.* ref. 20).

Although this controversy about the environmental hazards associated with the accumulation of chlorocarbons in the stratosphere is likely to continue for some time, it has been said[21] that the U.S. Federal Government is considering placing a temporary ban on aerosols employing chlorofluorocarbon propellants to ensure than further research can be carried out. Certainly there is much research needed in this area. For example, the stratospheric ClO concentrations are not known and yet, according to Nicholls,[22] they may be directly measurable from spectroscopic data already available but previously attributed to stratospheric NH. The primary manner in which the $ClO_x$ cycle is interrupted is probably through the reaction of Cl with various H-containing species, with the formation of HCl[18b,23] Methane is a likely reagent for this purpose, but the rate constants of the key processes are only now being made available. Thus Clyne and Watson[23] have investigated the kinetics of the reactions between ClO, $X^2\Pi$, $v = 0$, radicals and NO, CO, $N_2O$, $H_2$, $CH_4$, $C_2H_4$, $C_2H_2$, and $NH_3$ at 298 K. Both NO and $NO_2$ are believed to be closely involved in the $NO_x$ cycle, in determining the present level of $O_3$ concentration in the earth's atmosphere. Clyne and Watson conclude that the $ClO_x$ cycle is more effective than the $NO_x$ cycle and they point out that $ClO_x$ chemistry is strongly dependent on the concentrations of NO and of OH in the stratosphere; these quantities are not known with adequate accuracy. Some of their results are reproduced in Table 1.

Further examples of the selective isotopic enrichment of chlorine-containing molecules have been described, *e.g.* HCl,[24] ICl,[25] $SCCl_2$:[26] Moore and Young have now patented the use of laser irradiation of chlorine- and bromine-containing molecules for the purposes of isotopic enrichment.[27]

The pseudo-first-order rate constant for the reaction of atomic chlorine with $Br_2$ has been determined by means of measurements of the Cl resonance fluorescence.[28] Yellowish-orange $M^+Cl_2^-$ (M = Li, Na, K, Rb, Cs, or $\frac{1}{2}$Ba) species have been produced in low-temperature matrixes.[29] Resonance Raman spectra of

---

[18] M. J. Molina and F. S. Rowland, (*a*) *Nature*, 1974, **249**, 810; (*b*) *J. Phys. Chem.*, 1975, **79**, 667.
[19] S. C. Wofsy, M. B. McElroy, and N. D. Sze, *Science*, 1975, **187**, 535.
[20] R. J. Cicerone, S. Walters, and R. S. Stolarski, *Science*, 1975, **188**, 378.
[21] Editorial, *Chem. in Britain*, 1976, **12**, 2.
[22] R. W. Nicholls, *J. Atmos. Sci.*, 1975, **32**, 856.
[23] M. A. A. Clyne and R. T. Watson, *J.C.S. Faraday I*, 1974, **70**, 2250.
[24] D. Arnoldi, K. Kaufmann, and J. Wolfrum, *Phys. Rev. Letters*, 1975, **34**, 1597.
[25] D. D. S. Liu, S. Datta, and R. N. Zare, *J. Amer. Chem. Soc.*, 1975, **97**, 2557.
[26] M. Lamotte, H. J. Dewey, R. A. Keller, and J. J. Ritter, *Chem. Phys. Letters*, 1975, **30**, 165.
[27] C. B. Moore and E. S. Young, Ger. Offen. 2403580.
[28] P. P. Bemand and M. A. A. Clyne, *J.C.S. Faraday II*, 1975, **71**, 1132.
[29] W. F. Howard and L. Andrews, *Inorg. Chem.*, 1975, **14**, 767.

**Table 1** Calculated steady-state concentrations and relative reaction rates in the $ClO_x$ cycle

| Altitude/km | 15 | 25 | 35 | 45 |
|---|---|---|---|---|
| | \multicolumn{4}{c}{Concentration ratios} | | | |
| [HCl]/[Cl] | $0.52 \times 10^6$ | $0.63 \times 10^5$ | $3.3 \times 10^3$ | $1.7 \times 10^2$ |
| [ClO]/[Cl] | $9.2 \times 10^2$ | $3.3 \times 10^3$ | $1.2 \times 10^3$ | 18.2 |
| [ClO]/[HCl] | $17.3 \times 10^{-4}$ | $5.3 \times 10^{-2}$ | 0.36 | $10.5 \times 10^{-2}$ |
| | \multicolumn{4}{c}{Concentrations/molecule cm$^{-3}$} | | | |
| [HCl] | $4.3 \times 10^9$ | $8.4 \times 10^8$ | $1.4 \times 10^8$ | $3.8 \times 10^7$ |
| [ClO] | $7.4 \times 10^6$ | $4.4 \times 10^7$ | $6.0 \times 10^7$ | $4.4 \times 10^6$ |
| [Cl] | $7.9 \times 10^3$ | $13.1 \times 10^3$ | $4.3 \times 10^4$ | $2.2 \times 10^5$ |
| | \multicolumn{4}{c}{Rates of reaction/molecule cm$^{-3}$s$^{-1}$} | | | |
| $ClO + O \rightarrow Cl + O_2$ | 54.7 | $4.4 \times 10^4$ | $10.2 \times 10^5$ | $12.4 \times 10^5$ |
| $NO + ClO \rightarrow NO_2 + Cl$ | $17.6 \times 10^4$ | $10.5 \times 10^5$ | $11.5 \times 10^5$ | $2.2 \times 10^4$ |
| $O + NO_2 \rightarrow NO + O_2$ | $8.1 \times 10^2$ | $4.4 \times 10^5$ | $3.5 \times 10^6$ | $6.8 \times 10^5$ |
| $Cl + O_3 ClO + O_2$ | $17.4 \times 10^4$ | $10.3 \times 10^5$ | $18.6 \times 10^5$ | $11.5 \times 10^5$ |

the anion were obtained, and from the progression frequencies for the $Li^+$, $Rb^+$, and $Cs^+$ salts a dissociation energy, $D_e$, of $1.25 \pm 0.11$ eV was deduced, which is in agreement with the thermodynamic value. It has been reported[30] that $ClO_2$ centres, formed by $X$- or $\gamma$-irradiation of $Ca(ClO_3)_2,2H_2O$ at 298 K, are replaced by $Cl_2^-$ centres when irradiated by 360 nm light.

The polarity of $(Cl_2)_2$ has been determined in a molecular beam electric resonance spectrometer and is said[31] to be consistent with either an L- or a T-shaped structure. Fluorescence spectra have been obtained from both monomeric and dimeric $Cl_2$ and $Br_2$ molecules in inert-gas matrixes at 15 K.[32] The i.r. spectra of the molecular complexes of $Cl_2$ and $Br_2$ with $NH_3$ have been obtained at ca. 72 K:[33] at or above 150 K reactions occur with the formation of ammonium salts. Raman spectra have also been reported by the same group of workers[34,35] for $Cl_2$ and $Br_2$ complexes with $NH_3$, alkylamines, and pyridine. From the results, simple treatments of normal co-ordinates and of the extent of charge transfer were performed. Raman spectra in the range 20—600 cm$^{-1}$ of the hydrates of $Cl_2$ and $Br_2$, obtained at 77 K, show that the interaction between the donor and either acceptor is remarkably weak.[36]

Shapovalov[37] has reported the results of a study of the effects of continuous exposure to elemental bromine, bromides, and methyl bromide on the functions of liver and thyroid gland, and on haematological change, in 140 workers. Spectrophotometric methods[38] have been used to obtain the rate law and other parameters at 298.2 K for $Br_2 + HCN \rightarrow BrCN + H^+ + Br^-$. Sill[39] has discussed the

---

[30] D. Suryanarayana and J. Sobhanadri, *Mol. Phys.*, 1975, **29**, 1369.
[31] S. J. Harris, S. E. Novick, J. S. Winn, and W. Klemperer, *J. Chem. Phys.*, 1974, **61**, 3866.
[32] B. S. Ault, W. F. Howard, and L. Andrews, *J. Mol. Spectroscopy*, 1975, **55**, 217.
[33] E. V. Belousova, Ya. M. Kimel'feld, and A. P. Shvedchikov, *Zhur. fiz. Khim.*, 1975, **49**, 1075.
[34] Ya. M. Kimel'feld, A. B. Mostovoy, and L. M. Mostovaja, *Chem. Phys. Letters*, 1975, **33**, 114.
[35] Ya. M. Kimel'feld, A. B. Mostovoy, and L. M. Mostovaja, *Zhur. fiz. Khim.*, 1975, **49**, 284.
[36] J. W. Anthonsen, *Acta Chem. Scand.* (A), 1975, **29**, 175.
[37] Yu. D. Shapovalov, *Vrach. Delo*, 1974, 110.
[38] M. F. Nolan, J. N. Pendlebury, and R. H. Smith, *Internat. J. Chem. Kinetics*, 1975, **7**, 205.
[39] G. T. Sill, *J. Atmos. Sci.*, 1975, **32**, 1201.

composition of the u.v.-dark clouds on Venus and has proposed that they consist of $Br_2$ dissolved in droplets of aqueous HBr.

Comparative pulsed and standard neutron irradiations of $Br_2$ and $I_2$ in organic solvents have been carried out in both liquid and gaseous phases in order to study the chemical behaviour of halogen atoms after $(n, \gamma)$ reactions.[40] The results showed that a 2500-fold increase of the thermal neutron flux does not influence the organic yields, and it was concluded that in the liquid phase the stabilization time of the recoil halogen atoms is $10^{-7}$ s, of the same order as the mean lifetime of free radicals. According to a report by Bryant and Jones[41] the cumulative quantities of $^{129}I$ in the environment by the year 2000, based on predicted fuel-reprocessing programmes, are so low that this nuclide will pose no significant health problems this century. Whereas decontamination factors of 100 are sufficient for the $^3H$ and $^{85}Kr$ emissions from fuel-reprocessing plants, for $^{131}I$ from fast breeder reactor fuel a factor of greater than 2000 may be necessary.[42]

The use of a sodium hydroxide containment spray to prevent the accidental escape of radioactive iodine vapour has been discussed by a group from the Institute of Atomic Energy at Kyoto University.[43] The reaction between iodine (as $I_2$ and organic iodides) and hyperazeotropic nitric acid ($>16$ mol l$^{-1}$) is a promising technique for the retention of radioiodine, the Iodox process. Mailen and Tiffany[44] have investigated the rate of reaction and have found that the initial stage of the reaction produces $I^+$, which then undergoes further oxidation in a second slower stage to $IO_3^-$. The experimental data have been analysed in terms of the reactions (2) and (3), for which equilibrium constants (at 25 °C) were calculated.[44b] The efficiencies of absorption of radioiodine from an air stream by a solution of $Hg(NO_3)_2$ in concentrated $HNO_3$[45] and by 13X molecular sieves[46] have also been evaluated. Moravek and Hladky[47] have discussed the methods of determining radioiodine in the presence of large excesses of gaseous fission products, especially radioxenon.

$$I_2 + 4HNO_3 \rightleftharpoons 2I^+ + 2NO_3^- + N_2O_4 + 2H_2O \qquad (2)$$

$$I^+ + 3HNO_3 + NO_3^- \rightleftharpoons IO_3^- + H^+ + 2N_2O_4 + H_2O \qquad (3)$$

The recombination of iodine atoms in an excess of HCl, DCl, and HBr has been studied over a wide range of temperature, using flash photolysis techniques.[48] The results suggest that reaction is dominated at least at lower temperatures, by the radical–molecule complex mechanism. It appears that I—(HX) are the unstable intermediates, in which the iodine atom is interacting with both H and X, rather

---

[40] Y. Llabador and M. E. Halter, *Radiochem. Radioanalyt. Letters*, 1975, **20**, 241.
[41] P. M. Bryant and J. A. Jones, National Radiological Protection Board, (U.K.) (Report) 1972, NRPB-R8 (*Chem. Abs.*, 1975, **82**, 144454).
[42] H. Beaujean, J. Bohnenstingl, M. Laser, E. Merz, and H. Schnez, 'Environmental Behaviour of Radionuclides Released', Nuclear Industry Processing Symposium, IAEA, Vienna, 1973, p. 63.
[43] M. Adachi, W. Eguchi, F. Tohdo, and M. Yoneda, *J. Chem. Eng. Japan*, 1974, **7**, 360.
[44] (a) J. C. Mailen and T. O. Tiffany, *J. Inorg. Nuclear Chem.*, 1975, **37**, 127; (b) J. C. Mailen, *ibid.*, p. 1019.
[45] J. M. Scmitt and D. J. Crouse, U.S.P. 3585407.
[46] G. Collard, J. Broothaerts, W. Goossens, and J. Stevens, Centre Étude Energie Nucléaire [Rapp.], 1974, BLG 492.
[47] J. Moravek and E. Hladky, *Radioisotopy*, 1975, **16**, 63.
[48] H. W. Chang and G. Burns, *J. Chem. Phys.*, 1975, **62**, 2426.

than the stable hydrogen dihalide radicals IHX, observed earlier by the matrix-isolation technique. The reactions of $I_2$ and alkali-metal atoms in an Ar matrix have been examined by Raman spectroscopy;[49] using a Kr ion laser a resonance Raman progression was obtained for each of the $M^+I_2^-$ species (M = Li, Na, K, Rb, or Cs). The dissociation energy of $I_2^-$ was estimated to be 20 kcal mol$^{-1}$, which is in reasonable agreement with the value of $24.5 \pm 2$ kcal mol$^{-1}$ calculated by the same authors from thermochemical data.

The quality of $I_2$ crystals formed in reaction (4) does not depend on temperature (20—80 °C) or the salt concentration;[50] the impurities trapped in the crystals are

$$2KI + 2NaNO_2 + 2H_2SO_4 \rightarrow I_2 + Na_2SO_4 + K_2SO_4 + 2NO + 2H_2O \qquad (4)$$

non-isomorphic. Iodine adsorbed on silica has been investigated by Raman spectroscopy.[51] Shifts of the fundamental and overtones were observed to be small, thus indicating that only weak interactions are occurring, probably between surface OH groups and $I_2$ molecules.

A voltammetric method has been used to study the electrochemical behaviour of $I_2$ in pyridine solutions.[52] The electrochemical behaviour of $I_2$ in $AlCl_3$–NaCl melts is consistent with the production of a stable species of oxidation state +1, which is stabilized in basic rather than in acidic melts, presumably owing to the formation of $ICl_n^{1-n}$ anions.[53]

The $^{129}I$ Mössbauer spectra of some molecular complexes of $I_2$ at 77 K have been reported.[54] Based on the values of the hyperfine constants, the complexes may be divided into three main groups according to the nature of the organic (donor) solvent involved: in $\pi$-donor solvents the charge-transfer interaction is dominant, whereas in amine complexes electrostatic forces are more important. Bhat[55] has demonstrated that the absorption band due to free $I_2$ in $CS_2$ shifts from 517 to 498 nm on the addition of sulphur. The association constant of the complex between 2,5-dimethylpyrazine and $I_2$ varies between 280 and 420 (mol fraction)$^{-1}$ in the solvents $C_6H_{12}$, $CCl_4$, and $C_6H_6$.[56] A study of the complexes of $I_2$ with pyridine and quinoline derivatives shows[57] that a reasonably linear correlation exists between log $K_{ct}$ ($K_{ct}$ = equilibrium constant) and p$K_a$ for donors bearing no ortho-substituents. Sambhi and Khoo[58] have reported their estimates of the magnitude of the contribution of effects other than charge transfer to $\Delta H°$ of amine–$I_2$ complexes. The 1:1 and 1:2 adducts of hexamethylenetetramine with $I_2$ have been investigated by X-ray crystallography.[59] Both compounds belong to the class of $n-\sigma^*$ donor–acceptor complexes. The I—I distances are 2.830 in the orange 1:1 adduct and 2.791 and $2.771 \pm 0.002$ Å in the red-brown

---

[49] W. F. Howard and L. Andrews, *J. Amer. Chem. Soc.*, 1975, **97**, 2956.
[50] A. I. Goncharov, Yu. D. Nekrasov, B. M. Binshtok, and V. S. Koryakov, *Zhur. priklad. Khim.*, 1975, **48**, 895.
[51] T. Nagasao and H. Yamada, *J. Raman Spectroscopy*, 1975, **3**, 153.
[52] J. M. Nigretto and M. Josefowicz, *Electrochim. Acta*, 1974, **19**, 809.
[53] R. Marassi, G. Mamantov, and J. Q. Chambers, *Inorg. Nuclear Chem. Letters*, 1975, **11**, 245.
[54] S. Bukhshpan, C. Goldstein, T. Sonnino, L. May, and M. Pasternak, *J. Chem. Phys.*, 1975, **62**, 2606.
[55] S. N. Bhat, *J. Inorg. Nuclear Chem.*, 1975, **37**, 276.
[56] P. Boule, *Compt. rend.*, 1975, **280** C, 623.
[57] S. Sorriso, G. G. Aloisi, and S. Santini, *Z. phys. Chem.* (Frankfurt), 1975, **94**, 117.
[58] M. S. Sambhi and S. K. Khoo, *J. Phys. Chem.*, 1975, **79**, 666.
[59] H. Pritzkow, *Acta Cryst.*, 1975, **B31**, 1589.

1:2 adduct. The latter is converted slowly but spontaneously into the bis(hexamethylenetetramine) iodonium tri-iodide. A careful investigation of the nature of dialkyltellurium tetraiodides has shown that these compounds should actually be considered as adducts between dialkyltellurium di-iodides and $I_2$.[60] Preliminary results from the on-line separation of At from molten $ThF_4$,LiF in the 600 MeV C.E.R.N. synchrotron have been reported by Ravn et al.[61] The energies and intensities of γ-rays and internal-conversion electrons from the electron capture and from the particle decay of $^{211}$At have been measured.[62] The rates of exchange of iodine and astatine species between submonolayers on Pt and aqueous solution have been compared.[63]

**Halides.**—Mason[64] has shown that the n.m.r. shielding parameters for the (spin-paired) binary fluoride molecules and ions vary periodically with the atomic number of the central atom. The periodic correlation shows the basis of the familiar relation between fluorine shielding and electronegativity of the central atom, and also of the important exceptions to this relation. Fluorine-19 n.m.r. relaxation rates of ionic fluorides MF (M = Na, K, Rb, Cs, or $ND_4$) as a function of concentration have been interpreted as showing $F^-$—$F^-$ contacts.[65]

Estimates have been obtained of the $F^-$ ion affinities of $MF_5$ in the reaction $MF_5(c) + F^-(g) \rightarrow MF_6^-(g)$:[66] the values reported were −412 and −491 kJ mol$^{-1}$ for $MoF_5$ and $WF_5$, respectively. Workers at the Rennselaer Polytechnic Institute[67] have collected and evaluated electrical conductance, density, viscosity, and surface-tension data for 44 binary mixtures of fluoride salts.

A new type of ceramic membrane electrode that is selective for $F^-$ ions has been prepared by sintering $LaF_3$, $EuF_3$, and $CaF_2$ at above 1200 °C in an atmosphere of HF.[68] The approximate determination of the $F^-$ ion concentration in polluted waters can be made by means of a $F^-$ ion electrode in the presence of citrate ion;[69] however, for an accurate determination a photometric method was applied after separation of $F^-$ by distillation. The probable composition of the $La^{3+}$–alizarin complexone ($H_4A$)–$F^-$ complex is said to be $La(LaA)_4F_2^{3-}$.[70] Alizarin fluorine blue 5-sulphonate is more soluble in water than the unsulphonated reagent;[71] the ternary $La^{3+}$–$F^-$–complexone absorbs at 583 nm. The application of neutron activation analysis to the determination of fluorine has has been reviewed briefly by Verot and Jaumier.[72] X-Ray photoelectron spectroscopy (p.e.s.) has been used to identify and to provide a quantitative estimate of $CF_3$, $CF_2$, and CF groups on the surface of fluorinated diamond.[73] Proton and deuteron radiolysis of

---

[60] K. V. Smith and J. S. Thayer, *Inorg. Chem.*, 1974, **13**, 2021.
[61] H. L. Ravn, S. Sundell, and L. Westgaard, *Nuclear Instrum. Methods*, 1975, **123**, 131.
[62] L. J. Jardine, *Phys. Rev. (C)*, 1975, **11**, 1385.
[63] G. A. Nagy and V. A. Khalkin, *Magyar Kém. Folyóirat*, 1975, **81**, 33.
[64] J. Mason, *J.C.S. Dalton*, 1975, 1426.
[65] H. G. Hertz and C. Rädle, *Ber. Bunsengesellschaft phys. Chem.*, 1974, **78**, 509.
[66] J. Burgess, I. Haigh, R. D. Peacock, and P. Taylor, *J.C.S. Dalton*, 1974, 1064.
[67] G. J. Janz, G. L. Gardner, U. Krebs, and R. P. T. Tomkins, *J. Phys. Chem. Ref. Data*, 1974, **3**, 1.
[68] H. Hirata and M. Ayuzawa, *Chem. Letters*, 1974, 1451.
[69] A. Franke, *Vom Wasser*, 1974, **42**, 161.
[70] T. Anfaelt and D. Jagner, *Analyt. Chem. Acta*, 1974, **70**, 365.
[71] M. A. Leonard and G. T. Murray, *Analyst*, 1974, **99**, 645.
[72] J. L. Verot and J. J. Jaumier, *Inf. Chem.*, 1974, **137**, 179.
[73] P. Cadman, J. D. Scott, and J. M. Thomas, *J.C.S. Chem. Comm.*, 1975, 654.

chlorine-containing samples apparently produces high yields of $HCl_2$ and $DCl_2$, which are thought to be anionic species.[74] Rate constants and, in some instances, also activation-energy dependences have been measured for the capture of low-energy (thermal) electrons by molecules containing C—Cl, N—Cl, or Cl—Cl bonds.[75]

An inexpensive solid-state $Cl^-$-sensitive electrode formed from a mixture of HgS and $Hg_2Cl_2$ is reported to have better overall performance characteristics than commercially available AgCl electrodes.[76]

A quantitative separation of the halide ions $F^-$, $Cl^-$, $Br^-$, and $I^-$ has been obtained by means of a column of Sephadex G-15 gel.[77] Thin-layer voltammetric data for Pt electrodes indicate that these ions form chemisorbed layers which withstand rinsing with typical aqueous electrolytes.[78] The chemisorbed species are much less reactive towards electrochemical oxidation than the aqueous ions.

Crystals of hepta(tetrathiafulvalene) pentaiodide show a high electrical conductivity at room temperature. The structure of the $(hk0)$ projection has now been solved, and it has been shown that it contains infinite columns of tetrathiafulvalene groups and infinite rows of iodide, not polyiodide, ions.[79]

The removal of iodide as HI from $LiF-BeF_2$ melts by $HF-H_2$ sparging has been investigated to determine its applicability to molten-salt reactor fuels.[80] The iodide-removal mechanism is explained by a model which assumes that the rate-controlling step is the transport of $I^-$ from the bulk of the melt to the surface. By varying the temperature and the $H_2$ pressure, the nature of the iodine species displaced can be affected. The gravimetric determination of iodine in aqueous solution as $Ph_4AsICl_2$ has been described:[81] the iodine-containing solution in 6M-HCl saturated with $Cl_2$ is treated with excess $Ph_4AsCl$ to precipitate the dichloroiodate salt.

**Interhalogens and Related Species.**—The vibrational spectra of $ClONO_2$ and $FONO_2$ are said[82] to be consistent with a structure in which the halogen atom is perpendicular to the $ONO_2$ plane: this conclusion is contrary to previous assumptions, to new Raman data for $ClONO_2$,[83] and is also at variance with the known (planar) structures of $HONO_2$ and $MeONO_2$. Measurements of the equilibrium constant for the gas-phase reaction $F + Cl_2 \rightarrow ClF + Cl$, by a Cl-atom chemiluminescence technique, have yielded $\Delta H_0^\circ = 3.27 \pm 0.24$ kcal mol$^{-1}$;[84] this result settles the existing uncertainty in the spectroscopic determination of $D_0^\circ(ClF)$ in favour of the larger value, viz. 60.35 kcal mol$^{-1}$.

The $^{35}Cl$ n.q.r. frequencies have been reported for the two trichloride salts $Ph_4AsCl_3$ and $Et_4NCl_3$.[85] Using the Townes and Dailey model, the negative

---

[74] L. Andrews, B. S. Ault, J. M. Grzybowski, and R. O. Allen, *J. Chem. Phys.*, 1975, **62**, 2461.
[75] E. Schultes, A. A. Christodoulides, and R. N. Schindler, *Chem. Phys.*, 1975, **8**, 354.
[76] J. F. Lechner and I. Sekerka, *J. Electroanalyt. Chem. Interfacial Electrochem.*, 1974, **57**, 317.
[77] T. Deguchi, *J. Chromatog.*, 1975, **108**, 409.
[78] R. F. Lane and A. T. Hubbard, *J. Phys. Chem.*, 1975, **79**, 808.
[79] J. J. Daly and F. Sanz, *Acta Cryst.*, 1975, **B31**, 620.
[80] C. F. Baes, R. P. Wichner, C. E. Bamberger, and B. F. Freasier, *Nuclear Sci. Eng.*, 1975, **56**, 399.
[81] N. Ganchev and A. Kirev, *Doklady Bolg. Akad. Nauk*, 1975, **28**, 99.
[82] K. O. Christe, C. J. Schack, and R. D. Wilson, *Inorg. Chem.*, 1974, **13**, 2811.
[83] D. W. Amos and G. W. Flewett, *Spectrochim. Acta*, 1975, **31A**, 213.
[84] P. C. Nordine, *J. Chem. Phys.*, 1974, **61**, 224.
[85] E. F. Riedel and R. D. Willett, *J. Amer. Chem. Soc.*, 1975, **97**, 701.

charge on the $Cl_3^-$ ion is shared evenly between the two terminal atoms, while the central Cl bears a slight positive charge. The charge distribution was stated to indicate little or no $d$-orbital contribution to the bonding. Narain and Chandra[86] have given their interpretation of the n.q.r. coupling constants for $ClF_3$ and $BrF_3$, according to which there is approximately 14% and 6.2% $d$-character in the hybrid orbitals formed by Cl and Br, respectively.

A method for concentrating lean gold ores and flotation concentrates, involving fluorination with $ClF_3$, has been developed.[87] It has been tested on ores containing as little as 0.6 (g gold) ton$^{-1}$. The kinetics of the reaction between H atoms and $ClF_3$ have been studied directly by the discharge–flow e.p.r. technique.[88] Low-temperature Raman and i.r. spectra of $ClF_3$ have confirmed the existence of an intermediate solid phase (I), previously deduced from Grisard's calorimetric measurements.[89] It has also been suggested that the intermolecular forces in $ClF_3$ dimers are weak, and that no exchange of fluorine occurs as they dissociate.

The reaction between $ClF_3$ and $OF_2$ at high pressures (100—700 atm) and temperatures (200—250 °C) has been reported to yield $ClF_5$.[90] N.m.r. relaxation-time measurements have been used to derive more n.m.r. parameters for $ClF_5$ and $ClOF_3$.[91]

The previously unreported $ClF_6^-$ radical has now been produced by $\gamma$-irradiation of $ClF_5$ in an $SF_6$ matrix, and has been identified by means of its e.s.r. spectrum.[92] The unpaired electron in this radical evidently populates the chlorine $3s$ orbital (spin density 0.46). It is interesting to note that the lines in the e.s.r. spectrum are broader than those of the $ClF_4^-$ radical; it has been suggested that this broadening arises from the geometry of $ClF_6^-$ (which should be non-octahedral, for reasons similar to those discussed elsewhere for $XeF_6$).

It has been shown[93] that the spectroscopically determined dissociation energy of BrF is unreliable and that the best present estimate is in the range 20 540–21 860 cm$^{-1}$. Naumann and Lehmann[94a] have reinvestigated the low-temperature fluorination of $Br_2$ with elemental $F_2$. At $-78$ °C an orange-red solid is precipitated in $CFCl_3$ which appears to be BrF, as described by Ruff; however, it could not be isolated in a pure state, owing to the disproportionation into $Br_2$ and $BrF_3$. The same workers could isolate the 1:1 adduct of BrF and pyridine from the solution formed by the disproportionation of $BrF_3$ in $CFCl_3$ in the presence of CsF: the new compound decomposes above 60 °C. All attempts to prepare $CsBrF_2$, previously reported by Surles et al.,[94b] were unsuccessful.

Laser-induced fluorescence has been observed and attributed to isolated BrCl and ICl molecules:[95] in addition, resonance Raman series have been obtained from

---

[86] P. Narain and S. Chandra, *Indian J. Pure Appl. Phys.*, 1975, **13**, 496.
[87] A. A. Opalovskii, N. I. Tyuleneva, and S. V. Zemskov, *Zhur. priklad. Khim.*, 1974, **47**, 2157.
[88] S. J. Pak, R. H. Krech, D. L. McFadden, and D. I. MacLean, *J. Chem. Phys.*, 1975, **62**, 3419.
[89] R. Rousson and M. Drifford, *J. Chem. Phys.*, 1975, **62**, 1806.
[90] C. Merrill, U.S.P. 3843700 (*Chem. Abs.*, 1975, **82**, 19116).
[91] M. Alexandre and P. Rigny, *Canad. J. Chem.*, 1974, **52**, 3676.
[92] K. Nishikida, F. Williams, G. Mamantov, and N. Smyrl, *J. Amer. Chem. Soc.*, 1975, **97**, 3526.
[93] J. A. Coxon, *Chem. Phys. Letters*, 1975, **33**, 136.
[94] (a) D. Naumann and E. Lehmann, *J. Fluorine Chem.*, 1975, **5**, 307; (b) T. Surles, L. Quarterman, and H. Hyman, *J. Inorg. Nuclear Chem.*, 1973, **35**, 668.
[95] C. A. Wright, B. S. Ault, and L. Andrews, *J. Mol. Spectroscopy*, 1975, **56**, 239.

BrCl, ICl, and IBr in inert matrixes. The Raman spectrum of BrCl hydrate recorded at $-196\,°C$ shows a slight interaction between donor and acceptor.[36] Crystals of $NEt_4BrCl_2$ contain 2 independent $BrCl_2^-$ ions, one of which is symmetric whereas the other is significantly asymmetric.[96]

Bromine(III) fluoride has been shown to form a colourless 1:1 adduct with pyridine in $CFCl_3$ solution at $-78\,°C$; this new compound is stable up to *ca.* $110\,°C$ in the absence of water.[94a] Naumann and Lehmann have also demonstrated that $BrF_3$ disproportionates in $CFCl_3$ into $BrF_5$ and $BrF$ in the presence of a catalytic amount of CsF. The fluorination of $Br_2$ by $F_2$ in an electric glow discharge produces $BrF_5$ almost quantitatively.[97] When $NF_3$ was used as the fluorinating agent a mixture of $BrF_3$ and $BrF_5$ resulted, rather than bromodifluoroamine.

The $BrF_6^-$ radical has been identified from its e.s.r. spectrum as a product of the $\gamma$-irradiation at 77 K of a mixture of $BrF_5$ and $SF_6$.[98] Analysis of the coupling-constant data indicates that the $4s$ spin density is 0.54, as compared with 0.46 for the $3s$ orbital on Cl in $ClF_6^-$ radical. Christe and Wilson[99] have described the preparation of $BrF_6^+$ salts and have also obtained complete data for the vibrational spectra, from which they have calculated force constants for this cation: the Br—F stretching force constant, 4.9 mdyn $Å^{-1}$, is the highest yet reported for any BrF bond.

Emissions from both the $^3\Pi_0^+$ and the previously unreported $^3\Pi_1$ states of the IF molecule have been observed in the gas-phase reaction of $I_2$ with $F_2$ at low pressure;[100] a four-centre complex has been proposed as the reaction intermediate. A combined theoretical–experimental programme has been conducted to establish techniques for the study of excited-state transitions in $I_2$ and ICl.[101] Experimental techniques based on two-step excitation using two synchronized, tunable lasers have been developed, and successfully applied to excited-state fluorescence measurements on ICl. Iodine(I) chloride adsorbed on silica gives the same Raman spectrum as that obtained from adsorbed $I_2$.[51]

Resonance Raman series have been observed from samples of ICl and IBr in inert matrixes and assigned to these species.[95] A visible spectral method has been used to investigate the equilibrium $I_2 + Br_2 \rightleftharpoons 2IBr$ in $CCl_4$ solution.[102] The value obtained for the equilibrium constant, $495 \pm 42$, corresponding to a degree of dissociation of 8%, is intermediate between the two previously reported values.

The rate constant for the reaction between pyridine and ICl shows the expected dependence on the dielectric constant of the solvent.[103] Electrolysis of ICl in pyridine, $MeCONMe_2$, and $Et_2O$ has been studied, using a radiochemical modification of the Hittorf method;[104] it was observed that both I and Cl migrate

[96] W. Gabes and K. Olie, *Cryst. Struct. Comm.*, 1974, **3**, 753.
[97] I. V. Nikitin and V. Ya. Rosolovskii, *Zhur. neorg. Khim.*, 1975, **20**, 263.
[98] K. Nikishida, F. Williams, G. Mamantov, and N. Smyrl, *J. Chem. Phys.*, 1975, **63**, 1693.
[99] K. O. Christe and R. D. Wilson, *Inorg. Chem.*, 1975, **14**, 694.
[100] J. W. Birks, S. D. Gabelnick, and H. S. Johnston, *J. Mol. Spectroscopy*, 1975, **57**, 23.
[101] C. R. Claydon, R. H. Barnes, R. H. Kahn, and C. M. Verber, U.S. N.T.I.S., AD/A Rep. 1974, No. 00384/8GA.
[102] J. D. Childs, L.-N. Lin, and S. D. Christian, *J. Inorg. Nuclear Chem.*, 1975, **37**, 757.
[103] A. Das Gupta and R. Basu, *J. Chim. phys.*, 1975, **72**, 271.
[104] Yu. A. Karapetyan, *Ukrain khim. Zhur.*, 1974, **40**, 1205.

to the anode, and this was attributed to the presence of $ICl_2^-$. However, no migration to the cathode could be detected, and the author attributed this to the presence of a very slowly moving solvated $I^+$ ion. Ionic transport of ICl in $SbCl_5$ was also investigated, for which it was found that I migrates to the cathode and Cl to the anode.

Calorimetrically determined $\Delta H°$ values have been reported for some adducts of IX (X = I, Br, Cl, CN, or $CF_3$) with oxygen donors.[105] The following order of acceptor strengths was deduced: $ICF_3 < I_2 < ICN < IBr < ICl$; donor strengths followed the pattern $Ph_2CO < Ph_2SO < Ph_3PO < Ph_2SeO < Ph_3AsO$. Complexes of ICN and a number of pyridine derivatives have been investigated by i.r. spectroscopy in $CCl_4$, from which it was shown that the analogous $I_2$ complexes are more stable than those of ICN.[106] The charge-transfer complex $CF_3I,NMe_3$ has been studied by microwave spectroscopy and the N—I bond length was shown to be 2.93 Å.[107]

The oxidation of $I_2$ in acetic anhydride in the presence of phosphoric acid was reported earlier to yield $IPO_4$: however, the Mössbauer spectrum of the product clearly demonstrates that it should be treated as an iodine(I) compound.[108] The yields of organic iodides produced in the reaction at 77 K of saturated aliphatic hydrocarbons with $I^+$ and $I_2^+$, generated by electron bombardment of $I_2$ vapour, have been accounted for by a reaction mechanism involving cage effects.[109]

The reactions of $I_2$ with $ICl–AlCl_3$ mixtures have been investigated by thermal, microscopic and X-ray analysis;[110] the only products were $I_3^+ AlCl_4^-$ (dimorphic), $I_5^+ AlCl_4^-$, and $I_2Cl^+ AlCl_4^-$. Analogous derivatives containing the anions $FeCl_4^-$, $HfCl_6^{2-}$, and $TaCl_6^{2-}$ do not exist (see also Table 2). Crystals of these three $AlCl_4^-$ compounds decompose with loss of ICl and/or $I_2$ even in a dry atmosphere within a few minutes. However, they could be characterized by $^{35}Cl$, $^{127}I$ n.q.r., and $^{27}Al$ n.m.r. spectroscopy. The absorption band at 330 nm due to $I_5^+$ in sulphuric acid is enhanced by the addition of DMSO, and, in addition, a new band appears at ca. 350 nm whose intensity is dependent on the DMSO concentration.[111]

The 1:2 adduct of hexamethylenetetramine with $I_2$, which is a molecular donor–acceptor complex, has been shown[59] to change spontaneously into bis(hexamethylenetetramine)iodonium tri-iodide. The structure of the latter has also been

**Table 2** *Stability of some salts containing polyhalide cations*[*a]

| Anion | $I_3^+$ | $I_5^+$ | $I_7^+$ | $I_2Cl^+$ | $ICl_2^+$ | $Br_3^+$ |
|---|---|---|---|---|---|---|
| $AlCl_4^-$ | Yes | Yes | No | Yes | Yes | No |
| $FeCl_4^-$ | No | No | No | No | — | — |
| $HfCl_6^{2-}$ | No | No | No | No | — | — |
| $TaCl_6^{2-}$ | No | No | No | No | — | — |
| $SbCl_6^-$ | No | No | No | Yes | Yes | No |
| $SO_3F^-$ | Yes | No | Yes | Yes | Yes | No |

[a]* Based on ref. 110 and other works cited therein.

[105] H.-P. Sieper and R. Paetzold, *Z. phys. Chem.* (Leipzig), 1974, **255**, 1125.
[106] J. De Leeuw, M. Van Cauteren, and Th. Zeegers-Huyskens, *Spectroscopy Letters*, 1974, **7**, 607.
[107] A. C. Legon and D. J. Millen, *J.C.S. Chem. Comm.*, 1975, 580.
[108] C. H. W. Jones, *J. Chem. Phys.*, 1975, **62**, 4343.
[109] J. Cailleret, J. M. Paulus, and J. Ch. Abbe, *J.C.S. Faraday I*, 1975, **71**, 637.
[110] D. J. Merryman, J. D. Corbett, and P. A. Edwards, *Inorg. Chem.*, 1975, **14**, 428.
[111] S. N. Bhat, *Current Sci.*, 1975, **44**, 308.

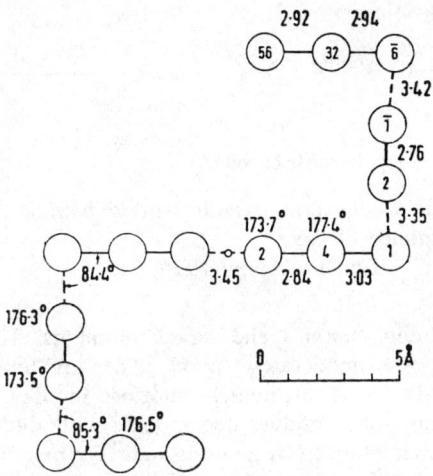

**Figure 1** The centrosymmetric $I_{16}^{4-}$ is shown in projection on (001) in the (theobromine)$_2$H$_2$I$_8$ structure. The best plane has been calculated through the central $I_2$--$I_3^-$--$I_3^-$--$I_2$ unit and the deviations of all the iodine atoms from this plane are given in units of $10^{-2}$ Å.

(Reproduced from *J.C.S. Chem. Comm.*, 1975, 677)

determined by X-ray methods and contains a nearly symmetrical $I_3^-$ ion;[112] the N—I distances in the cation are significantly longer than those in the corresponding bis(pyridine)iodonium cation.

The structure of (theobromine)$_2$H$_2$I$_8$ has revealed that this is a polyiodide salt containing protonated theobromine.[113] The anion $I_{16}^{4-}$ is the largest discrete polyiodide ion that has been characterized until now. A projection of this centrosymmetric anion is shown in Figure 1: it is evident that the ion is nearly planar and is best described as being $I_3^-$--$I_2$--$I_3^-$--$I_3^-$--$I_2$--$I_3^-$.

The increase in the molar conductivity, corrected for viscosity effects, with increasing I$_2$ content of MI (M = Me$_4$N, Et$_4$N, Bu$_4$N, or Na) in I$_2$–DMF solutions and of NaI in I$_2$–DMSO solutions has been attributed to the so-called relay conduction mechanism of I$^-$ ions in I$_2$ solutions.[114] These results support the proposal made earlier that the lower, but normal conductivities of NaI in liquid I$_2$ are due to the thermal instability of Na$^+$ I(I$_2$)$_n^-$. The formation of $I_3^-$ has been studied spectrophotometrically in MeOH, Pr$^i$OH, Bu$^t$OH, MeCN, and DMSO[115a] and in mixtures of these solvents both with and without water.[115b].

Rode[116] has reported the results of *ab initio* MO–SCF calculations on IF$_n$ ($n$ = 1, 3, or 5) and has used the results to discuss electronic structures and physical

---

[112] H. Pritzkow, *Acta Cryst.*, 1975, **B31**, 1505.
[113] F. H. Herbstein and M. Kapon, *J.C.S. Chem. Comm.*, 1975, 677.
[114] A. E. Gorenbein and E. Ya. Gorebein, *Zhur. fiz. Khim.*, 1975, **49**, 371.
[115] (*a*) A. A. Ramadan, P. K. Agasyan, and S. I. Petrov, *Zhur. obshchei Khim.*, 1974, **44**, 2299; (*b*) ibid., p. 983.
[116] B. M. Rode, *J.C.S. Faraday II*, 1975, **71**, 481.

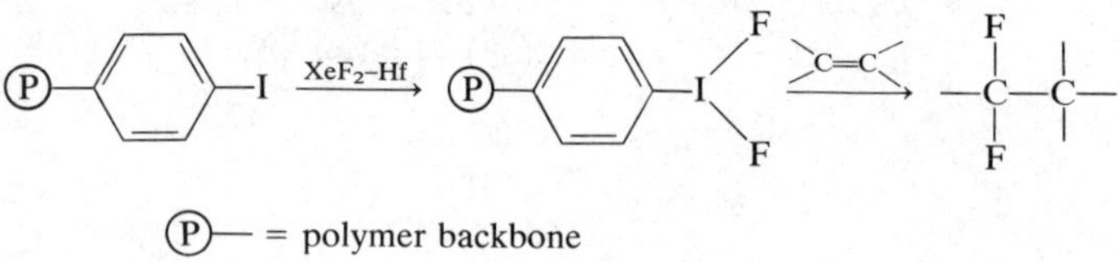

Figure 2 *Formation of a polymeric aryliodine(III) difluoride and its reaction with phenyl-substituted olefins.* (Reproduced from *J.C.S. Chem. Comm.*, 1975, 715)

properties, such as bond energies and dipole moments. The agreement with experimental values was moderately good. The low-temperature ($-78\,^{\circ}\text{C}$) fluorination of $(CF_2I)_2$ with elemental fluorine in $CFCl_3$ solution yields $(CF_2IF_2)_2$.[117] The white solid product decomposes with the formation of $I_2$ at $+28\,^{\circ}\text{C}$ but could be characterized by analysis and $^{19}F$ n.m.r. spectroscopy. Zupan and Pollak[118] have described the preparation of a polymeric aryliodine(III) difluoride by the reaction of iodinated 'pop-corn' polystyrene with $XeF_2$ in the presence of HF. They used the iodine(III)-containing polymer to fluorinate a variety of phenyl-substituted olefins to their geminal difluorides (see Figure 2 under mild conditions. The standard enthalpy of formation of $RbIF_4$ has been determined as $-191 \pm 4$ kJ mol$^{-1}$ from measurements of the heat of hydrolysis.[119]

A non-aqueous electrolyte battery using $ICl_3$ in $LiClO_4$–propylene carbonate solution, with Li as the anode and a graphite cathode, has been patented.[120] The high discharge current and voltage (up to 4.5 V) of such batteries are comparable to those of lithium–halogen batteries; moreover the $ICl_3$ battery does not require a halogen gas reservoir. Raman and i.r. spectra of $ICl_2^+ SbCl_6^-$ have been reported and assigned.[121] Force constants calculated for the cation of $C_{2v}$ symmetry are in good agreement with the corresponding data for other isoelectronic species. Aubke *et al.*[122] have investigated some reactions of triatomic iodine(III) cations of the types $IX_2^+$ and $I_2X^+$, as their fluorosulphates. Halogen substitution with more electronegative halogens (X') [reaction (5)] as well as halogen redistribution, *e.g.* reaction (6), have also been shown to occur. Anionic solvolysis was shown to occur with such compounds in excess $SbF_5$, *e.g.* as shown in reaction (7).

$$IX_2SO_3F + X_2' \rightarrow IX_2'SO_3F + X_2 \quad (5)$$

$$ICl_2SO_3F + IBr_2SO_3F \rightarrow 2IClBrSO_3F \quad (6)$$

$$ICl_2SO_3F + 3SbF_5 \rightarrow ICl_2Sb_2F_{11} + SbF_4SO_3F \quad (7)$$

---

[117] D. Naumann and L. Deneken, *J. Fluorine Chem.*, 1975, **5**, 443.
[118] M. Zupan and A. Pollak, *J.C.S. Chem. Comm.*, 1975, 715.
[119] A. Finch, P. N. Gates, and S. J. Peake, *Thermochim. Acta.* 1974, **10**, 203.
[120] H. Ikeda, T. Saito, H. Kato, H. Tamura, and Y. Matsuda, Japan Kokai, 74 105928 (*Chem. Abs.*, 1975, **82**, 158614).
[121] R. Forneris and Y. Tavares-Forneris, *J. Mol. Structure*, 1974, **23**, 241.
[122] W. W. Wilson, J. R. Dalziel, and F. Aubke, *J. Inorg. Nuclear Chem.*, 1975, **37**, 665.

According to Christe et al.[82] the vibrational spectrum of $I(NO_3)_3$ is consistent with the presence of predominantly covalent —$ONO_2$ ligands, although the structure is probably polymeric, with bridging nitrato-groups. The same workers have also reported that the unstable $CF_3I(NO_3)_2$ is produced in the reaction between $CF_3I$ and $ClONO_2$. The $^{129}I$ Mössbauer parameters of $I(OCOMe)_3$, as a frozen solution, are similar to those for $I_2Cl_6$, in which the co-ordination around iodine is square-planar.[108] An interesting intermediate, $[(R_f)_2I]^+[I(ClO_4)_4]^-$, has been isolated for $R_f = (CF_3)_2CF$ or n-$C_7F_{15}$, from the reaction between perfluoroalkyl iodides and $Cl_2O_4$;[123] under carefully controlled conditions this intermediate decomposes to give the perfluoroalkyl perchlorate.

A new graphite intercalation compound has been prepared using $IF_5$, with the approximate composition $C_{8.5}IF_5$.[124] The group who have produced and identified $ClF_6^-$ were unable to detect the formation of $IF_6^-$ under comparable conditions.[98]

Iodine(VII) fluoride and NOF react at below room temperature to form $NO^+$ $IF_8^-$:[125] since $BF_3$ displaces $IF_7$ quantitatively from this compound, the latter must be a weak Lewis acid. Although CsF and $IF_7$ were not found to combine at 150 °C, they do react in NOF (b.p. −56 °C) to form $Cs^+$ $IF_8^-$ in up to 90% yield. The $IF_8^-$ anion appears to be present in both the $Cs^+$ and the nitrosonium salts. Raman spectra and X-ray powder data (tetragonal cell) have been reported for $CsIF_8$. The 1:1 adduct between $SbF_5$ and $IF_7$ has now been prepared in a pure state from mixtures containing an excess of $IF_7$, and evidence has also been obtained for an antimony-rich adduct.[126] The 1:1 adduct has an ionic structure $IF_6^+$ $SbF_6^-$, with a cubic unit cell. The complex reacts rapidly with radon gas at ambient temperature, forming a non-volatile radon compound.

**Oxide Halides.**—An improved preparation of $FClO_2$ has been reported by Christe et al.[127] Equimolar quantities of $NaClO_3$ and $ClF_3$ react according to equation (8). Further examples of explosive reactions involving perchloryl fluoride have been reported:[128] the latest examples involved low temperatures and an organo-alkali-metal derivative as the reducing agent.

$$6NaClO_3 + 4ClF_3 \rightarrow 6NaF + 2Cl_2 + 3O_2 + 6FClO_2 \qquad (8)$$

Raman spectra of solid and liquid $FBrO_2$ have been obtained and indicate that this compound is monomeric and pyramidal, of $C_s$ symmetry, in both states,[129] and it thus resembles $FClO_2$. The iodine(V) analogue is polymeric. The BrO bonds are relatively strong, comparable with those in $BrO_3F$, whereas the BrF bond is weak; this has been rationalized in terms of an important contribution to the bonding from the ionic structure $BrO_2^+$ $F^-$. Rode[116] has calculated some properties of iodine oxide fluorides $IO_2F$, $IO_2F_3$, and $IO_3F$ and ions derived from these species. His results show that the energy necessary to eliminate elementary

---

[123] C. J. Schack, D. Pilipovich, and K. O. Christe, Inorg. Chem., 1975, **14**, 145.
[124] H. Selig and O. Gani, Inorg. Nuclear Chem. Letters, 1975, **11**, 75.
[125] C. J. Adams, Inorg. Nuclear Chem. Letters, 1974, **10**, 831.
[126] F. A. Hohorst, L. Stein, and E. Gebert, Inorg. Chem., 1975, **14**, 2233.
[127] K. O. Christe, R. D. Wilson, and C. J. Schack, Inorg. Nuclear Chem. Letters, 1975, **11**, 161.
[128] (a) J. H. J. Peet and B. W. Rockett, J. Organometallic Chem., 1974, **82**, C57; (b) W. Adcock and T. C. Khor, ibid., 1975, **91**, C20.
[129] R. J. Gillespie and P. Spekkens, J.C.S. Chem. Comm., 1975, 314.

oxygen decreases with the number of bonding O atoms. Naumann et al.[130] have reported the preparation of $CF_3IOF_2$ from reaction (9).

$$2CF_3IF_4 + SiO_2 \rightarrow 2CF_3IOF_2 + SiF_4 \quad (9)$$

$$CF_3IF_2 + O_3 \xrightarrow{CCl_3F, -78\,°C} CF_3IOF_2 + O_2 \quad (10)$$

$$IO_2F_2^- \xrightarrow{H_2O} IO_3^- + 2HF \quad (11)$$

The new compound is very readily decomposed by traces of water and should be stored at $-30\,°C$. It was also produced in reaction (10). The i.r. and Raman spectra of $CsIO_2F_2, \tfrac{1}{3}H_2O$, $Cs[H(IO_2F_2)_2], 2H_2O$, and $Co(NH_3)_6(IO_2F_2)_3, H_2O$ are consistent with these formulations;[131] more particularly, the water is present in these solids as water of hydration and is not involved in hydroxo-iodato-species or as $H_3O^+$. Milne and Moffett have also estimated the hydrolysis constant for reaction (11), from Raman measurements, to be $17 \pm 2\,m^2\,l^{-2}$. It has been claimed by Okrasinski et al.[132] that sparingly soluble $BaIO_3F$ may be prepared by the reaction of $Ba(OH)_2, H_2O$ with $KIO_2F_2$ in absolute EtOH: the $IO_3F^{2-}$ ion was reported to break up rapidly in aqueous solution to $IO_3^-$ and $F^-$.

Rode[133] has calculated the electronic structures for two monomeric forms of $IO_2F_3$, with $C_{2v}$ (trigonal bipyramidal) and $C_s$ (tetragonal pyramidal) symmetry. He reports that the $C_{2v}$ isomer is strongly favoured. Aubke et al.[134] have reported the preparation of $CsF, IO_2F_3$ and $IO_2F_3, SbF_5$. The Raman spectra of these new compounds have been interpreted in terms of an ionic formulation for $Cs^+\,IO_2F_4^-$, with $C_{2v}$, cis-octahedral anion symmetry, and of an O-donor interaction in the $SbF_5$ adduct.

**Oxides and Oxyanions.**—The u.v. photolysis of $CF_3OF$ in a dilute Ar matrix at 8 K has produced only $COF_2$;[135] no evidence was obtained for the formation of $CF_3OOCF_3$, which is the expected product. The analogous chlorine derivative, $CF_3OCl$, photodissociated at wavelengths $<280$ nm under similar conditions to give both $COF_2$ and $CF_3OOCF_3$ along with lesser amounts of other products.

The kinetics of the decay of ClO, prepared by the reaction of Cl with OClO or with $O_3$, have been followed by molecular beam mass spectrometry at 298 K.[136] The major reaction gives $Cl + ClOO$ and the minor reaction $Cl + OClO$. The heats of formation of ClO, OClO, and ClOO were estimated; the latter two compounds are unimportant under stratospheric conditions because they will probably be photolysed.

The kinetics and mechanism of the cathodic reduction of $OCl^-$ in the pH range 8—11 were found to be well represented by the equations. (12)—(14).[137] The crystal structure of $NaClO_2, 3H_2O$ has been determined by X-ray methods;[138] the

---

[130] D. Naumann, L. Deneken, and E. Renk, J. Fluorine Chem., 1975, **5**, 509.
[131] J. B. Milne and D. Moffett, Inorg. Chem., 1975, **14**, 1077.
[132] S. Okrasinski, R. Jost, R. Rakshapal, and G. Mitra, Inorg. Chim. Acta, 1975, **12**, 247.
[133] B. M. Rode, Monatsh., 1974, **105**, 807.
[134] H. A. Carter, J. N. Ruddick, J. R. Sams, and F. Aubke, Inorg. Nuclear Chem. Letters, 1975, **11**, 29.
[135] R. R. Smardzewski and W. B. Fox, J. Phys. Chem., 1975, **79**, 219.
[136] M. A. A. Clyne, D. J. McKenney, and R. T. Watson, J.C.S. Faraday I, 1975, **71**, 322.
[137] V. I. Skripchenko, E. P. Drozdetskaya, and K. G. Il'in, Zhur. priklad. Khim., 1975, **48**, 352.
[138] V. Tazzoli, V. Riganti, G. G. Guiseppetti, and A. Coda, Acta Cryst., 1975, **B31**, 1032.

dimensions of the chlorite ion were found to be: Cl—O 1.557 and 1.570(3) Å, ∠OClO 108.4°. The redox potential of the $ClO_2^-/Cl^-$ couple has been measured over the pH range 8—12.[139] The observed potential was shown to be dependent on pH and $[ClO_2^-]$, but to be independent of $[Cl^-]$.

$$ClO^- + e^- \xrightarrow{slow} ClO^{2-} \quad (12)$$

$$ClO^{2-} + H_2O \rightarrow Cl + 2OH^- \quad (13)$$

$$Cl + e^- \rightarrow Cl^- \quad (14)$$

The kinetics of the second-order gas-phase reaction (15) have been measured in the range 22—55 °C and 40—150 Torr.[140] An e.s.r. study of γ-irradiated $Ba(ClO_3)_2,H_2O$ has revealed that three main radical centres are formed: these are $O_3^-$ and $ClO_2$, as well as the novel $ClO_2,H_2O$ radical.[141] The latter breaks down above 120 °C, at which temperature the crystal loses its water of hydration. Exposure of the irradiated cystal to u.v. light bleaches the chlorine centres, with an increase in the concentration of $O_3$ centres. The thermal decomposition of $KClO_3$ is said[142] to produce an aerosol containing KCl, $KClO_3$, and $KClO_4$; in the presence of $MnO_2$ the formation of $KClO_4$ is suppressed.

$$2ClO_2 \rightarrow ClO + ClO_3 \quad (15)$$

The reactions of chlorine perchlorate, $Cl_2O_4$, with several fluoroalkyl halides have been examined by Schack et al.[123] Under carefully controlled conditions, high yields of the novel perfluoroalkyl perchlorates $R_fOClO_3$ ($R_f = CF_3$, $C_2F_5$, i-$C_3F_7$, or n-$C_7F_{15}$) have been produced from $R_fI$. Intermediates, whose likely composition is $[(R_f)_2I]^+[I(ClO_4)_4]^-$ {$R_f = (CF_3)_2CF$ or n-$C_7F_{15}$}, were isolated and it was considered that the reaction proceeds according to (16) and (17). At room temperature $[IClO_4]$, a polymeric material, decomposes to $Cl_2$, $Cl_2O_7$, and, eventually, $I_2O_5$.

$$2R_fI + 4Cl_2O_4 \rightarrow [(R_f)_2I]^+[I(ClO_4)_4]^- + 2Cl_2 \quad (16)$$

$$[(R_f)_2I]^+[I(ClO_4)_4]^- \rightarrow 2R_fOClO_3 + 2[IClO_4] \quad (17)$$

The crystal structure of $LiClO_4,3H_2O$ contains $Li^+$ ions, co-ordinated by an almost regular octahedron of $H_2O$ molecules, with each $H_2O$ forming a hydrogen bond with two $ClO_4^-$ ions.[143] In the sodium salt, a monohydrate, the $Na^+$ ions are also octahedrally co-ordinated by oxygens, four of which are from different $ClO_4$ groups:[144] nevertheless the $ClO_4^-$ ions are linked together via weak hydrogen bonds from the water molecules. The i.r. spectra of the vapours over alkali-metal perchlorates at 400—450 °C have been studied by isolation of molecular beams in

---

[139] K. Teruya and I. Nakamori, Kyushu Daigaku Kogaku Shuho, 1974, **47**, 775.
[140] V. I. Gritsan and V. N. Panfilov, Kinetika i Kataliz, 1975, **16**, 316.
[141] D. Suryanarayana and J. Sobhanadri, J. Chem. Phys., 1974, **61**, 2827.
[142] I. Cholakova, R. Proinova, and N. Koharov, God. Vissh. Khimikotekhnol. Inst., Sofia, 1974, **19**, 11.
[143] A. Sequeira, I. Bernal, I. D. Brown, and R. Faggiani, Acta Cryst., 1975, **B31**, 1735.
[144] B. Berglund, J. O. Thomas, and R. Tellgren, Acta Cryst., 1975, **B31**, 1842.

various matrixes at 12 K.[145] The data indicate that at these temperatures the dominant vapour species are the monomeric ion pairs $M^+$ $ClO_4^-$, although for $M = Li$ a large percentage of dimers was detected. The separation of the components of $\nu_3$ for $ClO_4^-$ was taken to indicate a bidentate interaction with the cation in the ion pairs. The average hydration number of 4 for $ClO_4^-$ in aqueous solutions has been inferred from vibrational spectra.[146]

According to Karelin et al.,[147] anhydrous $HClO_4$ undergoes autodissociation to $Cl_2O_7$ and $HClO_4,H_2O$ on melting; the rate of dissociation increases rapidly at temperatures above $-30\,°C$. An ion-selective electrode, based on a solution of niobium(v) oxinate in $CHCl_3$, has been devised[148] which is sensitive to pH and to certain anions, especially $ClO_4^-$: the electrode has been used to determine $10^{-4}$–$10^{-1}M$-$ClO_4^-$ in solutions containing an excess of $NO_3^-$, $Cl^-$, $SO_4^{2-}$, and $CH_3CO_2^-$.

The kinetics of the reactions $BrO + NO \rightarrow NO_2 + Br$, $2BrO \rightarrow 2Br + O_2$, and $Br + O_3 \rightarrow BrO + O_2$ have been measured at 298 K, using a discharge–flow system connected to a mass spectrometer, with collision-free sampling.[149] The Raman spectra of $Br_2O_3$ are consistent with the OBrOBrO structure.[150] This molecule reacts with $O_3$ at $-45\,°C$ to form $OBrOBrO_2$, which then decomposes to give the short-lived $BrO_2$ monomer. The latter can dimerize to either $O_2BrBrO_2$ or $OBrOBrO_2$; there was no evidence for the existence of $BrOBrO_3$. On the other hand, bromine perchlorate, $BrOClO_3$, reacts with perfluoroalkyl iodides in an analogous fashion to the reactions of $Cl_2O_4$ (q.v.).[123]

The ionization constant of $HBrO_3$ in formamide has been determined at 25, 30, and 35 °C from solubility measurements.[151] The polarographic behaviour of $BrO_3^-$ in mixed solvent systems containing water and an organic solvent (e.g. formamide, MeCN) is essentially the same as the behaviour in neutral or alkaline aqueous media.[152] Isotope exchange induced by $\gamma$-irradiation between $Br^-$ and $BrO_3^-$ in aqueous solution at room temperature is essentially complete.[153] The rate of exchange was found to be first-order with respect to both dose rate and $[BrO_3^-]$.

The photodecomposition mechanism of $BrO_4^-$ has been elucidated by flash photolytic studies in aqueous solutions containing NaOH and also from e.s.r. and optical absorption measurements in aqueous NaOH glasses after photolysis at 77 K.[154] The sole primary reaction was decomposition to $O(^1D)$ and $BrO_3^-$, and is therefore unlike the additive primary processes for the other halogen oxyanions. The radiolytic reduction of $KBrO_4$ in aerated and deaerated aqueous solutions has been investigated.[155] In acidic medium the only reaction is to give H (or $HO_2$) and $BrO_4^-$, with $G(-BrO_4^-) = 3.5$. In neutral solution $BrO_4^-$ is reduced by the

---

[145] G. Ritzhaupt and J. P. Devlin, J. Chem. Phys., 1975, **62**, 1982.
[146] M. C. R. Symons and D. Waddington, J.C.S. Faraday II, 1975, **71**, 22.
[147] A. I. Karelin, Z. I. Grigorovich, and V. Ya. Rosolovskii, Izvest. Akad. Nauk S.S.S.R., Ser. khim., 1975, 665.
[148] W. Szczepaniak and K. Ren, Chem. analit., 1975, **20**, 91.
[149] M. A. A. Clyne and R. T. Watson, J.C.S. Faraday I, 1975, **71**, 336.
[150] J. L. Pascal, A. C. Pavia, J. Potier, and A. Potier, Compt. rend., 1975, **280**, C, 661.
[151] U. N. Dash, Thermochim. Acta, 1975, **11**, 25.
[152] B. K. Gupta, D. S. Jain, and J. N. Gaur, J. Indian Chem. Soc., 1974, **51**, 777.
[153] H. J. Arnikar, S. F. Patil, P. P. Joshi, and S. D. Prasad, Radiochem. Radioanalyt. Letters, 1974, **19**, 303.
[154] U. K. Kläning, K. J. Olsen, and E. H. Appelman, J.C.S. Faraday I, 1975, **71**, 473.
[155] N. Hassan and C. Heitz, J. Chim. phys., 1975, **72**, 119.

**Table 3** Distribution of halogen recoil atoms following (n, γ) reactions in alkali-metal perhalates[156]

| Perhalate | Mass of nuclide | Distribution of activity (%) | | |
|---|---|---|---|---|
| $ClO_4^-$ [a] | 38 | 82.9 | 16.5 | ≤0.8 |
| $BrO_4^-$ | 80m | 73 | 25 | <1 |
|  | 80 | 75 | 24 | <0.4 |
|  | 82 | 74 | 24 | <0.1 |
| $IO_4^-$ [b] | 128 | 2.7 | 87 | 10.1 |
|  | 130 | 9.8 | 83.4 | 6.7 |

[a] G. E. Boyd and Q. V. Larson, *J. Amer. Chem. Soc.*, 1968, **90**, 5092;
[b] ibid., 1969, **91**, 4639.

solvated electron, although with $G = 5.3$ other species are also active. The behaviour of recoil Br atoms in $MBrO_4$ (M = Li, Na, K, Cs, or $NH_4$) has been studied in a thermal neutron flux.[156] The radiobromine distributions, as determined by high-voltage electrophoresis, are shown in Table 3. Bromide is oxidized to $BrO_3^-$ on thermal annealing, the $^{82}Br^-$ ions being oxidized more rapidly than $^{80m}Br^-$ ions. Table 3 provides a comparison of the behaviour of (n, γ) recoils in the three perhalate ions. The polarographic reduction of $BrO_4^-$ in dilute alkali-metal halide solutions has been studied;[157] the results are of interest for the information they provide on the double layer structure.

The bond energy of the iodosyl radical, IO, has been estimated to be 53 kcal mol$^{-1}$ from molecular beam reactive scattering measurements.[158] The reaction of $KIO_4$ with $HSO_3F$ at room temperature does not appear to produce $IO_3F$[159a] as claimed by Paul *et al.*[159b] According to Cernik and Krejcik, $O_2$ is evolved in the reaction whilst the $I^{VII}$ content diminishes. The Czech workers obtained a crystalline sample of $IO_2SO_3F$ from the reaction, for which they proposed the equation (18).

$$3KIO_4 + 9HSO_3F \rightarrow O_2 + 2IO_2SO_3F + HOIOF_4 + 4H_2SO_4 + 3KSO_3F \quad (18)$$

Naumann *et al.*[130] have prepared pure $CF_3IO$ by the action of $O_3$ on $CF_3I$; the same product could also be produced by the reaction of $CF_3IF_2$ and $SiO_2$ in MeCN between $-40$ and $0\,°C$. These workers also isolated the more stable iodine(v) derivative from the reaction (19); the oxide difluoride $CF_3IOF_2$ was shown to be an intermediate in this reaction.

$$CF_3IF_4 + SiO_2 \rightarrow CF_3IO_2 + SiF_4 \quad (19)$$

The preparations of carrier-free $^{131}IO_3^-$ and $^{131}IO_4^-$ have been described by Palagyi:[160] he has shown that pure radio-iodate solutions are stable in the pH range

[156] N. Hassan and C. Heitz, *J. Inorg. Nuclear Chem.*, 1975, **37**, 395.
[157] R. De Levie and M. Nemes, *J. Electroanalyt. Chem. Interfacial Electrochem.*, 1975, **58**, 123.
[158] D. St. A. G. Radleind, J. C. Whitehead, and R. Grice, *Nature*, 1975, **253**, 37.
[159] (a) M. Cernik and D. Krejcik, *Z. Chem.*, 1974, **14**, 491; (b) R. C. Paul, K. Krishnan, and K. C. Malhotra, *Indian J. Chem.*, 1970, **8**, 1030.
[160] S. Palagyi, Report, 1973 INIS-mf-1604, Internat. Atomic Energy Agency, Vienna, Austria.

3—11 for long periods, whereas the iodine(VII) oxo-anion is only stable at higher pH (12—13). The thermal decomposition reactions of metal iodates have been studied:[161] apart from $O_2$ and $I_2$ the products are: $O^{2-}$ for $Mg^{2+}$; $IO_6^{5-}$ for $Li^+$, $Ca^{2+}$, $Sr^{2+}$, and $Ba^{2+}$; $I^-$ and $O^{2-}$ for $Na^+$ and $K^+$; $I^-$ for $Rb^+$ and $Cs^+$. In the presence of $Na_2O$ or NaOH, iodate ion undergoes disproportionation according to reaction (20).[162] Calcium oxide or hydroxide caused some disproportionation to occur. Thus this behaviour of iodates in the presence of good oxide-ion donors is similar to that of chlorates. The reaction between $IO_3^-$ and $Br^-$ ions in the LiBr–KBr eutectic melt (temperature range 420—483 °C) proceeds according to reaction (21), and lithium orthoperiodate separates out from solution.[163] The initial rate of consumption of $IO_3^-$ is first-order in $[IO_3^-]$.

$$4IO_3^- + 6O^{2-} \rightarrow 3IO_6^{5-} + I^- \qquad (20)$$

$$10IO_3^- + 14Br^- \rightarrow 4IO_6^{5-} + 2O_2 + 2O^{2-} + 3I_2 + 7Br_2 \qquad (21)$$

The standard potential of the $IO_3^-/I_2$ electrode (1.1942 V) has been remeasured at 25 °C.[164] Dash has determined the ionization constant of $HIO_3$ in formamide at 25, 30, and 35 °C from solubility measurements.[151] The $^{127}I$ n.q.r. frequencies of twenty iodic acid derivatives have been measured at 77 K;[165] quadrupole coupling constants and asymmetry parameters were evaluated. The data are said to indicate the presence of covalently bonded $IO_3$ groups in iodates and tri-iodates. Crystal structures of $K_2H(SO_4)(IO_3)$[166] and $KIO_3, HIO_3$[167] have been determined by Vavilin et al. and have been shown to contain pyramidal $IO_3$ groups. In pyroelectric $Nd(IO_3)_3, H_2O$ there are three different $IO_3$ groups in the structure:[168] overall co-ordination numbers of iodine are 7, 7, and 6 for the three types of iodine. The kinetics of the reaction between $IO_3^-$ and $I^-$ in the presence of phenol [reaction (22)], have been determined at 35 °C.[169] The initial rate of the aqueous reaction in acetate buffer has been reinvestigated:[170] the reaction now appears to be first-order in $[IO_3^-]$ and second-order in $[H^+]$, whereas the apparent order in $[I^-]$ increases with temperature (293—308 K) in the concentration ranges considered.

$$IO_3^- + 2I^- + 2H^+ \rightarrow IO_2^- + I_2 + H_2O \qquad (22)$$

A d.t.a. study of $Me_4NIO_4$, which is isostructural with the perchlorate salt, has shown no phase changes up to 220 °C, above which temperature the compound exploded violently.[171] The kinetics of the solid-state reaction of $KIO_4$ in alkali-metal iodide matrices at 298, 328 and 378 K have been determined.[172] The rate

[161] Z. Gontarz and A. Gorski, *Roczniki Chem.*, 1974, **48**, 2091.
[162] Z. Gontarz and A. Gorski, *Roczhiki Chem.*, 1974, **48**, 2101.
[163] P. Pacak and I. Slama, *Coll. Czech. Chem. Comm.*, 1975, **40**, 254.
[164] R. D. Spitz and H. A. Liefbafsky, *J. Electrochem. Soc.*, 1975, **122**, 363.
[165] V. S. Grechishkin and V. A. Shishkin, *Zhur. strukt. Khim.*, 1974, **15**, 624.
[166] V. I. Vavilin, V. V. Ilyukhin, and N. V. Belov, *Doklady Akad. Nauk S.S.S.R.*, 1974, **219**, 1352.
[167] V. I. Vavilin, V. M. Ionov, V. V. Ilyukhin, and N. V. Belov, *Doklady Akad. Nauk S.S.S.R.*, 1974, **219**, 1108.
[168] R. Liminga, S. C. Abrahams, and J. L. Bernstein, *J. Chem. Phys.*, 1975, **62**, 755.
[169] D. N. Sharma and Y. K. Gupta, *Indian J. Chem.*, 1975, **13**, 56.
[170] S. M. Schildcrout and F. A. Fortunato, *J. Phys. Chem.*, 1975, **79**, 31.
[171] S. Okrasinski and G. Mitra, *J. Inorg. Nuclear Chem.*, 1975, **37**, 1315.
[172] H. S. Kimmel, J. P. Cusumano, and D. G. Lambert, *J. Solid State Chem.*, 1975, **12**, 110.

# The Halogens and Hydrogen

constant at 298 K in the RbI matrix was $2\frac{1}{2}$ times that in the KI matrix. Potentiometric studies of the reduction of $IO_4^-$ by $I^-$ at 25 °C in either aqueous HCl or $H_2SO_4$ have shown that the latter is preferred for quantitative work.[173] Trömel and Dölling have determined the crystal structures of $M_3IO_5$ (M = K,[174a] Rb, or Cs[174b]). The proposed structures are related to that of $(NH_4)_3FeF_6$ except that tetragonal-pyramidal $IO_5$ groups replace the octahedral $FeF_6$ units. However, in the lanthanide mesoperiodates $LnIO_5,4H_2O$ (Ln = Pr—Lu) the anionic species is $IO_4(OH)_2$, and these are bound to each other by the $LnO_8$ polyhedra.[175]

**Hydrogen Halides.**—The crystal structure of DF has been determined by neutron powder diffraction at both 4.2 and 85 K.[176] The orthorhombic structure contains 4 DF molecules per unit cell, forming zig-zag chains parallel to the $b$-axis; neighbouring chains are parallel. The FFF angle is 116° and the D—F distances are 0.95 and 1.56 ± 0.03 Å at 85 K. Johnson *et al.* have suggested that the parallel chain arrangement, with no disorder in the position of deuterium, will also be shown to be the structure of solid HF, in spite of the existing confusion over this matter in the literature. The Auger-electron spectrum and ionization potentials of the HF molecule have been measured.[177] The ionization potentials obtained for the $1\pi$, $3\sigma$ and $2\sigma$ configurations, *viz.* 16.05, 19.82, and 39.58 eV, are in much better agreement with the results of Banna *et al.*,[178] who used $X$-ray p.e.s., than with Berkowitz's data. The gas-phase ion chemistry of HF has been investigated, using the techniques of ion cyclotron resonance spectroscopy.[179] The only reaction observed for the parent ion was $HF^+ + HF \rightarrow H_2F^+ + F$. Proton-transfer reactions in mixtures of HF with $N_2$, $CH_4$, and $CO_2$ have also been examined, and it was shown that the proton affinity of HF is 112 ± 2 kcal mol$^{-1}$, comparable with that of $N_2$. This is interesting because the proton affinities of the other three hydrogen halides are all 25% higher than this value. The activity of $F^-$ in aqueous HF[180] and in anhydrous HF[181] has been measured potentiometrically by means of an electrode that is selective to $F^-$ ion, as a function of HF concentration.[180] The $F^-$ activity was found to increase very rapidly with increasing HF concentration up to a maximum at 2% HF; at higher HF concentrations the activity decreased steadily, presumably due to solvation of $F^-$ by HF molecules, until it reached an estimated value of $1 \times 10^{-13.9}$ in 0.1M-KF (in anhydrous HF). This result clearly shows that the solid-state $LaF_3$ electrode is not responding to $HF_2^-$, $H_2F_3^-$ $\cdots H_nF_{n+1}^-$ ions. Menard *et al.*[181] have also determined the acidity level of hydrofluoric acid (>70 wt. % HF) by means of a hydrogen electrode: for reference purposes they

---

[173] S. N. Prasad and S. C. Agrawal, *Trans. Soc. Adv. Electrochem. Sci. Technol.*, 1974, **9**, 120.
[174] (*a*) M. Trömel and H. Dölling, *Z. anorg. Chem.*, 1975, **411**, 41; (*b*) *ibid.*, p. 49.
[175] N. B. Shamrai, M. B. Varfolomeev, Yu. N. Saf'yanov. E. A. Kuz'min, and V. V. Ilyukhin, *Zhur. neorg. Khim.*, 1975, **20**, 57.
[176] M. W. Johnson, E. Sandor, and E. Arzi, *Acta Cryst.*, 1975, **B31**, 1998.
[177] R. W. Shaw and T. D. Thomas, *Phys. Rev. (A)*, 1975, **11**. 1491.
[178] M. S. Banna, B. E. Mills, D. W. Davis, and D. A. Shirley, *J. Chem. Phys.*, 1974, **61**, 4780.
[179] M. S. Foster and J. L. Beauchamp. *Inorg. Chem.*, 1975, **14**, 1229.
[180] A. Vaillant, J. Devynck, and B. Tremillon, *J. Electroanalyt. Chem. Interfacial Electrochem.*, 1974, **57**, 219.
[181] H. Menard, J. P. Masson, J. Devynck, and B. Tremillon, *J. Electroanalyt. Chem. Interfacial Electrochem.*, 1975, **63**, 163.

used the half-wave potentials for the ferrocene–ferricenium and perylene$^+$–perylene$^{2+}$ systems. Masson et al.[182] have studied the electrochemical behaviour of various quinones in anhydrous HF and also in HF–H$_2$O mixtures.[182a] In a second paper the use of these quinones as electrochemical indicators to follow acid–base titrations has been reported.[182b] One surprising conclusion was that SbF$_5$ is a weak acid in anhydrous HF. The anodic behaviour of polished Pt in liquid HF (0.1M-NaF) has been investigated at 0 °C.[183] Under these conditions Pt is oxidized to Pt$^{IV}$ before F$^-$ is oxidized to F$_2$. Thus Pt cannot be considered to be unattacked in liquid HF, especially under basic conditions, and its electrochemical use is limited to about +2.3 V versus the CuF(s)\Cu(s) reference electrode.

The addition of small amounts of water to HF–SbF$_5$ mixtures followed by the removal of all volatile material at 25 °C yields a white solid residue, according to Christe et al.,[184] who have shown the product to be H$_3$O$^+$ SbF$_6^-$. The same cation had been produced previously by the controlled hydrolysis of BrF$_4^+$ Sb$_2$F$_{11}^-$: since its i.r. spectrum closely resembles that reported by Couzi et al. for H$_2$F$^+$, the existence of the latter ion seems highly improbable. Christe et al. have also isolated a second stable hydroxonium salt, H$_3$O$^+$ AsF$_6^-$, and obtained evidence for the existence of the thermally less stable H$_3$O$^+$ BF$_4^-$:[184] they also suggested the use of SbF$_5$ or possibly BiF$_5$ as a means of drying liquid HF. In a second paper Christe[185] describes the preparation and characterization of H$_3$S$^+$ SbF$_6^-$; the fluoroarsenate analogue was shown to be unstable, and As$_2$S$_5$ was the only solid isolated from H$_2$S–HF–AsF$_5$ mixtures. No evidence was obtained for the protonation of either Xe or HCl by HF–SbF$_5$ mixtures.

The solvolysis of Te(OH)$_6$ in liquid HF has been investigated, mainly by $^{19}$F n.m.r. spectroscopy, and a range of products Te(OH)$_n$F$_{6-n}$ ($n$ = 0, 1, 2, or 3) have been identified in solution.[186] From the hydrolysis of TeF$_6$, products with $n$ = 4 or 5 were also characterized spectroscopically. On the other hand, solvolysis of selenic acid in HF leads to pentafluoro-orthoselenic acid (F$_5$SeOH), fluoroselenic acid (HSeO$_3$F), and another product. The solvolysis of the methyl ester of orthotelluric acid, Te(OMe)$_6$, in HF is a slow reaction at room temperature; $^{19}$F n.m.r. spectroscopy of the reaction mixture at different stages has shown that most of the possible species of the type Te(OMe)$_n$F$_{6-n}$ are formed.[187] Paine and Asprey have given details of the use of silicon powder or hydrogen as reducing agents in anhydrous HF for the preparation of the pentafluorides and tetra-fluorides of Re, Os, and Ir from their hexafluorides.[188a] The slow reaction between quartz wool and HF has been employed for the partial hydrolysis of UF$_6$ in order to produce crystals of UOF$_4$.[188b]

Ionic transference numbers in liquid mixtures of NH$_3$ and HF containing 80.85 wt % HF have been measured at room temperature;[189] the results for NH$_4^+$

---

[182] (a) J. P. Masson, J. Devynck, and B. Tremillon, J. Electroanalyt. Chem. Interfacial Electrochem., 1975, **64**, 175; (b) ibid., p. 193.
[183] A. Thiebault and M. Herlem, Compt. rend., 1974, **279**, C, 545.
[184] K. O. Christe, C. J. Schack, and R. D. Wilson, Inorg. Chem., 1975, **14**, 2224.
[185] K. O. Christe, Inorg. Chem., 1975, **14**, 2230.
[186] U. Elgad and H. Selig, Inorg. Chem., 1975, **14**, 140.
[187] I. Agranat, M. Rabinovitz, and H. Selig, Inorg. Nuclear Chem. Letters, 1975, **11**, 185.
[188] (a) R. T. Paine and L. B. Asprey, Inorg. Chem., 1975, **14**, 1111; (b) R. T. Paine, R. R. Ryan, and L. B. Asprey, ibid., p. 1113.
[189] G. Pourcelly, M. Rolin, and H. Pham, J. Chim. phys., 1974, **71**, 1347.

and $HF_2^-$ are 0.4 and 0.6, respectively. The temperature dependences of $^1H$ n.m.r. shifts indicate that $NOF,3HF$ should be described as the strongly hydrogen-bonded species $NO^+ F(HF)_3^-$.[190] The hydrogen bonds in $NO_2F,6HF$ are weaker than those in $NOF,3HF$.

Nikolaev et al.[191] have reviewed the thermodynamic properties of the alkali-metal and alkaline-earth-metal hydrofluorides. Boinon et al.[192] have reinvestigated the NaF–HF system for 30—45 wt. % NaF and discussed their results in the light of earlier work. Examination of the 10 °C isotherm of the RbF–HF–$H_2O$ system confirms the existence of a number of hydrates and hydrofluoride phases, and provides evidence for the formation of a new solvate $RbF,4HF$.[193]

The liquid–vapour equilibrium in the HF–$H_2SO_4$–$H_2O$ system at 22 °C has been determined by a dynamic method for the range of pressure 7.0—348.2 Torr.[194] The extraction of HF into di-isoamyl methylphosphonate from the HF–HCl–$H_2O$ system increases as the HF content rises;[195] on the other hand, the extraction of HCl decreases as the HF content rises.

The recombination of iodine atoms in an excess of HCl, DCl, or HBr has been investigated:[48] the results suggest that at lower temperatures the reaction proceeds by a radical molecule mechanism involving I—(HX) in which I is interacting with both H and X atoms. The reaction between hydrogen astatide and atomic chlorine has been investigated using $^{217}At$ ($t_{1/2} = 0.032$ s).[196] Angular distribution of the HAt, $Cl_2$, HCl, and At beams shows the expected general resemblance between this reaction and the HI+Cl reaction.

## 2 Hydrogen

**Hydrogen-bonding.**—The results of ab initio calculations on the hydrogen bond in $(HF)_2$ show that there is a relatively small net effect of the total correlation energy on the interaction energy, on the F(H)F distance, and on the intermolecular vibration.[197] N.m.r. studies ($^1H$ and $^{19}F$) of $HF_2^-$ in a single crystal of $KHF_2$ have confirmed that the proton in this linear anion is centred, to within ±0.025 Å.[198] The mean H—F bond length $\langle r^{-3}\rangle^{1/3}$ is 1.168±0.002 Å; the F—F distance was predicted to be 2.274±0.005 Å, in excellent agreement with the refined value obtained from neutron diffraction. Polarized i.r. and Raman spectra of $KHF_2$ and $KDF_2$ have been examined in connection with the isotope effect on the hydrogen-bond length:[199] a shift in the Raman-active symmetric stretching frequency was detected.

---

[190] A. V. Nikolaev, A. S. Nazorov, and V. V. Lisitsa, Izvest. sibirsk Otdel. Akad. Nauk S.S.S.R., Ser. khim. Nauk, 1974, 132.
[191] A. V. Nikolaev, A. A. Opalovskii, V. E. Fedorov, and T. D. Fedotova, J. Thermal Analysis, 1974, **6**, 461.
[192] B. Boinon, A. Marchard, and R. Cohen-Adad, Compt. rend., 1975, **280**, C, 1413.
[193] B. Boinon, Compt. rend., 1975, **280**, C, 657.
[194] S. V. Ostrovskii, S. A. Amirova, S. A. Nazarov, and T. V. Ostrovskaya, Zhur. priklad. Khim., 1975, **48**, 431.
[195] V. M. Vdovenko, L. S. Bulyanitsa, G. P. Savoskina, and E. N. Sventitskii, Kompleksoobrazovanie Ekstr. Aktinoidov Lantanoidov, 1974, 148.
[196] J. R. Grover and C. R. Iden, J. Chem. Phys., 1974, **61**, 2157.
[197] H. Lischka, J. Amer. Chem. Soc., 1974, **96**, 4761.
[198] J. C. Pratt and J. A. S. Smith, J.C.S. Faraday II, 1975, **71**, 596.
[199] P. Dawson, M. M. Hargreave, and G. R. Wilkinson, Spectrochim. Acta, 1975, **31A**, 1055.

Martin and Fujiwara[200] have shown that the H–F coupling ($^1$H and $^{19}$F n.m.r.) is resolved in solutions of HF in basic solvents, whereas in non-basic solvents it is averaged by exchange. They have therefore proposed that HF forms polymers in inert solvents, but forms 1:1 complexes in basic media. The same workers have also examined $FHX^-$ ions in aprotic solvents by n.m.r. spectroscopy.[201] They observed $J$(H—F) for all four bihalide ions (X = F, Cl, Br, or I). The formation and interconversion equilibria were interpreted by assuming quantitative H transfer to the most basic halide ion in the system. Consistent with this, they showed that the predominant complex in a solution of excess $Br^-$ in HCl is $ClHBr^-$, and not $HBr_2^-$, as suggested earlier.

It has been confirmed that the 1:1 adduct is the main component in $Me_2O$–HF mixtures containing an excess of $Me_2O$:[202] the equilibrium constant at 28 °C has been determined by i.r. spectroscopy to be 0.17 $cmHg^{-1}$.

Microwave spectroscopy of gaseous mixtures of HF with $MeCN^{203}$ or $H_2O^{204}$ has confirmed that hydrogen-bonded species are generated: this technique promises to yield useful structural information for such complexes. Thus the N—F distance in MeCN–HF has been estimated to be 2.741 Å,[203] and the O—F distance in $H_2O$–HF is 2.68 Å.[204] The i.r. spectrum of the latter complex in the vapour phase has been reported for the first time.[205] The enthalpy of association was estimated to be $-26$ kJ $mol^{-1}$ at 315 K. The i.r. spectra of mixtures of HCl and DCl or DBr cooled to 166 K have been shown to give additional lines in the HCl rotation—vibration band attributable to HCl,DCl and HCl,DBr molecules.[206]

Hydrogen-bonding between $ClO_4^-$ and water molecules has been shown to vary widely: in aqueous solution[146] and in the $LiClO_4,3H_2O$ lattice[143] all the oxygen atoms are hydrogen-bonded. However, in $NaClO_4,H_2O$ the $Li^+$ ion is six-coordinated by oxygen atoms, four of which are provided by four different $ClO_4^-$ ions.[144]

**Miscellaneous.**—It has been shown that pH measurements can be made in aerated, concentrated solutions of reactive electrolytes (e.g. aqueous $NH_4F$–HF solutions) with Pd electrodes cathodically charged with respect to palladium hydride.[207] Brown[208] has described the reactions of KH with a variety of primary amines; thus from trimethylenediamine, the product, potassium 3-aminopropylamide, is soluble in excess base and is claimed to be a novel alkylamide superbase of exceptional reactivity.

The ion equilibria $H_n^+ + H_2 \rightarrow H_{n+2}^+$ for $n = 3, 5, 7,$ and 9 have been determined in a pulsed-electron-beam high-ion-source mass spectrometer.[209] Hiraoka and Kebarle have measured the equilibrium constants at different temperatures and have hence estimated thermodynamic data for the formation of these cluster ions.

[200] J. S. Martin and F. Y. Fujiwara, J. Amer. Chem. Soc., 1974, **96**, 7632.
[201] F. Y. Fujiwara and J. S. Martin, J. Amer. Chem. Soc., 1974, **96**, 7625.
[202] T. Mejean and P. Pineau, J. Chim. phys., 1974, **71**, 955.
[203] J. W. Bevan, A. C. Legon, D. J. Millen, and S. C. Rogers, J.C.S. Chem. Comm., 1975, 130.
[204] J. W. Bevan, A. C. Legon, D. Millen, and S. C. Rogers, J.C.S. Chem. Comm., 1975, 341.
[205] R. K. Thomas, Proc. Roy. Soc., 1975, **A344**, 579.
[206] M. Larvor, J. P. Houdeau, and C. Haeusler, Compt. rend., 1974, **279**, B, 423.
[207] R. Jasinski, J. Electrochem. Soc., 1974, **121**, 1579.
[208] C. A. Brown, J.C.S. Chem. Comm., 1975, 222.
[209] K. Hiraoka and P. Kebarle, J. Chem. Phys., 1975, **62**, 2267.

# 8
# The Noble Gases

BY M. F. A. DOVE

## 1 The Elements

The preferred method of disposal of radioactive krypton isotopes, after being separated from other volatile fission products, is by dumping at sea as the compressed gas, confined in steel cylinders.[1] According to a report by Bryant and Jones[2] the cumulative quantities of $^{85}$Kr and $^{129}$I in the environment by the year 2000 are such that these nuclides will pose no significant health problem.

The first directly observed $^{129}$Xe ($I = \frac{1}{2}$, natural abundance 26.4%) n.m.r. spectra of xenon compounds have been reported by Seppelt and Rupp;[3] the key results are depicted in Table 1 and are discussed in later sections. Other workers have used pulsed F.T. n.m.r. spectroscopy to obtain $^{129}$Xe chemical shifts in the gaseous phase at lower pressures than was previously possible.[4] Accurate vapour-pressure measurements for Ar, Kr, and Xe in their normal liquid ranges have been determined and thermodynamic quantities calculated from them.[5]

The kinetics of the growth and decay of long-lived, excited $Ne_2$, $Ar_2$, and $Kr_2$ molecules have been investigated by means of fast absorption spectrophotometry in the 810—995 nm region;[6] the results indicate that the predominant decay mode at pressures in the range 200—1400 Torr is *via* a spontaneous radiative transition to the ground state. Deuteron radiolysis of Ar at 15 K generated a matrix which absorbs at 644 cm$^{-1}$ in the i.r. region: Andrews *et al.*[7] have assigned this band to a $DAr_n^+$ species. Vacuum-u.v. chemiluminescence from bound upper states of ArO, KrO, and ArCl has been observed from the quenching of excited inert-gas atoms by $N_2O$, $O_3$, $Cl_2$, and $CCl_4$.[8] In these systems the metastable $^3P_2$ states of Ar and Kr strongly resemble ground-state alkali-metal atoms in their chemical properties. Reactions of xenon in the $^3P_2$ state with halogen-containing molecules have been shown to give rise to XeX species (X = F, Cl, Br, or I) which

---

[1] M. Laser, H. Beaujean, T. Bohnenstingl, P. Filss, M. Heidendael, St. Mastera, E. Merz, and H. Vygen, Proceedings of the Symposium on the Management of Radioactive Wastes Fuel Reprocessing, 1972, OECD Publishing Center, Washington D.C., 1973, pp. 77—98.
[2] P. M. Bryant and J. A. Jones, National Radiological Protection Board, (U.K.), (Report) 1972, NRPB-R8 (*Chem. Abs.*, 1975, **82**, 144 454).
[3] K. Seppelt and H. H. Rupp, *Z. anorg. Chem.*, 1974, **409**, 331.
[4] C. J. Jameson, A. K. Jameson, and S. M. Cohen, *J. Chem. Phys.*, 1975, **62**, 4224.
[5] H. H. Chen, C. C. Lim, and R. A. Aziz, *J. Chem. Thermodynamics*, 1975, **7**, 191.
[6] T. Oka, K. V. S. R. Rao, J. L. Redpath, and R. F. Firestone, *J. Chem. Phys.*, 1974, **61**, 4740.
[7] L. Andrews, B. S. Ault, J. M. Grzybowski, and R. O. Allen, *J. Chem. Phys.*, 1975, **62**, 2461.
[8] M. F. Golde and B. A. Thrush, *Chem. Phys. Letters*, 1974, **29**, 486.

**Table 1** N.m.r. spectroscopic data[3,21]

| Compound | Solvent | δ/p.p.m. | $^1J(^{129}\text{Xe–F})$/Hz | $^3J(^{129}\text{Xe–F})$/Hz | $^2J(^{129}\text{Xe–}^{225}\text{Te})$/H |
|---|---|---|---|---|---|
| Xe | n-$C_6F_{14}$ | 0 | — | — | — |
| $XeF_2$ | MeCN | −3508 | 5550 | — | — |
| $FXeOTeF_5$ | $CFCl_3$ | −3264 | 5670 | 30 | 540 |
| $Xe(OTeF_5)_2$ | $CFCl_3$ | −2952 | — | 31 | 470 |
| $FXeOSeF_5$ | $CFCl_3$ | −3379 | 5630 | 37 | — |
| $Xe(OSeF_5)_2$ | $CFCl_3$ | −3131 | — | 37 | — |
| $Xe(OSeF_5)(OTeF_5)$ | $CFCl_3$ | −3042 | — | — | 480 |
| $XeF_4$ | $BrF_5$ | −5590 | 3900 | — | — |
| $XeF_6$ | $BrF_5$ | −5296 | — | — | — |
| $XeOF_4$ | $XeOF_4$ | −5331 | 1116 | — | — |
| $XeO_3$ | $H_2O$ | −5548 | — | — | — |

were characterized by emission spectral measurements.[9] The r.f. and microwave spectrum of ArClF has been measured by molecular beam electric resonance spectroscopy;[10] the atomic arrangement was shown to be Ar—Cl—F, with the Ar—Cl distance 3.33 Å and the Ar—Cl stretching frequency 47 cm$^{-1}$ in this linear molecule. Slivnik et al.[11] have now investigated the influence of $NiF_2$, $AgF_2$, $MgF_2$, $Ag_2O$, and $Ni_2O_3$ on the thermal reaction between Xe and fluorine (mole ratio 1:4.5, total pressure 36 atm at 22 °C). The reaction was found to be explosive in the presence of small amounts of $Ag_2O$ or $Ni_2O_3$ at 0 °C and sometimes in the presence of $NiF_2$ or $AgF_2$. Neiding and Sokolov[12] have written a review, with 400 references, of the chemistry of Kr, Xe, and Rn.

## 2 Krypton, Xenon, and Radon(II)

A new method for the preparation of $KrF_2$ has been reported by Slivnik et al.[13] It involves the irradiation of a liquefied mixture of fluorine and krypton with near-u.v. light at −196 °C and, compared with other methods, it may be used for larger quantities, e.g. 4.7 g of $KrF_2$ after 48 h irradiation. Russian workers have recorded the absorption spectrum of gaseous $KrF_2$ and have hence calculated the oscillator strengths.[14]

The 1:1 adduct $KrF_2,XeF_6$, prepared using either $BrF_5$ or HF as solvent, is said to be a molecular compound, on the basis of its i.r. spectrum.[15] The same workers have also proposed that the relative fluoride-ion donor strengths of the noble-gas fluorides decrease along the series $XeF_6 > XeF_2 > KrF_2 > XeF_4$. The new 1:1 adduct $KrF_2,VF_5$ has also been characterized by i.r. spectroscopy:[16] it is the only adduct formed from mixtures of the two binary fluorides in the range of composition 0—65 mole % $KrF_2$. The formidable fluorinating ability of $KrF_2$ has been clearly demonstrated by the reaction with gold powder in HF at 20 °C; the

---

[9] J. E. Velazco and D. W. Setser, J. Chem. Phys., 1975, **62**, 1990.
[10] S. J. Harris, S. E. Novick, W. Klemperer, and W. E. Falconer, J. Chem. Phys., 1974, **61**, 193.
[11] J. Levec, J. Slivnik, and B. Zemva, J. Inorg. Nuclear Chem., 1974, **36**, 997.
[12] A. B. Neiding and V. B. Sokolov, Uspekhi Khim., 1974, **43**, 2146.
[13] J. Slivnik, A. Smalc, K. Lutar, B. Zemva, and B. Frlec, J. Fluorine Chem., 1975, **5**, 273.
[14] G. N. Makeev, V. F. Sinyanskii, and B. M. Smirnov, Doklady Akad. Nauk S.S.S.R., 1975, **222**, 151.
[15] V. D. Klimov, V. N. Prusakov, and V. B. Sokolov, Doklady Akad. Nauk S.S.S.R., 1974, **217**, 1077.
[16] B. Zemva, J. Slivnik, and A. Smalc, J. Fluorine Chem., 1975, **6**, 191.

resulting yellow solid is $KrF^+AuF_6^-$, in which the $KrF^+$ cation is probably fluorine-bridged to the $AuF_6^-$ ion.[17] This adduct reacts with $O_2$ to give $O_2^+AuF_6^-$ and with Xe to give some $XeF_5^+AuF_6^-$; it decomposes at 60—65 °C to yield pure $AuF_5$. The same workers have also reinvestigated the reaction between $KrF^+SbF_5X^-$ (X = F or $SbF_6$) and $XeOF_4$, and have claimed that $XeOF_4,XeF_5^+SbF_5X^-$ is formed rather than a derivative of the $XeOF_5^+$ cation.

The radiolytic synthesis of $XeF_2$ from its elements has been investigated and the observed temperature dependence was found to be consistent with a chain-reaction mechanism.[18] γ-Radiolysis and photolysis of Xe and $F_2$ mixtures at 77 K have been investigated by Malkova et al. and shown to yield principally $XeF_2$, with yields of from 3 to 10 molecules per 100 eV of energy absorbed.[19] The product of a Xe–$Cl_2$ glow-discharge reaction was identified by Nelson and Pimentel in 1967 as $XeCl_2$ from its i.r. band at 313 cm$^{-1}$; Boal and Ozin subsequently reported the Raman band (253 cm$^{-1}$) of the same molecule. Howard and Andrews[20] have now shown that 5017 Å (30 mW) laser irradiation of Xe–$Cl_2$ mixtures yields the same product. They showed that this is the lowest energy laser beam capable of photoproducing $XeCl_2$. The 4765 Å laser irradiation of 2% ClF in Xe produced no $XeF_2$ but rather a mixture of products which were identified from their Raman spectra as $XeCl_2$ and also the novel XeClF ($\nu_3$ = 479.5, $\nu_1$ = 315.5 cm$^{-1}$); the same bands were seen in the i.r. spectrum of the product of photolysis using a mercury arc (see Figure 1).

The $^{129}$Xe n.m.r. chemical shifts of some xenon(II) derivatives are shown in Table 1;[21] in this oxidation state the shifts fall within the range $-2952$ to $-3508$ p.p.m. (relative to xenon), which is intermediate between the range for both xenon-(IV) and -(VI) species and for xenon itself. The observation of $J(^{129}Xe-F)$ as well as of $^2J(^{129}Xe-^{125}Te)$ coupling constants is of spectroscopic interest.

It has been inferred[22] from Raman and $^{19}F$ n.m.r. studies that the adducts $XeF_2,WOF_4$ and $XeF_2,2WOF_4$ are best formulated as covalent structures containing Xe—F · · · W bridges, both in the solid state and in solution. This could be confirmed for the 1:1 adduct by X-ray crystallography:[23] the angle at the bridging fluorine is 147° and the short Xe—F (bridging) bond length (2.04 Å) confirms that the structure of this complex is essentially covalent. N.m.r. evidence for the existence of an oxygen-bridged form of the 1:2 adduct has yet to be confirmed.[22]

Frlec and Holloway[24] have measured and discussed the vibrational spectra of the $XeF_2-MF_5$ (M = Sb, Ta, or Nb) systems in detail. They concluded that although the spectra are best interpreted in terms of ionic formulations involving $XeF^+$ and $Xe_2F_3^+$ there is increasing covalent character in the series $XeF_2,SbF_5 < XeF_2,TaF_5 < XeF_2,NbF_5$, $XeF_2,2SbF_5 < XeF_2,2TaF_5 < XeF_2,2NbF_5$, and $XeF_2,SbF_5 < 2XeF_2,TaF_5$. The same workers also obtained evidence for the

---

[17] J. H. Holloway and G. J. Schrobilgen, *J.C.S. Chem. Comm.*, 1975, 623.
[18] E. K. Il'in, *Khim. vysok. Energii*, 1974, **8**, 323.
[19] A. I. Malkova, V. I. Tupikov, I. V. Isakov, and V. Ya. Dudarev, *Zhur. neorg. Khim.*, 1974, **19**, 1729.
[20] W. F. Howard and L. Andrews, *J. Amer. Chem. Soc.*, 1974, **96**, 7864.
[21] K. Seppelt and H. H. Rupp, *Z. anorg. Chem.*, 1974, **409**, 338.
[22] J. H. Holloway, G. J. Schrobilgen, and P. Taylor, *J.C.S. Chem. Comm.*, 1975, 40.
[23] P. A. Tucker, P. A. Taylor, J. H. Holloway, and D. R. Russell, *Acta Cryst.*, 1975, **B31**, 906.
[24] B. Frlec and J. H. Holloway, *J.C.S. Dalton*, 1975, 535.

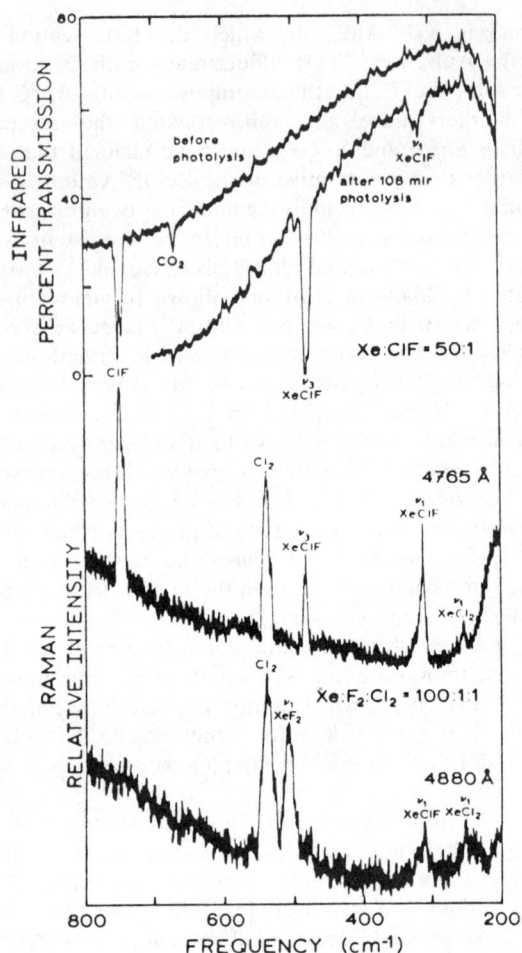

**Figure 1** *Infrared and Raman spectra of photolysed Xe–ClF samples, with the Raman spectrum of Xe–$F_2$–$Cl_2$ shown for comparison*
(Reproduced by permission from *J. Amer. Chem. Soc.*, 1974, **96**, 7864)

weak interactions of $2XeF_2$ with $Xe_2F_3^+$ and of $XeF_2$ with $XeF^+$ cations. On the basis of i.r. and $^{119}Sn$ Mössbauer spectra the lemon yellow adduct with $SnF_4$ has been formulated as $(XeF)_2SnF_6$.[25] A study of the $XeF_2$–HF system by d.t.a. has revealed the existence of a eutectic, at −92 °C, with mole fraction $XeF_2 < 0.2$, as the only significant feature.[26]

Wilson[27] has described the fluorination of $SO_2$ by $XeF_2$ as a base-catalysed process. The catalysts studied were of the type MX (M = $Me_4N$, Cs, or K; X = F or

[25] V. N. Zarubin and A. S. Marinin, *Zhur. neorg. Khim.*, 1974, **19**, 2925.
[26] A. S. Marinin, *Zhur. neorg. Khim.*, 1974, **19**, 1705.
[27] I. L. Wilson, *J. Fluorine Chem.*, 1975, **5**, 13.

Cl): in the reactions, $SO_2$ was oxidized to $SO_2F_2$ and $SO_3F^-$. The fluorination of iodinated 'pop-corn' polystyrene with $XeF_2$ in the presence of HF has been shown to yield a polymeric aryliodine(III) difluoride derivative.[28] In his review of the chemistry of radon Avrorin[29] agrees with the widely accepted view that $RnF_2$ is the product of the fluorination of this element. Relativistic quantum mechanics as applied to Rn (or element 118) fluoride structures indicate that ionic crystalline forms are probably more stable than the molecular form exhibited by $XeF_2$.[30] This is consistent with the properties reported by Stein for radon fluoride in that the radon compound does not vapourize as a molecular species but rather decomposes *in vacuo* above 250 °C to the elements. Liebman[31] proposes that $RnF_2$ is kinetically so stable that further fluorination to higher radon fluorides is blocked: he believes that $RnF_4$ will prove to be as stable as $XeF_4$, although $RnF_6$ will probably be less stable than $XeF_6$ owing to the large spin–orbit splitting of the $p_{3/2}$ and $p_{1/2}$ levels for $Rn^+$ (3.86 eV) than for $Xe^+$ (1.31 eV).

## 3 Xenon(IV)

High-voltage discharge in gaseous mixtures of xenon and fluorine at a nominal temperature of 193 K yielded $XeF_2$, $XeF_2,XeF_4$, and $XeF_6$ but no crystals of $XeF_4$.[19] Seppelt and Rupp[3] prepared pure $XeF_4$ by the u.v. irradiation of $XeF_6$ in $BrF_5$. The $^{129}Xe$ chemical shift of $XeF_4$ in $BrF_5$ (see Table 1) is surprisingly close to that for the xenon(VI) compounds they investigated.[3,21]

The thermal reaction between $XeF_4$ and rare-earth trifluorides succeeds only for $CeF_3$ and $TbF_3$, which are converted into the tetrafluorides.[32] Xenon(IV) fluoride was found to be more reactive than $XeF_2$ for these fluorination reactions.

## 4 Xenon(VI)

The absorption spectrum of gaseous $XeF_6$ has been investigated in the 50—170 eV region, using synchrotron radiation.[33] Absorption bands corresponding to $4p \rightarrow 5s$ transitions were not observed, thus implying that $XeF_6$ has a filled $5s$ orbital. This lends further support to the Bartell and Gavin model of a slightly distorted octahedral monomer geometry. Wang and Lohr have extended their calculations on the electronic structure of $XeF_6$, using the two-electron crystal-field model, to include oscillator strengths and band shapes.[34] The key experimental features of the vapour absorption spectrum were reproduced; these results thus provide further support for the pseudo-Jahn–Teller model for this molecule.

The yellow-green solutions of $XeF_6$ in $(F_5S)_2O$ at room temperature become colourless at lower temperatures, owing to association.[3,35] The $^{19}F$ and $^{129}Xe$ n.m.r. spectra at −118 °C consist of seven and thirteen lines, respectively, showing a multiplet spacing of 330 Hz. Seppelt and Ruff have interpreted these results in

---

[28] M. Zupan and A. Pollak, *J.C.S. Chem. Comm.*, 1975, 715.
[29] V. V. Avrorin, *Sov. rem. Probl. Khim.*, 1973, 21.
[30] K. S. Pitzer, *J.C.S. Chem. Comm.*, 1975, 760.
[31] J. F. Liebman, *Inorg. Nuclear Chem. Letters*, 1975, **11**, 683.
[32] V. I. Spitsyn, Yu. M. Kiselev, and L. I. Martynenko, *Zhur. neorg. Khim.*, 1974, **19**, 3194.
[33] U. Nielsen, R. Haensel, and W. H. E. Schwarz, *J. Chem. Phys.*, 1974, **61**, 3581.
[34] S. Y. Wang and L. L. Lohr, *J. Chem. Phys.*, 1974, **61**, 4110.
[35] H. H. Rupp and K. Seppelt, *Angew. Chem.*, 1974, **86**, 669.

**Table 2** Bond lengths and force constants for several xenon compounds

| Compound | Bond distance/Å | | Bond angle/° | | Force constant/ mdyn Å$^{-1}$ | | Comments |
|---|---|---|---|---|---|---|---|
| | Xe—F | Xe—O | FXeF | OXeO | XeF | XeO | |
| $XeF_2$[a] | 1.984 | — | 180 | — | 2.69 | — | |
| $XeF_4$[b] | 1.932 | — | 180 | — | 2.83 | — | |
| $XeOF_4$[c] | 1.900 | 1.703 | 172 | — | 3.28 | 7.09 | Bridging oxygens |
| $XeO_2F_2$ | 1.899 | 1.714 | 175 | 106 | 3.00 | 6.21 | |
| $XeO_3F_2$ | — | — | — | 120 | 3.38 | 6.16 | Bridging fluorine |
| $XeO_3F^-$[d] | 2.36 | 1.767 | — | 100 | 0.82 | 5.48 | |
| $XeO_3$[e] | — | 1.76 | — | 103 | — | 5.66 | |
| $XeO_4$[f] | — | 1.736 | — | 109 | — | 6.06 | |

[a] S. Reichman and F. J. Schreiner, *J. Chem. Phys.*, 1969, **51**, 2355; [b] R. K. Bohn *et al.*, 'Noble Gas Compounds', ed. H. H. Hyman, University of Chicago Press, Chicago, 1963; [c] J. Martins and E. B. Wilson, *J. Mol. Spectroscopy*, 1968, **26**, 410; [d] P. LaBonville *et al.*, *J. Chem. Phys.*, 1971, **55**, 631; [e] D. H. Templeton *et al.*, *J. Amer. Chem. Soc.*, 1963, **85**, 817; [f] R. S. McDowell and L. B. Asprey, *J. Chem. Phys.*, 1973, **57**, 3062; G. Gunderson *et al.*, *ibid.*, 1970, **52**, 812.

terms of tetrameric $Xe_4F_{24}$ in which each fluorine appears to be equally coupled to all four Xe atoms and each xenon is equally coupled to 24 F atoms: the expected multiplets would then contain 9 and 25 lines each. This interesting hypothesis is credible in the light of the crystal structure of $XeF_6$ discussed in last year's Report; however, the n.m.r. interactions are very unusual and, of course, $J(^{129}Xe—F)$ is uncommonly small (see Table 1).

The oxidation of elemental xenon to $XeF_5^+$ by $KrF^+ AuF_6^-$ has been mentioned above, as has the oxidation of $XeOF_4$ to a salt containing the $XeOF_4, XeF_5^+$ cation.[17] A new graphite intercalation compound has been prepared by Selig and Gani;[36] the reactivity of $XeOF_4$ intercalated in $C_{8.7}, XeOF_4$ was found to be markedly less than that of free $XeOF_4$, and preliminary experiments indicated that the new compound will be a useful mild fluorinating agent, e.g. towards readily oxidizable organic substances.

A normal-co-ordinate treatment of $XeO_2F_2$ using a modified Urey–Bradley force field has been carried out and the results have been compared with new treatments for $XeOF_4$ and $XeO_3F_2$, as well as with force constants reported for other xenon compounds.[37] The results have been set out in Table 2, from which it is evident that, as oxygen atoms are added to $XeF_2$, the Xe—F bond distance decreases while the Xe—O bond distance increases. Aqueous solutions of pure $XeO_3$ at concentrations up to 0.1 mol l$^{-1}$ can be prepared from sodium perxenate by means of a cation-exchange resin.[38]

## 5 Xenon(VIII)

Aleinikov *et al.*[39] have investigated the kinetics of the thermal decomposition of $(NH_4)_4XeO_6$, at 49—78 °C, and of $Na_4XeO_6$, at 329—364 °C: the solid decomposition product of the ammonium salt was shown to be $XeO_3$.

[36] H. Selig and O. Gani, *Inorg. Nuclear Chem. Letters*, 1975, **11**, 75.
[37] R. D. Willett, P. LaBonville, and J. R. Ferraro, *J. Chem. Phys.*, 1975, **63**, 1474.
[38] J. C. Nelapaty and B. Jaselskis, *Analyt. Chem.*, 1975, **47**, 354.
[39] N. N. Aleinikov, B. L. Korsunskii, V. K. Isupov, and I. S. Kirin, *Izvest. Akad. Nauk S.S.S.R., Ser. khim.*, 1974, 2423.

# Author Index

Abakumov, A. S., 393
Abakumov, G. A., 142
Abakumova, R. N., 130
Abalonin, B. E., 330, 331, 332
Abassalti, M., 157, 161
Abbas, L. L., 25
Abbaschian, G. J., 129
Abbe, J. Ch., 404
Abdel-Rahman, M. O., 301
Abdullaev, G. K., 98, 99
Abdullina, I. G., 367
Abel, E. W., 235
Abello, L., 373
Abenoza, M., 310
Ablov, A. V., 246
Abraham, F., 341
Abraham, K. M., 280, 293
Abramowitz, S., 291
Abrahams, S. C., 412
Abrahamson, H., 142
Abrouk, M., 33, 177, 377
Abts, L. M., 65
Ackerman, J. L., 169
Ackermann, M. N., 264
Adachi, M., 398
Adair, R., 150
Adams, C. J., 394, 407
Adams, J. M., 191
Adams, M. J., 8
Adams, P. F., 3
Adams, R. D., 235
Adcock, W., 407
Addison, C. C., 2
Adkins, R. R., 360
Adler, A. D., 228
Adler, O., 296
Adrian, F. J., 21
Adylova, M., 321
Afanas'ev, Yu. A., 98, 99
Afanas'eva, D. N., 324
Affrossman, S., 43
Afinogenov, Yu. P., 379
Agabaev, Ch., 44
Aganov, A. V., 220
Agarin, S. G., 260
Agasyan, P. K., 405

Agranat, I., 155, 329, 389, 414
Agrawal, S. C., 413
Aharoni, C., 171
Ahlenius, T., 157
Ahlrichs, R., 61, 62, 114, 157, 270
Ahmed, F. R., 306
Ahmed, N. N., 273
Ahnell, J. E., 161
Ahrland, S., 320, 337
Ai, M., 202
Aika, K., 33
Ainsley, R., 4
Akao, M., 174
Akat'eva, A. S., 210
Akesson, G., 385
Akhapina, N. A., 345
Akhmedov, Sh. T., 294
Akhmerov, T. G., 176
Akimoto, S., 197
Akitt, J. W., 75
Akmyradov, R., 137
Aknazarov, S. Kh., 383
Akopov, A. S., 228
Akopov, E. A., 368
Aksnes, D. W., 330
Alamichael, C., 160
Albert, H. J., 214
Albrand, J. P., 278
Aleinikov, N. N., 422
Albrecht, J.-M., 256
Albright, M. J., 233
Alcock, N. W., 109, 226
Aldridge, J. P., 171
Alekperov, A. I., 342
Aleksandrov, K. S., 382
Aleksandrova, I. P., 368
Alekseev, N. V., 215
Alekseeva, A. A., 144
Aleonard, S., 23
Aleshin, V. G., 262
Aleskovskii, V. B., 188, 293
Alexander, J. C., 17, 26
Alexandre, M., 402
Alexandrova, V. A., 197
Al'fonsov, V. A., 210

Alger, T. D., 119
Ali, S. Z., 384
Alian, S., 379
Alikhanyan, A. S., 322
Allais, G., 326
Allakhverdov, G. R., 390
Allavena, M., 366
Allen, R. O., 165, 343, 401, 417
Allibert, M., 114
Allmann, R., 185
Almazova, N. G., 7
Almlöf, J., 169
Aloisi, G. G., 399
Alonzo, G., 207, 223, 233
Althoff, W., 292
Alymov, A. M., 80
Amarat, J. E., 6
Amaudrut, J., 294
Ambe, S., 338
Amberger, E., 112, 113
Aminova, R. M., 284
Amiot, C., 160, 171
Amirova, S. A., 415
Amma, E. L., 250
Amos, D. W., 273, 401
Anacker-Eickhoff, H., 144
Ananthakrishnan, T. R., 291
Ancsin, J., 259
Andersen, A., 258
Anderson, A., 178
Anderson, F., 217
Anderson, G. A., 110
Anderson, J. W., 325
Anderson, S. P., 278, 283
Andersson, J., 242
Andersson, S., 379
Andersson, T., 279
Andose, J. D., 327
Andrä, K., 204
Andreev, I. F., 196, 197
Andreev, Yu. P., 170
Andrews, L., 22, 23, 59, 163, 165, 343, 395, 396, 397, 399, 401, 402, 417, 419
Andrianov, K. A., 211

423

Andrianov, V. I., 215
Andrusev, M. M., 345
Anfaelt, T., 400
Angenault, J., 300
Angus, P. C., 184
Anicich, V. G., 163
Anisimov, K. N., 235, 239, 240
Anistratov, A. T., 382
Ando, F., 341
Annarelli, D. C., 240
Annopol'skii, V. F., 122, 176, 199
Anorova, G. A., 81
Ansell, G. B., 197
Anthonsen, J. W., 219, 397
Antic, E., 123
Antill, J. E., 3
Antilla, R., 160
Antonetti, G., 318, 322
Antonik, St., 165
Anundskås, A., 127
Anzai, S., 373
Aomura, K., 124
Apollonov, V. N., 147
Appel, R., 285, 286, 291, 292, 299, 356, 357, 360, 365
Appell, J., 150
Appell, K., 265
Appelman, E. H., 394, 410
Appeloff, C. J., 100
Araki, T., 217
Arbuzov, B. A., 96, 337
Ariguib-Kbir, N., 34, 331
Aripov, E. A., 195
Armand, M., 155
Armas, B., 113
Armbruster, A. M., 263
Armstrong, O. R., 102, 282, 389
Arnau, J. L., 345
Arnikar, H. J., 410
Arnold, D. E. J., 298
Arnold, D. P., 231
Arnoldi, D., 396
Arnol'dov, M. N., 2
Aronson, J. R., 171
Arpe, R., 133
Arrington, D. E., 356, 357
Aruldhas, G., 291
Arzi, E., 413
Asbrink, L., 178
Asgarouladi, B., 105, 112, 387
Ashby, E. C., 11, 115, 116
Ashchyan, T. O., 99, 122
Ashe, A. J., tert., 323
Ashford, N. A., 263
Ashuiko, V. A., 35, 54
Ashurova, M., 243
Askarova, R. A., 287
Askog, J., 386
Asprey, L. B., 414

Astrom, A., 379
Atchekzal, H., 90
Atkinson, R., 163, 268, 343
Atovmyan, L. O., 120, 215
Atwood, J. L., 117, 126
Aubke, F., 210, 369, 406, 408
Aubry, M., 262, 347, 353
Audcenko, M. A., 147
Auel, Th., 250
Augustine, A., 328
Aulich, H., 9
Ault, B. S., 22, 23, 59, 165, 343, 397, 401, 402, 417
Aurivillius, K., 316
Austad, T., 181
Austin, J. A., 168
Avallet, M., 45
Avedi, H. F., 29
Averbuch-Pouchot, M.-T., 316
Aver'yanova, L. N., 245, 319
Avilova, I. M., 263
Avinens, C., 25
Avrorin, V. V., 421
Avvakumova, L. V., 329
Awasarkar, P. A., 212
Axente, D., 267
Axworthy, A. E., 395
Ayedi, M.-F., 367
Aymonino, P. J., 245
Ayuzawa, M., 400
Aziz, R. A., 417
Azizov, A. A., 209
Azman, A., 272
Azzaro, M., 96

Babayants, V. A., 383
Babich, S. A., 81
Babkina, N. A., 133
Baburina, V. A., 225, 294
Bach, B., 152, 153, 156
Bachelier, J., 245
Bachet, B., 138, 384
Bachhuber, H., 266
Bachmann, H. R., 62, 87
Bachmann, K., 388
Baddiel, C. B., 8
Badre, R., 161
Badachhape, R. B., 151
Baer, B. P., 161
Baer, T., 161
Bärnighausen, H., 226
Baes, C. F., 401
Baggio, R. F., 369
Baggio, S., 369
Bagieu-Beucher, M., 316
Bagnall, K. W., 342
Bagus, P. S., 169
Baine, R. T., 67
Baird, N. C., 262
Bak, B., 178, 179
Bakakin, V. V., 199, 224

Baker, E. H., 332, 384
Baker, R. T. K., 146, 150
Bakhitov, M. I., 287
Bakhru, H., 378
Bakum, S. I., 114, 116
Balabanova, E., 102
Balan, T. R., 112
Bald, J. F., 241
Balducci, A., 119
Baldwin, R. R., 267
Balej, J., 343, 368
Balesdent, D., 33, 153
Bali, A., 351
Balicheva, T. G., 210
Balko, V. P., 199
Ball, J. R., 99
Ballard, J. G., 245
Balzani, V., 141
Bamberger, C. E., 401
Bancroft, G. M., 204, 240
Band, Y. B., 179
Banks, E., 316
Banna, M. S., 160, 413
Bannister, A. J., 342
Bapanaiah, K. V., 181
Baptista, J. L., 165
Baran, E. J., 133, 245
Baranov, A. I., 268, 383
Barbe, G., 140
Barbieri, P., 335
Barbieri, R., 207, 220, 223, 233
Barchuk, V., 23
Bardin, B. A., 53
Bardolle, J., 376
Bardoux, Y., 382
Bares, J., 369
Baresel, D., 265
Bargaftik, M. N., 265
Barker, G. K., 86, 104
Barker, M. G., 26
Barkigia, K. M., 331
Barley, B. J., 123
Barminova, N. P., 209
Barnes, A. J., 145, 165, 213
Barnes, R. H., 403
Barnett, B. L., 18
Barrans, J., 315
Barrer, R. M., 98, 99, 124, 186
Barrett, R. P., 148
Barrie, A., 1, 358
Barshchevskii, I. N., 345, 348
Barsukov, A. V., 141
Bartell, L. S., 279, 283, 290, 327
Bart, J. C. J., 225, 391
Barthelat, M., 296, 315
Barthes, D., 161
Bartke, T. C., 117
Bartlet, B., 80
Barton, S. S., 148
Basalgina, T. A., 232
Basch, H., 167

# Author Index

Bashirova, L. A., 294
Bashkirov, S. S., 220
Bass, J. L., 124
Bassett, P. J., 1
Bassett, W. A., 174
Bassi, A. B. M. S., 167
Bastian, V., 299
Bastick, M., 149, 150
Basu, R., 403
Bates, J. B., 103, 174, 186
Batra, L. P., 167
Batsanov, S. S., 144
Battacharyya, P. K., 160
Battaglia, L. P., 340
Battu, R. S., 312
Baturina, L. S., 220, 226
Batyeva, E. S., 210
Bätzel, V., 92, 95
Bau, R., 299
Baudler, M., 323, 364
Bauer, G., 230
Bauer, P., 380
Bauer, S. H., 93, 353
Bäumer, G., 285
Baur, W. H., 174, 321
Baussart, H., 165
Bayandina, E. V., 387
Bayanov, A. P., 253, 254
Bayer, H., 266
Bayushkin, P. Ya., 142
Bazakutsa, V. A., 340, 372, 386
Beachley, O. T., jun., 89
Beagley, B., 88
Beall, H., 28
Beall, T. W., 333
Beauchamp, J. L., 163, 284, 413
Beaudet, R. A., 77, 90
Beaujean, H., 398, 417
Bebeshko, G. I., 150
Becher, H. J., 40, 107, 112
Beck, M. T., 270
Becker, G., 229, 289
Beddard, G. S., 271
Beddoes, R., 245
Bedon, P., 254
Beer, D. C., 80, 81
Begley, M. J., 308
Beguin, F., 154
Begun, G. M., 25, 125
Behar, D., 181
Beier, B. F., 65
Bekmuratov, A., 345
Bekowies, P. J., 343
Beletskaya, I. P., 81
Belevskii, V. N., 272
Belkin, Yu. V., 96
Bell, S., 167
Bellama, J. M., 214, 216
Bellet, J., 167
Belousov, M. V., 263
Belousova, E. M., 209, 219, 262, 397

Belov, N. V., 35, 41, 186, 196, 197, 412
Belov, V. V., 116
Belyaev, E. K., 176, 199
Belyaev, I. N., 99, 245, 319
Belyaeva, I. I., 245, 319
Belyakov, V. N., 186
Belyuga, Yu. V., 241
Bemand, P. P., 396
Bender, C. F., 150, 179
Bendtsen, J., 260
Benedict, J. T., 126
Benezech, G., 123
Beng-Tiong Yo, 265
Ben Lakhdar, T., 96
Bennett, R. A., 177
Benoit, R. L., 8
Bensari-zizi, N., 160
Beremzhanov, B. A., 322
Berenblut, B. J., 160
Berezin, B. D., 228
Berg, L. G., 35
Berg, M. E., 163
Berger, H.-O., 108
Berger, R., 278, 326
Bergesen, K., 324
Berglund, B., 409
Berman, H. M., 311
Bernal, I., 27, 409
Bernard, D., 314
Bernardi, F., 157, 167
Bernstein, E. R., 67
Bernstein, J. L., 412
Bernstein, L. S., 291
Bernstein, M. L., 271
Bernstein, R. B., 163
Berrada, A., 316
Berry, A. D., 349
Berry, F. J., 390
Bersohn, R., 165
Bertaud, M., 380
Bertazzi, N., 141, 207, 220, 223, 233, 333
Berthier, G., 157
Bertin, F., 38
Bertin, J., 156
Bertyakova, L. V., 8
Berul', S. I., 125, 340, 372, 386
Beruto, D., 176
Besenhard, J. O., 151
Besson, J., 29, 367
Betowski, D., 180
Betrencourt, M., 160
Beuerle, E., 293, 300
Beumel, O., 25
Bevan, J. W., 416
Beveridge, S. J., 251
Beznoshchenko, A. M., 11
Bez'yazychnyi, P. I., 268
Bezzebov, V. M., 296
Bhadra, J. K., 137
Bhaduri, A., 125
Bhaduri, S., 170

Bhardwaj, S. S., 148
Bhargava, M. K., 136, 256
Bhat, S. N., 399, 404
Bhattacharya, S. N., 223
Biagini, C. M., 44
Bianco, P., 132, 133
Bickley, D. G., 205
Biedermann, S., 160
Bienert, R., 188
Billaud, D., 32, 33, 152
Billy, M., 123, 187, 228
Bingham, R. C., 145
Binnewies, M., 42
Binshtok, B. M., 399
Biradar, N. S., 220
Birchall, J. M., 217
Birchall, T., 184, 245
Bird, J. M., 254
Birks, J. W., 403
Biryuk, E. A., 121
Bist, H. D., 368
Bitsoev, K. B., 22
Bitter, W., 227, 297, 329
Biswas, A. B., 125
Bjerrum, N. J., 126, 388
Björk, N.-O., 320
Blachnik, R., 333, 373, 392
Black, D. L., 94, 283
Black, G., 171
Blackman, G. L., 178
Blackborow, J. R., 91
Blagoveshchenskii, V. S., 323
Blaise, B., 376
Blake, D. M., 261
Blanchard, J.-M., 34, 35, 376
Blander, M., 25
Blaschette, A., 352
Blass, W. E., 160
Blaton, N., 373
Blecker, A., 225
Bleckmann, P., 225
Bleijerveld, R. H. T., 380
Bleiman, C., 195
Blessing, R., 320
Blick, K. E., 110
Blint, R. J., 150
Bludova, L. N., 34
Blums, A., 343
Blutau, Chr., 225, 373
Blyholder, G., 170, 171
Bobkova, L. T., 220
Bobrysheva, N. P., 125
Bochkarev, M. N., 233
Bock, H., 365
Bockelmann, W., 4, 254, 255
Bockrath, B., 9
Bodas, M. G., 125
Bodner, E. V., 80
Boenig, I. A., 108
Boer, F. P., 16, 44
Böwing, W. G., 361

Bogle, W., 388
Bogorodskii, M. M., 268
Bohme, D. K., 180, 263
Bohnenstingl, J., 398, 417
Boiko, A. A., 383
Boiko, A. P., 299
Boiko, D. N., 117, 127
Boileau, S., 18
Boinon, B., 415
Boinon, M.-J., 143, 322
Boivin, J.-C., 341
Bojes, J., 358
Bokii, N. G., 226
Boldog, I. I., 123, 319
Boldyreva, A. M., 188
Bol'shakova, P. K., 392
Bond, G. C., 170, 201
Bondar', I. A., 197
Bone, L. I., 375
Bonel, G., 320
Bonham, R. A., 161
Bonner, F. T., 30, 269
Bonnet, B., 96, 144, 334
Bonnetain, L., 155
Bonnier, E., 254
Boonstra, L. H., 303
Bordner, J., 333
Borel, M. M., 19, 54, 377
Borisenko, A. A., 324
Borisova, V. P., 388
Borovikov, N. F., 147
Borodulenko, G. P., 372
Borsese, A., 253
Bos, K. D., 235, 250
Botschwina, P., 167
Bouchama, M., 339
Bougeard, D., 41
Boudjouk, P., 217
Boudreau, J. A., 301
Bouix, J., 100
Boule, P., 399
Bounds, D. G., 179
Bournay, J., 167
Bousquet, J., 34, 35, 335, 376
Bovin, J.-O., 337
Bowen, L. H., 339
Bowers, V. A., 21
Boy, A., 241
Brach, B. Ya., 125
Brachet, D., 123
Bradley, C. H., 259
Bradshaw, A. M., 170
Brandon, J. K., 136, 251
Brandstätter, E., 230, 231
Brandt, D., 150
Brandt, G., 98
Brauer, G., 45
Brauman, J. I., 151, 155, 164, 269, 336
Braunagel, N., 214
Brassfield, H. A., 301
Bratt, P. J., 91
Brattås, L., 333, 340, 373
Brattsev, V. A., 80

Breakell, K. R., 131
Breckner, A., 326
Bren, V. A., 226
Brenac, A., 384
Bresadola, S., 85
Breunig, H. J., 234
Breusov, O. N., 185
Brianso-Perucaud, M.-C., 177, 377
Brice, V. T., 68
Brickmann, J., 291
Bridenbaugh, P. M., 190
Briend-Faure, M., 186, 187
Briggs, A. G., 104
Briggs, R. W., 142
Briggs, T. S., 65
Brindley, P. B., 111
Brion, C. E., 161, 171
Britnell, D., 372
Britton, D., 141
Brixner, L. H., 27
Broadley, E. B., 110
Brochu, R., 316
Brockner, W., 336
Brodbeck, C., 160
Brodersen, S., 160
Brodie, A. M., 329
Broida, H. P., 267
Brokko, A. V., 112
Brom, J. M., 37
Bronger, W., 36, 134, 279, 373, 385
Bronoel, G., 165
Brooker, M. H., 174
Brookes, A., 239
Brooks, P. R., 163
Broothaerts, J., 398
Bros, J.-P., 132
Brouty, C., 341, 372
Brown, C., 301
Brown, C. A., 32, 416
Brown, C. W., 160, 365, 367, 376
Brown, I. D., 27, 174, 409
Brown, J. M., jun., 267
Brown, K. G., 171
Brown, L. D., 71, 73
Brown, M. P., 91, 105
Brown, P. W., 136
Brown, R. D., 178
Brown, S. D., 353
Brown, W. E., 50, 103, 175
Brownstein, M., 292
Brulet, C. R., 261
Brun, G., 315
Bruna, P. J., 169, 394
Bruners, V., 20, 345
Bruno, P. 300
Bruns, R. E., 167, 282
Brusiloverts, A. I., 384
Bruvo, M., 208
Bruzzone, G., 52, 53, 56, 113, 114
Bryan, P. J., 91

Bryant, A. W., 254
Bryant, P. M., 398, 417
Brozek, V., 372
Buchler, J. W., 261
Buchner, W., 287
Buck, H. M., 288
Buckle, J., 220
Budarina, A. N., 23
Buder, W., 303
Buenker, R. J., 344
Bürger, H., 160, 216, 226, 242, 288
Bues, W., 25, 379
Bugaenko, L. T., 272
Bugerenko, E. F., 210
Bugg, C. E., 48
Bugli, G., 373
Bukhalova, G. A., 36, 368
Bukhvalova, N. V., 271
Bukonvec, P., 125
Bukshpan, S., 399
Bula, M. J., 101, 283
Bulakh, A. Ya., 165, 168
Bullen, G. J., 309
Bulloch, G., 300, 309
Bulten, E. J., 235, 242, 250
Bulyanitsa, L. S., 415
Bulychev, B. M., 22, 116
Bumazhnov, F. T., 196
Bundel', A. A., 176
Bunnell, R., 147
Bunting, R. K., 90
Buntova, M. A., 191
Bur-Bur, F., 357
Burch, D. S., 270
Burewicz, A., 125
Burg, A. B., 90, 229, 289
Burgada, R., 314
Burgeat, J., 384
Burger, K., 228, 357
Burgess, J., 342, 400
Burke, J. D., 273
Burkhardt, W. D., 300
Burlitch, J. M., 236
Burmistrova, N. P., 35
Burnham, R. A., 240
Burns, G., 398
Burrows, H. D., 165
Burshtein, I. F., 246
Burtseva, K. G., 98, 322
Burushkina, T. N., 186
Burylev, B. P., 372
Buscarlet, E., 155
Busch, B., 220
Buschow, K. H. J., 48, 114
Buslaev, Yu. A., 134, 135, 291
Busetto, C., 118
Butler, J. F., 171
Butler, K. D., 240
Butler, W. M., 233
Buttenshaw, A. J., 220
Butuzov, V. P., 147

# Author Index

Butyagin, P. Yu., 186
Bychkov, V. T., 235
Byerley, J. J., 370
Bykova, I. G., 127
Byler, D. M., 329
Bytchkov, V. T., 233

Cabana, A., 87
Cabello, P. A., 163
Cabicar, J., 345
Cabina, A., 160
Cadle, R. D., 395
Cadman, P., 148, 153, 156, 400
Caerveau, G., 238
Cahen, Y. M., 10, 17
Cailleret, J., 404
Calabrese, A., 173
Calabrese, J. C., 69
Calaway, W. F., 259
Calhoun, H. P., 307, 309
Callaway, B. W., 171, 290
Callahan, K. P., 82
Calleri, M., 133, 196
Calvert, J. G., 275
Calvo, C., 58, 97, 316, 321
Cambray, J., 169
Cameron, J. D., 180
Cameron, T. S., 303, 304
Campbell, I. M., 163
Camus, A., 284
Cannon, J. F., 114
Cano, F. H., 98
Capderroque, G., 80
Capelli, R., 253
Carberrý, E., 230
Carcaly, C., 257, 392
Carey, F. A., 163
Carihso, O. C., 375
Carlson, G. L., 347
Carlson, R. E., 368
Cargioli, J. D., 170, 171
Carmel, S. J., 21
Carnahan, B. L., 160
Caro, P., 123
Carpenter, C. D., 325
Carré, D., 138, 384
Carré, J., 90, 335
Carrell, H. L., 311
Carrick, M. T., 263, 272
Carson, A. A., 203
Carter, A. E., 171
Carter, H. A., 210, 369, 408
Carton, B., 153
Casabianca, F., 298
Caselli, M., 300
Cassaigne, A., 312
Cassaretto, F., 29
Cassidy, J. E., 126, 318
Cassoux, P., 91
Castan, P., 96
Cathonnet, M., 165
Caullet, P., 122

Caulton, K. G., 281
Cauquis, G., 181
Causley, G. C., 160
Cecal, I., 142
Cederbaum, L. S., 167, 259
Celotta, R. J., 171
Centofanti, L. F., 93, 282
Ceraso, J. M., 9
Cernia, E., 118, 119, 313
Cernick, M., 411
Cesari, M., 118, 119
Cavell, R. G., 287, 298, 324, 372
Cavero-Ghersi, C., 316
Cazzoli, G., 276
Cha, F. V., 135
Chadaeva, N. A., 330
Chadha, R. K., 222
Chadha, S. L., 338
Chaigneau, M., 161
Chakravorti, M. C., 376
Chambers, J. G., 145
Chambers, J. Q., 126, 348, 399
Champion, J. P., 160
Chan, L. Y. Y., 236
Chan, P. K. H., 394
Chand, M., 242
Chandra, S., 402
Chandrasekhar, J., 329
Chang, H. W., 398
Chang, J. W., 214
Chang, S. C., 31
Chao, L.-C., 139, 140
Chapelon, R., 343
Chaplygina, N. M., 125
Chapman, C. J., 272
Charbonnel, Y., 315
Charlot, J. P., 26
Chastagnier, M., 161
Chatt, J., 260, 261
Chatterjee, C., 133
Chattopadhyay, T., 136
Chatillon, C., 114
Chatillon-Colinet, C., 254
Chatterji, A., 158, 160
Chau, F. T., 160
Chaudhuri, T. R., 320
Chaus, I. S., 372
Chavant, C., 37
Cheburina, L. A., 210
Chekhun, A. L., 280
Chen, C. T., 164
Chen, H. H., 417
Chen, S., 177
Chenion, J., 150
Chernokal'skii, B. D., 327, 332
Chernorukov, N. G., 331
Chernova, N. A., 331
Chernyshev, E. A., 210, 211, 215
Chernylch, S. M., 388
Chernyuk, G. P., 166

Cheung, P. S. Y., 259
Chervkal'skii, B. D., 329
Chevrel, R., 134
Cheyne, R. M., 393
Chianelli, R., 316, 375
Chiarizia, R., 16
Chichagov, Yu. V., 156
Chieh, P. C., 136
Chiesi, V. A., 44
Chikanova, M. K., 294
Childs, J. D., 366, 403
Childs, M. E., 230
Childs, P., 9
Chipperfield, J. R., 238, 239
Chirkiu, G. K., 130
Chittenden, R. A., 313
Chivers, T., 358, 376
Cho, S.-A., 339
Chobanyan, P. M., 99, 245
Choi, B. C., 251
Choi, C. S., 264
Choisnet, J., 185
Cholakova, I., 409
Choo, K. Y., 182
Chow, Y. M., 141
Choy, C. K., 228
Choy, E. M., 18
Christian, S. D., 366, 403
Christie, K. O., 159, 160, 272, 273, 276, 346, 401, 403, 407, 414
Christodoulides, A. A., 164, 401
Christoffersen, R. E., 167
Christopher, R. E., 305
Christophides, A. G., 332
Chrobokova, E., 380
Chuchmarev, S. K., 165, 168
Chukalin, V. I., 279
Chukhlantsev, V. G., 196
Chumaevskii, N. A., 66
Chung, H. L., 136
Chun-Hsu, L., 155, 329
Churchill, M. R., 68, 83, 235
Chutjian, A., 167
Ciach, S., 128
Cicerone, R. J., 396
Cichon, J., 288
Ciric, B. E., 335
Claesson, S., 12
Clapp, C. H., 311
Clare, P., 306
Clark, C. D., 147
Clark, J. B., 254
Clark, R. J. H., 160
Claxton, T. A., 89, 278
Claydon, C. R., 403
Clayton, G. T., 212
Clayton, W. R., 63
Clearfield, A., 320
Cloyd, J. C., jun., 284
Clyne, M. A. A., 394, 396, 408, 410
Cochran, E. L., 21

Coda, A., 408
Codding, E. G., 159
Cody, I. A., 187, 190
Coffey, P., 290
Coffy, G., 143, 322
Coghi, L., 222
Cohen, E. A., 298
Cohen, S. M., 417
Cohen-Adad, R., 415
Coignac, J. P., 23
Coker, H., 66
Cole, A. R. H., 160
Cole, S. B., 348
Collard, G., 398
Collet, A., 18
Colley, I. D., 107
Collier, J. E., 314
Collin, J. E., 171
Collin, G., 373
Collin, R. L., 316
Collings, P., 357
Collins, M., 74
Collins, M. J., 3
Collons, P. H., 127
Colomer, E., 238
Colussi, A. J., 278, 291, 329
Conflant, P., 341
Conlin, R. T., 231, 241
Connes, P., 171
Connor, J. A., 342
Contour, J. P., 124
Contreras, J. G., 137
Connay, W. R., 107, 108
Cook, C. J., 163
Cook, R. L., 160
Cooper, D. B., 313
Cooper, J. W., 171
Cooper, M. J., 63
Cooper, P. J., 222
Corazza, E., 98
Corbellini, M., 119
Corbett, J. D., 127
Cornu, O., 186
Cornwell, A. B., 243, 244, 245
Corosine, M., 284
Corriu, R., 238
Corazza, E., 28
Corbett, J. D., 340, 404
Corcoran, W. H., 271
Corradi, A. B., 340
Corset, J., 1
Cosby, P. C., 177
Costa, D. J., 293
Costes, J.-P., 100
Cot, L., 25, 316
Cotton, F. A., 235, 298, 324, 328, 370
Coudurier, M., 190
Coulson, C. A., 158
Couret, C., 241, 289, 290
Couret, F., 289
Courtois, A., 27, 123, 136, 279, 333, 337

Coutures, J.-P., 123
Couturier, J.-C., 300
Covington, A. K., 8
Cowfer, J. A., 182
Cowley, A. H., 104, 284, 291, 297, 298
Cox, A. W., jun., 284
Cox, B., 240
Cox, R. A., 270
Coxon, J. A., 402
Crabtree, R. H., 260, 261
Cradock, S., 111, 171, 207, 214, 224, 232, 355
Cragg, R. H., 108, 110, 225
Craido, J. M., 174
Crane, G. R., 190
Crasnier, F., 284
Craven, B. M., 18
Creffield, G. K., 2
Cremaschi, P., 164
Cremlyn, R. J. W., 294
Cresswell, R. A., 94
Crews, J., 354
Cros, G., 100
Crouse, D. J., 398
Cruceanu, E., 42, 113
Cruse, H. W., 58, 119, 172
Csakvary, E., 220
Csizmadia, I. G., 61, 157, 167
Cucinella, S., 118
Cullen, W. R., 235, 240
Cunningham, D., 221
Curlee, R. M., 147
Curtis, A. B., 345
Curtis, E. C., 159, 160, 395
Curtis, M. D., 237
Curtiu, M., 226
Cushner, M., 298
Cussler, E. L., 18
Cusumano, J. P., 412
Cyvin, B. N., 22, 23, 117
Cyvin, S. J., 22, 23, 117, 336
Czako-Nagy, I., 220
Czarnecki, J., 367
Czybulka, A., 255

Dacre, P. D., 365
Dagdigian, P. J., 58, 119, 172
Dagron, C., 373
Dahl, A. R., 229, 230, 290, 311
Dailey, B. P., 160
Daineko, Z. G., 329
Dalgleish, W. H., 306
Dal Negro, A., 97
Daly, F. P., 160, 365, 376
Daly, J. J., 225, 283, 401
Dalziel, J. R., 406
Dance, N. S., 390
Dang-Nhu, M., 160
Danesi, P. R., 16
Dangoisse, D., 167
Danilin, V. N., 257

Danilov, V. P., 99, 122, 125
Dann, P. E., 309
Daoudi, A., 316
Darragh, J. I., 351
Darriet, B., 45
Darriet, J., 312
Darriet, M., 312
Das, M. K., 242
Das, N. S., 220
Das, S. K., 148
Das Gupta, A., 403
Dash, U. N., 410
Da Silva, C. P. G., 394
Datta, S., 396
Daubendick, R. L., 275
Daugherty, N. A., 251
Daran, J. C., 37
David, J., 26
Davidovich, R. L., 337
Davidson, E. R., 167
Davidson, I. M. T., 230
Davies, A. G., 288
Davies, B. W., 204
Davies, C. G., 245
Davies, D. W., 345
Davies, G., 147
Davies, G. J., 160
Davies, N. C., 153, 156
Davis, B. C., 165
Davis, D. D., 165, 271, 344
Davis, D. W., 160, 413
Davis, R., 111
Davis, T. D., 167
Davituliani, V. V., 176
Davranov, M., 243
Davydov, A. A., 201
Dawber, J. G., 271
Dawson, P., 160, 213, 370, 415
Dazord, J., 62, 90
Dean, C. R. S., 280
Dean, P. A. W., 290, 329
De Backer, M., 262
Debeau, M., 23
Debies, T. P., 157
De Boer, B. G., 83, 235
De Boer, J. J., 143
Debye, N. W. G., 228
Decius, J. C., 368
Deckler, J. J., 142
De Cooker, M. G. R. T., 212
Deev, A. N., 146
Degg, T., 135
De Groot, J. H., 21
Deguchi, T., 401
De Haas, N., 163
Dehnicke, K., 117, 220
Deichman, E. N., 137, 139, 321, 322, 331
Deitz, V. R., 148
de Ketelaere, R. F., 287
Dekker, E. H. L. J., 125
De Jaeger, R., 367

# Author Index

de Jong, J. W., 212
Delafosse, D., 124, 186, 187
Delahay, P., 9
Delaplane, R. G., 371
DeLeeuw, J., 404
DeLevie, R., 411
Delfino, S., 253
Dell'Anna, A., 128
Della Monica, M., 300
Deller, K., 55, 333
Delobel, R., 165
Del Piero, G., 118, 119
Delpuech, J.-J., 121, 312
DeLucas, L., 48
DeLucia, F. C., 178
Démaret, A., 331
Demarcq, M., 293
De Marco, R. A., 160, 166
Dembovskii, S. A., 257, 385
Demchenko, L. E., 372
Dementjev, A., 97
Demidova, T. A., 27
Demnynck, J., 119
Demortier, A., 262
Demuth, R., 229, 288
Denes, G., 55
Deneken, L., 406, 408
Deneuville, J.-L., 254
Denisov, Yu. N., 139
Denniston, M. L., 63, 100, 105, 326
Dent Glasser, F. P., 190
Denton, D. L., 90
de Pena, R. G., 367
de Perazzo, P. K., 369
Derbyshire, F. J., 147
Dergashev, Yu. M., 114
Dergacheva, M. B., 7
Dergunov, Yu. I., 203, 223
Dergunov, Yu. T., 224
Dermigny, B., 144
Deroche, J. C., 160
De Rumí, V. B., 121
de Sallier-Dupin, A., 321
Descamps, B., 166
Deschanvres, A., 185, 255, 326, 339
Desclaux, J. P., 157
Desgardin, G., 339
Des Marteau, D. D., 166
Desnoyers, J. E., 368
Desre, P., 254
Destrade, C., 312
Desyatnik, V. N., 243
Deutsch, P. W., 157
Devalette, M., 45
Devarajan, V., 185
Devaud, M., 245, 273
Devi, A., 188
De Villepin, J., 370
Devin, C., 294
Devlin, J. P., 10, 27, 410
De Vries, A. E., 150
Devynck, J., 413, 414

Dewald, R. R., 9
Dewan, J. C., 330
Dewar, M. J. S., 145, 284
Dewkett, W. J., 28
Dewey, H. J., 396
DeWith, G., 169
De Zwaan, J., 160
Dhamelincourt, M.-C., 104
Dhar, S. K., 209
Di Bianca, F., 207, 223
Dick, A. W. S., 231
Dice, D. R., 167
Dickens, B., 50, 103, 175, 316
Dickinson, D. A., 89
Dickinson, R. J., 240
Dickman, B., 213
Dieck, R. L., 308
Diehl, R., 98
Dietl, M., 363
Dietsch, W., 333
Dill, J. D., 60
Dilworth, J. R., 261
Dines, M., 375
Dines, M. B., 28
Diot, M., 34
Dipali, N. L., 30
DiPaolo, T., 161
Dirand, M., 136
Dittmar, G., 224, 373
Dirstine, R. T., 56
Dixon, D. A., 73
Dixon, M., 179
Djuric, S., 273
Dmitrieva, L. Yu., 196
Dmitrievskii, V. A., 276
Dmitrik, A. L., 197
Doak, G. O., 333
Dobbie, R. C., 280
Dobbs, C. G., 394
Dobbs, F. R., 11, 116
Dobler, W., 14, 15
Dobrogowska, C., 272
Dobrovinskaya, E. K., 223
Dobrynina, T. A., 345
Dobson, D. C., 383
Dodge, M. C., 277
Dölling, H., 413
Doggett, G., 179
Domashevskaya, E. P., 132
Domanskii, A. I., 197
Domcke, W., 167
Don, B. P., 77
Donaldson, J. D., 219, 243, 245, 251, 333, 384
Donovan, D. J., 264
Donovan, R. J., 171
Donnet, J. B., 190
Donohue, P. C., 279, 373
Dorfman, L. M., 9
Dornfeld, H., 376
Dorofeeva, O. V., 63, 79
Dorofeeva, L. I., 165
Dorokhova, N. I., 220

Doronin, V. N., 279
Dostal, K., 342
Doty, L. F., 324, 372
Dotzel, P., 392
Douchet, J., 160
Douek, I., 221
Dougherty, J. P., 139
Douglade, J., 337, 369
Douglas, W. E., 238
Down, M. G., 2, 5, 6
Downs, A. J., 394
Dozsa, L., 270
Dozzi, G., 118, 119
Drabowicz, J., 310
Dräger, M., 215, 287, 332, 364
Drafall, L. E., 373
Dragulescu, C., 336
Drake, J. E., 62, 93, 282, 325
Drapkina, N. E., 220
Drbal, K., 60
Dreher, H., 323
Dreux, M., 324
Drew, M. G. B., 41
Driessler, F., 61, 157
Drifford, M., 402
Drits, V. A., 197
Drivobok, V. I., 125
Drozdetskaya, E. P., 408
DuBois, D. L., 285
Dubovoi, P. G., 23
Dubrava, E. F., 122
Duc, G., 38
Duchen, M., 220
Ducourant, B., 25, 144, 334
Dudarev, V. Ya., 419
Dudareva, A. G., 139
Duff, E. J., 320
Dugleux, P., 321
Dumas, J.-M., 161
du Mont, W.-W., 229, 242, 289
Dunbar, R. C., 163
Duncan, I. A., 210
Duncan, W., 171
Dunitz, J. D., 14, 15, 50
Dunkin, D. B., 168
Dunmur, R. E., 305
Dunning, T. H., 344
DuPont, T. J., 281
Dupré, B., 43
Dupriez, G., 19
Duquenoy, G., 29, 337
Durand, J., 316
Durand, S., 202
Durif, A., 316
Durig, J. R., 92, 93, 94, 203, 215, 222, 275, 282, 284, 294
Durinin, E. V., 125
Dutasta, J. P., 293
Dutchak, Ya. I., 254, 257
Dutta, N. C., 261

Duval, X., 172
Dvdarev, V. Ya., 214
Dvsyannikov, N. N., 80
Dwight, A. E., 114
Dwyer, M., 271, 344
D'yachenko, O. A., 120, 215
Dyatkina, M. E., 133, 331
Dyatlova, N. M., 40, 315, 316
Dye, J. L., 9, 17, 18
Dyke, J. M., 347
Dykman, E., 372
Dymova, T. N., 114, 116
Dyson, J., 306
d'Yvoire, F., 331
Dzhafarov, G. G., 98
Dzhamarov, S. S., 112
Dzhambek, A. A., 219
Dzhaparidze, O. I., 385
Dzhuraev, T. D., 53
Dzikovskaya, L. M., 325
Dzvonar, V. G., 211

Eaborn, C., 233, 236
Ebbinghaus, G., 6
Ebert, L. B., 151, 155, 336
Ebert, M., 310
Ebsworth, E. A. V., 111, 207, 214, 224, 232, 342, 355
Edenharter, A., 339, 373
Edlund, O., 62
Edwards, J. D., 237, 342
Edwards, P. A., 127, 404
Edwards, T. H., 160
Eeckhaut, Z., 101, 283
Efanov, V. A., 281
Efimov, A. N., 271
Efremov. A. A., 348 379
Efremov, E. A., 348, 379
Efremov, V. A., 379
Efstifeev, E. N., 99, 245
Eggers, D. F., 347
Egorov, A. S., 220
Egorova, A. G., 138
Eguchi, W., 399
Ehlert, K., 243
Eholie, R., 134, 257, 386
Eiletz, H., 307
Einig, H., 285, 356
Eisenmann, B., 50, 54, 55, 144, 333, 392
Eiyen, R., 216
Elbert, S. T., 167
Elgad, V., 389, 414
Elieeva, N. G., 114
Eliseev, A. A., 372
Elizarova, G., 243
Ellasted, O. H., 160
Ellerman, J., 220, 288
Ellert, G. V., 372, 386, 392
Ellinger, Y., 157
Elliot, L. E., 241
Ellis, J. E., 235
Ellis, P. D., 88

El Maslout, A., 27, 45, 279
Elmes, P. S., 328
Elson, I. H., 291
El-Wakil, A., 369
Elzaro, R. A., 94
Emelina, A. V., 380
Emel'yanov, Yu. M., 268
Emerson, D., 158
Emmenegger, F. F., 128
Emül, R., 126, 230, 289
Eng, G., 220
Engel, G., 321
Engelbrecht, A., 342
Engelhardt, U., 302
Engelmann, G., 3
Engert, G., 125, 384
England, C., 271
Engler, R., 215, 335, 364
English, C. A., 259
Ennan, A. A., 7, 218
Enninga, G., 392
Enterling, D., 227
Epimakhov, V. N., 51
Epperlein, B. W., 97
Epstein, I. R., 63
Erb, A., 331
Erbacher, J. K., 251
Erdos, E., 369
Eremin, E. N., 150, 268
Eremin, V. P., 98, 99
Eremin, Yu. G., 140
Eriksen, R., 378
Ershova, Z. V., 393
Ertt, G., 139, 364
Escudié, J., 241, 289, 290
Esin, O. A., 185
Esin, Yu. O., 254
Esperås, S., 144, 187, 387, 392
Esparza, F., 283
Etienne, J., 138, 384
Etzrodt, G., 92
Eujen, R., 242
Euninga, G., 333
Evans, A. G., 128
Evans, D. F., 18
Evans, D. L., 206
Evans, E., 21
Evans, E. L., 152, 153, 156
Evans, H. T., 392
Evans, J. C., 128
Evans, M., 160
Evans, T., 147
Evans, W. J., 82, 84, 266
Evdokimo, A. A., 245
Evdokimov, D. Ya., 204
Evenson, K. M., 267
Evsikov, V. V., 97
Evstaf'ev, G. I., 284
Evstaf'eva, O. N., 36, 382
Evstigneeva, M. M., 176
Evzhanov, Kh. N., 4
Ewing, C. S., 290
Ewing, G. E., 259

Ewings, P. F. R., 204, 244, 245
Ewo, A., 170, 171
Eyring, E. M., 144, 370
Eysel, H. H., 20, 344
Eysseltová, J., 310
Ezhov, A. I., 273
Ezhora, Zh. A., 137, 139, 321, 331

Fabian, D. J., 43
Faggiani, R., 27, 58, 97, 316, 321, 409
Falardeau, E. R. 290, 296
Falconer, W. E., 173, 418
Falcinella, B., 141
Farber, M., 42, 58
Faris, J. F., 171
Farona, M. F., 219
Farrar, T. C., 214
Farrow, M. M., 144, 370
Faucher, J,-P., 305, 306
Faurskov Nielsen, O., 351
Fawcett, V., 380, 381
Feates, F. S., 150
Fedorenko, A. M., 143
Fedorenko, T. P., 368
Fedorov, N. F., 197
Fedorov, P. I., 134, 139
Fedorov, P. P., 134
Fedorov, S. G., 297
Fedorov, V. A., 129, 144, 251, 273, 340, 348, 379
Fedorov, V. E., 415
Fedorova, A. V., 251
Fedorova, L. A., 331
Fedorova, N. M., 372
Fedoruk, T. I., 256
Fedorushkov, B. G., 321
Fedoseev, D. V., 150
Fedotikov, S. A., 348, 379
Fedotova, T. D., 415
Fedulova, T. V., 54
Fehlner, T. P., 75
Fehsenfeld, F. C., 168, 172, 269
Feiccabrino, J. A., 224
Feil, D., 169
Feitsma, P. D., 6
Fel'dshtein, N. S., 211
Felgate, P. D., 141
Feldmann, D., 1
Felton, R. H., 261
Feltz, A., 249, 384
Feng, D.-F., 165
Fenton, D. E., 244
Feretti, E. L., 178
Ferguson, E. E., 168, 172
Ferguson, G., 328, 331, 333, 334, 340
Ferguson, J. E., 185
Ferguson, R. B., 185
Ferraro, J. R., 422
Ferris, J. P., 164
Ferro, R., 253
Ferroni, G., 318, 322

# Author Index

Feser, M., 354
Feshchenko, N. G., 295
Fes'kova, Zh. K., 213, 388
Fetter, K., 63
Fevrier, G., 161
Fidler, R. S., 3
Field, F. H., 172
Field, J. S., 236
Fieldhouse, S. A., 89
Filatov, E. Ya., 20, 268
Filatov, L. Ya., 245
Filatova, L. N., 137
Fild, M., 292, 298, 304
Filseth, S. V., 171
Filss, P., 417
Finch, A., 23, 104, 280, 295, 406
Finch, M. A., 225
Fink, D., 35, 133, 137
Finkelshtein, N. A., 368
Finn, E. J., 171
Finn, T. G., 160
Firestone, R. F., 417
Firl, J., 313
Firnau, G., 394
Firsova, T. P., 20, 26, 268
Fischer, D., 227
Fischer, E., 364
Fischer, K., 254
Fischer, M. B., 69
Fischer, P., 108
Fischler, I., 258
Flahaut, J., 134, 257, 372, 386, 392
Flanagan, M. J., 203
Fleet, M. E., 185
Fletcher, J., 172
Fletcher, J. W., 262
Flewett, G. W., 273, 401
Flick, W., 298
Flicker, H., 171
Flor, G., 144
Fluck, E., 290, 293, 300, 310, 311, 323, 364
Fojtik, A., 169, 177
Fokin, A. V., 369
Fokin, V. N., 119, 127
Fokina, E. E., 114
Fokina, T. V., 242
Fokina, Z. A., 379
Foley, P., 249
Folger, F., 391
Fong, M. Y., 158, 160
Foord, A., 88
Ford, J., 238, 239
Forel, M. T., 95, 100
Forder, R. A., 116
Foris, C. M., 27
Fornasini, M. L., 113
Forneris, R., 406
Forst, W., 166
Fortunato, F. A., 412
Fortunatov, N., 379
Foster, M. S., 413

Fouassier, M., 95, 100
Fouda, S. A., 370
Foulatier, P., 125
Fourcade, R., 25, 144, 334
Fouskova, A., 383
Fowles, G. W. A., 372
Fox, W. B., 160, 166, 349, 408
Francis, W. H. G., 394
Franco, D. W., 365
Francois, G., 165
Frank, A., 281
Frank, U., 5, 251
Franke, A., 400
Franke, E., 392
Frankiewicz, T. C., 163
Frankiss, S. G., 351
Franklin, J., 203
Franklin, J. L., 172
Frantsevich, I. N., 112
Fraser, C. J. W.,
Fraser, G. W., 389
Fratini, A. V., 63
Fray, D. J., 7
Frazer, M. J., 206, 221
Frazer, W. L., 143
Freasier, B. F., 401
Fredin, K. S., 390
Fredin, L., 171
Freed, K. F., 179
Freeman, A. G., 275
Freedman, P. A., 102, 160, 347
Freimanis, Ya. F., 220
Frenchko, V. S., 254, 257
Frenz, B. A., 324, 370
Freund, E. F., 98, 99, 186
Freund, F., 46
Freund, H.-R., 133
Freund, S. M., 60
Freundlich, W., 331
Freymann, R., 80
Fridh, C., 178
Fridland, D. V., 225, 295
Fridman, Ya. D., 242
Friedman, R. M., 340
Frit, B., 391
Fritchie, C. J., 248
Fritz, G., 94, 126, 214, 230, 281, 289
Fritz, H. P., 151
Frlec, B., 334, 418, 419
Froben, F. W., 262
Frolov, V. I., 146
Fromageau, R., 3
Fromm, E., 254
Frost, D. C., 265
Fu, L. K., 257
Füllgrabe, H.-J., 109
Fueno, T., 180
Fuggle, J. C., 43
Fuhrhop, J.-H., 141
Fujikawa, T., 160
Fujimura, T., 165

Fujinaga, T., 343
Fujino, O., 320
Fujita, J., 205
Fujita, K., 259
Fujiwara, F. Y., 416
Fujiwara, M., 178, 180
Fukuda, M., 315
Fukui, K., 162, 242
Fukushima, E., 67
Full, R., 102
Fullam, B. W., 278, 291, 310
Fuller, M. J., 170, 201
Funck, E., 185
Furdin, G., 153
Furuichi, R., 123
Furukawa, Y., 113, 295
Furuseth, S., 373
Futrell, J. H., 163

Gaballah, I., 123
Gabe, E. J., 306
Gabelica, Z., 369
Gabelnick, S. D., 403
Gabes, W., 403
Gadsby, G. R., 146
Gäthje, D., 225, 373
Gaidamaka, S. N., 299
Gaillard, F., 180
Gaines, D. F., 67, 68, 69, 234
Gainullina, R. G., 325
Gal'chenko, G. L., 79
Gale, R. J., 267
Galea, J., 318, 322
Gallay, J., 326
Gallezot, P., 124
Gallon, T. E., 1
Gallup, A., 267
Galigné, J. L., 316
Galy, J., 137, 379, 380
Gamayurova, V. S., 329
Gambaruto, M., 353
Gambiro, M., 132
Gamble, F. R., 28, 375
Gams, R. A., 133
Ganchenko, E. N., 253
Ganchev, N., 401
Ganelina, E. Sh., 382
Gani, O., 155, 407, 422
Gann, R. G., 163
Ganne, M., 143
Gans, P., 305
Gapeev, A. K., 257
Gar, T. K., 214
Garazova, V. S., 264
Garber, A. R., 63, 78, 80
Gárcia-Blanco, S., 98
Garcia-Fernandez, H., 358
Gard, G. L., 353
Gardner, J. L., 171, 343
Gardner, G. L., 400
Gardner, P. L., 23
Gardner, P. J., 104, 280, 295

Gareev, R. D., 210, 302
Garifdzhanova, N. M., 367
Garg, S. K., 148
Garito, A. F., 354
Garner, C. D., 208, 274, 330
Garnett, E. S., 394
Garrigou-Lagrange, C., 312
Garrison, B. J., 167
Gartland, G. L., 18
Gartzke, W., 105
Gas'kov, A. M., 257, 392
Gaspar, P. P., 163, 182, 231, 241
Gast, E., 100
Gasteiger, J., 169
Gates, P. N., 104, 295, 406
Gatehouse, B. M., 18, 328
Gatineau, L., 148
Gatilov, Yu. F., 327, 328, 330, 331, 332
Gaughan, A. P., 139
Gaumann, A., 392
Gaur, J. N., 410
Gavlas, J. F., 9
Gavrilenko, V. V., 129, 130
Gavrilova, L. A., 218
Gavryuchenkov, F. G., 137
Gawlowski, J., 350
Gay, R. S., 67
Gazzoni, G., 133, 196
Geanangel, R. A., 91, 92, 248
Gebert, E., 407
Gebert, W., 316
Gebhardt, E., 254
Gehlert, P., 306
Geizler, E. S., 3
Gel'd, P. V., 253, 254
Gellert, W., 265
Gence, G., 315
Gennick, I., 27
Gentile, P. S., 54
Genty, A., 376
Georgiew, G., 341
Gerads, H., 58
Gerega, V. F., 223, 224
Gerger, W., 95
Germain, J. E., 202
Gerratt, J., 61
Gertenbach, P. G., 18
Gerwarth, V. W., 111
Gethin, A., 267
Ghenassia, E., 202
Ghose, S., 197
Ghosh, M. R., 242
Giachardi, P. J., 271
Gibbins, S. G., 226
Gibbons, C. S., 46
Gibson, J. A., 292, 293, 298
Giering, W. P., 235
Gigauri, R. D., 327
Giguere, P. A., 21, 344, 345
Gilak, A., 291
Gilbert, B., 25, 125

Gilbert, J. R., 172
Gilbert, R., 160
Gill, J. T., 72
Gillbro, T., 165
Gillespie, H. M., 171
Gillespie, R. J., 95, 342, 407
Gillis, N. S., 259
Gillman, H. D., 312
Gilot, B., 262, 347
Gilyazov, M. M., 295
Gingerich, K. A., 30
Giniyatullin, R. Kh., 387
Ginsburg, S. I., 320
Giordano, N., 391
Giorgi, A. L., 113
Giongo, G. M., 313
Girgis, K., 135, 136
Gladis, K., 139
Gladkowski, D., 77
Gladyshev, E. N., 142
Gladyshevsky, E. I., 256
Glänzer, K., 266
Glasberg, B. R., 218
Glasser, F. P., 199
Glaunsinger, W. S., 8
Glavas, S., 344
Glazhova, T. I., 385
Glazov, V. M., 132
Gleisberg, F., 168
Gleitzer, C., 27, 45, 123, 333, 337
Glekel', F. L., 123
Glemser, O., 299, 305, 322, 349, 351, 354, 355, 359, 360, 363, 372
Glibert, J., 369, 373
Glick, M. D., 233
Glicker, S., 171
Glickson, J. D., 133
Glidewell, C., 26, 210, 250, 276, 346
Glockling, F., 236
Glonek, T., 311
Gnidash, N. I., 340, 372, 386
Gobom, S., 251
Goddard, R., 237
Godfrey, P. D., 178
Godneva, M. M., 3, 370
Godovikov, N. N., 313
Goel, A. B., 204
Goel, R. G., 334, 338
Götze, H. J., 227
Goetze, R., 95
Gol'danskii, V. I., 191
Goldbaum, R. H., 163
Goldberg, I. B., 346
Golde, M. F., 417
Gol'din, G. S., 220, 226, 297
Goldstein, C., 399
Goldstein, S., 213
Goldstein, M., 242, 243
Goldwhite, H., 178, 179, 278, 283, 303

Golenwsky, G. M., 110
Golic, L., 19
Golob, L., 347
Golodets, G. I., 263
Goloshchapov, M. V., 321
Golovanov, I. B., 260
Golovin, Yu. M., 373
Golovko, L. V., 191
Gol'toova, K. N., 127
Gol'tyapin, Yu. V., 79
Golub, A. M., 123, 135, 137, 319
Golubeva, N. D., 63, 115
Golubinskii, A. V., 226
Golubtsov, S. A., 211
Golutvin, Yu. M., 390
Gombler, W., 350, 353, 378
Gomez, M., 161, 254
Goncharov, A. I., 399
Goncharov, Yu. I., 197
Gontarz, Z., 412
Gonzalez-Urena, A., 163
Goodard, W. A., 344
Goodman, D. W., 284
Goodman, K., 163
Goossens, W., 398
Gopalakrishnan, P. S., 337
Gopinathan, C., 212
Gopinathan, M. S., 360
Gopinathan, S., 212
Gorbatenko, Zh. K., 295
Gorbunov, Yu. V., 254
Gorbunova, S. L., 271
Gordeev, A. D., 334
Gordeeva, N. V., 383
Gordetsov, A. S., 223
Gordienko, V. I., 134
Gordon, D. J., 36, 178
Gorelov, P., 44
Gorelov, I. P., 44
Gorenbein, A. E., 405
Gorenbein, E. Ya., 405
Gorgoraki, V. I., 322
Goricher, I. G., 370
Gorlov, Y. I., 291
Gorokhov, S. D., 242
Goroshchenko, Ya. G., 7
Gorshtein, G. I., 368
Gorski, A., 412
Gosling, P. D., 280
Gospoduiov, G. G., 249, 340, 386
Gottlieb, C. A., 168
Goubeau, J., 300
Gould, R. K., 149
Gould, R. O., 49
Gould, S. E. B., 49
Goulter, J. E., 41
Goursat, P., 123
Goypiron, A., 370
Grace, M., 28
Gracia, N. I., 161
Graf, E., 16
Graf, H. A., 34, 339, 373

# Author Index

Graffeuil, M., 284, 290
Graham, B. W. L., 235
Graham, R. E., 172
Graham, S. H., 191
Graham, S. C., 147
Graham, W. A. G., 235, 236, 239
Gramlich, V., 321
Graner, G., 160
Granier, W., 316
Grankin, V. N., 349
Grankina, Z. A., 218
Graver, W. M., 163
Graves, G. E., 297
Grebenshchikov, R. G., 196, 199
Grechishkin, V. S., 412
Grechkin, N. P., 325
Green, M., 86
Greenberg, M. S., 10
Greenwood, N. N., 72, 73, 222
Gregory, N. W., 126, 128
Greiss, G., 107
Gremmo, N., 9
Gress, M. E., 337
Greve, K. S., 129
Grey, R. A., 84
Grey, I. E., 373
Gribov, B. G., 130
Grice, R., 411
Grffin, A. M., 357, 361
Griffing, K. W., 61
Grigor'ev, A. I., 38, 131, 213, 273, 315, 316
Grigorovich, Z. I., 410
Grim, S. O., 324
Grimes, R. N., 79, 81
Grimvall, S., 245
Grinberg, A. N., 245
Grishina, N. I., 125
Grishina, O. N., 324
Gritsan, V. I., 409
Grizik, A. A., 367, 372
Groke, J., 214, 288, 326
Gropen, O., 95, 99, 117, 119
Grosse-Böwing, W., 363
Groth, W., 344
Groult, D., 339, 369
Grover, J. R., 415
Grow, R. T., 20
Gruba, A. I., 125
Grubev, L. I., 251
Grunfest, M. G., 220, 226
Grunwald, E., 7
Gryder, J. W., 268
Grynkewich, G. W., 235
Grzybowski, J. M., 165, 343, 401, 417
Gschneider, K. A., 140
Gsell, R., 244
Guadagno, J. R., 253
Guarnieri, A., 160, 351
Guastini, C., 44

Gudlin, D., 120
Guelachvili, G., 160, 171
Guemas, L., 41
Guemas-Brisseau, L., 41
Guen, L., 134, 257, 386
Guentherodt, H. J., 56
Guerard, D., 56, 152, 153
Guerchais, J. E., 220, 325, 334
Guérin, H., 331
Güttler, H.-J., 376
Guest, M. F., 275
Guillen, D., 161
Guillet, A., 190
Guillory, W. A., 271
Guiraud, F., 353
Guiraud, H., 345
Guiraud, R., 262, 367
Guiseppetti, G. G., 408
Guitel, J.-C., 316
Guittard, M., 134, 143, 341, 372, 392
Guggenberger, L. J., 70
Gullevic, G., 62
Gunawardane, R. P., 190, 199
Gunde, R., 272
Gundermann, K.-D., 343
Gundrizer, T. A., 201
Gunn, H. I., 178
Gunning, H. E., 172, 383
Gupta, B. K., 410
Gupta, K. S., 142
Gupta, M. K., 52
Gupta, M. P., 19, 50
Gupta, T. N. P., 19
Gupta, V. D., 204
Gupta, Y. K., 142, 412
Gur'yanova, E. N., 129, 208, 220
Gusakov, G. M., 130
Gutierrez-Losa, C., 161
Guth, J.-L., 122
Gutman, D., 172
Gutmann, V., 96
Guttormson, R. J., 19
Guyader, J., 229, 255, 326
Guzhavina, T. I., 1
Gvozd, V. F., 268
Gwyther, J. R., 146
Gynane, M. J. S., 139
Gysegem, P., 303
Gyunner, E. A., 143, 244, 245, 319
Gzizik, A. A., 373

Ha, T.-K., 159, 160
Haas, A., 357
Haas, C. H., 182
Haas, R. G., 335
Haaland, A., 117, 119
Haase, A., 45
Haase, W., 36, 336
Haberl, A., 168

Habeeb, J. J., 136, 138
Haber, J., 367
Habibi, N., 25, 334
Hack, W., 268
Hackbarth, J. J., 68
Hackett, P., 240
Haddad, S., 54
Haddon, R. C., 169
Hadek, V., 57
Hadenfeldt, C., 51, 279
Hadjikostas, C. C., 332
Hadjiminolis, S., 243
Hadzi, D., 272
Hägele, G., 300
Haensel, R., 421
Hänssgen, D., 227
Haeusler, C., 416
Hagen, A. P. 171, 290
Hagen, G. P., 235
Hahn, H., 385
Haiduc, I., 226
Haigh, I., 400
Haines, L. I. B., 206
Hajela, B. P., 220
Haladjian, J., 132, 133
Halgren, T. A., 73, 75, 102
Halin, K. E. J., 171
Hall, C. R., 313
Hall, H. T., 114
Hall, J. H., jun., 62, 73, 102
Hall, L. W., 88
Hall, R. J. B., 380
Hall, W. K., 187
Hallam, H. E., 165
Hallers, J. J., 6
Halmos, Z., 169
Halonin, A. S., 370
Halter, M. E., 398
Hamada, S., 187, 360
Hameed, A., 104, 295
Hamill, W. H., 160
Hamilton, W. C., 311
Hamon, C., 51, 279, 333
Hamon, M., 229, 255, 326
Hancock, K. G., 89
Handke, R., 294
Handy, P. R., 10
Hang N. T'Li, 137
Hanke, H. E., 224
Hanke, W., 188
Hanraham, R. J., 163, 165
Hansen, D. A., 163
Hanzlik, J., 97
Harada, T., 203
Hargittai, I., 117, 203, 226
Hargreave, M. M., 370, 415
Haring, A., 150
Harland, P. W., 163
Harman, T. C., 171
Harmon, K. M., 27
Harmony, M. D., 158, 160
Harris, M. S., 146, 150
Harris, R. K., 300, 305, 306

Harris, S. J., 172, 173, 397, 418
Harrison, B. H., 148
Harrison, D. J., 351
Harrison, L. G., 243
Harrison, P. G., 146, 175, 201, 204, 206, 220, 244, 245, 250
Hart, S. A., 25
Hartl, H., 276, 277, 302
Hartlib, L. P., 4
Hartman, J. S., 95, 101, 103, 283
Hartman, M., 176, 367
Hasegewa, A., 291
Hasegawa, K., 321
Hasegawa, Y., 320
Hass, D., 213
Hassan, N., 410, 411
Haszeldine, R. N., 217
Hatta, S., 217
Hata, T., 311
Hatzenbuhler, D. A., 163
Hauge, R. H., 22, 178, 179
Hauge, S., 36, 180, 386, 393
Haupt, H. J., 25, 140, 218, 235, 238
Hausen, H.-D., 111, 121, 138, 139, 310, 365
Hauser, E., 56
Hauser, P. J., 137
Hauswirth, W., 160
Hawkes, G. E., 259
Haworth, D. T., 110
Hawthorne, M. F., 61, 70, 78, 82, 84, 85, 106
Hay, P. J., 344
Hayakawa, T., 163
Hayaski, M., 184
Hayashi, T., 341
Hayes, D. M., 318
Hayes, R. G., 173
Haynes, W. M., 395
Hayon, E., 345
Hayter, A. C., 238
Heal, H. G., 358
Hearn, R. A., 48
Heath, G. A., 260, 261
Hebecker, C., 25
Heckel, E., 163
Hecht, H. J., 228
Hedman, J., 258
Hedwig, G. R., 56
Heeger, A. J., 354
Hehre, W. J., 158
Heicklen, J., 165, 344, 367
Heidendael, M., 417
Heijdenrijk, D., 120
Heil, C. A., 229, 290
Heiman, J. R., 142
Heimburger, R., 336
Heindl, R., 348
Heinicke, E., 1
Heintz, O., 163

Heitz, C., 410, 411
Hellams, K. L., 215
Helton, R. W., 163
Hemmings, R. T., 325
Hemsworth, R. S., 263
Hénault, C., 339
Hencher, J. L., 140
Henderson, J., 146
Hendrick, J., 275
Hendricks, R. W., 186
Hendriksen, P. A., 180
Hendriock, J., 214
Hengge, E., 230
Henis, J. M. S., 163
Hennig, H. J., 183
Henrick, K., 330
Henriksen, P. A., 386
Hentges, S. G., 235
Hepburn, D. R., jun., 81
Herbecker, Ch., 142
Herek, R., 273
Herber, B., 240, 243
Herber, R. H., 206, 223
Herberich, G. E., 106, 107
Herbs, E., 182
Herbstein, F. H., 158, 160, 405
Herlem, M., 414
Herman, J. A., 350
Herman, K., 21, 344
Herman, M. A., 312
Herman, R., 320
Herman, R. G., 320
Herman, Z. S., 346
Heřmánek, S., 61, 63, 74
Hérold, A., 32, 33, 56, 152, 153, 154, 155, 156, 335
Herpin, P., 41, 341, 372
Herrmann, D., 333, 387
Herron, J. T., 271
Hersh, K. A., 205
Hertenstein, U., 223
Hertz, H. G., 400
Hertz, J., 136
Hertz, R. K., 63
Hervé, G., 331
Herzog, J.-F., 96
Hess, D. W., 160
Hess, H., 90, 111, 365
Hesse, R., 144
Heubel, J., 367
Heuschmann, M., 301
Hewett, W. D., 37
Hewitson, B., 301
Hewitt, A. J., 334
Hewson, M. J. C., 304, 305
Heyding, R. D., 385
Heyns, A. M., 329
Hickam, C. W., 90
Hierl, P. M., 163
Highsmith, S., 7
Hikichi, Y., 320
Hildebrandt, R. L., 62, 234
Hill, H. H., 113

Hill, N., 23, 104
Hill, R., 236
Hill, W. E., 80
Hillier, I. H., 167, 171, 275
Hilpert, K., 58
Hinchcliffe, A., 179
Hinton, J. F., 142
Hinrichs, E. V.,
Hirai, A., 178, 180
Hirakawa, A. Y., 158, 160
Hiraoka, K., 163, 164, 167, 416
Hirata, H., 400
Hirota, F., 160
Harrison, A. G., 163
Hirsch, R. G., 171
Hisatsune, I. C., 277, 344, 367
Hishiyama, Y., 148
Hladik, J., 382
Hladky, E., 398
Ho, B. Y. K., 203, 208
Hobdell, M. R., 4, 146
Hocking, W. H., 167, 169
Hodges, H. L., 279
Höfer, R., 349, 354, 355
Höfler, F., 231, 288
Höfler, M., 240
Hoel, E. L., 61, 85
Hölderich, W., 229, 289
Hoffmann, H.-J., 130
Hoffman, M. Z., 177
Hoffman, P. R., 281
Hofstötter, H., 94
Hogans, R. J., 88
Hohlwein, D., 156
Hohorst, F. A., 407
Holland, R. F., 171
Holliday, A. K., 105, 231
Holloway, J. H., 334, 419
Holroyd, P. M., 4
Holtzmann, G., 265
Home, R. A., 210
Homer, J. B., 147
Honey, F. R., 160
Hopfgarten, F., 341
Hopkins, A. G., 160, 365
Hopkins, H. P., jun., 11, 116
Hoppe, R., 35, 42, 133, 137
Hopper, H. R., 204
Hopper, J. R., 266
Horie, Y., 9
Horland, A. K., 233
Hornemann, U., 122
Horner, B. L., 111
Horton, R. L., 172
Hoskins, B. F., 392
Hosmane, N. S., 214, 232
Hosokawa, K., 295
Hossain, S. F., 351
Houdeau, J. P., 416
Houk, L. W., 297, 333
Houlden, S. A., 61
Hovdan, H., 336

# Author Index

Howard, A. V., 230
Howard, J., 237
Howard, C. J., 267
Howard, J. A., 350
Howard, J. A. K., 237
Howard, W. F., 22, 23, 395, 396, 397, 399, 419
Howarka, F., 260
Howat, G., 179
Howell, J. M., 103, 270
Howells, J. D. R., 165
Howie, R. A., 41, 196, 245, 316
Howlett, K. D., 303, 304
Hoyermann, K., 268
Hrncir, D. C., 126
Hseu, T. H., 108
Hsieh, A. T. T., 114
Hsu, C.-Y., 240
Hu, M. G., 92
Huang, C.-H., 210, 316
Huang, R. J., 241
Huang, T., 160
Hubbard, A. T., 401
Hubbard, C. R., 21
Hubbard, W. N., 20
Huber, F., 25, 218, 235
Hubberstey, P., 3, 5, 6, 7, 146, 175
Hubin-Franskin, M. J., 171
Hubner, H.-J., 392
Hucke, E. E., 148
Hudgens, B. A., 93, 94
Hudson, A., 60
Hudson, J. W., 103
Hudson, R. F., 301
Hübner, H.-J., 138
Hünig, S., 223
Huetz, A., 51
Huff, L., 278
Huffman, J. C., 62, 63, 87, 234
Huffman, J. C., 234
Huggins, R. A., 151, 155, 336
Hughes, B., 208, 330
Hughes, M. N., 258
Huie, R. E., 271
Hukuo, K., 320
Hummers, W. S., 151
Hunt, R. H., 160
Hunter, D. L., 235, 324, 370
Hunter, F. D., 124
Huntress, W. T., jun., 163, 172, 260, 263
Hurdus, M. H., 378
Hurst, J. R., 335
Hursthouse, M. B., 258
Husebye, S., 144, 387, 392
Hussain, A., 273
Hussein, M. A., 380
Huster, J., 385
Huttner, G., 105, 281, 296, 327

Hutton, W. C., 129
Hvoslef, J., 48
Hyde, M. R., 274
Hyman, H., 402

Ibbott, D. G., 290
Ichiba, S., 225
Iden, C. R., 415
Ido, T., 394
Ievins, A., 35
Ievin'sh, A. F., 97
Igel, R., 373
Ihle, H. R., 26
Ijima, M. I., 174
Ikeda, H., 406
Ikeda, M., 179
Ikorskü, V. N., 125
Ikuta, S., 161
Il'enko, T. M., 215
Il'in, E. G., 291
Il'in, E. K., 276, 419
Il'in, K. G., 408
Il'yasov, I. I., 243
Ilyukhin, V. V., 35, 196, 412, 413
Imai, H., 149
Imai, J., 187
Imai, N., 268
Imelik, B., 124
Immirzi, A., 119
Imoto, H., 152
Inaba, S. I., 223
Inaba, T., 268
Inch, T. D., 313
Indelli, A., 321
Indzhiya, M. A., 327
Infante, A. J., 78
Inoue, T., 205
Ioffe, O. V., 390
Iofis, N. A., 176
Ionov, L. B., 327, 328
Ionov, V. M., 35, 412
Ipatova, E. N., 224
Ipser, H., 392
Irene, E. A., 249, 384
Irusteta, M. C., 123
Irving, R. J., 121
Isaacs, E. E., 235, 239
Isaacs, T. J., 373
Isaev, I. D., 144
Isaeva, A. P., 127
Isakov, I. V., 419
Ishibitsu, K., 214
Ishihara, H., 128, 334
Ishii, T., 123
Ishikawa, N., 311
Ishizuka, M., 33
Islam, N., 226
Ismail, Z. K., 22, 178, 179
Ismailov, V. M., 294
Ismailova, Z. M., 330, 331
Israiloff, P. 251
Issleib, K., 294
Isupov. V. K., 422

Itkina, L. S., 125
Ito, M., 174
Ivanciu, O., 42, 113
Ivanov, N. R., 383, 386
Ivanova, E. F., 11
Ivanova, N. T., 211
Ivanov-Emin, B. N., 132, 273
Ivanovskii, M. N., 2
Ivchev, R. S., 218
Iwachido, T., 16
Iwai, S., 174
Iwamoto, M., 394
Iwatani, K., 110
Izasimova, S. V., 332

Jackels, C. F., 167
Jäckh, C., 354
Jackson, A. J., 1
Jackson, D. A., 259
Jackson, R. A., 233
Jacobson, R. A., 301, 337
Jäger, A., 333, 392
Jagner, D., 400
Jain, D. S., 410
Jain, S. C., 220
Jain, S. R., 287
Jain, Y. S., 368
James, B. D., 66
James, D. W., 263, 272
James, F. C., 383
James, H., 165, 180
Jameson, A. K., 417
Jameson, C. J., 417
Janda, K. C., 172
Jander, J., 276
Janickis, V., 387
Janitsch, A., 213, 242
Jannach, R., 231
Jansen, M., 41
Jansen, P., 178, 179
Janssen, E., 305, 363
Janz, G. J., 400
Jardine, L. J., 400
Jarke, F. H., 263
Jaros, M., 380
Jarvis, J. A. J., 126, 318
Jarzembowski, F., 29
Jaselskis, B., 422
Jasinski, R., 416
Jasinski, T., 272
Jaulmes, S., 332
Jaumier, P. J., 400
Jaura, K. L., 222
Javora, P. H., 326
Jaworiwsky, I. S., 68
Jaymes, M., 391
Jeanjean, J., 180
Jeanne, G., 339
Jeannin, Y., 37, 200
Jeannot, F., 45, 123
Jefferson, D. A., 191
Jeffes, J. H. E., 322
Jeffrey, J. W., 124

# Author Index

Jeitschko, W., 256, 279, 325, 373
Jenkins, H. D. B., 21, 173
Jenkins, R. L., 241
Jenks, J. M., 59, 376
Jennings, D. E., 160
Jennische, P., 144
Jerschkewitz, H.-G., 188
Jespersen, N., 251
Jessen, E. B., 209
Jevcak, J., 262
Jiang, G. J., 171
Joly, R. D., 35
Johannesen, R. B., 292
Johansen, R., 95, 100, 119
Johansson, A., 146
Joh, K.-P., 291, 292
Johnson, A. D., 90
Johnson, B. F. G., 170
Johnson, C. S., 160
Johnson, D. J., 148
Johnson, D. K., 173, 274
Johnson, D. W., 165
Johnson, F. A., 80
Johnson, G. K., 20, 26
Johnson, H. D., 68
Johnson, J., 230
Johnson, L. J., 158, 160
Johnson, M. W., 413
Johnson, V., 255
Johnson, W., 150
Johnson, W. C., 167
Johnston, H. S., 403
Jolibois, B., 144
Jolly, W. L., 65, 145, 278
Jonathan, N., 347
Jones, C. E., 159
Jones, C. H. W., 393, 404
Jones, C. J., 84
Jones, G. E., 160
Jones, J. A., 398, 417
Jones, L. H., 77
Jones, R. J., 8
Jones, W. J., 102, 160, 347
Jonsson, B., 170
Jordan, K. W., 100
Jordan, T. H., 50, 103
Jornes, J., 12
Josefowicz, M., 399
Joshi, P. P., 410
Josien, F.-A., 337
Jost, J. W., 262
Jost, K.-H., 316
Jost, R., 59, 408
Jou, F. Y., 9
Jouini, J., 331
Jourdan, G., 25, 315
Joy, G., 139
Joyez, G., 171
Joyner, R. W., 170
Judge, D. L., 171
Julien-Pouzol, M., 143, 341, 372, 392
Jumas, J. C., 29, 373

Jung, J., 3
Jungk, E., 130
Jurkowitz, D., 67
Jutzi, P., 283

Kaae, J. L., 147
Kabachnik, M. I., 295, 313, 323
Kabre, M. S., 372, 392
Kabré, T. S., 143, 373
Kaczmarczyk, A., 74
Kadenatsi, B. M., 174
Kadlec, V., 97
Kadvte, T., 187
Kaempf, B., 18
Kaewchansilp, V., 279
Kagan, E. G., 81
Kagan, H. B., 156
Kagawa, N., 311
Kahn, R. H., 403
Kaidymov, B. I., 264
Kaiser, H. J., 1
Kaistha, B. C., 148
Kajimoto, O., 180
Kajiwara, M., 308
Kao, R. R., 232
Kalashnikov, Ya. A., 147
Kalasinsky, V. F., 92, 94, 203, 282
Kalina, D. G., 235
Kalinin, V. N., 85
Kalinina, G. S., 232
Kalinnikov, V. I., 383
Kalishevich, G. I., 254
Kalosh, T. N., 273, 340
Kamenar, B., 208
Kamigaito, O., 187
Kaminskas, A. Yu., 125
Kammer, W. E., 169
Kamyshev, V. A., 263
Kanamaru, F., 193
Kananovich, Yu. S., 132
Kaneda, K., 394
Kaneko, Y., 269
Kanishcheva, A. S., 30, 339
Kapila, V. P., 312, 371
Kapustnikov, A. I., 44
Kapon, M., 405
Kapoor, P., 354
Kapoor, P. N., 12
Kapoor, R., 354
Karapetyan, Yu. A., 403
Karayannis, N. M., 312
Kardanov, N. A., 220, 313
Karelin, A. I., 410
Karimov, Yu. S., 152
Karl, R. R., 352
Karlstrom, G., 170
Karnaukhov, A. S., 368
Karplus, M., 261
Karpas, Z., 168
Karpenko, N. V., 243, 340
Karpov, L. G., 277
Karyuk, G. G., 112

Kasai, M., 186
Kasatochkin, V. I., 152
Kaschuba, J., 264
Kasdan, A., 1, 182
Kasheva, T. N., 303
Kastner, P., 42
Katada, M., 225
Katerberg, J., 47
Kato, H., 162, 406
Kato, K., 113
Katori, A., 373
Kats, B. M., 218
Katsman, L. A., 265
Kaufmann, G., 37
Kaufmann, J., 329
Kaufmann, K., 396
Karaksin, Yu. N., 129
Kawada, I., 113
Kawai, S., 50, 113
Kawano, M., 360
Kawasaki, M., 165
Kawashima, N., 174
Kaya, K., 174
Kazak, V. F., 196
Kazakov, M. E., 152
Kazokov, M. M., 123
Kazarnovskii, I. A., 345
Keat, R., 300, 305, 306, 308
Kebarle, P., 163, 164, 167, 416
Kedrova, N. S., 66
Kedrowski, R. A., 241
Keene, T., 230
Keer, H. V., 125
Keiderling, T. A., 67
Keil, F., 114, 270, 278
Keller, P. C., 90, 101
Keller, R. A., 396
Kellett, S. C., 237
Kelly, H. C., 115
Kendrick, J., 167, 171
Kennedy, R. C., 166
Kennedy, S. W., 143
Kenney, M. E., 228
Kenyon, G. L., 318
Kermarac, M., 186, 187
Keropyan, V. V., 368
Kerridge, D. H., 273
Kevan, L., 165
Khaddar, M. R., 121, 312
Khadzhi, V. E., 186, 197
Khaikin, L. S., 342
Khakimova, D. K., 147
Khalifan, H., 379
Khalkin, V. A., 400
Khalyapina, O. B., 125, 186
Khan, A. U., 310
Khanafer, M., 372
Khandozhko, V. N., 235, 239, 240
Kharitonov, N. P., 203
Kharitonov, Yu. Ya., 66, 115, 116, 119, 127, 137, 320, 331, 337

# Author Index

Khmaruk, A. M., 249
Kholoclov, E. P., 259
Khoo, S. K., 399
Khor, T. C., 407
Khozhainov, Yu. M., 36, 382
Khrameeva, N. P., 331
Khramov, A. S., 220
Kibardina, L. K., 228
Kidd, R. G., 336
Kido, H., 133
Kiel, G., 335
Kijima, I., 245
Kim, J. J., 291
Kim, J. K., 163, 172, 260, 263
Kim, P., 163, 271, 344
Kimball, C. W., 114
Kimel'feld, Ya. M., 262, 397
Kimmel, H. S., 412
Kinberger, K., 100
King, G. W., 171
King, R. B., 284
King, T. J., 204, 206, 208, 220, 245, 250, 306
King, W. T., 347
Kinney, J. F., 126
Kinoshita, K., 156
Kipker, K., 323, 364
Kirchmeier, R. L., 395
Kireev, V. V., 308
Kireeva, A. Yu., 315
Kirev, A., 401
Kirchheim, R., 254
Kirichenko, I. N., 218
Kirilin, A. D., 210
Kirin, I. S., 422
Kirsanova, N. A., 299
Kirpichnikova, L. F., 383
Kiselev, A. V., 187
Kiselev, Yu. M., 421
Kishi, K., 170
Kishimoto, S., 170
Kishore, N., 294
Kistrup, C. J., 255
Kjaikin, L. S., 258
Kjekshus, A., 333, 340, 373, 384
Kjellevold, K. E., 48
Klaeboe, P., 25
Kläning, U. K., 410
Klar, G., 334
Kleir, D. A., 73, 102
Klein, F. S., 168
Kleinberg, J., 348
Kleinerman, T. V., 133
Klemin, R. B., 171, 172
Klemperer, W., 172, 173, 397, 418
Klimchuk, M. A., 11
Klimenko, M. A., 54
Klimov, E. S., 142
Klimov, V. D., 418
Klimova, T. P., 80
Klimova, T. V., 81

Klingebiel, U., 227, 299, 363
Kliñgen, T. J., 81
Klinkova, V. V., 66
Klopman, G., 264
Kloth, B., 364
Klotzbuecher, N. E., 260
Klüfers, P., 279
Klug, W., 357
Klushina, T. V., 36, 382
Klyuchnikov, N. G., 370
Kneidl, F., 327
Knigarko, I. P., 176
Kniglov, L. D., 254
Knobler, C. D., 82
Knop, O., 210, 316
Knox, K., 228
Knox, S. A. R., 236, 237, 239
Knyaznev, S. P., 80
Ko, D., 178, 179, 283
Ko, Y., 89
Kobayashi, A., 66
Kobayashi, H., 203
Kobayashi, M., 321
Kobayashi, Y., 321
Kober, F., 296, 329
Kobycheva, S. A., 370
Koch, D., 323, 364
Kocheskov, K. A., 208, 209
Kochina, T. A., 203
Kodama, G., 96
Kodema, G., 63
Köhler, F. H., 287
Köhler, H., 31, 321
Koehler, P., 216
Koehler, T. R., 259
Koenig, M., 315
Körner von Gustorf, E., 258
Köster, R., 62
Koga, Y., 157
Kogan, E. A., 204
Kogan, V. A., 220
Koharov, N., 409
Kohatsu, J. J., 248, 339
Kohrmann, H. J., 167, 168
Koizumi, M., 193
Kojić-Prodić, B., 316
Kokes, R. J., 268
Koketsu, J., 341
Kokat, Z., 204
Kokorina, L. G., 327, 328
Kokurin, N. I., 213
Kolb, J. R., 67
Kolbanev, I. V., 186
Kolditz, L., 336
Kolesnikova, N. A., 324
Kolesov, V. S., 130
Kolesova, V. A., 122
Koleva, N., 50
Kolitsch, W., 159
Kolli, I. D., 102
Kollmann, G., 230
Kollmann, K., 171
Kollman, P. A., 7, 146, 318

Kollmannsberger, M., 335
Kolodyazhnyi, Yu. V., 220, 226, 313
Kolobova, N. E., 235, 239, 240
Kolosov, E. N., 23
Kolosova, N. D., 209
Kolsky, V., 305
Kol'tsov, S. I., 188, 293
Komalenkova, N. G., 215
Komarek, K. L., 392
Komatsu, T., 291
Komissarova, L. N., 331
Komler, I. V., 323
Kompa, K. L., 62, 87
Kompanichenko, N. M., 372
Komura, K., 113
Kon, A. Yu., 246
Kondo, M., 165
Koniak, C., 368
Koniger-Ahlborn, E., 377
Konnen, G. P., 150
Konopka, A., 302
Konoplev, V. N., 66
Konoplina, A. A., 345
Konov, A. V., 104, 127, 293
Konovalova, I. V., 302, 315
Konrad, P., 131
Konyasheva, N. V., 51
Koop, H., 292
Kopeikina, A. N., 257
Kopf, J., 334
Korecz, L., 228, 351
Korenev, Yu. M., 23
Korenstein, R., 316
Kornev, Y. I., 141, 315
Korneva, S. P., 233
Korobanova, N. L., 99
Korobov, I. I., 115
Korol, E. N., 227
Korol'ko, V. V., 81
Korol'kov, V. A., 243
Korsakov, M. V., 203
Korshak, V. V., 308
Korshunov, B. G., 127, 388, 390, 391
Korshunov, I. A., 331
Korsunskii, B. L., 422
Koryakov, V. S., 399
Koshi, W. S., 164
Kosmodem'yanskaya, G. V., 98
Kosova, L. M., 324
Koster, A. S., 341
Koster, J. B., 127
Kostin, V. I., 337
Kostikov, A. I., 385
Kostina, V. G., 295
Kostiner, E., 316, 325, 385
Kosyleva, O. F., 99, 122
Kosyrkin, B. I., 130
Kotova, I. T., 125
Kovac, R. A., 90
Kovalev, V. V., 383

Kovaleva, T. V., 295
Kovalevskaya, I. P., 241
Koyano, I., 171
Kozachenko, E. L., 368
Kozhakova, A. A., 380
Kozin, L. F., 7, 138
Kozlov, E. S., 299
Kozlova, V. S., 81
Kozlovskii, E. V., 138
Kozlowska-Milner, E., 272
Koz'mina, M. L., 321
Kozyukov, V. P., 223
Kracht, D., 97
Krämer, V., 337, 341, 373
Kramer, J. D., 89
Kramer, L., 311
Krautwasser, H. P., 147
Kravchenko, O. V., 66, 115
Kravstsov, D. V., 226
Krebs, B., 225, 298, 306, 373, 376
Krebs, U., 400
Krech, R. H., 402
Kreevoy, M. M., 65
Krejcik, D., 411
Krenev, V. A., 105
Kress, J., 44
Krieger, W., 120
Kriegsman, H., 216
Krikorian, O. H., 2
Kripyakevich, P. I., 251, 256
Krishnamachari, N., 97, 321
Krishnamurthy, S. S., 103, 308
Krishnan, K., 411
Krist, R., 96
Krivtsov, N. V., 47
Krivy, I., 367
Krogh-Moe, J., 129
Krommer, H., 29
Krommes, P., 140
Kroner, J., 266
Kroon, P. A., 298
Kroth, H.-J., 289
Kroto, H. W., 102
Kruck, T., 243
Krüger, C., 127
Kruglaya, O. A., 232
Krupnyi, A. I., 382
Krylov, E. I., 209
Ksenzenko, V. I., 388
Kuan Fu, L., 392
Kubasova, L. V., 321
Kuchen, W., 312, 324
Kucherenko, E. S., 254
Kuchitsu, K., 161, 167
Kuczkowski, R. L., 91
Kudinova, A. A., 392
Kudo, Y., 186, 360
Kudryavtsev, N. T., 322
Kuenzi, H. U., 56
Kugler, E. L., 268
Kuhn, N., 297
Kuhn, W., 187
Kuhtz, B., 305, 363

Kukhar', V. P., 299, 303
Kuklina, N. F., 388
Kulagin, N. E., 130
Kulagina, N. G., 253
Kulakov, V. M., 147
Kulakova, I. I., 150
Kul'ba, F. Ya., 132, 133, 136, 137, 143
Kulikov, I. A., 273
Kulkarni, V. H., 220
Kumada, M., 230
Kumagi, N., 299
Kumar, N., 141
Kumevich, G. I., 204
Kummer, D., 219
Kuna, J., 227
Kung, R. T. V., 276
Kunin, L. L., 4
Kuntz, I. D., 7
Kunz, A. B., 157
Kunze, U., 210
Kupchik, E. J., 224
Kupčik, V., 341
Kuprii, V. Z., 345
Kuramshii, I. Ya., 219, 220
Kurbakova, A. P., 129
Kurbatov, N. N., 243
Kurcz, W. E., 209
Kurdyumov, A. V., 112
Kurdyumora, T. N., 137
Kurina, L. N., 165
Kuriyama, K., 4, 113
Kurnygin, A. S., 196
Kuroda, H., 160
Kuroda, R., 143
Kuropatova, A. A., 135
Kurosawa, H., 140
Kushnikov, Y. A., 348
Kustan, E. H., 390
Kutzelnigg, W., 61, 114, 157, 278
Kuyatt, C. E., 171
Kuzetsov, E. V., 287
Kuznesof, P. M., 216, 282
Kuznetsov, L. D., 369
Kuznetsov, N. T., 56
Kuznetsov, V. A., 63, 114, 115, 186, 233
Kuznetsov, V. G., 30, 339
Kuznetsova, A. A., 383
Kuznetsova, L. S., 254
Kuznetsova, Z. M., 156
Kuz'menkov, M. I., 322
Kuz'menkov, V. I., 125
Kuzmicheva, V. P., 382
Kuz'min, E. A., 413
Kuz'min, N, M., 129
Kuzyakov, Yu. Ya., 266
Kvlividze, V. I., 26
Kyuntsel', I. A., 334

Labarbe, P., 95
Labarre, J.-F., 109, 117, 158, 284, 290, 296, 305, 306

Laber, R.-A., 336, 338
LaBreque, J. J., 378
La Bonville, P., 422
Lacoste, G., 267
Ladd, J. A., 218
Ladd, M. F. C., 173
Lagrange, P., 154
Lagow, R. J., 77, 103, 395
Lagowski, J. J., 110
Lam, S. Y., 8
Lamanova, I. A., 329, 332
Lambert, D. G., 412
Lambert, L., 87
Lambrecht, V. G., 372
Lamotte, M., 396
Lampe, F. W., 182
Landreth, R., 367
Lane, R. F., 401
Lang, F.-M., 150
Lang, J., 26, 51, 77, 132, 229, 255, 279, 326
Lange, U., 31, 321
Langer, K., 122
Langford, G. R., 241, 288
Langhoff, S. R., 167
Langland, J. T., 65
Langlois, M.-Cl., 373
Lapitskii, A. N., 339
Lapkina, N. D., 152
Lappert, M. F., 102, 103, 104, 226, 290
Lapshin, S. A., 3
Lapshin, V. A., 7
Larina, R. A., 99
Larsen, L. A., 108
Larsen, I. K., 18
Laruelle, P., 332
Larvor, M., 416
Laser, M., 398, 417
Laskorin, B. N., 331
Latsou, A., 179
Latte, J., 348
Latyaeva, V. N., 233, 235
Lau, C., 379
Lau, K.-K., 90
Laügt, M., 316
Laughlin, D. R., 245
Lauglin, W. C., 126
Laurence, G. S., 141
Laurent, J.-P., 96, 100
Laurent, Y., 51, 255, 333, 326
Lautie, A., 370
La Villa, R. E., 375
Lavrent'ev, A. N., 281
Lavrova, O. A., 167
Lavut, E. A., 127
Lawesson, S.-O., 300
Lawson, D. F., 308
Lawton, S. L., 335
Laye, P. G., 203
Lazarev, V. B., 340, 372, 386
Lazarini, F., 19, 376
Leach, J. B., 75, 234.

# Author Index

Lebedev, E. P., 204, 225, 227
Lebedeva, N. V., 204
Le Blanc-Soreau, A., 52
LeCaer, C., 256
Lechert, H., 183
Lechner, J. F., 401
Leclerc, B., 143, 373
Ledesert, M., 19, 54, 377
Lee, J. S., 161
Lee, J. H., 163
Lee, L. C., 171
Lee, R. C., 297
Lee, R. G., 100
Lee, S. H., 25, 335
Lee, S. J., 165
Lee, S. T., 265
Lee, Y., 141
Lees, T. P., 368
Le Flem, G., 316
Le Geyt, M. R., 323, 364
Legon, A. C., 161, 404, 416
Legros, J. P., 200
Lehmann, E., 402
Lehn, J. M., 16, 18
Lehnig, M., 233
Leibovici, C., 158, 284, 290
Leites, L. A., 129
Leiva, A.-M., 323, 364
Lemenshko, O. V., 388
Lemesheva, D. G., 345
Lemley, J. T., 59, 376
Lemmen, P., 313
Lenglet, M., 125, 133
Leonard, M. A., 400
Leong, W. H., 263
Leonhardt, G., 364
Lepeshkov, I. N., 99, 122, 125
Lepoutre, G., 262
Lerf, A., 375
Le Roux, G., 384
Leroy, J.-M., 165
Leroy, M. J. F., 336
Lesnikovith, L. A., 321
Lessard, C. R., 167
Lester, W. A., 167
Lestera, T. M., 167
Letcher, S. V., 6
Letoffe, J. M., 34, 35, 376
Letsou, A., 178, 283
Leung, K. Y., 321
Levy, B., 157
Levy, D. H., 168, 270
Levy, J. B., 166
Levy, M. R., 163, 165
Levy, R., 375
Levin, Ya. A., 295
Levy Clement, Cl., 50
Levason, W., 333, 340
Levec, J., 418
Levenshtein, I. B., 332
Levin, E. M., 126
Levin, E. S., 253
Levin, G., 12

Levin, I. W., 291
Levina, N. P., 85
Levitskaya, Z. G., 246
Lewellyn, M., 305
Lewis, G. J., 313
Leyden, R. N., 70, 106
L'Haridon, P., 51, 255, 326, 333
Li, A. N., 390
Li, Y. S., 92, 93, 222
Licht, K., 216
Lieberman, M. L., 147
Liebman, J. F., 421
Lieder, C. A., 164
Liefbafsky, H. A., 412
Likforman, A., 138, 384
Likolaeva, L. G., 279
Lim, C. C., 417
Liminga, R., 316, 383, 412
Lin, L.-N., 403
Lin, M. C., 171, 172
Lin, M. J., 366
Lin, S. B., 123
Lin, S. F., 161
Lincoln, S. F., 134
Lindberg, B. J., 258
Lindberg, K. B., 316
Lindinges, W., 260
Lindner, E., 323
Lindner, P., 157
Lindqvist, O., 390
Lindstrom, R. H., 307
Lindoy, S., 101
Lineberger, W. C., 1, 182
Línek, A., 80
Lines, E. L., 93, 282
Linke, K. H., 260
Linowsky, L., 279
Lipscomb, W. N., 62, 71, 73, 75, 78, 102, 260
Lippard, S. J., 67, 72
Liquornik, M., 124
Lischka, H., 61, 114, 157, 415
Lisitsa, V. V., 151, 415
Liskow, D. H., 150
Lisov, N. I., 35
Liu, D. D. S., 396
Litvak, H. E., 163
Litvin, L. T., 147
Litvinov, Yu. G., 243
Litz, P. F., 107
Llabador, Y., 398
Ll'in, M. K., 56
Lloyd, D. J., 328
Lo, D. H., 145
Lo, F. Y., 82
Lobkov, E. U., 348, 349
Lobkovskii, E. B., 115
Lockhart, S. H., 231
Lockwood, D. J., 98
Loewenschuss, A., 104
Loginova, E. M., 373
Lohr, L. L., 421
Loireau-Lozac'h, A.-M., 134

Lombard, L., 147
Long, D. A., 380, 381
Long, G., 4
Long, G. G., 339
Long, J. R., 63
Longato, B., 85
Lopatin, S. N., 324
Lopatto, Yu. S., 147
Lopitaux, J., 133
Lopusiński, A., 325
Lorberth, J., 140
Lorenz, H., 281
Loriers, J., 348
Lortholary, P., 187, 228
Lory, E. R., 93
Loshin, A. F., 23
Lott, J. W., 69
Loucks, L. F., 171
Loupy, A., 11
Lowman, D. W., 88
Lovelock, J. E., 161, 395
Lovetskaya, G. A., 139
Luber, J., 302
Lucas, G., 276
Lucas, J., 55
Lucazeau, G., 41, 370
Luche, J.-L., 156
Lucken, E. A. C., 95
Lucquin, M., 165
Luczynski, Z., 164
Luger, P., 228
Lugorskaya, E. S., 99
Luiko, E. M., 204
Lunak, S., 369
Lund, A., 165
Lund, T., 106
Lundeen, J. W., 10
Lundgren, B., 12
Lundgren, J.-O., 371
Lunenok-Burmakina, V. A., 345
Lunsford, J. H., 366
Lupu, I., 336
Lur'e, E. A., 196
Lutar, K., 346, 418
Lutsenko, I. F., 280, 295
Lutz, O., 97
Lygin, V. I., 187
Lygina, I. A., 260
Lynch, A. W., 146
Lynch, K. P., 182
Lysenko, L. A., 256
Lysenko, Yu. A., 135
Lyskova, Yu. B., 55
Lyubimov, V. N., 349
Lyubosvetova, N. A., 331
Lyudomirskaya, A. P., 368
Lyutin, V. I., 196
Luzhnaya, N. P., 257

Ma, C.-B., 124, 196
Mabbs, F. E., 274
McAllister, T., 163
McAuley, A., 342
McAuliffe, C. A., 333, 344

McCarthy, G. J., 373
MacCordick, J., 37
McCullough, J. D., 390
McDermott, C. P., 104, 295
MacDiarmid, A. G., 354
McDowell, C. A., 160, 265
McDowell, D., 70
McElroy, M. B., 396
McFadden, D. L., 107, 402
McFarlane, H. C. E., 280
McFarlane, W., 280
McGee, H. A., 91
McGlinchey, M. L., 212
MacKay, G., 180
Mackay, K. M., 232, 235
Mackay, R. A., 148
McKelvey, J., 146
McKenney, D. J., 408
McKennon, D. W., 297
McKinney, R. J., 237
McKown, G. L., 77
McLaughlin, K. L., 335
MacLean, D. I., 402
McMasters, O. D., 140
MacMillan, D. T., 315
McMurdie, H. F., 21
MacNamee, R. W., 275
McNeil, E. A., 75, 77
MacPartlin, M., 221
McPhail, A. T., 107
McWhinnie, W.-R., 390
McWilliam, D. C., 231
Madan, H., 338
Maddren, P. S., 105
Madec, M. C., 245
Madsen, H. E. L., 320
Märkl, G., 327
Magee, C. P., 81
Maggiora, G. M., 167
Magnuson, J. A., 351
Magunov, R. L., 241
Mahadik, S. T., 227
Mahajan, O. P., 148
Mahajan, V. K., 151
Mahenc, J., 267
Mahlberg, G., 138, 373
Maier, M., 102
Mailen, J. C., 272, 398
Maillard, D., 366
Maiorova, L. P., 233
Majid, A., 217
Makarov, A. V., 56
Makarov, E. F., 191
Makarova, L. M., 131
Makatun, V. N., 383
Makeev, G. N., 418
Maki, A. G., 99
Maksimenko, A. A., 125
Maksimov, A. I., 262
Maksina, N. V., 368
Mala, J., 375
Malaman, B., 136, 256
Maleki, P., 120

Malhotra, K. C., 219, 220, 336, 351, 411
Malicki, W., 164
Malikova, E. D., 4
Malinovskii, Yu. A., 186
Maliovskii, T. I., 246
Malisch, W., 235
Malkova, A. I., 419
Mallah, K. A., 357
Mallinson, P. D., 160
Mallinson, P. R., 13
Malm, D. N., 171
Malov, Yu. I., 117, 127
Malova, N. S., 139
Mal'tsev, V. T., 99, 245
Mal'tseva, N. N., 66, 126
Mal'tseva, V. S., 140
Malygin, A. A., 188, 293
Malyutin, G. V., 257
Mamantov, G., 25, 125, 126, 348, 399, 402, 403
Mamedov, Kh. S., 98, 99
Manaktala, H. K., 140
Manatt, S. L., 298
Mandal, S. S., 320
Mandel, N. S., 316
Mangini, A., 157
Mangion, M. M., 63
Manapov, R. A., 219, 220
Mann, G., 132
Manne, R., 135
Manning, A. R., 240
Manojlovíc-Muir, L., 59, 304, 369
Manohar, H., 337
Manoharan, P. T., 329
Manoussakis, G. E., 332
Mantz, A. W., 171
Manuccia, T., 93
Maracek, J., 369
Maraschin, N. J., 77, 395
Marassi, R., 126, 348, 399
Maraud, J.-P., 137
Marcantonatos, M., 111
Marcati, F., 313
Marcelin, G., 163
March, F. C., 328, 331, 333, 340
Marchard, A., 415
Marchand, R., 33, 51, 132, 143, 174, 200, 279, 333
Marconi, G. C., 160
Marconi, W., 313
Marcus, L. H., 227
Mardivosova, I. V., 36
Marecek, J. F., 311, 313
Marechal, Y., 167
Mareev, Yu. M., 337
Margrave, J. L., 22, 103, 151, 178, 179
Margulis, M. A., 273
Maries, A., 190
Mariolacos, K., 341
Marinin, A. S., 218, 420

Mark, W., 390
Markheeva, D. M., 141
Markiv, V. Ya., 256
Markov, B. F., 23
Markov, V. S., 368
Markova, O. A., 59
Marks, T. J., 67, 235
Maroy, K., 390
Marsden, C. J., 388
Marsh, H., 150
Marshall, K., 190
Marsili, M., 327
Martell, A. E., 285, 315
Martin, D. R., 100, 105, 326
Martin, J. S., 416
Martin, L. R., 163
Martin, R. L., 61
Martinas, F., 226
Martinez-Ripoll, M., 45
Martin-Frère, J., 331
Martin-Garin, L., 254
Martynenko, B. V., 321
Martynenko, L. I., 141, 315, 421
Marumo, F., 174
Marynıck, D. S., 62
Marzi, W. B., 330
Masaki, N., 4, 113, 242
Masanet, J., 164
Mascherpa, G., 25, 96, 144, 334
Masdupuy, E., 202
Mase, H., 299
Mashcerpa-Corral, D., 135
Maslennikov, I. G., 281
Maslennikova, E. G., 390
Mason, J., 266, 268, 400
Massarotti, V., 144
Masse, R., 316
Massey, A. G., 104, 105
Masson, J. P., 413, 414
Mastera, St., 417
Mastryukov, V. S., 63, 79, 226
Mataitene, L. S., 125
Mataitis, A. I., 125
Mateeva, Zh. A., 167
Matheson, A. J., 8
Mathey, F., 80
Mathias, E., 344
Mathiasch, B., 225
Mathieu, Fr., 245
Mathieu, J.-C., 254
Mathis, F., 296, 313, 314
Mathis, R., 296, 315
Matousek, I., 169, 177
Matrosev, E. I., 323
Matschoss, V., 388
Matsuda, Y., 406
Matsui, Y., 197
Matsumoto, A., 163
Matsumoto, K. Y., 320
Matsuo, T., 321
Matsuura, N., 321

# Author Index

Matsuzawa, K., 133
Mattes, R., 40, 112, 377
Matthew, J. A. D., 1
Matthews, J. D., 221
Matthews, R. W., 336
Matvienko, L. G., 243
Matyas, P., 392
Mauer, F., 175
Maunaye, M., 132
Maurin, M., 25, 29, 225, 334, 373
Maxwell, I. E., 143
Maxwell, W. M., 79
May L., 399
May, N., 40
Maya, L., 229
Mayer, B., 342
Mayer, E., 94, 284, 289, 389
Mayer, T. M., 182
Mayer, U., 96
Mazanec, T. J., 110
Mazeau, J., 171
Mazolov, L. N., 1
Mazzei, A., 118
Mazzeo, B., 172
Meacher, J. F., 163
Mechkovskii, L. A., 257
Medeva, Z. S., 254
Medvedeva, M. D., 210, 228, 301
Medvedeva, Z. S., 326
Medvederskikh, Yu. G., 165, 168
Meek, D. W., 285
Meerschaut, A., 385
Meguro, K., 174
Mehetre, A. N., 266
Mehrotra, R. C., 12, 111, 204
Mehrotra, P. K., 329
Meider-Gorican, H., 16
Meikle, G. D., 224, 355, 389
Meisel, T., 169
Mejean, T., 416
Mekata, Y., 210
Melikhov, I. V., 368
Meliksetyan, N. S., 197
Melin, J., 155, 335
Meller, A., 95, 109, 227
Mel'nichenko, E. B., 271
Mel'nichenko, L. M., 244, 245, 319
Mel'nikov, N. N., 134, 135, 310
Mel'nikova, R. Ya., 383
Mel'nikova, S. V., 382
Menard, H., 413
Menchetti, S., 28, 98
Mengersen, C., 141
Meneghelli, B. J., 75
Mente, D. C., 93, 94, 282, 283
Menzebach, B., 225
Menzel, D., 170

Menziger, C., 111
Meot-Ner, M., 172
Merbach, A. E., 219
Mercer, G. D., 77, 78, 83
Mercey, B., 255
Mercier, J. P., 195
Mercier, R., 337, 367
Merer, A. J., 171
Merlo, F., 52, 56, 114
Merrill, C., 402
Merrill, L., 174
Merryman, D. J., 127, 404
Mertz, K., 138, 139
Merz, E., 398, 417
Meschede, W., 377
Meserole, F., 368
Mester, Z. C., 324
Metham, T. N., 236
Metz, W., 156
Meunier, G., 379, 380
Mewis, A., 279
Mews, R., 350, 351, 360
Meyer, E., 265
Meyer, K., 187
Meyer, R. T., 146
Mezentseva, L. P., 197
Mgaloblishvili, D. M., 348
Michael, J. V., 182
Michaelson, R. C., 171
Michalik, M., 364
Michalski, J., 325
Michaud, M., 157, 161, 294
Michel, A., 50
Michel, C., 339
Miedema, A. R., 253
Miehe, G., 341
Mielcarek, J. J., 101
Miftakhova, A. Kh., 315
Migachev, A. I., 276
Migeon, M., 96, 104, 294
Mignon, P., 161
Mihichuk, L., 235
Mijlhoff, F. C., 203, 380
Mikhailichenko, A. I., 54
Mikhailova, V., 50
Mikhailyuk, Yu. I., 134
Mikheev, N. V., 131
Mikheeva, L. M., 131
Mikheeva, V. I., 65, 66
Mikheeva, V. M., 126
Mikler, H., 392
Mikler, J., 213, 242, 392
Mikolajczyk, M., 310
Mikulski, C. M., 312, 354
Milhaud, J., 160
Milinski, N., 273
Milker, R., 286
Millard, M. M., 208
Millen, D. J., 161, 404, 416
Miller, F., 30
Miller, R. E., 294
Miller, V. R., 79
Millero, F. J., 368
Millington, D., 25, 342

Millie, P., 157
Mills, B. E., 160, 413
Mills, I. M., 179
Mills, J. L., 93, 94, 281, 282, 283, 310
Milne, C. A., 210, 369
Milne, J., 335
Milne, J. B., 391, 408
Milstein, R., 165
Minami, R., 21
Minkwitz, R., 265
Mirbabaeva, N. N., 196, 199
Mironenko, O. N., 102
Mironov, V. F., 210, 214, 223
Mirsaidov, U., 116
Misaki, T., 163
Misakian, M., 171
Mishenov, Yu. M., 51
Mishra, S. P., 276, 291
Mislow, K., 183, 327
Mitchell, J. D., 324
Mitchell, P. D., 160
Mitchell, R. E., 93, 282
Mitchell, T. N., 232
Mit'kin, V. N., 349
Mitra, G., 59, 408, 412
Miura, M., 291
Miyahara, K., 170
Miyazaki, T., 179
Mizoguchi, T., 370
Mizushima, M., 270
Mladenova-Bontschewa, Z., 341
Mocellin, A., 124, 186
Modinos, A., 105
Moedritzer. K., 294
Moffett, D., 391, 408
Moiseenko, Zh. G., 368
Mokhosoev, M. V., 125
Mokoulu, J. A. A., 180, 249
Molina, M. J., 396
Molleyre, F., 150
Molloy, L. R., 170, 201
Molls, W., 243
Molodkina, A. N., 26
Molyavko, L. I., 303
Momot, E., 165
Momot, O., 321
Monahan, K. M., 171
Mondal, P., 124
Mongeot, H., 62, 90
Monnaye, B., 199
Monn, E. H., 128
Monosova, A. Ya., 36, 382
Monteil, Y., 322, 323, 325, 372, 384
Montel, G., 320, 373
Montenarh, M., 356, 357, 360
Moody, D. C., 62, 63, 65, 234, 241, 288
Mooney, J. R., 228
Moore, C. B., 396

Moore, C. E., 29
Mooy, J. H. M., 120
Morandini, F., 85
Moravec, J., 367, 368, 398
Moraweck, B., 124
Moret, J., 390
Morgunova, M. M., 220
Morgunova, M. J., 226
Morillon, M., 160
Morillon-Chapey, M., 160
Morimoto, N., 197
Morimoto, T., 187
Moritani, I., 89
Moriyama, H., 66
Mornasini, M. L., 53
Morosin, B., 38
Morozova, M. P., 40
Morozova, S. S., 40
Morozova, T. G., 26
Morris, A., 347
Morris, J. H., 67, 108
Morris, H., 203
Morris, M. C., 21
Morrison, J. A., 157, 182, 214, 216
Morrow, B. A., 187, 188, 190
Morse, J. G., 280, 290, 296
Morse, K. W., 280, 290, 296
Morse, R. I., 378
Morss, L. R., 126
Mortier, W. J., 185
Morton, J. R., 278, 291, 329
Moruzzi, J. L., 172
Moseley, J. T., 177
Moskva, V. V., 294
Moskvitina, E. N., 266
Mostovaja, L. M., 397
Mostovoy, A. B., 397
Motl, A., 345
Motooka, I., 321
Motov, D. L., 370
Motte, J.-P., 27, 279
Moule, D. C., 167
Mrozek, J., 261
Mucka, V., 345
Mück, G., 126
Müller, A., 376, 377
Müller, H., 323, 364
Müller, H.-D., 281, 296
Mueller, W., 5, 7, 40
Müller-Buschbaum, H., 133
Muetterties, E. L., 61, 65, 266
Muir, K. W., 304
Muiry, I. B., 111
Mujtaba, S. Q., 331
Mukanov, I. P., 327, 328
Mulcahy, M. F. R., 150
Mulder, N., 149
Muller, A., 160
Muller, D., 385
Muller, M., 159
Muller, W., 251

Mulvihill, J. N., 26, 180
Mulyanov, P. V., 224
Mumaia, M. J., 171
Muminov, S. Z., 195
Munn, R. W., 179
Munoz, A., 315
Muradov, V. G., 253
Muradyan, L. A., 215
Muraev, V. A., 142
Murakhtanov, V. U., 1
Muralikrishna, U., 181
Muranaka, S., 50
Murase, I., 315
Murato, H., 173
Muratova, A. A., 219, 220, 222
Murphy, J. A., 335
Murphy, R. M., 25, 335
Murthy, A. R. V., 308, 366
Murray, G. T., 400
Murray, I. B., 207
Murray, R. M., 385
Murray, S. G., 340
Mushkov, S. V., 43
Musina, A. A., 220
Muylle, E., 101, 283
Myakishev, K. G., 63
Myers, E. A., 326
Myers, T. C., 311
Myers, W. H., 91, 285
Myl'nikova, L. N., 127
Mysin, N. I., 203

Nabatnikov, A. P., 146
Nabi, S. N., 273, 306
Nabier, M. N., 321
Nagai, Y., 223
Nagao, M., 187
Nagasao, T., 190, 399
Nagornyi, V. G., 146
Nagorsen, G., 51
Nagpal, K. C., 384
Nagy, G. A., 400
Nagy, J., 207
Nagy, S., 220
Nainan, K. C., 28
Nair, K. R., 366
Nakagawa, J., 184
Nakagawa, T., 364
Nakajima, M., 66
Nakajima, T., 21, 152
Nakajima, Y., 197
Nakamori, I., 409
Nakamura, T., 11
Nakamura, Y., 9
Nakanishi, K., 21
Nakayama, H., 143
Nalbandyan, A. B., 165
Nalewajeski, R., 261
Nambiar, P. R., 287
Nandi, R. N., 158, 160
Naono, H., 187
Naoumidis, A., 58
Narain, P., 402

Nardelli, M., 246
Nardin, G., 284
Naumann, D., 402, 406, 408
Naumov, A. D., 210
Naumov, V. A., 295, 296
Naumova, I. S., 197
Naumova, T. N., 127
Navaly, O. I., 219
Naveck, J., 313
Nazar, M. A., 277
Nazarenko, V. A., 121, 204
Nazarenko, Yu. P., 262
Nazarov, A. S., 151, 156, 335, 415
Nazarova, R. I., 370
Neal, H. G., 360
Nebylitsyn, B. D., 368
Neeley, C. M., 165
Nefedov, V. D., 203
Nefedov, V. I., 38, 132
Negita, H., 128, 225, 295, 334
Negro, A. D., 19
Neidhard, H., 40, 112
Neiding, A. B., 418
Neilson, R. H., 104, 291, 297
Neimeyer, S., 47
Nekrasov, Yu. D., 399
Nekrasov, Yu. V., 215
Nelander, B., 171
Nelapaty, J. C., 422
Nelson, D. J., 295
Nelson, D. L., 325
Nelson, J. F., 229
Nemee, L., 9
Nemes, M., 411
Nemoshkalenko, V. V., 262
Nepomnyashchii, L. B., 147
Nesmeyanov, A. N., 235, 239, 240
Nesterenko, A. I., 268
Neubert, A., 5
Neuman, M. A., 16, 44
Neumann, W. P., 214, 233, 249, 250
Neuzil, E., 312
Neves, E. A., 365
Nevett, B. A., 333
Newbery, W. R., 117, 126
Newham, R. J., 179
Newlands, M. J., 165
Newton, M. D., 150
Newton, M. G., 314
Neyrand, M., 225
Ng, H. N., 316
Ng, S., 161
Ng, T. L., 167
Nguyen-Van Thanh, 160
Ni, L. P., 125, 186
Nicholls, R. W., 396
Nicholson, D. G., 333, 334, 340, 384
Niculescu, D., 42

# Author Index 443

Niecke, E., 227, 297, 298, 329
Niedenzu, K., 107, 108, 110
Nielsen, O. F., 126, 160, 388
Nielsen, U., 421
Niendorf, K., 324
Nifant'ev, E. E., 323, 324
Nifant'eva, G. G., 251
Nigretto, J. M., 399
Nikishida, K., 403
Nikitin, O. T., 28, 56
Nikitin, I. V., 276, 403
Nikitina, A. S., 297
Nikitina, L. V., 40, 315, 316
Nikitina, V. K., 147
Nikitina, Z. K., 42
Nikolaeva, A. D., 369
Nikolaev, A. V., 99, 151, 415
Nikolaev, L. A., 345
Nikolina, G. P., 321
Nikol'skii, V. M., 44
Nikonorova, L. K., 325
Nikulenko, V. S., 391
Nill, K., 171
Nilsson, B. A., 316
Nimz, M., 213
Nisel'son, L. A., 127, 334, 388
Nishanov, I., 321
Nishikida, K., 402
Nishiyama, I., 161
Nitsch, G., 316
Nitsche, R., 325
Nitta, M., 124
Niwhi, F., 124
Niyazova, Z. U., 54
Noel, S., 369
Nöth, H., 62, 87, 95, 108, 109, 131, 303
Nofal, M., 379
Nofele, R. E., 354
Nolan, M. F., 397
Noles, G. T., 147
Noltes, J. G., 233, 235, 250
Nomura, S., 149
Nord, A. G., 316
Nord, G., 370
Nordine, P. C., 401
Nordmann, F., 3
Norman, A. D., 229, 230, 290, 311
Norman, A. H., 74
Noeman, J. G., 278
Norullaev, E., 147
Nosek, M. V., 257
Novak, A., 41, 44, 370
Novák, C., 80
Novak, D. P., 117
Novak, R. W., 80
Novick, S. E., 172, 173, 397, 418
Novikov, Yu. N., 152
Novikova, Z. S., 295
Norobilský, V., 305

Novoselova, A. V., 23, 249, 257, 340, 386, 392
Novotny, H., 113
Novr, M. M., 380
Novruzov, S. A., 294
Nowacki, W., 248, 339, 341, 373
Nowdjavan, I., 99, 367
Nozacki, T., 394
Nozik, Yu. Z., 383
Nowogrocki, G., 341
Nuber, B., 139
Nuffield, E. W., 248, 339, 373
Nuretdinov, I. A., 325, 387

Oakley, R. T., 301
Oates, G., 75, 217, 234
Ocheretnyi, V. A., 99
Ochiai, E. I., 343
Ochs, J., 207
Oddon, Y., 143, 322
Odin, I. N., 249, 340, 386
Odinets, Z. K., 273
Odom, J. D., 88, 92, 93, 94, 241, 282, 288
Oesterreicher, H., 114
Øye, H. A., 25, 127, 129
Offernan, R. E., 151
Ogawa, K., 124
Ogil, J., 96
Ogilvie, J. F., 165
Ogren, P. J., 129
Ogura, H., 165
O'Hare, P. A. G., 26
Ohashi, K., 341
Ohmasa, M., 341
Ohta, T., 160
Ojima, I., 223
Oka, T., 417
Okabe, T., 370
Okada, S., 163
Okamoto, Y., 315
Okawara, R., 130, 140
Okhrimenko, R. F., 370
Okita, T., 394
O'Konski, C. T. O., 262
Okransinski, S., 59, 408, 412
Okuda, T., 128, 295, 334
Olah, G. A., 264
Olapinski, H., 132
Oldershaw, G. A., 267
Olie, K., 403
Olin, A., 251
Oliver, J. P., 88, 116, 232, 233
Olivier-Fourcade, J., 29
Olofsson, G., 336
Olovsson, I., 174, 271, 272, 371
Olsen, K. J., 410
Omelanczuk, J., 310
Onak, T., 75, 234
Onaka, S., 235, 236
O'Neal, H. E., 183

O'Neill, P., 142
Onishchenko, M. K., 368
Onsan, Z., 202
Opalovskii, A. A., 156, 335, 348, 349, 402, 415
Opedal, D., 386
Opitz, J., 230
Oppermann, H., 388
Orchin, M., 240
Orlander, P., 214
Orlando, G., 160
Orshanskaya, Z. N., 127, 334
Orville-Thomas, W. J., 145, 213
Orzeszko, S., 148
Osetzky, D., 147
Oshima, C., 113
Oshman, V. A., 366
Osipov, O. A., 220, 226, 313
Ostapchuk, L. V., 218
Ostapenko, T. V., 196
Osteryoung, R. A., 267
Ostrovskaya, T. V., 415
Ostrovskii, S. V., 415
Oswald, H. R., 277
Othen, D. A., 210, 316
Otin, S., 161
Otsuji, Y., 342
Ott, R., 132
Ottinger, C., 150
Ovcharenko, F. D., 122, 191
Overend, J., 167
Overill, R. E., 89
Ovramenko, N. A., 122
Oyama, Y., 187
Oza, T. M., 30
Ozaki, A., 33
Ozawa, T., 373
Ozier, I., 160
Ozin, G. A., 22, 260

Pacak, P., 412
Pacansky, J., 169
Pace, S. C., 217, 395
Pack, S. P., 320
Paddock, N. L., 306, 307, 309, 323, 364
Padma, D. K., 348, 350
Paetzold, R., 324, 404
Page, D. I., 260
Pagès, M., 331
Paine, R. T., 93, 282, 414
Pak, S. J., 402
Pak, V. N., 188
Palagyi, I., 411
Palavit, G., 369
Palazzi, M., 332
Palie, M., 33, 154
Palkina, K., 316
Palladino, N., 313
Palvadeau, P., 41
Panderson, S., 178, 179
Pandey, J. D., 373

Pandit, S. K., 212
Pandyopadhyay, B., 137
Panfilov, V. N., 409
Pannan, C. D., 392
Pannetier, G., 373
Panetier, J., 55
Panov, E. M., 208
Panster, P., 235
Pantukh, B. I., 167
Pantzer, R., 300
Papatheodorou, G. N., 128
Papin, G., 161
Pardo, M. P., 372
Parent, C., 316
Paretzkin, B., 21
Parish, R. V., 240
Parkanyi, L., 207
Parker, A. J., 56
Parker, W. C., 394
Parkhomenko, N. G., 97
Parmentier, M., 333, 337
Parpiev, N. A., 123
Parrott, M. J., 288, 291
Parry, D. E., 152
Parry, G., 4
Parry, R. W., 93, 282
Parsons, D. G., 16
Parsons, G. H., 177
Parthe, E., 251
Pascal, J. L., 410
Pasinkiewicz, S., 120
Passmore, J., 342, 379
Pasternak, M., 399
Patel, M., 326
Patil, B. R., 220
Patil, K. C., 218
Patil, S. F., 410
Patmore, D. J., 131, 132
Patrick, J. W., 150
Pattison, P., 63
Pattoret, A., 114
Paul, R. C., 312, 338, 354, 371, 411
Paulus, J. M., 404
Pavars, I. A., 253
Pavia, A. C., 410
Pavlikov, V. N., 99
Pavlovich, V. K., 79
Pavlyuchenko, M. M., 321
Pawlak, Z., 272
Pawlikowska-Czubak, J., 367
Paxson, T. E., 71, 72
Payen, E., 96, 294
Payling, D. W., 280
Payne, N. C., 204
Payzant, J. D., 180, 263
Peacock, R. D., 400
Peakall, K. A., 3
Peake, S. C., 292
Peake, S. J., 406
Pearman, A. J., 261
Pearson, P. K., 179, 269
Pearson, W. B., 136, 251
Pebler, J., 220

Pechkovskii, V. V., 125, 322, 383
Pechurova, N. I., 141, 315
Pecul, K., 179
Pedersen, E. B., 300
Pedersen, S. E., 142
Pederson, L. G., 347
Pedley, J. B., 102, 103, 104, 226, 290
Pedregosa, J. C., 245
Peers, A. M., 160
Peet, J. H. J., 407
Peguy, A. A., 121, 312
Peica, Dz., 20, 345
Pelini, N., 165
Pelissier, J., 147
Pelissier, M., 117
Pelizzi, C., 222, 246
Pelizzi, G., 222, 231, 246, 340
Pellacani, A. C., 242
Pellerito, L., 141, 207, 220, 233, 333
Pelter, A., 104
Pelyukpashidi, R., 380
Penchina, C. M., 147
Pendlebury, J. N., 397
Penfold, B. R., 206
Penkovsky, V. V., 291
Penning, D., 47
Pennings, J. F. M., 288
Pépin, C., 87
Pepperberg, J. M., 75
Perachon, G., 35
Perchard, J. P., 366
Perche, A., 165
Pereira, A. R., 184
Perego, G., 118, 119
Perelygin, L. S., 11
Perez, R., 202
Perkins, P. G., 102
Perotti, A., 19, 97
Perregaard, J., 300
Perrin, C., 134
Perron, G., 368
Perry, A., 333
Perry, D., 248
Perry, W. B., 145, 278
Person, W. B., 171
Pertlik, F., 330
Peshikov, E. V., 383
Pessine. F. B. T., 282
Peter, J. R., jun., 333
Peter, W., 214, 217
Peters, J. W., 343
Peterson, J. R., 177
Petillon, F., 220, 334
Petit, J.-M., 294
Petov, G. M., 323
Petrenko, L. A., 11
Petrenko, Yu. A., 11
Petrini, G., 391
Petrosyants, S. P., 134, 135
Petrov, K. I., 373
Petrov, S. I., 405

Petrova, E. V., 125
Petrovici, E., 336
Petrovskaya, M. L., 321
Petter, W., 136
Peurichard, H., 161
Peyerimhoff, S. D., 344
Peyronel, G., 242
Pfeil, R., 92
Pham, H., 414
Philippot, E., 25, 29, 225, 334, 373
Philipps, M. J., 136
Philips, L. E., 180
Phillips, G. R., 163
Phillips, L. F., 26
Phillips, R. C., 204
Phillips, R. F., 171
Phizackerley, R. P., 14, 15
Pidcock, A., 236
Pierce, R. C., 102
Pierce-Butler, M., 226
Pierre, G., 181
Pierson, H. O., 147
Piffard, Y., 33, 143, 174, 200
Pignolet, L. 11, 142
Pilard, R., 132, 133
Pilipovich, D., 407
Pilyankevich, A. N., 112
Pimentel, G. C., 23, 163
Pinchuk, A. M., 249
Pinchuk, V. V., 135
Pine, A. S., 160
Pineau, P., 416
Pinnavaia, T. J., 195
Piskunov, I. N., 196
Pistorius, C. W. F. T., 254, 329
Pitner, T. P., 133
Pitts, J. N., jun., 163, 268, 343
Pitzer, K. S., 291, 421
Plachinda, A. S., 191
Plaistowe, J., 267
Plakhotnik, V. N., 97
Plakhov, G. F., 41, 196
Planet, W. G., 171
Platunova, N. B., 137
Platt, E., 278
Plavnik, G. M., 147
Plekhov, V. P., 222
Plešek, J., 61, 63, 74
Pleshivtsev, N. D., 2
Pletcher, J., 311
Pluth, J. J., 123, 124, 197
Plyler, E. K., 160
Plyshevskii, S. V., 125, 322
Plyushchev, V. E., 31, 368
Pobedimskaya, E. A., 186, 197, 295
Pocker, Y., 165
Podafa, B. P., 23
Podder, S. N., 220
Podol'skikh, L. D., 147
Pönicke, K., 208
Pogrebnaya, V. L., 268

# Author Index

Pogulyai, V. E., 161
Pohl, S., 298, 306
Pokorny, J., 19
Pokrovskaya, Yu. A., 322
Polanyi, J. C., 277
Polborn, K., 113
Poletaev, E. V., 322
Poletaev, I. F., 368
Polev, B. N., 54
Polezhaeva, N. A., 96
Politanska, U., 125
Polivanova, N. N., 368
Pollak, A., 406, 421
Pollock, R. J., 236
Polonskii, Yu. A., 3
Poltavtsev, Yu. G., 347, 378, 388
Poltmann, F. E., 385
Polyakov, A. M., 8
Polyakov, E. V., 176
Polyakova, V. B., 40, 66, 119, 127
Pomeroy, R. K., 235, 239, 240
Pomianowski, A., 367
Ponomarev, N. M., 373
Ponomarev, V. I., 215
Pons, J.-N., 177
Pool, M. J., 253
Poole, P. R., 163
Poonia, N. S., 16
Popa, M., 113
Pope, M. T., 331
Pople, J. A., 60, 158
Poppinger, D., 169
Popov, A. I., 10, 17, 18
Popova, L. N., 161
Popović, S., 316
Porte, A. L., 306, 322
Porter, D. A., 267
Porter, R. A., 351
Porter, R. F., 102, 110
Portheine, J. C., 248, 339, 373
Portnova, S. M., 125
Posnaya, I. S., 63
Posvic, H., 29
Potier, A., 135, 410
Potier, J., 99, 367, 410
Pott, G. T., 124
Potts, A. W., 22
Poulin, D. D., 298
Poulsen, F. W., 126, 388
Pour, V., 376
Pourcelly, G., 414
Poussigue, G., 166
Poutcharovsky, D. J., 251
Pouyet, B., 343
Povitskij, V. A., 152
Powles, J. G., 259, 260
Pozdeev, V. V., 214
Pozharitskii, A. F., 209
Pozhidaev, V. M., 302
Poziomek, E. J., 148

Prasad, H. S., 115
Prasad, R. N., 132
Prasad, S. D., 410
Prasad, S. M., 19, 50
Prasad, S. N., 413
Prasad, R., 242
Prasad, V. A. V., 313
Prask, H. J., 264
Pratt, J. C., 415
Pratt, J. N., 254
Pravilov, A. M., 277
Predvoditelev, D. A., 324
Preiss, I. L., 378
Preston, K. F., 278, 291, 329
Preston, R. S., 114
Pretzsch, J., 321
Preuss, H., 179
Preut, H., 25, 140, 238
Prewitt, C. T., 373
Price, W. C., 22
Prince, E., 264, 316
Pringle, W. C., 100
Prins, R., 149, 185, 318
Prishchenko, A. P., 295
Pritchard, M., 269
Pritzkow, H., 144, 277, 399, 405
Prodan, E. A., 321
Proinova, R., 409
Pronin, I. S., 272
Proshin, N. A., 369
Proskurnina, M. V., 280
Prost, M., 34, 376
Protas, J., 136
Prout, C. K., 303, 304
Prout, K., 116
Prusakov, V. N., 418
Prusazcyk, J., 271, 344
Puckvova, V. V., 129, 220
Puddephatt, R. J., 231
Pudovik, A. N., 210, 219, 220, 222, 228, 284, 296, 301, 302, 315
Pudovik, M. A., 210, 228, 296, 301
Pujol, R., 313
Pulham, R. J., 2, 3, 4, 5, 6, 7
Pullen, B. P., 171
Pullman, A., 263
Pungor, E., 169
Pupp, G., 331
Purcell, K. F., 61
Purdie, N., 144, 370
Puri, B. R., 148
Puri, J. K., 220
Purnell, H., 240
Pusatcioglu, S., 91
Pušelj, M., 136, 140
Pushkina, G. Ya., 331
Puskareva, K. S., 220
Puskaric, E., 367
Puxley, D. C., 251
Pyrkin, R. I., 295

Pytlewski, L. L., 312
Pyykko, P., 157

Quarterman, L., 402
Quast, H., 301
Quemeneur, E., 245
Quintana, J. G., 121
Quist, A. S., 103
Quicksall, C. O., 331

Rabbering, G., 369
Rabelais, J. W., 157
Rabilloud, G., 80
Rabinowitz, M., 155, 329, 389, 414
Rack, E. P., 163
Rackwitz, R., 1
Radbil, B. A., 276
Rademacher, P., 334
Radford, H. E., 168
Radleind, D. St. A. G., 411
Radom, L., 158, 169
Raede, H. S., 3
Rädle, C., 400
Rafaeloff, R., 134
Rahman, M. T., 233
Raich, J. C., 259
Rainey, P., 147
Raj, P., 223
Rajami, V., 373
Rajaram, P., 169
Rajković, L. M., 331
Rake, A. T., 240
Rakhimov, A. A., 195
Rakke, T., 373
Ra'kovskii, I. A., 54
Rakshapal, R., 59, 408
Ramadan, A. A., 405
Ramakrishna, B. L., 270
Raman, C. V., 101, 283
Ramanujam, V. M. S., 265
Ramaprasad, K. R., 183, 229
Ramey, R. R., 46
Ramirez, F., 311, 313
Ramondelli de Staricco, E., 346
Rand, Y., 269
Randaccio, L., 284
Randall, E. W., 259
Randles, J. E. B., 9
Ranganathan, T. N., 306
Range, K.-J., 125, 138, 373, 384, 392
Rankin, D. W. H., 224, 230, 236, 298, 355
Rao, C. N. R., 161
Rao, K. N., 178
Rao, K. V. S. R., 417
Rao, M. N. S., 308
Rao, P. S., 345
Rao, T. S., 266
Rapkin, A. I., 369
Rapoport, N. L., 390
Rapp, B., 62, 93, 282

Ramsey, R. J., 358, 360
Rannev, N. V., 27, 56
Raspopin, S. P., 243
Rassat, A., 157
Rathke, J., 63, 65
Ratkouskii, J. A., 35
Rau, H., 326
Rauh, P. A., 112
Rauk, A., 327
Raveau, B., 339
Ravid, B., 30
Ravn, H. L., 400
Rawat, J. P., 331
Ray, A. K., 214
Rayment, I., 361
Raynal, S., 18
Raynor, J. B., 106
Razumov, A. I., 294
Razuraev, G. A., 142, 232, 233, 235
Rea, J. R., 316
Reade, W., 88
Readon, E. J., 368
Reason, M. S., 104, 105
Rebsch, M., 353
Reck, G. P., 172
Redoules, G., 241
Redpath, J. L., 417
Reed, R., 77
Rees, D. A., 49
Reeves, L. W., 141
Regel, F., 100
Reikhsfel'd, V. O., 204, 225, 227
Rein, A. J., 206
Reinsborough, V. C., 46
Rellick, G. S., 150
Rembarz, G., 364
Rembaum, A., 51
Remizov, A. B., 219
Remmel, R. J., 68
Rempel, G. L., 370
Ren, K., 410
Renaud, J. P. P., 341
Renaud, R., 337
Renauprez, A., 124
Rendle, D. F., 131, 132
Renes, G., 203
Renk, E., 408
Rentzeperis, P. J., 242
Reshetova, L. N., 38
Rethfeld, H., 40, 112
Rettig, S. J., 108, 130, 131
Reuben, B. G., 150
Revel, M., 313
Reznichenko, V. N., 219
Ribar, B., 273
Ribbegard, G., 171
Ribeiro da Silva, M. A. V., 121
Ribes, M., 29, 225
Riccardi, R., 144
Ricci, J. S., jun., 311
Rice, D. A., 372

Richard, P., 23
Richards, J. A., 204, 206, 220, 250
Richards, R. L., 260, 261
Richards, W. G., 126, 179
Richardson, D., 119
Richardson, J. H., 269
Richardson, N. V., 347
Richardson, R. J., 170
Richter, W., 328
Ridley, D. R., 333, 334
Rie, J. E., 116
Rieck, G. D., 125, 341
Riedel, C., 373
Riedel, E. F., 401
Riedmann, W. D., 228, 357
Riegel, F., 101, 387
Rieke, R. D., 139, 140
Riera, V., 239
Riesel, L., 300, 301
Riess, J. G., 292, 293, 298, 395
Riethmiller, S., 92, 282
Rietz, R. R., 85
Riganti, V., 408
Riggin, M., 163
Rigny, P., 402
Rigo, A., 264
Rinaldi, R., 123, 124, 197
Rinck, R., 62, 87
Ring, M. A., 182, 183, 241
Rink, W., 144
Rinne, D., 352
Ritchie, D. I., 111
Ritter, G., 353, 366
Ritter, J. J., 60, 396
Ritzhaupt, G., 27, 410
Rivest, R., 48
Rivet, J., 267, 372, 392
Riviere, P., 241, 242
Riviere-Baudet, M., 241, 242
Roach, E. T., 10
Roaves, B., 256
Robaux, O., 167
Robbins, M., 372
Robert, D. U., 292, 293
Robert, J. B., 293
Roberts, B. P., 288, 291
Roberts, J. D., 259
Roberts, M. W., 170
Roberts, N., 23, 104
Roberts, P., 127
Roberts, R., 276
Robertson, A., 230, 236
Robertson, B. E., 19
Robertson, S. D., 149
Robineau, M., 177, 377
Robinet, G., 158. 290
Robinson, J. N., 148
Robinson, W. R., 142
Robison, D. H., 264
Robor, A. M., 144
Rochester, C. H., 190
Rockett, B. W., 407

Rockwood, S. D., 103
Roddy, J. W., 326
Rode, B. M., 405, 408
Rodesiler, P. F., 250
Rodicheva, G. V., 322
Rodriguez-Reinoso, F., 150
Röder, N., 117
Roeder, S. B. W., 67
Rösch, L., 289
Röschenthaler, G.-V., 292, 296
Roesky, H. W., 305, 358, 361, 363, 365
Roessel, K., 121, 310
Roest, E., 385
Rogacheva, E. I., 340, 372, 386
Roger, J., 177
Rogers, M. T., 160
Rogers, P. S., 190
Rogers, S. C., 161, 416
Rogl, P., 113
Rogstad, A., 347
Rohart, F., 160
Rokhlina, E. M., 226
Roland, A., 354
Rolandsen, A., 387
Rolin, M., 414
Romanetti, R., 318
Romanov, L. G., 8, 315
Romley, R. J., 235
Ron, A., 158, 160, 171
Roques, B., 136
Rose, F., 363
Rose, V. J., 335
Rosenberg, A., 160
Rosenfeld, J. L. J., 147
Roshchina, V. R., 254
Roslaya, T. L., 321
Rosmus, P., 365
Rosolovskii, V. Ya., 42, 47, 276, 345, 403, 410
Ross, B., 227
Ross, B. W., 330
Ross, L., 215, 364
Ross¡ S. D., 219, 243
Rossi, G., 19, 97
Rossi-Sonnichsen, I., 160
Rossiter, J. J., 335
Rotella, F. J., 235
Rothan, R. N., 126
Rothe, E. W., 172
Rothenberg, S., 146
Rothgery, E. F., 91
Rothon, R. N., 318
Rothschild, L., 300
Rothschild, W. G., 167
Rotzoll, G., 163
Rouanet, A., 123
Rouchy, J. P., 148
Roulleau, F., 45
Roussen, R., 402
Routledge, V. I., 274
Rouxel, J., 41, 52, 385

# Author Index

Rouxhet, P. G., 187
Rowbotham, P. J., 240
Rowe, K., 104
Rowland, F. S., 165, 396
Rozhnova, G. P., 368
Rozsondai, B., 226
Rozyev, N., 44
Rubini, P. R., 121, 312
Rubtsov, A. S., 257
Rubtsova, E. A., 268
Rudajevova, A., 376
Ruddick, J. N. R., 220, 333, 338, 408
Rudenko, A. P., 150
Rudin, V. N., 368
Rudolf, R. W., 61, 75, 290
Rudomino, M. V., 315
Ruf, W., 102
Rundqvist, S., 279
Ruoff, A., 160
Rupp, H. H., 417, 419, 421
Ruppert, I., 292, 299, 360
Rusakova, L. A., 150
Ruse, D., 226
Rush, J. J., 264
Russegger, P., 291
Russell, B. G., 187
Russell, B. R., 160
Russell, D. B., 19
Russell, D. R., 419
Russo, P. J., 354
Rustembekov, K. T., 348
Rutherford, J. S., 19
Rutledge, C. T., 212
Ruzicka, S. J., 219
Ružic-Toroš, Z., 316
Ryabinin, A. I., 98, 99
Ryabinina, A. F., 366
Ryabkin, V. S., 383
Ryan, R. R., 163, 414
Ryason, P. R., 187
Rybakova, G. A., 40
Ryer, M. N., 383
Rykov, A. N., 23
Ryschkewitsch, G. E., 28, 91
Rytter, E., 25, 129
Rytter, B. E. D., 129

Sabadash, N. G., 35
Sabelli, C., 28, 98
Sadakane, A., 16
Sadimenko, A. P., 313
Sado, A., 259
Sadykova, M. M., 98, 107
Saethre, L. J., 376
Safarik, I., 172, 383
Safiullin, R. K., 284
Safiullina, N. R., 222
Safonov, V. V., 104, 127, 213, 293, 388, 390, 391
Saf'yanov, Yu. N., 413
Saha, H. K., 320
Sahara, M., 160
Sahay, B. K., 183

Sahl, K., 124, 245
Saibova, M. T., 321
Saito, G., 143
Saito, H., 308, 322
Saito, K., 133, 205
Saito, T., 66, 406
Sakamoto, K., 113
Sakk, Zh. G., 47
Sakura, S., 343
Sakurai, H., 315
Salakhutdinov, R. A., 294
Salama, S. B., 353
Salares, V. R., 165
Salentine, C. A., 61, 82, 85
Salin, Ya. V., 132
Salinovich, O., 346
Salmaso, R., 264
Salmon, G. A., 165
Salmon, R., 316
Salov, A. V., 30, 339, 340, 372, 386
Salta, A., 20, 345
Salvatori, T., 118
Salyn, Ya. V., 38
Samanos, J., 65, 66
Samarai, L. I., 299
Sambe, H., 261
Sambhi, M. S., 399
Samitov, Yu. Yu., 220, 284, 330, 337
Samoilenko, V. N., 135
Samoilovich, M. I., 147, 186
Sams, J. R., 220, 240, 333, 338, 408
Sams, R. L., 99
Samson, J. A. R., 171, 343
Samson, S., 51
Sanad, W., 379
Sandercock, A. C., 134
Sandor, E., 413
Sandorfy, C., 160, 161
Sanhueza, E., 165, 344
Sankaranarayanan, V. N., 380, 381
Sano, H., 210, 235
Santarromana, M., 161
Santini, S., 399
Sanz, F., 226, 283, 401
Sara, V., 368
Saran, M. S., 354
Sarholz, N., 265
Sarig, S., 155, 329
Sasaki, K., 269
Sasaki, Y., 66, 149, 320
Sasnovskaya, V. D., 115
Sastri, M. V. C., 169
Sataty, Y. A., 158, 160, 171
Satgé, J., 241, 242, 289, 290
Sato, F., 73, 78
Sato, S., 143, 165, 322
Sato, T., 170
Sattler, E., 94, 281
Saturnino, D. J., 63
Sau, A. C., 308

Sauer, D. T., 351
Sauer, J. C., 265
Sauka, J., 343
Saunders, J. E., 294
Sauvage, M., 55
Sauvageau, P., 160
Saval'skii, S. L., 98, 322
Savel'ev, B. A., 105
Savignac, P., 324
Savin, V. I., 327
Savory, B., 7
Savory, C. G., 75
Savory, C. J., 170
Savoskina, G. P., 415
Sawyer, J. F., 109
Sax, M., 311
Scanlon, J. C., 375
Schack, C. J., 159, 160, 272, 273, 401, 407, 414
Schadow, H., 306, 308
Schaaf, T. F., 232, 233, 278
Schäfer, H., 5, 34, 40, 42, 50, 54, 113, 127, 144, 224, 251, 289, 316, 339, 373, 392
Schaefer, H. F., 150, 161, 167, 179, 269, 379
Schaeffer, R., 62, 63, 65, 229, 234
Schaffer, L. B., 186
Schaible, B., 121, 310
Schaper, K.-J., 105
Schaper, W., 363
Scheer, V. M., 110
Scheibe, R., 364
Scheler, H., 302, 306, 308
Scheltino, V., 265
Schempp, E., 31
Scherer, O. J., 297
Schermann, C., 171
Scherr, P. A., 88
Scherzer, J., 124
Scheuren, J., 240
Schiebel, H.-M., 305
Schiff, H. I., 263
Schildcrout, S. M., 412
Schindler, R. N., 164, 401
Schiwy, W., 225, 373, 377
Schlak, O., 305
Schlegel, H. B., 157, 167, 183
Schless, M., M., 243
Schleyer, P. von R., 60
Schlumper, H. U., 139
Schlyer, P. J., 241
Schmid, G., 92, 95, 353, 366
Schmid, H.-G., 327
Schmidbaur, H., 130, 287, 328
Schmidpeter, A., 302, 307, 315
Schmidt, A., 303, 336, 338
Schmidt, J. F., 165
Schmidt, K. H., 160
Schmidt, M., 100, 342

Schmidt, U., 281
Schmitz, D., 36, 134, 166, 373
Schmutzler, R., 291, 292, 293, 296, 305
Schneider, H., 120, 122
Schneider, I. A., 142
Schnez, H., 398
Schöllhorn, R., 373, 375
Schön, G., 144, 392
Schoen, P. E., 259
Schoener, J., 276
Schönig, G., 299
Scholer, F. R., 75, 77, 83
Schomburg, D., 226
Schoning, G., 363
Schow, S., 303
Schramm, W., 364
Schreiner, A. F., 137
Schrobilgen, G. J., 419
Schroeder, L. W., 50, 103, 316
Schroedter, K., 260
Schröer, U., 214, 250
Schubert, K., 136, 140, 251, 256
Schubert, W., 235
Schue, F., 18
Schugerl, K., 163
Schuijff, A., 369
Schulte-Frohlinde, D., 142
Schultes, E., 164, 401
Schultz, A., 58, 172
Schulz, H., 276
Schultz, S., 336
Schulze-Nahrup, M., 279
Schultze-Rhonhof, E., 196
Schumacher, H. J., 353, 394
Schumacher, R., 164
Schumann, H., 229, 230, 234, 242, 289
Schurath, U., 344
Schussler, D. P., 142
Schuster, H.-U., 4, 254, 255, 279, 333
Schuster, P., 7
Schutte, W. C., 66
Schwarz, A., 249
Schwarz, H., 213
Schwarz, W., 111, 138, 139, 365
Schwarz, W. H. E., 421
Schweiger, J. R., 298
Schwendeman, R. H., 94, 159
Schwenk, A., 97
Schwenzer, G. M., 179, 378
Schwering, H.-U., 118, 130
Schwing-Weill, M. J., 120
Scibona, G., 16
Scmitt, J. M., 398
Scott, J. D., 148, 400
Scott, W. E., 49
Scotti, M., 110, 243
Searle, H. T., 306

Searcy, A. W., 176
Sears, W. M., 182
Secco, E. A., 218
Secker, J. A., 223
Seddon, K. R., 91
Seddon, W. A., 262
Sedel'nikova, N. D., 3
Seefurth, R. N., 21
Seel, F., 264, 350, 376, 378
Sefuk, M. D., 163
Segal, J. A., 170
Sehgal, S. M., 219, 336
Seifullina, I. I., 209
Seiler, P., 14, 15, 50
Seip, H. M., 99, 100, 101
Sekerka, I., 401
Sekine, M., 311
Selig, H., 155, 329, 389, 407, 414, 422
Selim, M., 80
Selinova, G. A., 322
Selivanov, V. D., 224
Selivanova, N. M., 36, 382
Selph, C., 395
Selvoski, M. A., 308
Semenenko, K. N., 22, 40, 63, 66, 114, 115, 116, 117, 119, 127
Semenova, A. I., 348
Sementsova, D. V., 368
Semiokhin, I. A., 170
Sempels, R. E., 187
Senegas, J., 137
Senenii, V. Ya., 299
Senf, L., 249
Sengupta, K. K., 104, 295
Seppelt, K., 379, 388, 417, 419, 421
Septsova, N. M., 31
Sequeira, A., 27, 409
Sera, K., 216
Serafini, A., 109
Serazetdinov, D. Z., 322
Serebryakova, G. V., 40
Serezhkina, L. B., 273
Sergent, M., 134, 326
Sergienko, V. I., 337
Sernetz, F., 375
Serp, R., 236
Serpone, N., 205
Seshadri, T., 219
Sethi, D. S., 171
Setser, D. W., 418
Setton, R., 154, 156
Seybold, K., 169
Seyferth, D., 240
Serastjanova, T. G., 107
Serastyanov, V. G., 372, 386
Sevcuk, V. G., 368
Shabanova, I. N., 114, 251
Shabanova, L. A., 382
Shafer, M. W., 375
Shagidullin, R. R., 329, 332
Shaidarbekova, Zh. K., 322

Shaidulin, S. A., 295
Shakirova, L. I., 165
Shalimov, M. D., 147
Sham, T. K., 204
Shamir, J., 294
Shamrai, N. B., 413
Shamsvidinov, T. M., 161
Shapovalov, Yu. D., 397
Sharma, D. N., 412
Sharma, K. K., 122, 222
Sharma, R. A., 21
Sharma, R. K., 141
Sharma, S. K., 312, 371
Sharov, V. A., 209
Sharp, D. W. A., 351
Sharp, G. J., 104, 226
Sharp, K. G., 214, 216, 241
Sharpless, R. L., 171
Sharrocks, D. N., 104
Shashkin, D. P., 196
Shaw, A. W., 269
Shaw, R. A., 305, 306, 308
Shaw, R. W., 413
Shchegolev, B. F., 133, 331
Shcherbakova, M. N., 291
Shcherbina, K. G., 125
Shearer, H. M. M., 361
Sheets, R. W., 171
Sheka, I. A., 372
Shekhtman, V. Sh., 383
Sheludyakov, V. D., 210
Sheldrick, B., 48
Sheldrick, G. M., 141, 208, 223, 226, 357, 361
Sheldrick, W. S., 291, 292, 293, 294, 304, 315
Shemyakin, N. F., 87
Shen, J., 264
Shen, Q., 140
Shen, S. S., 253
Shenar, H., 229
Shendrik, A. V., 321
Shepard, W. N., 351
Sherwood, R. C., 190
Shevchenko, V. E., 28
Shevchuk, V. G., 125, 368
Shevelev, N. P., 370
Shibanov, E. V., 196
Shibata, M., 179
Shieh, C.-F., 128
Shih, S., 344
Shihada, A.-F., 300
Shilkin, S. P., 40, 66, 115, 117
Shilov, I. V., 323
Shimoji, M., 9
Shin, H. K., 163
Shiokawa, T., 161
Shipkov, N. N., 147
Shirley, D. A., 160, 413
Shick, J. S., 125
Shiro, Y., 173
Shirokov, A. M., 383
Shishkin, V. A., 412
Shitora, V. I., 196, 199

# Author Index

Shkodina, T. B., 209
Shkuryaeva, V. A., 368
Shmatko, B. A., 2
Shinyd'ko, L. I., 273, 340
Shokol, V. A., 303
Shol'ts, V. B., 23
Shore, S. G., 63, 68, 90
Shortridge, R. G., 172
Shpakova, V. M., 36
Shreeve, J. M., 351, 395
Shriver, D. F., 139, 282, 329
Shtain, A. V., 382
Shtanòv, V. I., 257
Shternberg, I. Ya., 220
Shugal, N. F., 316
Shuklina, N. P., 331
Shultin, A. A., 363
Shulz-Ekloff, G., 265
Shurai, P. E., 257
Shurvell, H. F., 272
Shuvalov, L. A., 383
Shvarts, E. M., 97
Shvedchikov, A. P., 262, 397
Sicre, J. E., 353, 394
Sidamashvili, Ts. A., 327
Siddiqi, K. S., 219, 226
Sidebottom, D. P., 370
Siderer, Y., 165
Sideswaren, P., 165, 168
Sidorov, L. N., 23, 28
Siebert, W., 100, 101, 102, 105, 112, 342, 387
Siedle, A. R., 70, 78, 80, 85
Siemers, P. A., 249, 384
Sienko, M. J., 8
Sieper, H.-P., 404
Šiftar, J., 125
Signorelli, A. J., 183
Silber, H., 321
Silberstein, A., 294
Silber, P., 177, 377
Silbernagel, B. G., 28
Silberstein-Hirsch, A., 134
Silina, T. A., 66
Sill, G. T., 397
Sillion, B., 80
Silver, J., 219, 243
Silvestri, A., 141, 207, 220
Simon, A., 6
Simon, G., 202
Simon, K., 207
Simonaitis, R., 344
Simonetta, M., 164
Simmons, J. D., 171
Simonov, M. A., 41, 196
Simonov, V. D., 161
Simons, J., 61
Simons, J. P., 163, 165
Sims, A. L., 82
Singh, A., 111, 204
Singh, B., 222
Singh, O. P., 220
Sinha, R. P., 50
Sinotova, E. N., 203

Sinyaev, V. A., 35
Sinyanskii, V. F., 418
Sipachev, A. V., 38
Sipe, C. A., 373
Sirazhiddinov, N. A., 196, 199
Sirko, V. N., 131
Sizareva, A. S., 65
Sizova, N. I., 220
Skabichevskii, P. A., 10
Skibsted, L. H., 370
Skillern, K. R., 115
Sklarz, E. G., 113
Skoropanov, A. S., 257
Skowrońska-Ptasińska, M., 120
Skralec, W. J., 277
Skripchenko, V. I., 408
Skvortsov, V. G., 98, 99, 368
Sladky, F., 342, 389
Slagle, I. R., 172
Slagter, G. K., 6
Slama, I., 375, 412
Slanger, T. G., 171
Sleziona, J., 293
Slivnik, J., 346, 418
Slogvik, S., 390
Slovyanskikh, V. K., 372, 386, 392
Smalc, A., 346, 353, 418
Smalley, R. E., 270
Smardzewski, R. R., 16, 160, 408
Smart, E. F., 3
Šmied, R., 310
Smirnov, B. M., 418
Smirnova, N. A., 297
Smith, B. C., 390
Smith, B. E., 66
Smith, D. F., 36, 178
Smith, G. R., 172
Smith, H. D., jun., 78
Smith, J. A. S., 415
Smith, J. L., 113
Smith, J. V., 123, 124, 197
Smith, K., 104
Smith, K. M., 141
Smith, K. V., 390, 400
Smith, L. R., 281
Smith, P. D., 261
Smith, R. H., 397
Smolikov, V. V., 187
Smyrl, N., 10, 402, 403
Smyth, N. P. A., 163
Sneddon, L. G., 81
Snow, M. R., 143
Snelson, A., 22
Snelson, A., 23
Snyder, S. C., 163
So, S. P., 126, 179
Sobolev, V. M., 260
Sobczak, R., 113
Sobhanadri, J. 397, 409

Sochilin, F. G., 281
Sogabe, K., 291
Sohma, J., 62
Sokolov, V. B., 418
Sokolov, A. N., 196
Sokolov, E. I., 81
Sokolova, I. D., 322
Sokolova, N. P., 170
Sokolova, V. R., 243
Solan, D., 213
Soledova, Yu. P., 147
Solenko, T. V., 125, 186
Solka, B. H., 163
Sollberger, F., 348
Solmajer, T., 272
Solntsev, K. A., 56
Solodushchenko, G. F., 299
Solomon, R. E., 265
Solouki, B., 365
Solov'eva, L. P., 199, 224
Solov'eva, T. V., 174
Solov'eva, Yu. I., 394
Somieski, R., 300, 301
Somoano, R. B., 51
Sonnino, T., 399
Sonsale, A. Y., 212
Sopchyshyn, F. C., 262
Soria, D., 346
Sorokina, S. F., 324
Sorrell, C. A., 46
Sorriso, S., 399
Sosedov, V. P., 146
Sosinsky, B. A., 239
Soto, M. R. C., 279
Southern, J. T., 333, 340, 384
Souza, G. G. B., 212
Sowerby, D. B., 306, 308
Speakman, J. C., 180, 249
Spek, A. L., 235, 250
Spekkens, P., 407
Spelbos, A., 203
Spencer, J. L., 86
Spencer, P. J., 253
Spilker, D., 240
Spinat, P., 341, 372
Spinner, E., 167, 169
Spiridonov, V. P., 117
Spiridonova, N. N., 235
Spiro, M., 370
Spiro, T. G., 67
Spitsyn, V. I., 98, 102, 421
Spitz, R. D., 412
Spitzin, V. I., 107
Spratley, R. D., 355
Sprenger, G. H., 395
Srikrishnan, T., 341, 373
Srinivasan, V., 169
Srivastava, K., 222
Srivastava, O. N., 242
Srivastava, P. C., 222
Srivastava, R. C., 223
Srivastava, R. D., 42, 58
Srivastava, T. N., 222
Srylalin, I. T., 257

Stadelhofer, J., 117
Stadelmaier, H. H., 140
Staemmler, V., 61, 114, 157
Stafast, H., 178, 179
Stafford, S., 218
Stahl-Brasse, R., 34, 331
Staley, R. H., 163, 284
Stam, C. H., 120
Stankiewicz, T., 304
Stanko, V. I., 79, 80, 81
Starer, H., 171
Staricco, E. H., 346
Starowieysky, K. B., 120
Stas', M. N., 144
Steblevskii, A. V., 322
Stec, N. J., 302, 303, 325
Stedman, G., 357
Steele, B. R., 236
Steenhoek, L., 129
Steer, I. A., 88, 91
Steer, R. P., 167
Steger, J., 325, 385
Steggall, M., 144
Stein, L., 407
Steinberger, H., 312, 324
Steinbeisser, H., 351
Steiner, E. C., 16, 44
Steinfatt, M., 343
Steinfink, H., 59, 376
Steinkilberg, M., 170
Steinmetz, J., 256
Steinmetz, W. E., 264
Stelzer, O., 290
Stephens, F. S., 238
Stephens, M., 104, 295
Stephenson, L. M., 269
Stepin, B. D., 127, 390
Stepina, S. B., 31
Stern, M., 177
Steudel, R., 347, 348, 353, 363, 376
Stevens, J., 398
Stewart, A. W., 163
Stewart, J. M., 38, 335
Štíbr, B., 84
Stief, L. J., 171, 172
Stilbs, P., 103, 336
Stobart, S. R., 184, 235, 240
Stocco, G. C., 141, 207, 220, 223, 333
Stockdale, J. A. D., 171
Stoeckli, F., 348
Stoékli-Evans, H., 213
Stokhuyzen, R., 136
Stokkeland, O., 119
Stolarski, R. S., 396
Stone, F. G. A., 86, 237, 239
Stone, J. M. R., 179
Storek, W., 364
Storhoff, B. N., 78
Stork, W., 377
Stork, W. H. J., 124
Storr, A., 130, 131, 132
Stotskaya, N. M., 271

Stotskii, A. A., 271
Stover, G., 388
Strandberg, R., 320
Stranks, D. R., 134
Strattan, L. W., 163
Stratton, C., 306
Strausz, O. P., 172, 383
Street, F. J., 1
Streiff, R., 43
Streitwiesser, A., 157, 169
Strelko, V. V., 186
Streltsova, Yu. F., 372
Stremovsov, V. I., 257
Strey, G., 179
Strezhneva, I. I., 99
Stroganov, E. V., 196, 199
Stromme, K. O., 174
Strouse, C. E., 82
Struchkov, Yu. T., 152, 226
Strüver, W., 285
Studnev, Yu. N., 369
Stukan, R. A., 152
Stul, M. S., 193
Su, L. S., 279
Su, Y. Y., 241
Subba Rao, G. V., 375
Subbotin, V. I., 2
Subra, R., 157
Subramanian, C. R., 167
Subramanian, J., 141
Subramanian, S., 329
Šubrtová, V., 80
Sudakova, T. M., 284
Sudol, T., 303
Sugai, R., 394
Sugi, Y., 170
Sujishi, S., 235
Suhara, M., 259
Sukhenko, V. D., 372
Sukhorukova, L. N., 372, 386
Sullivan, B. P., 70
Sullivan, G. W., 63
Sun, T. S., 178
Sundaram, S., 265
Sundell, S., 400
Sundermeyer, W., 354
Sunthanker, S. V., 227
Sumarokova, T. N., 271
Surikov, V. I., 254
Surkin, Ya. K., 265
Surles, T., 402
Suryanarayana, D., 397, 409
Sutor, P. A., 214
Suvorov, A. V., 128, 130, 294
Suyunova, Z. E., 191
Suzdalev, I. P., 191
Suzuki, H., 245
Suzuki, I., 269
Svanström, P., 251
Svensson, C., 337
Sventitskii, E. N., 415
Svettsov, V. I., 262

Swanson, N., 171
Swyke, C., 303
Syano, Y., 197
Syutkina, O. P., 208
Symons, M. C. R., 89, 276, 278, 291, 295, 310, 410
Szczepaniak, W., 410
Sze, N. D., 396
Szilassy, I., 270
Szymanski, J. T., 373
Szwarc, M., 12

Tabachik, V., 310
Tabayashi, K., 180
Tachiyashiki, S., 143
Tadasa, K., 268
Taesler, I., 271, 272
Taillandier, E., 96
Tait, J. C., 350
Takagi, J., 249
Takagik, J., 373
Takéuchi, Y., 124, 249, 373
Takusaka, T., 110
Taldenko, Yu. D., 262
Tamagake, K., 158, 160
Tamai, Y., 149
Tamao, K., 230
Tamres, M., 204
Tamm, N. S., 273
Tamura, H., 406
Tan, T. S., 212
Tana, S.-Y., 367
Tanabe, K., 160
Tanaka, K., 162, 167, 170
Tanaka, S., 113
Tanaka, T., 117
Tananaev, I. V., 137, 139, 321, 322, 331
Tandon, J. P., 132, 220
Taneja, R. P., 114
Tang, S., 75, 234
Tang, S. Y., 172
Tang, Y. N., 241
Tani, M. E. V., 340
Taniguchi, S., 163
Tanttila, W. H., 270
Tappen, D. C., 241
Taqui Khan, M. M., 285
Tarasevich, Yu. I., 191
Tarasov, V. P., 134
Tarasova, A. I., 131
Tarimci, C., 31
Tarnorutskii, M. M., 51
Tarrago, G., 160
Tarte, P., 185
Tasaka, A., 322
Tasker, A., 372
Tate, C., 1
Tatsii, V. F., 185
Tattershall, B. W., 180
Taugbol, K., 209
Tavares-Forneris, Y., 406
Taylor, A., 225
Taylor, B., 325, 385

# Author Index

Taylor, B. F., 86
Taylor, P., 400, 419
Taylor, R., 208, 223, 226
Taylor, R. G., 91, 94, 283
Tazeeva, N. K., 330
Tazzoli, V., 408
Tchir, P. O., 355
Tebbe, K.-F., 40, 112
Tedesco, P. H., 121
Tegenfeldt, J., 169
Tehan, F., 18
Teichner, S. J., 65, 66
Teitelboum, M. A., 161
Tellgren, R., 169, 174, 383, 409
Tellier, J.-C., 125
Tempere, J.-F., 124
Tenisheva, N. Kh., 327
Tennakoon, D. T. B., 191
Tennent, N. H., 300
Teplyakova, N. V., 384
Terekhov, V. A., 132
Terent'eva, S. A., 296
Terry, J. H., 220
Terry, S., 146
Terunuma, D., 217
Teruya, K., 409
Terzieff, P., 392
Terzis, A., 48
Teterin, E. G., 368
Tevault, D. E., 163, 165
Thain, J. M., 8
Thakur, K. P., 344, 373
Thayer, J. S., 223, 390, 400
The, K. I., 287
Theard, L. P., 163, 172, 260, 263
Theobald, F., 337, 367
Thevenot, F., 112
Thevet, F., 373
Thewalt, U., 226
Thibault, R. M., 81
Thiebault, A., 414
Thiebault, C., 134, 257, 386
Thiele, G., 144
Thierling, M., 220, 288
Thirase, G., 32, 183
Thistlethwaite, P. J., 128
Thomas, B., 306, 308
Thomas, B. S., 73
Thomas, D., 341
Thomas, J. M., 148, 152, 191, 202, 400
Thomas, J. O., 169, 174, 409
Thomas, R. B., 146, 150
Thomas, R. K., 416
Thomas, T. D., 413
Thomas, T. E., 145
Thomas-David, G., 38
Thompson, D. A., 61
Thomson, C., 95, 99, 179, 283
Thornton, E. W., 201
Thourey, J., 35

Thrower, P. A., 150
Thrush, B. A., 417
Thulstrup, E. W., 258
Thumova, M., 343
Thym, S., 20, 344
Tiffany, T. O., 272, 398
Tikavyi, V. F., 339
Tilford, S. G., 171
Tiller, H.-J., 187
Tillmanns, E., 124, 316
Tillott, R. J., 308
Timmons, R. B., 163, 172
Timms, P. L., 103, 105
Timoshchenko, N. I., 379
Timoshenko, N. J., 259
Timoshin, A. S., 132
Tinelli, C., 3
Ting-Po, I., 7
Tinhof, W., 109, 303
Tipping, A. E., 217
Tishura, T. A., 23
Titenko, Yu. V., 347
Titov, L. V., 115
Titova, K. V., 66
Tiwari, P., 242
Tkachev, A. S., 223
Tkachev, K. V., 99
Tkalenko, V. G., 220, 313
Tobias, C., 12
Tobias, R. S., 10
Toby, S., 344
Todd, L. J., 63, 70, 73, 78, 80
Todd, S. M., 307
Todo, N., 170
Toei, K., 16
Tofield, B. C., 190
Tohdo, F., 398
Tok, G. C., 243
Tokahashi, T., 230
Tokue, I., 161
Tolls, E., 323, 364
Tolmacheva, L. N., 65, 66
Tolokonnikova, L. I., 174
Tomarchio, C., 223
Tomat, R., 264
Tomita, A., 149
Tomita, I., 320
Tomizuka, I., 148
Tomkins, R. P. T., 400
Tomlinson, L., 378
Tong, D. A., 306
Topor, N. D., 174
Torbov, V. I., 279
Tordjman, I., 316
Toropova, V. F., 315
Torre, J.-P., 124, 186
Torri, J., 96
Tossell, J. A., 122, 185
Totani, T., 110
Toth, R. A., 171
Tournoux, M., 33, 143, 174, 200, 339
Touzain, Ph., 29, 155, 367
Touzin, J., 380

Tozer, D. J. N., 251
Tranquard, A., 143, 322
Tran Qui Duc, 23
Trapeznikov, V. A., 114, 251
Tremillon, B., 413, 414
Trenkel, M., 127
Tresvyatskii, S. G., 99
Tret'yakova, K. K., 127, 334, 388
Trevino, S. F., 264
Treweek, R. F., 60
Tribo, M., 83
Tricker, M. J., 153, 156, 191, 202, 251
Tridot, G., 341
Trigunayat, G. C., 242
Trillo, J. M., 174
Trimm, D. L., 147, 202
Triplett, K., 237
Tripodi, K., 163
Troe, J., 266
Trömel, M., 413
Trombe, J.-C., 320, 373
Trombe, F., 113
Trotter, J., 108, 130, 131, 132, 307
Troup, J. M., 298
Trunov, V. K., 245
Trusov, V. I., 128, 130, 294
Truter, M. R., 16
Tsai, B. P., 161
Tsang, J. C., 375
Tsapkova, N. N., 220
Tsay, Y.-H., 127
Tsemekhman, L. Sh., 372
Tsiklis, D. S., 348
Tsimmerman, S. S., 262
Tsin, T. B., 240
Tsinober, L. I., 186
Tsipis, C. A., 332
Tsuboi, M., 158, 160
Tsuchida, T., 123
Tsuhako, M., 321
Tsunashima, S., 172
Tsyganok, L. P., 133
Tuck, D. G., 136, 137, 138, 140
Tucker, P. A., 303, 364, 419
Tudo, J., 144
Tupikov, V. I., 419
Turban, K., 54, 113
Turbini, L. J., 110
Turetskaya, R. A., 211
Turner, A. G., 360
Turner, D. J., 370
Turner, J. J., 342
Tur'yan, Ya. I., 131
Tusek, L., 16
Tuvaeva, T. N., 23
Tychinskaya, I. I., 218
Tyfield, S. P., 146
Tyler, B. J., 217
Tyuleneva, N. I., 402
Tzschach, A., 208

Uchida, I., 242
Udagawa, Y., 174
Uebel, R., 31, 321
Ufintsev, V. B., 132
Ugai, Ya. A., 132
Ugi, I., 311, 313
Ugorets, M. Z., 348, 385
Ugulava, M. M., 327
Uhlenhaut, H., 365
Ukhova, T. V., 188
Ul'chenko, N. I., 263
Ul-Hasan, M., 306
Ullrich, D., 277
Ulman, J., 68, 234
Uminskii, A. A., 156, 335
Ungemach, S. R., 161
Unger, E., 290
Ungurenasu, C., 33, 154
Unnai, T., 165
Urchsovskaya, L. N., 54
Urikh, V. A., 35
Urry, G., 208
Usanovich, M. I., 271
Ushakova, V. G., 143
Uskova, S. L., 137
Usov, A. P., 268
Uspenskaya, K. S., 150
Ustynyuk, Yu. A., 209
Utida, T., 173
Uytterhoeven, J. B., 193
Uznański, B., 302, 303

Vahrenkamp. H., 342
Vaillant, A., 413
Vaisfeld, M. I., 368
Vakhidov, Sh. A., 147
Vakhobov, A. V., 53, 54, 55
Vakratsas, Th., 323, 364
Valentine, J. S., 345
Valimukhametova, R. G., 227
Van Brunt, R. J., 171
Van Cauteren, M., 404
Van den Berg, P. J., 212
Van den Hof, R. P. A., 212
Vanderschmitt, A., 366
van der Helm, D., 366
Van der Grampel, J. C., 303, 364
van der Kelen, G. P., 101, 240, 283, 287
Van der Lugt, W., 6
van der Veken, B. J., 312
Van der Voet, A., 22
Vanderwielen, A. J., 183
Van Deventer, E. H., 25
van Dijk, J. M. F., 288
Van Dyke, C. H., 225, 227
Vangelisti, R., 153, 156
Vankateswarlu, P., 368
Van Paaschen, J. M., 91
Van Remoortere, F. P., 16, 44
van Tamelen, E. E., 261

Van Wazer, J. R., 103, 270, 280, 290, 293, 311
van Zee, R. J., 310
Van Zytveld, J. B., 47
Varchuk, K. I., 219
Varfan-Jarec, M., 376
Varfolomeev, M. B., 413
Varma, R., 183, 214, 229
Varnek, V. A., 243
Varney, R. N., 260
Varwig, J., 351
Vashman, A. A., 272
Vasilenko, G. I., 332
Vasil'ev, N. G., 191
Vasil'ev, V. G., 368
Vasil'ev, V. P., 138, 213
Vasil'eva, M. B., 340
Vasil'eva, M. P., 386
Vasil'eva, N. P., 132
Vasil'eva, V. N., 213
Vasilyeva, G. A., 233
Vas'kov, A. P., 325
Vassilyev, A. M., 80
Vasudeva Murthy, A. R., 348
Vasudevan, K., 169
Vavilin, V. I., 35, 412
Volkov, A. F., 129
Vdovenko, V. M., 415
Vdovina, T. Z., 254, 326
Vecher, A. A., 257
Vedeneev, V. I., 161
Vegas, A., 98
Vegnere, A. A., 97
Veigel, E., 139
Veillard, A., 119
Veillard, H., 119
Veith, M., 226, 227
Velazco, J. E., 418
Veleckis, E., 25
Vel'mozhnyi, I. S., 244, 245, 319
Venables, J. A., 259
Venetopoulos, C. C., 242
Veniamov, N. N., 215
Venkatasubramanian, N., 265
Vepřek, S., 277
Vepřek-Šiška, J., 369
Verall, R. E., 263
Verber, C. M., 403
Verbetskii, V. N., 22
Verdier, P., 132
Vereshchagin, L. F., 147
Vergamini, P. J., 77
Verkade, J. G., 301
Verma, R. D., 354
Vermeil, C., 164
Vermeulen, L. A., 147
Vermor-Gaud-Daniel, F., 29
Vernet, J.-L., 161
Verot, J. L., 400
Vertes, A., 220
Vetter, A., 334

Viard, B., 294
Viard, R., 163
Vicat, J., 23
Vicentini, G., 300
Vidal, M., 102
Viebahn, W., 134
Vigdorovich, V. N., 53
Vignoles, C., 320
Vikane, O., 390, 393
Vikhreva, L. A., 313
Vilesov, F. I., 277, 324
Vilkov, L. V., 63, 79, 226, 258, 296, 342
Vinarov, I. V., 245
Vinarova, L. I., 204
Vincent, H., 323, 372, 384
Vinogradov, L. I., 302
Vinogradova, G. Z., 257
Vishinskaya, L. I., 233
Vishnevskii, V. B., 204
Visser, J. W., 369
Viswanathan, B., 169
Viterbo, D., 44
Vitse, P., 99, 367
Vlačil, 60
Vladimirova, M. V., 273
Vlasov, O. N., 310
Vlasov, V. S., 129
Vlasova, I. V., 31
Vlasse, M., 316
Völke, H. P., 210
Vogel, P. L., 235
Vojtech, O., 367, 368
Vol'fkovich, L. A., 321
Volk, K., 7
Volkov, V. L., 99, 245
Volkova, A. N., 188, 293
Vollenkle, H., 251
Vol'nov, I. I., 345
Volodin, A. A., 308
Vol'pin, M. E., 152
Volz, P., 299
von Niessen, W., 167, 259
von Schenk, R., 133
von Schnering, H. G., 279
von Seyerl, J., 327
Vorob'ev, P. N., 213
Voronezheva, N. I., 38
Voronin, A. I., 161
Voronin, G. F., 34
Voronkov, Yu. M., 170
Voronkov, V. V., 310
Voronova, E. I., 257
Vorontsov, E. S., 294
Vorontsova, N. A., 310
Vorontsova, N. V., 165
Vosper, A. J., 258, 269
Voss, G., 364
Vostokov, I. A., 223
Vovna, V. I., 324
Voznyak, O. M., 254
Vtkina, O. N., 368
Vvanova, N. A., 220
Vyazankin, N. S., 232, 233

# Author Index

Vyaznev, M. Ya., 43
Vyatkin, I. P., 43
Vygen, H., 417
Vyshinskaya, L. I., 235

Waddington, D., 410
Waddington, T. C., 21, 173
Wade, K., 226
Wafflart, J., 382
Wagner, D.-L., 359, 360
Wagner, H., 359
Wagner, H. G., 265, 268
Wagner, R. I., 276
Wagner, S., 302
Wahlgren, U., 169, 179
Waite, D. W., 73
Waite, R. J., 146
Wakatsuki, K., 117
Wakeshima, I., 245
Walden, R. T., 160
Waldvogle, G. G., 90
Walker, A., 150
Walker, J., 147
Walker, M. L., 310
Walker, N., 166
Walker, P. L., 150, 375
Walker, R. W., 267
Walker, W. C., 171
Walker, W. R., 251
Wallbridge, M. G. H., 25, 63, 342
Wallenwein, G., 323, 364
Walsh, A. D., 167
Walter, E., 300
Walter, R. H., 170
Walter-Levy, L., 369
Walters, M. J., 191
Walters, S., 396
Wan, C., 197
Waněk, W., 305
Wang, S. Y., 421
Wanklyn, B., 197
Wanrooy, J., 369
Warner, A. E. M., 322
Warning, K., 285, 286
Warwick, M. E., 201
Washburn, B., 89
Wasif, S., 353, 380
Wasson, J. R., 173, 274
Watanabe, H., 110
Watanabe, M., 322
Watanabe, N., 21, 152
Watanabe, O., 148
Watari, F., 331
Waterworth, L. G., 139
Watkins, J. J., 115
Watkiss, P. J., 219
Watson, L. M., 43
Watson, R. T., 396, 408, 410
Watt, W., 150
Watts, R. O'B., 217
Wauchop, T. S., 171
Waugh, J. S., 169
Way, G. M., 105

Wayne, R. P., 271, 272
Wazeer, M. I. M., 300, 305, 306
Webb, J., 133
Webb, T. R., 328
Weber, D., 290, 310, 311
Weber, J., 149
Weber, J. Q., 395
Weber, K., 344
Weber, L., 114
Weber, N. P., 230
Weber, W., 111
Webster, D. E., 238, 239
Webster, M., 220
Weed, H. C., 2
Wegner, P. A., 87
Wehner, E., 365
Wehner, P., 321
Wehrer, A., 172
Wehrer, P., 172
Wei, C. N., 172
Wei, L. Y., 375
Weibel, A. T., 88
Weiden, N., 122
Weidenbruch, M., 214, 217
Weidlein, J., 108, 117, 121, 130, 132, 139, 310
Weinberger, P., 347
Weinmaier, J. H., 315
Weis, J., 144, 392
Weiss, A., 122, 135
Weiss, E., 32, 183
Weiss, J., 139, 364
Weiss, K., 346
Weiss, S., 29
Welch, A. J., 86, 87
Welge, K. H., 171
Welk, E., 255
Wells, M., 150
Wells, P. R., 231
Wells, R. D., 126
Wells, W. C., 160
Weltner, W., 37, 172
Welty, P. K., 195
Wennerstrom, H., 170
Wenschuh, E., 228, 357
Werff, F. B. D., 6
Werner, A. S., 161
Werner, F., 233
Werner, H., 110, 240, 243
West, A. R., 41, 196
West, B. O., 328
Westenberg, A. A., 163
Westgaard, L., 400
Westheimer, F. H., 311
Weston, A. F., 108, 110
Westwood, N. P. C., 104, 265
Wey, R., 122
Whaley, T. P., 25
Whangbo, M.-H., 157
Wharf, I., 139
Wharton, L., 270
Whitaker, A., 316

White, A. H., 330
White, W. B., 373
Whitehead, J. C., 411
Whitehead, M. A., 360
Whitehead, W. D., 171
Whitla, W. A., 46
Whittaker, E. J. W., 123
Whittingham, A. C., 4
Whittingham, M. S., 28, 375
Wiberg, N., 266
Wichelhaus, W., 279
Wichner, R. P., 401
Widler, H. J., 139
Wiedemeier, H., 249, 384
Wiegel, K., 226
Wieser, D., 62
Wieser, J. D., 212, 234
Wiezer, H., 358
Wiffen, J. T., 60
Wight, G. R., 161, 171
Wild, S. B., 330
Wilkins, B. T., 290
Wilkins, C. J., 329
Wilkins, R. A., jun., 277
Wilkinson, G. R., 370, 415
Wilkinson, J. R., 73
Wilkinson, M., 135
Willemot, E., 167
Willemsen, H.-G., 62
Willett, R. D., 401, 422
Willey, G. R., 226
Williams, E. A., 170, 171
Williams, F., 402, 403
Williams, F. W., 162
Williams, J. E., 157
Williams, J. O., 191
Williams, T. A., 22
Willson, M., 314
Wilson, I. L., 366, 420
Wilson, R. D., 272, 276, 299, 346, 401, 403, 407, 414
Wilson, S., 61
Wilson, W. W., 406
Wincel, H., 164
Windus, C., 235
Winer, A. M., 343
Winfield, J. M., 25, 217, 345
Wingfield, J. N., 16, 309
Winn, J. S., 397
Winnewisser, B. P., 178, 179
Winnewisser, G., 167, 169
Winnewisser, M., 178, 179
Winsor, A., 21
Winter, N. W., 267
Winterton, R. C., 236
Winther, F., 160, 351
Wishart, B. J., 95
Wisløff Nilsen, E., 99
Wittel, K., 135
Wittman, A., 251
Wofsy, S. C., 396
Wold, A., 325, 385
Wolf, A. P., 182

Wolf, E., 388
Wolf, R., 314, 315
Wolfe, S., 157, 167, 183
Wolfes, N., 238
Wolff, G., 58
Wolfrum, J., 396
Wolmershäuser, G., 361
Wong, H. S., 78
Wong, K. F., 161
Wong, T. C., 161, 283, 327
Wood, M. K., 311
Woodhams, F. W. D., 210, 316
Woods, M., 308
Woodward, P., 105, 237
Woolf, A. A., 329
Woollam, J. A., 51
Wooten, J. B., 339
Wopersnow, W., 251
Woplin, J. P., 305
Worrall, I. J., 135, 139
Worzala, F. J., 136
Wozniak, W. T., 67
Wright, C. A., 402
Wright, D. A., 160
Wu, C. H., 26
Wu, K. K. 174
Wuensch, B. J., 248, 339
Wulff, C. A., 25, 335
Wuyts, L. F., 240
Wyatt, J. R., 163

Yablokov, V. A., 232
Yablokova, N. V., 232
Yakel, H. L., 112
Yakovlev, Yu. B., 136, 143
Yakovleva, A. F., 127
Yakshin, V. V., 331
Yakubchik, V. P., 253
Yamabe, T., 162
Yamada, H., 190, 199, 399
Yamada, K., 128, 167, 179, 295, 334
Yamamoto, A., 173
Yamamoto, O., 116
Yamamoto, T., 73
Yamamoto, Y., 89, 328
Yamanaka, S., 193
Yamanaka, T., 249, 273
Yamaoka, S., 59, 376
Yambushev, F. D., 327
Yamnov, A. L., 259
Yanaka, T., 394
Yang, K. T., 148
Yang, Y. S., 125

Yankelevich, R. G., 219
Yarkova, E. G., 220, 222
Yasuno, H., 242
Yatsimirskii, K. B., 262
Yaws, C. L., 266
Yedunov, N. F., 218
Yesinowski, J. P., 141
Yokata, T., 172
Yoneda, M., 398
York, R. J., 178
Yoshida, T., 373
Yoshihara, K., 161
Yoshino, N., 219
Yoshino, T., 219
Youll, B., 72, 222
Young, B. C., 150
Young, E. S., 396
Young, R. A., 259
Yow, H. Y., 290
Yu, M. W., 248
Yu, S.-L., 351
Yudin, B. F., 3
Yudin, D. M., 321
Yuen, P.-S., 316
Yurchenko, E. N., 243
Yurchenko, V. A., 99
Yushchenko, S. F., 273
Yust, S. C., 147
Yuzhelevskii, Yu. A., 81
Yuz'ko, M. I., 320
Yuzvak, V. I., 368
Yvon, K., 251

Zabel, M., 125, 384
Zademidko, G. A., 53
Zahradnik, R., 169, 177
Zaidi, S. A. A., 219, 226
Zaitsev, B. E., 132, 273
Zaitseva, I. S., 125
Zaitseva, M. P., 368
Zaitov, M. M., 186
Zakhar'ev, Yu. V., 349
Zakharkhin, L. I., 81, 85, 87, 129, 130
Zakharov, P. I., 209
Zakolodyazhnaya, O. V., 241
Zalkin, V. M., 254
Zamaletdinova, G. U., 210
Zamanskii, V. M., 266
Zanne, M., 45
Zare, R. N., 58, 119, 172, 396
Zaripov, M. M., 186
Zarubin, V. N., 218, 420
Zavalishina, A. I., 324

Zavodnik, V. E., 27
Zazzetta, A., 119
Zdamovskii, A. B., 99
Zeck, O. F., 241
Zeegers-Huyskens, Th., 404
Zeil, W., 167, 168
Zeldin, M., 207, 213, 244, 249
Zelenetskii, S. N., 308
Zelionkaite, V., 387
Zeller, C., 153
Zemlyanskii, N. I., 325
Zemlyanskii, N. N., 209
Zemnukhova, L. A., 337
Zemskov, S. V., 402
Zemva, B., 418
Zenaidi, N., 337
Zenchenko, D. A., 136
Zernyakova, N. B., 99
Zettler, F., 90, 138
Zhaikin, L. S., 296
Zharkov, A. P., 132, 133
Zhdanov, S. P., 98
Zhelnin, B. I., 368
Zherebtsova, L. I., 368, 382
Zhikharev, M. I., 271
Zhilinskaya, V. V., 262
Zhinkin, D. Ya., 226
Zhirnov, Yu. P., 271
Zhuk, M. I., 331
Zhukov, E. G., 385
Zhuravlev, E. Z., 224
Ziegler, M. L., 139
Ziehn, K. D., 285
Ziemann, H., 379
Zilber, R., 316
Zimin, M. G., 302
Zimina, I. D., 262
Zimmermann, H., 325
Zingaro, R. A., 326
Zink, J. I., 142
Zinner, L. B., 300
Zins, D., 177, 377
Zipf, E. C., 161
Zlomanov, V. P., 257, 392
Zolotov, Yu. A., 129
Zorina, E. N., 130
Zmbov, K. F., 5
Zubieta, J. A., 208
Zuckerman, J. J., 203, 208
Zunger, A., 258
Zupan, M., 406, 421
Zur, Z., 372
Zvarikina, A. V., 152
Zviedra, I., 35
Zykova, T. V., 294